Vickery

SPECIATION AND ITS CONSEQUENCES

SPECIATION
and Its Consequences

Edited by Daniel Otte
Academy of Natural Sciences of Philadelphia
and John A. Endler
University of California, Santa Barbara

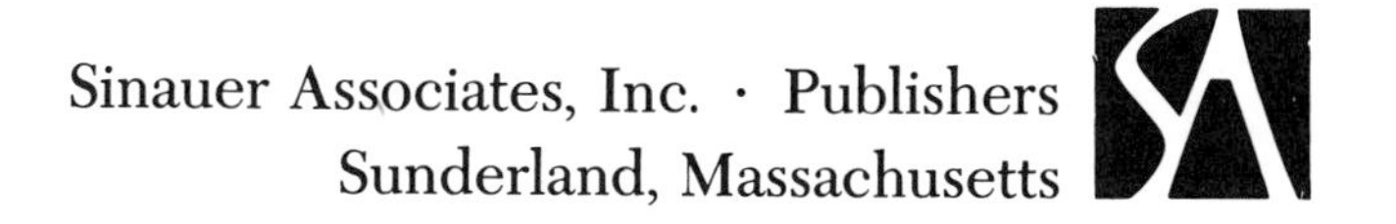

ABOUT THE COVER

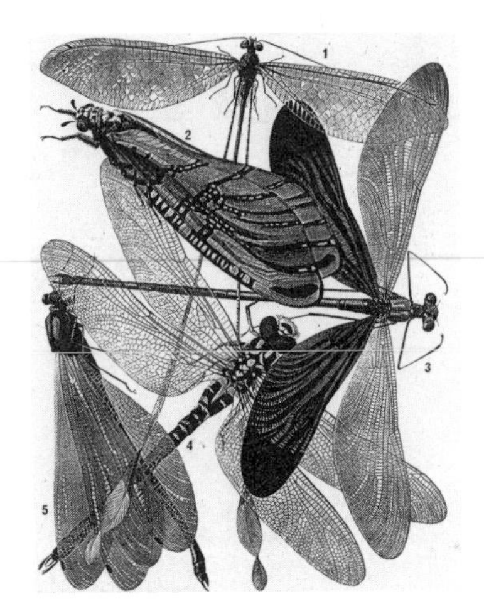

The cover drawing is by the French designer E. A. Seguy from a republication of his portfolio *Insectes,* originally published in Paris during the 1920s. Species depicted in this composition—which Seguy created based on scientific publications, paying strict attention to accuracy of form and color—include: 1, *Nemopistha imperatrix* (West Africa); 2, *Tomatares citrinus* (South Africa); 3, *Neurolasis chinensis* (Asia); 4, *Aeschna cyanea* (Europe); and 5, *Mnais earnshawi* (Indochina). (From *Seguy's Decorative Butterflies and Insects in Full Color,* published in 1977 by Dover Publications Inc., New York, N.Y.)

SPECIATION AND ITS CONSEQUENCES

Library of Congress Cataloging-in-Publication Data

Speciation and its consequences/edited by Daniel Otte and John A. Endler.
p. cm.
Includes indexes and bibliographies.
ISBN 0-87893-657-2 —ISBN 0-87893-658-0 (pbk.)
1. Species. I. Otte, Daniel. II. Endler, John A., 1947-
QH83. S685 1989
575—dc19 89-5867
CIP

Printed in U.S.A.

3 2 1

Contents

Contributors

SPENCER C. H. BARRETT, Department of Botany, University of Toronto, Toronto, Ontario M5S 1A1, Canada

N. H. BARTON, Department of Genetics and Biometry, University College London, 4 Stephenson Way, London NW1 2HE, U.K.

KURT BENIRSCHKE, Center for Reproduction of Endangered Species, Zoological Society of San Diego, P.O. Box 551, San Diego, CA 92112, U.S.A.

GUY L. BUSH, Department of Zoology, Michigan State University, East Lansing, MI 48824, U.S.A.

ROGER BUTLIN, School of Pure and Applied Biology, University of Wales, Cardiff CF1 3TL, Wales, U.K.

JERRY A. COYNE, Department of Ecology and Evolution, University of Chicago, 1103 E. 57th Street, Chicago, IL 60637, U.S.A.

JOEL CRACRAFT, Department of Anatomy and Cell Biology, University of Illinois at Chicago, Chicago, IL 60605, U.S.A.

SCOTT R. DIEHL, Department of Human Genetics and Psychiatry, Box 710, Medical College of Virginia, Richmond, VA 23298, U.S.A.

BARBARA S. DURRANT, Center for Reproduction of Endangered Species, Zoological Society of San Diego, P.O. Box 551, San Diego, CA 92112, U.S.A.

JOHN A. ENDLER, Department of Biological Sciences, University of California, Santa Barbara, CA 93106, U.S.A.

MONICA M. FRELOW, Museum of Vertebrate Zoology, University of California, Berkeley, CA 94720, U.S.A.

DOUGLAS J. FUTUYMA, Department of Ecology and Evolution, State University of New York at Stony Brook, Stony Brook, NY 11794, U.S.A.

DOUGLAS E. GILL, Department of Zoology, University of Maryland, College Park, MD 20742, U.S.A.

B. ROSEMARY GRANT, Biology Department, Princeton University, Princeton, NJ 08544, U.S.A.

PETER R. GRANT, Biology Department, Princeton University, Princeton, NJ 08544, U.S.A.

RICHARD G. HARRISON, Section of Ecology and Systematics, Corson Hall, Cornell University, Ithaca, NY 14853, U.S.A.

GODFREY M. HEWITT, School of Biological Sciences, University of East Anglia, Norwich NR4 7TJ, U.K.

ARLENE T. KUMAMOTO, Center for Reproduction of Endangered Species, Zoological Society of San Diego, P.O. Box 551, San Diego, CA 92112, U.S.A.

ALLAN LARSON, Department of Biology, Washington University, St. Louis, MO 63110, U.S.A.

JOHN D. LYNCH, School of Biological Sciences, University of Nebraska, Lincoln, NE 68588, U.S.A.

GARETH NELSON, Department of Herpetology and Ichthyology, American Museum of Natural History, New York, NY 10024, U.S.A.

H. ALLEN ORR, Department of Ecology and Evolution, University of Chicago, 1103 E. 57th Street, Chicago, IL 60637, U.S.A.

DANIEL OTTE, Academy of Natural Sciences of Philadelphia, 19th and The Parkway, Philadelphia, PA 19103, U.S.A.

STEPHEN PACALA, Ecology Section, Biological Sciences Group, University of Connecticut, Storrs, CT 06268, U.S.A.

JAMES L. PATTON, Museum of Vertebrate Zoology, University of California, Berkeley, CA 94720, U.S.A.

DAVID M. RAND, Museum of Comparative Zoology, Harvard University, Cambridge, MA 02138, U.S.A.

ROBERT E. RICKLEFS, Department of Biology, University of Pennsylvania, Philadelphia, PA 19104, U.S.A.

JONATHAN ROUGHGARDEN, Department of Biological Sciences, Stanford University, Stanford, CA 94305, U.S.A.

OLIVER A. RYDER, Center for Reproduction of Endangered Species, Zoological Society of San Diego, P.O. Box 551, San Diego, CA 92112, U.S.A.

MARGARET F. SMITH, Museum of Vertebrate Zoology, University of California, Berkeley, CA 94720, U.S.A.

CATHERINE A. TAUBER, Department of Entomology, Comstock Hall, Cornell University, Ithaca, NY 14853, U.S.A.

MAURICE J. TAUBER, Department of Entomology, Comstock Hall, Cornell University, Ithaca, NY 14853, U.S.A.

ALAN R. TEMPLETON, Department of Biology, Washington University, St. Louis, MO 63110, U.S.A.

ROBERT C. VRIJENHOEK, Center for Theoretical and Applied Genetics, New Jersey Agricultural Experiment Station, Rutgers University CN 231, New Brunswick, NJ 08903, U.S.A.

DAVID B. WAKE, Museum of Vertebrate Zoology, University of California, Berkeley, CA 94720, U.S.A.

David Sloan Wilson, Department of Biological Sciences, State University of New York, Binghamton, NY 13901, U.S.A.

Kay P. Yanev, Museum of Vertebrate Zoology, University of California, Berkeley, CA 94720, U.S.A.

Preface

[There] is the tendency in organic beings descended from the same stock to diverge in character as they become modified. That they have diverged greatly is obvious from the manner in which species of all kinds can be classed under genera, genera under families, families under sub-orders and so forth; and I can remember the very spot in the road, whilst in my carriage, when to my joy the solution occurred to me; and this was long after I had come to Down. The solution, as I believe, is that the modified offspring of all dominant and increasing forms tend to become adapted to many and highly diversified places in the economy of nature.

Charles Darwin[1]

. . . the principle of divergence, plays, I believe, an important part in the origin of species. The same spot will support more life if occupied by very diverse forms: we see this in the many generic forms in a square yard of turf. . . , or in the plants and insects on any little uniform islet. . . . Now every single organic being, by propagating rapidly, may be said to be striving its utmost to increase in numbers. So it will be with the offspring of any species after it has broken into varieties, or sub-species, or true species. And it follows, I think, from the foregoing facts, that the varying offspring of each species will try . . . to seize on as many and as diverse places in the economy of nature as possible. Each new variety or species when formed will generally take the place of, and so exterminate its less well-fitted parent. This, I believe, to be the origin of the classification or arrangement of all organic beings at all times. These always *seem* to branch and sub-branch like a tree from a common trunk; the flourishing twigs destroying the less vigorous—the dead and lost branches rudely representing extinct genera and families.

Charles Darwin[2]

Darwin's ideas on the origin of species radically changed our view of the world. The splitting and divergence of lineages is the most important phenomenon in biological science because it is the source of the earth's kaleidoscopically varying forms. The two quotations illustrate Darwin's disposition to view the multiplication of species in the broadest possible context. Although he could see that lineages were splitting, he did not know the underlying mechanisms, and his emphasis was therefore principally on divergence and incompatibility among lineages, ecological specialization, and various other consequences of the splitting of lineages.

[1]Autobiography, in *The Life and Letters of Charles Darwin,* 1959, edited by Francis Darwin.
[2]Letter to Asa Gray, 5 Sept. 1857, in *The Life and Letters of Charles Darwin,* 1959, edited by Francis Darwin.

Speciation is the process of becoming a species. It is a subject of great interest for a variety of reasons. Speciation represents the formation of the units of evolution, which need to be named, systematically arranged, and used to reconstruct the history of life. It also represents the formation and interaction of lineages, and the connection between microevolution and macroevolution. Speciation is common, widespread, and continuous, but in all but a relatively few instances the process does not go to completion.

This book has three purposes. First, it illustrates the inhomogeneity among diverse taxa in their patterns and processes of speciation and differentiation. Second, it considers some of the ecological, paleontological, and systematic consequences of different patterns of speciation. Third, it presents, in a single place, the extreme diversity of concepts and viewpoints on the subject. The field is in a state of rapid evolutionary radiation. Thus, in different chapters the reader will find examples of ideas and interpretations which are mutually contradictory. We have not attempted to provide readers with a (false) consensus, but rather with an overview of the field as a whole. We hope that this will encourage reassessment of both data and theory at all levels, and ultimately contribute to a new synthesis of evolutionary ideas.

We have organized the book into six Parts and a Conclusion. The Parts are somewhat arbitrary in that the authors wrote what they thought would be most interesting and we arranged them into groups afterwards; the quality of this classification reflects the complexity of evolution and of the species problem.

Part One summarizes many of the radically different views about the nature and reality of species. These contrasting views partially explain some of the contentious arguments that have occurred over speciation, and also bring out the different purposes of diverse research programs. For this reason the reader should not expect consistent use of any one species concept throughout the book; this is discussed in the concluding chapter.

Part Two summarizes what we know of the fascinating genetic phenomena that distinguish species, and that occur when species come into contact and hybridize. Some of these phenomena can also generate (biological) species. Part Three also emphasizes genetics, but concentrates on how the genetic consequences of population structure and mating systems can affect or lead to speciation.

Parts Four and Five are concerned with the biogeographical and ecological causes and correlates of speciation and their interactions. These chapters range from strong advocacy of sympatric speciation through assumption of allopatric speciation; from consistency arguments through testing of hypotheses; and from mostly theoretical through strongly empirical approaches. As in the case of species concepts, this is a very contentious area.

Part Six discusses how speciation can affect the generation of diversity and macroevolution, as well as effecting the evolution of ecological communities and ecosystems.

The concluding chapter draws some of the ideas together, tries to put them in a broad perspective, and discusses some of the major unresolved issues in the study of species and speciation.

This book evolved from an international symposium on speciation held in Philadelphia to celebrate the 175th anniversary of the founding of the Academy of Natural Sciences of Philadelphia. A committee consisting of Academy members B. Chernoff, G. Davis, C. Gouldon, and D. Otte chose the topic and, with the help of colleagues throughout the country, selected a group of speakers who could lead us into a broad range of subjects related to speciation and its consequences for all aspects of biology. Not all symposium papers are represented in this book, and a few new ones have been added.

We are pleased to thank the 70 external reviewers who very kindly made extensive criticisms and suggestions about each manuscript, and two very helpful anonymous reviewers of the entire manuscript. We are also pleased that the authors were very good about utilizing virtually all of the external reviewers' comments.

We are grateful to the Academy of Natural Sciences, in particular Keith Thomson (President) and Frank B. Gill (Vice President of Systematics and Evolutionary Biology) for their support and encouragement of the book and symposium. We are especially indebted to the Women's Committee of the Academy for their generous financial support. Without the tremendous efforts of the organizing committee (George M. Davis, chair, B. Chernoff, C. Goulden, J. Silver, B. Jenkins, and J. Wheeler) and the expert technical assistance of Richard Clark, such a successful symposium would not have been possible.

DANIEL OTTE, *Academy of Natural Sciences of Philadelphia*
JOHN A. ENDLER, *University of California, Santa Barbara*

PART ONE

CONCEPTS OF SPECIES

CHAPTER 1

THE MEANING OF SPECIES AND SPECIATION: A Genetic Perspective

Alan R. Templeton

INTRODUCTION

What is a species? This fundamental question must be answered before the process of species formation can be investigated. As any survey of the evolutionary literature will quickly reveal, there are many definitions of species already in existence. These different definitions reflect the diverse types of evolutionary questions and/or organisms with which their authors were primarily concerned. Consequently, a species concept can be evaluated only in terms of a particular goal or purpose. My goal is to understand speciation as an evolutionary genetic process. A fundamental assumption behind this goal is that speciation, regardless of the precise definition of species, is best approached mechanistically by examining the evolutionary forces operating on individuals within populations or subpopulations and tracing their effects upward until they ultimately cause all of the members of that population or subpopulation to acquire phenotypic attributes conferring species status on the group.

This emphasis on the evolutionary genetic mechanisms operating within populations of individuals places speciation fully within the province of population genetics. Accordingly, what is needed is a concept of species that can be directly related to the mechanistic framework of population genetics. To achieve this goal, I will first review three species concepts that have strong supporters in the current literature: the evolutionary species concept, the biological species concept, and the recognition species concept. All of these species concepts treat species as real biological entities and attempt to define species in terms of some fundamental biological property. In this regard, all of these definitions are biological species concepts, although one of them is often referred to as "*the* biological species concept." Since "the biological species concept" defines species in terms of isolating mechanisms, it is more accurately known as the isolation concept (Paterson 1985). Paterson's terminology will be used in the remainder of this chapter.

After reviewing the strengths and weaknesses of these three concepts, I will propose a fourth biological species concept, the cohesion concept, which attempts to utilize the strengths of the other three while avoiding their weakness with respect to the goal of defining species in a way that is compatible with a mechanistic population genetic framework. In this manner, a definition of species can be achieved that illuminates, rather than obscures or misleads, the mechanisms of speciation and their genetic consequences.

THREE BIOLOGICAL SPECIES CONCEPTS

The evolutionary species concept

Under this definition, a species consists of a population or group of populations that shares a common evolutionary fate through time. This definition has the advantage of being applicable to both living and extinct groups and to sexual and asexual organisms. Moreover, it emphasizes the fact that a species unit can be held together not only through gene flow but also through developmental, genetic, and ecological constraints. Finally, this concept is useful because it is close to the operational species definition used by most practicing taxonomists and paleontologists. Decisions as to species status are usually made on the basis of patterns of phenotypic cohesion within a group of organisms versus phenotypic discontinuity between groups. However, when a variety of phenotypes are studied, it is often discovered that the patterns of cohesion/discontinuity vary as a function of the phenotype being measured. One fault of the evolutionary species concept is that it provides little or no guidance as to which traits are the more important ones in defining species.

There are two other principal difficulties with this concept. First, there is the problem of judging what constitutes a "common" evolutionary fate. Obviously, polymorphisms can exist even within local populations, and many species are polytypic. Therefore, "common" does not mean "identical" evolutionary fates, so some judgment must be made as to just how much diversity is allowed within a "common" evolutionary fate. Finally, and most importantly with regard to the goal of this chapter, the evolutionary species concept is not a mechanistic definition. It deals only with the manifestation of cohesion rather than the evolutionary mechanisms responsible for cohesion. Hence it does not provide an adequate framework for integrating population genetic factors into the species concept.

The isolation species concept

The species concept that is dominant in much of the evolutionary literature is popularly known as the biological species concept. Mayr (1963) defined the isolation species concept as "groups of actually or potentially interbreeding natural populations which are reproductively isolated from other such groups." Similarly, Dobzhansky (1970) states that "Species are systems of populations: the gene exchange between these systems is limited or prevented by a reproductive isolating mechanism or perhaps by a combination of several such mechanisms." As White (1978) has emphasized, the isolation concept species "is at the same time a reproductive community, a gene pool, and a genetic system." It is these later two attributes that make this concept of species particularly useful for integrating population genetic considerations into the problem of the origin of species. Population genetics is concerned with the evolutionary forces operating on gene pools and with the types of genetic systems that arise from the operation of these forces. The isolation species concept is therefore potentially useful in analyzing speciation from a population genetic perspective, but it unfortunately has some serious difficulties that must be rectified before this potential can be realized.

The difficulties stem from the fact that this species concept is defined in terms of isolating mechanisms. Table 1 presents a brief classification of the types of isolating barriers, and similar tables can be found in any of the books on speciation by Mayr or Dobzhansky. Under the isolation species concept, these isolating barriers define the boundaries of the reproductive community and gene pool and preserve the integrity of the genetic system of the species.

Paterson (1985) has pointed out that a fundamental difficulty with the isolation concept of species is that it is misleading when thinking about the process of speciation. For example, under the classic allopatric model of speciation, speciation occurs when populations are totally separated from each other by geographical barriers. The intrinsic isolating mechanisms given in Table 1 are obviously irrelevant as isolating barriers during speciation because they cannot function as isolating mechanisms in allopatry. Hence, the evolutionary forces responsible for this allopatric speciation process have nothing to do with "isolation." This is true for other speciation mechanisms as well (Templeton 1981). This is not to say that isolation is not

TABLE 1. Classification of isolating mechanisms.

1. Premating mechanisms that prevent interpopulational crosses
   a. Ecological or habitat isolation: the populations mate in different habitats in the same general region, or use diferent pollinators, etc.
   b. Temporal isolation: the populations mate at different times of the year
   c. Ethological isolation: potential interpopulational mates meet but do not mate
2. Postmating but prezygotic isolation
   a. Mechanical isolation: interpopulational matings occur but no transfer of sperm takes place
   b. Gametic mortality or incompatibility: sperm transfer occurs but the egg is not fertilized
3. Postzygotic isolation
   a. F_1 inviability: hybrid zygotes have a reduced viability
   b. F_1 sterility: hybrid adults have a reduced fertility
   c. Hybrid breakdown: the F_2 or backcross hybrids have reduced viability or fertility
   d. Coevolutionary or cytoplasmic interactions: individuals from a population infected by an endoparasite or with a particular cytoplasmic element are fertile with each other, but fertility and/or viability break down when matings occur between infected and uninfected individuals

a product of the speciation process in some cases, but the product (i.e., isolation) should not be confused with the process (i.e., speciation). The isolation concept has been detrimental to studies of speciation precisely because it has fostered that confusion (Paterson 1985).

The recognition species concept

Paterson (1985) has argued strongly that this confusion can be avoided by looking at the so-called isolating mechanisms from a different perspective. For example, consider the premating isolation mechanisms listed in Table 1. It is commonplace in the evolutionary literature to find statements that complex courtship rituals, mating signals, etc. function as premating isolating barriers that exist to prevent hybridization with other species. The works of Dobzhansky (1970) indicate how dominant this idea was in the thinking of one of the principal architects and proponents of the biological species concept. Yet, as Tinbergen (1953) has pointed out, such premating mechanisms have several functions in addition to isolation: the suppression of escape or aggressive behavior in a courted animal, the synchronization of mating activities, the persuasion of a potential mate to continue courtship, the coordination in time and space of the pattern of mating, the orientation of the potential mates for copulation, and, finally, fertilization itself. The importance of these other functions of premating behavior is illustrated by the work of Crews (1983) on pseudomale courtship and copulatory behavior in the all-female parthenogenetic lizard, *Cnemidophorus uniparens.* In these lizards, insemination and premating isolation are totally irrelevant since reproduction is strictly parthenogenetic. Yet females show elaborate courting behaviors that mimic male courtship in closely related species. These behaviors serve as a neuroendocrine primer that coordinates reproductive events. Obviously, mating behavior in these lizards facilitates reproduction, but isolation is irrelevant.

The critical question then becomes, which of these many functions (or which combination) is important in the process of speciation? Paterson (1985) has argued that isolation is an irrelevant function in the process of speciation. Consequently, to examine the reason why a premating "isolating" barrier arose, it is necessary to focus attention on the other functions of these premating mechanisms and to examine the evolutionary forces operating on these functions (Paterson 1985). In this regard, all the other functions of these premating behaviors can be thought of as facilitating reproduction, not hindering it as in the isolation function. The isolation function can indeed arise as a by-product of the evolution of the other functions, but in general it is not an active part of the process of speciation.

Consequently, isolating mechanisms are a misleading way of thinking about the process of speciation. Although all of the mechanisms listed in Table 1 are defined in terms of preventing reproduction between populations, they can also be thought of in an intraspecific fashion as facilitating reproduction within populations. In general, it is this positive inverse of the functions given in Table 1 that plays the major role in speciation. Paterson (1985) has focused upon the positive function of these mechanisms in facilitating reproduction among members of a certain population. Accordingly, Paterson accepts the premise, shared by the isolation concept, that a species is a field for gene recombination. Unlike the isolation concept, which defines the limits of this field in a negative sense through isolating mechanisms, Paterson defines the limits of this field in a positive sense through fertilization mechanisms, that is, adaptations that assist the processes of meiosis and fertilization. Species are defined as the most inclusive population of individual biparental organisms which share a common fertilization system.

In a sense, the isolation and recognition concepts of species are two sides of the same coin. Flipping the coin is worthwhile because the recognition concept yields a clearer vision of evolutionary process versus pattern, whereas the isolation concept is actively misleading. Hence, given the goal of defining species in such a manner that it facilitates the study of speciation as an evolutionary process, the recognition concept is clearly superior to the isolation concept.

Paterson (1985) has burdened the recognition concept with several restrictions that do not necessarily follow from his primary definition. The most serious of these is his exclusive use of fertilization mechanisms to define a species. Obviously, a field of genetic recombination requires more than fertilization; it requires a complete life cycle in which the products of fertilization are viable and fertile. Moreover, the so called "fertilization" mechanisms of Paterson have other evolutionary functions that he ignores, as is well illustrated by the courtship behavior of the parthenogenetic lizards previously discussed. Hence, just as Paterson criticized isolation mechanisms because they may evolve for reasons other than isolation, his "fertilization" mechanisms may likewise evolve for reasons other than fertilization.

Other minor criticisms of Paterson's concept can be made (Templeton 1987), but I want to concentrate on two serious and fundamental difficulties that are shared by both the isolation and recognition concepts. As with many other problems in the biological world, these problems are caused by sex—either too little or too much.

SEXUAL HANGUPS OF THE ISOLATION AND RECOGNITION CONCEPTS

Too little sex

Both the isolation and recognition concepts of species are applicable only to sexually reproducing organisms (Vrba 1985). Accordingly, large portions of the organic world are outside the logical domain of these species definitions. This is a serious difficulty to people who work with parthenogenetic or asexual organisms.

One particular troublesome aspect of excluding nonsexual species is that most parthenogenetic "species" display the same patterns of phenotypic cohesion within and discontinuity between as do sexual species. For example, Holman (1987) examined the recognizability of sexual and asexual species of rotifers. Contrary to the predictions made by the isolation concept, he discovered that species in the asexual taxa are actually more consistently recognized than those from the sexual taxa. Thus, he concluded that for asexual rotifers "species are real and can be maintained by nonreproductive factors." As this example illustrates, the asexual world is for the most part just as well (or even better) subdivided into easily defined biological taxa as is the sexual world. This biological reality should not be ignored.

Ignoring nonsexual taxa is a major failure of the isolation and recognition concepts, but this failure is actually more extensive than many people realize. For example, the evolutionary genetics of self-mating populations is simply a special case of automictic parthenogenetic populations (e.g., Templeton 1974a). Hence, self-mating sexual species are also outside the logical domain of the isolation and recognition concepts. But the problem does not stop with self-mating sexual species. For example, many species of wasps have mandatory sib mating (Karlin and Lessard 1986). Such a system of mating, as well as any other closed system of mating, will display evolutionary dynamics that can be regarded as a special case of automixis, just as self-mating can. Hence, all sexual taxa with a closed system of mating are outside the logical domain of the isolation and recognition concepts.

The problem does not stop here, however. Models for analyzing multilocus selection in automictic and self-mating populations were very successfully applied to a barley population that was 99.43% self-mating (Templeton 1974b). The reason for this success is straightforward: with this much selfing, the evolutionary dynamics of the population closely approximate that of a 100% selfing population. When outbreeding is at such a low level, its primary role is to introduce genetic variability into the population. Once introduced, the evolutionary fate of that variation is more like that of a selfing population than that of an outcrossing population. Moreover, the genetic impact of the occasional outbreeding is further reduced by isolation by distance, which causes most outbreeding to be between nearly genetically identical individuals. Consequently, from a population genetic perspective, this barley population could not be regarded in any meaningful way as a "field for genetic recombination," and accordingly it lies outside the logical domain of both the isolation and recognition concepts.

The problem of isolation by distance previously mentioned creates a further restriction on the logical domain of the isolation and recognition concepts. An outcrossing population characterized by very limited gene flow and small local effective sizes has much the same genetic consequences and evolutionary dynamics as a predominantly selfing population. Ehrlich and Raven (1969) were among the first to point out in strong terms that many animal and plant species cannot be regarded as fields of genetic recombination in any meaningful sense with respect to basic evolutionary mechanisms, and therefore are also outside the logical domain of the isolation and recognition concepts.

The barley example leads to an interesting question. If a 99.47% selfing population is outside the logical domain of the isolation and recognition concepts, what about a 99% selfing population or a 95% selfing population? Ehrlich and Raven's (1969) work leads to a similar set of questions. At what point is isolation by distance and population subdivision sufficiently weak to bring a taxa into the logical domain of the isolation and recognition concepts? Although this is not an easy question to answer, the problem of genetically closed taxa is usually dismissed in a sentence or two, with sexual and genetically closed taxa being treated as distinct categorical types (e.g., Mayr 1970; Vrba 1985). However, from the viewpoint of evolutionary mechanisms (and, hence, from the viewpoint of speciation as an evolutionary process), there is a continuum from panmictic evolutionary dynamics to genetically closed evolutionary dynamics. Consequently, the logical domain of the isolation and recognition concepts is not at all clear or well defined. The only thing that is certain is that this domain is much more restrictive and limited than is generally perceived.

Too much sex

As discussed, genetically closed reproductive systems cause serious difficulties for the isolation and recognition concepts, but so do genetically open systems. For example, Grant (1957), one of the stronger proponents among botanists of the isolation concept, concluded that less than 50% of the outcrossing species in 11 genera of Californian plants were well delimited by isolation from other species. Again and again in plants, taxonomists have defined species that exist in larger units known as syngameons that are characterized by natural hybridization and limited gene exchange. Grant (1981) defines the syngameon as "the most inclusive unit of interbreeding in a hybridizing species group." The frequent occurrence of syngameons in plants creates serious difficulties for both the isolation and recognition concepts because the field of genetic recombination is obviously broader than the taxonomic species and the groups that are behaving as evolutionarily independent entities. One solution is simply to deny the species status of the members of the syngameon. For example, Grant (1981) refers to the members of a syngameon as "semispecies." Under the recognition concept, the syngameon itself would be the species, since Grant's definition of syngameon is virtually identical to Paterson's (1985) definition of species. However, botanists have not made these taxonomic decisions arbitrarily. The species within a syngameon are often real units in terms of morphology, ecology, genetics, and evolution. For example, the fossil record indicates that balsam poplars and cottonwoods (both from the genus *Populus*) have been distinct for at least 12 million years and have generated hybrids throughout this period (Eckenwalder 1984). Even though the hybrids are widespread, fertile, and ancient, these tree species have and are maintaining genetic, phenotypic, and ecological cohesion within and distinction between and have maintained themselves as distinct evolutionary lineages for at least 12 million years (Eckenwalder 1984). Hence, cottonwoods and poplars are real biological units that should not be ignored.

It is commonplace for zoologists to acknowledge that the isolation concept runs into serious difficulties when it is applied to outbreeding, higher plants, but then to argue that the isolation concept works reasonably well for sexually reproducing, multicellular animals. However, this view is no longer tenable with the increased resolution that recombinant DNA techniques provide. For example, in mammals, studies are being carried out in my laboratory on baboons, wild cattle, canids, and gophers and cotton rats, examples, respectively, of primates, ungulates, carnivores, and rodents—the four major mammalian groups. In every case, there is evidence for naturally occurring interspecific hybridization (Baker et al. 1989; Davis et al. 1988; unpublished data). In spite of hybridization, many of the taxonomic units within these groups represent real biological units in a morphological, ecological, genetic, and evolutionary sense. For example, wolves and coyotes can and do hybridize. Yet, they are morphologically quite distinct from each other, have extremely different behaviors in terms of social structure and hunting, and represent distinct evolutionary lineages with diagnostic genetic differences (Figure 1). Moreover, the fossil record indicates that they have evolved as distinct and continuous lineages for at least 0.5 million years (Hall 1978) and perhaps for as much as 2 million years (Nowak 1978). Although these taxa do not satisfy the criterion of the isolation species concept, Hall (1978) argues that these are biologically real groups and that species status is clearly appropriate.

Animal syngameons are by no means limited to mammals. *Drosophila heteroneura* and *D. silvestris* are two Hawaiian *Drosophila* species on which we have worked. Although they are phylogenetically very close and broadly sympatric on the Island of Hawaii (Carson 1978), they are morphologically extremely distinct, with the most dramatic difference

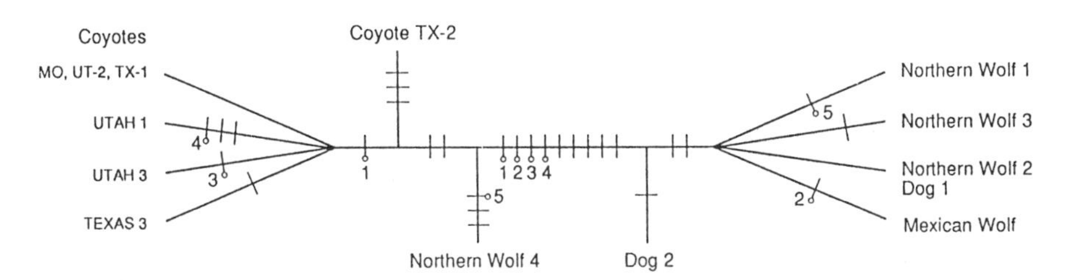

FIGURE 1. Unrooted cladogram of North American coyotes, wolves, and dogs as constructed by maximum parsimony. The cladogram is based upon restriction endonuclease site mapping of mitochondrial DNA. Each line that crosses a segment of the cladogram indicates a single evolutionary change in the map within that evolutionary segment. Five sites were inferred to have changed twice, and their two inferred positions within the cladogram are indicated by the numbered crosslines.

being that *silvestris* has a round head and *heteroneura* a hammer-shaped head (Val 1977). They can be hybridized in the laboratory, and the hybrids and subsequent F_2 and backcrosses are completely fertile and viable (Val 1977; Templeton 1977; Ahearn and Templeton 1989). Because the morphology of hybrids is known from these laboratory studies, Kaneshiro and Val (1977) were able to discover that interspecific hybridization occurs in nature. Our molecular studies (DeSalle and Templeton 1987) confirm that hybrids are indeed formed in nature, and, moreover, that these hybrids can and do backcross to such an extent that a *heteroneura* mitochondrial haplotype can occasionally be overlaid on a normal-looking *silvestris* morphology. In spite of this natural hybridization, the species can and do maintain their very distinct, genetically based morphologies (Templeton 1977; Val 1977) and have distinct nuclear DNA phylogenies (Hunt and Carson 1983; Hunt et al. 1984) in spite of the limited introgression observed with mitochondrial DNA (DeSalle et al. 1986). Hence, both morphology and molecules define these taxa as real, evolutionarily distinct lineages.

As these and other studies illustrate, animal taxa frequently display natural hybridization that yields fertile and viable hybrids. These taxa have often been recognized as species because of their distinct morphologies and ecologies and because modern molecular studies have revealed that they are behaving as independent evolutionary lineages, at least with respect to their nuclear genomes. In other words, many animal species are members of syngameons, just as plants are. Hence, the problem of syngameons is a widespread one for the isolation and recognition concepts.

THE COHESION SPECIES CONCEPT

Another biological definition of species is now possible, which I call the cohesion concept of species. The cohesion concept species is the most inclusive population of individuals having the potential for phenotypic cohesion through intrinsic cohesion mechanisms (Table 2). I will now elaborate on the meaning of this species concept, showing how it borrows parts of the evolutionary, isolation, and recognition concepts, while it avoids their serious defects.

As with the evolutionary species concept, the cohesion species concept defines species in terms of genetic and phenotypic cohesion. As a consequence, the cohesion concept shares with the evolutionary concept the strengths of being applicable to taxa reproducing asexually (or by some other closed or nearly closed breeding system) and to taxa belonging in syngameons. Unlike the evolutionary species concept, the cohesion concept defines species in terms of the mechanisms yielding cohesion rather than the manifestation of cohesion over evolutionary time. This is a

TABLE 2. Classification of cohesion mechanisms.

- I. Genetic exchangeability: the factors that define the limits of spread of new genetic variants through *gene flow*
  - A. Mechanisms promoting genetic identity through *gene flow*
    - 1. Fertilization system: the organisms are capable of exchanging gametes leading to successful fertilization
    - 2. Developmental system: the products of fertilization are capable of giving rise to viable and fertile adults
  - B. Isolating mechanisms: genetic identity is preserved by the lack of *gene flow* with other groups
- II. Demographic exchangeability: the factors that define the fundamental niche and the limits of spread of new genetic variants through *genetic drift* and *natural selection*
  - A. Replaceability: *genetic drift* (descent from a common ancestor) promotes genetic identity
  - B. Displaceability
    - 1. Selective fixation: *natural selection* promotes genetic identity by favoring the fixation of a genetic variant
    - 2. Adaptive transitions: *natural selection* favors adaptations that directly alter demographic exchangeability. The transition is constrained by:
      - a. Mutational constraints on the origin of heritable phenotypic variation
      - b. Constraints on the fate of heritable variation
        - i. Ecological constraints
        - ii. Developmental constraints
        - iii. Historical constraints
        - iv. Population genetic constraints

mechanistic focus similar to that taken by the isolation concept, although in this case the focus is on cohesion mechanisms rather than isolation mechanisms. By defining a species in terms of cohesion mechanisms, the cohesion concept can easily be related to a mechanistic population genetic framework and can provide guidance in understanding speciation as an evolutionary process. In particular, speciation is now regarded as the evolution of cohesion mechanisms (as opposed to isolation mechanisms). This also means that the cohesion concept focuses primarily on living taxa rather than fossil taxa.

As pointed out by Paterson (1985), it is useful to define the mechanisms underlying species status in such a way that the definitions reflect the most likely evolutionary function of the mechanisms during the process of speciation. Accordingly, cohesion mechanisms will be defined to reflect their most likely evolutionary function. The basic task is to identify those cohesion mechanisms that help maintain a group as an evolutionary lineage. The very essence of an evolutionary lineage from a population genetic perspective is that new genetic variants can arise in it, spread, and replace old variants. These events occur through standard microevolutionary forces such as gene flow, genetic drift, and/or natural selection. The fact that the genetic variants present in an evolutionary lineage can be traced back to a common ancestor also means that the individuals that comprise this lineage must show a high degree of genetic relatedness. The cohesion mechanisms that define species status are therefore those that promote genetic relatedness and that determine the populational boundaries for the actions of microevolutionary forces.

The isolation and recognition concepts are exclusively concerned with genetic relatedness promoted through the exchange of genes via sexual reproduction. These definitions have elevated a single microevolutionary force—gene flow—into the conclusive and exclusive criterion for species status. There is no doubt that gene flow is a major microevolutionary force, and hence the factors that define the limits of spread of new genetic variants through gene flow are valid criteria for species status. Accordingly, genetic exchangeability is included in Table 2 as a major class of cohesion mechanisms. Genetic exchangeability simply refers to the ability to exchange genes via sexual reproduction. This implies a shared fertilization system in the sense of Paterson (1985). Effective exchange of genes also demands that the products of fertilization be both potentially viable and fertile (Templeton 1987). As shown in Table 2, the role of gene flow in determining species status can be defined in either a positive (I.A in Table 2) or a negative (I.B in Table 2) sense. As stated earlier, the positive sense generally provides a more accurate view of the evolutionary processes involved in speciation.

Gene flow is not the only microevolutionary force that defines the boundaries of an evolutionary lineage. Indeed, genetic drift and natural selection play a far more potent and universal role because these two classes of microevolutionary forces are applicable to all organisms, not just outcrossing sexual species. An important question is, therefore, what factors define the limits of spread of new genetic variants through genetic drift and natural selection? Since these forces can operate in asexual populations, it is obvious that the factors that limit the field of action of drift and selection are not necessarily the same as those limiting the actions of gene flow. As seen, gene flow requires genetic exchangeability, that is, the ability to exchange genes during sexual reproduction. For genetic drift and natural selection to operate, another type of exchangeability is required: demographic exchangeability (Table 2).

From an ecological perspective, members of a demographically exchangeable population share the same fundamental niche (Hutchinson 1965), although they need not be identical in their abilities to exploit that niche. The fundamental niche is defined by the intrinsic (i.e., genetic) tolerances of the individuals to various environmental factors that determine the range of environments in which the individuals are potentially capable of surviving and reproducing. The realized niche (Hutchinson, 1965) refers to that subset of the fundamental niche that is actually occupied by a species. The realized niche is usually a proper subset of the fundamental niche because of the lack of opportunity to occupy certain portions of the fundamental niche (e.g., the environmental ranges might be within the tolerance limits in some locality, but geographical barriers prevent the colonization of that locality) or because of interactions with other species that prevent the exploitation of the entire range of ecological tolerance. Hence, the realized niche is influenced by many extrinsic factors, but demographic exchangeability depends only on the intrinsic ecological tolerances.

To the extent that individuals share the same fundamental niche, they are interchangeable with one another with respect to the factors that control and regulate population growth and other demographic attributes. It is demographic exchangeability that is used to define populations in most models of population and community ecology. Indeed, most models from these ecological disciplines do not even specify the mode of reproduction, so genetic exchangeability is not used to define a population.

From a genetic perspective, the chances of a neutral or selectively favorable mutation going to fixation in a demographically exchangeable population are nonzero regardless of the particular individual in which the mutation occurred. In other words, every individual in a demographically exchangeable population is a potential common ancestor to the entire population at some point in the future. Ancestor-descendant relationships can be defined just as readily in asexual populations as in sexual populations. Hence, demographic exchangeability does not require genetic exchangeability and is a distinct biological attribute at the population level.

Just as genetic exchangeability can vary in strength, so can demographic exchangeability. From an ecological perspective, complete demographic exchangeability occurs when all individuals in a population display exactly the same ranges and abilities of tolerance to all relevant ecological variables. Demographic exchangeability is weakened as individuals begin to differ in their tolerance ranges or abilities. From a genetic perspective, a population is completely demographically exchangeable if the probability of a neutral or selectively favorable mutation going to fixation is exactly the same regardless of the individual in which it occurs. A weakly demographically exchangeable population would consist of members who display very different (but still nonzero) fixation probabilities.

Demographic exchangeability allows us to readily incorporate microevolutionary forces other than gene flow as being important in defining an evolutionary lineage. One such microevolutionary force is genetic drift, which promotes genetic cohesion through ancestor-descendant relationships (i.e., the concept of identity-by-descent in population genetics). For the special case of neutral alleles (alleles that have no selective importance), the rate at which genetic drift promotes identity-by-descent depends only on the neutral mutation rate and is therefore equally important in both large and small populations. Interestingly, this prediction about the neutral rate of evolution and the other basic predictions of the standard neutral theory do not depend upon the assumption of sexual reproduction—these predictions are equally applicable to asexual organisms. Although the neutral theory does not require genetic exchangeability, demographic exchangeability is a critical and necessary assumption (e.g., Rothman and Templeton 1980). Making *only* the assumption of demographic exchangeability, it is inevitable that at some point in the future all the alleles will be descended from one allele that presently exists. It makes no difference for the operation of genetic drift whether alleles or the individuals carrying the alleles are exchangeable. Hence, demographic exchangeability must be regarded as a major cohesion mechanism because it defines the populational limits for the action of genetic drift. This aspect of demographic exchangeability is called "replaceability" in Table 2.

Natural selection is another powerful force that can help define an evolutionary lineage. The concept of natural selection does not require genetic exchangeability because selection models are as easily formulated for genetically closed populations as for genetically open ones (e.g., Templeton, 1974a, 1974b). As pointed out by Darwin, natural selection requires two demographic conditions: (1) that organisms can produce more offspring than are needed for strict replacement, and (2) that unlimited population growth cannot be sustained indefinitely. When these demographic conditions are coupled with heritable variation in traits influencing survival and reproduction, the logical consequence is that the offspring of some individuals will displace those of others within the population. This aspect of demographic exchangeability is called "displaceability" in Table 2.

Natural selection promotes cohesion both through favoring genetic relatedness and through affecting the limits of demographic exchangeability itself. Whenever natural selection causes a new, favorable mutation to go to fixation, genetic relatedness at that locus is obviously a direct consequence. Moreover, as this mutation goes to fixation, that subset of the species' genetic variation that remains linked to the new mutation likewise goes to fixation. This is known as the hitchhiking effect, and it is important to note that as genetic exchangeability declines in importance, hitchhiking effects increase in importance, for the simple reason that genetic recombination is less effective in breaking down the initial linkage states that were created at the moment of mutation. Hence, selective fixation of one allele by another is an extremely powerful cohesion mechanism in populations with genetically closed systems of reproduction (Levin 1981). As an example, Figure 2 shows the results of selection in a parthenogenetic strain of *D. mercatorum* (Annest and Templeton 1978). As can be seen from that figure, the population rapidly converged to a single genotype for all the marker loci being examined. The dynamics of this convergence indicated that very strong selective forces were operating (Annest and Templeton 1978). Other replicates of this same population, all subject to genetic recombination during the first parthenogenetic generation, selectively converged to other genotypic states at the marker loci, thereby indicating that the marker loci were not being selected directly. Thus, selection at perhaps a few loci promoted genetic identity at all loci in these parthenogenetic populations.

The extent of demographic exchangeability is intimately intertwined with the ecological niche requirements of the organisms and the habitats that are available for satisfying those requirements. It is these very same

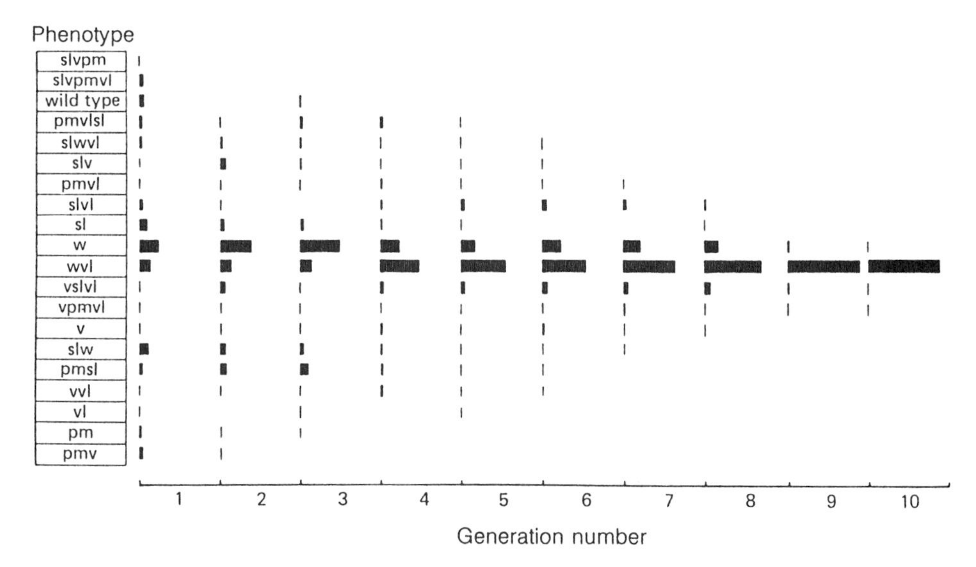

FIGURE 2. Clonal selection in a parthenogenetic population of *Drosophila mercatorum*. The initial generation was heterozygous for several visible markers on all the major chromosomes (sl, w, v, pm, and vl). Parthenogenetic reproduction during this initial generation creates a large number of genotypes since meiotic recombination and assortment occur in these automictic strains. After the first generation genetic recombination is irrelevent since virtually all flies are totally homozygous. From Annest and Templeton (1978).

ecological requirements and available habitats that provide many of the selective forces that drive the process of adaptation. Hence, the process of adaptation by natural selection can directly alter the traits that determine the extent of demographic exchangeability. Adaptive transitions therefore play a direct role in defining demographically exchangeable groups of organisms.

The importance of adaptive transitions in defining demographic exchangeability opens up a whole new set of cohesion mechanisms that constrain the possible courses of adaptive transitions, as shown in Table 2 (II.B.2). The first is mutational constraints that limit the types of phenotypic variants that are likely to be produced. Such constraints make it difficult to alter some aspects of the existing genetic/developmental system, but facilitate evolutionary change along other lines. For example, the genus *Drosophila* consists of some flies that have pigmented spots, clouds, or patterns on their wings, such as the Hawaiian "picture-wings," and others that have clear wings, such as *D. melanogaster.* Yet, as Basden (1984) points out, no picture-winged *Drosophila* has ever produced a clear-winged mutant, nor has a clear-winged species produced a picture-winged mutant. This negative result is of biological significance for *D. melanogaster,* for probably no other higher eukaryote has been examined more thoroughly for visible mutations. Thus, Basden concluded that at the species level there is a block to certain types of mutations. This is simply another way of stating that constraints exist that make certain types of mutations impossible or highly improbable.

Given that phenotypic variation has been produced by the mutational process, there are constraints that influence the selective fate of that variation (Table 2, II.B.2.b). First, there are ecological constraints that select against certain phenotypes and that restrict the range of environmental variability experienced by the species. Moreover, for an adaptive transition to persist, a niche must be available for the organisms with the new adaptation. Ecological constraints are undoubtedly one of the more important cohesion mechanisms maintaining species within syngameons, as is demonstrated by what happens within syngameons when the constraints are altered. For example, under most environmental conditions, red and black oaks live together in the same woods and cross-pollinate. Nevertheless, they remain two distinct, cohesive populations because the F_1 hybrid acorns do not germinate well under the dark, cool conditions of a mature forest. When a forest is partially cleared and thinned (mostly by humans), the black oak and red oak acorns germinate poorly, whereas the hybrid acorns do very well. As a result, many current woods consist of a continuous intergradation between black and red oaks. Hence, the normal cohesion of red and black oak populations is lost when the ecological constraints are altered.

Ecological constraints are also important in asexual taxa because these constraints often determine the populational limits of selective fixation, which, as previously mentioned, is a major cohesion mechanism in taxa with closed systems of reproduction. Moreover, the work of Roughgarden (1972) predicts that asexual populations can evolve more sharply delimited niche widths than can otherwise equivalent sexual populations. This property may help explain the greater recognizability of asexual species over sexual species (Holman 1987).

Developmental constraints constitute the second class of cohesion mechanisms related to the fate of heritable variation in adaptive transitions. When there is strong selection on one trait, pleiotropy (a form of developmental constraint) ensures that other traits will evolve as well. Hence, pleiotropy can facilitate evolutionary changes that would otherwise not occur. Although many people have emphasized the nonadaptive, even maladaptive nature of these pleiotropic-induced changes, Wagner (1988) has shown that pleiotropy is essential for the evolution of complex adaptive traits. He examined a model in which fitness depends on the simultaneous states of several traits and then contrasted models of adaptive evolution in which all traits are genetically independent (no pleiotropy or developmental constraints) with a model in which developmental constraints were imposed. He found that, when there are no developmental constraints, the rate of adaptive evolution decreases dramatically as the number of characters involved in functional integration increases. Hence, developmental constraints and pleiotropy seem to be necessary for the evolution of functionally integrated phenotypes.

Further adaptive evolution can be facilated even when the primary adaptation induces pleiotropic effects that are maladaptive. This phenomenon can be illustrated by malarial adaptations in humans (Templeton 1982). The primary malarial adaptations (such as sickle cell) often induce highly deleterious pleiotropic effects (such as anemia), which, in turn, generate secondary adaptive processes on modifiers to diminish or eliminate the deleterious effects (such as persistence of fetal hemoglobin to suppress anemia). In this manner a single adaptive transition can trigger a cascade of secondary transitions, which cumulatively can have a large impact on demographic exchangeability.

Another cohesion mechanism that constrains the selective fate of phenotypic variability is historical constraint. Evolution is an historical process, and, consequently, the evolutionary potential of a lineage is shaped by its past adaptive transitions. For example, a prerequisite for the evolution of aposematic coloration in insects with gregarious larvae is the evolution of unpalatability. Without the prior existence of distastefulness, there is no selective force for warning coloration within the broods (Templeton 1979). Hence, the adaptation of distastefulness is an historical constraint on the evolution of aposematic coloration and gregarious larvae. This prediction was recently tested by Sillen-Tullberg (1988), who showed through a phylogenetic analysis that in every case in which resolution was possible, distastefulness evolved prior to the evolution of gregarious, aposematic larvae. As shown by this example, one adaptation can make a second one more likely, thus reinforcing the cohesion of the lineage that shares these adaptive transitions.

Population genetic constraints also limit the selective fate of new phenotypic variability. These constraints arise from the interaction of population structure (system of mating, population size, population subdivision) with the genetic architecture underlying selected traits (the genotype–phenotype relationship, number of loci, linkage relationships, etc.). For example, in 1924 Haldane showed that selectively favorable dominant genes are much more likely to be fixed than selectively favorable recessive genes in randomly mating populations. However, this constraint disappears if the system of mating is changed from random mating to inbreeding (Templeton 1982). Thus, an alteration of system of mating can alter the phenotypic and genetic cohesion of a population by making whole new classes of genetic variability responsive to natural selection.

ADVANTAGES OF THE COHESION CONCEPT OF SPECIES

The cohesion concept of species defines a species as an evolutionary lineage through the mechanisms that limit the populational boundaries for the action of such basic microevolutionary forces as gene flow, natural selection, and genetic drift. The genetic essence of an evolutionary lineage is that a new mutation can go to fixation within it; and genetic drift and natural selection as well as gene flow are powerful forces that can cause such fixations. Hence, there is no good rationale for why gene flow should be the only microevolutionary mechanism that is used to define an evolutionary lineage; yet this is precisely what the isolation and recognition concepts do.

Under the cohesion concept, many genetically based cohesion mechanisms (Table 2) can play a role in defining a species. Not all species will be maintained by the same cohesion mechanism or mixture of cohesion mechanisms, just as proponents of the isolation concept acknowledge that not all isolating mechanisms are equally important in every case. By adjusting the mixture of cohesion mechanisms, it is possible to take into account under a single species concept asexual taxa, the taxa that fall within the domain of the isolation and recognition concepts, and the members of syngameons.

Figure 3 gives a simplified graphic portrayal of the relative importance of genetic versus demographic exchangeability in defining species over

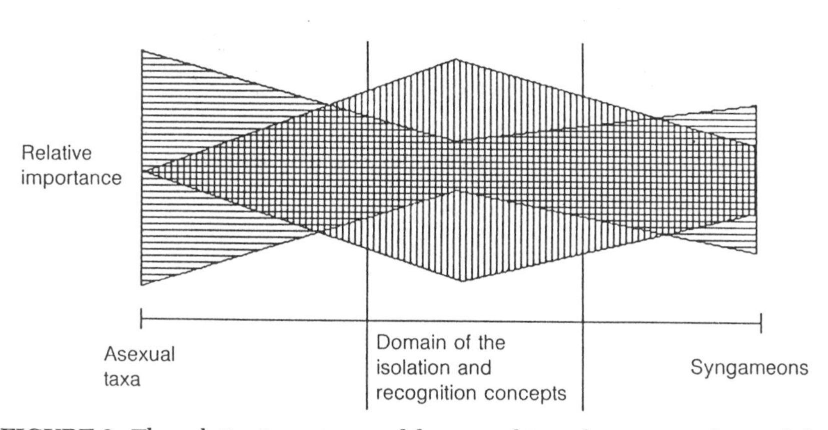

FIGURE 3. The relative importance of demographic and genetic exchangeability over the reproductive continuum. The areas marked by vertical lines indicate the importance of genetic exchangeability, with the width of that area at any particular point in the reproductive continuum indicating its importance in defining species. Similarly, the areas marked by horizontal lines are used to indicate the importance of demographic exchangeability. The diagram gives only the general trend in relative importance. Because the strength of both genetic and demographic exchangeability can vary continuously, the relative importance can be altered from that shown in the diagram at virtually any point in the reproductive continuum, except for asexual taxa.

the entire reproductive continuum. For asexual taxa, genetic exchangeability has no relevance, and species status is determined exclusively by demographic exchangeability. As the reproductive system becomes more open, not only does genetic exchangeability become a factor, but demographic exchangeability is diminished in importance because selective replacement becomes increasingly less effective in promoting genetic relatedness. In the middle range, genetic exchangeability dominates because the factors determining the limits of gene flow also limit the actions of drift and selection in outbreeding Mendelian populations. In this domain, the recognition and isolation concepts are valid, and hence, both are special cases of the more general cohesion concept of species. Finally, in moving toward the syngameon end of the continuum, genetic exchangeability decreases in importance relative to the ecological constraints that define demographic exchangeability.

This continuity of applicability of the cohesion concept is consistent with the biological reality that there is a continuum in the degree of genetic openness of reproductive systems found in the organic world. This is a tremendous advantage over the recognition or isolation concepts that are applicable only to the middle range of this reproductive continuum and that deal with the remainder of the range either by denying the existence of species outside this range (e.g., Vrba 1985) or by using qualitatively different species concepts (e.g., Mayr 1970) to impose an artificial discreteness on the reproductive continuum.

Another strength of the cohesion concept is that it clarifies what is meant by a "good species" and the nature of the difficulties that can occur with the isolation and recognition concepts. "Good species" are generally regarded as geographically cohesive taxa that can coexist for long periods of time without any breakdown in genetic integrity. The fact that there is no breakdown in genetic integrity in spite of sympatry implies the lack of genetic exchangeability between the taxa. However, the condition of prolonged coexistence also implies that they have distinct ecological niches (Mayr 1970). Hence, "good species" are those that are well defined both by genetic and demographic exchangeability. (Similarly, members of a "good" higher taxa lack both genetic and demographic exchangeability.) Given this definition of a "good species," there are two principal ways to deviate from this ideal. One occurs when the population boundaries defined by genetic exchangeability are more narrow than those defined by demographic exchangeability. This is precisely the problem of asexual taxa previously discussed. The other mode of deviation occurs when the boundaries defined by genetic exchangeability are broader than those defined by demographic exchangeability—in other words, the problem posed by the syngameon. Hence, these two seemingly very disparate problems with the isolation and recognition concepts actually have a common underlying cause: the boundaries defined by demographic exchangeability are different than those defined by genetic exchangeability.

Speciation is generally a process, not an event (Templeton 1981). While the process is still occurring, the tendency is to have "bad" species. Although the taxa associated with these incomplete speciation processes are the bane of the taxonomist, they provide the most insight into speciation. By providing a precise definition of "bad species" (the conflict between genetic and demographic exchangeability), the cohesion concept is a useful tool for gaining insight into the process of speciation. "Bad species" need no longer be regarded as a diverse set of special cases; rather, the cohesion concept provides the means for seeing the patterns found in these troublesome taxa. For example, Levene (1953) long ago postulated a model in which different genotypes display different fitnesses in niches that are demographically independent. However, in this model, there is complete genetic exchangeability and there is still sufficient demographic exchangeability among all the genotypes within the various realized niches (through within-niche selective displacement) that this is clearly a model of intraspecific polymorphism. The situation modeled by Levene (1953) bears some resemblance to the syngameon examples discussed earlier in that a conflict arises between genetic and demographic exchangeability (through adaptation to different realized ecological niches that alter the intrinsic tolerances that define the fundamental niche). Hence, there can be a continuum in relative strength between these conflicting species boundary criteria. Interestingly, there has been an implicit acknowledgment of this tension in the speciation literature. Most models of sympatric speciation start with a Levene-type model, with the model of Wilson (this volume) being an example (also see Maynard Smith 1966). Although these models differ greatly in detail, the cohesion concept clarifies the evolutionary significance of this entire class of speciation models: it is the evolution of demographic *non*exchangeability that triggers the speciation process in these cases, and speciation proceeds through shifts in the relative importance of demographic and genetic exchangeability within and between populations adapting to different realized niches. Thus, a seemingly diverse set of speciation models all have a common theme, and the cohesion concept allows that theme to be clearly discerned.

Note also that natural selection is the driving force of speciation in all of these sympatric speciation models, with the effects on gene flow being secondary. Because the cohesion concept explicitly incorporates a broad set of microevolutionary forces as being important in speciation, we can deal directly with natural selection as being the primary trigger of speciation in these models rather than having to constantly rephrase the evolutionary significance of natural selection in terms of its secondary effects on gene flow. The cohesion concept therefore facilitates the study of speciation as an evolutionary process by making explicit the role played by a broad array of evolutionary forces that includes, but is not limited to, gene flow.

As illustrated by the Levene-type speciation models, one of the evolutionary forces important in speciation is natural selection. Natural selection is important in defining a species under the cohesion concept in part because of the impact of adaptive transitions on demographic exchangeability. Interestingly, Mayr (1970) argues that most species have distinct ecological niches (that is, they are not demographically exchangeable), and that this ecological distinctiveness is the "keystone of evolution" because it serves as the basis of diversification of the organic world, adaptive radiation, and evolutionary progress. Although Mayr therefore concludes that the "evolutionary significance of species" lies in their ecological distinctiveness, he still argues that adaptive transitions and natural selection generally play no direct role in speciation and contribute to defining a species only through the "incidental by-product" of isolating mechanisms. Mayr does allow for selective pressures to reinforce isolating mechanisms and to accentuate ecological exclusion if sympatry has been established, but he emphasizes that this occurs only after the process of speciation has been basically completed. Hence, under the isolation concept, the factors responsible for the "evolutionary significance of species" play no direct role in defining species. Under the cohesion concept, the evolutionary significance of a species can arise directly out of its defining attributes.

SPECIATION

Now that species has been defined, what is speciation? Speciation is the process by which new genetic systems of cohesion mechanisms evolve within a population. This process can be thought of as being analogous to the process of genetic assimilation of individual phenotypes. Genetic assimilation is a process discussed by Waddington (1957) in light of his work with the fruit fly, *Drosophila melanogaster.* For example, he discovered that by subjecting strains of this fly to a heat shock, many of the flies would express the phenotype of lacking a certain vein on their wings. Initially, this "crossveinless" phenotype appeared to be purely environmental. By artificially selecting those flies expressing the phenotype, Waddington discovered he was selecting for the genetic predisposition to express this phenotype as well. Therefore, over several generations this "environmental" phenotype acquired a genetic basis to such an extent that the phenotype eventually came to be expressed even in the absence of the heat shock. Similarly, a purely environmental alteration in the manifestation of cohesion can lead to evolutionary conditions that favor the assimilation of the new pattern of cohesion into the gene pool. For example, consider the case of allopatric speciation in which an ancestral taxa that was continuously distributed in a region is now, by the erection of some geographical barrier, split in two totally isolated subpopulations. The erection of the geographical barrier potentially alters the manifestation of several cohesion mechanisms. For sexual taxa, genetic relatedness through gene flow has been altered, and for both sexual and asexual taxa, the potential for genetic relatedness through genetic drift and natural selection is altered as soon as the populations become demographically independent due to geographical separation. Moreover, if the geographical barrier is associated with altered environments and/or altered breeding systems, alterations in the constraints on adaptive transitions could be directly induced and a new realized niche may be occupied. However, none of this constitutes speciation until these alterations in the manifestation of genetic and demographic exchangeability are genetically assimilated into the gene pool as new cohesion mechanisms. Thus, speciation is the genetic assimilation of altered patterns of genetic and demographic exchangeability into intrinsic cohesion mechanisms.

This is a simple definition of speciation, but because of the breadth of the cohesion species concept, this definition can be used to study a wide variety of evolutionary processes that contribute to the formation of a new species within a single mechanistic framework. This is an exciting prospect, and one that I hope will result in a deeper application of evolutionary genetics to the problem of the origin of species.

SUMMARY

The "biological species concept" defines species as reproductive communities that are separated from other similar communities by intrinsic isolating barriers. However, there are other "biological" concepts of species, so the classic biological species concept is more accurately described as the "isolation" species concept. The purpose is this chapter was to provide a biological definition of species that follows directly from the evolutionary mechanisms responsible for speciation and their genetic consequences.

The strengths and weaknesses of the evolutionary, isolation, and recognition concepts were reviewed and all three were judged to be inadequate for this purpose. As an alternative, I proposed the cohesion concept that defines a species as the most inclusive group of organisms having the potential for genetic and/or demographic exchangeability. This concept borrows from all three biological species concepts. Unlike the isolation and recognition concepts, it is applicable to the entire continuum of reproductive systems observed in the organic world. Unlike the evolutionary concept, it identifies specific mechanisms that drive the evolutionary process of speciation. The cohesion concept both facilitates the study of speciation as an evolutionary process and is compatible with the genetic consequence of that process.

ACKNOWLEDGMENTS

The ideas in this chapter were greatly influenced by my discussion with the other participants at the symposium and with the people working in my laboratory, and I thank them all for the challenging intellectual stimulation that they have provided. Special thanks go to Allan Larson and John Endler for their helpful comments on an earlier version of this manuscript. This work was supported by NIH Grant R01 GM31571.

LITERATURE CITED

Ahearn, J. N., and A. R. Templeton. 1989. Interspecific hybrids of *Drosophila heteroneura* and *D. silvestris*. I. Courtship success. Evolution 43:347–361.

Annest, L., and A. R. Templeton. 1978. Genetic recombination and clonal selection in *Drosophila mercatorum*. Genetics 89:193–210.

Baker, R. J., S. K. Davis, R. D. Bradley, M. J. Hamilton, and R. A. Van Den Bussche. 1989. Ribosomal DNA, mitochondrial DNA, chromosomal and electrophoretic studies on a contact zone in the pocket gopher, *Geomys*. Evolution, in press.

Basden, E. B. 1984. The species as a block to mutations. Drosophila Inform. Serv. 60:57.

Carson, H. L. 1978. Speciation and sexual selection in Hawaiian *Drosophila*. Pp. 93–107 in: P. F. Brussard (ed.), *Ecological Genetics: The Interface*. Springer-Verlag, New York.

Crews, D. 1983. Alternative reproductive tactics in reptiles. BioScience 33:562–566.

Davis, S. K., B. Read, and J. Balke. 1988. Protein electrophoresis as a management tool: Detection of hybridization between Banteng (*Bos javanicus* d'Alton) and domestic cattle. Zoo Biol. 7:155–164.

DeSalle, R., L. V. Giddings, and A. R. Templeton. 1986. Mitochondrial DNA variability in natural populations of Hawaiian *Drosophila*. I. Methods and levels of variability in *D. silvestris* and *D. heteroneura* populations. Heredity 56:75–85.

DeSalle, R., and A. R. Templeton. 1987. Comments on "The Significance of Asymmetrical Sexual Isolation." Evolution. Biol. 21:21–27.

Dobzhansky, Th. 1970. *Genetics of the Evolutionary Process*. Columbia University Press, New York.

Eckenwalder, J. E. 1984. Natural intersectional hybridization between North American Species of *Populus* (Salicaceae) in sections *Aigeiros* and *Tacamahaca*. III. Paleobotany and evolution. Can. J. Bot. 62:336–342.

Ehrlich, P., and P. Raven. 1969. Differentiation of populations. Science 165:1228–1232.

Grant, V. 1957. The plant species in theory and practice. Pp. 39–80 in: E. Mayr (ed.), *The Species Problem*. American Association for the Advancement of Science, Publication No. 50, Washington, D. C.

Grant, V. 1981. *Plant Speciation*, 2nd ed. Columbia University Press, New York.

Haldane, J. B. S. 1924. A mathematical theory of natural and artificial selection. Part 1. Trans. Cambridge Philos. Soc. 23:19–41.

Hall, R. L. 1978. Variability and speciation in canids and hominids. Pp. 153–177 in: R. L. Hall and H. S. Sharp (eds.), *Wolf and Man: Evolution in Parallel*. Academic Press, New York.

Holman, E. W. 1987. Recognizability of sexual and asexual species of rotifers. System. Zool. 36:381–386.

Hunt, J. A., J. G. Bishop III, and H. L. Carson. 1984. Chromosomal mapping of a middle-repetitive DNA sequence in a cluster of five species of Hawaiian *Drosophila*. Proc. Natl. Acad. Sci. U.S.A. 81:7146–7150.

Hunt, J. A., and H. L. Carson. 1983. Evolutionary relationships of four species of Hawaiian *Drosophila* as measured by DNA reassociation. Genetics 104:353–364.

Hutchinson, G. E. 1965. The niche: An abstractly inhabited hypervolume. Pp. 26–78 in: *The Ecological Theatre and the Evolutionary Play*. Yale University Press, New Haven.

Kaneshiro, K., and F. C. Val. 1977. Natural hybridization between a sympatric pair of Hawaiian *Drosophila*. Am. Natur. 111:897–902.

Karlin, S., and S. Lessard. 1986. *Theoretical Studies on Sex Ratio Evolution*. Princeton University Press, Princeton, N.J.

Levene, H. 1953. Genetic equilibrium when more than one ecological niche is available. Am. Natur. 87:311–313.

Levin, B. R. 1981. Periodic selection, infectious gene exchange and the genetic structure of *E. coli* populations. Genetics 99:1–23.

Maynard Smith, J. 1966. Sympatric speciation. Am. Natur. 100:637–650.

Mayr, E. 1963. *Animal Species and Evolution*. Harvard University Press, Cambridge, MA.

Mayr, E. 1970. *Populations, Species, and Evolution*. Belknap Press, Cambridge, MA.

Nowak, R. M. 1978. Evolution and taxonomy of coyotes and related *Canis*. Pp. 3-16 in: M. Bekoff (ed.), *Coyotes: Biology, Behavior, and Management*. Academic Press, New York.
Paterson, H. E. H. 1985. The recognition concept of species. Pp. 21-29 in: E. S. Vrba (ed.), *Species and Speciation*. Transvaal Museum Monograph No. 4, Pretoria.
Rothman, E. D., and A. R. Templeton. 1980. A class of models of selectively neutral alleles. Theor. Pop. Biol. 18:135-150.
Roughgarden, J. 1972. Evolution of niche width. Am. Natur. 106:683-718.
Sillen-Tullberg, B. 1988. Evolution of gregariousness in aposematic butterfly larvae: A phylogenetic analysis. Evolution 42:293-305.
Templeton, A. R. 1974a. Density dependent selection in parthenogenetic and self-mating populations. Theor. Pop. Biol. 5:229-250.
Templeton, A. R. 1974b. Analysis of selection in populations observed over a sequence of consecutive generations. I. Some one locus models with a single, constant fitness component per genotype. Theor. Appl. Genet. 45:179-191.
Templeton, A. R. 1977. Analysis of head shape differences between two interfertile species of Hawaiian *Drosophila*. Evolution 31:630-642.
Templeton, A. R. 1979. A frequency-dependent model of brood selection. Am. Natur. 114:515-524.
Templeton, A. R. 1981. Mechanisms of speciation—a population genetic approach. Annu. Rev. Ecol. System. 12:23-48.
Templeton, A. R. 1982. Adaptation and the integration of evolutionary forces. Pp. 15-31 in: R. Milkman (ed.), *Perspectives on Evolution*. Sinauer, Sunderland, MA.
Templeton, A. R. 1987. Species and speciation. Evolution 41:233-235.
Tinbergen, N. 1953. *Social Behaviour in Animals*. Methuen, London.
Val, F. C. 1977. Genetic analysis of the morphological differences between two interfertile species of Hawaiian *Drosophila*. Evolution 31:611-629.
Vrba, E. S. 1985. Introductory comments on species and speciation. Pp. ix-xviii in: E. S. Vrba (ed.), *Species and Speciation*. Transvaal Museum Monograph No. 4, Pretoria.
Waddington, C. H. 1957. *The Strategy of the Genes*. Allen & Unwin, London.
Wagner, G. 1988. The influence of variation and of developmental constraints on the rate of multivariate phenotypic evolution. J. Evol. Biol. 1:45-66.
White, M. J. D. 1978. *Modes of Speciation*. Freeman, San Francisco.

CHAPTER 2

SPECIATION AND ITS ONTOLOGY: The Empirical Consequences of Alternative Species Concepts for Understanding Patterns and Processes of Differentiation

Joel Cracraft

INTRODUCTION

Biology has long endured arguments over species concepts. The reasons for this contentiousness are not simple, but the willingness of biologists to engage in continued intellectual debate underscores the critical importance placed on species and the fact that no solution to the "species problem" has been generally accepted. Species concepts have assumed this role as instigators of debate largely because biologists have used species with two different objectives in mind. First, species have served as the basis for describing and cataloging biotic diversity and for our attempts to represent the historical relationships of that diversity in a hierarchical manner. Species are thus taken to be the primary taxa of systematic biology. Second, species have also been employed as basic entities of evolutionary theory. They are said to be the things produced by the process called speciation and the entity that speciates.

It may seem that these two roles are not disparate enough to require different conceptions of species, but biologists have found it difficult to reconcile them. At the center of this dialogue is the biological species concept (Mayr 1942, 1963, 1969), which has a wide following within evolutionary biology (Dobzhansky et al. 1977; Futuyma 1986), particularly in those disciplines concerned with processes operating at the microgeographic level of demes and populations. Thus, population geneticists and ecologists have generally embraced the biological species concept (BSC) even though most of their studies are not directly concerned with species-level taxa (e.g., Lewontin 1974; Endler 1977; Nei 1987).

In spite of this general acceptance, controversy over species concepts has arisen because many evolutionary biologists and systematists have found the BSC to be untenable in theory and unworkable in practice. The list of critiques of the BSC is long indeed, emanating from both botanists (Cronquist 1978; Levin 1979; Raven 1980; Mishler and Donoghue 1982; Donoghue 1985; Ehrendorfer 1984) and zoologists (Ehrlich 1961; Sokal and Crovello 1970; Rosen 1978, 1979; Cracraft 1983, 1987, 1988; McKitrick and Zink 1988). The majority of this criticism, however, has gone unanswered by supporters of the BSC and has been overlooked or discounted by those many evolutionary biologists who endorse biological species in their work. This is perplexing because most botanists, and an increasing number of zoologists, have chosen to abandon the BSC (Donoghue 1985:173).

One theme of this chapter is that systematic and evolutionary biologists can no longer afford to ignore this body of criticism. Use of the BSC presents fundamental obstacles to describing and interpreting patterns and processes of evolutionary differentiation. At present, many biologists have not realized the depth of these difficulties, which suggests that further dialogue is desirable. The ensuing discussion will deemphasize the theoretical reasons for abandoning the BSC as these have been treated in detail elsewhere (see especially Rosen 1978, 1979; Cracraft 1983, 1987; Rosenberg 1985; Donoghue 1985; McKitrick and Zink 1988). Instead, the focus will be on the empirical consequences of using the BSC in evolutionary studies. A major conclusion will be that evolutionary biologists should abandon the BSC. No doubt this will be a radical proposal to some, yet if the BSC obscures our ability to reconstruct evolutionary history accurately and to investigate the processes responsible for evolutionary differentiation, then abandoning that definition is a decision having strong scientific justification.

A decision such as this necessitates adopting an alternative definition that does not possess the difficulties of the BSC. One viable candidate is the phylogenetic species concept (Rosen 1978, 1979; Nelson and Platnick 1981; Cracraft 1983, 1987, 1988; McKitrick and Zink 1988). It will be argued that, compared to the BSC, the phylogenetic species concept is more successful at unifying the two roles of species, namely serving as the entity of evolutionary theory and as a basis for describing the historical pattern of taxonomic diversity and reflecting that pattern in biological classifications.

BIOLOGICAL AND PHYLOGENETIC SPECIES: CONCEPTUAL CONTRASTS

Speciation is the process whereby new species originate. Species are speciated; they are effects of lower level processes. From this perspective, a theory of speciation has as its units entities we call species. Species must be treated as discrete real entities because all theories require this (Hull 1976, 1977, 1978; Gaukroger 1978): empirical theories cannot be about unreal, arbitrarily delimited entities and it would be biological nonsense to say "species are speciated" if they were not real things. Entities chosen to function in a particular theory, moreover, must be irreducible to other units that serve a similar role within the domain of that theory. This means that if we are to compare these entities, or use them in descriptions of pattern, or perhaps as participants in processes specified by some theory, then they must be the same kind of entity (Gaukroger 1978; Cracraft 1987). These simple requirements of every theory have been largely overlooked by evolutionary biologists. "Biological species" are often delimited subjectively, many are clearly reducible to other discrete entities, and they frequently lack comparability among themselves as basal evolutionary taxa (Rosen 1978, 1979; Cracraft 1983, 1987). In numerous instances, moreover, "biological species" are not the entities that result from "speciation" (taxonomic differentiation).

These introductory remarks emphasize the utmost importance of a species definition within evolutionary theory: that definition establishes a particular ontology for the theory itself. If the ontology specified by the definition fails to match the real entities participating in or produced by natural processes, then not only will nature be described incorrectly, but it will be difficult to evaluate any theory that makes use of those entities. Such is the effect of the BSC, for it does not describe a correct ontology for theories about species origins. To appreciate this more fully, we need to identify the characteristics of entities presumably produced by the process of speciation (hereafter, these entities will be termed *evolutionary taxa*).

First and foremost, the speciation process produces differentiated taxa, that is, populations of interbreeding (reproductively cohesive) organisms having one or more evolutionary novelties distinguishing this new unit from all other similar units. These novelties could be any intrinsic attribute, from fixed differences at the genomic level to new morphological, biochemical, or behavioral characters. Whatever the novelty, populations are delineated as new taxa when they are, in principle, 100% diagnosable (see below). Many populations exhibit quantitative differences from other populations, but recognizing them as a taxon means that a biologist would have to apply some subjective criterion to subdivide continuous variation. In these instances, however, the taxa of our theories would be individuated arbitrarily and would not have ontological status as discrete entities.

Evolutionary taxa (as discrete entities) are also basal in the sense that none of them can be further subdivided into smaller populations that are themselves discrete, diagnosable units. Speciation theory does not specify the sizes of differentiated populations, and some can clearly be quite small (many founder populations, for example). Thus, size is irrelevant to the question of whether a population is a basal, differentiated taxon; what is important is whether that population can be recognized as being distinct.

Finally, the entities of speciation theory would be expected to have some degree of geographic integrity if reproductive cohesion is to be maintained over time. An essential element of current speciation theory is that an incipient new taxon will have some degree of spatial disjunction from its ancestral population (or sister taxon) so that reproductive cohesion can be disrupted, thereby allowing differentiation to take place. Note that this does not necessarily imply *reproductive isolation* (in the sense of the BSC), only that the loss of cohesion through spatial isolation leads to differentiation.

These are the minimal characteristics seemingly possessed by entities that are produced by processes of differentiation. If this is an accurate representation of current knowledge about the origin of evolutionary taxa, how well do existing definitions of species facilitate our understanding of these events?

The biological species concept

As noted earlier, a major contributor to controversies over species concepts has been the conceptual antagonism between seeing species as taxonomic entities or as evolutionary entities. The biological species concept has exacerbated that antagonism through the use of a plethora of subsidiary concepts, including nondimensional, multidimensional, and polytypic species, which ostensibly enable application of the BSC to biologically disparate situations. Of these, the polytypic species concept has had the widest influence.

Polytypic species are taxonomic, not evolutionary, constructs, and gained broad acceptance in the 1930s and 1940s as a response to what was perceived to be a pernicious trend within taxonomy: the tendency to name differentiated isolates as distinct species (what Mayr calls "typological-morphological species," e.g., 1963:338). As an alternative, it was proposed that diagnosable, or nearly diagnosable, taxa which replace one another spatially be united into a single "polytypic species." Historically, the main function of this procedure was to simplify classification and reduce the number of species names (Mayr 1942:126; 1969:38). As Mayr (1942:127) noted about the advantage of polytypic species within birds: "The total number of species to be memorized by the taxonomist has thus been cut by two-thirds."

Although the notion of polytypic species may be thought beneficial for taxonomists' memories, it unfortunately has canalized thinking about the historical pattern of speciation. Species are envisioned to be subdivided into numerous geographic races or subspecies. These taxonomic units exhibit a broad range of phenotypic differentiation, from being barely distinct quantitatively from their neighbors to being diagnosably distinct populations. The glue uniting all these units into a single polytypic species is the presumption of potential interbreeding; that is, it is assumed they are not reproductively isolated, no matter what their degree of differentiation. If a population is so markedly different that a taxonomist would judge it could not interbreed with other closely related taxa if they were in sympatry, then it is treated as a separate biological species. Speciation, under this ontology, is a process whereby these populations become more and more differentiated, so much so that they eventually cross the line of reproductive compatibility with the other populations of the same polytypic species.

The fact that the biological polytypic species concept was created to solve what was felt to be an undesirable glut of species names has had a profound affect on our attempts to describe and explain evolutionary pattern and process. Given this ontology, what is the unit of evolution? What entity speciates? Certainly it is not the polytypic species itself. Most workers seem to think it is the subspecies that is differentiating and becoming a new taxonomic unit, and Mayr himself held this view until recently. Now he considers subspecies as merely "pigeon-holing" devices for taxonomists (Mayr 1982:289). Unfortunately, subspecies names are frequently applied to a population showing any degree of differentiation deemed worthy of recognition by a taxonomist. This contributes to a classificatory system that is meaningless for describing evolutionarily relevant variation (see below). Given that many subspecies are themselves arbitrary subdivisions of continuous variation, it seems inescapably clear that many subspecies could hardly be considered units of evolution inasmuch as they do not qualify as objective units in the first place. Indeed, it can only be concluded that under the biological polytypic species concept there is no consistent, objective unit of evolution and, as such, the concept does not provide an empirically sound ontology for studying the origin of species.

It is essential for evolutionary biology to have an objective ontology with which to individuate the entities relevant to our theories. The biological species concept is frequently characterized as an objective or nonarbitrary descriptor of taxonomic diversity (Mayr 1969:27; Sudhaus 1984; Willmann 1987). In fact, in all situations that are critical for evolutionary analysis, the biological species concept can be applied only subjectively. The sole case in which the BSC can be said to be objective is in spatially restricted areas in which two diagnosably distinct taxa are in sympatry and reproductively isolated. But this is also a case in which all other species concepts currently in use would treat this situation in precisely the same manner. Cases of sympatry, therefore, do not speak for the objectivity of the BSC any more than for other concepts of species.

There are, however, two situations in which we would expect different species concepts to provide alternative interpretations of species limits. In both the BSC can be applied only subjectively and inconsistently. The first case involves two largely allopatric taxa that hybridize in a zone of contact. A decision as to whether a systematist recognizes one or two species depends on a personal assessment of the amount of hybridization and the width of the hybrid zone. Cases of this kind cannot be resolved in a straightforward manner using the BSC, as even proponents of that concept readily admit.

The second situation involves decisions about the specific status of differentiated populations that are entirely allopatric. As Mayr notes (1969:196), the criterion of reproductive isolation cannot be directly applied to these taxa, and all surrogate criteria used to evaluate potential reproductive isolation also fail the test of objectivity. These solutions generally include extrapolations from comparisons of morphological, behavioral, and ecological differences in closely related species that are sympatric or parapatric (e.g., Mayr 1963:31). But morphological differentiation among populations is not always closely correlated with their genetic compatibility, even among closely related taxa, therefore a solution such as this would require acceptance of numerous untested and untestable assumptions. Perhaps the most important requirement for this methodology is approval of the proposition that morphological, behavioral, and genetic rates of differentiation have been constant across the taxa being compared. Only if rates are constant can assessments between the amount of morphological differentiation and reproductive isolation in one group be used to guess at this relationship in allopatric taxa in other groups. Ironically, the hypothesis of constant rate cannot be examined *unless* the biological species concept is abandoned and a corroborated phylogenetic hypothesis for all the differentiated taxa is obtained.

Decisions about the species status of hybridizing taxa or differentiated allopatric taxa are critical in any analysis of speciation. Because the criterion of reproductive isolation cannot be applied in a uniform and objective manner, the ontological status of the units identified by the biological species concept must always be in doubt. This point requires emphasis: the ability to interbreed—*reproductive cohesion*—by itself cannot establish those populations as a discrete, evolutionary unit (contra Ayala 1981:46), because reproductive cohesion, which manifests a primitive morphogenetic organization, can transcend well-defined species boundaries. *Reproductive disjunction*, on the other hand, reflects evolutionary changes that signify the presence of discrete taxonomic entities, but so too do other changes that fail to affect reproductive isolation. This is one reason why the BSC cannot establish a consistent ontology for evolutionary theory. Similarly the recognition concept of species (Paterson 1981, 1982, 1985) possesses many of these same difficulties. Although the evolution of a new "specific-mate recognition system" could be used to diagnose an evolutionary unit, many evolutionary taxa obviously are capable of sharing these systems. Thus, the recognition concept of species, like the BSC, has the potential to confound the historical analysis of taxonomic diversification.

Many investigators have stressed that the biological species concept confounds the causal analysis of differentiation because the historical pattern of differentiation will not always be congruent with the historical pattern of reproductive isolation. Worse still, the BSC can lead to incorrect conclusions about the causal relationship between morphological (and genetic) differentiation and reproductive disjunction. Thus, Mayr (1963:31) claims that "The degree of morphological difference displayed by a natural population is a by-product of the genetic discontinuity resulting from reproductive isolation." In fact, however, reproductive isolation is itself a by-product of genetic differentiation following isolation and not causal of any accompanying morphological differentiation (e.g., Levin 1979:383). Reproductive isolation, therefore, is merely a subset of the numerous possible consequences of the more general process of differentiation, as many biologists have realized.

Phylogenetic species concept

The preceding section outlines some conceptual difficulties of the biological species concept that impair its effectiveness in evolutionary analysis. This raises the question of whether an alternative species concept can provide a better ontological foundation for evolutionary theory and at the same time satisfy the needs of systematic biology. In this regard, it has been suggested that a phylogenetic species concept constitutes a solution to the problems presented by the BSC (see discussions of Rosen 1978, 1979; Nelson and Platnick 1980; Cracraft 1983, 1987; McKitrick and Zink 1988; Zink 1988).

A phylogenetic species is an irreducible (basal) cluster of organisms, diagnosably distinct from other such clusters, and within which there is a parental pattern of ancestry and descent (Cracraft 1983, 1987; see also Rosen 1978, 1979; Nelson and Platnick 1980). The phylogenetic species concept circumvents all of the difficulties of the BSC because, as defined here, species are equivalent to evolutionary taxa in the sense they were portrayed earlier. Phylogenetic species are, therefore, basal, differentiated, evolutionary taxa. In the majority of cases, phylogenetic species will be demonstrably monophyletic; they will never be nonmonophyletic, except through error. Some phylogenetic species may be diagnosably distinct from other such units and yet not possess characters that can be hypothesized to be derived. In some instances, these species may be truly monophyletic but evidence of that fact has remained undiscovered. Or their historical status may be unresolved because relative to their sister species they are primitive in all respects. Whether they might be the "ancestor" of that species, and therefore be truly paraphyletic with respect to the historical structure of their populations, is probably unresolvable.

The phylogenetic species concept emphasizes the most general aspect of taxonomic diversification, namely differentiation. Some differentiation results in reproductive isolation and some does not. By relying solely on reproductive isolation as the central criterion for species status, the BSC precludes recognition of a very large class of evolutionary taxa, namely that constituting all instances of diagnostically distinct populations that are not reproductively isolated from other such populations. The BSC relegates this ubiquitous phenomenon to a position of secondary importance, and this is a primary reason why so many botanists in particular have not found the BSC useful in evolutionary analysis (Cronquist 1978; Levin 1979; Raven 1980; Donoghue 1985; Ehrendorfer 1984). The phylogenetic species concept, in contrast, views reproductive isolation as an important, but not necessarily predominant subset of effects produced by the process of differentiation. Reproductive isolation signifies the evolution of diagnostic characters, but not all newly evolved characters necessarily affect reproductive isolation.

The phylogenetic species concept, as all species concepts must, recognizes the critical importance of reproductive cohesion. This component of the definition is required if we wish to avoid assigning species status to individual organisms, to different sexes and morphs, or to developmental stages. In this sense, then, reproductive cohesion is a trivial component of all species concepts, including those that are purely morphological.

Because phylogenetic species are equivalent to basal evolutionary taxa, their use at once unifies the notion of species as it is applied to evolutionary theory, and as it is used in taxonomic practice. This concept thus provides a theoretically coherent ontology for systematic and evolutionary biology. The phylogenetic species concept emphasizes *diagnostic* character variation for individuating basal evolutionary taxa, thereby allowing non-diagnostic character variation to be partitioned into its evolutionary relevant intra- and interspecific components. Because the biological species concept does not necessarily divide nature at its true historical "joints" (Rosenberg 1985:197), there will always be the possibility of confounding within- and among-taxon patterns of variation.

The phylogenetic species concept is not a resurrected version of the so-called morphological species concept. Diagnostic characters can be represented by any intrinsic attribute of organisms, from the genome level on up. Determining whether character variation is diagnostic or not is decidedly more objective than assessing whether allopatric populations might hybridize or whether hybridization is sufficiently extensive to recognize a single biological species (contra Coyne and Barton 1988). In principle, populations should be 100% diagnosable, that is, all of the individuals will have the relevant diagnostic character(s). But realistically, the biological situation will almost always call for deeper analysis (see McKitrick and Zink 1988). Diagnostic characters may be restricted to males, to females, or perhaps to a particular ontogenetic stage; some individuals, moreover, may exhibit variation that obscures recognition of diagnostic characters. It is because of situations such as these that understanding the reproductive relationships of individuals within populations is often critical for delineating species correctly. In spite of potential difficulties, assigning a differentiated population to species rank under the phylogenetic species concept is still a hypothesis whose verification or rejection will always be dependent on the data available and the thoroughness with which they are interpreted.

It was noted earlier that some diagnosable populations may be quite small. This is not an arbitrary artifact of the phylogenetic species concept (Coyne and Barton 1988) but simply a reflection of natural processes of taxonomic differentiation: populations of all sizes can become isolated and then differentiate. If the delineation of species were based on the relative degree of similarity or dissimilarity, as is effectively the case with the BSC (direct application of reproductive disjunction being relatively rare in practice), then species limits inevitably will be arbitrary in many cases. The polytypic species concept has sought to unite many small differentiated populations into larger ones to "simplify" taxonomy. Yet, this too is arbitrary, and it confounds an accurate description and causal analysis of evolutionary diversity. Because the phylogenetic species concept recognizes the evolutionary singularity of diagnosably distinct populations—of whatever size—it offers the ontological foundation on which we can begin to understand any historical pattern underlying population differentiation.

The preceding sections indicate that the conceptual differences between biological and phylogenetic species are profound. It remains to be seen how these concepts have different consequences for interpretation of real-world data.

EMPIRICAL CONSEQUENCES OF ALTERNATIVE SPECIES CONCEPTS

The analysis of evolutionary pattern and process begins with an established species ontology derived from a theoretical expectation of how nature is organized and from previous empirical experience. In speciation analysis, for example, observational data on patterns of character variation and spatial distribution, along with perhaps a theoretical model of speciation, are used to individuate species-level taxa *prior* to subsequent investigations into their history. Within this context, then, species might be considered units of evolution and might be expected to exhibit a pattern of phylogenetic deployment through space and time. These elementary precepts can be used to compare the empirical consequences of applying biological and phylogenetic species concepts.

The historical pattern of taxonomic diversification

Different concepts of species influence analysis in several ways. The most important of these will be considered first: different species concepts often imply different ontologies and this results in misinterpretations of the historical pattern of differentiation. The following examples illustrate this influence.

Example 1. Speciation in *Cinclosoma.* The quail-thrushes (*Cinclosoma*) of Australia consist of six well-defined taxa distributed allopatrically and parapatrically in central and southern Australia (Figure 1). Considerable controversy regarding species limits has arisen because various authors have not been able to apply the biological species concept in a consistent manner. Table 1 summarizes ornithologists' attempts to assign species status to these taxa. Within the context of the BSC, conspecificity is typically decided on the basis of relative phenotypic similarity, which functions as a surrogate for a direct measure of reproductive compatibility (Mayr 1942, 1963, 1969). In the quail-thrushes, this procedure has failed because estimating relative similarity has been subjective at best.

These different judgments of species limits within *Cinclosoma* have obvious implications for speciation analysis. Taken at face value, each implies a different history for the pattern of differentiation. More importantly, each of these different estimates of species limits influences our ability to reconstruct that historical pattern accurately, because if biological species are real entities in nature and these are our best interpretations of those

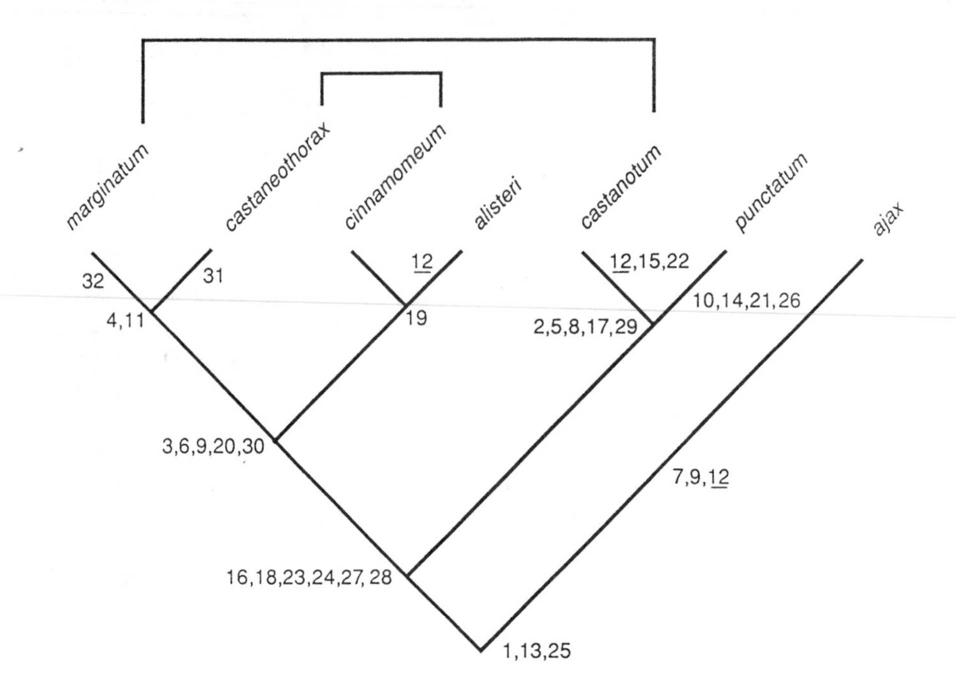

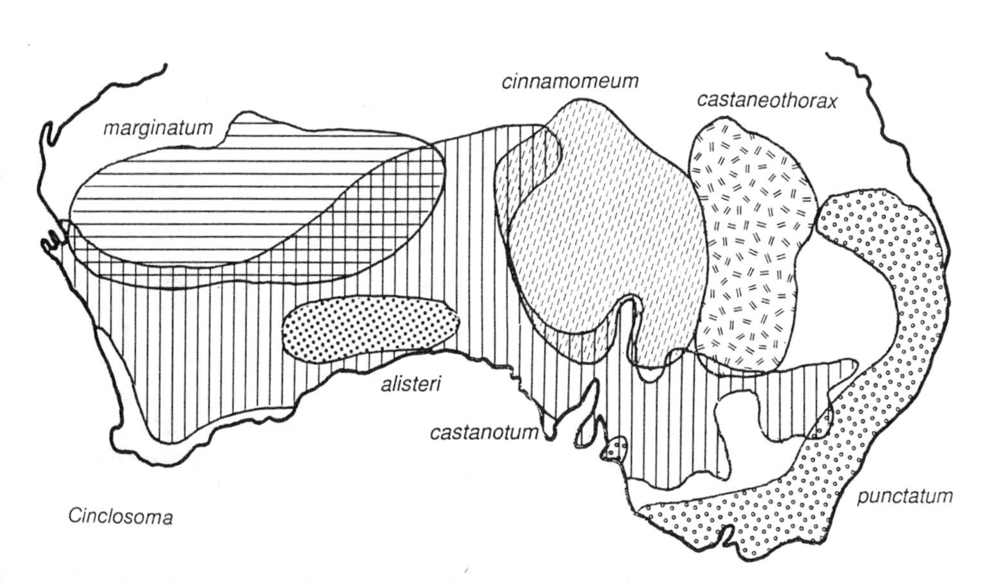

FIGURE 1. Distribution and phylogenetic hypothesis for the phylogenetic species of Australian quail-thrushes (*Cinclosoma*). An analysis of 32 ordered characters (Table 2) produced a best-fit tree of 35 steps (consistency index = 0.914). The tree was rooted using the outgroup taxa, *Ptilorrhoa castanota* and *P. leucosticta*. Underlined characters identify parallelisms and brackets link taxa that are known to hybridize. Distributions are after Ford (1983).

TABLE 1. Historical summary of species limits among the quail-thrushes (*Cinclosoma*) of Australia.

Campbell and Campbell (1926)	Condon (1962)	Ford (1974, 1976)	Ford (1983)
1. *punctatum*	1. *punctatum*	1. *punctatum*	1. *punctatum*
2. *castanotum*	2. *castanotum*	2. *castanotum*	2. *castanotum*
3. *alisteri*	3. *alisteri*	3. *cinnamomeum*	3. *cinnamomeum*
4. *cinnamomeum*	4. *cinnamomeum*	(*marginatum*)	(*alisteri*)
5. *marginatum*	5. *marginatum*	(*castaneothorax*)	4. *castaneothorax*
6. *castaneothorax*	6. *castaneothorax*		(*marginatum*)

MacDonald (1968, 1973)	Deignan (1964)	Pizzey (1980)	Wolstenholme (1926)
1. *punctatum*	1. *punctatum*	1. *punctatum*	1. *punctatum*
2. *castanotum*	2. *castanotum*	2. *castanotum*	2. *castanotum*
(*alisteri*)	3. *cinnamomeum*	3. *cinnamomeum*	3. *cinnamomeum*
3. *cinnamomeum*	(*castaneothorax*)	(*castaneothorax*)	4. *castaneothorax*
(*castaneothorax*)	(*alisteri*)	(*marginatum*)	(*marginatum*)
(*marginatum*)	(*marginatum*)	4. *alisteri*	5. *alisteri*

entities, then clearly many of these sets of postulated species limits will lead us astray as we attempt to recover the one true history. Reproductive isolation is not an intrinsic attribute, but a relational concept, and thus does not constrain biological species to be strictly monophyletic. By definition, nonmonophyletic species imply history has been misrepresented.

These problems do not exist with the phylogenetic species concept. When this concept is applied, minimally six phylogenetic species are recognized in Australia (diagnoses are contained in the data of Table 2). A hypothesis of their phylogenetic relationships can then be generated by cladistic analysis of a set of discrete character data derived from external morphology (Table 2). The hypothesis of Figure 1, for example, is the most parsimonious tree for the data (length = 35 steps; consistency index = 0.914) and represents our current best estimate of the historical pattern of taxonomic differentiation (see also Cracraft 1986).

This phylogenetic hypothesis also permits analysis of patterns of hybridization in a way that is not possible when employing the BSC. The species pairs, *castanotum–marginatum* and *cinnamomeum–castaneothorax*, hybridize sporadically in zones of overlap (Ford 1983). Hybridization is apparently not extensive because of habitat segregation of the

TABLE 2. Character-state data for the genus *Cinclosoma*.[a]

Taxa	Characters																														
	1	2	3	4	5	6	7	8	9	10	11	12	13	14	15	16	17	18	19	20	21	22	23	24	25	26	27	28	29	30	31	32
P. leucosticta	0	0	0	0	0	0	0	0	0	0	0	0	0	0	0	0	0	0	0	0	0	0	0	0	0	0	0	0	0	0	0	0
P. castanota	0	0	0	0	0	0	0	0	0	0	0	0	0	0	0	0	0	0	0	0	0	0	0	0	0	0	0	0	0	0	0	0
C. ajax	1	0	0	0	0	0	1	0	1	0	0	1	1	0	0	0	0	0	0	0	0	0	0	0	1	0	0	0	0	0	0	0
C. punctatum	1	1	0	0	1	0	0	1	0	1	0	0	1	1	0	1	1	1	0	0	1	0	1	1	1	1	1	1	1	0	0	0
C. castanotum	1	1	0	0	1	0	0	1	0	0	0	1	1	0	1	1	1	1	0	0	0	0	1	1	1	0	1	1	1	0	0	0
C. castaneothorax	1	0	1	1	0	1	0	0	1	0	1	0	1	0	0	1	0	1	0	1	0	1	1	1	1	1	1	1	0	1	1	0
C. cinnamomeum	1	0	1	0	0	1	0	0	1	0	0	0	1	0	0	1	0	1	1	1	0	0	1	1	1	0	1	1	0	1	0	0
C. marginatum	1	0	1	1	0	1	0	0	1	0	1	0	1	0	0	1	0	1	0	1	0	0	1	1	1	0	1	1	0	1	0	1
C. alisteri	1	0	1	0	0	1	0	0	1	0	0	1	1	0	0	1	0	1	1	1	0	0	1	1	1	0	1	1	0	1	0	0

[a]Outgroups include *Ptilorrhoa castanota, P. leucosticta,* and C. ajax, all distributed in New Guinea. Key for characters (0, primitive and absent; 1, derived and present, in all cases): 1, male throat blue-black; 2, female throat gray; 3, female throat buff or cream; 5, female breast solid gray; 6, female breast light brown to brown; 7, female breast reddish chestnut; 8, feathers at sides of breast and upper throat solid gray; 9, feathers at sides of breast brown to rufous; 10, male breast solid gray; 11 male breast with extensive chestnut or rust-red patch; 12, male breast extensively blue-black; 13, male flanks with spotting; 14, female flanks with spotting; 15, male flanks gray-brown; 16, male with white spots on lesser primary coverts; 17, crown and forehead gray; 18, crown light brown; 19, crown light cinnamon or rufous cinnamon; 20 upperparts cinnamon to light rufous cinnamon; 21, upperparts heavily streaked; 22, male back and rump deep rufous; 23, primaries light brown; 24, tertials rufous to cinnamon with dark central streak; 25, male white malar streak not extending onto throat; 26, male malar streak reduced anteriorly; 27, male light eye stripe; 28, relative bill size decidedly reduced; 29, ear coverts gray-brown; 30, ear coverts rufous or cinnamon; 31, male upper breast rich rust-red; 32, male upper breast pale chestnut.

parental forms. It might be imagined, however, that if climate changed, and these habitat differences were abolished, then hybridization could become rampant. Given that this might occur, use of the BSC would imply these species pairs are conspecific. The problem, however, is that these differentiated evolutionary taxa are not sister groups but are each separated by three and four speciation events. In fact, sister species within this complex are typically widely separated geographically. Application of the BSC has created paraphyletic taxa and has hindered recognition of this historical pattern, but it is easily revealed when the phylogenetic species concept is employed.

Example 2. Speciation in *Daphoenositta.* A classic and frequently mentioned example of avian speciation used to illustrate the biological species concept involves the sittellas of Australia (e.g., Keast 1961; Mayr 1963:372–373). Five well-differentiated taxa are distributed peripherally around northern, eastern, and southern Australia (Figure 2). At one time the five forms were thought to be essentially allopatric, and some authors (e.g., McGill 1948) considered them to be distinct species. With more extensive collecting and examination of specimens, it soon was realized that these taxa have much larger ranges than previously thought. Individuals of these taxa exhibit a substantial capacity to disperse, and the result has been hybridization in zones of contact. Indeed, there is evidence for hybridization among all the forms in parts of Queensland (Ford and Parker 1974; Ford 1980; Short et al. 1983a, 1983b).

Documentation of hybridization has led most workers to recognize a single polytypic biological species (Mayr 1950; Keast 1961; Ford 1980; Short et al. 1983a, 1983b), with each form being ranked as a subspecies or even something called a "megasubspecies" (Short et al. 1983a). The conventional story about speciation is that the ancestor of the five parental forms was once distributed broadly across Australia, with the five populations becoming isolated in relatively more humid refuges during a time of increased aridity. With amelioration of the climate, these populations, now differentiated, spread outward from these refuges to eventually hybridize in areas of overlap (Keast 1961; Mayr 1963:372; Ford 1980).

At first it might appear that the sittellas provide a clear example of the success of the biological species concept. Yet, application of the BSC in this and many other similar cases raises some serious difficulties. The first is ontological in nature: what is the unit of evolution in a situation such as this? The single global biological species is certainly not the unit that was speciated: it possesses no singularity as a differentiated taxon that is not shared by any monophyletic group. Instead, virtually all workers have treated the subspecies as the unit of evolution, because each of these populations is interpretable as having evolved its own characters in isolation.

Ironically, if subspecies are only "a unit of convenience for the tax-

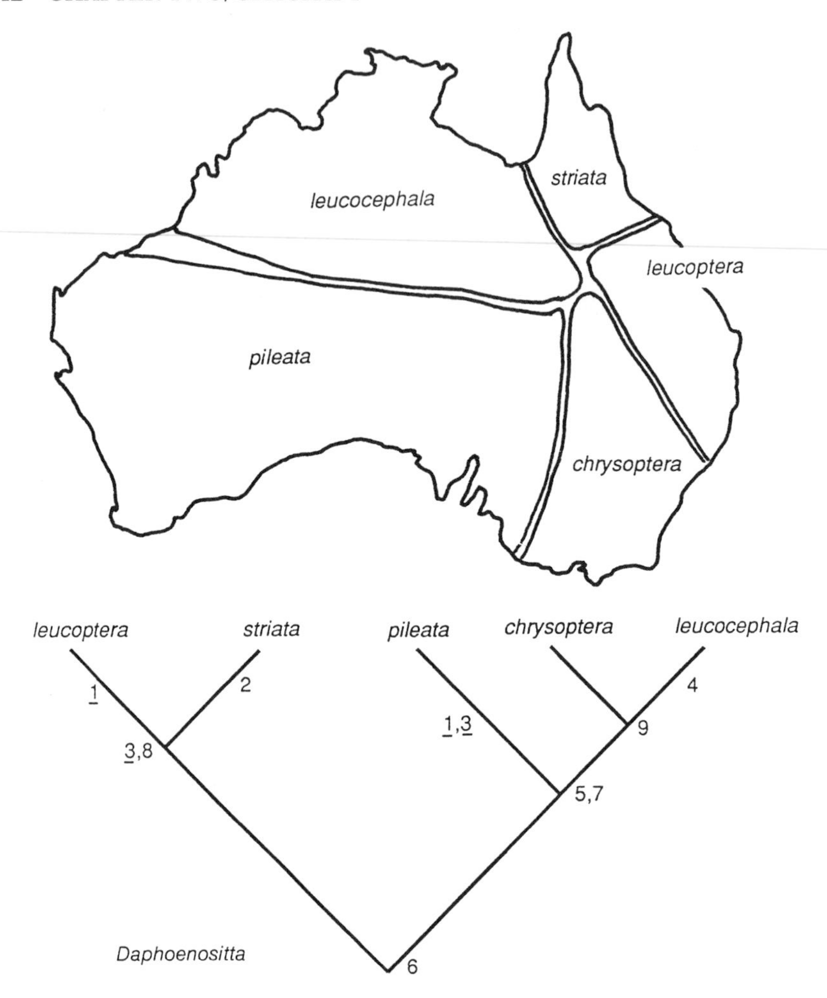

FIGURE 2. Distribution and phylogenetic hypothesis for the phylogenetic species of Australian sittellas (*Daphoenositta*). An analysis of nine ordered characters (Table 3) produced a best-fit tree of 11 steps (consistency index = 0.818). The tree was rooted using the outgroup taxon, *Daphoenositta papuensis*. Underlined characters signify parallelisms. Distributions after Ford (1980) and Short et al. (1983a).

onomist, but not a unit of evolution," as Mayr has stressed (1982:289), then seemingly we are left without taxa that could be called evolutionary units. To deny these well-defined entities a unitary evolutionary role merely to satisfy the sanctity of a particular species definition is to cast doubt on the usefulness of that definition. If evolutionary theory is supposed to be generally applicable across organisms, there cannot be ontological confusion about the identity of its units. Are they species, subspecies, megasubspecies, or what? The biological species concept does not make this clear nor is it capable of a consistent answer.

The sittellas raise another problem for the biological species concept. If the five forms are considered to be phylogenetic species (diagnoses in Table 3) and their character variation is tabulated, then cladistic procedures can be used to generate a hypothesis of their history. This hypothesis (Figure 2) proposes a very specific pattern of relationships with the northeastern form, *striata*, being the sister species of the northwestern form, *leucoptera*. These, in turn, are the sister group of the other species. This latter group includes the sister pair, *chrysoptera* (Southeast) and *leucocephala* (East), and their sister species, *pileata* of southwestern Australia. This pattern of historical and spatial relationships is markedly congruent with speciation patterns in other Australian birds (Cracraft 1982, 1986).

This hypothesis indicates a complex history of differentiation within sittellas. The forms are not of the same ages, as is often implied in the scenarios associated with the biological species concept. Placing all these evolutionary taxa in a single biological species obfuscates this historical pattern, because the ability to hybridize merely signifies the retention of a primitive morphogenetic milieu.

TABLE 3. Character-state data for the genus *Daphoensitta*.[a]

	Characters								
Taxa	**1**	**2**	**3**	**4**	**5**	**6**	**7**	**8**	**9**
D. albifrons	0	0	0	0	0	0	0	0	0
D. papuensis	0	0	0	0	0	0	0	0	0
D. pileata	1	0	1	0	1	1	1	0	0
D. chrysoptera	0	0	0	0	1	1	1	0	1
D. leucocephala	0	0	0	1	1	1	1	0	1
D. striata	0	1	1	0	0	1	0	1	0
D. leucoptera	1	0	1	0	0	1	0	1	0

[a]Outgroups include two taxa in New Guinea (*albifrons, papuensis*), both usually placed in a single species, *D. papuensis*. Key for characters (0, primitive and absent; 1, derived and present): 1, male with white breast and belly (unstreaked); 2, female throat black; 3, crown jet glack; 4, male with white crown; 5, reduction in melanin deposition in feathers and upperparts (streaks and background less dense); 6, wing patch on primaries; 7, cinnamon wing patch on primaries and secondaries; 8, white wing patch on primaries; 9, bill virtually all black (without extensive yellow).

Example 3. Speciation in the *Pipilo fuscus* complex. The brown towhees of the southwestern United States and Mexico include four differentiated taxa (Figure 3). Two of these, *P. aberti* and *P. albicollis,* are moderately distinct, whereas *P. fuscus* and *P. crissalis* are very similar, differing primarily in the presence (*fuscus*) or absence (*crissalis*) of a breast spot. Virtually all workers have united these latter two taxa into the same biological species on the assumption that if they were in contact, they would probably interbreed (Davis 1951; Marshall 1960, 1964; Mayr and Short 1970; AOU 1983). Eight subspecies have been described within the *crissalis* group and 11 within the *fuscus* group (Davis 1951). Populations of *crissalis* from southern Baja California are said to bridge the morphological gap between more typical *crissalis* and populations in the *fuscus* group. None of these subspecies is apparently diagnosably distinct.

Placing *fuscus* and *crissalis* in the same biological species, even though they are diagnosably different, has clearly seemed like a reasonable decision to many biologists. Yet, it constrains our view of speciation within the brown towhees. Such a decision implies *fuscus* and *crissalis* are more closely related to each other than either is to *albicollis* or *aberti.* The analysis of speciation has effectively been reduced to a three-taxon statement.

The history of these taxa has recently been reevaluated by Zink (1988). Using variation at 39 presumptive loci and 29 skeletal measurements, he was able to demonstrate, first, that *P. fuscus* and *P. crissalis* are not closest phenetically given either data set. Furthermore, cladistic analysis of Rogers' genetic distances (Zink 1988:76, Table 3) suggests that *P. crissalis* is the sister species of *P. aberti,* and *P. fuscus* is the sister species of *P. albicollis* (Figure 3), although the data are by no means unambiguous.

Assuming that this historical hypothesis will be substantiated by future work, the similarities that led workers to unite *fuscus* and *crissalis* in the same biological species are easily interpretable as retentions of primitive characters. Even if *fuscus* and *crissalis* are eventually shown to be each other's closest relative, they still remain diagnosably distinct, evolutionary taxa. Research undertaken within the context of the biological species concept has led to misunderstandings about the evolutionary roles of these taxa and their history, whereas use of the phylogenetic species concept can avoid these misinterpretations.

Example 4. Speciation in the *Thomomys umbrinus* complex. Use of the biological species concept often results in species taxa that lack a unitary historical role. This is illustrated by the pocket gophers of central Mexico that are currently placed in the biological species, *Thomomys umbrinus.* Patton and Feder (1978) and Hafner et al. (1987) have shown that at the gross level of chromosomal variation, this species can be subdivided into

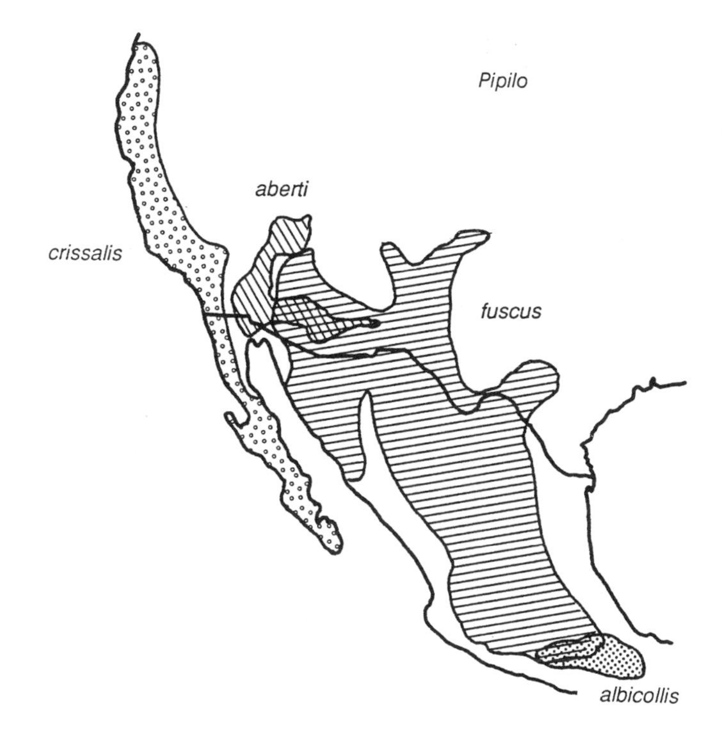

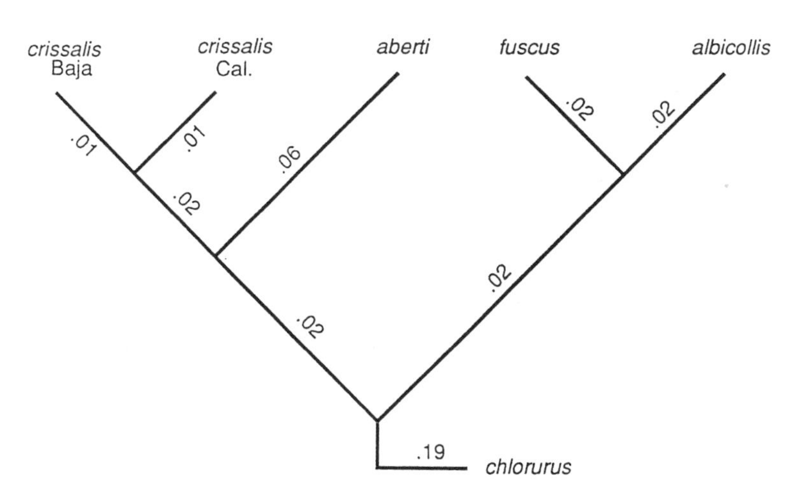

FIGURE 3. Distribution and phylogenetic hypothesis for four phylogenetic species in the North America sparrow genus *Pipilo*. The Fitch–Margoliash tree based on Rogers' genetic distances (data from Zink 1988) is rooted using the outgroup species *P. chlorurus*. Distributions are from Hubbard (1973).

two karyotypic groups, one of $2N = 76$ and the other $2N = 78$. Hafner et al. (1987) present arguments and evidence to suggest that populations in each group are reproductively isolated from populations in the other. Taken at face value, these observations would seem to suggest that minimally two biological species are involved instead of one. The situation is more complex, however, for each of these karyotypic entities is paraphyletic and is comprised of sets of populations having closer phyletic relationships to populations with a different karyotype than to populations within the same karyotypic group.

The biological species concept cannot be applied to the *Thomomys umbrinus* complex unless one is willing to accept paraphyletic species, and to do that would be a de facto admission that biological species are not units of evolution. Hafner et al. (1987:18) conclude that "paraphyletic species may be common, perhaps the rule, in naturally occurring organisms" (see also Patton 1981, and Patton and Smith 1981, for similar observations). To the extent this is true, the biological species concept will fail to provide an accurate trace of the history of taxonomic differentiation. The phylogenetic species concept, on the other hand, does not present this difficulty, because it identifies the unit of evolution to be differentiated evolutionary taxa.

Species concepts and the analysis of geographic variation

Evolutionary biologists have long known that the study of geographic variation is central to understanding the patterns and processes of speciation (e.g., Dobzhansky 1937; Mayr 1942), and countless studies have been undertaken over the last half century (for reviews of only a small portion of this extensive literature see Mayr 1963; Gould and Johnston 1972; Endler 1977; Zink and Remsen 1986). Much of this research has been motivated by two important goals: first, to elucidate the causal agents responsible for spatial patterns of phenotypic or genotypic variation, and second, to identify those microevolutionary processes by which new species arise. It is clear the second problem cannot be solved without having answers to the first, yet the observation that variation can often cut across species boundaries indicates that an investigator could pursue facets of the first problem without necessarily directly being concerned with the second. Thus, although some evolutionary biologists have treated these two problems as inseparable, this need not be so.

There has been very little discussion about the influence that different species concepts might have on the analysis of geographic variation (but see Zink and Remsen 1986). For certain kinds of questions, it may not matter how species are defined or how that concept is applied to natural situations. Thus, a causal analysis of a well-defined cline that trends across two taxa may be of interest regardless of whether the two taxa are considered subspecies of a single biological species or two sister species. Nevertheless, an interpretation of any pattern of variation is predicated upon a correct description of the pattern itself, and sometimes the latter is influenced by the choice of species boundaries. Sensitivity to this influence is particularly important when the impetus for an analysis of geographic variation is the study of speciation. In this case, the investigator is presumably interested in the processes that apportion variation within an ancestral population into variation between descendant populations. If this is the case, then application of the biological species concept to this problem would seem to require an emphasis on variation that is considered of importance for reproductive isolation (e.g., Frost and Platz 1983; Nevo et al. 1987; Zink and Remsen 1986:33), whereas use of a phylogenetic species concept, in contrast, would focus attention on those aspects of variation that are relevant to the origin of any evolutionary novelty (including those affecting reproductive isolation). Of more importance for the present discussion is a consideration of the ways in which alternative species concepts might cause us to resolve patterns in different ways.

Some of the potential influences of species concepts can be illustrated by the following hypothetical situation. Consider three parapatric populations (call them A, B, C) distributed along a latitudinal transect. Assume that each is weakly differentiated but still diagnosably distinct and that some hybridization is observed between each adjacent pair of populations (A × B, B × C). Finally, assume that the population means for some attribute, say body size, exhibit a pattern of clinal variation, with size increasing with latitude. Many such cases have been described in the literature. The standard interpretation would be to treat these populations as subspecies of a single biological species (e.g., Mayr 1942, 1963; Endler 1977). Defined as a single species, the task at hand would be to find an explanation for the trend in size, employing perhaps correlation analysis with a suite of environmental variables. In this case, the biological species concept predisposes us to look at and attempt to explain an *intraspecific* pattern of variation, and, as Endler (1977:7) noted for similar situations, this type of variation is generally not deemed important from the standpoint of speciation because the populations in these cases are neither strongly allopatric nor strongly differentiated.

The use of a phylogenetic species concept compels us to investigate this same problem in a different way. Even though weakly differentiated, the presence of diagnostic characters implies that we are dealing with three evolutionary taxa. As such, both intra-and interspecific patterns of variation are potentially of interest and therefore must be defined. In this example, it is entirely possible that intraspecific size patterns will *not* be concordant with the interspecific pattern: variation within each species might show a decrease in body size as latitude increases even as the population means themselves increase. Such an observation might caution us against accepting an explanation for the interspecific trend when that explanation is based only on a correlation with some environmental variable that also happens to show a trend across the distributions of the species.

This example can be made still more complex, and once again species concepts are important in the resolution of pattern. An historical analysis of geographic variation requires a phylogenetic hypothesis for the entities under study (Straney and Patton 1980). If Mayr's (1982:289) assessment that subspecies should not be interpreted as units of evolution is correct, then application of the biological species concept to situations such as this implies that historical analysis will be at an impasse. In contrast, the phylogenetic species concept compels us to ask how the three evolutionary taxa—A, B, C—might be interrelated. If there is reason to believe they form a monophyletic group, then there are three relevant phylogenetic hypotheses that could be considered. Two of these hypotheses, in which the centrally distributed taxon is more closely related to either adjacent taxon [i.e., (A + B) + C and A + (B + C)], would imply concordance between the historical pattern of speciation and the size trend. As a consequence, either (or both) environmental causation and historical constraint could be important in explaining the observed trend. The third hypothesis, in which the northernmost and southernmost taxa are sister groups [i.e., (A + C) + B], would suggest that differentiation associated with body size is not related to phylogenetic history, which strengthens the case for seeking a causal relationship with some environmental factor (see Straney and Patton 1980; and below).

Although the preceding example is hypothetical, many published studies of geographic variation illustrate instances in which the description of intra-and interspecific patterns of variation could potentially be misinterpreted by application of the BSC. Some aspects of the problem will be illustrated by the following examples using data from natural situations.

Geographic patterns of morphological differentiation. One important component of the study of geographic variation is the most elementary: to describe and explain spatial patterns of differentiation. In this type of analysis, it is of interest to explore whether the spatial patterns of variation are correlated with spatial patterns of environmental variation or with the phylogenetic pattern of the entities being studied. Each type of correlation implies a different underlying causal fabric for the observed pattern of differentiation (Straney and Patton 1980). Resolution of the phylogenetic pattern itself is strongly influenced by decisions about species limits. This influence can be illustrated by again considering variation within the brown towhee complex (Zink 1988).

The question to be investigated is whether patterns of morphological variation are congruent with phylogenetic pattern. If they are, this suggests differentiation has a strong historical component that is related to divergence following cladogenesis; if patterns of variation exhibit congruence with environmental patterns but not with phylogenetic history, then adaptive or epigenetic determinants on variation might be interpreted as being more important (see Straney and Patton 1980, for detailed discussion).

The best available estimate of phylogenetic pattern for the brown towhees is a cladistic analysis of Rogers' genetic distances (data from Zink 1988; Figure 3). These results can be compared to an estimate of phenetic relationships that has been calculated using the standardized group means of six external dimensions for each of 46 populations (data taken from Davis 1951). A phenogram was generated using UPGMA on a matrix of taxonomic distances (Sneath and Sokal 1973), and this has been mapped onto the phylogenetic hypothesis (Figure 4, left). Given these data and the simple approach used to analyze them, patterns of morphological variation across space are not very congruent with the phylogenetic pattern. Thus, *Pipilo crissalis* is generally phenetically more similar to *P. albicollis* and *P. fuscus* than to *P. aberti.*

This relationship between the two patterns would be altered if the species limits themselves were modified. Previous workers, using phenetic assessments of plumage, ecology, and behavior to judge the extent of reproductive isolation, have considered *crissalis* and *fuscus* to be parts of a single biological species. Rearranging the phyletic pattern to reflect this conception of species limits, and then juxtaposing that with the phenetic pattern, reveals a new set of relationships (Figure 4, right). Now there is somewhat more congruence between the two patterns, particularly as they are expressed between *crissalis* and *fuscus.*

The purpose of this exercise is to show that the perceived pattern of geographic variation is in part dependent on prior judgments about species limits. The decision to place *P. crissalis* and *P. fuscus* in the same biological species has been based on overall resemblance. If, hypothetically, they happened to be sympatric and hybridize, this would seemingly strengthen and justify this interpretation. Yet, they are apparently distantly related geographically and do not comprise an evolutionary unit with a singular history. Only be treating them as if they were phylogenetic species can the historical pattern be resolved. And if it is necessary to treat them as phylogenetic species for analytical purposes, then it makes scientific sense to call them phylogenetic species. The biological species concept is superfluous in this case.

Clinal variation across evolutionary taxa. Uniting evolutionary taxa into a single biological species may cause us to describe a pattern of clinal variation where one does not exist. Two evolutionary taxa within the galliform

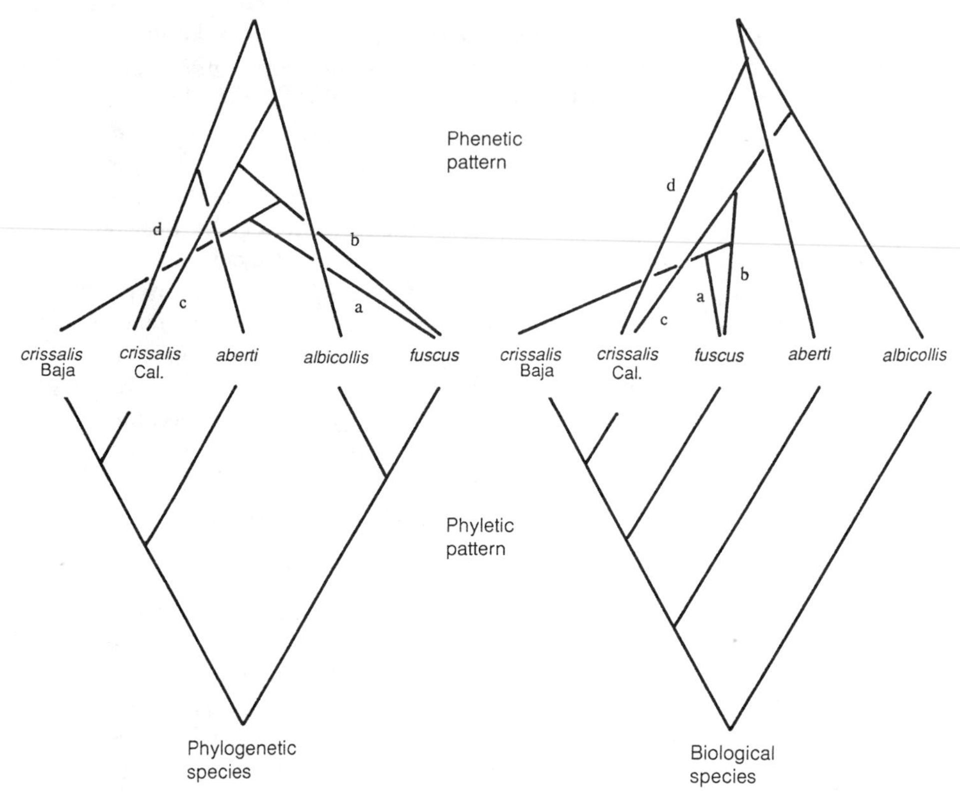

FIGURE 4. Comparison of the congruence between phenetic and phylogenetic patterns using phylogenetic species (left) and biological species (right). The phenetic pattern was generated as discussed in the text.

family Cracidae, *Ortalis cinereiceps* and *O. garrula,* are often placed in a single biological species on the basis of their potential to hybridize (Vaurie 1965, 1968). Although recognizing these two taxa as phylogenetic units, Vaurie (1965) described a size cline across them. *Ortalis cinereiceps* exhibits a cline of decreasing size from Nicaragua south to northwestern Colombia. *Ortalis garrula,* which is distributed across northern Colombia to the east of *O. cinereiceps,* is smaller still but shows no noticeable size variation across its range. It is doubtful that there is any significant gene flow between these two forms: only one specimen seems to be intermediate, and the taxa themselves are allopatric (Delacour and Amadon 1973:95). Within the context of a single biological species, there is a strong predilection to describe and seek a causal explanation for intraspecific clinal variation, even when it extends across two or more evolutionary taxa.

Because these two forms of *Ortalis* are diagnostically distinct, however, the transspecific trend may be spurious. There exists the possibility that *O. garrula* may be more closely related to a third species of *Ortalis* than it is to *O. cinereiceps*. If such is the case, the small size of *O. garrula* might be a primitive character shared with that other species. Use of phylogenetic species and a search for their relationships could help solve this problem.

Geographic variation and the historical analysis of hybridization. Within the context of a biological species concept investigators rarely make a distinction between diagnostic and nondiagnostic components of variation. The history of speciation can thereby be obscured. In those cases in which differentiated isolates have presumably come into geographic contact and hybridize, the usual procedure is to unite these taxa into a single biological species. From this perspective, subsequent analysis of geographic variation focuses almost entirely upon nondiagnostic variation because the problem of interest is to document character variation through the zone of intergradation. Indeed, under the aegis of the biological species concept, the study of hybridization has taken center stage in the analysis of geographic variation in contrast to those investigations that emphasize the historical analysis of taxonomic diversification. Without an understanding of the historical pattern of taxic origins, however, it is very likely that the causal dynamics of hybridization itself will be misconstrued. This general point can be illustrated by one example from Haffer's (1974) classic study of geographic variation and hybridization in the toucans of Amazonia.

Haffer's (1974:265–284) study included an analysis of variation within an entity said to be a biological species. Thus, *Ramphastos vitellinus* is distributed across much of Amazonia, but is subdivided into three well-marked evolutionary taxa (assigned subspecific rank by Haffer) that broadly hybridize north and south of the Amazon, although apparently not across the lower Amazon (Figure 5). Haffer examined patterns of variation in a number of variables, only one of which will be considered here. Haffer (1974:274, Figure 16.34) demonstrated a zig-zag cline in the width of the red breast band, first decreasing from the population in the Guianas and Venezuela (*R. v. vitellinus*) toward that in eastern Ecuador and Peru (*R. v. culminatus*) and then increasing again toward the population of central and southeastern Brazil (*R. v. ariel*). These and other clines are interpreted by Haffer (1974) as the result of gene flow within hybridizing populations that have come into secondary contact.

Haffer's general interpretation appears correct. Yet, interpreting patterns of variation *within* what is taken to be a single biological species has made it conceptually more difficult to examine that same variation from

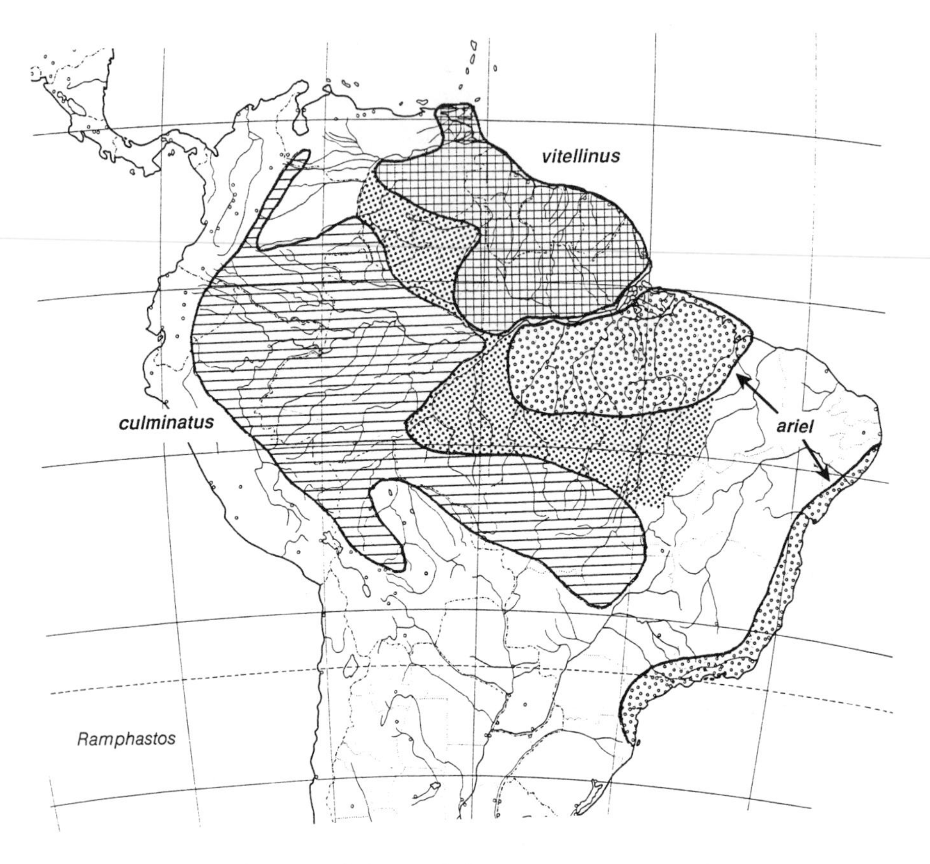

FIGURE 5. Distribution of three phylogenetic species in the toucan genus *Ramphastos.* Distributions of hybrids between *R. vitellinus* and *R. culminatus* and between *R. culminatus* and *R. ariel* are shown by black dots. *Ramphastos vitellinus* and *R. ariel* are apparently sister species (Prum 1982). Distributions from Haffer (1974). See text.

the perspective of historical patterns that may have arisen *among* differentiated taxa. The three taxa of toucans in this example are well-defined evolutionary units, or phylogenetic species. Prum (1982, 1988) has proposed that *vitellinus* of the Guianas is the sister species of the south Amazon form, *ariel,* and that both comprise the sister group of *culminatus* from upper Amazonia. Consequently, historical analysis indicates that the sister species of this clade do not hybridize and that all hybridization is taking place among taxa separated by at least two speciation events. Whether the sister species *vitellinus* and *ariel* are capable of hybridizing is conjecture at this point, for in spite of having an ability to fly long distances, they are isolated by the Amazon.

Two important conclusions can be drawn from this historical hypothesis. First, the patterns of variation resulting from gene flow must be the result of secondary contact and hybridization and not primary intergradation because the hybridizing taxa are not sister groups. Second, the clines themselves must be younger than the speciation event that gave rise to *vitellinus* and *ariel*. If that event could be dated in some manner, either by a geological or biological clock, then it would establish a time constraint on any analysis that attempted to describe or reconstruct the dynamics of this hybridization.

The preceding example is also similar in some ways to the situation described earlier for the Australian sittellas (*Daphoenositta*). Using hybridization as a criterion for uniting all these taxa into a single biological species clearly impairs the historical analysis of hybridization itself: even though two differentiated taxa hybridize, they may not be sister species (Rosen 1978, 1979). In the sittellas (Figure 2), patterns of hybridization are expressed across a complex pattern of historical interrelationships. These examples illustrate the point that as long as hybridization is used as a criterion to define species limits, those limits cannot then be used to study the degree of concordance between phylogenetic history and the ability to hybridize. This circularity is not confronted when using phylogenetic species.

It might be argued that these observations and conclusions could be studied within the context of the biological species concept. One response to this is that if investigations of history were an integral part of the application of the biological species concept to speciation analysis, then we might expect to see many examples. We do not, however, and it is only necessary to inspect the major texts on speciation to document what little role historical analysis has played (e.g., Mayr 1942, 1963, 1969; Endler 1977; White 1978). This situation has begun to change, primarily because workers are seeing the benefits of treating differentiated taxa as evolutionary units having a unique historical pattern of interrelationships. As this work continues, more and more patterns of variation and hybridization will be found to transcend the boundaries defined by speciation events. When this occurs, the biological species concept not only becomes inapplicable, it carries with it the potential to complicate evolutionary analysis.

Species concepts and the genetics of speciation

Evolutionary geneticists, operating within the framework of the biological species concept, have viewed the genetic analysis of speciation and differentiation as two separate problems, what Templeton (1981:25) calls the genetics of speciation versus the genetics of species differences. To most evolutionary geneticists, probably, the genetics of speciation is essentially equivalent to the genetics of reproductive isolation (Ayala 1975; Bush

1975; Avise 1976; Templeton 1981, 1982; Nei et al. 1983; Rose and Doolittle 1983; Barton and Charlesworth 1984; Barton and Hewitt 1985; Krieber and Rose 1986), and the research protocol is one of searching for the kinds of genetic changes that bring about reproductive disjunction. As Templeton (1981, 1982) notes, however, generalities have failed to emerge from this work.

Acceptance of a phylogenetic species concept implies a shift in perspective: the genetics of speciation now becomes one component of a causal chain that seeks to explain the origin of evolutionary novelties in populations. Once again, we assume these novelties can be any intrinsic attribute as long as it characterizes the existence of an evolutionary taxonomic unit. To simplify the problem, we can envision the contribution of evolutionary genetics to be twofold:

1. to propose causal explanations for the origin of these novelties within the ontogenies of individual organisms within a population, and
2. to propose causal explanations for the spread and fixation of those novelties in the population, thus characterizing the latter as a new, differentiated taxon.

The first contribution appears to reside within the realm of developmental molecular genetics and the second within population genetics and ecology.

Seeing speciation from the standpoint of phylogenetic species thus calls for a new emphasis on the genetics of species differences, but not in the way this subject has sometimes been studied in the past. Attempts to relate genetic distance to taxonomic rank (Ayala et al. 1974; Ayala 1975; Avise 1976; Zimmerman et al. 1978) are unlikely to lead to useful generalizations because within the context of current widely accepted methods of classification, any correlations might well be spurious (see also Patton 1981:286–287). Given the methods of evolutionary classification, in which taxonomic rank is often a subjective assessment of the degree of phenotypic divergence, there is no reason to expect that taxa assigned to ranks such as subspecies, semispecies, sibling species, or nonsibling species will be internally homogeneous with respect to age or be comparable phylogenetically. Many so-called subspecies or semispecies, for example, might be as old as, if not older than, entire clades of species.

This cautionary note implies that studies on species differences should have a strong measure of phylogenetic control. By definition, sister taxa, whatever taxonomic rank they might be given, are of the same age, and comparisons between sister groups offer a wealth of opportunities to explore the relationships between genetic differences, on the one hand, and phenotypic divergence or relative differences in diversity, on the other.

In one sense, it is possible to remain pessimistic regarding the possibility of finding general explanatory laws relating genetic divergence and speciation. Many different kinds of genetic change appear responsible for perturbations of developmental pathways and hence contribute to the origin of novelties. Moreover, many of these processes have a strong stochastic component, which contributes even more to the state of pessimism about the generation of deterministic laws. Yet, at an idiographic level, at least, population genetics has contributed substantially to our understanding of differentiation, including the origin of reproductive isolation. From a systematic viewpoint, however, there seems to be a need for more discussion about the exact nature of the problems facing those biologists interested in the genetics of speciation and what the major questions might be. Only by defining the problems more precisely are we likely to develop major explanatory generalizations.

CONCLUSIONS

Carson (1985:380) has recently argued that evolutionary biology suffers from "a regrettable lack of unification of theory relating to the modes or processes involved in the origin of new species." One primary reason for this, he contends, is the biological species concept, which Carson sees as having been particularly unsuccessful in plants. The examples discussed in this chapter demonstrate that the difficulties of the biological species concept are much more general and extend to animals as well. Even in the two groups most often associated with the biological species concept, birds and mammals, these difficulties are commonplace. As Hafner et al. (1987) note, many biological species are likely to be paraphyletic. If so, then it will not be possible to recover the true history of speciation, and all results based on these taxa will be misleading. Moreover, even so-called "monophyletic" biological species will obscure historical reconstruction because they mask the true number of differentiated evolutionary taxa and their genealogical relationships. Given these difficulties, nothing is served by using the biological species concept.

Equating differentiated evolutionary taxa with species at once provides a basis for unifying the description of evolutionary pattern in both plants and animals. Phylogenetic species are basal taxonomic units, and as such are broadly comparable. Although many different processes may underlie evolutionary change, a constant outcome is the origin of evolutionary taxa. The phylogenetic species concept recognizes this, and consequently provides a powerful ontological framework for systematic and evolutionary biology.

ACKNOWLEDGMENTS

I want to thank the authorities of the American Museum of Natural History, New York, and the Museum of Zoology, University of Michigan, for permitting me to study specimens in their care. I am grateful to Drs. Thomas E. Dowling, Robert M. Zink, and Mary C. McKitrick for their comments on the manuscript. This work was supported by a grant from the National Science Foundation.

LITERATURE CITED

American Ornithologists' Union. 1983. *Check-list of North American Birds,* 6th ed. American Ornithologists' Union, Lawrence, Kansas.

Avise, J. C. 1976. Genetic differentiation during speciation. Pp. 106–122 in: F. J. Ayala (ed.), *Molecular Evolution.* Sinauer Associates, Sunderland, Massachusetts.

Ayala, F. J. 1975. Genetic differentiation during the speciation process. Evolution. Biol. 8:1–78.

Ayala, F. J. 1981. Speciation: Stages, modes and genetic analysis. Pp. 45–67 in: O. A. Reig (ed.), *Ecology and Genetics of Animal Speciation.* Univ. Simon Bolivar, Decanato Invest., Caracas, Venezuela.

Ayala, F. J., M. L. Tracey, D. Hedgecock, and R. C. Richmond. 1974. Genetic differentiation during the speciation process in *Drosophila.* Evolution 28:576–592.

Barton, N. H., and B. Charlesworth. 1984. Genetic revolutions, founder effects, and speciation. Annu. Rev. Ecol. System. 15:133–164.

Barton, N. H., and G. M. Hewitt. 1985. Analysis of hybrid zones. Annu. Rev. Ecol. System. 16:113–148.

Bush, G. L. 1975. Modes of animal speciation. Annu. Rev. Ecol. System. 6:339–364.

Campbell, A. J., and A. G. Campbell. 1926. A review of the genus *Cinclosoma.* Emu 26:26–40.

Carson, H. L. 1985. Unification of speciation theory in plants and animals. System. Bot. 10:380–390.

Condon, H. T. 1962. Australian quail-thrushes of the genus *Cinclosoma.* Records South Aust. Mus. 14:337–370.

Coyne, J. A., and N. H. Barton. 1988. What do we know about speciation? Nature (London) 331:485–486.

Cracraft, J. 1982. Geographic differentiation, cladistics, and vicariance biogeography: Reconstructing the tempo and mode of evolution. Am. Zool. 22:411–424.

Cracraft, J. 1983. Species concepts and speciation analysis. Curr. Ornithol. 1:159–187.

Cracraft, J. 1986. Origin and evolution of continental biotas: Speciation and historical congruence within the Australian avifauna. Evolution 40:977–996.

Cracraft, J. 1987. Species concepts and the ontology of evolution. Biol. Philos. 2:63–80.

Cracraft, J. 1988. Species as entities of biological theory. In M. Ruse (ed.), *What the Philosophy of Biology Is.* D. Reidel, Dordrecht. In press.

Cronquist, A. 1978. Once again, what is a species? Beltsville Symp. Agricult. Res. 2:3–20.

Davis, J. 1951. Distribution and variation of the brown towhees. Univ. Calif. Publ. Zool. 52:1–120.

Delacour, J., and D. Amadon. 1973. *Curassows and Related Birds.* American Museum of Natural History, New York.

Diegnan, H. G. 1964. Subfamily Orthonychinae. Pp. 228–240 in: E. Mayr and R. A. Paynter, Jr. (eds.), *Check-list of Birds of the World.* Museum of Comparative Zoology, Harvard University.

Dobzhansky, Th. 1937. *Genetics and the Origin of Species.* Columbia University Press, New York.

Dobzhansky, Th., F. J. Ayala, G. L. Stebbins, and J. W. Valentine. 1977. *Evolution*. W. H. Freeman, San Francisco.
Donoghue, M. J. 1985. A critique of the biological species concept and recommendations for a phylogenetic alternative. The Bryologist 88:172–181.
Ehrendorfer, F. 1984. Artbegriff und Artbildung in botanischer Sicht. Z. zool. System. Evolution. 22:234–263.
Ehrlich, P. R. 1961. Has the biological species concept outlived its usefulness? System. Zool. 10:167–176.
Endler, J. A. 1977. *Geographic Variation, Speciation, and Clines*. Princeton University Press, Princeton, N.J.
Ford, J. 1974. Taxonomic significance of some hybrid and aberrant-plumaged quail-thrushes. Emu 74:80–90.
Ford, J. 1976. Systematics and speciation in the quail-thrushes of Australia and New Guinea. Proc. 16th Int. Ornithol. Congr. 542–556.
Ford, J. 1980. Hybridization between contiguous subspecies of the Varied Sittella in Queensland. Emu 80:1–12.
Ford, J. 1983. Evolutionary and ecological relationships between quail-thrushes. Emu 81:57–81.
Ford, J., and S. A. Parker. 1974. Distribution and taxonomy of some birds from south-western Queensland. Emu 74:177–194.
Frost, J. S., and J. E. Platz. 1983. Comparative assessment of modes of reproductive isolation among four species of leopard frogs (*Rana pipiens* complex). Evolution 37:66–78.
Futuyma, D. J. 1986. *Evolutionary Biology*. Sinauer Associates, Sunderland, Massachusetts.
Gaukroger, S. 1978. *Explanatory Structures*. Humanities Press, Atlantic Highlands, N.J.
Gould, S. J., and R. F. Johnston. 1972. Geographic variation. Annu. Rev. Ecol. System. 3:457–498.
Haffer, J. 1974. Avian speciation in tropical South America. Publ. Nuttall Ornithol. Club 14:1–390.
Hafner, M. S., J. C. Hafner, J. L. Patton, and M. F. Smith. 1987. Macrogeographic patterns of genetic differentiation in the pocket gopher *Thomomys umbrinus*. System. Zool. 36: 18–34.
Hubbard, J. P. 1973. Avian evolution in the aridlands of North America. Living Bird 12:155–196.
Hull, D. L. 1976. Are species really individuals? System. Zool. 25:174–191.
Hull, D. L. 1977. The ontological status of species as evolutionary units. Pp. 91–102 in: R. Butts and J. Hintikka (eds.), *Foundational Problems in the Special Sciences*. D. Reidel, Dordrecht-Holland.
Hull, D. L. 1978. A matter of individuality. Philos. Sci. 45:335–360.
Keast, A. 1961. Bird speciation on the Australian continent. Bull. Mus. Comp. Zool. 123:303–495.
Krieber, M., and M. R. Rose. 1986. Molecular aspects of the species barrier. Annu. Rev. Ecol. System. 17:465–485.
Levin, D. A. 1979. The nature of plant species. Science 204:381–384.
Lewontin, R. C. 1974. *The Genetic Basis of Evolutionary Change*. Columbia University Press, New York.
MacDonald, J. D. 1968. Notes on the genus *Cinclosoma*. Emu 67:283–289.
MacDonald, J. D. 1973. *Birds of Australia*. A. H. & A. W. Reed, Sidney, Australia.
Marshall, J. T., Jr. 1960. Interrelations of Abert and Brown towhees. Condor 62:49–64.
Marshall, J. T., Jr. 1964. Voice in communication and relationships among brown towhees. Condor 66:345–356.
Mayr, E. 1942. *Systematics and the Origin of Species*. Columbia University Press, New York.
Mayr, E. 1950. Taxonomic notes on the genus *Neositta*. Emu 49:282–291.
Mayr, E. 1963. *Animal Species and Evolution*. Harvard University Press, Cambridge.
Mayr, E. 1969. *Principles of Systematic Zoology*. McGraw-Hill, New York.
Mayr, E. 1982. *The Growth of Biological Thought*. Harvard University Press, Cambridge.

Mayr, E., and L. L. Short. 1970. *Species Taxa of North American Birds.* Publications of the Nuttall Ornithological Club No. 9, Museum of Comparative Zoology, Harvard University, Cambridge.
McGill, A. R. 1948. A distributional review of the genus *Neositta.* Emu 48:33–52.
McKitrick, M. C., and R. M. Zink. 1988. Species concepts in ornithology. Condor 90:1–14.
Mishler, B. D., and M. J. Donoghue. 1982. Species concepts: A case for pluralism. System. Zool. 31:491–503.
Nei, M. 1987. *Molecular Evolutionary Genetics.* Columbia University Press, New York.
Nei, M., T. Maruyama, and C. Wu. 1983. Models of evolution of reproductive isolation. Genetics 103:557–579.
Nelson, G. J., and N. I. Platnick. 1981. *Systematics and Biogeography: Cladistics and Vicariance.* Columbia University Press, New York.
Nevo, E., G. Heth, A. Beiles, and E. Frankenberg. 1987. Geographic dialects in blind mole rats: Role of vocal communication in active speciation. Proc. Natl. Acad. Sci. U.S.A. 84:3312–3315.
Paterson, H. E. H. 1981. The continuing search for the unknown and unknowable: A critique of contemporary ideas on speciation. South African J. Sci. 77:113–119.
Paterson, H. E. H. 1982. Perspective on speciation by reinforcement. South African J. Sci. 78:53–57.
Paterson, H. E. H. 1985. The recognition concept of species. Transvaal Mus. Monograph No. 4:21–29.
Patton, J. L. 1981. Chromosomal and genic divergence, population structure, and speciation potential in *Thomomys bottae* pocket gophers. Pp. 255–295 in: O. A. Reig (ed.), *Ecology and Genetics of Animal Speciation.* Univ. Simon Bolivar Decanato Investigaciones, Caracas, Venezuela.
Patton, J. L., and J. H. Feder. 1978. Genetic divergence between populations of the pocket gopher, *Thomomys umbrinus* (Richardson). Z. Saugetier. 43:17–30.
Patton, J. L., and M. F. Smith. 1981. Molecular evolution in *Thomomys* pocket gophers: phyletic systematics, paraphyly, and rates of evolution. J. Mammal. 62:493–500.
Pizzey, G. 1980. *A Field Guide to the Birds of Australia.* Princeton University Press, Princeton, New Jersey.
Prum, R. O. 1982. Systematics and biogeography of the family Ramphastidae (Aves). Unpublished Senior Honors Thesis, Harvard University, Cambridge, Massachusetts.
Prum, R. O. 1988. Historical relationships among avian forest areas of endemism in the neotropics. Proc. 19th Int. Ornithol. Congr., Ottawa, Canada.
Raven, P. H. 1980. Hybridization and the nature of species in higher plants. Can. Bot. Assoc. Bull. Suppl. 13:3–10.
Rose, M. R., and W. F. Doolittle. 1983. Molecular biological mechanisms of speciation. Science 220:157–162.
Rosen, D. E. 1978. Vicariant patterns and historical explanation in biogeography. System. Zool. 27:159–188.
Rosen, D. E. 1979. Fishes from the uplands and intermontane basins of Guatemala: Revisionary studies and comparative geography. Bull. Am. Mus. Nat. Hist. 162: 267–376.
Rosenberg, A. 1985. *The Structure of Biological Science.* Cambridge University Press, New York.
Short, L. L., R. Schodde, and J. F. M. Horne. 1983a. Five-way hybridization of Varied Sittellas *Daphoenositta chrysoptera* (Aves: Neosittidae) in central Queensland. Aust. J. Zool. 31:499–516.
Short, L. L., R. Schodde, R. A. Noske, and J. F. M. Horne. 1983b. Hybridization of 'White-headed' and 'Orange-winged' Varied Sittellas, *Daphoenositta chrysoptera leucocephala* and *D. c. chrysoptera* (Aves: Neosittidae), in eastern Australia. Aust. J. Zool. 31:517–531.
Sneath, P. H. A., and R. R. Sokal. 1973. *Numerical Taxonomy.* W. H. Freeman, San Francisco.
Sokal, R. R., and T. J. Crovello. 1970. The biological species concept: A critical evaluation. Am. Natur. 104:127–153.

Straney, D. O., and J. L. Patton. 1980. Phylogenetic and environmental determinants of geographic variation of the pocket mouse *Perognathus goldmani* Osgood. Evolution 34:888–903.

Sudhaus, W. 1984. Artbegriff und Artbildung in zoologischer Sicht. Z. zool. System. Evolution. 22:183–211.

Templeton, A. R. 1981. Mechanisms of speciation—a population genetic approach. Annu. Rev. Ecol. System. 12:23–48.

Templeton, A. R. 1982. Genetic architectures of speciation. Pp. 105–121 in: D. Barigozzi (ed.), *Mechanisms of Speciation.* A. R. Liss, New York.

Vaurie, C. 1965. Systematic notes on the bird family Cracidae. No. 4. *Ortalis garrula* and *Ortalis ruficauda.* Am. Mus. Novitates 2237:1–16.

Vaurie, C. 1968. Taxonomy of the Cracidae (Aves). Bull. Am. Mus. Nat. Hist. 138: 131–260.

White, M. J. D. 1978. *Modes of Speciation.* W. H. Freeman, San Francisco.

Willmann, R. 1987. Missverständnisse um das biologische Artkonzept. Paläontol. Z. 61:3–15.

Wolstenholme, H. 1926. *Official Checklist of the Birds of Australia,* 2nd ed. H. J. Green, Melbourne.

Zimmerman, E. G., C. W. Kilpatrick, and B. J. Hart. 1978. The genetics of speciation in the rodent genus *Peromyscus.* Evolution 32:565–579.

Zink, R. M. 1988. Evolution of brown towhees: Allozymes, morphometrics and species limits. Condor 90:72–82.

Zink, R. M., and J. V. Remsen, Jr. 1986. Evolutionary processes and patterns of geographic variation in birds. Curr. Ornithol. 4:1–69.

CHAPTER 3

SPECIES AND TAXA:

Systematics and Evolution

Gareth Nelson

Were classical taxonomy a science it would have its basic unit—"the species"—adequately defined.—P. Raven, 1974:167, from W. H. Camp, 1951:118

The manifest tendency of life toward formation of discrete arrays is not deducible from any a priori considerations. It is simply a fact to be reckoned with. —Th. Dobzhansky, 1935:347.

This fact shows that the title of Darwin's book is a misnomer, it should have been called "On the Origin of Taxa."—S. Løvtrup, 1987c:9

In 1812 Napoleon invaded Russia, Beethoven composed his seventh and eighth symphonies, Cuvier published his *Recherches sur les Ossemens Fossiles de Quadrupèdes*, the United States declared war on England, and in January of the year in Philadelphia, six gentlemen, assembled in an apothecary shop at the corner of Market and Second Streets, founded the Academy of Natural Sciences of Philadelphia. We are here today to celebrate that event and the subsequent history of this Academy by addressing the topic of speciation.

At the outset I confess a disbelief in species, as that word is commonly understood to refer to the basic taxonomic unit or to the taxonomic unit of evolution (Gingerich 1985; Hecht and Hoffman 1986). There seem to be no basic taxonomic unit and no particular taxonomic unit of evolution. Still, taxa are real. Among taxa there are some that arbitrarily are termed families, there are others termed genera, and there are others termed species. But as Louis Agassiz said long ago (1859:8), "species do not exist in nature in a different way from the higher groups"—a view expressed nearly a century later by Carl Hubbs (1943:110) and more recently by others (Mishler and Donoghue 1982; Donoghue 1985; Løvtrup 1987a, 1987b, 1987d, 1987e). Indeed, taxa are all much the same, even if some taxa include others. I hesitate to suggest that if there are taxonomic units of evolution, the units are taxa generally, for the suggestion contradicts all that I have been taught by my teachers in school and by most of my colleagues since, who apparently have adopted the biological species doctrine of the "new systematics" and of the "modern synthesis" of the late 1930s (e.g., White 1978; Grant 1985).

The suggestion is not radical because certain aspects of evolution are seen as processes occurring not within taxa but within populations, so much so that one occasionally reads that "there is no such thing as evolution of *species*, but only evolution of *populations* of organisms" (Bunge 1981:284; Levin 1979; Carson 1985; Raven 1986). Nor is the suggestion at odds with the consensus among botanists that species are not fundamental units in any case (e.g., Backmann 1987). The suggestion is at odds with the view that species, and taxa generally, are arbitrary constructs—that they are units of convenience but not of evolution (e.g., Sylvester-Bradley 1956; cf., Bonde 1981)—and with the biological species doctrine, whether isolationist or recognitional (Lambert et al. 1987; Paterson 1987), according to which a species is, or should be, or should be regarded as, one integrated population or suprapopulation (Bush 1981), so much so that one occasionally reads that "what really evolve are species: they speciate and become transformed. Genera and higher taxa do nothing whatever" (Ghiselin 1984:213). Counterintuitively, the suggestion affirms that in some factual sense taxa—such as Animalia, Vertebrata, Amniota, *Homo sapiens*—evolve in the same way, that they all differentiate and diversify (cf. Mayr 1987b:308: "Every deme evolves and so does every subspecies as well as every higher taxon"). The suggestion is really all that I have to offer on the narrow subject of speciation. If there is some meaning to the suggestion I suppose that it is taught by nature, although I am skeptical of a nature that teaches only the few and not the many.

What of taxa, then, and their diversification? I approach this question from the field of systematics and, more particularly, from the context of the last 25 years—really a short time in the history of the discipline—in which the subjects of cladistics and vicariance have been extensively discussed and developed. Some persons see the value of these recent developments as revolutionary, liberating us from the burdens of the past, and particularly from the necessity to understand the work and thought of our predecessors. I see recent developments as embedded in, and shedding light on, an old tradition, dating from the days of Cuvier and his colleagues, if not from the time of Linnaeus, or for that matter from the time of Aristotle (e.g., Nelson and Platnick 1981; Bonde and Stangerup 1985; Papavero and Balsa 1986). This view is sometimes portrayed as "dull, atheoretical, and forever committed mindlessly to dusting off, correcting, and adding to the accomplishments of our forerunners" (Eldredge and Novacek 1985:65).

The early nineteenth century gave birth not merely to this Academy but, because of its development of experimental procedure, to much of biology as we know it today. The topics of species and speciation were extensively discussed in those years, and the topics were ably reviewed in 1859 by Isidore Geoffroy Saint-Hilaire, the only son of his illustrious father Etienne (Appel 1987). He commented (1859:349, translated):

> The word species is, of all the words, the one most often encountered in the books of natural history. On what page is it not found? And what notion could be said to be fundamental with more right than that which this word expresses? It is the first and the last word in natural history, and the day when we become complete masters of it we would be very close to the millennium of science in general.

In reading Geoffroy's account of these early days, one is struck by the many themes enduring to our own time (e.g., Mayr 1957; Slobodchikoff 1976; Bocquet et al. 1976, 1977; Barigozzi 1982; Sober 1984). Another commentator of that time and place is Pierre Trémaux, who also considered the problem of the origin of species. Taking his cues from Geoffroy's compilation he wrote (1865:133–134, translated):

> Two principal notions have served to define the species: resemblance between individuals and ability to reproduce. To the extent that the first of these conditions, which is only the consequence of the second, can be abandoned in order to focus on the second, one approaches a solution to the problem. For Laurent de Jussieu the species is a succession of individuals entirely similar, perpetuated through reproduction. For Buffon the species is the continual succession of similar individuals that reproduces itself, and the character of the species is common fecundity. Blainville describes the species thus: the individual repeated in time and space. According to Lamarck the species is a collection of similar individuals, which reproduction perpetuates in the same state, as long as the circumstances of their situation do not change enough to cause variation in their habits, their character, and their form. Of definitions of species there are as many as there are naturalists.

Trémaux's problem is the origin of species, and his solution to it is the same as that known today as punctuated equilibrium. It has been said, although I have lost track of the botanical source, that "for want of knowing the work of our predecessors we grind over and over again the same grist." No doubt. But I suspect that our notions—our theories—are the millstones that carry us round in circles. I have in mind evolutionary theory in particular and its demand for a particular taxonomic unit—the species as opposed to taxa generally—that participates in evolutionary processes. Perhaps the demand is to conflate two notions of species that are best kept separate, the one from systematics and the other from genetics (Paterson 1981; Donoghue 1985). A third, and ecological, notion of species is sometimes added to the conflation (Bock 1986). In any case, the conflated unit (the species) is seen, incredibly, as real, whereas other taxa (genera, families, orders) are seen as not real, or at least not so real, and sometimes progressively less real in the degree to which they are remote from the species. Thus, other taxa are seen as more or less arbitrary aggregates of related species (Mayr 1969). Even when all taxa are seen as real, a view common among cladists, species are seen as units of evolution, and other taxa as by-products of evolution of species (Hennig 1950; Ax 1984, 1985a, 1985b, 1987). Or species are seen as individuals and other taxa are not (Eldredge and Cracraft 1980; cf. Cracraft 1983). Or other taxa are seen as individuals and species are not (von Vaupel Klein 1987). Or species are seen as units of evolution and other taxa as units of history (Wiley 1981b). Or species are seen as the terminal, and other taxa as the nonterminal, entities of evolutionary history (Løvtrup 1979; Wiley 1980), the former making history and the latter having made it, the former in a world of becoming and the latter in a world of being (Patterson 1988). However intuitively appealing, distinctions between species and other taxa always encounter empirical difficulties. They are a perennial, if not eternal, topic of discussion and dispute about whose distinctions will prevail in the science, and in the educational curriculum, of tomorrow. They are like selfish genes. They are the stuff of theory and sociology.

To the founders of the Academy sociology was unknown and theory seemingly was a bad word. In the first volume of the Academy's journal the editors stated the following hope (pp. 1–2):

> to publish a few pages whenever it appears to them that materials worthy of publication have been put in their possession. In so doing, they propose to exclude entirely all papers of mere theory—to confine their communications as much as possible to facts—and by abridging papers too long for publication in their original state, to present the facts thus published, clothed in as few words as are consistent with perspicuous description.

In keeping with this hope, I take as fact that evolutionary theory is in a period of revolving (I cannot say of revolution)—a period of cycles and epicycles of intepretation of theoretical notions now centuries old of species, speciation, and the like. Trémaux and punctuated equilibrium are a case in point. In this particular case, matters are bleaker still if we consider empirical claims, such as those regarding stasis in the fossil record. Here the findings of Heinrich Bronn (1858) are touted as modern discoveries of great significance for evolutionary theory (e.g., Wake et al. 1983; Eldredge and Stanley 1984; cf. Janvier 1983). Bronn's survey of the fossil record was prepared in response to an open proposal by the Academy of Sciences in Paris in 1850 (Anon. 1850), and in 1857 Bronn was given an award by that institution (Anon. 1857). The proposal (Bronn 1859:81) was as follows:

Study the laws of the distribution of fossil organic bodies in the different sedimentary strata according to the order of superposition. Discuss the question of their successive or simultaneous appearance and disappearance. Investigate the nature of the relations existing between the present state of the organic kingdom and its former states.

Bronn's survey was done late in life and summarized much of his life's work. There is no reason today to ignore such a document (Stanley 1979; Valentine 1982). Trémaux's writings were never well known. Perhaps they deserve the oblivion into which they have fallen, but their general theme of earth and life evolving together accords well with the spirit of our modern concerns with vicariance. For reasons not altogether clear Tremaux's work was of interest for a time to Karl Marx, who felt that it was "a very important advance" over the state of darwinism in the late 1860s (Cohen 1985:346). He corresponded with Frederick Engels about it (Lecourt, 1983).

What of taxa, then, and their diversification? A few years ago Norman Platnick and I addressed this question for a chapter in a proposed book entitled *Beyond Neodarwinism.* Our initial problem was to see that darwinism, or neodarwinism, has an identity within the area of systematics. This was no small chore, but after some agonizing we wrote (Nelson and Platnick 1984:143–144):

We believe that darwinism has an identity within the area of biological systematics, that it has a history within that discipline, that it is, in short, a theory that has been put to the test and found false. We believe further that the process of test and falsification has changed the nature of systematics, so that the discipline is already "beyond darwinism" in a significant sense.

These are strong words, but they have some sense, which I will try once again to address. I do so in spite of the reaction that this sense, or the lack of it, has provoked in some quarters, and may provoke again. An example of the reaction is furnished by Richard Dawkins (1986:284) (also Ridley 1986):

False, note well, is precisely the word Nelson and Platnick used. Needless to say, their words have been picked up by the sensitive microphones that I mentioned in the previous chapter, and the result has been considerable publicity. They have earned themselves a place of honour in fundamentalist, creationist literature.

Even if true, which these remarks are not, they do not address the point at issue, and what I believe is the result of our experience with cladistics.

Dawkins (1986:284) continued:

As Mark Ridley more mildly said [Ridley 1985], in a review of the book in which Nelson and Platnick made that remark about darwinism being false, who would have guessed that all they really meant was that ancestral species are tricky to represent in cladistic classification?

Again these remarks do not address the point at issue nor even express the truth, but this sort of commentary has been repeated elsewhere, for example, in a review of Dawkins' own book by Douglas Futuyma (1987:36):

> According to the philosophy of classification that transformed cladists have adopted ancestral forms (such as the therapsid ancestors of mammals) should not be recognized as formal, named categories in classification. They deduce by a curious logic that Dawkins analyzes as well as anyone bound by real logic can, that if ancestors are not formally recognized by names, they cannot be real.

Once identified, or assumed to be identified, ancestral taxa may be placed in a classification, as they always have been, in the phylogenetic order of their appearance. Naming and placing them in a classification have never been a problem (cf. Wiley 1987). Their identification, however, was and remains a problem, and it is an empirical one. The alleged therapsid ancestors of mammals have already been dealt with in existing literature (Gardiner 1982; Kemp 1982, 1988; Hopson and Barghusen 1986; McKenna 1987). Their fate is much the same as that of the alleged thecodont ancestors of birds, which I will mention in due course.

I pass over the more florid remarks of these and similar commentators—I could natter on at some length about the remarks of Steve Gould, for example—because their remarks do, and my candid responses to them might, fall beyond the bounds of gentlemanly deportment as viewed by the founders of this Academy (Phillips 1953:266):

> If therefore any person after being a member shall discover himself to be so insensitive to propriety at any time in the Society as to insult another, much less to strike, or in or out of the Society, by any act or language to aim disgrace upon the Society, or to reflect disrespect upon any member to strangers, such individual is by this fact itself expelled without debate.

I pass over also any further reference to the piece by Platnick and myself in *Beyond Neodarwinism*. I will attempt to deal with the problem afresh. I begin by a statement, attributed to Richard Lewontin, of impeccable logic but of "very weak" (Lewontin in Bethell 1988:205) empirical content (Futuyma 1983:161, from Lewontin 1981):

> It is time for students of the evolutionary process, especially those who have been misquoted and used by the creationists, to state clearly that evolution is fact, not theory. . . . Birds arose from nonbirds and humans from nonhumans. No person who pretends to any understanding of the natural world can deny these facts any more than she or he can deny that the earth is round, rotates on its axis, and revolves around the sun.

Birds arose from nonbirds. Nonbirds are devoid of empirical content.

This truth is not an original discovery of systematics. Parmenides long ago quipped that "being is, non-being is not" (Bahm 1970:267). What systematics has discovered is that this quality of nonness inheres in all of those taxa that were thought, with some reason, to be ancestral. The reason is the nonness. Indeed, nonness is their hallmark, which has frequently been presented as empirical evidence of plausible ancestry. Systematists discovered otherwise not because they adopted some arcane philosophical stance or peculiar logic or because they wanted to make the discovery. Quite the contrary. They dearly wanted ancestral taxa with empirical content, and they went to tortuous lengths to try and find them in ever lower levels of their classifications. They were in fact dismayed, and I and all of my cladistic colleagues among them. We were dismayed to find that such taxa as were proposed always dissolved with further empirical study. It was left to Willi Hennig (1950, 1966) in the 1950s and later in the 1960s, when his work was translated into English, to formalize this phenomenon of dissolution under the name of paraphyly, and to Colin Patterson (1977, 1981a, 1981b) later to generalize the concept to all fossil groups thought to be ancestral. As a result, paleontologists involved in these investigations came to change their expectations and quickly reached the consensus that ancestral taxa cannot be empirically demonstrated (Schoch 1986). The older ancestral groups were abandoned first piecemeal and then wholesale. There was a last-ditch hope that ancestral taxa at the species level could be saved, at least in theory, and this hope persists.

Birds arose from nonbirds. An oak is not a bird. Did birds evolve from oaks? Obviously not. Birds evolved from nonbird animals. They evolved from nonbird vertebrates. They evolved from nonbird amniotes. We accept these as facts because we accept birds as members of these taxa—Animalia, Vertebrata, Amniota. It might seem that this taxonomic progression, from animal to vertebrate to amniote, is progress toward increasing empirical content—ancestry. Really all that the progression accomplishes is to transfer to lower and lower taxonomic levels a notion (nonbird) itself devoid of empirical content. What meaning we might see in these more precise statements, and surely they are more precise, resides only in the structure of the classification.

Alfred Romer (1966:166) asserted:

> Surely the birds arose independently of either dinosaurian stock, from Triassic thecodonts, and there is little in the structure in some of the small bipedal thecodonts to debar them from being avian ancestors.

Here at last we might believe there is some empirical content—the small bipedal thecodonts in the rocks. Alas, not only are the thecodonts nonbird archosaurs, but they are, on the authority of Romer's classification of reptiles, also noncrocodilian, nonpterosaur, nonsaurischian, and nonornithischian archosaurs as well. They are truly an impressive archosaurian nonness (Charig 1983; Carroll 1988). It might seem that the more things that they are not, the more that they are. But a string of zeros, however long, sums to zero.

Everett Olson (1971:347) commented that the thecodonts include "an array of primitive archosaurian reptiles" that

> fits well into the "horizontal" pattern of classification that is often imposed on a "basal complex" in which lines are difficult to sort out. With increased knowledge, this sorting may take place, and if fully successful, the primitive complex may disappear from classification.

Hennig's "paraphyly" is another name for the syndrome of the disappearing basal complex. True to form, thecodonts subsequently were empirically found (Gauthier and Padian 1985; Gauthier 1986; Padian 1986) to be another of those paraphyletic groups that Patterson (1982:64) has portrayed with his usual clarity and eloquence:

> Is it not strange that the justification of phylogeny, as something beyond systematics, resides in extinct paraphyletic groups? For those groups are the inventions of evolutionists, those who appeal to them as demonstrating the path of descent. So far as I know, such groups did not exist in pre-Darwinian taxonomy, for paleontologists were then preoccupied with the real problem of allocating fossils to recent groups (Patterson, 1977:596). Nor do I find any extinct paraphyletic groups in Haeckel's (1866) trees [e.g., Patterson 1983; Oppenheimer 1987]. Such groups are therefore a later invention, imagined by evolutionists committed to the confirmation of Darwin's views.

To ardent darwinians these, incredibly, are fighting words, but in themselves the words are calm, clear, and reasonable. They bring us to a point at which I can easily summarize. Yes, we have knowledge of evolution. Yes, we have knowledge of systematics. If we ask how these two areas of knowledge interrelate we find that they do so by way of a third and intermediate area of knowledge—of phylogeny. How we come to know phylogeny is variously viewed. Some persons believe and argue, on grounds approaching faith it seems to me, that phylogeny comes from our knowledge of evolution. Others have found to their surprise, and sometimes dismay, that phylogeny comes from our knowledge of systematics. If we analyze these conflicting claims we find that each has its own way of seeing and portraying phylogeny. Evolutionists, or better, darwinians, see taxa in relation as ancestor and descendant and necessarily so—never mind the syndrome of the disappearing basal complex. Cladists see taxa as descendants only, necessarily without empirical claims as to direct ancestry, because the syndrome by its very nature is chronic and incurable. Interestingly, these two ways of seeing are not diametrically opposed, but rather they overlap because all ancestors are also descendants. Here the cladists have the advantage of greater generality. The darwinians (Szalay 1977; Ridley 1986) have the potential advantage of more empirical content in the form of ancestral taxa—if only there were an ancestral taxon with an identity so that specimens of it could be recognized as such—if only there were an ancestral taxon immune to the infectious agent causing the syndrome. Alas, the infectious agent is the darwinian himself, when he invents the ancestral "taxon" and then appeals to it as an empirical demonstration of the path of descent toward adaptation, speciation, adaptive radiation, extinction, and the sundry other byways of evolution.

The conflicting claims are opposed on the matter of ancestral taxa, and ultimately on whether alleged "taxa" such as nonbirds—or nonbird animals, or nonbird vertebrates, or nonbird amniotes—have an empirical identity. Cladists claim not. On this matter I side, of course, with the cladists. Through their empirical and theoretical work they have amply demonstrated, at least in my judgment, that claims to the contrary are false. I choose the word "false" with all due regard to its meaning, implications, and empirical requirements (cf. Kemp 1985). In sum, darwinians are in the "ludicrous" situation imagined by Kitcher (1985:64):

> . . . there are global constraints on the collection of Darwinian histories. Pleas of bad luck in the face of absent fossils look suspect when we are considering a single Darwinian history. They would take on a ludicrous appearance if we were to find that the entire fossil record was marked by missing ancestors.

How important are these matters, for us, for biology, and for science? I do not know how important they are. Perhaps they are not important at all. It is possible that future discovery of evolutionary principles and laws may occur without reference to phylogeny. But we have not yet achieved the millennium, when taxa—or species if you like—are irrelevant to these concerns (Rosen 1984). If we are to judge the importance of these matters, be the importance nil, small, or large, we may first try to understand them clearly. It is not realistic to expect clear and widespread understanding to spring effortlessly into existence. I personally regret that it must be so, and I beg your indulgence for these few minutes, indeed for the entire period of what Gould calls the "cladist wars"—of what might more accurately be called the darwinian hundred-year war on systematics.

Is it fair to Darwin, or to darwinians, to see a darwinian presence, be it benign or otherwise, within the area of systematics? And to characterize that presence as theory by the expectation that taxa stand in relation as ancestor and descendant? On reconsideration I answer yes, why not? The characterization is historically accurate. Besides, the characterization is interesting, for it places in perspective certain recent discussions concerning individuality of species (Ghiselin 1981, 1987b; Fink 1981; Hull 1981; Wiley 1981a; Rosenberg 1985; Vrba 1985; Mayr 1987a; Zandee and

Geesink 1987), punctuated equilibrium (Stanley 1981; Gould 1984; Eldredge 1985a), species selection (Hoffman 1984; Eldredge and Gould 1988), nomothetic paleontology (Raup and Gould 1974; Gould and Eldredge 1977; Gould 1980a; Gilinsky and Bambach 1986; Sepkoski and Raup 1986; Erwin et al. 1987), hierarchical theories of evolution (Gould 1982a, 1982b; Salthe 1985; Eldredge 1985b), evolution as entropy (Rifkin 1980; Brooks and Wiley 1986; Wicken 1987; Weber et al. 1988), and who knows what future efforts along these lines.

These discussions of individuality, equilibrium, selection, nomothesia, hierarchy, and entropy indicate to me that in spite of the progress in cladistic systematics, there is a lingering problem with the species. Ancestral groups might all be paraphyletic, but according to these modern darwinians ancestral species are not paraphyletic, or even if they are it does not matter (de Queiroz and Donoghue 1988). Given the consensus about ancestral groups, the darwinian presence in systematics thus retreats to the species level, where perhaps it always was most comfortably entrenched—with species as the alleged units of evolution after all. I personally view these discussion with dismay, for in a way they carry us full circle. Some of my aging cladist colleagues now stand fast in the darwinian trench. Their colors, so far as I can discern, are a newer "new systematics" and a more modern "modern synthesis" (Grene 1983; Ghiselin 1987a; Hull 1987; cf. Kitcher 1984a, 1984b, 1987, 1988). Well, such are the vicissitudes of life on a millstone, or if not on a millstone, then perhaps on a random walk (Bookstein 1987).

Are ancestral species paraphyletic? One can argue about theory and hypothetical examples, and here again my sympathies lie with the cladists, not with the darwinians: of course, ancestral species are paraphyletic. Robert Schoch commented (1986:164):

> From the discussion above it should be clear that, by definition, ancestors at the species level are not recognized by unique positive occurrences of characters; they do not have unique identities, individualities, or histories (Forey, 1982). At best an ancestor can be recognized by what it is not; it is not the descendant. In other words, even species-level ancestors are paraphyletic, not monophyletic, and suffer from all the problems inherent to nonmonophyletic taxa (see section on monophyly, polyphyly, and paraphyly). This problem is not solved by adopting a convention that all species (whether ancestors or not) will, by definition, be considered individuals and monophyletic. Even at the species level, ancestors may be taxonomic artifacts or arbitrary conventions that do not correspond to real groups in nature.

But what about the facts, the alleged ancestral species known from specimens and the results of further study of them? There is virtually nothing that can be said unless some significant number of them is found—or alleged to have been found so that they might be studied with significant result. Recent interest in punctuated equilibrium has produced many schemes of ancestor-descendant relationships among species (e.g., Cheetham 1987). These alleged ancestral species provide raw material so that the matter might be put to the test and laid to rest. Will the ancestral species exhibit the syndrome of the disappearing basal complex? Or will they, because they are the ultimately simple, immune, and perfect units, defy the probes of further systematic investigation?—as if there is a last, and yet unopened, Chinese box that *must* contain the pearl of great price. Well, is there? And does it? Ledyard Stebbins (1987:200) has answered in a related context:

> Again, the answer is "No." I know of no instance in which a species known from the fossil record has, as a single unit, evolved into another species.

If Stebbins' is the correct result, will the last trench fall or stand firm? Will the hundred-year war ever be over? Or, with the species level, have we reached the trench of ultimate insignificance? Does anyone care what really happens there?

Here the answer apparently is yes (Avise et al. 1987:518):

> No longer will it be defensible to consider species as phylogenetically monolithic entities in scenarios of speciation or macroevolution. Phylogenetic differences within species are qualitatively of the same kind as, though often smaller in magnitude than, those normally pictured in higher order phylogeny reconstructions.

Is phylogeny relevant to evolution? I suppose it is and always will be, but maybe not. If knowledge of phylogeny really comes from systematic endeavor, not from knowledge or theory of evolution, is that circumstance a bad thing? Perhaps it might help us better to understand the proper domain of evolutionary theory. If darwinism had at one time intruded into the area of systematics and proved ineffective, or even false in its expectations, would that be a calamitous development? I hardly think so, for darwinism was not originally a theory of systematics anyway. If there were no taxonomic units of evolution beyond taxa generally, would that circumstance be destructive to our endeavors? Surely we would continue on in much the same way as before, hopefully having learned a lesson along the way.

Whatever else might be said of neodarwinism, I think in retrospect that it was flawed in its handling of systematics and phylogeny (cf. Antonovics 1987 vs. Futuyma 1988). Ernst Mayr (1980:37) asserted that "more than anyone else, George Gaylord Simpson was responsible for bringing paleontology and macroevolution into the Synthesis." And Michael Ghiselin (1986:462) concurred, stating that in these respects Mayr's "views . . . have been largely derivative of Simpson." So with paleontology and macroevolution came a certain view of systematics and phylogeny. Ghiselin (1987a:109) asserted that Simpson's "taxonomic philosophy was one long series of blunders." Simpson's attitude toward darwinism is well known (1961:238): "a quiet but firm pronouncement by which we orient our lives" (cf. Gould 1980b, 1983 vs. Løvtrup 1987). His attitude toward cladistics (cf. Ghiselin, 1984) is exemplified in the following remarks, delivered at a scientific meeting in 1974 (Hull 1988:169):

> Many people—some people anyway—have said that if you start applying Hennigian taxonomy to a group of organisms, you are likely to wind up with nonsense. And the reply—the only reply that I have seen—a general reply, a reply on matter of principle rather than attacking an example, has been that the system cannot generate nonsense because everything that it generates is consistent with its premises. Well now, that sounds like a marvelous argument. But you know, if your premises are wrong and are idiosyncratic, you are insane if you argue this way. That is known as paranoia in psychiatric terms. So I think literally this system is a paranoiac system. Sure, it is logical if you accept its premises, but its premises are wrong and they are idiosyncratic. That is certainly a very violent, strong criticism to make and it does require some backing up. You may say that I have invested my life in a different system of taxonomy and so, of course, I'm going to think that anybody who does not agree with me is paranoid. That may or may not be true. But even if it is true, I do not believe it.

I have mentioned elsewhere that cladistics began as a reform of paleontology as represented, for example, in the "modern synthesis" (Nelson and Platnick 1984:156). To some extent this reform has been successful (cf. Campbell and Barwick 1988), even though it is retrograde in its rejection of darwinian, and neodarwinian, doctrine; and even though (Patterson 1987:8) its

> style of analysis was seen by some paleontologists as a threat or insult to their subject. By about 1960 paleontology had achieved such a hold on phylogeny reconstruction that there was a commonplace belief that if a group had no fossil record its phylogeny was totally unknown and unknowable.

The "modern synthesis" meanwhile has accommodated itself to punctuated equilibrium (Dawkins 1985; Gould and Eldredge 1986; Mayr 1988) and other such nomothetic aspirations. Thus, another cycle of reform—more retrograde motion, more perception of threat and insult—may prove inevitable at the species level (Janvier 1986; Rieppel 1986; cf. Rieppel 1984; Gould 1985; Laws and Fastowsky 1987). Symptomatic are the remarks of Olson and Brunk (1986:353; cf. Eldredge 1985c):

> A landmark in the trend to "trust the fossil record" was the publication of the model of punctuated equilibria (Eldredge and Gould, 1972), elaborating the concept developed by Eldredge (1971) and, before him, Simpson (1944).

What does it mean, "to trust the fossil record?" Unless I misread history there once was an expectation that the fossil record would reveal phylogeny in the form of taxa standing in relation as ancestor and descendant. By the 1930s this expectation was taken for granted and was merged into the "synthesis" without murmur or demur.

Mayr remarked (1980:42–43):

It was not that the synthesis was hammered out during the Princeton conference—rather, the conference constitutes the most convincing documentation that a synthesis occurred during the preceding decade.

With respect to the conference itself Mayr remarked (1980:42):

it was almost impossible to get a controversy going, so far-reaching was the basic agreement among the participants.

The Princeton conference of 1947 took place under the auspices of the "Committee on Common Problems of Genetics, Paleontology, and Systematics" of the National Research Council. The proceedings were published in book form under the title *Genetics, Paleontology, and Evolution* (Jepsen et al. 1949; cf. Levinton 1988). The transmutation of title itself suggests that paleontology had long been entrusted to deliver the goods of phylogeny (cf. Rachootin and Thomson 1981). Again, in retrospect, I think that neodarwinism was flawed in its handling of systematics and phylogeny, and that the flaw was born of this trust tacitly given to the fossil record, and received—I will not say abused—by the paleontology of that time and place. The development of cladistics was in effect a reclamation of the trust (e.g., Patterson and Smith 1987, 1988; Smith and Patterson 1988), which we may hope will never be given again—even at the species level—in spite of continuing and tacit claims for it made by "modern," or "nomothetic," paleontology (e.g., Gould 1985; Tiffney 1985; Raup 1986, 1987; Raup and Jablonski 1986; Sepkoski 1987; Raup and Boyagian 1988; Vrba 1988).

Among other issues is the evidence (more paraphyletic taxa, e.g. Boucot 1988) in favor of periodic extinction and its cause (as if artifacts were phenomena needing explanation). One favored cause is a hypothetical star, companion to our sun, that periodically spawns comet showers that punctuate the evolution of life on earth here below. Gould (1985:11) commented:

if the solar companion [star] exists, paleontology shall be the impetus for the greatest revision of cosmology, at least for our corner of the heavens, since Galileo.

I would have thought that rather than poor Galileo, the appropriate reference is Immanuel Velikovsky (1950). What was it that Thomas Huxley said of Herbert Spencer? Oh yes, that his only sense of tragedy is the "slaying of a beautiful deduction by an ugly fact" (Kennedy 1978:16). Indeed, the more things change, the more they do remain the same (Valentine 1982).

In commenting on the Chicago conference of 1980, Hampton Carson remarked (1981:773; also, 1987):

> Forty years ago, the modern followers of Darwin (Fisher, Haldane, Wright, Dobzhansky, and Mayr) stole the evolutionary spotlight from the paleontologists. This conference saw an attempt by a few fossil zealots who are able to charm reporters to regain attention. Most unfortunately, the ideas they used have neither data base nor innovation.

Phylogeny as a phenomenon has long since passed through the limelight of scientific discovery. It would be a mistake to think that cladistics, like genetics, has stolen rather than reclaimed something from "the paleontologists." Quite the contrary, for paleontologists—and geneticists, too—have significantly contributed to the development of cladistics, which has rendered phylogeny, in particular the analytical means required for its discovery, once again accessible to us all.

In a review of *Beyond Neodarwinism* Joel Cracraft stated (1985: 302):

> Whereas it is certainly true that many Darwinians have recognized higher taxa as ancestors and continue to do so, the argument of Nelson and Platnick remains a strawman. If on the other hand, they had argued that species-level hypotheses of ancestry and descent are artifacts (of something), then it would be back to the drawing board for Darwinism. No such argument was put forth, however. Foolish statements by some advocates of a theoretical worldview, in this case Darwinism, do not mean logically that the worldview is incorrect.

Well, now, do foolish statements mean that the worldview is correct? The statements are too concerted and too numerous to mean nothing. Strawmen never have stuffing so stern and solemn.

In truth I never thought that a special argument for species-level taxa would be necessary. Only an entrenched darwinian would think that. Duly admonished, nevertheless, I hope that this omission regarding species-level hypotheses has now been rectified to the satisfaction of all darwinians, so that in contentment they may honorably surrender their trench and go back to the drawing board, to confront the ugly fact of paraphyly and to leave the world of systematics in peace.

SUMMARY

The "species problem" is perennial (Howard, 1988), and "speciation remains as much a black box as ever" (Jackson, 1988). If we examine these problems we find a spectrum of solutions: some writers claim that everything, or everything important, is known; others claim that nothing, or nothing important, is known (Hull, 1988). I claim that the problems are insoluble, for they stem from a false assumption: that there is an empirical difference between species and other taxa, such that species evolve through speciation of other species.

My claim is based on the findings of cladistic systematics that ancestral taxa, regardless of whether they are species or higher taxa, are paraphyletic and, therefore, artifactual both in theory and in practice. If so, taxa give rise to other taxa only in the sense that ancestral taxa differentiate: taxa give rise to subtaxa, and all taxa evolve. Evolution of taxa is not a phenomenon confined to the species level except in neodarwinian theory, which in this respect is simply false.

ACKNOWLEDGMENTS

I am grateful to many persons for reading and commenting on one or another draft of this manuscript: S. Coats, M. Donoghue, N. Eldredge, J. Farris, P. Janvier, J. Kwok, S. Løvtrup, N. Macbeth, M. Novacek, C. Patterson, S. Salthe, and R. Schoch. I am particularly grateful to Philippe Janvier for suggesting to me long ago that fossil species are inherently problematic when considered from a cladistic point of view, and to Colin Patterson and Søren Løvtrup for their efforts, above and beyond the call of duty, to understand the relation between evolution and systematics.

LITERATURE CITED

Agassiz, L. 1859. *An Essay on Classification.* Longman, London. Reprint 1962. Harvard University Press, Cambridge.

Anon. 1850. Grand prix des sciences physiques. Compt. Rend. Séanc. Acad. Sci. Paris 30(9):257–260.

Anon. 1857. Rapport sur le concours pour le grand prix des sciences physiques. Compt. Rend. Séanc. Acad. Sci. Paris, 44(5):167–169.

Antonovics, J. 1987. The evolutionary dys-synthesis: Which bottles for which wine? Am. Nat. 129(3):321–331.

Appel, T. A. 1987. *The Cuvier–Geoffroy Debate: French Biology in the Decades before Darwin.* Oxford University Press, New York.

Avise, J. C., J. Arnold, R. M. Ball, E. Bermingham, T. Lamb, J. E. Neigel, C. Roeb,and N. C. Saunders. 1987. Intraspecific phylogeography: The mitochondrial DNA bridge between population genetics and systematics. Annu. Rev. Ecol. System. 18:489–522.

Ax, P. 1984. *Das phylogenetische System: Systematiesierung der lebenden Natur aufgrund ihrer Phylogenese.* Fischer, Stuttgart.

Ax, P. 1985a. Stem species and the stem lineage concept. Cladistics 1(3):179–287.

Ax, P. 1985b. Die stammesgeschichtliche Ordnung in der Natur. Abhand. Matemat. Naturwissenschaft. Klasse, Akad. Wissenschaft. Literat., Mainz, 1985(4):1–27.

Ax, P. 1987. *The Phylogenetic System: The Systematization of Organisms on the Basis of Their Phylogenies.* Wiley, Chichester.

Backmann, K. 1987. Speciation in *Microseris.* P. 67–80 in: P. Hovenkamp E. Grittenberger, E. Hennipman, R. de Jong, M. C. Roos, R. Sluys, and M. Zandee (eds.), *Systematics and Evolution: A Matter of Diversity.* Utrecht University, Utrecht.

Bahm, A. J. 1970. *Polarity, Dialectic, and Organicity*. Thomas, Springfield.
Barigozzi, C. (ed.). 1982. Mechanisms of speciation: Proceedings from the international meeting on mechanisms of speciation sponsored by the Academia Nazionale dei Lincei. May 4–8 1981, Rome, Italy. *Progress in Clinical and Biological Research*, Vol. 96. Liss, New York.
Bethell, T. 1988. *The Electric Windmill: An Inadvertent Autobiography*. Regnery Gateway, Washington, D.C.
Bock, W. J. 1986. Species concepts, speciation, and macromutation. Pp. 31–57 in: K. Iwatsuki, P. H. Raven, and W. J. Bock (eds.), *Modern Aspects of Species*. University of Tokyo Press, Tokyo.
Bocquet, C., J. Génermont, and M. Lamotte (eds.). 1977. Les problèmes de l'espéce dans le règne animal, Tome I. Mém. Soc. Zool. France 38:1–407.
Bocquet, C., J. Génermont, and M. Lamotte (eds.). 1976. Les problèmes de l'espéce dans le règne animal, Tome II. Mém. Soc. Zool. France 39:1–381.
Bonde, N. 1981. Problems of species concepts in paleontology. Pp. 19–34 in: J. Martinell (ed.), *International Symposium on "Concept and Method in Paleontology."* Universitat de Barcelona, Barcelona.
Bonde, N., and H. Stangerup. 1985. *Naturens historie fortaellere, 1: udviklingsideens historie fra Platon til Darwin*. G.E.C. Gad, Copenhagen.
Bookstein, F. L. 1987. Random walk and the existence of evolutionary rates. Paleobiology 13(4):446–464.
Boucot, A. J. 1988. Periodic extinctions within the Cenozoic. Nature 331(6155):395–396.
Bronn, H. G. 1858. *Untersuchungen über die Entwicklungs-Gesetze der organischen Welt während der Bildungs-Zeit unserer Erd-Oberfläche*. Schweitzerbart, Stuttgart.
Bronn, H. G. 1859. On the laws of evolution of the organic world during the formation of the crust of the earth. Ann. Mag. Nat. Hist. Ser. 3, 4(20):81–90, 4(21):175–184.
Brooks, D. R., and E. O. Wiley. 1986. *Evolution as Entropy: Toward a Unified Theory of Biology*. University of Chicago Press, Chicago.
Bunge, M. 1981. Biopopulations, not biospecies, are individuals and evolve. Behav. Brain Sci. 4(2):284–285.
Bush, G. L. 1981. Stasipatric speciation and rapid evolution in animals. Pp. 201–218 in: W. R. Atchley and D. S. Woodruff (eds.), *Evolution and Speciation: Essays in Honor of M. J. D. White*. Cambridge University Press, Cambridge.
Camp, W. H. 1951. Biosystematy. Brittonia 7(2):113–127.
Campbell, K. S. W., and R. E. Barwick. 1988. Geological and palaeontological information and phylogenetic hypotheses. Geol. Mag. 125(3):207–227.
Carroll, R. L. 1988. *Vertebrate Paleontology and Evolution*. Freeman, New York.
Carson, H. L. 1981. [Letter to the editor]. Science 211(4484):773.
Carson, H. L. 1982. Speciation as a major reorganization of polygenic balances. Pp. 411–433 in: C. Barigozzi (ed.), Mechanisms of Speciation: Proceedings from the International Meeting of Mechanisms of Speciation Sponsored by the Accademia Nazionale dei Lincei. May 4–8 1981, Rome, Italy. *Progress in Clinical and Biological Research*, Vol. 96. Liss, New York.
Carson, H. L. 1985. Unification of speciation theory in plants and animals. System. Bot. 10(4):380–390.
Carson, H. L. 1987. Population genetics, evolutionary rates and neo-Darwinism. Pp. 209–219 in: K. S. W. Campbell and M. F. Day (eds.), *Rates of Evolution*. Allen and Unwin, London.
Charig, A. 1983. *A New Look at the Dinosaurs*. Facts on File, New York.
Cheetham, A. H. 1987. Tempo of evolution in a Neogene bryozoan: Are trends in single morphologic characters misleading? Paleobiology 13(3):286–296.
Cohen, I. B. 1985. *Revolution in Science*. Harvard University Press, Cambridge.
Cracraft, J. 1983. The significance of phylogenetic classifications for systematic and evolutionary biology. Pp. 1–17 in: J. Felsenstein (ed.), *Numerical Taxonomy*. Springer, Berlin.
Cracraft, J. 1985. [Review of] Beyond neodarwinism: An introduction to the new evolutionary paradigm. Cladistics 1(3):300–303.

Dawkins, R. 1985. What was all the fuss about? Nature 316(6030):683–684.
Dawkins, R. 1986. *The Blind Watchmaker*. Norton, New York.
de Queiroz, K., and M. J. Donoghue. 1988. Phylogenetic systematics and the species problem. Cladistics 4(4):317–338.
Dobzhansky, Th. 1935. A critique of the species concept in biology. Philos. Sci. 2(3): 344–355.
Donoghue, M. J. 1985. A critique of the biological species concept and recommendations for a phylogenetic alternative. The Bryologist 88(3):172–181.
Eldredge, N. E. 1971. The allopatric model and phylogeny in Paleozoic invertebrates. Evolution 15(1):156–167.
Eldredge, N. 1985a. *Time Frames: The Rethinking of Darwinian Evolution and the Theory of Punctuated Equilibria*. Simon and Schuster, New York.
Eldredge, N. 1985b. *Unfinished Synthesis: Biological Hierarchies and Modern Evolutionary Thought*. Oxford University Press, New York.
Eldredge, N. 1985c. Evolutionary tempos and modes: A paleontological perspective. Pp. 113–137 in: L. R. Godfrey (ed.), *What Darwin Began: Modern Darwinian and Non-Darwinian Perspectives on Evolution*. Allyn and Bacon, Boston.
Eldredge, N., and J. Cracraft. 1980. *Phylogenetic Patterns and the Evolutionary Process: Method and Theory in Comparative Biology*. Columbia University Press, New York.
Eldredge, N., and S. J. Gould. 1972. Punctuated equilibria: an alternative to phyletic gradualism. Pp. 82–115 in: J. M. Schopf (ed.), *Models in Paleobiology*. Freeman, Cooper, San Francisco.
Eldredge, N., and S. J. Gould. 1988. Punctuated equilibrium prevails. Nature 332(6161): 211–212.
Eldredge, N., and M. J. Novacek. 1985. Systematics and paleobiology. Paleobiology 11 (1):65–74.
Eldredge, N., and S. M. Stanley. 1984. *Living Fossils*. Springer, New York.
Erwin, D. H., J. W. Valentine, and J. J. Sepkoski, Jr. 1987. A comparative study of diversification events: The early Paleozoic versus the Mesozoic. Evolution 41(6): 1177–1186.
Fink, W. L. 1981. Individuality and comparative biology. Behav. Brain Sci. 4(2):288–289.
Forey, P. L. 1982. Neontological analysis versus palaeontological stories. Pp. 119–157 in: K. A. Joysey and A. E. Friday (eds.), *Problems of Phylogenetic Reconstruction*. Academic Press, London.
Futuyma, D. J. 1983. *Science on Trial: The Case for Evolution*. Pantheon, New York.
Futuyma, D. J. 1987. World without design. Nat. Hist. 96(3):34–36.
Futuyma, D. J. 1988. Sturm und Drang and the evolutionary synthesis. Evolution 42(2): 217–226.
Gardiner, B. G. 1982. Tetrapod classification. Zool. J. Linnean Soc. 74(3):207–232.
Gauthier, J. 1986. Saurischian monophyly and the origin of birds. Pp. 1–55 in: K. Padian (ed.), *The Origin of Birds and the Evolution of Flight*. California Academy of Sciences, Memoir 8.
Gauthier, J., and K. Padian. 1985. Phylogenetic, functional, and aerodynamic analyses of the origin of birds and their flight. Pp. 185–1987 in: M. K. Hecht, J. H. Ostrom, G. Viohl, and P. Wellnhofer (eds.), *The Beginning of Birds: Proceedings of the International Archaeopteryx Conference, Eichstätt, 1984*. Freunde des Jura-Museums Eichstätt, Willibaldsburg.
Geoffroy Saint-Hilaire, I. 1859oire Naturelle Générale des Règnes Organiques, Principalement Etudiée Chez l'Homme et Les Animaux, Vol. 2. Masson, Paris.
Ghiselin, M. T. 1981. Categories, life, and thinking [including 18 commentaries and author's response]. Behav. Brain Sci. 4(2):269–313.
Ghiselin, M. T. 1984. Narrow approaches to phylogeny: A review of nine books on cladism. Pp. 209–222 in: R. Dawkins and M. Ridley (eds.), *Oxford Surveys in Evolutionary Biology*, Vol. 1. Oxford University Press, Oxford.
Ghiselin, M. T. 1986. Mayr versus Darwin on paraphyletic taxa. System. Zool. 34(4): 460–462.
Ghiselin, M. T. 1987a. Hierarchies and their components. Paleobiology 13(1):108–111.

Ghiselin, M. T. 1987b. Species concepts, individuality, and objectivity [including response to six commentaries]. Biol. Philos. 2(2):127–143, 207–212.
Gilinsky, N. L., and R. K. Bambach. 1986. The evolutionary bootstrap: A new approach to the study of taxonomic diversity. Paleobiology 12(3):251–268.
Gingerich, P. D. 1985. Species in the fossil record: Concepts, trends, and transitions. Paleobiology 11(1):27–41.
Gould, S. J. 1980a. The promise of paleobiology as a nomothetic, evolutionary discipline. Paleobiology 6(1):96–118.
Gould, S. J. 1980b. G. G. Simpson, paleontology, and the modern synthesis. Pp. 153–172 in: E. Mayr and W. B. Provine (eds.), *The Evolutionary Synthesis: Perspectives in the Unification of Biology*. Harvard University Press, Cambridge.
Gould S. J. 1982a. The meaning of punctuated equilibrium and its role in validating a hierarchical approach to macroevolution. Pp. 83–104 in: R. Milkman (ed.), *Perspectives on Evolution*. Sinauer, Sunderland, MA.
Gould, S. J. 1982b. Darwinism and the expansion of evolutionary theory. Science 216 (4544):380–387.
Gould, S. J. 1983. In praise of Charles Darwin. Pp. 1–10 in: C. L. Hamrum (ed.), *Darwin's Legacy*. Harper & Row, San Francisco.
Gould, S. J. 1984. Toward the vindication of punctuational change. Pp. 9–34 in: W. A. Berggren and J. A. Van Couvering (eds.), *Catastrophes and Earth History: The New Uniformitarianism*. Princeton University Press, Princeton.
Gould, S. J. 1985. The paradox of the first tier: An agenda for paleontology. Paleobiology 11(1):2–12.
Gould, S. J., and N. Eldredge. 1977. Punctuated equilibria: The tempo and mode of evolution reconsidered. Paleobiology 3(2):115–151.
Gould, S. J., and N. Eldredge. 1986. Punctuated equilibrium at the third stage. System. Zool. 35(1):143–148.
Grant, V. 1985. *The Evolutionary Process: A Critical Review of Evolutionary Theory*. Columbia University Press, New York.
Grene, M. 1983. Introduction. Pp. 1–15 in: M. Grene (ed.), *Dimensions of Darwinism: Themes and Counterthemes in Twentieth-Century Evolutionary Theory*. Cambridge University Press, Cambridge.
Haeckel, E. 1866. *Generalle Morphologie der Organismen: allgemeine Grundzüge der organischen Formen-Wissenschaft mechanisch begründet dur die von Charles Darwin reformirte Descendenz-Theorie*. Reimer, Berlin.
Hecht, M. K., and A. Hoffman. 1986. Why not neo-darwinism? A critique of paleontological challenges. Pp. 1–47 in: R. Dawkins and M. Ridley (eds.), *Oxford Surveys in Evolutionary Biology*, Vol. 3. Oxford University Press, Oxford.
Hennig, W. 1950. *Grundzüge einer Theorie der phylogenetischen Systematik*. Deutscher Zentralverlag, Berlin.
Hennig, W. 1966. *Phylogenetic Systematics*. University of Illinois Press, Urbana.
Hoffman, A. 1984. Species selection. Pp. 1–20 in: M. K. Hecht, B. Wallace, and G. T. Prance (eds.), *Evolutionary Biology*, Vol. 18. Plenum, New York.
Hopson, J. A., and H. R. Barghusen. 1986. An analysis of therapsid relationships. Pp. 83–106 in: N. Hotton III, P. D. MacLean, J. J. Roth, and E. C. Roth (eds.), *The Ecology and Biology of Mammal-like Reptiles*. Smithsonian Institution, Washington, D.C.
Howard, D. J. 1988. The species problem. Evolution 42(5):1111–1112.
Hubbs, C. L. 1943. Criteria for subspecies, species and genera, as determined by researches on fishes. Ann. N.Y. Acad. Sci. 44(2):108–121.
Hull, D. L. 1981. Metaphysics and common usage. Behav. Brain Sci. 4(2):290–291.
Hull, D. L. 1987. Genealogical actors in ecological roles. Biol Philos. 2(2):168–184.
Hull, D. L. 1988. *Science as a Process: An Evolutionary Account of the Social and Conceptual Development of Science*. University of Chicago Press, Chicago.
Jackson, J. 1988 Does ecology matter? Paleobiology 14(3):307–312.
Janvier, P. 1983. Groupes panchroniques, "fossiles vivants" et systematique: l'Exemple des "Crossopterygii" et des Petromyzontida. Bull. Soc. Zool. France 108(4):609–616.

Janvier, P. 1986. L'Impact du cladisme sur la recherche dans les sciences de la vie et de la terre. Pp. 101–120 in: P. Tassy (ed.), *L'Ordre et la Diversité du Vivant: Quel Statut Scientifique pour les Classifications Biologiques*? Fayard, Paris.
Jepsen, G. L., E. Mayr, and G. G. Simpson (eds.). 1949. *Genetics, Paleontology, and Evolution*. Princeton University Press, Princeton.
Kemp, T. S. 1982. *Mammal-Like Reptiles and the Origin of Mammals*. Academic Press, London.
Kemp, T. S. 1985. Models of diversity and phylogenetic reconstruction. Pp. 135–158 in: R. Dawkins and M. Ridley (eds.), *Oxford Surveys in Evolutionary Biology*, Vol. 2. Oxford University Press, Oxford.
Kemp, T. S. 1988. Haemothermia or Archosauria? The interrelationships of mammals, birds and crocodiles. Zool. J. Linnean Soc. 92(1):67–104.
Kennedy, J. G. 1978. *Herbert Spencer*. Twayne, Boston.
Kitcher, P. 1984a. Species. Philos. Sci. 51(2):308–333.
Kitcher, P. 1984b. Against the monism of the moment: A reply to Elliott Sober. Philos. Sci. 51(4):616–630.
Kitcher, P. 1985. *Vaulting Ambition: Sociobiology and the Quest for Human Nature*. MIT Press, Cambridge.
Kitcher, P. 1987. Ghostly whispers: Mayr, Ghiselin, and the "philosophers" on the ontological status of species. Biol. Philos. 2(2):184–192.
Kitcher, P. 1988. Some puzzles about species. In press.
Lambert, D. M., B. Michaux, and C. S. White. 1987. Are species self-defining? System. Zool. 36(2):196–205.
Laws, R. A., and D. E. Fastovsky. 1987. Characters, stratigraphy, and "depopperate" logic: An essay on phylogenetic reconstruction. PaleoBios 44:1–9.
Lecourt, D. 1983. Marx au crible de Darwin. Pp. 227–249 in: Y. Conry (ed.), *De Darwin au Darwinisme: Science et Ideologie*. Vrin, Paris.
Levin, D. A. 1979. The nature of plant species. Science 204(4391):381–384.
Levinton, J. 1988. *Genetics, Paleontology, and Macroevolution*. Cambridge University Press, New York.
Lewontin, R. 1981. Evolution/creation debate: A time for truth. BioScience 31(8):559.
Løvtrup, S. 1979. The evolutionary species: Fact or fiction? System. Zool. 28(3):386–392.
Løvtrup, S. 1987a. On species and other taxa. Cladistics 3(2):157–177.
Løvtrup, S. 1987b. On the species problem and some other taxonomic issues. Environ. Biol. Fishes 29(1):3–9.
Løvtrup, S. 1987c. *Darwinism: The Refutation of a Myth*. Croom Helm, London.
Løvtrup, S. 1987d. The theoretical basis of evolutionary thought. Ann. Sci. Nat. Zool. Paris [13] 8(4):219–236.
Løvtrup, S. 1987e. Phylogenesis, ontogenesis and evolution,. Boll. Zool. 54(3):199–208.
Mayr, E. (ed.). 1957. *The Species Problem: A Symposium Presented at the Atlanta Meeting of the American Association for the Advancement of Science, December 28–29, 1955*. American Association for the Advancement of Science, Washington, D.C.
Mayr, E. 1969. *Principles of Systematic Zoology*. McGraw-Hill, New York.
Mayr, E. 1980. Prologue: Some thoughts on the history of the evolutionary synthesis. Pp. 1–48 in: E. Mayr and W. B. Provine (eds.), *The Evolutionary Synthesis: Perspectives in the Unification of Biology*. Harvard University Press, Cambridge.
Mayr, E. 1987a. The ontological status of species: Scientific progress and philosophical terminology [including response to six commentaries]. Biol. Philos. 2(2):145–166, 212–220.
Mayr, E. 1987b. The species as category, taxon and population. Pp. 303–320 in: Anon. (ed.), *Histoire du Concept d'Espèce dans les Sciences de la Vie*. Fondation Singer-Polignac, Paris.
Mayr, E. 1988. *Towards a New Philosophy of Biology: Observations of an Evolutionist*. Harvard University Press, Cambridge.
McKenna, M. C. 1987. Molecular and morphological analysis of high-level mammalian relationships. Pp. 55–93 in: C. Patterson (ed.), *Molecules and Morphology: Conflict or Compromise*? Cambridge University Press, Cambridge.

Mishler, B. D., and M. J. Donoghue. 1982. Species concepts: A case for pluralism. System. Zool. 31(4):491-503.

Nelson, G., and N. Platnick. 1981. *Systematics and Biogeography: Cladistics and Vicariance*. Columbia University Press, New York.

Nelson, G., and N. Platnick. 1984. Systematics and evolution. Pp. 143-158 in: M. -W. Ho and P. T. Saunders (eds.), *Beyond Neo-darwinism*. Academic Press, London.

Olson, E. C. 1971. *Vertebrate Paleozoology*. Wiley-Interscience, New York.

Olson, E. C., and C. F. Brunk. 1986. The evolutionary synthesis today: An essay on paleontology and molecular evolution. Pp. 351-361 in: K. M. Flanagan and J. A. Lillegraven (eds.), *Vertebrates, Phylogeny, and Philosophy. Contributions to Geology*. University of Wyoming, Special Paper 3.

Oppenheimer, J. M. 1987. Haeckel's variations on Darwin. Pp. 123-135 in: H. M. Hoenigswald and L. F. Wiener (eds.), *Biological Metaphor and Cladistic Classification: An Interdisciplinary Perspective*. University of Pennsylvania Press, Philadelphia.

Padian, K. 1986. Introduction. Pp. 1-7 in: K. Padian (ed.), *The Beginning of the Age of Dinosaurs: Faunal Change across the Triassic-Jurassic Boundary*. Cambridge University Press, Cambridge.

Papavero, N., and J. Balsa. 1986. *Introdução histórica et epistemológica à biologia comparada, com especial referência à biogeografia, 1: do Gênesis ao fim do Império Romano do Ocidente*. Sociedade Brasileira de Zoologia, Belo Horizonte.

Paterson, H. E. H. 1981. The continuing search for the unknown and unknowable: A critique of contemporary ideas on speciation. South African J. Sci. 77(3):113-119.

Paterson, H. E. 1987. A view of species. Pp. 211-215 in: G. Sermonti (ed.), Proceedings of the International Workshop on Structuralism in Biology, Osaka (Japan), December 7-11, 1986. Riv. Biol 80(2):157-268, 269-338.

Patterson, C. 1977. The contribution of paleontology to teleostean phylogeny. Pp. 579-643 in: M. K. Hecht, P. C. Goody, and B. M. Hecht (eds.), *Major Patterns in Vertebrate Evolution*. Plenum, New York.

Patterson, C. 1981a. Significance of fossils in determining evolutionary relationships. Annu. Rev. Ecol. System. 12:195-223.

Patterson, C. 1981b. Agassiz, Darwin, Huxley, and the fossil record of teleostean fishes. Bull. Br. Mus. Nat. Hist. (Geol.) 35(3):213-224.

Patterson, C. 1982. Morphological characters and homology. Pp. 21-74 in: K. A. Joysey and A. E. Friday (eds.), *Problems of Phylogenetic Reconstruction*. Academic Press, London.

Patterson, C. 1983. How does phylogeny differ form ontogeny? Pp. 1-31 in: B. C. Goodwin, N. Holder, and C. C. Wylie (eds.), *Development and Evolution*. Cambridge University Press, Cambridge.

Patterson, C. 1987. Introduction. Pp. 1-22 in: C. Patterson (ed.), *Molecules and Morphology in Evolution: Conflict or Compromise*? Cambridge University Press, Cambridge.

Patterson, C. 1988. The impact of evolutionary theories on systematics. Pp. 59-91 in: D. L. Hawksworth (ed.), *Prospects in Systematics*. Systematics Association Special Volume, 36. Oxford University Press, Oxford.

Patterson, C., and A. B. Smith. 1987. Is the periodicity of extinctions a taxonomic artefact? Nature (London) 330(6145):248-251.

Patterson, C., and A. B. Smith. 1988. Periodicity in extinction: The role of systematics. Ecology, in press.

Phillips, M. E. 1953. The Academy of Natural Sciences of Philadelphia. Trans. Am. Philos. Soc. New Ser. 43(1):366-274.

Rachootin, S. P., and K. S. Thomson. 1981. Epigenetics, paleontology, and evolution. Pp. 181-193 in: G. G. E. Scudder and J. L. Reveal (eds.), *Evolution Today: Proceedings of the Second International Congress of Systematic and Evolutionary Biology*. Hunt Institute, Pittsburgh.

Raup, D. M. 1986. *The Nemesis Affair: A Story of Dinosaurs and the Ways of Science*. Norton, New York.

Raup, D. M. 1987. Major features of the fossil record and their implications for evolutionary rate studies. Pp. 1-14 in: K. S. W. Campbell and M. F. Day (eds.), *Rates of Evolution*. Allen and Unwin, London.

Raup, D. M., and G. E. Boyajian. 1988. Patterns of generic extinction in the fossil record. Paleobiology 14(2):109–125.

Raup, D. M., and S. J. Gould. 1974. Stochastic simulation and evolution of morphology—towards a nomothetic paleontology. System. Zool. 23(3):305–322.

Raup, D. M., and D. Jablonski (eds.). 1986. *Patterns and Processes in the History of Life*. Springer, Berlin.

Raven, P. H. 1974. Plant systematics: 1947–1974. Ann Missouri Bot. Garden 61(1): 166–178.

Raven, P. H. 1986. Modern aspects of the biological species in plants. Pp. 11–29 in: K. Iwatsuki, P. H. Raven, and W. J. Bock (eds.), *Modern Aspects of Species*. University of Tokyo Press, Tokyo.

Ridley, M. 1985. More of the same please. Nature (London) 313(6005):823–824.

Ridley, M. 1986. *Evolution and Classification: The Reformation of Cladism*. Longman, London.

Rieppel, O. 1984. Atomism, transformism and the fossil record. Zool. J. Linnean Soc. 82(1–2):17–32.

Rieppel, O. 1986. Species are individuals: A review and critique of the argument. Pp. 283–317 in: M. Hecht, B. Wallace, and G. T. Prance (eds.), *Evolutionary Biology*, Vol. 20. Plenum, New York.

Rifkin J. 1980. *Entropy: A New World View*. Viking, New York. Reprint 1981. Bantam, New York.

Romer A. S. 1966. *Vertebrate Paleontology*. University of Chicago Press, Chicago.

Rosen, D. E. 1984. Hierarchies and history. Pp. 77–97 in: J. W. Pollard (ed.), *Evolutionary Theory: Paths into the Future*. Wiley, Chichester.

Rosenberg, A. 1985. *The Structure of Biological Science*. Cambridge University Press, Cambridge.

Salthe, S. N. 1985. *Evolving Hierarchical Systems: Their Structure and Representation*. Columbia University Press, New York.

Schoch, R. M. 1986. *Phylogeny Reconstruction in Paleontology*. Van Nostrand Reinhold, New York.

Sepkoski, J. J., Jr., 1987. Sepkoski replies. Nature (London) 330(6145):251–252.

Sepkoski, J. J., Jr., and D. M. Raup. 1986. Periodicity in marine extinction events. Pp. 3–36 in: D. K. Elliott (ed.), *Dynamics of Extinction*. Wiley, New York.

Simpson, G. G. 1944. *Tempo and Mode in Evolution*. Columbia University Press, New York.

Simpson, G. G. 1961. Lamarck, Darwin and Butler: Three approaches to evolution. Am. Schol. 30(2):238–249.

Slobodchikoff, C. N. (ed.). 1976. *Concepts of Species. Benchmark Papers in systematic and Evolutionary Biology*, Vol. 3. Dowden, Hutchinson, Ross; Stroudsburg.

Smith, A. B., and C. Patterson. 1988. The influence of taxonomic method on patterns of evolution. Pp. 127–216 in: M. K. Hecht and B. Wallace (eds.), *Evolutionary Biology*, Vol. 23. Plenum, New York.

Sober, E. (ed.). 1984. *Conceptual Issues in Evolutionary Biology: An Anthology*. MIT Press, Cambridge.

Stanley, S. 1979. *Macroevolution: Pattern and Process*. Freeman, San Francisco.

Stanley, S. 1981. *The New Evolutionary Timetable: Fossils, Genes and the Origin of Species*. Basic Books, New York.

Stebbins, G. L. 1987. Species concepts: Semantics and actual situations. Biol. Philos. 2(2):198–203.

Sylvester-Bradley, P. C. (ed.). 1956. *The Species Concept in Palaeontology*. Systematics Association Publication No. 3. Sytematics Association, London.

Szalay, F. S. 1977. Ancestors, descendants, sister groups and testing of phylogenetic hypotheses. System. Zool. 26(1):12–18.

Tiffney, B. H. (ed.). 1985. *Geological Factors and the Evolution of Plants*. Yale University Press, New Haven.

Trémaux, P. 1865. *Origine et Transformations de l'Homme et des Autres Etres*. Hachette, Paris.

Valentine, J. W. 1982. Darwin's impact on paleontology. BioScience 32(6):513–518.
Velikovsky, I. 1950. *Worlds in Collision*. Doubleday, Garden City.
von Vaupel Klein, J. C. 1987. Phylogenetic analysis and its foundations. Pp. 159–172 in: P. Hovenkamp, E. Grittenberger, E. Hennipman, R. de Jong, M. C. Roos, R. Sluys, and M. Zandee (eds.), *Systematics and Evolution: A Matter of Diversity*. Utrecht University, Utrecht.
Vrba, E. S. 1985. Introductory comments on species and speciation. Pp. ix–xviii in: E. S. Vrba (ed.), *Species and Speciation*. Transvaal Museum Monograph no. 4. Transvaal Museum, Pretoria.
Vrba, E. 1988. Ecological predictions for macroevolutionary patterns in the fossil record. In: N. C. Stenreth (ed.), *Coevolution in Ecosystems and the Red Queen Hypothesis*. Cambridge University Press, Cambridge. In press.
Wake, D. B., G. Roth, and M. H. Wake. 1983. On the problem of stasis in organismal evolution. J. Theor. Biol. 101(2):211–224.
Weber, B. H., D. J. Depew, and J. D. Smith (eds.). 1988. *Entropy, Information, and Evolution: New Perspectives in Physical and Biological Evolution*. MIT Press, Cambridge.
White, M. J. D. 1978. *Modes of Speciation*. Freeman, San Francisco.
Wicken, J. S. 1987. *Evolution, Thermodynamics, and Information: Extending the Darwinian Program*. Oxford University Press, New York.
Wiley, E. O. 1980. Is the evolutionary species fiction?—A consideration of classes, individuals and historical entities. System. Zool. 29(1):76–80.
Wiley, E. O. 1981a. The metaphysics of individuality and its consequences for systematic biology. Behav. Brain Sci. 4(2):302–303.
Wiley, E. O. 1981b. *Phylogenetics: The Theory and Practice of Phylogenetic Systematics*. Wiley, New York.
Wiley, E. O. 1987. Evolution and classification [review of Ridley, 1986]. Quart. Rev. Biol. 62(3):293–295.
Zandee, M., and R. Geesink. 1987. Phylogenetics and legumes: A desire for the impossible? Pp. 131–167 in: C. H. Stirton (ed.), *Advances in Legume Systematics*, Part 3. Royal Botanic Gardens, Kew.

PART TWO

GENETIC STRUCTURE OF SPECIES BOUNDARIES

CHAPTER 4

THE SUBDIVISION OF SPECIES BY HYBRID ZONES

Godfrey M. Hewitt

INTRODUCTION

To understand speciation it would seem useful to have as much information as possible on the spatial genetic structure of taxa that are considered good species. This reveals to us what we have to explain, suggests hypotheses, and can provide suitable material for experimentation. It is becoming increasingly apparent that species across their range are often divided into patchworks of parapatric subspecies and races and where two forms meet, mate, and hybridize, a hybrid zone occurs. A review of the literature (Hewitt 1985) yielded over 150 clear examples of such hybridization and this number is rising steadily. Indeed a perusal of the distributions of well-documented organisms such as African birds (cf. Hall and Moreau 1970; Snow 1978) suggests that one-third to one-half of the species may be subdivided by hybrid zones. There are similar indications in butterflies, grasshoppers, lizards, and amphibians in which there has been a moderate level of investigation. Thus, the phenomenon is common in some groups, and may represent a much more frequent species substructure than previously imagined.

While these hybrid zones have often been recognized on morphological criteria they can involve behavioral, chromosomal, or molecular differences, and detailed examination frequently reveals a combination of these. Some zones are narrow, just a few meters wide, and some are broad clines many kilometers wide. Some are associated with geographic boundaries or ecotones and some would seem to be affected by man's agricultural activities. This variety of subdivision may have several origins and proximate causes.

CAUSES OF ZONES

It is perhaps worth emphasizing the usefulness of thinking in terms of genes as well as genomes in this context; hybridizing genomes will tend to be broken down by segregation and recombination, and the selection acting on each gene will be different and produce various responses. In the middle of most hybrid zones there is little chance of finding a "pure" racial genotype and less chance of an F_1 hybrid genome; it can be misleading to think simply of parentals and hybrids when the gene is the basic unit of currency. Of course, individual fitness is the product of many genes and it is necessary also to consider the affects of epistasis and recombination between loci within the genome.

The genetic differences between the parapatric forms may have arisen by selection or drift. This could have occurred either primarily, with gene flow possible, or in allopatry, with secondary contact forming the hybrid zone. It is difficult, if not impossible, to decide between these historical possibilities (Endler 1982a; Barton and Hewitt 1985). However, if we first consider the current structure of hybrid zones, the available data indicate several modes of selection that may act on allelic differences between the two hybridizing races.

A cline between two races fixed for different alleles at a locus could be the result of (1) heterozygote disadvantage, (2) differential environmental selection on the two homozygotes, (3) selective equality of homozygotes and heterozygotes, (4) frequency-dependent selection on homozygotes, (5) superiority of one homozygote over the other, or (6) superiority of heterozygote over homozygote in the particular environment of the zone. Where several genes differ between the races the recombinants may be disadvantaged if there is coadaptive epistasis between alleles of the same race, which will produce clines of type 1. In practice there is little strong evidence for the fifth or sixth possibilities in hybrid zones (Barton and Hewitt 1985) while the first four can each explain a number of known examples.

Firstly, the evidence from studies on some 170 zones shows that many of them involve some form of hybrid unfitness, which may be due to heterozygote or recombinant disadvantages (Table 1). This reduction in reproduction of heterozygotes and hybrid genotypes will be balanced by gene flow due to dispersal and mating from the opposing racial homozygote and together determine the width of the cline (Endler 1977). Such clines have been aptly called tension zones (Key 1968) since they will tend to minimize their length. They are not tied to an ecological transition because they are due to internal genetic incompatabilities. Particularly clear examples are in *Podisma* (Barton and Hewitt 1981), *Rana* (Kocher and Sage 1986), and *Caledia* (Shaw et al. 1982).

TABLE 1. Summary of the attributes of reasonably well demonstrated hybrid zones from a survey of the literature.[a]

1. Showing differences in morphology 126, allozymes 61, chromosomes 54, behavior 28
2. Position near ecotone 45, at barrier 26, no environmental change 10, stable 20, unstable 7
3. Clear hybrid unfitness: *Acomys, Allonemobius, Bufo, Caledia, Chilocorus, Chorthippus, Didymuria, Erebia, Geomys, Gryllus, Hyalophora, Hyla, Keyacris, Litoria, Mus, Notropis, Peromyscus, Phlox, Phytomyza, Pinus, Podisma, Pseudophyrne, Rhus, Sceloporus, Sturnella, Triturus, Uroderma, Warramaba*
4. Ecologically differentiated: *Campylorhyncus, Gaillardia, Geomys, Icterus, Parus, Pholidobous, Piplio, Quiscalus, Rhus, Sceloporus, Spalax, Stenella, Sturnella, Thomomys, Triturus*
5. Probably neutral mixing: *Dendroica, Lepomis, Passerina, Peromyscus, Vermivora*
6. Noncoincident: *Ambystoma* (LdH-2?), *Caledia* (mtDNA), *Clarkia* (chrom. trans.), *Gaillardia* (ME, Pgm), *Mus* (mtDNA), *Peromyscus* (morph?)
7. In density troughs or through inhospitable places: *Bufo, Campylorhyncus, Chilocorus, Chorthippus, Clarkia, Cnemidophorus, Crotaphytus, Dendroica, Erebia, Gerbillus, Hyalophora, Perognathus, Podisma, Spalax, Spermophilus, Uroderma*

[a]There are many more putative zones. The examples chosen are the better established ones and given by generic name for simplicity. References to most are given in Barton and Hewitt (1985) and Hewitt (1985); additional ones are referenced in this paper. Total zones surveyed, 170.

Second, quite a number of zones are broadly associated with an environmental transition and may include clines for alleles suited to different conditions (Table 1). An environmental cline would be expected to follow closely the course and shape of the ecological change, while clines for other loci determining different adaptations need not occur in the same place. At present only a few zones provide evidence supporting these expectations (e.g., *Gaillardia*, Heywood 1986; *Parus*, Dixon 1955; *Sturnella*, Rohwer 1972). Where distinct environments intermingle as large patches, the differentially adapted genomes may exist as a mosaic, and there is some recent positive evidence for this in crickets (Howard and Harrison 1984; Harrison 1986; Harrison and Rand, this volume; Howard 1986). Overall the evidence indicates that while the general position of many hybrid zones may be determined by environmental gradients, they do not seem to be the only or main factor maintaining them.

Third, if the allelic differences between the two races and their recombinants are selectively equal they will gradually diffuse into each other, producing increasingly shallow clines—most probably following secondary contact. The width of such a cline will depend on the dispersal rate and the number of generations since contact (Endler 1977). Both parameters are difficult to ascertain but some zones are so wide in comparison to the species dispersal that this seems a likely explanation, e.g., *Dendroica* (Barrowclough 1980), *Lepomis* (Avise et al. 1984), and *Peromyscus* (Baker et al. 1983). It is quite possible that many other racial and geographic differences are of a similar nature.

MANY CONCORDANT DIFFERENCES

A widespread and striking feature of hybrid zones is the coincidence and often concordance of multiple clines. This is true of the hybrid zones in *Podisma pedestris* and *Chorthippus parallelus* on which I have worked (Barton and Hewitt 1981; Hewitt et al. 1988). In fact, in very few cases is there evidence that genes and characters are not changing in the same place (Table 1). While being coincident, the various clines may have different widths, i.e., nonconcordant, e.g., *C. parallelus* (Table 2), and this argues for a different mode or intensity of selection at each locus. There are a number of possible explanations of this coincidence. Often the first to be suggested is an ecological transition, in which a number of clines due to an environmental gradient occur together. However, there are problems with this simple view since the null points of selection for different genes on an environmental gradient are unlikely to be the same. These separate clines would not be expected to be exactly coincidental unless there is

TABLE 2. Widths (1/max slope) of coincident clines for different characters in three organisms of differing mobility.[a]

Chorthippus parallelus **(flightless grasshopper)**	***Bombina bombina/variegata*** **(fire-bellied toad)**	***Mus musculus/domesticus*** **(commensal mouse)**
σ = 30–100 m	σ = 430–890 m	σ = 100 m–3 km
Esterase-2, 15–20 km	CK, 7.3 km	Hybrid index, 20–40 km
Song syllable length, ±19 km	Ldh, 6.3 km	α-Globin locus, ±20 km
Stridulatory pegs, 4.1 km	Ak, 6.1 km	β-Globin locus, ±20 km
Courtship song, 2.1 km	mtDNA, similar	Four allozyme loci, 20 km
Song echeme interval, 1.4 km	Mdh-1, 5.7 km	mtDNA, 20–20 km
Nucleolus organizer, 0.6 km	Gpi, 5.5 km	Y sequence, ±2 km

[a]The first dispersal (σ) figure is measured in the experiment; the second is deduced from zone width.

strong positive epistasis among the alleles within each race. If this is the case then recombinants between the two races should be less fit than the parentals and the zone will have the properties of a tension zone. Indeed, it is possible to envisage the evolution of an environmental cline into a tension zone. This argument emphasizes that environmental and tension zones are not mutually exclusive.

Where tension zones involve a large epistatic environmental component we might expect them to lie where a patch of one habitat type changes to the other. For a mosaic to develop the habitat patches will need to be relatively large compared with the dispersal rate of the species, to obviate mixing. The zones between such large patches should comprise sets of coincident clines. The situation in *Gryllus* (Harrison and Rand, this volume) provides the best evidence for a mosaic and it will be interesting to follow its detailed investigation.

Now, tension zones themselves have properties in addition to epistasis that will tend to hold clines together. Individual clines produced by heterozygote disadvantage will tend to move together when they overlap due to linkage disequilibria and dispersal into the hybrid sink. Tension zones coalesce (Hewitt 1975; Hewitt and Barton 1980; Barton and Hewitt, 1981, 1985). The heterozygote and recombinant disadvantage can form steep clines for the genes concerned and act as a barrier to gene flow at other loci; the permeability of this barrier depends on the level of selection and recombination (Barton and Hewitt, 1983). Advantageous alleles that are advancing through the range will be slowed significantly only if they are tightly linked to a locus with considerable heterozygote disadvantage, while the diffusion through of effectively neutral alleles will be more affected. There will be some tendency for differences to accumulate and be held at a tension zone.

Two further causes of coincidence and concordance—population structure, and range changes with secondary contact—have several ramifications and will be dealt with in more detail.

DENSITY TROUGHS

The property of a tension zone that has particularly interesting consequences is the strong tendency to come to rest in density troughs (Hewitt 1975; Barton and Hewitt 1985). This is because there will be a flux of genes due to dispersal from high-density to low-density areas and this will push the tension zone into the low-density region from either side (Figure 1). The density gradient need not be very steep to trap zones that would otherwise move under a strong advantage to one homozygote or race (Barton 1979; Hewitt and Barton 1980). Density troughs due to stable features such as valleys, ridges, or rivers can be very strong traps and pull together all the

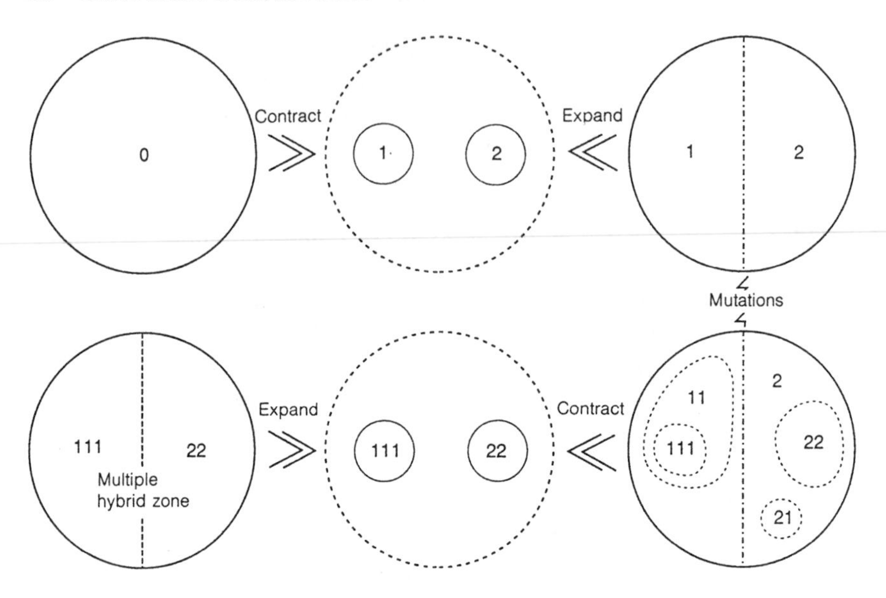

FIGURE 1. Major change in density of an organism across a part of its range. The distance scale might be in kilometers for grasshoppers, 10x for mice and 100x for some birds. The net dispersal (m) will be greater from the high density side at A; a zone there will tend to move down the density slope until it reaches B, where it will rest, since dispersal from both sides will be the same. Even though there is a lower density at position C, the zone will not reach it without a catastrophic change in distribution of the species.

tension zones in their catchment area, providing another cause of coincidence of clines. Periodic changes in density distribution may well cause tension zones to move from one place to another, but larger, more permanent troughs can be expected to hold zones for much longer. Quite a few hybrid zones appear to coincide with less hospitable regions (Table 1), but detailed sampling and analysis are necessary to substantiate this relationship. We have firm data on this from two distant parts of the alpine hybrid zone in *Podisma pedestris* (Hewitt and Barton 1980; Nichols and Hewitt 1986). This hybrid zone is not associated with an environmental change along its course as has been shown by detailed vegetational analysis (Nichols 1985; Nichols and Hewitt 1988). On the other hand, detailed density distributions show that the zone is held for large stretches by regions of low density. Where it runs between such stretches through higher density areas, then computer modeling using the real local insect distributions along with measured estimates of dispersal and selection shows that the zone course is predictable. The detailed patchiness on the scale of individual dispersal distances determines the exact position of the zone, and not any racial differences in ecology or environment. It is possible to be misled without this information. It would be reassuring to have more data of this type for other zones.

RANGE CONTRACTION AND EXPANSION

Given the heterogeneity of environment and patchy population structure, an individual tension zone will not travel far from its origins before being trapped by a density trough; this will depend, of course, on the grain of the organism's environment. The large areas covered by many parapatric races, dissected as they often are by major density troughs, are better explained by range contraction and expansion with secondary contact producing hybrid zones (Hewitt 1979, 1985; Hewitt and Barton 1980). A cycle of contraction and expansion will reduce individual tension zones down to multiple allelic differences between isolates so that they expand out together to form a set of coincident clines on secondary contact, and repeated cycles will accumulate more (Figure 2). Not only does this model produce coincident multiple tension clines but it also will include in the hybrid zone those ecologically dependent and neutral alleles that differed in the isolates. It is a very powerful explanation; the other reasons for coincidence still apply but they are subsequent and secondary to this major population restructuring process. More generally, many of the local genetic adaptations and divergences will also be lost in this process of contraction and expansion; those adaptations will become extinct and be replaced by one of the surviving genomes and its accumulated adaptation.

The recent ice ages provide a time scale for major range change in the temperate regions and most zones are best explained by secondary contact following the last ice age, even if the story in the tropics (Endler 1982b; Mayr and O'Hara 1986) is more difficult to unravel. Quite a number of

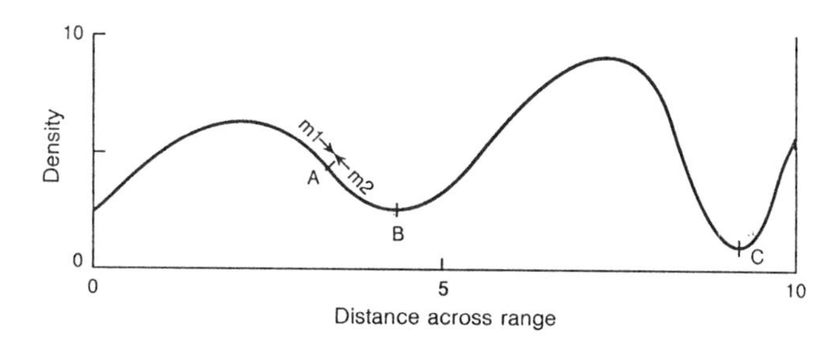

FIGURE 2. A simple diagram of range expansion and contraction to show how this process tends to homogenize the species range for a series of different mutations (digits). When two forms distinguished by negatively heterotic alleles meet they form a hybrid zone. Ecological and neutral differences will also be coincident.

TABLE 3. Examples of hybrid zones in Europe and North America.[a]

Genus	Area	Reference
GLACIATED MOUNTAINS		
Chilocorus tricyclus/ hexadactylus	Canadian Rockies	Smith and Virkhi (1978)
Chorthippus parallelus	Pyrenees	Hewitt et al. (1987)
Corvus corone	Scotland	Cook (1975)
Dendroica coronata	Canadian Rockies	Barrowclough (1980)
Erebia cassoides/ tyndarus/hispania	Alps, Pyrenees	Lorkovic (1958)
Hyalophora euryalis/ gloveri	Sierra Nevada, Calif.	Collins (1984)
Podisma pedestris	Alps	Hewitt (1975)
Sorex araneus, several	Great Britain, Scandinavia	Searle (1984)
Spermophilus richardsonii	Rockies, Montana	Nadler et al. (1971)
Thomomystalpoides	Rockies	Nevo et al. (1974)
Zygaena ephialtes	Alps	Turner (1971)
TUNDRA/STEPPE		
Allonemobius fasciatus/ socius	Northeast/southeast USA	Howard (1986)
Ambystoma tigrinum	Colorado	Kocher (personal communication)
Bombina bombina/ variegata	Central Europe	Arntzen (1978)
Bufo americanus/ hemiophyrs	Manitoba	Green (1983)
Ceratophyllus styx	Northwest Europe	Riddoch et al. (1984)
Chilocorus bipustulatus/ geminus	Turkmenia USSR	Zaslavskii (1967)
Clarkia nitens/speciosa	Sierra Nevada, Calif.	Hauber and Bloom (1983)
Colaptes auratus/cafer	Western USA	Moore and Buchanan (1985)
Corvus corone	North and central Europe	Cook (1975)
Cryptocephalus hypochoeridis	Slovakia	Barabas (1978)
Desmognathus fuscus/ ochrophaes	Appalachians	Karlin and Guttman (1981)
Ensatina eschscholtzii	Sierra Nevada, Calif.	Wake and Yanev (1986)
Gasterosteus aculeatus	Washington	Hagen and Moodie (1979)
Geomys bursarius/ lutescens	Nebraska	Heaney and Timm (1985)
Gryllus pennsylvanicus/ firmus	Northeast USA	Harrison (1986)
Icterus galbula	North and central USA	Rising (1983)
Mus domesticus/ musculus	North and central Europe	Sage et al. (1986)

TABLE 3. (*Continued*)

Genus	Area	Reference
Natrix natrix	North and central Europe	Thorpe (1984)
Peromyscus californicus	California	Smith (1979)
Peromyscus leucopus	Oklahoma	Stangl (1986)
Pheucticus ludovicianus/ melanocephalus	Central North America	West (1962)
Plethodon cinereus	Massachusetts	Highton (1977)
Plethodon jordani	Appalachians	Peabody (1978)
Plethodon ouachitae	Ouachita	Duncan and Highton (1979)
Prosimulium magnum	Great Lakes	Rothfels (personal communication)
Rana pipiens	Several, Dakota–Texas	Kruse and Dunlap (1976)
Rumex acetosa	Massif, France	Wilby (1987)
Sturnella magna/neglecta	Great Plains	Rohwer (1976)
Thomomys bottae/ townsendii	California	Thaeler (1968)
Thomomys talpoides	Central North America	Thaeler (1974)
Triturus cristatus/ marmoratus	France	Schoorl and Zuiderwijk (1981)
Vermivora pinus/ chrysoptera	Northeast USA	Ficken and Ficken (1968)

[a]In mountains that were glaciated during the last ice age and in periglacial tundra and steppe that would have been inhospitable to these species. Many other probable cases of range changes occur south of these (e.g., Texas, Mexico, Mediterranean) and in South America, Africa, and Australia.

ranges have been found in the mountains and northern regions of Europe and North America that would not have supported the species during the glaciations (Table 3), and a similar argument would hold for many Australian and South American examples. This does not mean that any or all of the racial differences arose during the last ice age; they could have arisen by various means through the Pleistocene and more distant times, but they were brought together as a hybrid zone in the present postglacial period. Evidence on the age of these zonal differences from molecular studies is beginning to appear. The differences in mitochondrial DNA, immunological and biochemical markers between *Bombina bombina* the fire bellied toad and *B. variegata* the yellow-bellied toad that form a complex zone through central Europe, indicate that they diverged some 5 million years ago (Szymura et al. 1985). In the much-studied case of the mice *Mus musculus* and *Mus domesticus*, which also form a zone in central Europe, it is estimated from protein and mitochondrial DNA divergence patterns that there has been little genetic exchange for 1 million years (Wilson et al.

1985; Sage et al. 1986b). Immunological distances between the call races of the Australian frog *Limnodynastes tasmaniensis* are placed in a Pleistocene time scale (Roberts and Maxson 1986). We anticipate with interest more evidence on this topic, which will, in turn, be useful in the debate on the apparently slow rates of evolutionary change from biochemical and chromosomal characters (Bush et al. 1977; Kimura 1983; Wilson 1985; Chambers 1987).

UNCOUPLING CLINES AND ZONES

The coincidence of many tension clines and ecological clines can be explained by contraction–expansion cycles with secondary contact and they can be held together by heterozygote or recombinant disadvantage. How then do we explain the few cases of zones with noncoincident clines (Table 1). In principle it could be that (1) the separate differences have arisen relatively recently, probably since the formation of the zone, and spread through to their present position near a hybrid zone, or (2) they have become uncoupled from the hybrid zone. Such uncoupling seems to be a possibility in several cases. In the classic *Warramaba viatica* complex that is subdivided into a series of chromosomal races (White 1978), it seems necessary to postulate an uncoupling and reassortment of chromosome differences to give the present distributions (Hewitt 1979). It was suggested that this was most likely achieved in a period of contraction when a series of small surviving isolates containing mixtures of karyotypes may have become fixed for different chromosomal combinations, and then spread out as reassorted races on expansion. In *Mus musculus/domesticus* the mitochondrial DNA differences are not coincident with the nuclear differences in Scandinavia, but they are coincident in other parts of the zone in Germany (Sage et al. 1986b; Vanlerberghe et al. 1986). The favored explanation for the noncoincidence is that a reassorted female from the hybrid zone in North Germany colonized Sweden first during the postglacial expansion (Gyllensten and Wilson 1987). In *Caledia captiva* the mitochondrial DNA of the Moreton race extends several hundred kilometers north of the nuclear hybrid zone with the Torresian race (Arnold et al. 1988), and a similar combination of population bottlenecking and reassortment followed by expansion could have been involved. The population structure at the trailing-leading edge of a species range in a contraction–expansion cycle seems to be a likely setting for reassortment, and also perhaps for the establishment of new mutations, recombinants, and rearrangements.

POSITION AND STABILITY

Given that hybrid zones tend to rest in density troughs as predicted by theory and supported by detailed data in *Podisma pedestris* (Barton and Hewitt 1981; Nichols and Hewitt 1986), then following expansion and secondary contact the zone should stay in the same general position until the next major change in its environment that modifies population structure sufficiently. Local change may cause the zone to bend and bulge as the result of population restructuring (cf. *Podisma*), but general or drastic changes are necessary to alter the position of a long stretch of zone. The zones on Kangaroo Island between chromosome races of *W. viatica* (White 1978) are where they would be in relation to the same races on the South Australian mainland if the sea had not risen after the last ice age (Hewitt 1979). The same is true of the mountain "island" near Seyne-les-Alpes and the hybrid zone of *Podisma*. Such examples argue that these zones have been essentially immobile since the last postglacial contact. In *Podisma* at Seyne, where we can predict the density and distribution of individuals by altitude and vegetation (Nichols and Hewitt 1986), the zone has probably disappeared during the postglacial optima, when the two races of alpine grasshopper were restricted to two higher mountain refuges and reformed in the same place. We can see here on a small scale what is envisaged to occur on a continental scale with the ice ages.

Man's more recent activities provide evidence of ecological changes creating, modifying, or moving hybrid zones (Table 4). A particularly striking example is that of the lizard *Pholidobolus* in Ecuador (Hillis and Simmons 1986) where xeric changes brought about by agricultural practices seem to be the cause of a 30-km movement of the zone in less than 12 years. Several of the bird zones in the central United States (Rising 1983) have been caused or modified greatly by modern agricultural pursuits and provide evidence of movement, but this also makes their dynamics difficult to interpret. Man may also have created a hybrid zone in the fire ants *Solenopsis* that have been introduced in Mississippi (Vander Meer et al. 1985).

CHORTHIPPUS PARALLELUS IN THE POSTGLACIAL

Several hybrid zones occur along mountain ranges, including those in *Podisma pedestris* and *Chorthippus parallelus*, and these are clearly the result of secondary contact. In *C. parallelus* the French *C. p. parallelus* and Spanish *C. p. erythropus* subspecies (formerly designated species) meet and form a multiple zone along the Pyrenees (Butlin and Hewitt 1985a, 1985b). Its altitude limit is around 2000 m so that the subspecies are separated by the higher mountain ridge and meet through the cols and around the eastern and western ends of the range (Figure 3). The zone does not follow the political boundary, but runs between the highest peaks. In the ice ages the Pyrenees were covered with an ice sheet and *Chorthip-*

TABLE 4. Examples of hybrid zones in which there is evidence that man's activities seem to have caused a change in composition or position of the contact.

Genus	Change	Reference
Anser caerulescens (snow geese), south USA and north Canada	Since 1925—due to agriculture in winter feeding grounds	Cooke (personal communication)
Bolitoglossa franklini/ resplendens (salamanders), southwest Guatemala	Due to recent disturbance and road building	Wake et al. (1980)
Geomys bursarius (pocket gophers), central USA	Cropland changes and roads	Tucker and Schmidly (1981)
Gryllus pennsylvanicus/ firmus (crickets), northeast USA	Loam and sand habitats opened up by man	Harrison (1986)
Passerina cyanea/ amoena (buntings), central North America	Recent contact in Great Plains; history of invasions	Sibley and Short (1959)
Pholidobolus montium/ affinis (lizards), Ecuador	Field clearance allowed more xeric *montium* to advance	Hillis and Simmons (1986)
Pipilo erythrophthalmus/ocai (towhees), Mexico	Mixing due to forest clearing	Sibley and West (1958)
Quiscalus quiscula (grackles), Louisiana	Moved north due to disturbance of hardwood forest	Yang and Selander (1968)
Vermivora pinus/chrysoptera (warblers), northeast USA	Blue displacing golden, changing farming practice since 1860	Gill (1980)

pus would have retreated southwest into Spain and east, probably through lowland Provence, and south into Italy. As the climate warmed in the postglacial period the two races would have expanded from their refugia to meet, first low down at both ends of the Pyrenees and later in the higher cols. The clines are coincident at these cols, and the width of the clines for different loci and characters differ consistently (Table 2) (Hewitt et al. 1988).

The cline for X chromosome nucleolar organizer is very narrow (0.6 km), indicating significant hybrid disadvantage. The clines for some other

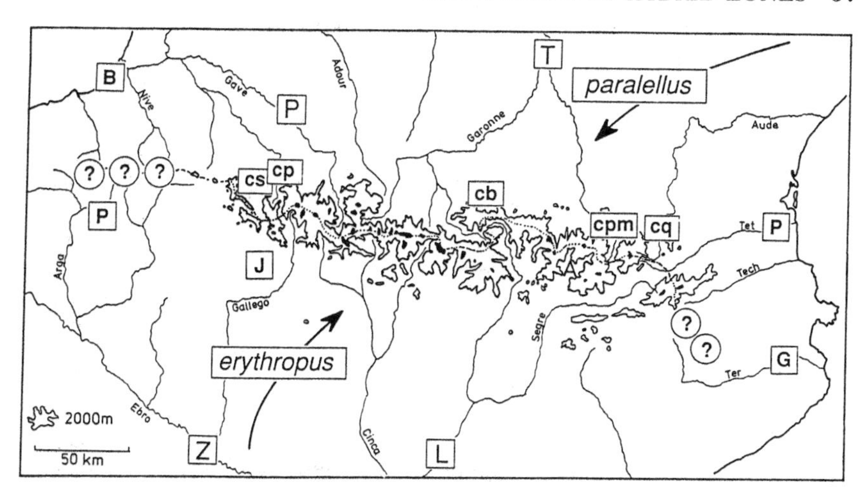

FIGURE 3. The Pyrenees range with the 2000-m contour marked to show the upper altitudinal limit for *Chorthippus parallelus*. The major cols through the mountains are indicated; cs, Col du Somport; cp, Col du Pourtalet; cb, Col de Beret; cpm, Col de Puy Morens; cq, Col de la Quillane. Major towns are shown by their initial letters. The course of the ridge and zone between *C. p. parallelus* and *C. p. erthropus* is given as a dotted line.

characters are much wider (10–20 km) and indicate diffusion with little, if any, selection. Individual dispersal (σ) has been measured by simple mark-recapture experiments in a uniform natural habitat at some 20–30 m/generation. So using Endler's (1977) formula for neutral diffusion following secondary contact, $T = 0.35(w/\sigma)^2$, a cline 15 km wide (w) would take 87,500 years (T) to form. Since the zone is thought to have formed in these high cols some 8000 years ago, it would need an average dispersal of around 100 m/generation to produce a 15-km cline. Something is wrong. Indeed, because we found testicular dysgenesis in F_1 hybrids between parental *C. p. erythropus* and *C. p. parallelus* (Hewitt et al. 1987), the genes for which will act as a barrier to neutral gene flow along with other negatively heterotic loci, these are underestimates of dispersal. Now, while a rare individual of this small flightless grasshopper might manage 100 m in its lifetime, for the whole population to have a standard deviation of parent–offspring distances of this magnitude is most unlikely. Except, that is, at times when the population structure is patchy and is subject to local extinction and recolonization (cf. Lande 1979). Under these circumstances long distance migrants can colonize areas with their alleles only. Our computer simulations of secondary contact show that zones for several clines become much wider with low population sizes (*Ne*) than with high ones. There is also evidence from *Podisma* at Seyne that variation in the width of the zone is due to present differences in population structure (Nichols and Hewitt 1986), low effective population size allowing greater interpenetration of alleles. Detailed vegetation analysis with measures of grasshopper density, distribution, and dispersal through the hybrid zone is necessary to model and understand the level and pattern of gene flow in such situations, and these are in hand for two of the Pyrenean cols.

The historical aspect should not be overlooked in resolving this type of dilemma, and we are fortunate that both the palynology and archaeology of this region are of growing interest (Jalut 1977; Bahn 1985). It appears that pastoral activity that favors *C. parallelus* was well established at high altitude in the Pyrenees in the early neolithic. There were considerable grazing resources above about 1900 m and there is even evidence of cereals on one of the cols we study in the High Aude around 6200 BP! Lower down there was thick forest from early postglacial times until the seventeenth century that is inhospitable to *C. parallelus*. There could have been an early postglacial broad mixing of races in the high pastures with little dispersal from lower down through the forests. We have observed grasshoppers to be carried some distance in the fleeces of mountain sheep, and there is a Paleolithic tradition of transhumance over the Pyrenees.

In spite of this genetic evidence of considerable dispersal and mixing, the introgression of even neutral alleles is no more than a few kilometers, and the same is true for other zones such as *Bombina*, *Caledia*, *Litoria*, *Ranidella*, *Sceloporus*, *Thomomys*, and *Warramaba*. This means that there is virtually no genetic exchange between the main bodies of the races except for alleles that are advantageous in both genomes, which will pass through the zone with only moderate hindrance. Consequently races separated by hybrid zones could be fairly autonomous in their evolution unless reorganized at times of major range contraction and expansion.

Several authors (e.g., Grant 1980; Slatkin 1987) have argued that the spread of alleles through a species by dispersal and selection may be very slow, and therefore immigration over larger distances may explain the relative similarity of allele frequencies between populations. The data from hybrid zones indicate that this may be true for a substantial part of the genome and operate at the level of subspecies and races as well as smaller groups of populations. Grant (1980) also emphasizes the differences in mobility between species (cf. Table 2). Thus *C. parallelus*, a flightless grasshopper, has a field-measured dispersal of the order of 30 m/generation and its widest clines are around 15 km. Since these are probably selectively equal differences, this argues for an overall dispersal during the last 8000 years of 100 m/generation from Endler's (1977) neutral diffusion model. Similar data and reasoning for both the toad *Bombina bombina* and the commensal mouse *Mus domesticus* show an increasing dispersal ability and wider zones. Even so, the exchange of genetic material is limited compared with the range of the genomes and species. Much wider zones exist for several bird species, which in general are more mobile. This variation in dispersal along with the extent of suitable environment will determine how discrete is the dissection of the species range.

POSTGLACIAL EXPANSION

A growing body of information from various sources permits a more precise description of the biotic changes since the last glaciation. In western Europe the use of fossil insect distributions (Coope 1970, 1977) and pollen analysis (Huntley and Birks 1983) is particularly valuable. These indicate that the present warm Holocene period began following a final cold spell about 10,000 BP. Much of France down to the Pyrenees was tundra around 11,000 BP, although birch and conifers were established in the east and Germany. The English Channel, an effective barrier to orthopteran dispersal, was cut around 8000 BP, so that the expansion of *C. parallelus* to reach England before this must have been very rapid—1000 km in 2000 years requires 500 m/year—an almost unbelievable figure for an unaided, flightless, fairly heavy insect! An alternative explanation comes from a consideration of the details of the postglacial warming period as revealed by radiocarbon dating of fossil *Coleoptera* in Britain (Atkinson et al. 1987). About 13,000 BP there was a rapid warming to present-day temperatures followed by a slow cooling and cold spell until 10,500 BP with temperatures nearly as cold as 13,500 BP (Figure 4). The tundra was much more extensive across western Europe around 13,000 BP than around 11,000 BP, so that during the warm spell around 12,000 BP many plant and animal species expanded north and some, including *C. parallelus*, may have survived the 10,500 BP cold spell in sheltered refugia north of the Alps. However, even with the most favorable assumptions (13,000–8000 BP = 5000 years to cover 1000 km), this still requires an average range expansion of 200 m/year, as compared with field dispersal estimates of less than 30 m/year.

Such rapid expansion carries certain genetic implications. Given the rapid warming over 500 years, many species would have been expanding into unoccupied favorable territory. Consequently, those populations at the edge of the distribution would contribute most to the expansion, since they would arrive first in greater numbers and colonize a location; subsequent migrants would contribute a much smaller proportion to the gene pool. Those genomes at the edge of a refugium would spread to fill the new territory; internal genomes behind hybrid zones would be unlikely to spread. In addition, this type of population structure at the expanding edge would be more favorable for the establishment of new chromosomal

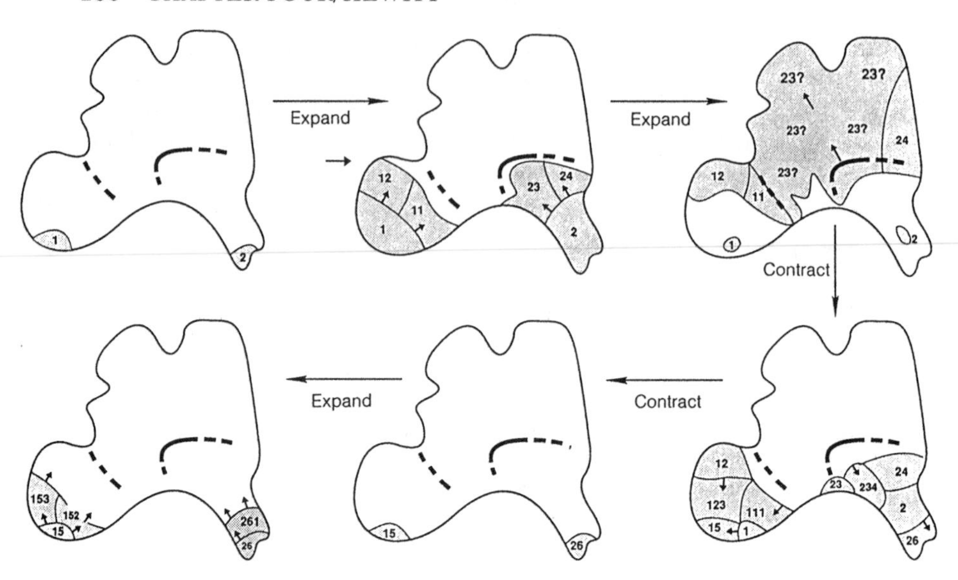

FIGURE 4. A suggested sequence of distribution for the *Chorthippus parallelus* complex from the pleniglacial through the initial warming spell (to 12,000 BP), the subsequent cooling (to 10,500 BP) and the last warming immediately afterward (e.g., 8000 BP). This is reconstructed from information in West (1977), Huntley and Birks (1983), and Atkinson et al. (1987) along with a knowledge of the current habitats and locations of the species (Reynolds 1980).

rearrangements or other negatively heterotic mutants. Computer simulations are being used to investigate these ideas further.

Genetic drift contributes to the differentiation among populations at the edge of the expanding range. The study of genetic drift has a long history stretching back to the ideas of Wright (1931, 1969, 1978) and which have been developed in many areas of genetics, from the origin and establishment of new genes and genomes to the founder flush and peripatric models of speciation (e.g., Mayr 1954, 1963; Carson 1975, 1982; Powell 1978; Templeton 1980; Carson and Templeton 1984; Barton and Charlesworth 1984; Barton, this volume). The work of Endler (1973, 1977) is particularly comprehensive and relevant. He ran a series of two-dimensional simulations to see if stepped clines could form regularly as a result of genetic drift with gene flow between demes. Steep clines developed, particularly at intermediate levels of gene flow, but they were not stable, even though local patches of high and low gene frequency persisted for hundreds of generations. In one series he examined the effects of barriers to dispersal caused by regions of small population size, and noted that areal differentiation was greater and persisted longer around these barriers.

Even under optimal conditions of size and flow, strongly differentiated areas did not grow larger than 15 demes. However, genetic drift at the expanding edge of a species range has an extra dimension, since the novel fixed genomes for negatively heterotic mutants can colonize the newly available territories. A stable and immobile tension zone would form near the population in which a heterotic mutant became established. This would separate the older allele from the advancing wave front at which the new allele was spreading into unoccupied territory.

Following the last ice age Europe had two such major rapid expansions with a partial contraction in between, which was overlaid with smaller cycles. The foregoing considerations suggest that the chromosomal races in *Mus* and *Sorex* originated at the expanding-contracting edge, and did not reach their present positions as hybrid zones spreading through the species range. The same may apply to parapatric races for other characters and in other continents, e.g., *Warramaba viatica* in Australia.

A geographic model for *C. parallelus* in Western Europe through two ice ages is given in Figure 5, which incorporates these ideas. They are further developed in a stylized manner in Figure 6, which includes more mutations, edge expansion, and mountain and lowland refugia, and demonstrates how a variety of racial distributions may have developed (see legend for explanation). The details of the genomic boundaries will be provided by further research and this will allow testing and improvement of this type of explanation.

In addition to the spread of genomes and subdivision by hybrid zones proposed, glaciation and range changes probably had other effects on the distribution of species and their genetic composition. Several have been discussed by various authors over the years. This is particularly evident in earlier plant cytogenetic literature (Darlington 1963; Stebbins 1971). Darlington presents a variety of pleistocene scenarios for the distribution of polyploid complexes and Stebbins emphasizes that empty habitats would have become available around the advancing and retreating margins of the ice sheet into which the best adapted hybrid and polyploid genomes could have spread. In animals Key (1968) proposed peripheral isolates for the origin of differentiating traits in *W. viatica*, and the wide distribution of parthenogens such as the orthopterans *Saga pedo* and *Moraba virgo* may well be the product of postglacial expansion (Hewitt 1979). In attempting to explain levels of chromosome variation and allozyme similarity in the pocket gophers *Thomomys* from the southwestern United States, Patton and Yang (1977) suggest that major shifts in vegetational zones and greater glacial fragmentation of northern populations were formative factors (see also Patton and Smith, this volume). Sage and Wolff (1986) have examined the protein heterozygosity in 18 specimens of the Dall sheep, *Ovis dalli dalli*, from three mountain ranges in Northern Alaska and found it to be

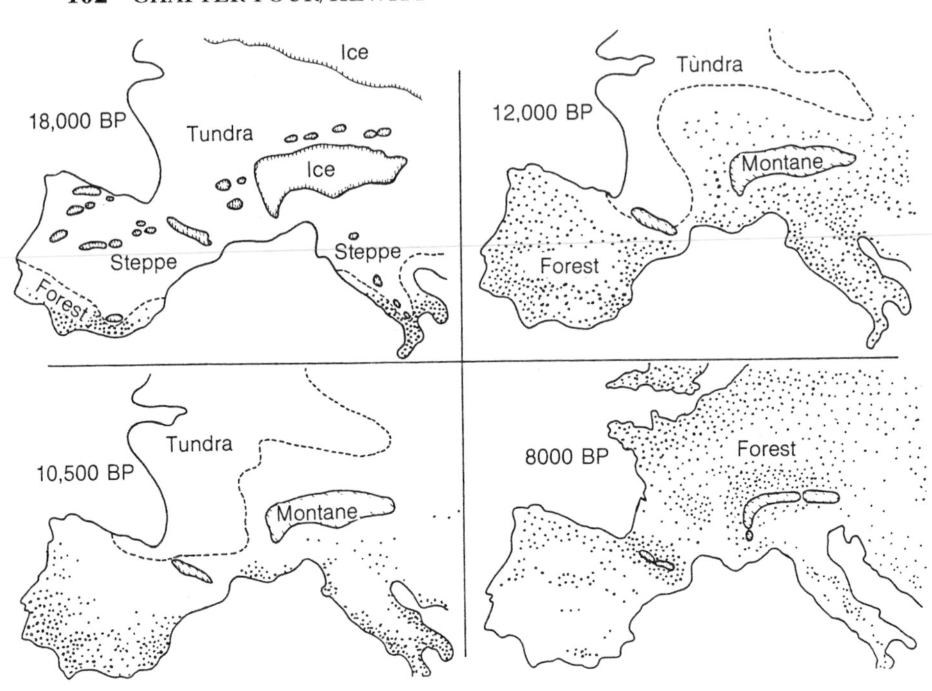

FIGURE 5. A stylized map of western Europe including the Pyrenees and Alps in which a cycle of expansion contraction and expansion is followed for *Chorthippus parallelus*. It shows glacial refugia in southern Spain and Italy with range change northward in the interglacial, leaving isolates in the high Iberian and Italian mountains and forming hybrid zones in the Pyrenees and possibly the Alps and other places. Mutations at the edge of the expanding range that become established are shown as arrows and an additional digit on the identifying number of the genome. Following a contraction the races in refugia contain different mutations; genomes to the north go extinct. The different mutations then all expand out together as a front to form a zone of coincident clines on contact with another race.

remarkably low ($H_{\text{Nei}} = 0.015$) as compared with the average for 184 mammalian taxa ($H_{\text{Nei}} = 0.041$). They argue cogently that repeated fluctuation of ice fronts will have shifted its periglacial range several times, thereby providing repeated opportunities for advancing founder individuals to occupy new patches of suitable environment and generating a serial loss of alleles from oldest to youngest populations. This could clearly have produced the observed low heterozygosity in these sheep and is supported by a generally lower heterozygosity in northern species from higher latitudes as compared with those from lower latitudes (Sage and Wolff 1986). They propose that species in previously glaciated regions should have lost genic variability due to these range changes. This might be extended to predict lower heterozygosity at high latitude within subdivided species with extensive north–south ranges. Such a loss of heterozygosity would be concomitant with the production of the suggested pattern of species subdivision by tension zones of various kinds and strengths. Both would depend on the life-style and population dynamics of each species, and genetic data on distribution and frequency are needed from a variety of organisms and ranges.

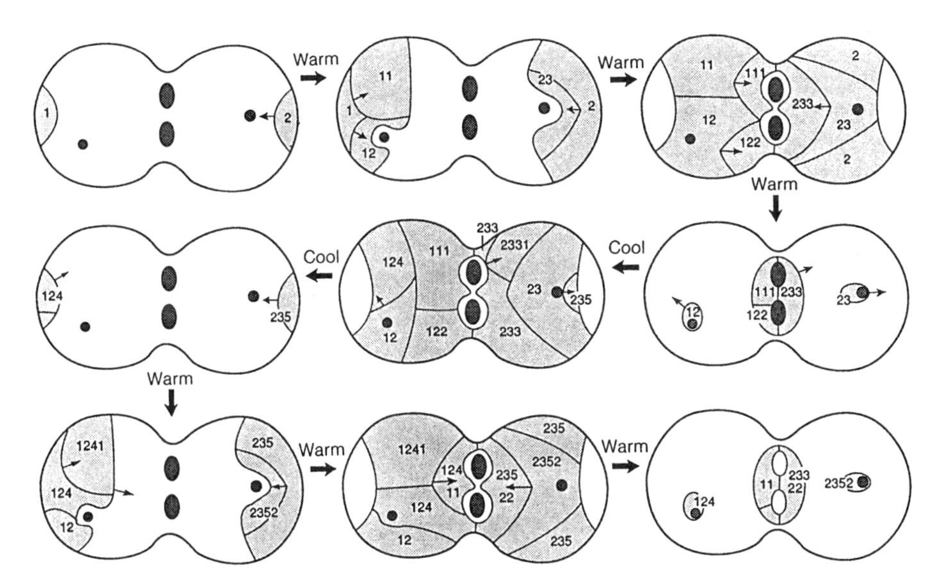

FIGURE 6. A diagram of range expansion–contraction cycles to illustrate how races may accumulate genomic changes and form stable hybrid zones comprising many coincident clines. The dumbell species range contains a refugium at each end (during an ice age), two small isolated blocks of inaccessible high mountains, and a larger central ridge of such high mountains. Genomes 1 and 2 could be in Spain and Italy with the Pyrenees forming the central ridge. As the climate warms the genomes expand their range and mutations establish themselves at the edge and spread (arrows and digits). If these mutants are negatively heterotic then a tension zone forms between them and the parent homozygote. In a warm climate the species is restricted to mountains and forms a multiple hybrid zone in the central ridge of mountains. Subsequent cooling causes the expansion of the ranges and then contraction into the refugia with a further accumulation of genetic differences between the races. These then expand again to form an even more complex hybrid zone.

CONCLUSIONS

There is growing evidence that a large number of species are subdivided by hybrid zones, which usually comprise multiple coincident clines for genes determining a whole range of characters. They are most often caused by hybrid unfitness, but ecological differences may also be integrated into the racial genomes. A consideration of these properties and population dynamics suggests that hybrid zones once formed do not move far, and the present positioning of several zones supports this. It is proposed that these multiple racial differences are brought together in a zone by cycles of range contraction and expansion; the details of the current postglacial period in Europe suggest that large areas could be covered by an expanding genome before contact with another genome produces a stable hybrid zone. The population structure at the edge of distribution during range changes may be particularly suitable for the establishment of new negatively heterotic mutants. Thus, the major substructure of the species may be put in place after major environmental changes such as the postglacial amelioration. Smaller changes of a similar nature may produce a lower substructure, and in the relatively stable periods genomes may adapt to local conditions.

These hybrid zones are semipermeable; genes showing negative heterosis and those positively epistatic with them will form the zone. Neutral genes closely linked to such zonal loci will be retarded in their spread, while advantageous alleles may pass through relatively unhindered. Significantly hybrid zones show that the diffusion of apparently neutral alleles is often for small distances compared with the range of the races. This is more pronounced for organisms with low dispersal. However, the particular population structure in which this dispersal occurs is critical and this will vary from time to time.

Thus, many species will be divided into compartments between which there is little gene flow; the range and content of these genomic compartments are modified by major changes in their environment. There may well be a substructure below this since some zones are apparently simple, involving few characters. These will be more difficult to find, but the coming of more rapid molecular techniques offers an exciting prospect.

Hybrid zones can provide information and insight into many aspects of speciation in addition to species substructure, including important areas such as the origin and establishment of differences, the nature of subspecific and specific differences, and models of speciation such as reinforcement. Several of these are dealt with by other speakers. Together they show what a valuable probe this phenomenon is.

SUMMARY

Many species are subdivided by hybrid zones comprising coincident clines for many characters. Most of these zones involve hybrid unfitness, although the genetic divergence of the races may have included adaptation to different environments that is now integrated into their genomes. The population dynamics of such zones implies that once formed they do not move far. It is proposed that range contraction–expansion cycles forge zones comprising multiple racial differences.

Postglacial Europe provides a particularly well-studied example of large-scale range changes. During this period large areas were colonized by an expanding genome before meeting another. During range changes the population structure at the edges seems particularly suitable to establish new mutations, and the expansion into virgin territory provides a rare opportunity for the spread of such genetic novelties. Thus, the gross substructure of a species may be put in place by major environmental changes such as the postglacial amelioration. Smaller fluctuations may produce a lesser substructure, along with adaptation to local conditions in stable periods.

Advantageous mutations may pass through hybrid zones, but most alleles diffuse through for only small distances compared to the range of the races. This is more pronounced for species with low dispersal, and their population structure, which may vary temporally, is critical. Thus, many species may be divided into compartments among which gene flow is greatly reduced.

ACKNOWLEDGMENTS

I am grateful to Nick Barton, Roger Butlin, John Endler, Richard Nichols, Michael Ritchie, and Richard Sage for helpful comments. The work on *Podisma* and *Chorthippus* was supported by grants from the NERC and SERC.

LITERATURE CITED

Arntzen, J. W. 1978. Some hypotheses on postglacial migrations of the fire bellied toad *Bombina bombina* (L) and the yellow bellied toad *Bombina variegata* (L). J. Biogeog. 5:339–395.

Atkinson, T. C., K. R. Briffa and G. R. Coope. 1987. Seasonal temperatures in Britain during the past 22,000 years, reconstructed using beetle remains. Nature (London) 325:587–592.

Avise, J. C., E. Bermingham, L. G. Kessler, and N. C. Saunders. 1984. Characterization of mitochondrial DNA variability in a hybrid swarm between subspecies of bluegill sunfish (*Lepomis mactochirus*). Evolution 38:931–941.

Bahn, P. G. 1985. *Pyrenean Prehistory*. Aris & Phillips Ltd., Warminster, UK.
Baker, R. J., L. W. Robbins, F. B. Stangl, and E. C. Burney. 1983. Chromosomal evidence for a major subdivision in *Peromyscus leucopus*. J. Mammal. 64:356–359.
Barabas, L. 1978. Studium zony kontaktu arealov *Crypocephalus hypochocridis* u *Cryptocephalus transiens*. Biologia (Bratislava) 33:407–412.
Barrowclough, G. F. 1980. Genetic and phenotypic differentiation in a wood warbler (genus *Dendroica*) hybrid zone. The Auk 97:655–668.
Barton, N. H. 1979. The dynamics of hybrid zones. Heredity 43:341–359.
Barton, N. H., and B. Charlesworth. 1984. Genetic revolutions, founder effects and speciation. Annu. Rev. Ecol. System. 15: 133–164.
Barton, N. H., and G. M. Hewitt. 1981. Hybrid zones and speciation. Pp. 109–145 in W. R. Atchley and D. S. Woodruff (eds.), *Evolution and Speciation: Essays in Honour of M. J. D. White*. Cambridge University Press, Cambridge.
Barton, N. H. , and G. M. Hewitt. 1983. Hybrid zones as barriers to gene flow. Pp. 341–359 in: G. S. Oxford and D. Rollinson (eds.), *Protein Polymorphism: Adaptive and Taxonomic Significance*. Blackwell, Oxford.
Barton, N. H., and G. M. Hewitt. 1985. Analysis of hybrid zones. Annu. Rev. Ecol. System. 16:113–148.
Bush, G. L., S. M. Case, A. C. Wilson and J. L. Patton. 1977. Rapid speciation and chromosomal evolution in mammals. Proc. Natl. Acad. Sci. U.S.A. 74:3942–3946.
Butlin, R. K., and G. M. Hewitt. 1985a. A hybrid zone between *Chorthippus parallelus parallelus* and *C.p. erythropus* (Orthoptera: Acrididae): Morphological and electrophoretic characters. Biol. J. Linnean Soc. 26:269–285.
Butlin, R. K., and G. M. Hewitt. 1985b. A hybrid zone between *Chorthippus parallelus parallelus* and *C.p. erythropus* (Orthoptera: Acrididae): Behavioural characters. Biol. J. Linnean Soc. 26:287–299.
Carson, H. L. 1975. The genetics of speciation at the diploid level. Am. Natur. 109:83–92.
Carson, H:. L. 1982. Speciation as a major reorganisation of polygenic balances. Pp. 411–433 in: C. Barigozzi (ed.), *Mechanisms of Speciation*. Alan R. Liss, New York.
Carson, H. L., and A. R. Templeton. 1984. Genetic revolutions in relation to speciation phenomena: The founding of new populations. Annu. Rev. Ecol. System. 15:97–131.
Chambers, S. M. 1987. Rates of evolutionary change in chromosome numbers in snails and vertebrates. Evolution 41:166–175.
Collins, M. M. 1984. Genetics and ecology of a hybrid zone in *Hyalophora* (Lepidoptera: Saturniidae). Univ. Calif. Pub. Entomol. 104:1–93.
Cook, A. 1975. Changes in the carrion/hooded crow hybrid zone and the possible importance of climate. Bird Study 22:165–168.
Coope, G. R. 1970. Interpretations of Quaternary insect fossils. Annu. Rev. Entomol. 15: 97–120.
Coope, G. R. 1977. Fossil coleopteran assemblages as sensitive indicators of climatic changes during the Devensian (last) Cold Stage. Philos. Trans. R. Soc. London, Ser. B 280:313–340.
Darlington, C. D. 1963. *Chromosome Botany and the Origin of Cultivated Plants*. George Allen and Unwin, London.
Dixon, K. L. 1955. An ecological analysis of the interbreeding of crested titmice in Texas. Univ. Calif. Pub. Zool. 54:125–206.
Duncan, R., and R. Highton. 1979. Genetic relationships of the eastern large Plethodon of the Oachita mountains. Copeia 1979: 95–110.
Endler, J. A. 1973. Gene flow and population differentiation. Science 179:243–250.
Endler, J. A. 1977 *Geographic Variation, Speciation and Clines*. Princeton University Press, Princeton.
Endler, J. A. 1982a. Problems in distinguishing historical from ecological factors in biogeography. Am. Zool. 22:441–452.
Endler, J. A. 1982b. Pleistocene forest refuges: fact or fancy? Pp. 641–657 in: G. T. Prance (ed.), *Biological Diversification in the Tropics*. Columbia University Press, New York.
Ficken, M. S., and R. W. Ficken. 1968. Reproductive isolating mechanisms in the blue-winged and golden-winged warbler complex. Evolution 22:166–179.

Gill, F. B. 1980. Historical aspects of hybridization between blue-winged and golden-winged warblers. The Auk 97:1–18.
Grant, V. 1980. Gene flow and the homogeneity of species populations. Biol. Zentralb. 99: 157–169.
Green, D. M. 1983. Allozyme variation through a clinal hybrid zone between the toads *Bufo americanus* and *B. hemiophrys* in south eastern Manitoba. Herpetologica 39:28–40.
Gyllensten, U., and A. C. Wilson. 1987. Interspecific mitochondrial DNA transfer and the colonization of Scandinavia by mice. Genet. Res. Cambridge 49:25–29.
Hagen, D. W., and G. E. E. Moodie, 1979. Polymorphism for breeding colors in *Gasterosteus aculeatus*. I. Their genetics and geographic distribution. Evolution 33:641–648.
Hall, B. P., and R. E. Moreau. 1970. *Atlas of Speciation in African Passerine Birds*. British Museum (Natural History), London.
Harrison, R. G. 1986. Pattern and process in a narrow hybrid zone. Heredity 56:337–350.
Hauber, D. P., and W. L. Bloom. 1983. Stability of a chromosomal hybrid zone in the *Clarkia nitens* and *Clarkia speciosa* spp *polyantha* complex (Onagracae). Am. J. Bot. 70:1454–1459.
Heaney, L. R., and R. M. Timm. 1985. Morphology, genetics and ecology of pocket gophers (genus *Geomys*) in a narrow hybrid zone in Nebraska. Biol. J. Linnean Soc. 25:301–317.
Hewitt, G. M. 1975. A sex-chromosome hybrid zone in the grasshopper *Podisma pedestris* (Orthoptera: Acrididae). Heredity 35:375–385.
Hewitt, G. M. 1979. *Animal Cytogenetics. III. Orthoptera*. Gebruder Borntraeger, Stuttgart.
Hewitt, G. M. 1985. The structure and maintenance of hybrid zones—with some lessons to be learned from alpine grasshoppers. Pp. 15–54 in: J. Gosalvez, C. Lopez-Fernandez, and C. Garcia de la Vega (eds.), *Orthoptera*. Fundación Ramón Areces, Madrid.
Hewitt, G. M., and N. H. Barton. 1980. The structure and maintenance of hybrid zones as exemplified by *Podisma pedestris*. Pp. 149–169 in: R. L. Blackman, G. M. Hewitt, and M. Ashburner (eds.), *Insect Cytogenetics* (Royal Entomological Society of London Symposia 10). Blackwell, Oxford.
Hewitt, G. M., R. K. Butlin, and T. M. East. 1987 Testicular dysfunction in hybrids between parapatric subspecies of the grasshopper *Chorthippus parallelus*. Biol. J. Linnean Soc. 31:25–34.
Hewitt, G. M., J. Gosalvez, C. Lopez-Fernandez, M. G. Ritchie, W. Nichols, and R. K. Butlin. 1988. Differences in the nucleolar organisers, sex chromosomes and Haldane's Rule in a hybrid zone. Pp. 109–119 in: P.E. Brandham (ed.), *Kew Chromosome Conference III*. HMSO, London.
Heywood, J. S. 1986. Clinal variation associated with edaphic ecotones in hybrid populations of *Gaillardia pulchella*. Evolution 40:1132–1140.
Highton, R. 1977. Comparison of microgeographic variation in morphological and electrophoretic traits. Evol. Biol. 10:397– 436.
Hillis, D. M., and J. E. Simmons. 1986. Dynamic change of a zone of parapatry between two species of *Pholidobolus* (Sauria: Gymnophthalmidae). J. Herpetol. 20:85–87.
Howard, D. J. 1986. A zone of overlap and hybridization between two ground cricket species. Evolution 40:34–43.
Howard, D. J., and R. G. Harrison, 1984. Habitat segregation in ground crickets: Experimental studies of adult survival, reproductive success and oviposition preference. Ecology 65:61–68.
Huntley, B., and H. J. B. Birks. 1983. *An Atlas of Past and Present Pollen Maps for Europe*. Cambridge University Press, Cambridge.
Jalut, G. 1977. *Vegetation et Climat des Pyrenees Mediterraneennes depuis Quinze Mille Ans*. Archives d'Ecologie Prehistorique No. 2, Toulouse, 2 vols.
Karlin, A. A., and S. I. Guttman. 1981. Hybridization between *Desmognathus fuscus* and *D. ochrophaeus* (Amphibia: Urodela: Plethodontidae) in north eastern Ohio and north western Pennsylvania. Copia 1981:371–377.
Key, K. H. L. 1968. The concept of stasipatric speciation. System. Zool. 17:14–22.
Kimura, M. 1983 *The Neutral Theory of Molecular Evolution*. Cambridge University Press, Cambridge.

Kocher, T. D., and R. D. Sage. 1986. Further genetic analyses of a hybrid zone between Leopard frogs (*Rana pipiens* complex) in central Texas. Evolution 40:21–33.

Kruse, K. C., and D. G. Dunlop. 1976. Serum albumins and hybridization in two species of the *Rana pipiens* complex in the north central U. S. Copeia 1976:394–396.

Lande, R. 1979. Effective deme sizes during long-term evolution estimated from rates of chromosomal rearrangement. Evolution 33:234–251.

Lorkovic, Z. 1958. Some peculiarities of spatially and sexually restricted gene exchange in the *Erebia tyndarus* group. Cold Spring Harbor Symp. Quant. Biol. 23:319–325.

Mayr, E. 1954. Change of genetic environment and evolution. Pp. 157–180 in: J. Huxley, A. C. Hardy, and E. B. Ford (eds), *Evolution as a Process*. Macmillan, New York.

Mayr, E. 1963. *Animal Species and Evolution*. Belknap Press of Harvard University Press, Cambridge.

Mayr, E., and R. J. O'Hara. 1986. The biogeographic evidence supporting the Pleistocene forest refuge hypothesis. Evolution 40:55–67.

Moore, W. S., and D. B. Buchanan. 1985. Stability of the Northern Flicker hybrid zone in historical times: Implications for adaptive speciation theory. Evolution 39:135–151.

Nadler, C. F., R. S. Hoffman, and K. R. Greer. 1971. Chromosomal divergence during evolution of ground squirrel populations. Genetics 20:298–305.

Nevo, E., Y. J. Kim, C. R. Shaw, and C. S. Thaeler. 1974. Genetic variation, selection and speciation in *Thomomys talpoides* pocket gophers. Evolution 28:1–23.

Nichols, R. A. 1985. Genetical and ecological differentiation across a hybrid zone. Pp. 55–83 in: J. Gosalvez, C. Lopez-Fernandez, and C. Garcia de la Vega (eds.), *Orthoptera*. Fundación Ramón Areces, Madrid.

Nichols, R. A., and G. M. Hewitt. 1986. Population structure and the shape of a chromosomal cline between two races of *Podisma pedestris* (Orthoptera: Acrididae). Biol. J. Linnean Soc. 29:301–316.

Nichols, R. A., and G. M. Hewitt. 1988 Genetical and ecological differentiation across a hybrid zone. Ecol. Entomol. 13:39–49.

Patton, J. L., and S. Y. Yang. 1977. Genetic variation in *Thomomys bottae* pocket gophers: Macrogeographic patterns. Evolution 31:697–720.

Peabody, R. B. 1978. Electrophoretic analysis of geographic variation and hybridization of two Appalachian salamanders, *Plethodon jordani* and *P. glutinosus*. Ph.D. thesis, University of Maryland.

Powell, J. R. 1978. The founder-flush speciation theory: An experimental approach. Evolution 32:465–474.

Reynolds, W. J. 1980. A re-examination of the characters separating *Chorthippus montanus* and *C. parallelus* (Orthoptera: Acrididae). J. Nat. Hist. 14:283–303.

Riddoch, B. J., M. T. Greenwood, and R. D. Ward. 1984. Aspects of the population structure of the sand martin flea, *Ceratophyllus styx*, in Britain. J. Nat. Hist. 18:475–484.

Rising, J. D. 1983. The Great Plains hybrid zones. Pp. 131–157 in: R. F. Johnston (ed.), *Current Ornithology*, Vol. 1. Plenum, New York.

Roberts, J. D., and L. R. Maxson. 1986. Phytogenetic relationships in the genus *Limnodynastes* (Anura: Myobatrachidae): A molecular perspective. Aust. J. Zool. 34:561–573.

Rohwer, S. A. 1972. Distribution of meadow larks in the central and southern Great Plains and desert grasslands of eastern New Mexico and Texas. Transact. Kansas Acad. Sci. 75:1–19.

Rohwer, S. A. 1976. Specific distinctness and adaptive differences in south western meadow larks. Occasional papers Mus. Nat. Hist. Univ. Kansas 44:1–14.

Sage, R. D., and J. O. Wolff. 1986. Pleistocene glaciations, fluctuating ranges, and low genetic variability in a large mammal (*Ovis dalli*). Evolution 40:1092–1095.

Sage, R. D., D. Heyneman, K-C. Lim, and A. C. Wilson. 1986a. Wormy mice in a hybrid zone. Nature (London) 324:60–63.

Sage, R. D., J. B. Whitney III, and A. C. Wilson. 1986b. Genetic analysis of a hybrid zone between domesticus and musculus mice (*Mus musculus* complex): Hemoglobin polymorphisms. Pp. 75–85 in: *Current Topics in Microbiology and Immunology*, Vol. 127. Springer-Verlag, Berlin, Heidelberg.

Schoorl, J., and A. Zuiderwijk. 1981. Ecological isolation in *Triturus cristatus* and *Triturus marmoratus* (Amphibia: Salamandridae). Amphibia/Reptilia 3/4:235–252.
Searle, J. B. 1984. Three new karyotypic races of the common shrew *Sorex araneus* (Mammalia: Insectivora) and a phylogeny. System. Zool. 33:184–194.
Shaw, D., M. Arnold, A. Marchant, and N. Contreras. 1988. Chromosomal rearrangements, ribosomal genes and mitochondrial DNA: Contrasting patterns of introgression across a narrow hybrid zone. Pp. 121–129 in P.E. Brandham (ed.), *Kew Chromosome Conference III.* HMSO, London.
Shaw, D. D., P. Wilkinson, and D. J. Coates. 1982. The chromosomal component of reproductive isolation in the grasshopper *Caledia captiva* II. The relative viabilities of recombinant and nonrecombinant chromosomes during embryogenesis. Chromosoma 86:533–547.
Sibley, C. G., and L. L. Short. 1959. Hybridization in the buntings (*Passerina*) of the Great Plains. The Auk 76:443–463.
Sibley, C. G., and D. A. West. 1958. Hybridization in the red-eyed towhees of Mexico: The eastern plateau populations. Condor 60:85–104.
Slatkin, M. 1987. Gene flow and the geographic structure of natural populations. Science 236:787–792.
Smith, M. F. 1979. Geographic variation in genic and morphological characters in *Peromyscus californicus*. J. Mammal. 60: 705–722.
Smith, S. G., and N. Virkhi. 1978. *Animal Cytogenetics III. Coleoptera*. Gebruder Borntraeger, Stuttgart.
Snow, D. W. 1978. *An Atlas of Speciation in African Non-passerine Birds*. British Museum (Natural History), London.
Stangl, F. B. 1986. Aspects of a contact zone between two chromosomal races of *Peromyscus leucopus* (Rodentia: Cricetidae). J. Mammal. 67:465–473.
Stebbins, G. L. 1971. *Chromosomal Evolution in Higher Plants*. Edward Arnold, London.
Szymura, J. M., C. Spolsky, and T. Uzzell. 1985. Concordant change in mitochondrial and nuclear genes in a hybrid zone between two frog species (genus *Bombina*). Experientia 41:1469–1470.
Templeton, A. R. 1980. The theory of speciation via the founder principle. Genetics 94:1011–1038.
Thaeler, C. S. 1968. An analysis of three hybrid populations of pocket gophers (genus *Thomomys*). Evolution 22:543–555.
Thaeler, C. S. 1974. Four contacts between ranges of different chromosomal forms of the *Thomomys talpoides* complex (Rodentia: Geomyidae). System. Zool. 23:343–354.
Thorpe, R. S. 1984. Primary and secondary transition zones in speciation and population differentiation: A phytogenetic analysis of range expansion. Evolution 38:233–243.
Tucker, P. K., and D. J. Schmidly. 1981. Studies of a contact zone among three chromosomal races of *Geomys bursarius* in East Texas. J. Mammal. 62:258–272.
Turner, J. R. G. 1971. Two thousand generations of hybridization in a *Heliconius* butterfly. Evolution 25:471–482.
Vander Meer, R. K., C. S. Lofgren, and F. M. Alvarez. 1985. Biochemical evidence for hybridization in fire ants. Florida Entomol. 68:501–506.
Vanlerberghe, F., B. Dod, P. Boursot, M. Bellis, and F. Bonhomme. 1986. Absence of Y chromosome introgression across the hybrid zone between *Mus musculus domesticus* and *Mus musculus musculus*. Genet. Res., Cambridge 48:191–197.
Wake, D. B., and K. P. Yanev. 1986. Geographic variation in allozymes in a "ring species", the plethodontid salamander *Ensatina eschscholtzii* of western North America. Evolution 40:702–715.
Wake, D. B., S. Y. Yang, and T. J. Papenfuss. 1980. Natural hybridization and its evolutionary implications in Guatemalan plethodontid salamanders, genus *Bolitoglossa*. Herpetologica 36:335–345.
West, D. A. 1962. Hybridization in grosbeaks (*Pheucticus*) of the Great Plains. The Auk 79:399–424.
West, R. G. 1977. *Pleistocene Geology and Biology*. Longman, London.

White, M. J. D. 1978. *Modes of Speciation*. Freeman, San Francisco.

Wilby, A. S. 1987. Population cytology of *Rumex acetosa*. Ph.D. thesis, University of London.

Wilson, A. C. 1985. The molecular basis of evolution. Sci. Am. 253:148–157.

Wilson, A. C., R. L. Cann, S. M. Carr, M. George, U. B. Gyllensten, K. M. Helm-Bychowski, R. G. Higuchi, S. R. Palumbi, E. M. Prager, R. D. Sage, and M. Stoneking. 1985. Mitochondrial DNA and two perspectives on evolutionary genetics. Biol. J. Linnean Soc. 26:375–400.

Wright, S. 1931. Evolution in Mendelian populations. Genetics 16:97–159.

Wright, S. 1969. *Evolution and the Genetics of Populations*. Volume 2, *The Theory of Gene Frequencies*. University of Chicago Press, Chicago.

Wright, S. 1978. *Evolution and the Genetics of Populations*. Volume 4, *Variability within and among Natural Populations*. University of Chicago Press, Chicago.

Yang, S. Y., and R. K. Selander. 1968. Hybridization in the grackle *Quiscalus quiscula* in Louisiana. System. Zool. 17:107–143.

Zaslavskii, V. A. 1967. Reproduktivnoe samounichtozhenie ekologicheskii faktor (ekologicheskie posledstviya geneticheskogo vzaimodeistviya populyatsii). Z. obshchei biol. 28:3–11.

CHAPTER 5

MOSAIC HYBRID ZONES AND THE NATURE OF SPECIES BOUNDARIES

Richard G. Harrison and David M. Rand

HYBRID ZONES AND SPECIATION

Hybrid zones have figured prominently in two persistent and fundamental debates about the process of speciation. One debate concerns the origin of hybrid zones and the geographic context in which differentiation occurs. Mayr (1942) distinguished zones of primary and secondary intergradation, the former representing clines that develop along a series of populations in continuous contact and the latter resulting from the joining of populations that have differentiated in allopatry. Most narrow hybrid zones were interpreted as zones of secondary contact because of doubts that such abrupt discontinuities could represent a direct response to the environment. However, Clarke (1966), Slatkin (1973), and Endler (1973, 1977) provided explicit models demonstrating that even very steep clines can develop along an environmental gradient in the absence of geographic isolation. Endler (1977) argued persuasively that it is extremely difficult to distinguish primary and secondary intergradation, since both processes result in the same pattern of variation. Only if we encounter populations within a few hundred generations of secondary contact will we be able to decipher their recent history. The debate over hybrid zone origins mirrors a similar debate about speciation—between proponents of parapatric models of speciation and an opposition that contends that geographic isolation is a prerequisite for the evolution of intrinsic barriers to gene exchange.

One focus of this chapter will be to examine the consequences of a mosaic hybrid zone model for resolving the issue of hybrid zone origins. I will argue that in the context of a mosaic hybrid zone, the issue of origins can often be resolved, but that this does not necessarily provide information about the process of speciation. As Barton and Hewitt (1981) emphasized: "the way in which the present zone was formed may be quite unrelated to the way in which the differentiation itself arose."

Not only are hybrid zone origins uncertain, but the eventual fate of such zones has also proved to be a contentious issue and one that has important implications for competing hypotheses of allopatric speciation. The central debate is whether prezygotic barriers to gene exchange are a fortuitous byproduct of differentiation in allopatry or whether they arise within areas of secondary contact (hybrid zones) in response to selection against individuals of mixed ancestry. Do hybrid zone interactions lead to speciation by reinforcement or are fusion or extinction more likely outcomes? Alternatively, are hybrid zones in fact stable configurations? The current consensus is that there is little convincing evidence of reinforcement (see Littlejohn 1981; Barton and Hewitt 1985; Bush and Howard 1986; Butlin 1987), but there is some disagreement about whether the lack of evidence reflects the improbability of such an outcome, the restricted conditions under which it might be expected, or the difficulty in providing convincing evidence that it has occurred. The fate of hybrid zone interactions is obviously dependent on the structure of the zone. As we shall see, most hybrid zones are interpreted as simple clines and this interpretation is clearly appropriate in many situations. However, in some cases hybrid zones are more complex, and the complexity will have important consequences for predicting the outcome of the interaction.

Many recent hybrid zone papers have avoided both the issue of origins and the question of whether hybrid zones play a direct role in the speciation process. Instead, they have focused on the structure of hybrid zones and on patterns of introgression and linkage disequilibrium (Szymura and Barton 1986; Kocher and Sage 1986; Harrison 1986). Regardless of their possible role in the speciation process, natural hybridization events are excellent situations for investigating the genetic architecture of species differences and the nature of the interaction between two differentiated genomes. The term genetic architecture refers to the number, effect (large or small), and chromosomal distribution (linkage relationships) of the genes that are responsible for species differences. Of particular interest is the nature of the "species boundary," the sum of all allelic differences that contribute to genetic (reproductive) isolation. Patterns of linkage disequilibrium within hybrid zones and the nature and extent of introgression at marker loci are a consequence of many generations of recombination between differentiated genomes and, therefore, can provide information about the number and distribution of genes affecting hybrid fitness and/or assortative mating. Again, the structure of the hybrid zone must provide the context for interpreting the significance of observed patterns of variation and association.

TRADITIONAL HYBRID ZONE MODELS

There is no clear consensus definition for a hybrid zone, and the term has been used in a variety of ways. Endler (1977) defines hybrid zones as "narrow belts (clines) with greatly increased variability in fitness and morphology . . . , separating distinct groups of relatively uniform sets of populations." Barton and Hewitt (1981, 1985) argue that common usage simply equates a hybrid zone with a cline, although they would prefer to restrict the term to clines that are maintained by a balance between dispersal and selection against hybrids. Moore (1977) considers the term hybrid zone to be interchangeable with zone of secondary intergradation. These authors invoke different models to explain the existence of hybrid zones or clines. Both Endler (1977) and Barton and Hewitt (1981, 1985) believe that hybrid zones are maintained by a balance between dispersal and selection, but their models differ in the role played by selection. Endler emphasizes genotype-specific responses to environmental factors that are geographically variable and, therefore, views clines as a balance between dispersal and selection along environmental gradients or across discontinuities. Barton and Hewitt claim that the dynamic equilibrium is a balance between dispersal and hybrid unfitness and assume that many hybrid zone interactions are independent of the environment (except insofar as environment affects dispersal). Moore (1977) stresses the importance of ecological factors, but argues that hybrid zones persist because hybrids are actually more fit within the regions in which they occur (narrow ecotones).

A common theme running through much of the hybrid zone literature is that hybrid zones can be interpreted as clines and that they can be characterized using samples collected along linear transects across the zone. Indeed, many hybrid zone studies rely on data from a single transect (e.g., McDonnell et al. 1978; Sage and Selander 1979; Gartside 1980). Cline models do appear to be appropriate for many hybrid zones reported in the literature, in which "transects . . . follow either the smooth, sigmoid curves expected from single-locus models, or the stepped clines expected from genetic barriers" (Barton and Hewitt 1985:129). Thus, Szymura and Barton (1986) provide an elegant analysis of patterns of variation across a hybrid zone between two species of toads (*Bombina*). We wish to draw attention to examples of natural hybridization in which simple monotonic clines are not a good representation of observed patterns of variation.

HYBRID ZONES IN A PATCHY ENVIRONMENT

A fundamental question to be asked about any hybrid zone is the role of environmental heterogeneity in determining its structure and position. The prevailing models of spatial variation in selection coefficients are smooth environmental gradients or abrupt discontinuities (see Slatkin 1973; Endler 1977, for examples). In both cases, selection coefficients are described along a one-dimensional transect that parallels the direction of presumed change.

Hybrid zones (clines) in the real world are often associated with climatic gradients [e.g. the *Mus musculus/M. domesticus* hybrid zone in Denmark (Hunt and Selander 1973), many avian hybrid zones across the Great Plains (Rising 1983)]. However, it is difficult to demonstrate that an environmental gradient is, in fact, the cause of observed clinal variation. Barton and Hewitt (1985:131) suggest that association of clines with environmental gradients might occur under a number of different hybrid zone models and that "one would expect a more broken pattern if the distribution directly reflects environmental heterogeneity." In fact, some hybrid zones exhibit just such a "broken pattern"—a pattern that almost certainly reflects a direct response to the environment. Before discussing examples, it is useful to consider an alternative view of environmental heterogeneity.

Hybrid zones may occur in an ecological setting far more complex than implied by gradient models. Many habitats, resources, and physical environmental factors are distributed in relatively discrete patches, so that an organism's environment can be viewed as a shifting mosaic of patches that differ in quality (Pickett and White 1985). Therefore, a transition from environment A to environment B does not necessarily involve a series of intermediate environments (with genotypes having continuously varying selection coefficients across the region). Instead, it may represent changing proportions of two patch types (A and B), from 100% A on one side of the zone to 100% B on the other (Figure 1). If closely related species differ in characteristics that affect their fitness in or preference for alternative patch types, then the mosaic nature of the environment may have a profound influence on the interaction between these taxa and on the structure of a hybrid zone between them. The patch model (like the gradient model) is an oversimplification, since the world is not divided into two discrete patch types. Nevertheless, a two-patch mosaic model is a useful context for discussing hybrid zone dynamics.

In fact, whether a mosaic model or a gradient model is appropriate depends on the dispersal ability of the organism relative to the spatial scale of the environmental patches—or the "grain" size (Levins 1968). In coarse-grained environments, dispersal distances are small compared with

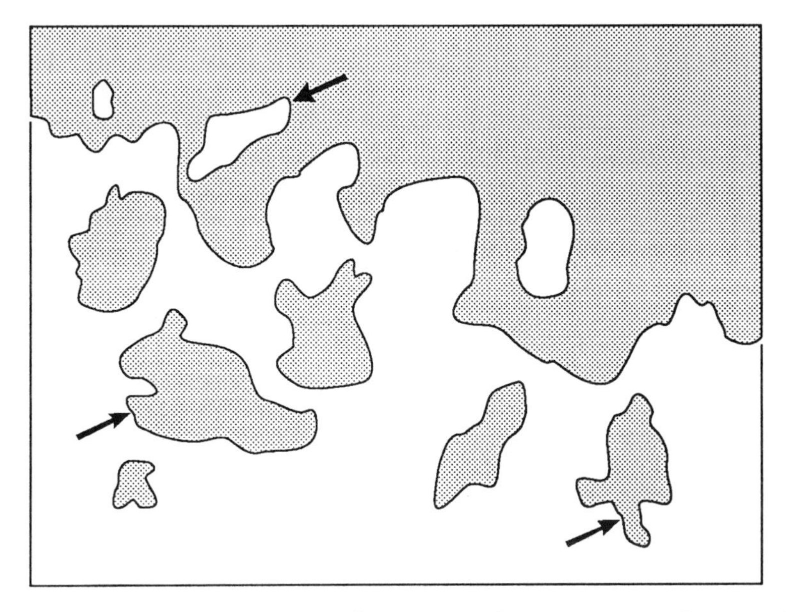

FIGURE 1. Mosaic environment with a transition between two patch types. The structure of a hybrid zone occurring in this environment will depend on the scale of the mosaic relative to the dispersal ability of the hybridizing taxa and the species' relative fitness in and/or preference for the two patch types. Mosaic hybrid zones may consist of many, effectively independent, encounters between the species involved; the arrows mark three such sites of interactions.

the average patch size and a hybrid zone will reflect the mosaic nature of the environment. In fine-grained environments, dispersal distances are large relative to patch size and the mosaic model reduces to a gradient model. Under these circumstances, the appropriate selection coefficient at any point across the zone will be a function of the proportions of patch types A and B encountered by an individual organism. These proportions may change continuously across the zone.

When dispersal distances are much smaller than patch size, a mosaic hybrid zone may contain many local gradients at the patch boundaries. In this situation local dynamics can be analyzed in terms of traditional cline models. However, on a larger scale the hybrid zone is still a mosaic and it is not possible to predict global outcomes from the dynamics of local interactions.

Slatkin (1973) has modeled the effects of a "pocket" in a one-dimensional environment. The model assumes a single locus with two alleles, such that one homozygote is favored within the pocket and the alternative homozygote elsewhere. Slatkin shows that only if the size of the pocket is greater than a characteristic length ($l_c = l/s$, *where* l is average dispersal distance and s is a measure of the strength of selection) will variation at the locus reflect the presence of the pocket. If we know the size of local patches occupied by individuals of two different genotypes and the dispersal distance of these individuals, we can use this relationship to estimate the strength of selection needed to maintain the patchy distribution. A similar model will be appropriate for understanding the distributions of two populations or species that interact in a patchy environment. In this case, selection differentials may depend on allelic differences at many loci (e.g., in instances of secondary contact between entities that have differentiated in allopatry). Hybridization will lead to decay in disequilibrium and correspondingly will reduce the variance in fitness relative to the two patch types. Note that if the two species exhibit strong preferences for alternative patch types, it may not be necessary to invoke current selection to explain a mosaic pattern.

Not only is the environment a mosaic of patches, but the distribution of these patches may change over time. For species that utilize ephemeral resources or occupy disturbed habitats, the turnover rate of patches will be high. Since human disturbance has been a major influence in many areas over the past several hundred (or more) years, the nature of the environmental mosaic has no doubt been dramatically altered. For some groups of hybridizing species, this has resulted in the creation of many new patches of suitable habitat. Furthermore, many patches may not be suitable for either of the interacting taxa (e.g., forests for denizens of old fields and pastures). These gaps can act as major barriers to dispersal, perhaps isolating fragments of the hybrid zone. The mosaic model, therefore, recognizes that because of a need for specific resources or habitats and/or limited environmental tolerance, species are often distributed discontinuously with many independent encounters between local populations (Figure 1). Furthermore, the dynamics of patch origination and extinction may be such that many suitable patches are empty. The existence of empty patches has important consequences for hybrid zone dynamics.

ANALYSIS OF A MOSAIC HYBRID ZONE

The field crickets *Gryllus pennsylvanicus* and *G. firmus* interact along an extensive hybrid zone in the eastern United States (Harrison and Arnold 1982). *G. pennsylvanicus* occurs in inland localities throughout the northeastern United States and extends south in the mountains to northern Georgia. *G. firmus* [described as the "beach cricket" by Fulton (1952)] is found in coastal and lowland areas from Florida to New England. Individuals of mixed ancestry have been found at many sites along the hybrid zone, from the Blue Ridge and Shenandoah Valley in Virginia to southern

Connecticut (Figure 2). In Pennsylvania, the hybrid zone appears to follow the eastern front of the Appalachian mountains (Harrison, unpublished data).

Within the group of eastern North American field crickets, *G. pennsylvanicus* and *G. firmus* are sister taxa [Harrison 1979b; unpublished mitochondrial DNA (mtDNA) data]. In the northeast both species overwinter in the egg stage, with adults appearing in late summer and early fall.

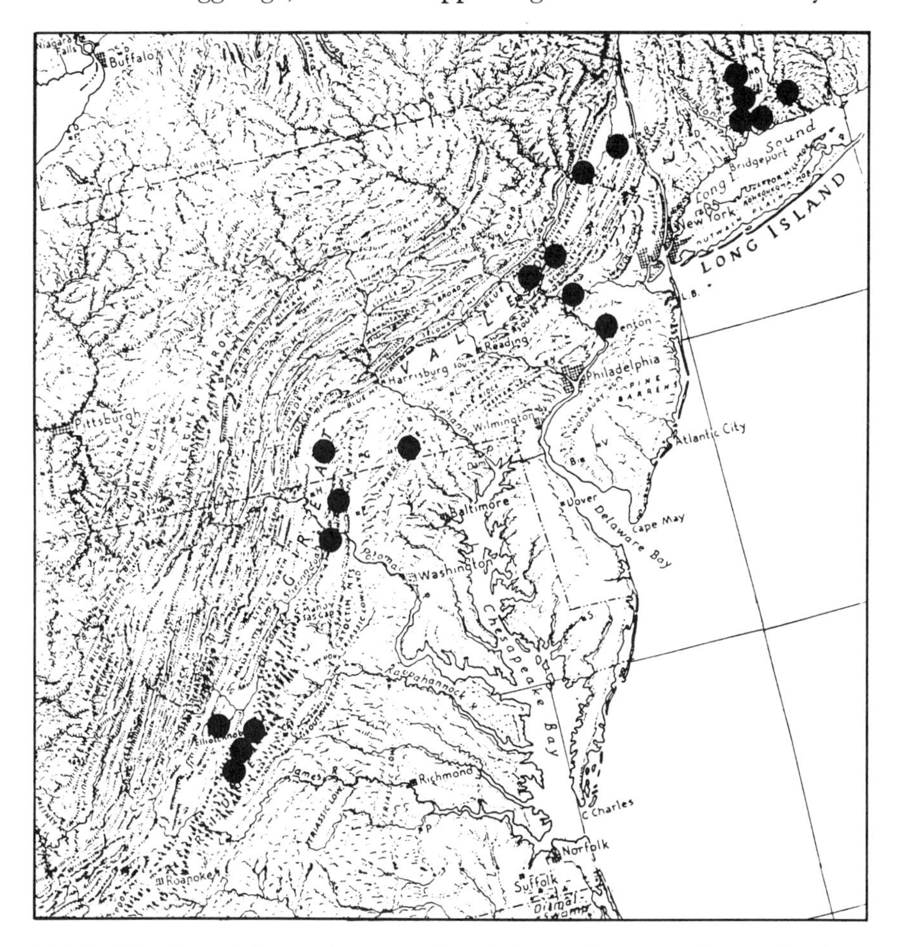

FIGURE 2. Map of the northeastern United States, showing collecting localities within the *Gryllus firmus/Gryllus pennsylvanicus* hybrid zone. These are sites in which both parental species and/or individuals of mixed ancestry have been collected. In Connecticut, the sites shown on this map represent a small subset of the hybrid zone populations that have been identified. Note that the hybrid zone appears to parallel the eastern front of the Blue Ridge and the Appalachian mountains.

The two cricket species differ in a number of morphological characters, including body size, ovipositor length, and tegmina color (tegmina are the modified forewings used in song production). They also differ in allele frequencies at several loci assayed by standard protein electrophoretic techniques. However, no single morphological character is diagnostic, and there are no fixed allelic differences between the species. Based on allozyme comparisons, the genetic distance (Nei 1972) between *G. firmus* and *G. pennsylvanicus* is only 0.03. Alexander (1957) documented differences in calling song between the two species. Although the songs of males from "pure" populations adjacent to the hybrid zone differ significantly in chirp rate and interpulse interval, there is considerable overlap in these song parameters (Harrison, unpublished data). Recently, we have shown that crickets in Connecticut belong to two distinct mtDNA lineages and that outside the hybrid zone the association between mtDNA genotype and species designation is nearly perfect (Harrison et al. 1987). The two lineages differ in the presence/absence of six restriction sites (out of a total of 44 mapped sites). We refer to the two lineages simply as A and B, the A lineage corresponding to *G. pennsylvanicus* and the B lineage to *G. firmus*. In fact, each of these mtDNA lineages consists of a single widely distributed genotype together with several rare genotypes, each of which differs from the common genotype by one restriction site loss and occurs in only a single local population (Harrison et al. 1987; Rand, unpublished data). In the southeastern United States, the mtDNA genotypes of the two species differ by only a single restriction site, even when crickets are collected far from the hybrid zone, along the coast (*G. firmus*) and in the mountains (*G. pennsylvanicus*) of North Carolina (Harrison, unpublished data).

Both premating and postmating barriers operate to limit genetic exchange between the two cricket species. Although crosses between *G. firmus* males and *G. pennsylvanicus* females give rise to viable and fertile F_1, the reciprocal cross consistently fails to produce offspring (Harrison 1983). The same asymmetric outcome is observed in interspecific crosses of crickets from Connecticut and Virginia, whereas crosses within species, between localities (Connecticut and Virginia) consistently produce normal F_1. In Virginia, where the two species interact along a steep environmental gradient, temporal isolation of adults is a significant barrier to gene exchange (Harrison 1985). In Connecticut, where adults appear synchronously, there is strong positive assortative mating in which nearly "pure" individuals of both species occur together within the hybrid zone (Harrison 1986). In spite of these barriers, hybridization and introgression have been clearly documented, using allozymes and mtDNA as genetic markers (Harrison 1986; Harrison et al. 1987).

The influence of habitat mosaics of different spatial scales will depend on the dispersal ability of the crickets. Because the vast majority of adult G.

firmus and *G. pennsylvanicus* are flightless, individual dispersal distances are usually on the order of tens or hundreds of meters per generation. From mark-release-recapture studies of *G. pennsylvanicus* within a large area of favorable habitat we know that greater than 70% of marked adults are recaptured at least once, and often many times, within 20 m of the release point (Harrison, unpublished). Continued sampling beyond the marking period revealed no decline in the proportion of marked crickets, suggesting little mixing with crickets from surrounding areas. Further studies in the same area documented that there is very little exchange of marked individuals between adjacent patches of favorable habitat (quadrats laid out within a continuous area of pasture). Where unsuitable habitat (forests) intervene, dispersal is presumably reduced further. On the other hand, occasional long distance dispersal events cannot be ruled out by these mark- recapture studies, since a fraction of the marked crickets are never recaptured.

Occasional long-winged individuals capable of flight do appear (Alexander 1968; Harrison 1979a). The frequency of long-winged crickets is higher in *G. firmus* than in *G. pennsylvanicus*, higher in the south than in the north, and variable from year to year (Harrison 1979b, unpublished data). In 1978, 25–75% of the *G. firmus* in the Shenandoah Valley and along the crest of the Blue Ridge were long-winged, perhaps in response to the very high population densities. However, crickets were not observed to fly, and the impact of such episodes on gene flow and colonization remains unclear. The occurrence of large numbers of individuals capable of flight is certainly a rare event, restricted temporally and spatially.

Here we wish to focus attention primarily on the structure and dynamics of the hybrid zone in Connecticut. The earliest samples from Connecticut were collected from a series of populations along a transect from the coast to the northwest hills. These samples revealed a transition from "pure" *G. firmus* along the coast to "pure" *G. pennsylvanicus* inland and suggested simple clinal variation of morphological characters (ovipositor and hind wing length) and allele frequencies at an *Esterase* locus (see Figure 8 in Harrison and Arnold 1982). More extensive sampling showed that patterns of variation across the zone are not consistent with simple cline models and led to the suggestion that the hybrid zone is a patchwork of populations (Harrison 1986; Rand 1987). Crickets from adjacent sites within the zone often differ significantly in morphology, allele frequency, and mtDNA genotype. Within 10 km of the coast are pockets of "pure" *G. pennsylvanicus* and 100 km inland is a population of mixed ancestry in which some crickets are morphologically indistinguishable from *G. firmus*.

Collecting localities within Connecticut and patterns of variation for a single morphological character (ovipositor length), three enzyme loci, and mtDNA are shown in Figures 3A–F. The figures are based on all previously

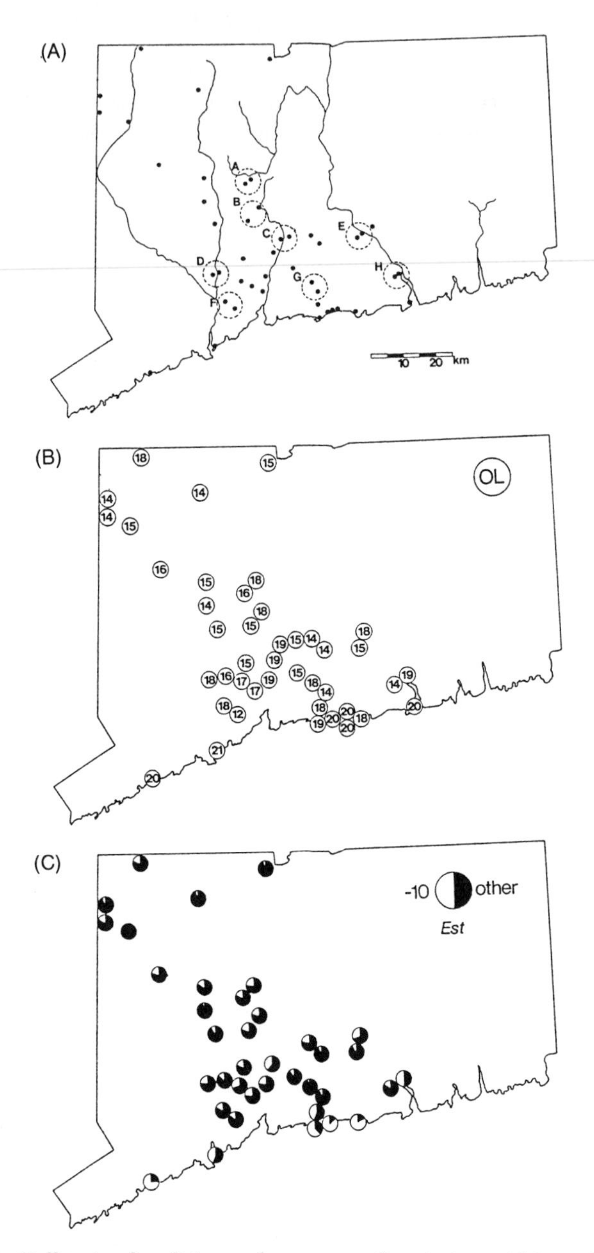

FIGURE 3. Collecting localities and patterns of variation within and adjacent to the Connecticut hybrid zone. (A) Sites from which crickets have been collected, including eight sets of paired populations within the hybrid zone (A–H). Comparisons of these paired populations appear in Table 1. Complete data (morphological measurements, allozyme frequencies, mtDNA genotype frequencies) are available only for about half of the sites. (B) Mean ovipositor length (in mm) for females from each of the populations. (C) Frequency of the Est^{-10} allele (which oc-

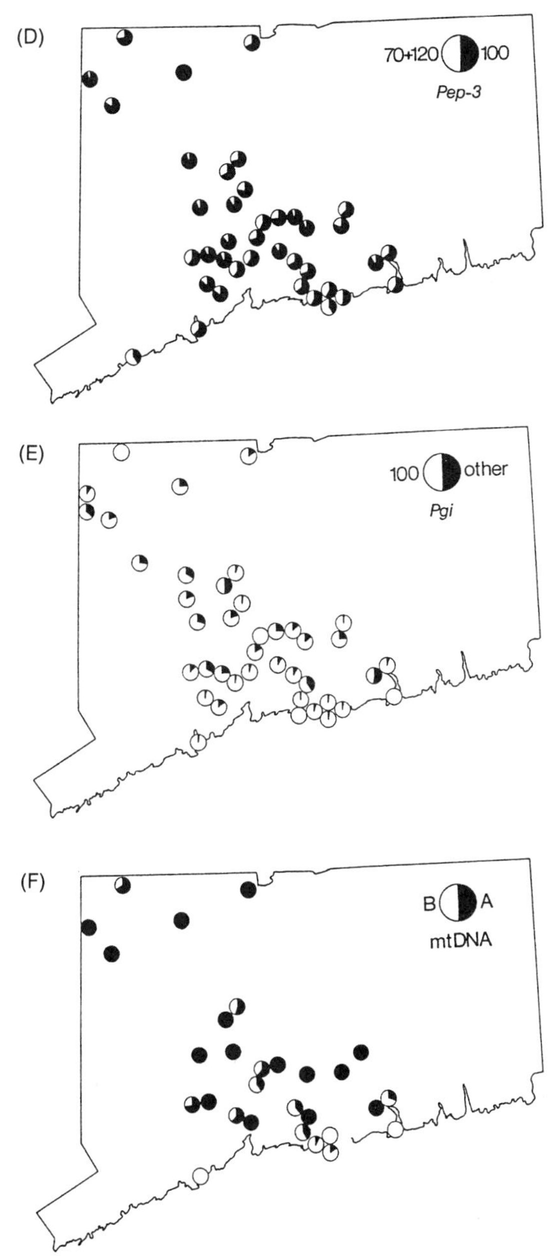

curs at high frequency in *Gryllus firmus* and low frequency in *G. pennsylvanicus*). (D) Combined frequencies of the *Pep*-3^{70} and *Pep*-3^{120} alleles, both of which occur at higher frequency in *G. firmus*. (E) Frequency of the *Pgi*^{100} allele, which occurs at consistently higher frequency in *G. firmus* populations. (F) Frequencies of the *A* and *B* mtDNA genotypes, which characterize allopatric populations of *G. pennsylvanicus* and *G. firmus*, respectively.

published accounts together with new data from several populations. Each of the five characters reveals the mosaic structure of the cricket hybrid zone. Furthermore, the patterns of the individual character mosaics are generally concordant. Eight pairs of populations within the hybrid zone are used to examine patterns of differentiation for the five characters (Figure 3A). Each pair represents collections from two neighboring sites and includes one population in which crickets are morphologically like *G. firmus* and a second population in which crickets are morphologically like *G. pennsylvanicus*. Differences in ovipositor length between the members of the paired populations are evidence of the morphological differentiation (Table 1). Distances between paired sites range from only 200 m (pair H) to 6 km. Clearly, there are many areas within the hybrid zone in which morphologically distinct populations abut or occur in close proximity to one another (relative to the width of the hybrid zone).

The paired populations exhibit significant differences in the frequencies of mtDNA genotypes and alleles coding for soluble enzymes. mtDNA data are available for five of the paired sites. In each case, the population of crickets with short ovipositors is fixed for the *A* mtDNA genotype whereas in the neighboring population, in which females have longer ovipositors, the *B* mtDNA genotype occurs at frequencies ranging from 0.27 to 0.72

TABLE 1. Comparison of paired populations.[a]

	Ovipositor length (mm)	Est^{-10}	$Pep\text{-}3^{70}$ + $Pep\text{-}3^{120}$	Pgi^{100}	mtDNA
A	17.7/16.2	0.24/0.17	0.30/0.33	0.94/0.50	0.44/0.00
B	18.4/14.7	0.19/0.21	0.23/0.10	0.97/0.81	
C	19.0/14.9		0.40/0.25	1.0/0.75	0.37/0.00
D	17.9/15.8	0.24/0.12	0.42/0.08	0.87/0.66	0.27/0.00
E	17.5/15.1	0.29/0.08	0.41/0.21	0.98/0.75	
F	18.2/12.3	0.16/0.14	0.14/0.14	0.97/0.86	
G	17.6/14.0	0.03/0.08	0.32/0.31	0.89/0.62	0.63/0.00
H	18.5/14.4	0.50/0.16	0.39/0.11	0.94/0.47	0.72/0.00
	20.5/14.8	0.87/0.03	0.41/0.03	0.97/0.80	1.0/0.00

[a]The values shown in the table are for eight paired populations from within the *Gryllus firmus/Gryllus pennsylvanicus* hybrid zone in Connecticut. The two numbers represent ovipositor length (in mm), allele frequency, or mtDNA genotype frequency (the B genotype) for the *firmus*-like and *pennsylvanicus*-like populations, respectively. The numbers at the bottom of the table provide a comparison of allopatric populations, a coastal *G. firmus* population and a G. *pennsylvanicus* population from the northwest hills.

(Table 1). The members of one population of each pair are descended entirely from *G. pennsylvanicus* females, whereas maternal lineages of both parental species persist in the neighboring population.

Frequencies of the Est^{-10}, $Pep\text{-}3^{70}$, $Pep\text{-}3^{120}$, and Pgi^{100} alleles are generally higher in populations of crickets that are similar to *G. firmus* in morphology (Table 1). Differences between adjacent paired populations within the hybrid zone are in the same direction as differences between "pure" parental populations on either side of the zone. However, the absence of allozyme differentiation at certain sites (notably at the *Est* locus) and evidence of some discordance among characters provide evidence of past hybridization and differential introgression. For example, the paired populations from Guilford (site G in Figure 3A) do not differ in *Est* or *Pep* allele frequencies. Curiously, both populations have *Pep* allele frequencies characteristic of *G. firmus* and *Est* allele frequencies characteristic of *G. pennsylvanicus*. Several other paired populations also show little or no differentiation at *Est* and/or *Pep*. In contrast, patterns of variation at the *Pgi* locus are generally concordant with the morphological data.

Although at low resolution the hybrid zone represents a transition from "pure" *G. firmus* along the coast to "pure" *G. pennsylvanicus* inland, detailed sampling reveals the hybrid zone to be a complex mosaic of populations. Such a mosaic might simply reflect stochastic colonization and extinction coupled with strong barriers to gene exchange (rivers and patches of unsuitable habitat may be major dispersal barriers for crickets). However, preliminary observations suggested that the pattern of the mosaic might be determined by the patchy distribution of soil types in Connecticut. *G. firmus* (the "beach cricket") appeared to occur primarily on sandy, well-drained soils, near the coast and inland along river valleys. In contrast, *G. pennsylvanicus* was more abundant on loam soils, on traprock ridges in the Central Valley and in upland areas in the eastern and western parts of the state. To test the hypothesis that soil type determines the distribution of crickets within the hybrid zone, we identified pairs of neighboring sites within the zone, one site of each pair having sandy soil and the other loam. These paired localities had not been sampled before. These sites are, in fact, among the eight paired sites previously discussed (A, C, D, H). In each case, the sandy site is occupied by crickets that are similar in most respects to *G. firmus*, whereas the loam sites have populations that are nearly "pure" *G. pennsylvanicus* (Rand 1987; Rand and Harrison 1989). Therefore, there is little doubt that the mosaic structure of the hybrid zone is determined by the patchy nature of the environment. Whether habitats select crickets or crickets select habitats is not entirely clear.

Our data clearly indicate that the hybrid zone is a mosaic. However, given uncertainties about the relative magnitudes of dispersal distance and patch size, it is not known whether the environmental grain is sufficiently coarse that gradient hybrid zones characterize patch boundaries. Our impression from sites in which differentiated populations occur in close proximity is that these boundaries do not reveal evidence of clinal variation. At other sites, however, clines may characterize the regions between patches. More detailed sampling will be needed to resolve this issue.

Having provided evidence that morphologically and genetically distinct cricket populations persist in different habitats in spite of some hybridization and introgression, we now show that both premating and postmating barriers to gene exchange also persist within the hybrid zone. The existence of an asymmetric postmating barrier to gene exchange was originally demonstrated in laboratory crosses of crickets from allopatric populations of *G. pennsylvanicus* and *G. firmus*. A similar asymmetric barrier also exists within the hybrid zone (Rand 1987; Rand and Harrison 1989). Using males and virgin females from three of the paired sites, we demonstrated that crosses involving females from the *firmus*-like populations and males from the *pennsylvanicus*-like populations produce offspring only rarely, whereas the reciprocal cross resulted in hatching of progeny in a majority of the trials. Thus, this component of the species boundary appears to remain intact in spite of introgression of mtDNA and alleles at loci coding for soluble enzymes.

Premating barriers to gene exchange are also evident when crickets from allopatric populations are tested in the laboratory. Mate selection experiments demonstrate that given a choice between males of the two cricket species, females of *G. pennsylvanicus* mate preferentially with conspecific males. Males from coastal Connecticut and the northwest hills were matched by size, introduced in pairs into large screened cages, and allowed to interact for 24–48 hours. A single *G. pennsylvanicus* female was then placed with the pair of males and allowed to mate over a 2- or 3-day period. At the end of each mating trial, the males were frozen and the female was allowed to lay eggs. Comparisons of enzyme genotypes of the two males, the female, and a sample of the offspring provided an estimate of the proportion of offspring sired by each male. Of 30 females tested, 16 produced offspring sired only by the conspecific male and 2 produced offspring sired only by the heterospecific male. (In one of the latter two cases, the conspecific male died during the experiment.) The remaining females produced offspring sired by both males, but in every case a majority of these were progeny of the conspecific male (Figure 4). Although differences between males of the two species in effectiveness of sperm transfer or sperm competition may contribute to the asymmetry (Harrison, unpublished), they cannot alone account for the biased patterns of progeny production. The results of these experiments argue strongly for nonrandom mating within laboratory cages (in the absence of environmental heterogeneity).

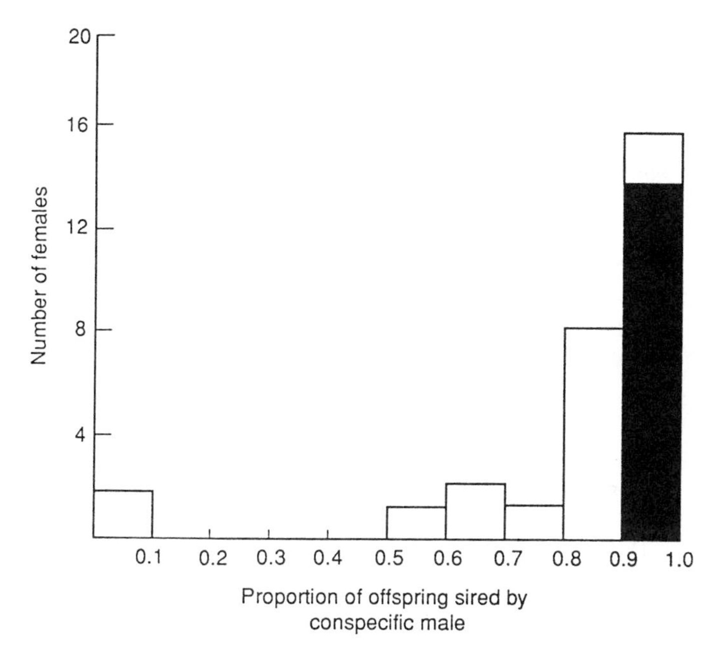

FIGURE 4. Proportion of offspring sired by conspecific males in laboratory mate-preference experiments. In each experiment, a *Gryllus pennsylvanicus* female was placed in a cage with a *G. pennsylvanicus* male and a *G. firmus* male (matched by size). After 48–72 hours, the female was removed and allowed to deposit eggs. Analysis of enzyme phenotypes (genotypes) of parents and offspring allowed us to identify the proportion of offspring sired by each male. The histogram shows that nearly half of the 30 females tested produced offspring sired only by the conspecific male (black bar) and that nearly all of the females produced a majority of offspring sired by the conspecific male.

No comparable laboratory experiments have been done using crickets from hybrid zone populations. However, *firmus*-like and *pennsylvanicus*-like crickets occur together at a few sites within the hybrid zone. Analysis of offspring produced by field-inseminated females from one of these populations (HN2) demonstrated strong (but not complete) assortative mating, using ovipositor length and alleles at the *Pep–3* locus as markers (Harrison 1986). In this case, assortative mating persists within the population in spite of dedifferentiation at the *Est* locus and clear introgression of mtDNA.

OTHER MOSAIC HYBRID ZONES

Descriptions of mosaic hybrid zones are not uncommon in the literature on hybridization in animals and plants. An excellent recent example involves two species of North American ground crickets (genus *Allonemobius*). *Allonemobius fasciatus* and *A. socius* meet and hybridize along an extensive east–west hybrid zone (Howard 1986). A transect from southern Ohio through the mountains of West Virginia demonstrates a clear transition from *A. fasciatus* in the north to *A. socius* in the south across a very broad hybrid zone. Along the transect, however, there are clear reversals in character index score (based on four allozyme loci). *A. fasciatus* populations appear to occur at higher elevations and *A. socius* populations in the valleys. Additional collections have substantiated the notion that within this segment of the hybrid zone cricket distributions closely parallel the topography (Howard, personal communication). Howard (1986) suggested that this pattern reflects adaptation of the two cricket species to different environments (climatic regimes). The patches in this mosaic would appear to be far larger than those defined by soil type in the *Gryllus* hybrid zone in Connecticut, although the dispersal distances for crickets in the two genera are likely to be comparable.

Descriptions of other presumed mosaic hybrid zones are consistent with the view that the interacting taxa are ecologically and/or behaviorally distinct and that the structure of the environment determines the structure of the hybrid zone. In describing interactions between taxa in the *Pseudacris nigrita* complex, Gartside (1980) argued that "The existence of a patchy environment in the east may have allowed a macro-geographic overlap . . . with a minimum of reproductive and ecological interaction. Sharp boundaries between the patches would provide little or no intermediate environment." Sperling (1987) indicated that differences between species or subspecies in resource (host plant) utilization patterns and the patchy distribution of these resources impose a mosaic structure on hybrid zones in the *Papilio machaon* complex (swallowtail butterflies).

The plant hybrid zone literature also contains references to mosaic hybrid zones, in which closely related but ecologically differentiated species occupy a patchy environment. Four broadly sympatric species of *Baptisia* hybridize where they are found together (Alston and Turner 1963). Because of ecological differences, the species often occur alone or in pairs and all four species are rarely present at one site. Sampling along a highway east of Houston, Texas revealed a "linear mosaic of *Baptisia* populations." In Australian *Ranunculus*, species adapted to different microsites occupy a mosaic environment and hybridize at the boundaries between local habitat patches (Briggs 1962). Individuals of mixed ancestry appear limited to narrow ecotones.

Clearly, a common theme is that a mosaic hybrid zone structure occurs when closely related species that differ in habitat or resource utilization patterns occupy a patchy environment. The complex internal structure of these zones is a reflection of an underlying environmental mosaic in which habitats and resources are found in discrete patches. How do such zones arise and how does the complex structure affect their fate?

THE ORIGIN OF MOSAIC HYBRID ZONES

Attempts to resolve hybrid zone origins must focus on factors determining the current distribution of the hybridizing taxa and not on the geographic context in which differences initially arose. A nonallopatric speciation event can give rise to species that later colonize new habitats and come together in a zone of secondary contact (Figure 5). The process of differentiation (speciation) may well be obscured by subsequent dispersal, colonization, and local extinction.

In a mosaic hybrid zone, any explanation for the origin of the zone (not the origin of the differentiated taxa) must account for the patchwork population structure and for association of the taxa with an underlying environmental mosaic. In a scenario for hybrid zone origin based on in situ selection (primary intergradation), colonization by a single (polymorphic) lineage is followed by local differentiation. Genetic variation is initially all within populations and local differences in selection regimes promote differentiation (variation among populations). If steep clines or abrupt discontinuities occur independently at many different sites, a model of in situ selection requires parallel changes at each of these sites for each character

FIGURE 5. Two scenarios for the origin of species differences and the formation of a hybrid zone. In (A), the speciation event occurs in the absence of geographic isolation, and the species then colonize new habitats in which they meet in an extensive zone of secondary contact. In (B), differentiation occurs in allopatry and subsequent spread also leads to formation of an extensive hybrid zone. Note that current patterns of variation are identical in the two cases. Although the geographic context in which speciation occurred is different, both hybrid zones represent secondary contact.

(locus) that varies among them. In contrast, a model of secondary contact suggests that two already distinct lineages colonize a hybrid zone (much of the variation is between populations). In a patchy environment, the resulting distribution of genotypes will depend on both stochastic colonization and extinction events and on the ecological and behavioral differences that have arisen prior to secondary contact. If there is a clear nonrandom mapping of genotypes onto environmental patches, then the initial colonists presumably differed in their fitness in and/or preference for alternative patch types. Sorting of lineages into alternative patch types may result from (1) local competition leading to exclusion, (2) differences in absolute tolerance, or (3) differences in patch preference.

What is the most likely explanation for patterns of variation within the field cricket hybrid zone in Connecticut? Local adaptation to the soil substrate has been proposed as a general explanation for the occurrence of crickets with long ovipositors and light colored tegmina in areas of sandy soil (Lutz 1908; Alexander 1968). Selection for these characteristics within local patches of sandy soil in Connecticut could account for the observed mosaic in morphological traits. Although in situ selection on morphological traits provides a plausible explanation for patterns of morphological differentiation, it fails to account for patterns of variation in allozymes and mtDNA. Concordance among morphological characters, allozymes, mtDNA genotypes, and the asymmetric postmating barrier to gene exchange argues strongly for secondary contact. The alternative is that natural selection, acting independently on each character or gene (including mtDNA), has produced the same or similar results at many sites within the hybrid zone. The probability of such an outcome seems remote.

We suggest that following the retreat of the glaciers, Connecticut was colonized by two distinct cricket lineages. These lineages were at least partially genetically isolated and there were already clear associations among morphology, allozymes, and mtDNA. Either because the two lineages were each better adapted to different local environments and/or because they exhibited strong habitat preferences, the distributions of these crickets came to reflect the underlying environmental mosaic. Of course, this mosaic has changed dramatically, particularly as a result of recent human disturbance. The cricket populations have closely tracked these changes. Because of initial linkage disequilibrium and the existence of both pre- and postmating barriers to gene exchange between the newly sympatric populations, selection on a subset of character differences could lead to the current distributions of character states and to a mosaic pattern for characters not under direct selection. Since the species boundary is semipermeable, however, hybridization has resulted in dedifferentiation for some fraction of the genome.

This interpretation of current patterns of variation does not confront the question of how the two cricket lineages came to be differentiated in the first place. Thus, the geographic context in which the pre-and postmating barriers arose remains unclear. Solving the riddle of hybrid zone origins has not provided direct information about how speciation occurred.

PROPERTIES OF MOSAIC HYBRID ZONES

Mosaic hybrid zones often appear to represent secondary contact between populations/species that differ in habitat utilization or in response to physical environmental factors. The ecological or behavioral differentiation causes a pattern of "curious interlacing of ranges with hybridization at the borders of habitats" (Mayr 1942). In a simple two-patch, two-species situation (with species A and B having greater fitness in or preference for patches a and b, respectively), the two patch types serve as refugia for alternative alleles at loci that affect fitness or preference. A patchy environment will tend to maintain diversity, so that multilocus genotypes characteristic of the two parental species can persist within the hybrid zone. Maynard Smith (1966) points out the relevance to cases of secondary contact of sympatric speciation models that involve adaptation of two populations to two different habitats or niches. If previously isolated populations are ecologically distinct, such that each is well adapted to only one of the patch types within the hybrid zone, the result may be a stable polymorphism (i.e., maintenance of diversity).

In situations in which hybrids are less fit than either parental type (heterozygotes less fit than either homozygote), deterministic population genetics models predict that the outcome will be extinction of the rarer species or of the rarer alleles at loci contributing to hybrid unfitness (e.g., see Paterson 1978; Templeton 1981). However, such models restrict attention to interactions at a single site, focusing on the local not the global picture. Mosaic hybrid zones, with complex internal structure, involve many independent contacts (Figure 1), each with a potentially unique evolutionary trajectory. The outcome of any local interaction will be a consequence of the relative abundance of the two taxa, selection pressures within the local patch, random drift, and the genetic architecture of the species boundary. If proportions of the two species vary from site to site and if local selection pressures also vary, global extinction of either species (or allele) is unlikely.

For species that occupy disturbed habitats (and mosaic hybrid zones may be especially common in such groups of organisms), the environmental context will be a shifting mosaic of patches. Thus, the dynamics of species interactions must be placed in the context of patch distributions that change over time. Patches will have characteristic origination and extinction rates. Depending on the dispersal ability of the organisms, the cost of dispersal, and the distance between patches, a proportion of suitable patches may, in fact, be empty. If hybrids are less fit, the availability of empty patches will favor early colonists, since first possession of a patch may provide a significant advantage. In the extreme case when hybrids are much less fit than either parental type and there is little or no positive assortative mating, a species less well adapted to a particular patch type may be able to resist invasion by a better adapted species. Introgression of alleles conferring high patch-specific fitness is expected, but will be retarded to the extent that these loci are linked to those involved in hybrid unfitness.

Mosaic hybrid zones are more favorable sites for reinforcement than are narrow tension zones. Obviously, factors that promote the persistence of two distinct entities also enhance the opportunity for reinforcement, since they reduce the probability of the alternative outcomes, fusion and extinction. The existence of a heterogeneous environment impedes fusion (by favoring alternative types in different patches) and complex internal population structure decreases the chances of global extinction. Because mosaic hybrid zones are broad, the center of these zones is often far removed from "pure" populations of either parental type. This minimizes the influx of genes from outside the zone, which otherwise would tend to overwhelm weak selection against hybridization within the zone (Howard 1986). Furthermore, the existence of many independent contacts provides multiple opportunities for local reinforcement. Littlejohn (1981) argues that with many "trials" there will be "a reasonable chance that homogamic mechanisms could develop." Favorable changes occurring at one site may then spread throughout the broad zone in which encounters between the hybridizing taxa occur. Such an argument adopts a Wrightian view of many semiindependent subdivisions—specifically, many localized episodes of hybridization and recombination. This discussion should not be taken to imply that reinforcement is the probable outcome in mosaic hybrid zones, only that conditions within such zones make such an outcome more likely.

Within a mosaic hybrid zone, hybridization and subsequent recombination will lead to gene flow across the species boundary, but chromosomal segments surrounding genes affecting fitness in or preference for patches of different types may remain differentiated (see Barton 1979; Barton and Hewitt 1981 for a model of the time-dependent decay in chromosomal differentiation when individual loci contribute to hybrid unfitness). In secondary contact there will initially be significant linkage disequilibrium between genes causing differential adaptation and those causing assortative mating. This can effectively retard the flow of neutral alleles across the species boundary (maintain linkage disequilibrium). Obviously, these effects will vary among loci, i.e., the species boundary will be semipermeable

(Key 1968; Harrison 1986). The *Gryllus* hybrid zone in Connecticut demonstrates this quite clearly. Furthermore, genetic isolation can arise (and decay) piecemeal within the genome. Certain portions of the genome may be effectively isolated by differences arising in allopatry, whereas other segments may become isolated following secondary contact. Because of recombination, genomes (and species) are not coherent entities, and therefore the consequences of hybridization in a mosaic environment must be considered at the level of individual genes or chromosomal segments.

Studies of linkage disequilibrium and patterns of introgression within hybrid zones can provide insights into the genetic architecture of species boundaries (Barton 1983; Szymura and Barton 1986). Because mosaic hybrid zones offer an opportunity to observe many independent interactions between hybridizing taxa, we can ask whether these patterns are consistent among sites (i.e., do "replicates" of the same interaction lead to the same end result?). If linkage disequilibrium between the same pairs of alleles persists at all sites and/or if patterns of variation in the extent of introgression are the same at many localities, deterministic forces (i.e., selection) must be invoked to explain such patterns. The availability of restriction fragment length polymorphisms (RFLPs) as markers of variation in DNA sequence will make possible detailed analysis of genetic architecture.

CONCLUSIONS

Mosaic hybrid zones exhibit patterns of variation that appear fundamentally different from the steep multilocus clines that have received so much attention in the hybrid zone literature. The complex structure of the environment appears to be a major factor determining the nature of these zones. In several well-studied examples, the interacting species are ecologically and/or behaviorally distinct. They differ in their fitness in and/or preference for alternative patch types. The patchy nature of the environment and the mosaic patterns of variation that result when closely related species interact in this context can have profound consequences for understanding the origin of hybrid zones and for predicting their eventual fate. Most mosaic zones appear to represent secondary contact between taxa that have differentiated in allopatry (perhaps through local adaptation to different environments or resources). Both the heterogeneity of the environment in which they occur and the complex internal structure of these hybrid zones promote the maintenance of diversity (species diversity or allelic diversity) and increase the probability of reinforcement. Many such zones may have gone unnoticed, because the patterns of variation are not as striking as the steep clines and abrupt discontinuities that characterize more traditional hybrid zones. The literature on mosaic hybrid zones is expanding, however, and promises to provide important insights into both the origin of barriers to gene exchange and the genetic architecture of speciation.

ACKNOWLEDGMENTS

Many individuals (students and research assistants) have contributed to the work on cricket hybrid zones. We thank Dan Howard and an anonymous reviewer for their comments. Research on the *Gryllus* hybrid zone has been supported for many years by NSF, most recently by Grant BSR 84-07474.

LITERATURE CITED

Alston, R. E., and B. L. Turner. 1963. Natural hybridization among four species of *Baptisia* (Leguminosae). Am. J. Bot. 50:159–173.

Alexander, R. D. 1957. The taxomony of the field crickets of the eastern United States (Orthoptera:Gryllidae:*Acheta*). Ann. Entomol. Soc. Am. 50:584–602.

Alexander, R. D. 1968. Life cycle origins, speciation and related phenomena in crickets. Quart. Rev. Biol. 43:1–41.

Barton, N. H. 1979. Gene flow past a cline. Heredity 43:333–339.

Barton, N. H. 1983. Multilocus clines. Evolution 37:454–471.

Barton, N. H., and G. M. Hewitt. 1981. Hybrid zones and speciation. In: W. R. Atchley and D. S. Woodruff (eds.), *Evolution and Speciation*, Cambridge University Press, Cambridge.

Barton, N. H., and G. M. Hewitt. 1985. Analysis of hybrid zones. Annu. Rev. Ecol. System. 16:113–148.

Briggs, B. G. 1962. Interspecific hybridization in the *Ranunculus lappaceus* group. Evolution 16:372–390.

Bush, G. L., and D. J. Howard. 1986. Allopatric and non-allopatric speciation: Assumptions and evidence. In: S. Karlin and E. Nevo (eds.), *Evolutionary Processes and Theory*. Academic Press, New York.

Butlin, R. 1987. Speciation by reinforcement. Trends Ecol. Evol. 2:8–13.

Clarke, B. C. 1966. The evolution of morph-ratio clines. Am. Natur. 100:389–402.

Endler, J. A. 1973. Gene flow and population differentiation. Science 179:243–250.

Endler, J. A. 1977. *Geographic Variation, Speciation and Clines*. Princeton University Press, Princeton.

Fulton, B. B. 1952. Evolution in the field cricket. Evolution 6:283–295.

Gartside, D. F. 1980. Analysis of a hybrid zone between chorus frogs of the *Pseudacris nigrita* complex in the southern United States. Copeia 1980:56–66.

Harrison, R. G. 1979a. Flight polymorphism in the field cricket *Gryllus pennsylvanicus*. Oecologia 40:125–132.

Harrison, R. G. 1979b. Speciation in North American field crickets: Evidence from electrophoretic comparisons. Evolution 33:1009–1023.

Harrison, R. G. 1983. Barriers to gene exchange between closely related cricket species. I. Laboratory hybridization studies. Evolution 37:245–251.

Harrison, R. G. 1985. Barriers to gene exchange between closely related cricket species. II. Life cycle variation and temporal isolation. Evolution 39:244–259.

Harrison, R. G. 1986. Pattern and process in a narrow hybrid zone. Heredity 56:337–349.

Harrison, R. G. and J. Arnold. 1982. A narrow hybrid zone between closely related cricket species. Evolution 36:535–552.

Harrison, R. G., D. M. Rand, and W. C. Wheeler. 1987. Mitochondrial DNA variation in field crickets across a narrow hybrid zone. Mol. Biol. Evol. 4:144–158.
Howard, D. J. 1986. A zone of overlap and hybridization between two ground cricket species. Evolution 40:34–43.
Hunt, W. G., and R. K. Selander. 1973. Biochemical genetics of hybridization in European house mice. Heredity 31:11–33.
Key, K. H. L. 1968. The concept of stasipatric speciation. System. Zool. 17:14–22.
Kocher, T. D., and R. D. Sage. 1986. Further genetic analysis of a hybrid zone between Leopard frogs (*Rana pipiens* complex) in central Texas. Evolution 40:21–33.
Levins, R. 1968. *Evolution in Changing Environments*. Princeton University Press, Princeton, New Jersey.
Littlejohn, M. J. 1981. Reproductive isolation: A critical review. In: W. R. Atchley and D. S. Woodruff (eds.), *Evolution and Speciation*. Cambridge University Press, Cambridge.
Lutz, F. E. 1908. The variation and correlation of certain taxonomic characters of *Gryllus*. Carnegie Institute of Washington Publication Number 101, 3–63.
Maynard Smith, J. 1966. Sympatric speciation. Am. Natur. 100:637–650.
Mayr, E. 1942. *Systematics and the Origin of Species*. Columbia University Press, New York.
McDonnell, L. J., D. F. Gartside, and M. J. Littlejohn. 1978. Analysis of a narrow hybrid zone between two species of *Pseudophryne* (Anura:Leptodactylidae) in southeastern Australia. Evolution 32:602–612.
Moore, W. S. 1977. An analysis of narrow hybrid zones in vertebrates. Quart. Rev. Biol. 52:263–278.
Nei, M. 1972. Genetic distance between populations. Am. Natur. 106:283–292.
Paterson, H. E. H. 1978. More evidence against speciation by reinforcement. South African J. Sci. 74:369–371.
Pickett, S. T. A., and P. S. White. 1985. *The Ecology of Natural Disturbance and Patch Dynamics*. Academic Press, New York.
Rand, D. M. 1987. Population biology of mitochondrial DNA in the crickets, *Gryllus pennsylvanicus* and *Gryllus firmus*. Ph.D. Thesis, Yale University, New Haven.
Rand, D. M., and R. G. Harrison. 1989. Ecological genetics of a mosaic hybrid zone: Mitochondrial, nuclear and reproductive differentiation by soil type in crickets. Evolution, in press.
Rising, J. D. 1983. The Great Plains hybrid zones. Curr. Ornithol. 1:131–157.
Sage, R. D., and R. K. Selander. 1979. Hybridization between species of the *Rana pipiens* complex in central Texas. Evolution 33:1069–1088.
Slatkin, M. 1973. Gene flow and selection in a cline. Genetics 75:733–756.
Sperling, F. A. H. 1987. Evolution of the *Papilio machaon* species group in western Canada (Lepidoptera:Papilionidae). Quaest. Entomol. 23:198–315.
Szymura, J. M., and N. H. Barton. 1986. Genetic analysis of a hybrid zone between the fire-bellied toads, *Bombina bombina* and *Bombina variegata*, near Cracow in southern Poland. Evolution 40:1141–1159.
Templeton, A. R. 1981. Mechanisms of speciation—a population genetic approach. Annu. Rev. Ecol. System. 12:23–48.

CHAPTER 6

SYMPATRY AND HYBRIDIZATION IN A "RING SPECIES": the Plethodontid Salamander *Ensatina eschscholtzii*

David B. Wake, Kay P. Yanev, and Monica M. Frelow

A central question in evolutionary biology relates to mode of speciation. Do species arise gradually, by the divergence of geographically disjunct or distant populations, or do species arise suddenly, on the margins of ranges, in newly occupied habitats? Although few would question that species can arise by other means as well, much controversy over modes of speciation relates to this question. Furthermore, most workers would accept that the answer to the question asked is neither one nor the other, but rather a mixture of these and other modes. But there have been many recent discussions of so-called "founder-effect" or "peripatric" speciation, and this mode seems to have been on the ascendency in recent years.

Speciation by geographic subdivision remains in the lexicon of speciation modes (Endler 1977; Templeton 1981; Mayr 1982; Woodruff 1981), but somewhat forgotten are the cases of isolation, and resulting speciation, by distance, best illustrated by "ring species." Such species played a role in the formulation of the geographic model of speciation (Mayr 1942, 1963), but their numbers have been eroded as various problems have arisen with different specific cases (Mayr 1970). In this chapter, we present a progress report of our continuing study of one of the best known "ring species," the lungless salamander *Ensatina eschscholtzii* of California.

The salamander *E. eschscholtzii* is a strictly terrestrial member of the lungless family Plethodontidae. It has seven subspecies, most of which are wrapped in a ring-like fashion around the Central Valley of California (Stebbins 1949; Wake and Yanev 1986). Occupancy of California has been from the North, via southward migration through the Coast Ranges and the Sierra Nevada and other interior mountain ranges; no *Ensatina* occupy the Central Valley currently. Stebbins (1949) believed that the subspecies intergrade at the north end of the valley. Midway along the length of the valley there has been a "transvalley leak," and a population from the coastal region has established itself in the foothills of the Sierra Nevada. Two subspecies with strikingly distinct color patterns meet along a front more than 100 km in length and everywhere they meet they hybridize (Brown 1974). In southern California two subspecies with even more strikingly distinct color patterns meet. Sympatry has been found in four mountain ranges (Wake et al. 1986). In three of these hybridization occurs, but in the fourth (Cuyamaca Mountains of San Diego County) no hybridization is found and the two morphs appear to be distinct biological species.

We are studying genetic differentiation throughout the range of *Ensatina* throughout western North America, with a concentration on zones of intergradation and hybridization. In this chapter we focus principally on a zone between about 900 and 1200 m elevation in the foothills of the central Sierra Nevada in southern Calaveras County, California. This region was studied by Brown (1974), who reported hybridization between *Ensatina eschscholtzii platensis* (a form with a spotted or blotched color pattern, consisting of red-orange spots or blotches on a dark brown background), a widespread Sierran subspecies, and *E. e. xanthoptica* (a form with a uniform, lively orange color pattern), a coastal subspecies with an outpost in the Sierran foothills. Although the area in which parental types cooccurred was restricted, Brown thought that the zone of hybrid influence might be relatively wide, on the order of tens of kilometers. We used electrophoretic analysis of variable electromorphic loci to investigate interactions in the hybrid zone. Our results are compared with our findings (Wake et al. 1986) from the contact zones in southern California.

MATERIALS AND METHODS

We established a study zone in southern Calaveras County, California, lying between two relatively large water barriers, San Antonio Creek (the major tributary of the Calaveras River system) and the Stanislaus River (Figure 1). Most salamanders were captured through the use of pitfall traps. From 20 to 50 such traps were placed in restricted areas of a favorable habitat in the vicinity of the village of Avery, at elevations between 975 and 1200 m. Two additional sites were located at higher elevations to the northeast, near the town of Arnold and the village of Camp Connell.

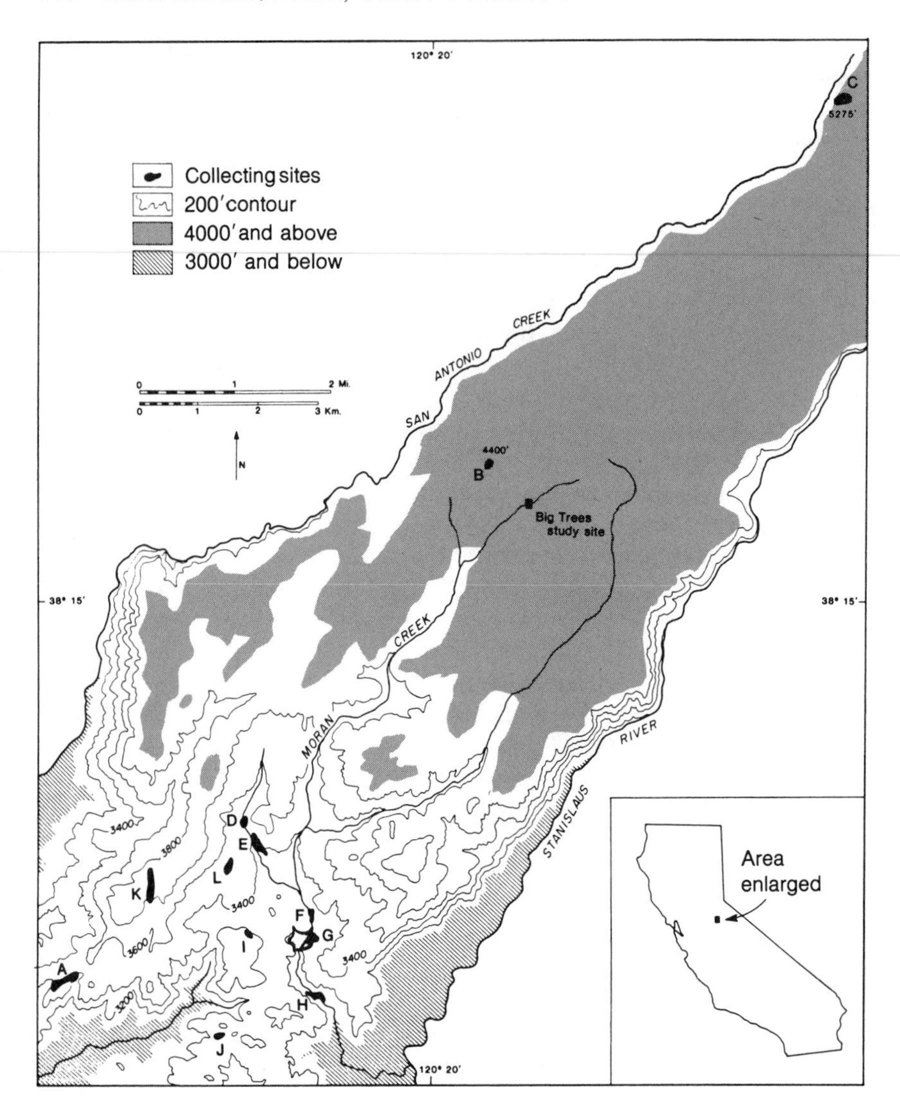

FIGURE 1. Map of portion of southern Calaveras County, California, indicating locations of populations (Table 1) used in this study. The only contour maps available are in English units (1 foot = 0.3048 m).

Details will be published elsewhere. For the purposes of the present study all sites at which 10 or more specimens were obtained were used (two sites thus were excluded from detailed analysis), and some sites that lay in very close proximity to each other were combined. The sites were no more than about 200 m in diameter, and we believe that we are justified in treating them as local populations on the basis of population studies of the species (Stebbins 1954). We have 12 sites in this region (Table 1).

These salamanders use underground retreats by day and during the extended dry season, and they are relatively difficult to find on the surface. Pitfall trapping was the only effective means of collection. Specimens encountered by turning cover objects in the study sites were included, but these were rare (less than 5% of the total). The period of collection extended over 5 years, and we have combined years. These animals are known to have very localized movements and are known to live for many years, and females are thought to reproduce every other year following sexual maturity at about age 4 years (Stebbins 1954), so we do not believe that combining years compromised our study.

Specimens were returned living to the laboratory and were scored for color pattern using a modification of the hybrid index devised by Brown (1974), who scored five separate color traits ranging from 1 (theoretical "pure" *platensis*) to 10 (theoretical "pure" *xanthoptica*). He presented an average score between 1 and 10 for each animal. Our study of coloration is incomplete at this time, and we present here information obtained from three traits, scored according to the index established by Brown. These traits are degree of iris iridophore development, degree of erythrophore development on ventral surfaces, and degree of melanophore development on ventral surfaces. We note here that Brown's Figure 3 has

TABLE 1. Populations used in study and sample size.

Designation	Locality (Calaveras Co., California)	Sample size
A	4 km WSW Avery	27
B	1 km NE Arnold	11
C	1 km NE Camp Connell	22
D	1.7 km N Avery	32
E	1.3 km NNE Avery	25
F	1.3 km E Avery	11
G	1.4 km ESE Avery	41
H	2.3 km SE Avery	26
I	0.5 km S Avery	12
J	2.6 km SSW Avery	26
K	2 km WNW Avery	53
L	0.9 km NNW Avery	21

transposed standard 1 and 4 for ventral melanophores, and we corrected this error prior to using the index. Furthermore, instead of averaging the scores for the three traits we simply summed them; our index ranges from 3 through 30.

Following scoring for color the specimens often were photographed (mainly suspected hybrids) and prepared for electrophoretic analysis (see Wake and Yanev 1986, for details). An initial survey of two populations (plus some additional material from nearby areas) as a part of our study of geographic variation in the species included 26 scorable proteins. We found fixed or nearly fixed differences between reference populations of *platensis* and *xanthoptica* in eight proteins (Nei $D = 0.419$), and for purposes of analysis in this chapter we treat these as genetic loci; the electromorphs were treated as alleles. We established a genetic hybrid index on the basis of these eight loci ranging from 0 (theoretical "pure" *platensis*) to 16 (theoretical "pure" *xanthoptica*). These alleles are indicated in Table 2. There are a few low-frequency alleles, and for purposes of initial analysis each of these was assigned either to *plantensis* or *xanthoptica*. This assignment was based on several factors. We used geographically remote samples of *xanthoptica* and *platensis* for comparative purposes. Our outside sample of *xanthoptica* is a combination of several sites in western Calaveras County below 750 m elevation (total of 25 specimens). The outside sample of *platensis* is a combination of a sample from Blodgett, El Dorado County, California (total of 10 specimens), and of samples B and C (Figure 1,

TABLE 2. Allele assignments for the eight loci used in this study.

Locus	Symbol	EC Number	*platensis* alleles	*xanthoptica* alleles
Phosphogluconate dehydrogenase	Pgd-A	1.1.1.44	*f*	*g*
Malate dehydrogenase (mitochondrial)	M-Mdh-A	1.1.1.37	*c*	*b*
Isocitrate dehydrogenase (cytosolic)	S-Icdh-A	1.1.1.42	*g*	*d*
Proline dipeptidase	Pep-D	3.4.13.9	*d*	*f*
Aspartate aminotransferase (cytosolic)	S-Aat-A	2.6.1.1	*b*	*a*
Tripeptide aminopeptidase	Pep-B	3.4.11.4	*h*	*g*
Glycerol-3-phosphate dehydrogenase	Gpd	1.1.1.8	*a*	*c*
L-Iditol dehydrogenase	Iddh-A	1.1.1.14	*d*	*b*

together totalling 32 specimens). For the outside samples of *xanthoptica* the alleles listed in Table 2 had a frequency of 1. Because each sample was relatively small, we probably missed rare alleles, such as an allele found in a heterozygous state in single individuals in both populations A and J. Because these populations were like *xanthoptica* in almost every other respect, and because these rare alleles otherwise were not found, they were counted as if they were *xanthoptica* allele *f*. When low-frequency alleles were encountered only in the hybrid zone and they did not appear in either set of outside samples they were assigned to one form or the other on the basis of cooccurrence with other alleles, and in every case assignment was unambiguous. At one locus (*PAP*) we had difficulty separating two fast-migrating bands in *platensis*, but since the *xanthoptica* allele was very slow in its migration, the separation of the two, and the identification of heterozygotes, was unambiguous, so we scored the two fast-migrating alleles as if they are one allele (designated g here; cf. Wake and Yanev 1986).

Coordination of allele designations throughout all of our studies on this species will be difficult because of the extraordinary number of alleles involved, but as far as possible we have used the allele designations of Wake and Yanev (1986). Alleles thought to be newly discovered in this study are given arbitrary designations that are simply alphabetical extensions of our earlier study, and they do not relate to mobility. Details can be obtained from D. B. Wake.

We restricted our study to the loci that had fixed or nearly fixed differences and that were readily scorable. We found 24 alleles for these 8 loci, but the 16 listed in Table 2 strongly dominated and the remaining 8 are absent in most populations and generally have a frequency of less than 0.05 when present. For only one locus does such an allele reach a moderate frequency. LGG allele *k* (assigned to *xanthoptica* on the basis of its presence in four heterozygotes in population J and its absence in all *platensis* populations) reaches a frequency of 0.19 in population H and 0.16 in population G. Several of these eight alleles are "rare," and three are found only in single heterozygotes. The greatest number of low-frequency alleles are found in populations G and K, both in the heart of the hybrid zone, but both also the largest samples.

Although scores of 0 and 16 in our allozyme hybrid index unambiguously identify parental *platensis* and *xanthoptica* individuals, respectively, the presence of certain alleles of one form in the other complicates the situation somewhat. We used population A as a reference for *xanthoptica* and populations B and C as reference populations for *platensis*. The profiles of these populations relative to outside comparative populations (Figures 2 and 3) show that such a procedure is justifiable. We assume that there is residual background variation in the form of low-frequency alleles

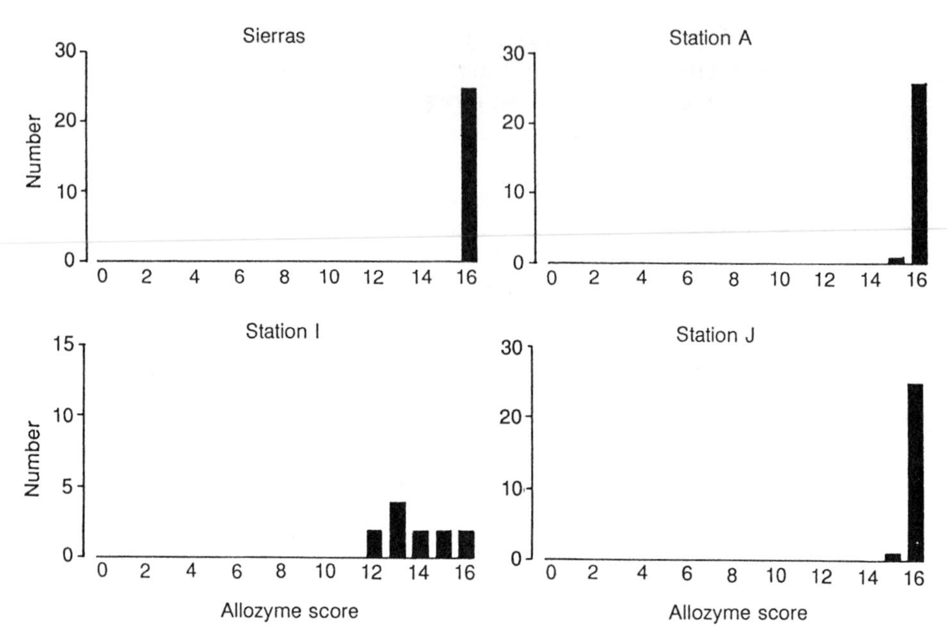

FIGURE 2. Frequency histograms for individuals in four samples classified according to allozyme hybrid index. The four samples include salamanders that have the general coloration of *xanthoptica*. The Sierras sample is used as an allozyme outgroup for this study. Samples from populations A, I, and J lie along the western edge of the hybrid zone. A score of 16 is considered to be "pure" *xanthoptica*, whereas a score of 15 is considered to be ambiguous (based on the presence of the same low-frequency allele in populations A and J, suggesting that this might be a low-frequency "background" allele in *xanthoptica* rather than a *platensis* allele). All scores lower than 15, and some scores of 15, represent incorporation of *platensis* alleles.

in both parental stocks (e.g., our Blodgett sample contains a few alleles in low frequency that we assigned to *xanthoptica*). A score of 15, although slightly ambiguous, is treated as a parental *xanthoptica* pattern. Similarly, the somewhat more variable *platensis* has hybrid index values of 1 and 2 assigned to it. Only by making this compromise can we deal effectively with all eight loci simultaneously.

RESULTS

Based on analysis of genotypes and allozyme hybrid index, as well as the color hybrid index, we found both parental types and hybrids in three pop-

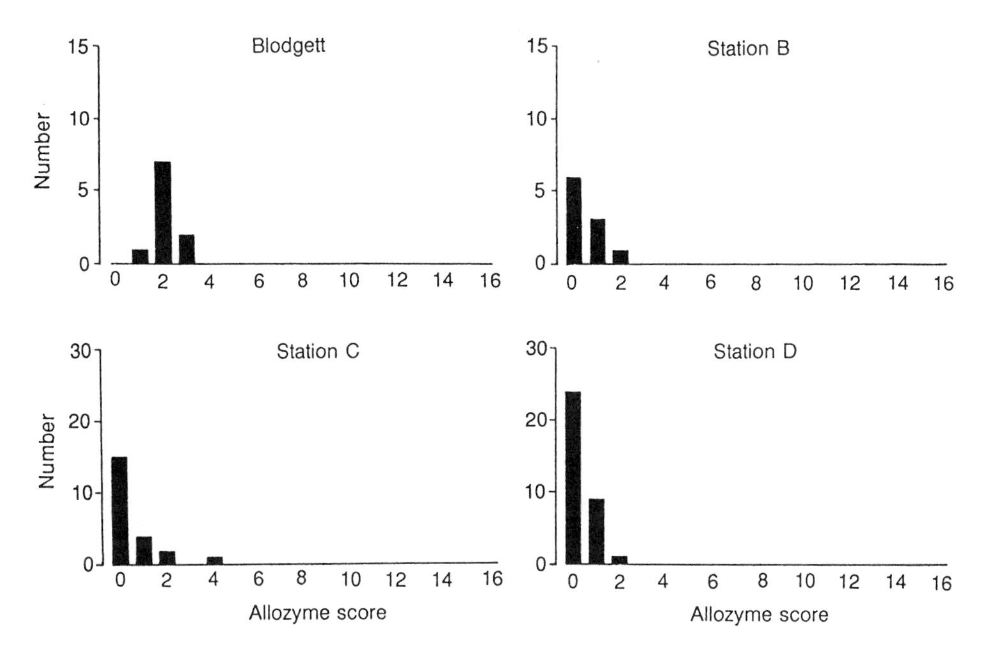

FIGURE 3. Frequency histograms for individuals in four samples classified according to allozyme hybrid index. The four samples include salamanders that have the general coloration of *platensis*. Sample D lies on the eastern and northern edge of the hybrid zone, whereas the other three samples are outgroups for the present study. A score of 0 is an unambiguous "pure" *platensis*. There is more background variation in *platensis* than in *xanthoptica*, and we have found no case of a population having an overall score of 0.

ulations: G, K, and L (Figure 4). Population K (our largest sample) contains one unambiguous F_1 hybrid, but the only other unambiguous F_1 we have found is in population F, which otherwise has only *platensis* genotypes. Individuals with scores of 8 elsewhere in the sample cannot be F_1.

The hybrid zone is very narrow. Endler (1977) has recommended estimating the width of hybrid zones as that distance between frequencies of 0.8 and 0.2 for a given allele. The distance between localities along Mill Creek and a tributary (Figures 1, 5, and 6) that have average gene frequencies (based on *platensis*) dropping from 0.88 to 0.2 (Table 3) is approximately 1.4 km.

The hybrid zone also can be crossed indirectly by a transect extending along a mainly south-facing slope between sites for samples A and B (Figure 1). Here the samples are less regular in relation to each other than those lying along Mill Creek. Both samples K and L lie on the hybrid zone,

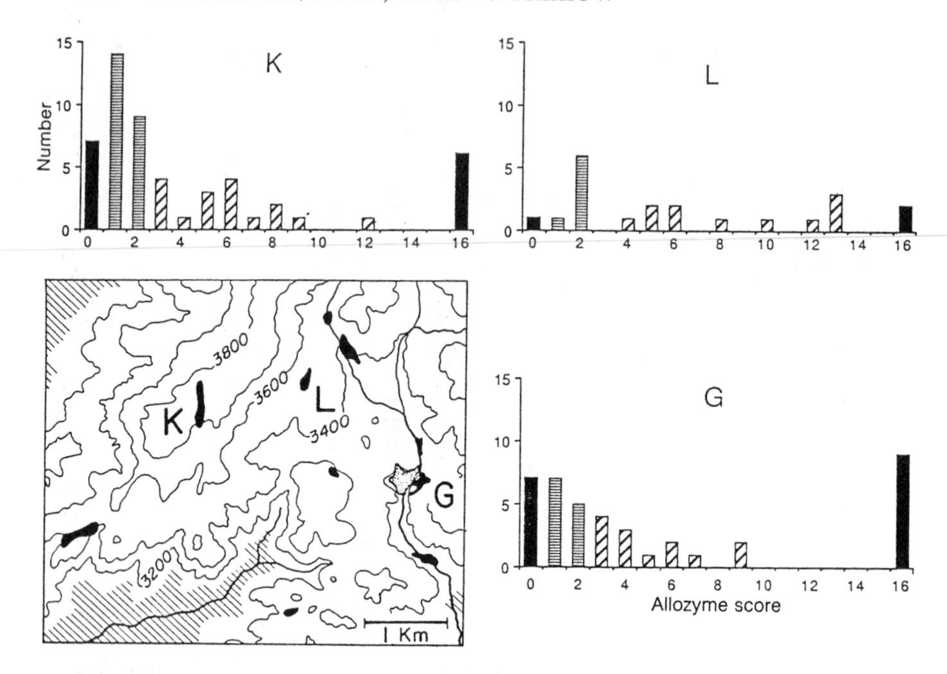

FIGURE 4. Location of the three sample sites near Avery, Calaveras County, California, at which "pure" parental *platensis* and *xanthoptica* are present. These samples lie along the middle of the narrow hybrid zone. Frequency histograms are based on the allozyme hybrid index. A single F_1 (sample K) was found. Scores of 1 and 2 are differentiated from scores of 0 or of 3 and higher to indicate their somewhat ambiguous nature (see text). The only contour maps available are in English units (1 foot = 0.3048 m).

which accordingly is even narrower than is apparent from Figure 7. Furthermore, we have a small sample (three specimens) from a site 500 m west of the western edge of K that has allozyme scores of 16, 16, and 15 (this last score is based on the same allele for 6-phosphogluconic dehyrogenase (6-PGD) that gives single scores of 15 in samples A and J), suggesting that the hybrid zone is entered abruptly.

We consider scores on the allozyme hybrid index from 3 through 14 to be unambiguous hybrids. Such scores are absent from samples immediately adjacent to the hybrid zone: A, D, and J. Percentage of hybrid individuals in samples in and on the margins of the hybrid zone are indicated in Table

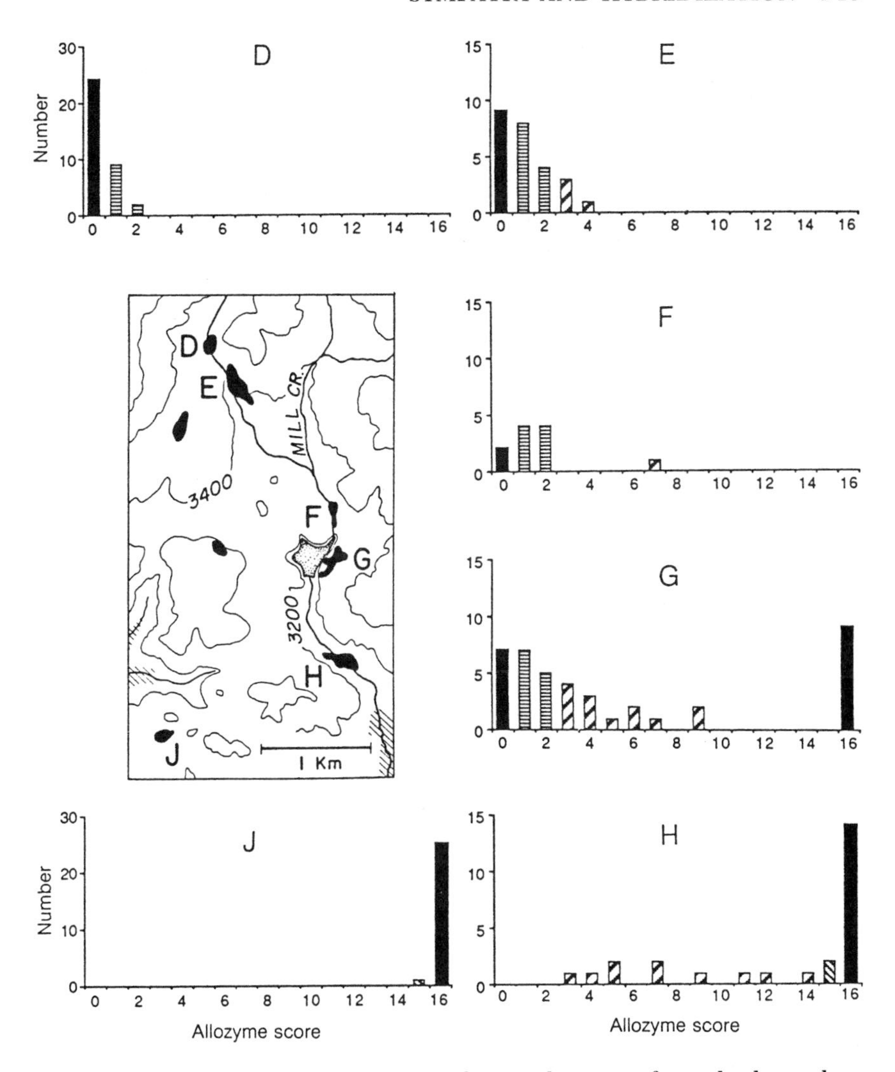

FIGURE 5. Section of the study zone indicating location of samples lying along and near Mill Creek and a tributary near Avery, Calaveras County, California. Frequency histograms based on the allozyme hybrid index are presented for samples extending through the hybrid zone. There is only a single F_1 (sample F), and only a single sample (G) at which "pure" parental *platensis* and *xanthoptica* are present. Scores of 1 and 2 are differentiated from scores of 0 or of 3 and higher to indicate their somewhat ambiguous nature (see text). The only contour maps available are in English units (1 foot = 0.3048 m).

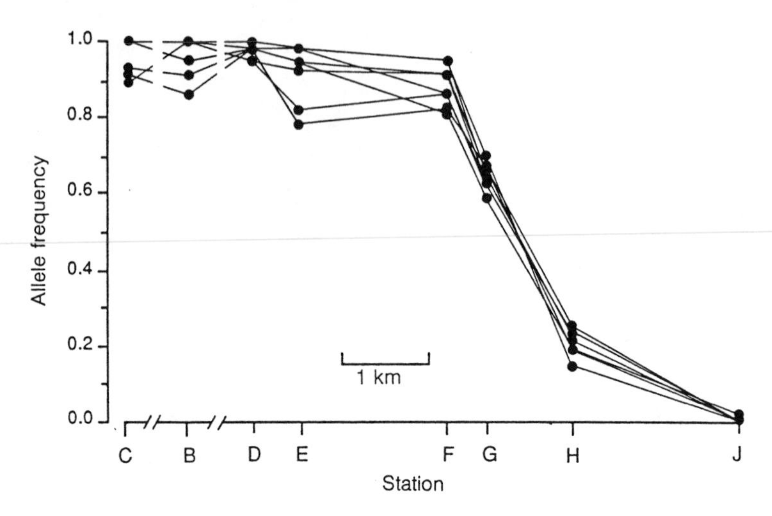

FIGURE 6. Gene frequency cline (based on *platensis* alleles) for eight loci in the samples from Figure 5, plus two relatively remote samples of *platensis* that show the background variation in that form (see also Table 3).

TABLE 3. Average frequency (and standard deviation) of *platensis* alleles for eight loci.

Population	Frequency	Number
A	0.003 (±0.01)	27
B	0.97 (±0.05)	11
C	0.97 (±0.05)	22
D	0.98 (±0.02)	32
E	0.92 (±0.08)	25
F	0.88 (±0.05)	11
G	0.64 (±0.04)	41
H	0.20 (±0.03)	26
I	0.13 (±0.24)	12
J	0.003 (±0.01)	26
K	0.74 (±0.03)	53
L	0.59 (±0.10)	21
	Total	307

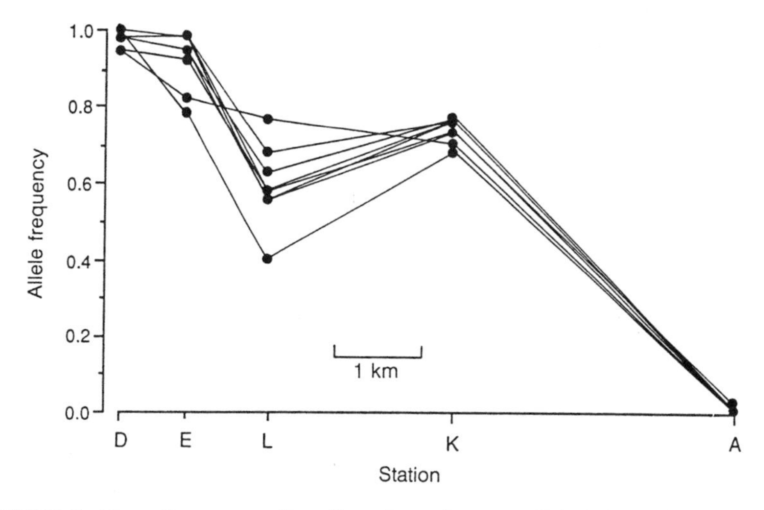

FIGURE 7. Gene frequency cline (based on *platensis* alleles) for eight loci in samples lying along a mainly south-facing slope between samples A and B (cf. Figure 1 and Table 3).

TABLE 4. Percentage individuals in different samples counted as hybrids.

Sample	Operational hybrids (%)	Sample	Operational hybrids
A	0	G	32
B	0	H	42
C	0[a]	I	67
D	0	J	0
E	16	K	30
F	9	L	52

[a]One individual in this population has a score of 4, but the alleles contributing to this score have a high probability of being *platensis* alleles; see text.

4. Two samples, I (N = 12) and L (N = 21) have more than 50% hybrids. In sample I there are no individual scores lower than 12 and the population is dominated by *xanthoptica* alleles (average frequency of *platensis* alleles in only 0.13 (SD = 0.24; Table 3). In contrast, in sample L the distribution of genotypes approximates that expected in a hybrid swarm (Figure 4), and only three scores of 0 or 16 are recorded.

The color hybrid index modified from Brown (1974) was established without knowledge of genetic information. The relationship between the two indices is illustrated in Figures 8, 9, and 10. Boxed values in these figures represent coincident occurrences in populations taken as representing parental conditions (A for *xanthoptica*; B and C for *platensis*). The *platensis* color pattern is much more variable than that of *xanthoptica*, and the color index ranges from 3 through 10. That for *xanthoptica* ranges only from 28 through 30. The two F_1 individuals found in this study have color scores of 15 and 16. Accordingly, we believe that the color index is suitable for comparison with the allozyme index.

Those populations in the heart of the hybrid zone (G, K, and L) have both color and hybrid scores skewed in the direction of *platensis* (Figure 8). No allozyme scores from 0 through 8 are missing, and no color scores from 3 through 15 are missing. In contrast, two allozyme scores from 8 through 16 are missing, and five color scores from 15 through 30 are missing-

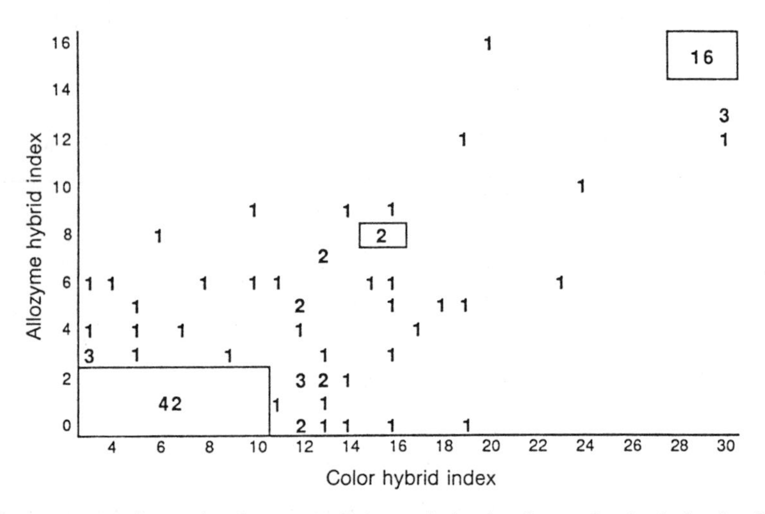

FIGURE 8. Relationship between allozyme hybrid index and color hybrid index in three populations in which both parental genotypes and hybrid genotypes are present. Color scores of 0, 1, and 2 cannot exist using our methods (see Materials and Methods). The box in the lower left outlines the coincidence between allozyme scores of 0 to 2 and color scores in populations B and C, which, for purposes of analysis, are considered to be typical parental *platensis*. The box in the upper right outlines the coincidence between allozyme scores of 15 and 16, and color scores in population A, which, for purposes of analysis, are considered to be typical parental *xanthoptica*. The box in the center surrounds the combined scores of the two F_1 individuals encountered in this study. The numbers identify individuals from samples G, K, and L for which information both on allozymes and color is available.

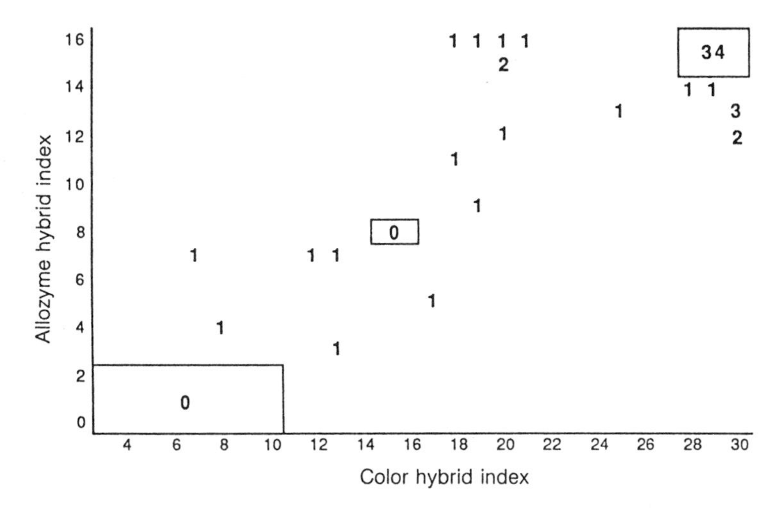

FIGURE 9. Relationship between allozyme hybrid index and color hybrid index for three samples (H, I, and J) on the *xanthoptica* side of the hybrid zone. See Figure 8 for further explanation.

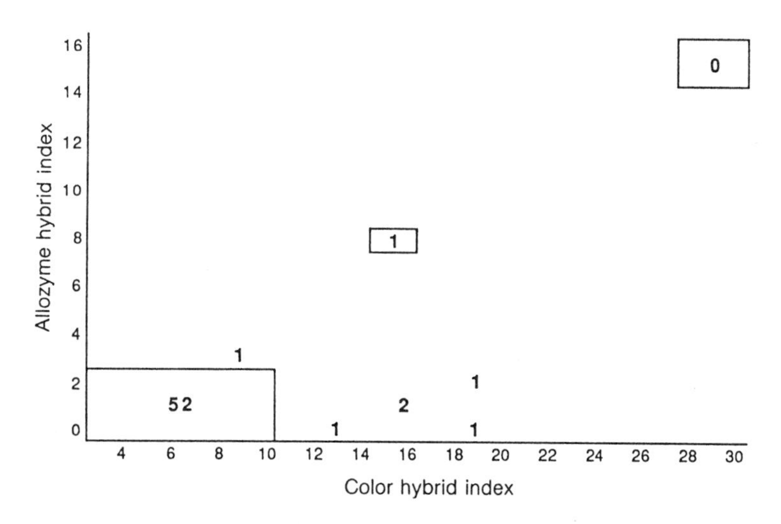

FIGURE 10. Relationship between allozyme hybrid index and color hybrid index for three samples (D, E, and F) on the *platensis* side of the hybrid zone. See Figure 8 for further explanation.

ing. Only nine individuals have scores from 8 through 16 for allozymes and simultaneously from 15 through 30 for color (excluding those in the parental box). In contrast, there are 35 individuals with scores from 0 through 8 for allozymes and simultaneously from 3 through 15 for color (again, excluding those in the parental box). Only two individuals have allozyme scores above 8 and color scores below 15, but nine individuals have allozyme scores below 8 and color scores above 15.

Those populations on the *xanthoptica* side of the hybrid zone (H, I, and J) have a strong predominance of coincident color and allozyme scores that are those identified as parental *xanthoptica* (Figure 9). There are 19 individuals with allozyme scores between 8 and 16 and color scores between 15 and 30 (excluding those in the parental box), but there are only 5 with allozyme scores from 0 through 8 and simultaneously with color scores from 3 through 15 (and there are no individuals in the parental box).

Those populations on the *platensis* side of the hybrid zone (D, E, and F) have a strong predominance of coincident color and allozyme scores that are those identified as parental *platensis* (Figure 10). No individuals have allozyme scores above 8, and only four individuals have color scores above 15 (19 is the highest).

DISCUSSION

Our analysis of allozyme variation in the study area has shown that hybridization between *platensis* and *xanthoptica* occurs in a geographically narrow zone in the foothills of the Sierra Nevada in Calaveras County, California. This area was chosen for study because it was known that a contact zone between *platensis* and *xanthoptica* existed there, and hybridization between the two had been demonstrated based on analysis of color pattern variation (Brown 1974). We located the exact sites used by Brown and added some additional ones. His series were as follows (our population designation follows in parentheses): (1) Dorrington (C); (2) Arnold (B); (3) Mill Creek (D); (4) Avery and Hunter Reservoir (G); (5) Indian Creek (A, our site is located a few hundred meters SE of his).

Brown (1974) estimated that the distance of overlap between the two parental types was about 0.3 miles (approximately 480 m), but stated that more intensive work would probably show the zone to be wider than he had observed. He thought that he detected genetic influence of *xanthoptica* on the coloration of *platensis* some distance from the zone of overlap (e.g., in the Arnold series). Introgression was believed to be taking place in both directions, and its effect was increased variability in color pattern in the parental forms (*platensis* was thought to be more affected than *xanthoptica*).

Using criteria from his color analysis, Brown (1974) constructed a hybrid index that gave an estimate of 8% hybrids in this entire region. In his Avery and Hunter Reservoir site (our population G), 22% (from his Figure 8) were hybrids. In the present study percentage hybrid individuals in population G by the operational definition using our allozyme and color indices is 32 and 38, respectively.

We have extended Brown's study in two ways: by adding more sites in the immediate vicinity of the hybrid zone near Avery, and by surveying variation in eight polymorphic proteins. Our results confirm Brown's identification of hybridization in a narrow zone. Along Mill Creek and its tributary the hybrid zone is only a little over 1 km in width, and both parental forms occur only in one site over a linear distance of about 300 m or less. The two other areas of overlap are equally narrow. We find no evidence of pervasive introgression, and only ambiguous evidence of gene flow beyond the immediate limits of the hybrid zone.

Prior work (Wake and Yanev 1986) showed that *platensis* has considerably more polymorphic proteins than does *xanthoptica* (1.38 to 1.69 alleles per locus, versus 1.12 for *xanthoptica*; heterozygosities for *plantensis* ranged from 6.9 to 12.3%, versus 1.9% for *xanthoptica*). Thus, although our hybrid index based on eight proteins is relatively certain for *xanthoptica*, the potential for overlooking rare *xanthoptica* alleles in the low-frequency background variation of *platensis* is relatively greater. For example, examination of allozyme hybrid index scores above zero in population D discloses that 5 of the 10 individual variants contributing such scores involve alleles of three proteins that also are present in reference populations B and C. Furthermore, these 10 variants occur in only 6 of the 8 surveyed proteins, whereas in the adjacent population E (which is a few hundred meters closer to the zone of overlap) the 32 variants contributing to scores above 0 are found in all eight proteins. Thus, little gene flow (possibly involving as few as 5 out of a total of 512 alleles) can be measured across the hybrid zone from *xanthoptica* through the zone of overlap to population D. Even less gene flow is occurring in the opposite direction. In population A only 1 of 430 alleles and in population J only 1 of 414 alleles (a single protein in one individual each in these two populations could not be scored) could be considered a *platensis* allele.

We do not believe that gene flow from *xanthoptica* is responsible for allozyme hybrid index scores greater than 0 in populations B and C. First, the Blodgett sample (Figure 3) is from an area approximately 75 km from B, far north of the Sierran range of *xanthoptica*, so scores greater than zero in that sample cannot be attributed to gene flow. Second, the one score in the hybrid range (a score of 4 in an individual in population C) is based on four variants that are present in Blodgett.

Population D has even fewer potentially *xanthoptica* variants than do more remote populations of *platensis*, in spite of the fact that it is a larger sample (N = 32) than the others and lies very near the zone of overlap (Figure 3). This suggests not only that there is little gene flow, but also that some of the low-frequency background alleles might have been lost. This population is the last of the "pure" *platensis* populations, and is at the extreme margin of the range of the subspecies, both in terms of elevation and in habitat. At lower elevations, and only a few meters away from the stream, as well as farther to the west generally, the forest thins and becomes interspersed with areas of brush and grasslands. Typically *platensis* lives in closed canopy, relative dense forest at higher elevation, whereas *xanthoptica* lives in lower, more open forest and brush mosaic habitats. Thus, it is possible that combinations of inbreeding and drift might increase homozygosity and reduce variability in the finger-like projections of the population of *platensis*, causing it to break up into relatively small and semi-isolated demes at the population front.

We found only 2 F_1 hybrids; all others are backcrosses. Our samples are too small to permit a detailed analysis of the pattern of backcrossing, but it appears to be restricted to the hybrid zone and to involve mainly other hybrid individuals. If there were more backcrossing into parental stocks we would find more evidence of gene flow. Thus, we conclude that hybridization between sympatric parental genotypes is rare.

Inspection of the allozyme hybrid index profile in population G (Figure 4) suggests that the *xanthoptica* individuals present in the population have not been outbreeding, for all classes from 9 through 16 are absent. There is only one individual in this range in station K, but there are several in Station L (Figure 4). Furthermore, in both station H (Figure 5) and I (Figure 2) there are numbers of individuals with scores between those of F_1 and "pure" *xanthoptica*. Thus, it seems unlikely that the absence of a series of potential genotypes from population G is the result of inviability or poor performance. However, on both sides of the contact zone, and even in the zone itself, the pattern of skew encountered suggests an absence of free interbreeding.

We believe that our hybrid zone arose from secondary contact resulting from the "transvalley leak" of *xanthoptica* from the west, probably during Late Pleistocene times (see Stebbins 1949, for a detailed consideration of this issue). The close genetic similarity of Sierran *xanthoptica* to populations in the hills just east of San Francisco Bay (Nei genetic distance is 0.021) argues in favor of this hypothesis, and the very low variability of Sierran *xanthoptica* compared with that of other populations of *Ensatina* is consistent with establishment of these populations by recent founder events (data in Wake and Yanev 1986). Further evidence in favor of the hypothesis of secondary contact is the fact that 8 of 26 proteins sampled show drops in frequency from greater than 0.95 to less than 0.05 in a distance just over 5 km, and they change in a highly concordant manner. The hybrid zone has been stable for at least 20 years.

A factor contributing to the narrowness of the hybrid zone is the tendency of terrestrial plethodontid salamanders to have small home ranges and to have no special pattern of dispersal. These salamanders lay direct-developing eggs on land and do not migrate to water. Stebbins (1954) studied *xanthoptica* in the East Bay region over approximately 4 years, and based on a mark–recapture study he reported on density (about 600–700 per acre), home range, and movements. Home ranges were small, with "greatest width" averaging 10.1 m for females and 19.6 m for males. The widest home range recorded was 41.4 m, and this also was the greatest movement. Sexual maturation does not occur until the fourth year, and individuals may live to be 10 to 15 years of age, so movements of single individuals over their lifetime will be difficult to follow. Dispersal distances are expected to be greater than home range sizes, but as yet little information of dispersal in *Ensatina*, or other terrestrial salamanders, exists. A mark–recapture study on a 100 by 300 m plot in Calaveras Big Trees State Park (Figure 1) is designed to find potential dispersing individuals, and preliminary results suggest that Sierran *platensis* move greater distances (at least one individual has moved more than 100 m) than do the *xanthoptica* Stebbins (1954) studied. Possibly his plot size was too small to record dispersers. Although the species is a relatively sedentary one, narrowness and persistence of the hybrid zone cannot be attributed to neutral diffusion based on lack of movement by the animals.

Much attention has been given to hybrid zones recently (e.g., Barton and Hewitt 1981, 1983, 1985; Woodruff 1981), and there have been some important theoretical advances. For example, Barton and Bengtsson (1986) concluded that strong selection against hybrids at numerous genetic loci is required to create a genetic barrier at linked neutral loci. We believe that hybrid individuals are at a disadvantage in the hybrid zone, and that the allozyme patterns we report are a consequence of that selection, even though the allozymes themselves may be selectively neutral. There are three main candidates for selective agents: habitat choice, predation, and mating behavior.

In general *xanthoptica* occurs at lower elevations than *platensis*, in warmer and drier environments, and in open pine to pine–oak forest, and mosaics of forest and grassland–brush habitats. Where *xanthoptica* approaches the range of *platensis* it reaches its highest elevational limit on south-facing slopes with relatively open forest, whereas *platensis* reaches its lowest elevational limit on north-facing slopes with more closed forest or in stream bottoms. For example, along Mill Creek and its tributaries *platensis* has been found as low as 975 m (station G), whereas just a short distance away *xanthoptica* has been taken as high as 1190 m (station K). Station K can be subdivided at the ridge-top into north and south halves, the latter south facing and covered by a relatively more open forest. Using our operational values with the allozyme hybrid index, the ratio of *xanthop-*

tica to *platensis* is 5:11 on the south half, but 1:16 on the north half, and the single F_1 from the site is from the south half. We do not know what sets the limits of either form, but both to the north and to the south of the range of *xanthoptica* in the Sierra Nevada, *platensis* reaches lower elevations than in the Avery area. Perhaps the combination of deteriorating habitat and competition with *xanthoptica* sets the lower elevational limits of *platensis* in this area. There is no apparent reason why *xanthoptica* could not live at slightly higher elevations. The winter snow zone commences between 1000 and 1200 m, but the environment is relatively mild. However, densities of *platensis* are relatively high immediately at the edge of the hybrid zone, and it seems likely that competition by preemptive occupancy of space could limit incursion of *xanthoptica*. The zone of overlap is surprisingly narrow given the known ability of single individuals to undertake movements equal to more than one-third the width of the known zone of overlap.

Several vertebrates are known or suspected to prey on *Ensatina* (Stebbins 1954). Raccoons regularly disturb pitfall traps and eat salamanders, and jays and other birds also feed on *Ensatina*. The cryptic (in natural habitat) color pattern of *platensis* contrasts with the vivid, conspicuous coloration of *xanthoptica* (Stebbins 1949; Brown 1974). Both Stebbins (1949) and Brown (1974) have argued that *xanthoptica* is a mimic of the highly poisonous salamanders of the genus *Taricha*, which have only one known predator (a garter snake). The robust tail of *Ensatina* is richly supplied with poison glands and it is an effective deterrent to at least some snake predators (Hubbard 1903). Furthermore, the tail can be autotomized. On several occasions of raccoon depredation of can traps, only the detached tail of the *Ensatina* has been left behind. Sierran populations of *xanthoptica* are an especially vivid yellow-orange, much brighter than the East Bay populations, and the Sierran populations of *Taricha torosa* are also the most brightly colored in that species (Riemer 1958). Hybrids between *platensis* and *xanthoptica* are neither mimics nor are they cryptic. They stand out, but do not look like anything else. Increased predation on them (for which we have no evidence) could contribute greatly to selection against hybrids and backcrosses. Presumably the mimics could ascend to higher elevations and the cryptic forms to lower elevations, so the position of the hybrid zone is unlikely to be set by increased predation on hybrids. Cryptic forms are less cryptic at lower elevations, because of the more open nature of the low elevation forests, and mimics move out of the range of *Taricha* at higher elevations, reducing the effectiveness of their mimicry. Selection related to habitat choice along the elevational gradient, in combination with increased predation on hybrids, could help define the elevational limits of the zone.

Mate recognition systems also may be important in the vicinity of the hybrid zone. Plethodontid salamanders have an involved courtship behavior, and in *Ensatina* it is a very lengthy process with many components (Stebbins 1954). We have no evidence as yet concerning the possibility of isolating mechanisms, or differences in mate recognition signals, but the existence of a courtship that is dependent on chemical and tactile cures contributes to the likelihood that such may be the case. F_1 individuals are rare, and all genotypic classes recorded could be the result of relatively limited mating among the hybrids themselves. We suspect on the basis of the genic patterns recorded that there is little mating of parental types with F_1s or other hybrids.

Parapatric distributions, in which two genetically distinct forms have closely abutting geographic ranges, are common in salamanders (see review by Larson 1984; Good et al. 1987), and frequently there is no hybridization between the taxa. Elevational replacement is also a well-known phenomenon in salamanders (Hairston 1951; Wake and Lynch 1976), and competition between closely related species of plethodontids has been demonstrated (see review by Hairston 1988). No direct evidence of competition in *Ensatina* exists, but arguing by analogy from studies on other species of salamanders, competition between *xanthoptica* and *platensis* may contribute to the location of the contact zone. We know that the zone varies in elevation from place to place in the Sierra Nevada, as Hairston (1951) demonstrated for *Plethodon jordani* and *P. glutinosis* in North Carolina, and this suggests that different combinations of factors are acting in concert to set the location of the zone in different areas. If we are correct in our argument that *xanthoptica* has invaded the Sierra Nevada, it probably became established first in regions of low elevation and open habitat, unoccupied by *platensis*. As it spread eastward and into higher elevations, it came into contact with *platensis*, and the amount of geographic overlap between the two has remained remarkably slight. We postulate that a combination of preemptive occupancy of space by *platensis*, competition, and different patterns of environmental adaptation has created a standoff between the two forms.

Comparisons with other contact zones in *Ensatina* show that although gene flow is low and very limited in geographic extent in Calaveras County, there is even less genic exchange in southern California. There are 10 nearly fixed genetic loci (out of 26 loci sampled) differentiating the blotched *klauberi* (the morphological analog of *platensis*) and unblotched *eschscholtzii*, and with this increased genetic differentiation there also is an apparent decrease in the number of hybrids found and the extent of backcrossing (Wake et al. 1986). A hybrid index based on fixed differences in 10 proteins can be used to display variation in the four main contact zones in southern California (Figure 11). Our samples for southern California reported here are regional ones, including sites in which both parental

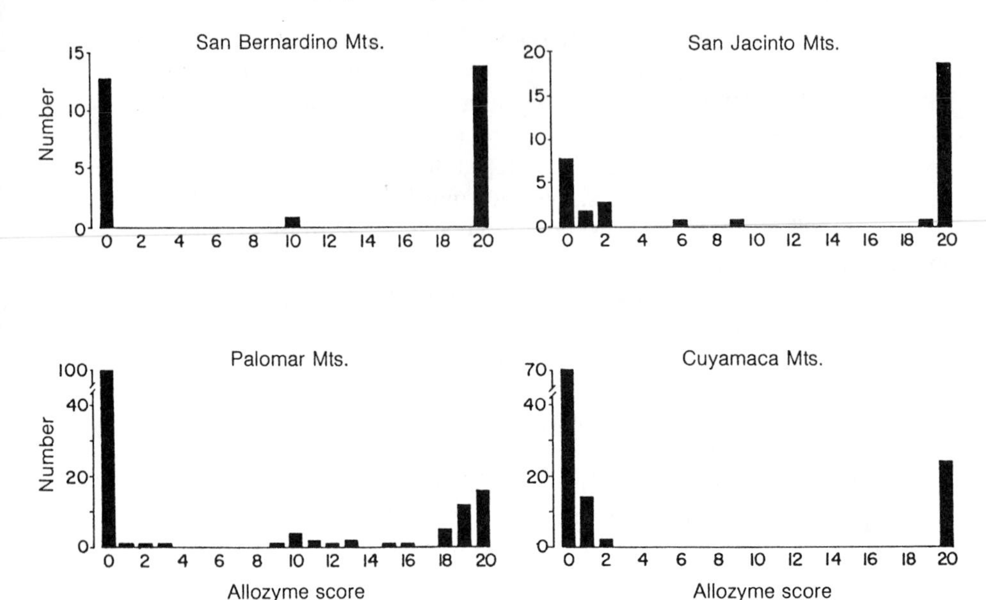

FIGURE 11. Frequency histograms for four samples of *Ensatina* in southern California. These samples represent amalgamated local samples in four discrete mountain regions. A score of 0 is unambiguous "pure" *klauberi*, whereas 20 is unambiguous "pure" *eschscholtzii*. Scores of 1 and 2 are also considered to be *klauberi*, and scores of 19 are accepted as *eschscholtzii*. All other scores are considered to be hybrids. Sympatry with no hybridization occurs in the Cuyamaca Mountains.

forms occur as well as areas in which only one occurs. In this figure *klauberi* has scores of 0 to 2, and *eschscholtzii* has scores of 19 and 20. We found only two hybrids in the San Bernardino Mountain region, both F_1 taken in close proximity to both parental types. In the San Jacinto Mountains we found only two hybrids, neither an F_1. In the Palomar Mountains there are more hybrids, including a potential F_1, and there is an apparent skew in the direction of *eschscholtzii*. Finally, in the Cuyamaca Mountains, our southernmost sample, there is complete sympatry with no evidence of hybridization; however, *eschscholtzii* is relatively uncommon (only 8 have been found, compared to more than 50 *klauberi*). Thus, in the zone of contact in southern California there is substantially less hybridization than in the Sierran zone of contact, and there even is local sympatry with no hybridization.

In both the southern California and Sierra Nevada hybrid zones the interacting units appear to be at the species level of differentiation. However, because of the complexities associated with the possibility of indirect connections between the units, we recommend no taxonomic changes as yet. Studies of genic variation, morphological variation, and ecology are in progress, and we are giving special attention to areas that previous workers have identified as zones of intergradation. The general pattern that is emerging is not one of gradual differentiation, but rather one in which there is much local and regional differentiation with occasional sharp zones of change. Speciation seems to have progressed in part by allopatric adaptive divergence and in part by essentially stochastic isolation by distance. We tentatively accept the general scenario of Stebbins (1949) that *Ensatina* moved southward and wrapped around the Central Valley. However, this appears to have been a much more ancient event than envisioned by him. There have been many opportunities for range expansion and contraction, and for adaptive morphological and ecological differentiation. We find genetic differentiation manifest at many levels. Thus, within what Stebbins identified as subspecies, there is substantial geographic differentiation. The amount of genetic differentiation within the blotched and unblotched groups of subspecies is as great as that between these two groups. Intergradation between coastal and inland (montane) groups north of the Sacramento Valley was postulated on the basis of an analysis of coloration (Stebbins 1949). This region must be studied carefully to determine if there is a substantial difference from the kinds of interactions that we have recorded in the Sierran foothills. The apparent natural experiment provided by the transvalley leak, which resulted in the secondary contact with hybridization of *xanthoptica* and *platensis*, may be just a more dramatic form of the secondary meeting of groups that has led to the abrupt change that we find elsewhere in this complex group. The cooccurrence of two biological species in the Cuyamaca Mountains is readily apparent because of the great differences in coloration, and the absence of hybrids. Our discovery of regions of abrupt changes in gene frequencies suggests that it is still too early to determine if *Ensatina* presents several stages in the process of speciation, or if the genus already has speciated extensively and some borders are more readily apparent than are others. Work in progress will contribute to the solution of this problem.

SUMMARY

A complex of morphologically differentiated populations of salamanders of the plethodontid genus *Ensatina* surround the Central Valley of California in ring-like fashion. At the bottom of the ring, in southern California, there is overlap, and sympatry occurs with limited or no hybridization in different locations. At about the midpoint in the ring there has been a "transvalley leak" of coastal populations into the foothills of the Central Sierra Nevada, where coastal unblotched and inland blotched populations meet along a long, narrow hybrid zone. There are either fixed or nearly fixed allozymic differences between the two interacting population groups, and these were used to study interactions in the hybrid zone. Local sympatry has been found in three places, with both parental types as well as a variety of hybrids being present. F_1s are rare, but there are backcrosses and perhaps multigenerational hybrids present in the sympatric sites and on their borders. The hybrid zone is very narrow; the average gene frequencies (eight loci) drop from 0.88 to 0.2 in about 1400 m, and the width of the zone of overlap between parental types is about 300 m. Gene flow is low, but apparently symmetrical. The two interacting groups appear to have achieved the status of biological species, with strong genetic and morphological, and moderate ecological differentiation. The groups are differently adapted, and there appears to be selection against hybrids. Nevertheless, the groups are linked by a series of intermediate populations around the north end of the Central Valley. This ring species displays several stages of speciation in what appears to be a continuous process of gradual allopatric, adaptive divergence.

ACKNOWLEDGMENTS

The research reported in this chapter has required substantial field work and we are grateful to many colleagues for their assistance in installing and monitoring the pitfall traps over a multiyear period. We especially thank Theodore J. Papenfuss, Robert Macey, Nancy Staub, John Cadle, Marvalee Wake, Thomas A. Wake, Aaron Bauer, David Good, David Darda, Kevin de Queiroz, and Stan Sessions. Chuck Brown has assisted in many ways, and has generously offered his advice and contributed specimens and information. We appreciate comments on the manuscript by James Patton, Marvalee Wake, David Good, Montgomery Slatkin, John Endler, and an anonymous reviewer. This research has been supported by the National Science Foundation (current Grant BSR 8619630) and the Museum of Vertebrate Zoology.

LITERATURE CITED

Barton, N., and B. O. Bengtsson. 1986. The barrier to genetic exchange between hybridising populations. Heredity 56:357–376.

Barton, N. H., and G. M. Hewitt. 1981. Hybrid zones and speciation. Pp. 111–145 in: W. R. Atchley, and D. Woodruff (eds.), *Evolution and Speciation, Essays in Honor of M. J. D. White*. Cambridge University Press, Cambridge.

Barton, N. H., and G. M. Hewitt. 1983. Hybrid zones as barriers to gene flow. Pp. 341–359 in: G. S. Oxford, and D. Rollinson (eds.), *Protein Polymorphism: Adaptive and Taxonomic Significance*. Blackwells, Oxford.

Barton, N. H., and G. M. Hewitt. 1985. Analysis of hybrid zones. Annu. Rev. Ecol. System. 16:113–148.

Brown, C. W. 1974. Hybridization among the subspecies of the plethodontid salamander *Ensatina eschscholtzi*. Univ. Calif. Pub. Zool. 94:1–57, Pl. 1–4.

Endler, J. A. 1977. *Geographic Variation, Speciation, and Clines*. Princeton University Press, Princeton.

Good, D. A., G. Z. Wurst, and D. B. Wake. 1987. Patterns of geographic variation in allozymes of the Olympic Salamander, *Rhyacotriton olympicus* (Caudata: Dicamptodontidae). Fieldiana: Zool., N.S. 32:1–15.

Hairston, N. G. 1951. Interspecies competition and its probable influence upon the vertical distribution of Appalachian salamanders of the genus *Plethodon*. Ecology 32:266–274.

Hairston, N. G. 1988. *Community Ecology and Salamander Guilds*. Cambridge University Press, Cambridge.

Hubbard, M. E. 1903. Correlated protective devices in some California salamanders. Univ. Calif. Pub. Zool. 1:157–170, pl. 16.

Larson, A. 1984. Neontological inferences of evolutionary pattern and process in the salamander family Plethodontidae. Pp. 119–217 in: M. K. Hecht, B. Wallace, and G. T. Prance (eds.), *Evolutionary Biology*, Vol. 17 Plenum, New York.

Mayr, E. 1942. *Systematics and the Origin of Species*. Columbia University Press, New York.

Mayr, E. 1963. *Animal Species and Evolution*. Belknap Press of Harvard University Press, Cambridge.

Mayr, E. 1970. *Populations, Species and Evolution*. Belknap Press of Harvard University Press, Cambridge.

Mayr, E. 1982. Processes of speciation in animals. Pp. 1–19 in: C. Barigozzi (ed.), *Mechanisms of Speciation*. Alan R. Liss, New York.

Riemer, W. 1958. Variation and systematic relationships within the salamander genus *Taricha*. Univ. Calif. Pub. Zool. 56:301–390.

Stebbins, R. C. 1949. Speciation in salamanders of the plethodontid genus *Ensatina*. Univ. Calif. Pub. Zool. 48:377–526, Pl. 11–16.

Stebbins, R. C. 1954. Natural history of the salamanders of the plethodontid genus *Ensatina*. Univ. Calif. Pub. Zool. 54:47–124.

Templeton, A. 1981. Mechanisms of speciation—a population genetic approach. Annu. Rev. Ecol. System. 12:23–48.

Wake, D. B., and J. F. Lynch. 1976. The distribution, ecology, and evolutionary history of plethodontid salamanders in tropical America. Sci. Bull. Nat. Hist. Mus. Los Angeles Co. 25:1–65.

Wake, D. B., and K. P. Yanev. 1986. Geographic variation in allozymes in a "ring species," the plethodontid salamander *Ensatina eschscholtzii* of western North America. Evolution 40:702–715.

Wake, D. B., K. P. Yanev, and C. W. Brown. 1986. Intraspecific sympatry in a "ring species," the plethodontid salamander *Ensatina eschscholtzii* in southern California. Evolution 40:866–868.

Woodruff, D. S. 1981. Toward a genodynamics of hybrid zones: Studies of Australian frogs and West Indian land snails. Pp. 171–197 in: W. R. Atchley and D. S. Woodruff (eds.), *Evolution and Speciation, Essays in Honor of M. J. D. White*. Cambridge University Press, Cambridge.

CHAPTER 7

REINFORCEMENT OF PREMATING ISOLATION

Roger Butlin

INTRODUCTION

The mechanism of speciation, now known as speciation by reinforcement, was proposed in its present form by Dobzhansky (1940) following a suggestion by Fisher (1930) and ultimately deriving from Alfred Russel Wallace's ideas. The term itself was first used by Blair (1955) in relation to a hybrid zone in the frog, *Gastrophryne*. In this chapter I will examine, briefly, some of the theoretical difficulties with reinforcement and some of the widely quoted examples. I will then describe progress to date in attempts to test for reinforcement in a grasshopper hybrid zone.

Perhaps the best way to introduce the process is with a quote by Dobzhansky (1951):

> Assume that incipient species, A and B, are in contact in a certain territory. Mutations arise in either or in both species which make their carriers less likely to mate with the other species. The non-mutant individuals of A which cross to B will produce a progeny which is adaptively inferior to the pure species. Since the mutants breed only or mostly within the species, their progeny will be adaptively superior to that of the non-mutants. Consequently, natural selection will favour the spread and establishment of the mutant condition.

Dobzhansky envisaged that this mechanism would "strengthen . . . these [isolating] mechanisms until the possibility of gene exchange . . . is severely limited or stopped."

This description clearly implies a progression from a pair of populations between which gene flow is possible, although it is restricted by selection against hybrids, to a pair of genetically isolated populations. Blair (1955) envisaged a similar progression when he coined the term "reinforcement." However, the laboratory example used by Dobzhansky to support his argument involved an increase in assortative mating between *Drosophila pseudoobscura* and *D. persimilis* (Koopman 1950). This experiment simulated a qualitatively different situation because no gene flow was possible at any stage. Selection cannot be said to have increased reproductive isolation between the populations since it was complete at the outset. If a species is defined as "a group of natural populations that is reproductively isolated from other such groups, but in which the component populations are not reproductively isolated from one another" (Bigelow 1965), and reproductive isolation is considered in terms of gene flow (Bigelow 1965; Barton and Hewitt 1985), then it is apparent that reinforcement is a process of speciation, but that Koopman's experiment is an interaction between species. Confusion of these two evolutionary scenarios hampers understanding of the evolution of mate recognition systems and the development of premating isolation. I have suggested (Butlin 1985, 1987a, 1987b) that the term "reinforcement" should be reserved for the evolution of increased isolation resulting from selection against hybrids and that interactions between species causing divergence in mate recognition systems should be called "reproductive character displacement." This phrase is appropriate by analogy with "ecological character displacement" due to competitive interactions between species (Brown and Wilson 1956; Grant 1975). Character displacement has been used to describe both patterns of geographic variation and the evolutionary origins of the patterns. Here I use reproductive character displacement, like reinforcement, to describe a process. The pattern of greater divergence in mate recognition systems in sympatry than in allopatry may be accounted for by either process (Butlin 1985, 1987a, and see below, "Supposed examples of reinforcement").

Under Paterson's "recognition concept" of species (Paterson 1986) this distinction may be viewed differently (Butlin 1987b; Spencer et al. 1987), although Paterson himself has distinguished "reinforcement within species" (reproductive character displacement) from "speciation by reinforcement" (Paterson 1982). Where species are defined as groups of organisms sharing a common fertilization system, the process that I call reproductive character displacement *is* part of speciation, since two populations that interbreed are considered conspecific, even when no gene exchange results. This is in direct opposition to Bigelow's (1965) conclusion that gene flow, not interbreeding, is the crucial criterion in species definition, and fails to recognize that the significance of species is their evolutionary independence (Butlin 1987b).

The distinction between reinforcement and reproductive character displacement does not depend on the mode of selection against hybrids but only on their total fitness. Selection against hybrids may be due to incompatibility of divergent genomes or to maladaptation of intermediate phenotypes, and may take the form of reduced viability, mating success, or fertility. Whatever the form of selection, if some F_1 individuals are able to survive, reproduce, and produce viable and fertile F_2 or backcross progeny, then continued production of hybrids will generate a wide range of genotypes in which parental combinations of alleles are disrupted by recombination. By contrast if F_1 individuals fail to survive or reproduce then continued interbreeding of the two populations produces no such "hybrid swarm." Reinforcement and reproductive character displacement are not extremes of a continuum since even when F_1 fitness is very low this mixing of genotypes will occur.

PROBLEMS WITH THE REINFORCEMENT MODEL

In principle reinforcement can occur in any of three evolutionary settings: in a tension zone in which populations that have diverged genetically in allopatry make contact and produce unfit hybrids (Barton and Hewitt 1985), in an environmentally-determined cline in which the direction of selection varies geographically (Endler 1977), or in sympatry in which a polymorphism is maintained in a heterogeneous environment (Maynard Smith 1966; Wilson, this volume). I will concentrate on reinforcement in tension zones, which was the situation envisaged by Dobzhansky (1951), but many of the conclusions apply equally in the other settings.

Dobzhansky's (1951) description of reinforcement involved individuals of incipient species A and B choosing mates from among a mixture of A and B individuals in a contact zone. However, this will happen only when the populations' ranges first meet. As soon as some individuals have mated "incorrectly" F_1 individuals will be generated, and then F_2 and backcross individuals as well. A series of coincident clines will be established for the loci that differ between A and B, with widths dependent on the dispersal distance and the selection against heterozygotes (Barton and Hewitt 1985). Only when selection is extremely strong and clines are, therefore, very narrow, will pure A individuals ever meet pure B individuals. Recombination within the tension zone breaks up associations between alleles derived from the parental races, producing a wide variety of genotypes and reducing linkage disequilibrium. This wide variety of genotypes is maintained in the populations of the tension zone by a balance between gene flow into the zone and removal of hybrid genotypes by selection.

There may be circumstances in which more extensive contact between parental types persists over many generations. Some authors distinguish "true hybrid zones" from "zones of overlap with hybridization" (Short

1969; Littlejohn and Watson 1985). True hybrid zones are parapatric interactions in which the parental types are separated by a hybrid swarm and include tension zones in the sense of Barton and Hewitt (1985) as well as zones maintained by bounded hybrid superiority in which hybrids have an adaptive advantage in the intermediate habitats in which they occur (Littlejohn and Watson 1985; Moore and Buchanan 1985). Reinforcement is not expected in zones of the latter type (if they exist; see Hewitt, this volume), since there is no fitness disadvantage to mismating. Zones of overlap with hybridization are parapatric interactions in which both parental types coexist with hybrids. Littlejohn and Watson (1985) consider reinforcement more likely in these conditions than in tension zones. However, the distinction between the two types of interaction is problematic since it rests on the use of the term "hybrid." In practice a zone is characterized as "overlap with hybridization" if a significant proportion of individuals in the zone falls within the range of variation of the parental types for one or a few characters. This ignores the fact that these "parental" individuals may well be "hybrid" with respect to many other loci. The distinction depends mainly on the genetics underlying the particular characters chosen to define parentals and hybrids—if few loci influence these characters then parental phenotypes are expected within a tension zone, but if many loci are involved parental phenotypes will be rare or absent in the center of the zone. True "overlap with hybridization" probably occurs only where hybrids are completely sterile, as in some cases in the *Thomomys bottae* complex (Patton and Smith, this volume).

Environmental patchiness may increase the frequency of contact between parental individuals (Harrison, this volume). The influence of patchiness is likely to depend critically on the scale of the patch size in relation to the dispersal of an organism: if patch size is large then typical tension zones may exist at patch boundaries; if it is small it may have little effect on zone dynamics compared to a uniform environment. At intermediate scales patchiness may result in a broader zone of contact in which some parental combinations of alleles are maintained in patches in which they confer increased fitness (Harrison, this volume). This could increase the probability of reinforcement and is a situation that needs to be examined further.

Where patches are small and isolated, the parapatric interaction may be dominated by the pattern of extinction and recolonization of local populations. Reinforcement could occur in these conditions if colonization of patches by individuals with more divergent mate recognition systems led to production of fewer hybrids and thus a lower chance of extinction. This possibility also deserves further study.

On present evidence, however, the great majority of parapatric interactions in which there is some gene flow between the two forms can be interpreted as "clines maintained by a balance between dispersal and selection," or tension zones (Barton and Hewitt 1985).

In a tension zone, selection cannot act to increase assortative mating between the A and B races. Instead reinforcement must operate to produce an association between genes for characters involved in mate choice and genes causing reduced hybrid fitness such that mate choice leads to the production of fewer unfit hybrid genotypes. Suppose the races differ at a locus *S* such that S_AS_B heterozygotes have reduced survival. Another locus, *M*, affects mating such that M_AM_A and M_BM_B individuals mate assortatively. Selection will favor linkage disequilibrium between the *S* and *M* loci and divergence between races at the *M* locus, although this selection will occur only within the tension zone and new alleles at the *M* locus will be at a disadvantage while rare because of difficulty in finding mates (Moore 1979).

A simulation model devised by Crosby (1970) is helpful in understanding how reinforcement may operate in a tension zone. In this model two races of plants met in a linear habitat. They had fixed differences at eight unlinked loci affecting fertility such that F_1 hybrids had a relative fertility of 0.25. Both populations varied at three unlinked loci affecting the time of commencement of flowering and two unlinked loci affecting the duration of the flowering period. At the start of the simulation the two populations had the same duration of flowering but slightly different dates of onset of flowering. Seed and pollen dispersal were limited and plants could be pollinated only by other plants in flower at the same time.

In the early generations of the simulation a narrow hybrid zone formed at the interface between the two races. However, as time progressed the frequencies of the two parental combinations of fertility alleles increased in the hybrid zone, the overlap between these types increased, and the frequency of hybrids decreased. This came about because an association developed between fertility alleles and flowering time and duration alleles such that individuals with one parental combination of fertility alleles had alleles for early flowering and vice versa. Duration of flowering decreased in both races. There was no direct selection on flowering time or duration, so that the observed changes were a result of selection against hybrids. They resulted in an increase in isolation and so this is a genuine model of reinforcement. Note, however, that some divergence in flowering time may result from differential migration (Stam 1983) and that selection acted to maintain the most fit combinations of fertility alleles; the fate of other characters that may have differed between the races was not considered.

This model provides some support for the possibility of reinforcement. Isolation increased in spite of maximum recombination among fertility loci and between fertility loci and mating loci, whereas other models show that the probability of reinforcement is highest with close linkage (Maynard

Smith 1966; Caisse and Antonovics 1978; Felsenstein 1981). At the same time the model indicates the weakness of reinforcement relative to direct selection pressures on the mate recognition system. The changes described took place over a large number of generations and only a very small selective advantage to longer duration of flowering was enough to prevent divergence completely. This is broadly equivalent to stabilizing selection that is expected to be a pervasive feature of mate recognition systems (Paterson 1982, 1986) and for which there is some evidence (Butlin et al. 1985; Gerhardt 1982). In addition to direct opposition to reinforcement, stabilizing selection may reduce the genetic variation available within each of the interacting populations, whereas in Crosby's model substantial variation was present.

Paterson (1978, 1982) considers that an alternative outcome, elimination of one population or of one set of alleles responsible for heterozygote advantage, is more likely in situations such as that modeled by Crosby. This was prevented in Crosby's simulation by the addition of parental genotype individuals at the ends of the linear habitat that, Paterson believes, biases the results in favor of reinforcement. However, tension zones may be stabilized by a variety of factors, particularly gene flow into a density trough (Barton and Hewitt 1985; Hewitt, this volume) and so this objection has limited force. Gene flow into a zone from populations that do not experience reinforcing selection, but may experience other selection pressures on the relevant characters, is certainly expected to retard reinforcement but, as in Crosby's model and others (e.g., Caisse and Antonovics 1978), it may not prevent it.

In Crosby's simulation eight loci generate a total selection against F_1 hybrids of 75%. This is not consistent with data on the best studied examples of tension zones, such as between the chromosomal races of the grasshopper *Podisma pedestris* (Barton and Hewitt 1981) and the toads *Bombina bombina* and *B. variagata* (Szymura and Barton 1986), the total selection against hybrids is strong (~0.5) but the cline widths and dispersal estimates suggest that many genes (~150 in *Podisma*, ~300 in *Bombina* are involved and that each individual locus experiences only weak selection (<1%). Reinforcement is less likely in these circumstances because within a hybrid zone population most individuals have similar proportions of alleles from the two parental races and so the potential benefits of assortative mating are reduced. The indirect selection operating on mating loci will be very weak. Note that even when the total selection against hybrids is very strong, if it is dispersed over many loci the tension zone will be wide and parental genotypes will not come into contact. This supports the argument that reinforcement and reproductive character displacement do not form a continuum.

Conversely, as the number of loci responsible for hybrid disadvantage decreases, the tension zone becomes narrower (for the same total selection against hybrids) and other factors operate more strongly against reinforcement. It would require a fortuitous close linkage between the rare selected loci and loci capable of producing significant assortment, probably also rare. In addition, selection for reinforcement occurs only within the zone. Cline widths would be narrow so that the available genetic variation would be limited. Immigration would be important and would be from areas in which the mating system experiences different selection pressures. New assortative mating alleles must appear in the small zonal populations to be available for reinforcement.

The generality of Crosby's result is also limited by the use of flowering time as the parameter influencing mating pattern and by the genetics of this character. Flowering time is undoubtedly an important part of the mate recognition system of many plants, but it is atypical of mate recognition systems in general because the single character influences both male and female components. The mate recognition system, in animals at least, more often consists of a sequence of signals and responses between sexes, with the male and female components under separate genetic control—although exceptions certainly exist, such as the direction of coiling in the shell of the snail, *Partula* (Johnson 1982), and possibly some acoustic signaling systems (Doherty and Hoy 1985). Involvement of two or more genetically distinct characters in the mate recognition system is likely to make reinforcement more difficult.

The numbers of genes controlling variation in mate recognition systems is largely unknown. Many loci are implicated in some cases (such as *Drosophila paulistorum*, Ehrman 1961), but in others apparently very few loci are involved (for example in *Ostrinia nubilalis*, Roelofs et al. 1987). In models in which single loci produce assortative mating (Caisse and Antonovics 1978; Felsenstein 1981), reinforcement is more likely if the level of assortment is high. This may be feasible for a case such as coil direction in *Partula*, but it does not address a situation such as that in *Ostrinia*, in which one or a few loci affects each of the male and female characters.

Where many genes are involved, quantitative genetic models may be more appropriate (Lande 1981, 1982). These suggest that the mate recognition system may be inherently labile due to the interaction between male and female components. However, these models assume stabilizing viability selection on the male character and no direct selection on the female preference, neither of which may be typical of mate recognition signals. It could also be argued that the models do not represent reinforcement because the divergence is not due to selection against hybrids but is initiated by adaptation of the male character to alternate environments and accentuated by directional sexual selection. Any isolation produced could not be said to be the evolutionary function of the divergence.

Nevertheless, the central point remains that male signal–female response systems may have the potential to evolve rapidly.

These theoretical considerations do not preclude the possibility of reinforcement, but they do indicate that it requires conditions that are likely to be very rare, if they occur at all. Reinforcement is favored by strong selection against hybrids, the availability of genetic variation in characters capable of producing substantial assortative mating, and close linkage between mating and selected loci. It is opposed by stabilizing selection on the mate recognition system, gene flow into the tension zone, recombination, and lack of suitable genetic variation. However, the weight of these theoretical conclusions is restricted by our limited understanding of tension zones and, particularly, of the genetics and evolution of mate recognition systems. In this light it is necessary to examine supposed examples of reinforcement with great care. Since theory suggests that reinforcement is improbable, the evidence must be very strong before a putative example can be accepted.

Before considering some of these cases the distinction between reinforcement and reproductive character displacement must be reemphasized. Much of the difficulty with reinforcement stems from recombination between genes from the two populations, either among genes influencing hybrid fitness or between these genes and loci influencing mating pattern. In the absence of gene flow between the populations (for example, when hybrids are completely sterile), these difficulties are removed and it becomes much easier to envisage divergence in mate recognition systems. However, the selection pressure here cannot be described as selection for isolation. The interaction is between, rather than within, species and it is reproductive character displacement, not reinforcement. The major difficulty here is the ecological question of whether the two species can coexist in sympatry or parapatry during the evolution of divergent recognition systems (Paterson 1978; Spencer et al. 1986, 1987). This is not relevant to the question of reinforcement in which coexistence of the interacting genotypes is maintained by the dynamics of the tension zone, geographically varying selection, or selection in a heterogeneous environment.

SUPPOSED EXAMPLES OF REINFORCEMENT

The central prediction of the reinforcement model is that where there is selection against hybrids, the mating pattern will evolve in a way that reduces the production of unfit genotypes. There has been no published attempt to test this directly; instead research has concentrated on patterns of geographic variation in male signal characters, comparisons of assortative mating between allopatric and sympatric pairs of populations, and measurements of assortative mating within clinal populations. Some of the most widely quoted examples are listed in Table 1. For these approaches to provide convincing evidence of reinforcement they must demonstrate (1) that gene flow occurs between the taxa, or did occur when they originally met, (2) that components of the mate recognition system have diverged in

TABLE 1. Proposed examples of reinforcement.

Species	Context	Type of evidence	Reference
Agrostis tenuis *Anthoxanthum odoratum*	Environmental cline	Inverse clines in flowering time, some evidence of genetic determination	McNeilly and Antonovics (1968)
Bufo woodhouseii and *B. americanus*	Range overlap	Decline in frequency of hybrids	Jones (1973)
Chauliognathus pennsylvanicus	Environmental cline	Assortative mating between phenotypes in cline but not between populations from outside the cline	McLain (1985, 1986)
Drosophila mojavensis and *D. arizonensis*	Range overlap	Greater premating isolation between sympatric than allopatric populations in laboratory tests	Wasserman and Koepfer (1977); Koepfer (1987)
D. paulistorum races	Complex of overlapping ranges	"	Dobzhansky et al. (1969, 1976)
Gastrophyrne olivacea and *G. carolinensis*	Tension zone	Divergence in male signal characters in area of hybridization	Blair (1955)
Icterus galbula	Tension zone	Bimodal phenotype distribution within zone	Corbin and Sibley (1977)
Litoria ewingi and *L. paraewingi*	Tension zone	Inverse cline in male signal characters	Littlejohn and Watson (1983)
L. ewingi and *L. verreauxi*	Range overlap	Divergence in male signal characters in broad area of sympatry	Littlejohn (1965)
Partula suturalis	Complex of overlapping ranges	Shell coil varies to be opposite to sympatric *Partula* species; coil types mate assortatively	Murray and Clarke (1980); Johnson (1982); Johnson et al. (1987)
Pinus muricata	Tension zone	Divergence in flowering time between populations	Millar (1983)

the area of contact and in the time since contact was established, (3) that this divergence is sufficient to alter the pattern of mating in a way that decreases the frequency of production of unfit hybrid genotypes, and, ideally, (4) that the divergence is not a result of other selection pressures on the mate recognition system.

Several of the examples involve greater premating isolation or greater signal divergence in areas where species' ranges overlap without exchanging genes (*Drosophila paulistorum*, *D. mojavensis*, and *D. arizonensis*; *Litoria ewingi* and *L. verreauxi*; and see Coyne and Orr, this volume). In cases such as these neither the discovery of occasional putative hybrids nor the viability and fertility of laboratory hybrids is sufficient evidence for past or present gene flow in the field. Reproductive character displacement cannot be excluded and, in view of the arguments outlined, is the more likely course of events. Another possibility is that variation in mate recognition systems predated contact, and sympatry is only possible where there is premating isolation. These historical alternatives are very hard to distinguish. In the case of Coyne's comparative data on *Drosophila*, and the *D. mojavenis/D. arizonensis* example in particular, there is the additional problem that the laboratory hybrid fitnesses are known to be high, whereas reinforcement is expected only where hybrid fitness is low. This makes the explanation of prior allopatric divergence more plausible than reinforcement. The reduction in occurrence of hybrids in *Bufo*, if it is real (Loftus-Hills 1975), may also be a case of reproductive character displacement as there is no evidence that the hybrids are fertile.

In the case of coil direction in *Partula*, gene flow does occur between the species in some localities, favoring the possibility of reinforcement. Coil direction is known to have a strong effect on mating pattern, but the fitness of hybrids in unclear. *Partula* is probably the most convincing example currently available but the present situation could have arisen from fixation of alternative coil directions in allopatric populations of *Partula suturalis* followed by increases in range to the present distributions and overlap with other species only where coil direction differs. It is also important to note that the assortative mating produced by differences in the direction of coiling is ineffective as a barrier to gene flow within *P. suturalis* (Johnson et al. 1987). The steep cline in coil type is stable rather than evolving toward speciation, and it is unlikely that the dextral form will spread away from the areas of sympatry with sinistral species.

The remaining examples all involve tension zones or environmentally determined clines and there is, therefore, no question that some gene flow occurs. Two classes of problems are common in these studies: inadequate mapping of transitions, and lack of information on the functions of signal characteristics. Where an inverse cline is to be used as evidence for reinforcement it is essential to have adequate mapping to exclude the possibility that an oblique transect has produced the appearance of greater divergence in close populations. Mapping in two dimensions rather than isolated transects (e.g., *Agrostis, Anthoxanthum, Chauliognathus, Litoria ewingi/L. paraewingi*) is needed. For some characters, such as flowering time, adaptation to local environments must be excluded and this may also be achieved by adequate mapping.

If patterns of variation in male signal characteristics are to be used as evidence for reinforcement, it is important to understand the functions of the specific parameters involved. Ideally, female preferences should also be studied in the tension zone or cline. For example, the significance of the weak inverse cline in number of notes per call and number of pulses per note in the *L. ewingi/L. paraewingi* tension zone is unknown because female responses to these characters have not been tested. They are not characters normally associated with species discrimination in anurans (Watson et al. 1971). If female rejection of heterospecifics, as opposed to recognition of conspecific males occurs (as in *Conocephalus nigropleurum*, Gwynne and Morris 1986), studies of female behavior will be essential to demonstrate reinforcement.

An additional form of evidence is the observation of mating pairs in natural populations. Rising (1983) failed to find evidence of assortative mating in pairs of orioles (*Icterus*). It seems likely the bimodal distribution of phenotypes observed by Corbin and Sibley (1977) in the oriole hybrid zone was due to patterns of colonization of disturbed habitats rather than to reinforcement. Similarly a large sample of mating pairs from a hybrid zone in the northern flicker (*Colaptes auratus)* showed no assortative mating (Moore 1987). McLain (1985) has found strong evidence for assortative mating in a soldier beetle cline that could be a result of reinforcement. However, much more background information is needed in this case—for example, only a single transect has been reported, the selection pressures are undefined, and the cline is very wide suggesting only weak selection. Current work by Kochmer (personal communication) may help to elucidate this situation.

REINFORCEMENT IN THE *CHORTHIPPUS PARALLELUS* TENSION ZONE

Against this background I will describe briefly my studies on reinforcement in the meadow grasshopper, *Chorthippus parallelus*, carried out in collaboration with Godfrey Hewitt and Michael Ritchie. Our intention was to look for evidence of reinforcement in male signal characters and female preferences in a well-described tension zone.

C. parallelus has two subspecies in western Europe: *C. p. parallelus*, whose range extends from southern France to Scotland and east into the

USSR, and *C. p. erythropus*, which is restricted to the Iberian Peninsula (Reynolds 1980). The subspecies probably diverged during the Pleistocene when the Pyrenees were glaciated and the grasshoppers were confined to refugia in southern Europe. They now differ in a wide range of characters: morphological, electrophoretic, behavioral and chromosomal (Butlin and Hewitt 1985a, 1985b; Hewitt et al. 1988). In the Pyrenees the grasshoppers are restricted to moist meadow habitats below about 2000 m and so the two subspecies now meet and hybridize in a series of cols. The hybrid zone in two of these cols has been mapped in detail using the most distinctive morphological character, the number of stridulatory pegs. In each case the zone is positioned close to the narrowest point of the grasshopper's range, as would be expected for a tension zone. As these localities are near the altitude limit of the species, the zones may also be located in density troughs (Barton and Hewitt 1985). For our main transect at Mont Louis in the eastern Pyrenees (Figure 1) some of the data on peg number are given in Figure 2.

Selection against hybrids takes the form of sterility of F_1 males and reduced fertility of backcross males (Hewitt et al. 1987b). Testes of F_1 males are greatly reduced in size and rarely contain any sperm, whereas backcross males have testes of intermediate size and do produce some sperm. The widths of the clines for loci influencing sterility have not yet been measured, but we would expect them to be much less than the width of the cline in peg number. The dispersal distance of *C. parallelus* is on the order of 20 m generation$^{-1/2}$, so that the cline width of 4 km for peg number at Mont Louis suggests selection of about 0.02% or 14,500 generations of neutral introgression (longer than the time since the melting of the last Pyrenean ice cap; see Hewitt, this volume).

A major component of the mate recognition system in grasshoppers is the male stridulation, or song, produced by rubbing the stridulatory pegs on the insides of the hind femora across specialized wing veins (Perdeck 1958; Otte 1977; von Helversen and von Helversen 1975). *C. parallelus* produces two principal song types, designated "calling" and "courtship" song. Calling song is made by isolated males and is generally supposed to attract and stimulate females, although there is only weak phonotaxis by *C. parallelus* females (Butlin and Hewitt 1986). Courtship song is usually produced by males in close proximity to females.

The two subspecies differ in several features of their songs (Figure 3). *C. p. erythropus* produce much less calling song. Individual echemes (discrete sequences of syllables about 2 seconds in duration) have similar structures in the two subspecies but *C. p. erythropus* produces longer syllables and longer intervals between echemes. The courtship song of *C. p. parallelus* is a short, unstructured series of sound pulses, whereas *C. p. erythropus* courtship song may be much longer and has a distinct

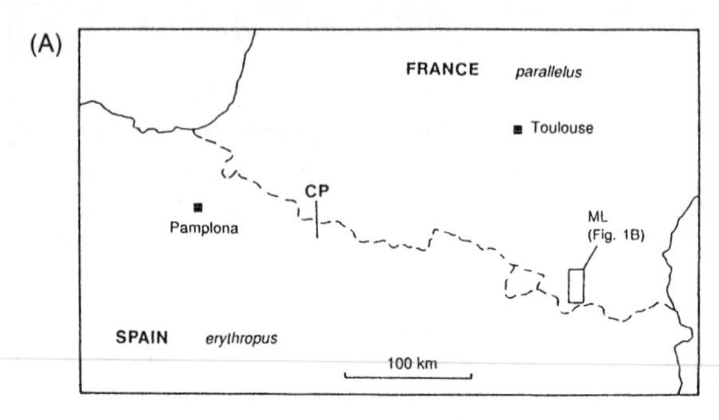

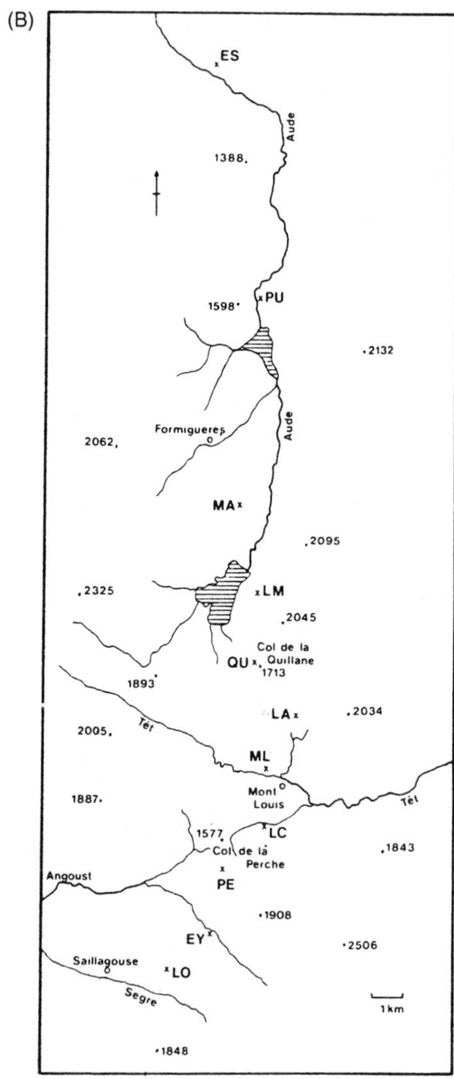

FIGURE 1. Location of the *Chorthippus parallelus* hybrid zone. (A) The zone runs close to the French–Spanish border and has been studied in detail at the Col du Portalet (CP) and near Mont Louis (ML). (B) At Mont Louis the zone center is close to the Col de la Quillane. Another contact exists to the east, in the Tet valley.

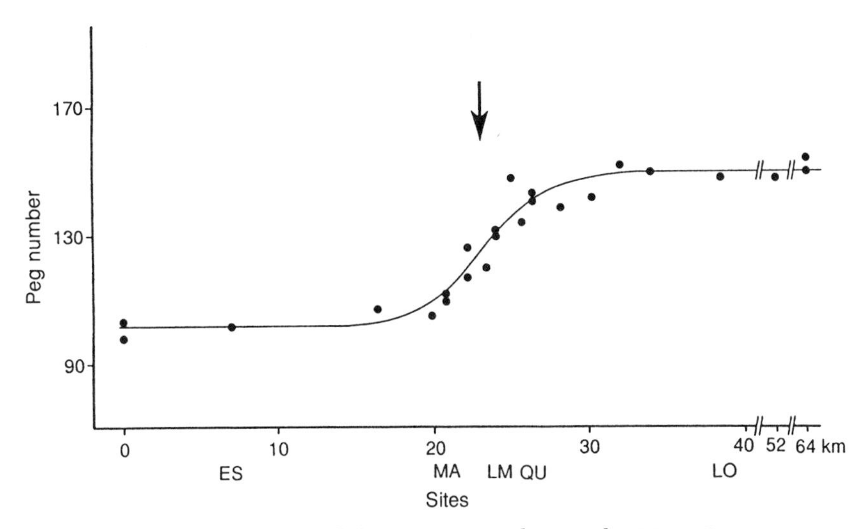

FIGURE 2. The cline in stridulatory peg number in the Mont Louis transect. Points represent samples of 7 to 35 individuals. The line is the best-fitting tanh curve. The estimated cline center is indicated by the arrow and the estimated cline width is 4.09 ± 0.56 km.

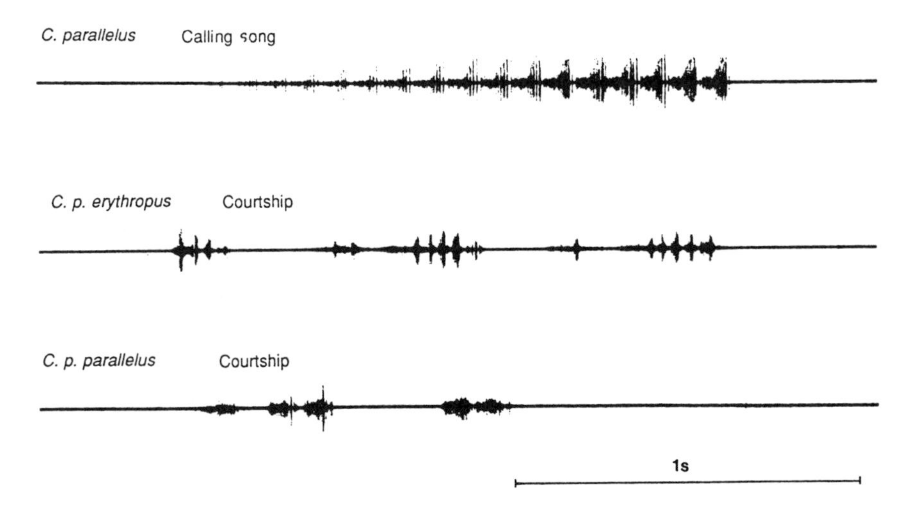

FIGURE 3. Oscillograms of *Chorthippus parallelus* songs. See text for further explanation.

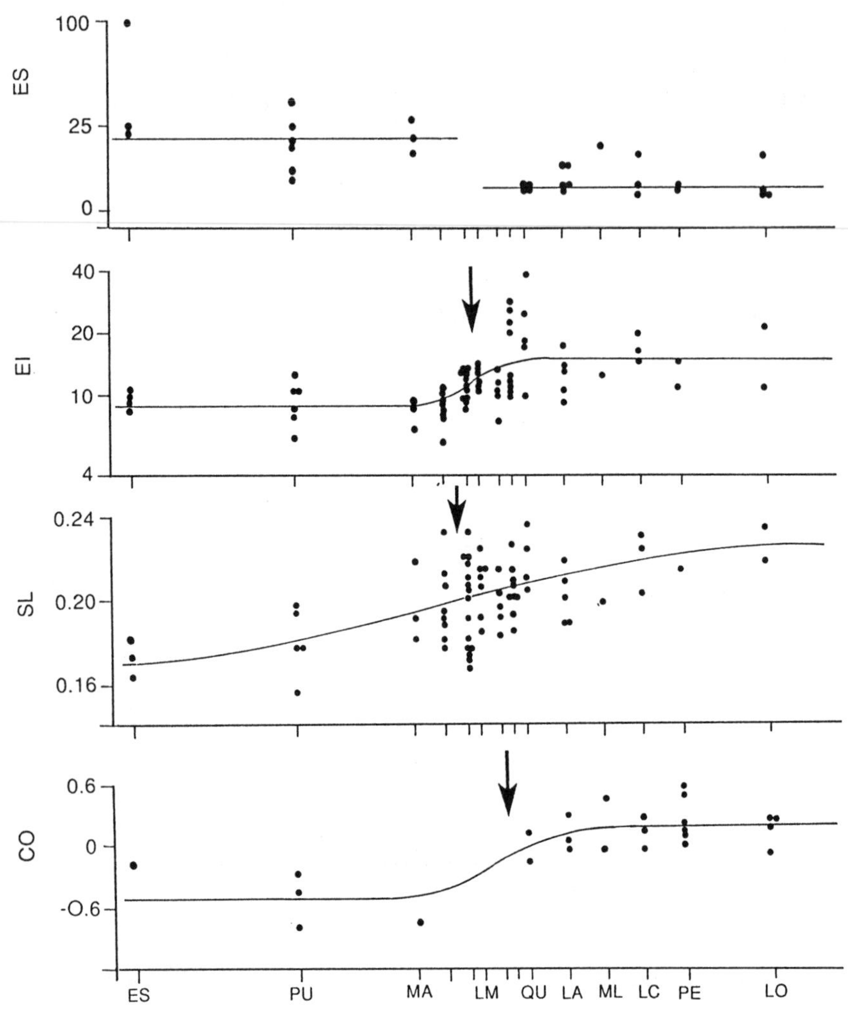

FIGURE 4. Clines in song characters in the Mont Louis transect. Points are individual mean values and lines are the best fitting tanh curves. ES, echemes per sequence (square root scale): insufficient data are available to estimate cline width or position. EI, echeme interval in seconds (log scale): cline width 1.41 ± 0.65 km. SL, syllable length in seconds: cline width 19.5 ± 32.9 km. CO, courtship index (see Butlin and Hewitt 1985b): cline width 2.08 ± 2.28 km. Arrows indicate cline centers.

periodicity at about 15 pulses/second. The pattern of change in these characters in one or our transects is shown in Figure 4. All of the characters for which we have sufficient data appear to change in a simple clinal pattern with no indication of greater divergence in the area of hybridization.

The cline widths estimated from these limited data are variable, but even the steepest, for echeme interval, suggests only weak selection (~0.16%) on male song genes. The numbers of genes involved and the total selection pressures are unknown. Preliminary data indicate a similar pattern in our other transect. It is difficult with this type of data to exclude the possibility of an inverse cline on some scale, either narrow or wide relative to our sampling interval. However, there is nothing in these data to suggest that such a pattern exists.

Looking for inverse clines in male signal characters has been the most common test for reinforcement. However, the absence of an inverse cline in male signal characters is only weak evidence against reinforcement for several reasons: reinforcement may produce only a transient inverse cline, relevant components of the mate recognition system may not have been studied, and female preferences may have changed in mean or variance without an associated change in male characters. Our next step, therefore, has been to measure female preferences in the Mont Louis transect. Female preferences could be studied in three ways: (1) by measuring assortative mating within populations, (2) by measuring assortative mating between pairs of populations at different distances on either side of the contact zone, or (3) by measuring preferences of females from transect populations against external standards. Approaches (1) and (2) have some advantages and have been attempted by Michael Ritchie (1988), but I will describe here results from approach (3), which is the only method that does not confound male variation in either mating "vigor" or components of the mate recognition system, with variation in female preference.

Females from each of a series of populations from the Mont Louis transect were allowed to mate with males from reference populations outside the transect. This was intended to assess female preference for characteristics of males of the two subspecies rather than males of their own versus another population. As males from the same reference populations were used with females from several transect populations, male vigor effects could be eliminated. Full experimental details will be published elsewhere (Ritchie, Butlin and Hewitt, in preparation). Briefly, in each trial two females from a test population were enclosed in a mating arena with one *C. p. parallelus* reference population male and one *C. p. erythropus* reference population male. Only the first mating in such a trial was scored. Only matings by virgin females will be considered here because there is evidence that females are less selective in subsequent matings (Ritchie 1988). The experiment was carried out in two parts. In the first part the reference populations were from the western Pyrenees but, unfortunately, these populations were not available for the second part. Reference populations from the eastern Pyrenees, outside the female transect, were used. Two female populations were included in both parts and these showed no significant differences between the two parts ($G_1 = 0.90$). The two sets of trials will, therefore, be considered together.

There was significant variation among populations in the proportion of females mating with the *C. p. parallelus* male ($G_8 = 17.4, p < 0.05$). The pattern of variation along the transect was surprising (Figure 5). There is a suggestion of an inverse cline centered about the middle of the morphological cline, with little preference expressed by the extreme populations. This pattern should be interpreted with care because the appearance of an inverse cline relies primarily on the data for just two populations: Escouloubre and Matemale. The Escouloubre result is surprising since mass mating tests with pairs of populations remote from the hybrid zone show significant assortative mating (Ritchie 1988). Nevertheless the pattern may represent evidence for reinforcement. On present data this interpretation is only one of several alternatives; for example, the observed variability of female preference may be typical of variation on this scale throughout the species range, the correspondence with the zone being fortuitous, or it may reflect the ecologically marginal nature of zonal populations, or there may be other selection pressures operating on female preference.

Accepting for the present that the pattern of female preference is a result of reinforcement, one can ask whether this is consistent with the other information available about this hybrid zone. An inverse cline produced by reinforcement should have a width approximately equal to the widths of the clines for loci responsible for hybrid disadvantage, in this case F_1 male sterility. If selection on these loci is strong enough for reinforcement to be a realistic possibility then the cline width should be on the order

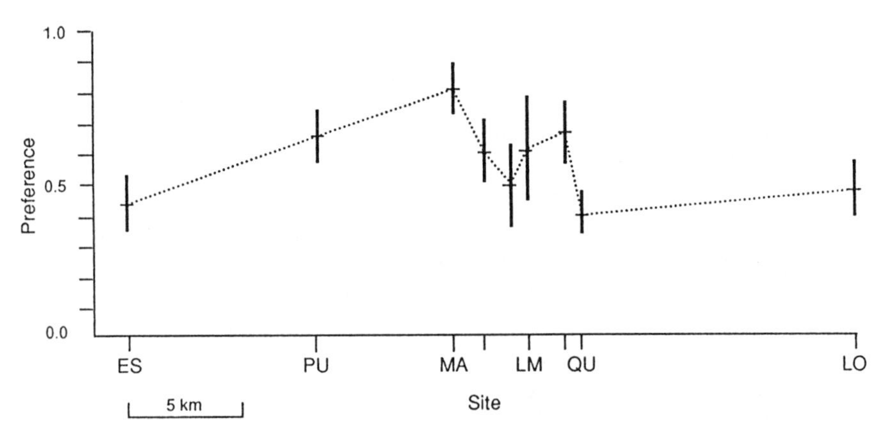

FIGURE 5. Female preference in the Mont Louis transect expressed as the proportion of virgin females mating with a *C. p. parallelus* reference male (± standard error).

of 250 to 800 m. (With dispersal of 20 m generation$^{-1/2}$ and total selection of 0.5, this corresponds to 10–100 loci contributing to hybrid sterility; Barton and Hewitt 1981.) The greatest selection for assortative mating would be on the shoulders of such a zone because linkage disequilibrium among sterility loci, and between sterility loci and mating loci, declines toward the center of the zone, whereas the frequency of "foreign" alleles declines toward the edges. The apparent inverse cline in female mating preference has a width between peaks of 5 and 10 km. A cline of this width can be accounted for by reinforcement only if the extra divergence initially produced on the shoulders of the sterility cline has spread back into the parental populations. However, this would imply a region of assortative mating between Matemale and Quillane, which would produce bimodal distributions of morphological and song characters. Such bimodality has not been observed. There should also be less sterility in the field than would be expected from random mating. This cannot be tested until the clines for sterility alleles have been accurately mapped.

Clearly the pattern of female preferences needs to be studied further. A finer scale transect is needed, spanning the clines for sterility alleles, and a similar transect in the western Pyrenees should also be examined. The technique for studying variation in female preference has considerable potential and hopefully it will be applied to other proposed cases of reinforcement. It could usefully be applied in other circumstances too, because the pattern of geographic variation in female preference within species is largely unknown.

At present the *C. parallelus* tension zone has produced no unequivocal evidence that reinforcement is occurring or has occurred. This is in spite of the considerable amount of background information available and the clear fit of this situation to the prerequisites of the reinforcement model. It is still necessary to qualify this conclusion because the predictions of the reinforcement model that are generally tested, particularly the existence of inverse clines in male signal characters, are indirect and may not even be necessary consequences of the process. Probably the most direct prediction of the model is that within a population in which reinforcement has occurred, mate choice will result in offspring of higher fitness on average than forced random mating. I know of no published attempt to test this prediction, but an appropriate experiment on *C. parallelus* is nearing completion and preliminary results do not support the reinforcement model (M. G. Ritchie, personal communication).

CONCLUSIONS

To date, studies on natural populations have produced no convincing example of reinforcement, and our study of *C. parallelus* has not altered this situation. Certainly it is difficult to provide evidence that clearly distinguishes reinforcement from reproductive character displacement or divergence in mate recognition systems in allopatry. It must remain possible, therefore, that the lack of examples is due to the limitations of our experiments rather than to the rarity of the phenomenon. It is also important to bear in mind that a process such as reinforcement can be rare in our experience and yet be of great importance in evolution. Nevertheless, theoretical considerations of reinforcement tend to confirm the impression that it is unlikely to be a major mode of evolution. Gene flow, recombination, and stabilizing selection on mate recognition systems combine to oppose the weak indirect selection due to production of unfit hybrid progeny.

Laboratory simulations, such as Koopman's (1950) experiments, quoted by Dobzhansky (1951), have almost always modeled reproductive character displacement rather than reinforcement (e.g., Ehrman 1973; Crossley 1974). One exception to this is the disruptive selection experiment of Thoday and Gibson (1962) that did produce a marked increase in premating isolation in spite of the opportunity for maximum gene flow. This experiment has been very difficult to repeat (Spiess and Wilke 1984) but it remains an important demonstration of the potential for selection against "hybrids" to increase isolation.

In spite of theoretical arguments against reinforcement and the paucity of examples, it remains a popular idea, at least in part because it gives a role to natural selection in the process of speciation. Some authors continue to refer to it as a likely consequence of selection against hybrids (e.g., Kat 1985; Heaney and Tim 1985; Howard 1986; Moore and Buchanan 1986). I think reinforcement should be treated with more caution. We need to improve our poor understanding of the genetic basis and evolution of mate recognition systems in general before we can expect to understand their behavior in the complex context of tension zones.

SUMMARY

The reinforcement model of speciation was promoted by Dobzhansky and has since been widely supported. However, there are several theoretical difficulties with the model. In the context of tension zones the probability of reinforcement is reduced by (1) recombination among genes influencing hybrid fitness, and between these genes and genes for assortative mating, (2) gene flow from outside the tension zone, and (3) stabilizing selection on the mate recognition system. There are no unequivocal examples of reinforcement in natural populations and there is little experimental support.

A tension zone in the grasshopper, *Chorthippus parallelus*, appears to satisfy all the conditions for the reinforcement model. Nevertheless, investigations of male signal characters and female mating preferences in this zone have failed to provide convincing evidence for reinforcement.

LITERATURE CITED

Barton, N. H., and G. M. Hewitt. 1981. A chromosomal cline in the grasshopper, *Podisma pedestris*. Evolution 35:1008–1018.

Barton, N. H., and G. M. Hewitt. 1985. Analysis of hybrid zones. Annu. Rev. Ecol. System. 16:113–148.

Bigelow, R. S. 1965. Hybrid zones and reproductive isolation. Evolution 19:449–458.

Blair, W. F. 1955. Mating call and stage of speciation in the *Microhyla olivacea-M. carolinensis* complex. Evolution 9:469–480.

Brown, W. L., and E. O. Wilson. 1956. Character displacement. System. Zool. 5:49–64.

Butlin, R. K. 1985. Speciation by reinforcement. Pp. 84–113 in: J. Gosalvez, C. Lopez Fernandez, and C. Garcia de la Vega (eds.), *Orthoptera* 1. Fundación Ramón Areces, Madrid.

Butlin, R. K. 1987a. Speciation by reinforcement. Trends Ecol. Evol. 2:8–13.

Butlin, R. K. 1987b. Species, speciation and reinforcement. Am. Natur. 130:461–464.

Butlin, R. K., and G. M. Hewitt. 1985a. A hybrid zone between *Chorthippus paralleius parallelus* and *Chorthippus parallelus erythropus* (Orthoptera:Acrididae): Morphological and electrophoretic characters. Biol. J. Linnean Soc. 26:269–285.

Butlin, R. K., and G. M. Hewitt. 1985b. A hybrid zone between *Chorthippus paralleus parallelus* and *Chorthippus parallelus erythropus* (Orthoptera:Acrididae): Behavioural characters. Biol. J. Linnean Soc. 26:287–299.

Butlin, R. K., and G. M. Hewitt. 1986. The response of female grasshoppers to male song. Animal Behav. 34:1896–1899.

Butlin, R. K., G. M. Hewitt, and S. F. Webb. 1985. Sexual selection for intermediate optimum in *Chorthippus brunneus*. (Orthoptera:Acrididae). Animal Behav. 33:1281–1292.

Caisse, M., and J. Antonovics. 1978. Evolution in closely adjacent plant populations. IX. Evolution of reproductive isolation in clinal populations. Heredity 40:371–384.

Corbin, K. W., and C. G. Sibley. 1977. Rapid evolution in orioles in the genus *Icterus*. Condor 79:335–342.

Crosby, J. L. 1970. The evolution of genetic discontinuity: Computer models of the selection of barriers to interbreeding between species. Heredity 25:253–297.

Crossley, S. A. 1974. Changes in mating behaviour produced by selection for ethological isolation between ebony and vestigial mutants of *Drosophila melanogaster*. Evolution 28:631–647.

Dobzhanzky, Th. 1940. Speciation as a stage in evolutionary divergence. Am. Natur. 74: 312–321.

Dobzhansky, Th. 1951. *Genetics and the Origin of Species*, 3rd ed. Columbia University Press, New York.

Dobzhansky, Th., O. Pavlovsky, and L. Ehrman. 1969. Transitional populations of *Drosophila paulistorum*. Evolution 23:482–492.

Dobzhansky, Th., O. Pavlovsky, and J. R. Rowell. 1976. Partially successful attempt to enhance reproductive isolation between semispecies of *Drosophila paulistorum*. Evolution 30:201–212.

Doherty, J. A., and R. R. Hoy. 1985. Communication in insects. III. The auditory behaviour of crickets: Some views of genetic coupling, song recognition, and predator detection. Quart. Rev. Biol. 60:457–472.

Ehrman, L. 1961. The genetics of sexual isolation in *Drosophila paulistorum*. Genetics 46:1025–1038.

Ehrman, L. 1973. More on natural selection for the origin of reproductive isolation. Am. Natur. 107:318–319.

Endler, J. A. 1977. *Geographic Variation, Speciation and Clines*. Princeton University Press, Princeton.
Felsenstein, J. 1981. Skepticism toward Santa Rosalia, or why are there so few kinds of animals? Evolution 35:124–138.
Fisher, R. A. 1930. *The Genetical Theory of Natural Selection*. Oxford University Press, Oxford.
Gerhardt, H. C. 1982. Sound pattern recognition in some North American treefrogs (Anura:Hylidae): Implications for mate choice. Am. Zool. 22:581–595.
Grant, P. R. 1975. The classical case of character displacement. Evol. Biol. 8:237–337.
Gwynne, D. T., and G. K. Morris. 1986. Heterospecific recognition and behavioural isolation in acoustic Orthoptera (Insecta). Evol. Theory 8:33–38.
Heaney, L. R., and R. M. Timm. 1985. Morphology, genetics and ecology of pocket gophers (genus *Geomys*) in a narrow hybrid zone. Biol. J. Linnean Soc. 25:301–317.
Hewitt, G. M., R. K. Butlin, and T. M. East. 1987. Testicular dysfunction in hybrids between parapatric subspecies of the grasshopper, *Chorthippus parallelus*. Biol. J. Linnean Soc. 31:25–34.
Hewitt, G. M., J. Gosalvez, C. Lopez-Fernandez, M. G. Ritchie, W. Nichols, and R. K. Butlin. 1988. Differences in the nucleolar organisers, on sex chromosomes and Haldane's Rule in a hybrid zone. Pp. 109–119, Kew Chromosome Conference III, HMSO, London.
Howard, D. J. 1986. A zone of overlap and hybridization between two ground cricket species. Evolution 40:34–43.
Johnson, M. G. 1982. Polymorphism for direction of coil in *Partula suturalis*: Behavioural isolation and positive frequency dependent selection. Heredity 49:145–151.
Johnson, M. S., J. Murray, and B. Clarke. 1987. Independence of genetic subdivision and variation for coil in *Partula suturalis*. Heredity 58:307–313.
Jones, J. M. 1973. Effects of thirty years hybridization on the toads *Bufo americanus* and *Bufo woodhouseii fowleri* at Bloomington, Indiana. Evolution 27:435–448.
Kat, P. W. 1985. Historical evidence for fluctuations in levels of hybridization. Evolution 39:1164–1169.
Koepfer, H. R. 1987. Selection for sexual isolation between geographic forms of *Drosophila mojavensis*. I. Interactions between the selected forms. Evolution 41:37–48.
Koopman, K. F. 1950. Natural selection for reproductive isolation between *Drosophila pseudoobscura* and *Drosophila persimilis*. Evolution 4:135–148.
Lande, R. 1981. Models of speciation by sexual selection on polygenic traits. Proc. Natl. Acad. Sci. U.S.A. 78:3721–3725.
Lande, R. 1982. Rapid origin of sexual isolation and character divergence in a cline. Evolution 36:213–223.
Littlejohn M. J. 1965. Premating isolation in the *Hyla ewingi* complex (Anura:Hylidae). Evolution 19:234–243.
Littlejohn M. J., and G. F. Watson. 1983. The *Litoria ewingi* complex (Anura:Hylidae) in south eastern Australia. VII. Mating call structure and genetic compatibility across a narrow hybrid zone between *L. ewingi* and *L. paraewingi*. Aust. J. Zool. 31:193–204.
Littlejohn, M. J., and G. F. Watson. 1985. Hybrid zones and homogamy in Australian frogs. Annu. Rev. Ecol. System. 16:85–112.
Loftus-Hills, J. J. 1975. The evidence for reproductive character displacement between the toads *Bufo americanus* and *B. woodhouseii fowleri*. Evolution 29:368–369.
Maynard Smith, J. 1966. Sympatric speciation. Am. Natur. 100:637–650.
Mayr, E. 1963. *Animal Species and Evolution*. Harvard University Press, Cambridge.
McLain, D. K. 1985. Clinal variation in morphology and assortative mating in the soldier beetle *Chauliognathus pennsylvanicus* (Coleoptera:Cantharidae). Biol. J. Linnean Soc. 25:105–117.
McLain D. K. 1986. Niche differentiation and the evolution of ethological isolation in a soldier beetle hybrid zone. Oikos 47:159–167.
McNeilly, T., and J. Antonovics. 1968. Evolution in closely adjacent plant populations. IV. Barriers to gene flow. Heredity 23:205–218.
Millar, C. I. 1983. A steep cline in *Pinus muricata*. Evolution 37:311–319.

Moore, W. S. 1979. A single locus mass action model of assortative mating, with comments on the process of speciation. Heredity 42:173–186.
Moore, W. S. 1987. Random mating in the Northern Flicker hybrid zone: Implications for the evolution of bright and contrasting plumage patterns in birds. Evolution 41:539–546.
Moore, W. S., and D. B. Buchanan. 1985. Stability of the Northern Flicker hybrid zone in historical times: Implications for adaptive speciation theory. Evolution 39:135–151.
Murray, J., and B. Clarke. 1980. The genus *Partula* on Moorea: Speciation in progress. Proc. R. Soc. London, Ser. B 211:83–117.
Otte, D. 1977. Communication in Orthoptera. Pp. 334–361 in: T. A. Sebeok (ed.), *How Animals Communicate*. Indiana University Press, Bloomington.
Paterson, H. E. H. 1978. More evidence against speciation by reinforcement. South African J. Sci. 74:369–371.
Paterson H. E. H. 1982. Perspective on speciation by reinforcement. South African J. Sci. 78:53–57.
Paterson, H. E. H. 1986. The recognition concept of species. Pp. 21–29 in: E. S. Vrba (ed.), *Species and Speciation*. Transvaal Museum, Pretoria.
Perdeck, A. C. 1958. The isolating value of specific song patterns in two sibling species of grasshoppers (*Chorthippus brunneus* Thunb. and *C. biguttulus* L). Behaviour 12:1–75.
Reynolds, W. J. 1980. A reexamination of the characters separating *Chorthippus montanus* and *C. parallelus* (Orthoptera:Acrididae). J. Nat. Hist. 14:283–303.
Rising, J. D. 1983. The progress of oriole hybridization in Kansas. Auk 100:885–897.
Ritchie, M. G. 1988. A Pyrenean hybrid zone in the grasshopper *Chorthippus parallelus* (Orthoptera:Acrididae). Ph.D. Thesis, University of East Anglia, Norwich, U.K.
Roelofs, W., T. Glover, X-H. Tang, I. Spreng, P. Robbins, C. Eckenrode, C. Lofstedt, B. S. Hansson, and B. O. Bengtsson. 1987. Sex pheromone production and perception in European corn borer moths is determined by both autosomal and sex-linked genes. Proc. Natl. Acad. Sci. U.S.A. 84:7585–7589.
Short, L. L. 1969. Taxonomic aspects of avian hybridization. Auk 86:84–105.
Spencer, H. G., B. H. McArdle, and D. M. Lambert. 1986. A theoretical investigation of speciation by reinforcement. Am. Natur. 128:241–262.
Spencer, H. G., D. M. Lambert, and B. H. McArdle. 1987. Reinforcement, species and speciation: A replay to Butlin. Am. Natur. 130:958–962.
Spiess, E. B., and C. M. Wilke. 1984. Still another attempt to achieve assortative mating by disruptive selection in *Drosophila* Evolution 38:505–515.
Stam, P. 1983. The evolution of reproductive isolation in closely adjacent plant populations through differential flowering time. Heredity 50:105–118.
Szymura, J. M., and N. H. Barton. 1986. Genetic analysis of a hybrid zone between the fire-bellied toads, *Bombina bombina* and *B. variegata*, near Cracow in Southern Poland. Evolution 40:1141–1159.
Thoday, J. M., and J. B. Gibson. 1962. Isolation by disruptive selection. Nature (London) 193:1164–1166.
von Helversen, D., and O. von Helversen. 1977. Verhaltensgenetische Untersuchungen am akustischen Kommunikationssystem der Feldheuschrecken. J. Comp. Physiol. 104: 273–323.
Wasserman, M., and H. R. Koepfer. 1977. Character displacement for sexual isolation between *Drosophila mojavensis* and *Drosophila arizonensis*. Evolution 31:812–823.
Watson, G. F., J. J. Loftus-Hills, and M. J. Littlejohn 1971. The *Litoria ewingi* complex (Anura:Hylidae) in south eastern Australia. I. A new species from Victoria. Aust. J. Zool. 19:401–416.

CHAPTER 8

TWO RULES OF SPECIATION

Jerry A. Coyne and H. Allen Orr

The importance of speciation in the modern synthetic theory of evolution is evident from the title of the book that launched this synthesis: Dobzhansky's *Genetics and the Origin of Species* (1937). Yet the past half century has produced only two real advances in our understanding of speciation. The first is Dobzhansky (1937) and Mayr's (1940) biological species concept. By showing that species were reproductively distinct communities, they narrowed the problem of the origin of species to the origin of reproductive isolating mechanisms. The second advance, also due largely to Mayr (1942), is the idea that the evolution of these mechanisms usually requires geographic isolation.

The importance of allopatry is probably the only widely accepted "rule" of speciation. Although this lack of generalization is not surprising in an enterprise so dependent on reconstructing history, it has also produced the view that students of speciation are evolutionary biologists' poor cousins, doomed to eternal speculation about untestable theories. In this chapter we would like to call attention to two frequently overlooked "rules" of speciation—empirical generalizations that are at least as compelling as any other in evolutionary biology. We will describe these rules, show how they are connected, and consider a number of possible genetic and evolutionary explanations. Although the correct explanation is not yet known, we believe it will represent an important advance in speciation theory.

The first generalization, named "Haldane's rule" after its discoverer, has actually been known since 1922 (Haldane 1922:101):

> When in the F_1 offspring of two different animal races one sex is absent, rare, or sterile, that sex is the heterozygous [heterogametic] sex.

Haldane cites many crosses between "races" (usually distinct species) that produce inviability or sterility in only the heterogametic offspring. Since 1922 many more interspecific hybridizations have been studied, but the rule remains intact. Table 1 shows the striking adherence of several groups to this pattern.

It has been suggested that Haldane's rule merely reflects greater male than female sensitivity to hybridization. But this cannot be true because the rule also holds in species having heterogametic *females*, including birds and butterflies (Haldane 1933; Gray 1958). The explanation of the pattern must therefore be connected not with sex, but with sex *chromosomes*. Genetic studies of hybrid sterility and inviability have confirmed this, yielding a second rule about postzygotic reproductive isolation: *The genes having the greatest effect on hybrid sterility and inviability are X-linked.*

Such analyses are usually performed on a pair of species whose hybridization produces one viable and fertile and one sterile (or inviable) sex, because there must be a phenomenon to study (sterility/inviability) and yet some viable and fertile F_1s are required for genetic analysis. A species carrying recessive mutations on all of its chromosomes is crossed to a second species lacking genetic markers. Homogametic F_1 hybrids are then backcrossed to the marked parental species. This cross (Figure 1) produces a number of genotypic classes that are identified by the marker alleles. The fertility and viability of these classes indicate which chromosomes or chromosome arms carry genes causing postzygotic isolation.

An example of this analysis is Orr's (1987) study of male sterility in hybrids between *Drosophila pseudoobscura* and *Drosophila persimilis*. Hybridization of these species produces completely sterile F_1 males and fairly fertile F_1 females. Backcrossing the latter to *D. pseudoobscura* males gives

TABLE 1. Conformity of sp‹ cies hybridizations to Haldane's rule.[a]

Group	Trait	Hybridizations with asymmetry	Number obeying Haldane's rule
Mammals	Fertility	20	19
Birds	Fertility	43	40
	Viability	18	18
Drosophila	Fertility and viability	145	141

[a]In addition, many species of Lepidoptera (Haldane 1922) and mosquitoes (Davidson, 1974) obey the rule. Source: Mammals and birds (Gray 1954, 1958) and *Drosophila* (Bock 1984; Orr and Coyne 1988).

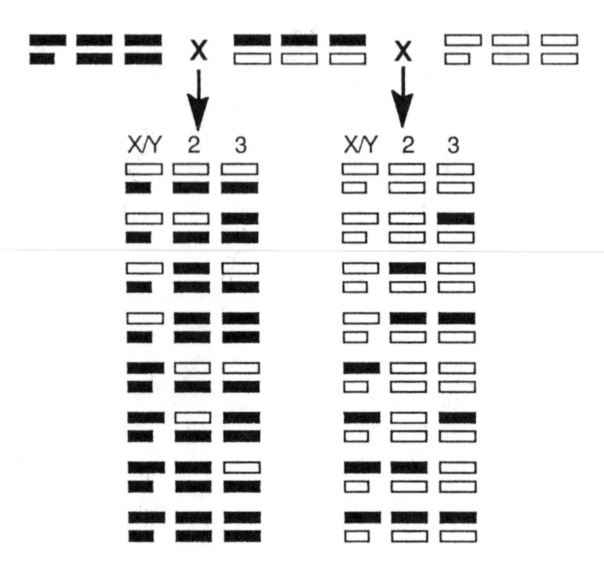

FIGURE 1. Genotypes produced in typical backcross analysis. Hybrid F_1 females (at top center) are separately backcrossed to males of each of the two parental species, producing the backcross male genotypes shown.

both fertile and sterile male offspring. Using a marked stock of *D. pseudoobscura*, Orr showed that each of the four major chromosomes has a significant effect on the fertility (measured as sperm motility) of backcross males. Figure 2 reveals that substitution of any *D. persimilis* chromosome into a largely *D. pseudoobscura* genetic background reduces the incidence of motile sperm. The X chromosome, however, has by far the largest effect. Almost no males carrying the *D. persimilis* X have motile sperm, whereas those with the *D. pseudoobscura* X usually have motile sperm. Further analysis also revealed a substantial effect of the Y chromosome on sterility (Orr 1987).

Similar analyses have been carried out in several species (Table 2). It is remarkable that *all* of these studies show predominant effects of the sex chromosomes on postzygotic isolation. When the X and Y chromosomes are studied separately, the X chromosome always has a large effect, whereas that of the Y varies from cross to cross. Because most of this work was done on species pairs obeying Haldane's rule, abnormalities were studied in the heterogametic sex. It is notable, however, that the X has the largest effect of any chromosome in all existing studies of *homogametic* sterility or inviability (see below). The large X effects in *female* as well as male hybrids will become important when we consider genetic explanations for this rule. The distribution of sex chromosomes across hybrid

FIGURE 2. Results from a typical backcross analysis of hybrid sterility. Backcross males were produced from the backcross of *D. pseudoobscura–D. persimilis* hybrid F_1 females to *D. pseudoobscura* males (from Orr 1987). *D. pseudoobscura* chromosomes are shown in white and *D. persimilis* in black. Substitution of a *D. persimilis* X chromosome causes almost complete sterility.

zones can also provide evidence of their fitness effects. Tucker et al. (1988) report a much steeper cline for X- and Y-linked DNA markers than for autosomal markers in transects across the European *Mus domesticus/ Mus musculus* hybrid zone, implying strong selection against hetero-specific sex chromosomes.

TABLE 2. Hybridizations showing large effects of sex chromosomes on hybrid traits.

Cross	Trait affected	Reference
Drosophila species		
pseudoobscura/persimilis	Male and female fertility	Dobzhansky (1936); Orr (1987)
pseudoobscura USA/Bogota	Male fertility	Dobzhansky (1974); Orr (1989)
simulans/mauritiana	Male fertility	Coyne and Kreitman (1984)
simulans/sechellia	Male fertility	Coyne (1986)
mohavensis/arizonensis	Male fertility	Zouros et al. (1988)
micromelanica A/B	Male fertility	Sturtevant and Novitski (1941)
littoralis/virilis	Male fertility	Orr and Coyne (1989)
novamexicana/virilis	Male and female fertility	Orr and Coyne (1989)
texana/virilis	Male and female fertility	Orr and Coyne (1989)
lummei/virilis	Female sterility	Orr and Coyne (1989)
buzzattii/serido	Female sterility	Naviera and Fontdevila (1986)
mulleri/aldrichi-2	Female viability	Crow (1942)
texana/montana	Female viability	Patterson and Griffen (1944)
Glossina morsitans subspp. 1/2	Fertility or mating ability	Curtis (1982)
Anopheles arabiensis/gambiae	Fertility or mating ability	Curtis (1982)
Colias eurytrheme/C. philodice	Female viability, fertility, mating vigor	Grula and Taylor (1980)
Cavia rufescens/C. porcellus	Male sterility	Detlefsen (1914)

How many genes contribute to these X effects? It is difficult to get accurate estimates with classic genetic techniques. A minimum of four (and probably five) X-linked loci cause male sterility in *Drosophila pseudoobscura/persimilis* hybrids (Lancefield 1929; Wu and Beckenbach 1983), whereas at least three X-linked genes are involved in the pairs *D. simulans/D. mauritiana*, *D. simulans/D. sechellia*, and *D. buzzattii/D. serido* (Naviera and Fontedevila 1986; Coyne and Charlesworth 1986, 1988). In at least two hybridizations, however, the large effect of the X is attributable to a relatively small section of the chromosome, implying a small number of loci affecting sterility (Orr 1989; Orr and Coyne, 1989). This suggests that postzygotic isolation is sometimes caused by substitutions at very few loci.

The X chromosome cannot, of course, cause sterility or inviability by itself. These phenomena occur only in hybrids, and must therefore involve epistatic *interactions* between X-linked genes and autosomal genes or cytoplasmic factors from another species. The nature of these other factors varies from hybridization to hybridization: they include the autosomes (Coyne 1984; Orr 1987, 1989; Zouros et al. 1988), the Y chromosome (Coyne 1985a; Orr 1987), and, in female hybrids, the cytoplasm (Crow 1942; Patterson and Griffen 1944; Orr 1987).

The pronounced effect of the X chromosome is not usually seen in genetic analyses of morphological or behavioral differences between species (data summarized in Charlesworth et al. 1987). The genes causing such differences are usually spread evenly among the chromosomes, with no disproportionately large effects of the X or Y. Table 3 shows the relative effects of the major chromosomes on differentiation of two morphological characters in the *D. melanogaster* group.

Why does the X chromosome play such a large role in the genetics of postzygotic reproductive isolation? Haldane's original explanation (1922) was that hybrid viability and fertility require a complete haploid genome from each species. Homogametic F_1 hybrids satisfy this requirement, but heterogametic hybrids lack an X chromosome to complement one haploid set of autosomes. This was once the commonly accepted explanation of Haldane's rule (e.g., Muller 1940; White 1973). Muller (1940) made this explanation more genetic by positing that hybrid sterility and inviability are predominantly caused by the expression of recessive alleles. Such alleles are expressed only in heterogametic hybrids having a single sex chromosome. This explanation was disproved, however, by Coyne (1985a) and Orr (1987, 1989), who showed that in some *Drosophila* species obeying Haldane's rule, hybrid *females* having the same X/autosome imbalance as sterile F_1 males are nevertheless completely fertile. In the Appendix we describe other problems with Muller's explanation of Haldane's rule.

Haldane later (1932:75) offered another explanation based on translocations between X and Y chromosomes. In crosses between a pair of

TABLE 3. Effects of chromosomes on character differences between *Drosophila* species.[a]

Character	Species pair	Proportionate contribution		
		X	2	3
Sex comb teeth	*mauritiana/simulans*	0.11	0.39	0.50
Genital size	*mauritiana/simulans*	0.08	0.36	0.56
Genital size	*sechellia/simulans*	0.30	0.52	0.18

[a]From Coyne (1982, 1985) and Coyne and Kreitman (1986).

species differing by such a translocation, heterogametic but not homogametic hybrids suffer from duplication or deletion of part of the Y chromosome. This explanation is probably wrong because pairs of *Drosophila* species that have not experienced such translocations (as well as orthopteran species having XO males) also obey Haldane's rule (Tan 1935; Bigelow 1960; Ohmachi and Masaki 1964; Lemeunier and Ashburner 1976).

Dobzhansky (1937:252–253) suggested that Haldane's rule results from reciprocal translocations between X chromosomes and *autosomes*, such that some genes that are X-linked in one species would be autosomal in the other. A simple crossing diagram shows that this explanation is incorrect, because heterogametic hybrids do not suffer from any duplications or deficiencies.

These early explanations of sex chromosome effects have been supplanted by a number of others, most of which fall into three categories. The first proposed that such effects are artifacts of the types of species available for genetic analysis. The second proposed that the effects are artifacts of the method of genetic analysis itself. The third proposed that the effects are not artifacts, but reflections of real biological phenomena involving the X chromosome. We consider these explanations in turn and conclude that the third explanation is probably correct.

A POTENTIAL OBSERVATIONAL BIAS

Does the large effect of the X chromosome result from an empirical bias? A possible bias arises merely because genetic analysis is limited to certain types of hybridizations. Genetic study of postzygotic isolation requires use of some viable, fertile F_1 hybrids, and therefore rules out species pairs that produce inviable or sterile hybrids of both sexes. Because sterility of both sexes may be caused more often by *autosomal* alleles than by sex-linked alleles (recessive X-linked genes can affect only heterogametic hybrids), autosomal incompatibilites may yield genetically unanalyzable species pairs more often than incompatibilities involving the X. As a result, X-linked incompatibilities could be grossly overrepresented among analyzable hybridizations. Therefore, the fact that a hybridization produces one viable, fertile sex and one inviable or sterile sex may a priori implicate the sex chromosomes.

These "analyzable" cases, then, may not be representative of taxa in the first steps of speciation: there may be many young species pairs that—because of autosomal incompatibilities—produce no viable, fertile hybrids. This "bias argument" therefore implies the existence of *two* genetic pathways to speciation, one obeying Haldane's rule (due to X effects) and the other producing hybrid effects in both sexes (due to autosomal genes).

Several lines of evidence militate against this explanation. First, the X chromosome plays a large role in *female* as well as male hybrid sterility and inviability. The X has the greatest effect of any chromosome in *all four* genetically analyzed cases of hybrid female sterility or inviability between closely related *Drosophila* species [*D. montana* × *D. americana texana* (Patterson and Griffen 1944), *D. mulleri* × *D. aldrichi*-2 (Crow 1942), *D. pseudoobscura* × *D. persimilis* (Orr 1987), *D. virilis* × other members of *virilis* phylad (Orr and Coyne, 1989)]. This pattern is unexpected if the large X effect is simply an artifact of analyzing cases of Haldane's rule.

Second, we have devised a method that allows analysis of the genetic basis of isolation between species pairs producing sterile or inviable hybrids of *both* sexes. This method takes advantage of the fact that two "uncrossable" species may both cross successfully with a third species, which can then be used to pass chromosomes between the uncrossable species. In this way we can produce hybrids differing only in the species origin of the X chromosome. We used this method to analyze the basis of isolation between *D. americana* and *D. montana*, whose hybridization produces completely sterile hybrids of both sexes (Throckmorton 1982). *D. montana* X chromosomes were transferred into a largely *D. americana* genome via *D. virilis* (Figure 3). The results of this study demonstrate that the X chromosome has a very large effect on both male and female hybrid sterility (Orr and Coyne, 1989). In a predominantly *americana* genome, males carrying a *montana* X chromosome almost never produce sperm (thus, the unmarked autosomes cannot have much effect on whether sperm are present). On the other hand, control males carrying a *virilis* X chromosome usually possess mature sperm. Similarly, hybrid females carrying a *montana* X chromosome are usually sterile, whereas control females are usually fertile.

We can also show that sterility and inviability of *both* sexes does not occur early in speciation, as is predicted by the bias hypothesis. To understand how reproductive isolation increases with time, we gathered literature data on the amount and type of postzygotic isolation between pairs of species, and correlated this isolation with a biochemical measure of divergence time, Nei's (1972) electrophoretic genetic distance *D* (Coyne and Orr 1989). Our index of postzygotic isolation between a pair of species came from published information about the sterility and inviability of hybrids in reciprocal crosses. Because each of the two reciprocal hybridizations produces males and females, the index of postzygotic isolation was the proportion of these four "sexes" that were either completely sterile or completely inviable as adults. This index thus takes the values of 0 (no sexes completely inviable or sterile), 0.25, 0.5, 0.75, or 1.00 (both sexes sterile or inviable in both reciprocal crosses). When *any* individuals of a class were viable or fertile, we counted the entire class as viable or fertile, so that our index underestimates the true amount of postzygotic isolation.

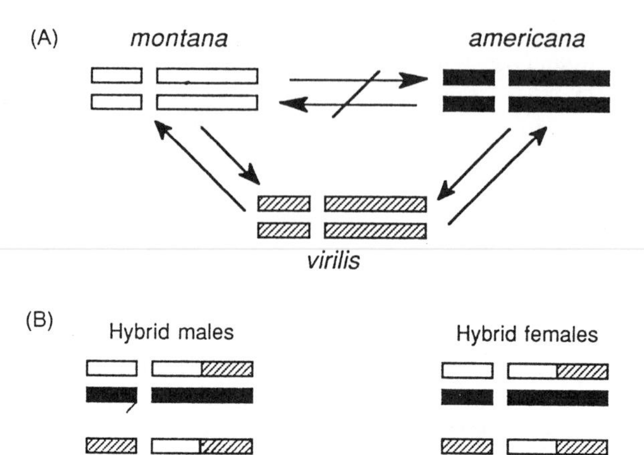

FIGURE 3. (A) *D. montana* and *D. americana* produce sterile hybrids of both sexes, barring direct study of the genetics of hybrid sterility. Each species, however, produces fertile hybrids when crossed to *D. virilis*. (B) *D. virilis* was used to pass *D. montana* chromosomes into a *D. americana* background, producing the male and female genotypes shown (see text). Fertility differences among the two male classes and the two female classes showed that the X has a large effect on both hybrid male and female fertility. (From Orr and Coyne, 1989.)

The index combines hybrid inviability and sterility because they appear to evolve at the same rate. Coyne and Orr (1989) give additional details, assumptions, and justifications of the analysis.

Figure 4 shows the time course for the evolution of postzygotic isolation, represented by the correlation between Nei's *D* and our index of sterility and inviability. Each point represents one of the 119 hybridizations for which we could find data on both reproductive isolation and genetic distance (some species take part in several hybridizations). Obviously, hybrid sterility and inviability increase with the time of divergence between the parental species. More important, however, is the *pattern* of divergence. Classes 0.25 and 0.5—those crosses in which only one or two of the four "sexes" are sterile or inviable—consist almost entirely of hybridizations obeying Haldane's rule (37 of 43 have sterility or inviability limited to males). This indicates that Haldane's rule does not result from a pattern of speciation in which male hybrids from one reciprocal cross become sterile or inviable, followed by females from that same cross. Instead, males from *both* reciprocal crosses become sterile or inviable before any female effects arise. This conclusion cannot be inferred from Haldane's rule alone.

In addition, there are very few cases in which *both* male and female offspring of the two reciprocal crosses are sterile or inviable early in the process of speciation. Of the 47 pairs of species separated by genetic distance smaller than 0.5, only four belong to the postzygotic isolation classes 0.75 or 1.00, the classes that include female effects. This means that Haldane's rule does not result from two pathways of speciation, one in which sterility or inviability occurs first in male hybrids and the other in which hybrids of both sexes become sterile or inviable simultaneously. Instead, sterility or inviability of all hybrids is nearly always preceded by sterility or inviability of males alone.

Haldane's rule therefore represents a nearly obligatory first step in the evolution of postzygotic isolation in Drosophila. Genetic studies of such cases therefore provide useful information about processes occurring early in speciation. We conclude that the large effects of sex chromosomes found in such studies are not artifacts of the restricted types of hybridization available for genetic analysis.

The question remains, however, whether postzygotic isolation is really an important cause of speciation. If, for example, complete mating discrimination between species always evolves before substantial hybrid sterility or inviability, then postzygotic isolation would be irrelevant to speciation. This is not the case, however. We can plot the evolution of *premating* isolation by correlating Nei's electrophoretic genetic distance between a pair of species with the strength of mating discrimination observed in the laboratory (Coyne and Orr 1989). We find that, among geographically isolated species, the restriction of gene flow due to assortative mating evolves at approximately the same rate as the restriction of gene

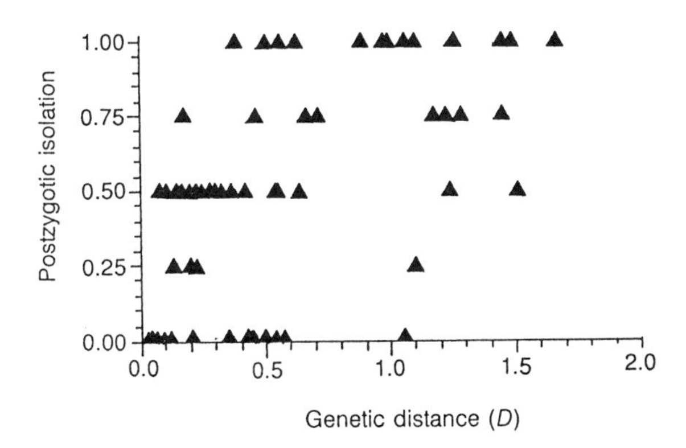

FIGURE 4. Postzygotic isolation versus genetic distance (Nei's D) between species pairs. Isolation obviously increases with the time since divergence. (From Coyne and Orr 1989.)

flow due to postzygotic isolation. Hybrid sterility and inviability must therefore be considered major contributors to reproductive isolation during allopatric speciation.

It has been suggested that large effects of the X chromosome on postzygotic isolation merely reflect the chromosome's large size. If genes producing such isolation were distributed throughout the genome and diverged at roughly equal rates, the largest chromosome would cause the most sterility or inviability. This cannot explain the data, however, because most pairs of species showing large X effects do not have disproportionately large X chromosomes (in the *D. melanogaster* group, for example, the X is the *smallest* major chromosome). In addition, the effects of even large X chromosomes are far greater than can be explained by their size (see Figure 2, for example, in which the *D. pseudoobscura* X chromosome is about twice the size of each of the three major autosomes).

Although large X effects cannot be explained by chromosome size, perhaps they could be explained by an unusually high *concentration* of fertility and viability genes on the X. This hypothesis itself requires some evolutionary explanation, for it is not obvious why genes affecting reproduction *and viability* should preferentially reside on sex chromosomes. The single piece of evidence addressing this hypothesis is that male-sterile mutations induced by chemical mutagenesis are not disproportionately concentrated on the X chromosome (Lindsley and Lifschytz 1972). It is of course possible that these loci are not representative of genes whose divergence among species causes hybrid sterility, so we cannot completely exclude the "concentration" hypothesis.

IS THE X EFFECT AN ARTIFACT OF ITS HEMIZYGOSITY?

The second group of hypotheses posit that the large effect of the X merely reflects its hemizygosity among experimental males in backcrosses. Such hemizygosity could inflate the effect of X-linked alleles in two ways. First, X-linked genes are dosage compensated in *Drosophila* males (Muller 1950). Such genes are effectively present in double dose in backcross males, and substitution of such an allele will have twice the effect of substituting an autosomal allele. Second, most of the alleles causing hybrid sterility or inviability could be recessive. Whereas X-linked recessive alleles are expressed in hybrid backcross males, autosomal recessive alleles from one species are not. Hence *all* X-linked genes but only partially dominant autosomal alleles are expressed in hybrid males. Muller (1940:203–204) believed this explained both Haldane's rule and the known large effect of the X chromosome among sterile male hybrids (e.g., Dobzhansky 1936).

Although attractive, these explanations are probably ruled out by two observations. First, the X chromosome also plays a large role in hybrid *female* sterility and inviability in *Drosophila* (see previous discussion). Females, of course, are not hemizygous for the X chromosome. X-linked alleles are thus present in the same dose as autosomal alleles, and X-linked recessive alleles (with h, the dominance coefficient, less than 0.5) have no more expression than autosomal recessives. Second, both of these hypotheses predict that X-linked alleles affecting *any* character should have a greater effect than autosomal alleles. Yet the X plays a large role only in hybrid sterility and inviability: genetic analysis of morphological or physiological differences among species reveals no disproportionate X effects (Charlesworth et al. 1987).

Several additional pieces of evidence militate against the "double dose" explanation. First, it predicts that X-linked alleles will show only twice the effect of additive genes. As previously noted, however, the effect of the X is often far greater than this. Second, large X effects have been found in the single genetic analysis of hybrid inviability/sterility in butterflies (Grula and Taylor 1980), a group that lacks dosage compensation (Johnson and Turner 1979).

In defense of Muller's recessivity argument, it could be claimed that the autosomes would display a large effect if homozygous autosomes from species A were forced to interact with homozygous autosomes from species B. In this genotype (*A/A B/B*), the recessive alleles from one species would interact deleteriously with those from the other. The autosomes, so this argument goes, ordinarily fail to show a large effect only because such genotypes are not produced in backcrosses.

This argument fails on two counts. First, it cannot explain the phenomenon of real interest, F_1 hybrid sterility or inviability. F_1 hybrids are heterozygous for *all* autosomes. Second, this argument implies that the X has a large effect only because recessive X-linked alleles from one species (e.g., from A) interact with "foreign" autosomal recessives (*B/B*) among backcross males. In reality, however, the X has a large effect even when it interacts with *heterozygous* autosomes (e.g., see Figure 2, class 9). A homozygous autosome, however, does not cause sterility when interacting with heterozygous autosomes (e.g., see Figure 2, classes 5, 6, 7).

Maynard Smith (personal communication) has suggested that both Haldane's rule and the large effect of sex chromosomes could result from a modification of Haldane's "balance" hypothesis: hybrids are fertile or viable only if they carry a complete conspecific set of autosomes and sex chromosomes (including the Y among males). Homogametic F_1 hybrids would satisfy this requirement, but heterogametic hybrids would be sterile or inviable because the X and Y come from different species. Under this hypothesis, a large X effect is detected among backcross males merely because substitution of a foreign X—but not of an autosome—leaves a hybrid with an incomplete X–Y autosome set. This explanation predicts that a large effect of autosomes would be observed if somehow hybrids could be produced that lack a conspecific set of autosomes.

This hypothesis fails on three counts. First, it cannot account for the large effects of sex chromosomes among *homogametic* hybrids, as previously discussed: sterile or inviable F_1 females obviously carry a *complete* haploid set of chromosomes from each species. Second, at least two *Drosophila* hybridizations show that the specific origin of the Y chromosome has no effect on sterility (Orr 1989; Orr and Coyne, 1989), though a strong effect is predicted by Maynard Smith's hypothesis. Finally, both Haldane's rule and strong X chromosomal effects on hybrid fitness have been in hybrids between orthopteran species having XO males (Bigelow 1960; Ohmachi and Masaki 1964; Shaw et al. 1979; Butlin and Hewitt 1988). In these hybridizations males are sterile or inviable in spite of their possession of a complete haploid genome from one species.

It therefore appears that large X effects cannot be explained away as artifacts of genetic analysis. Such effects appear instead to reflect some real biological process of evolutionary significance.

NONARTIFACTUAL CAUSES OF LARGE X EFFECTS

Two biological phenomena could explain the large effect of the X chromosome on postzygotic isolation. The effect may result from improper dosage compensation of X-linked genes in hybrids, or it may arise from differences in the rate of evolution of sex chromosomes versus autosomes.

Dosage compensation

In *Drosophila*, compensation is achieved by doubling the gene products of most X-linked loci in males (Lucchesi 1978). Mutants interfering with normal dosage compensation are lethal (Baker and Belote 1983). It is important to note that improper dosage compensation would adversely affect male (i.e., heterogametic) more than female hybrids: lack of normal compensation in males is equivalent to a deficiency, whereas unneeded hyperactivation of genes in females is equivalent to a duplication. Deficiencies are typically more detrimental than duplications (Lindsley et al. 1972), presumably because a decrease in enzyme activity has a greater effect on enzyme flux (and, by assumption, fitness) than an increase (Kacser and Burns 1981).

Although dosage appears to be regulated locus-by-locus via cis-acting sequences scattered along the *Drosophila* X chromosome (Jaffe and Laird 1986), the X-autosomal ratio determines whether compensation will occur (X/A = 1 in normal females and ½ in normal males). In *Drosophila*, there is evidence that individuals assess their X/A ratio by summing the number of X-linked and autosomal "counter genes" present; two apparent X-linked counter genes have been found (see Jaffe and Laird 1986).

Dosage compensation in hybrids could be disrupted in several ways: (1) the cis-acting sequences adjacent to most X-linked genes could diverge between two species. As a result, X-linked genes from species A would not be "recognized" by the regulatory signals produced in part by species B genes. (2) Similarly, counter genes could diverge between two species. In this case, X counters from A may not "recognize" autosomal counters from B. The apparent abnormal X/A ratio would then disrupt normal compensation.

Although we cannot completely discount the possibility that the X effect results from a breakdown in dosage compensation, several lines of evidence argue against this hypothesis. First, birds and butterflies—taxa that appear to lack dosage compensation (Baverstock et al. 1982; Johnson and Turner 1979)—obey Haldane's rule. Moreover, as noted, the X has a large effect in the one genetically analyzed butterfly hybridization that obeys Haldane's rule (Grula and Taylor 1980). Second, this hypothesis implies that *any* anomaly (not just sterility and inviability) appearing in hybrids results from a disturbance in dosage compensation; it therefore predicts that morphological or physiological anomalies among hybrids should obey Haldane's rule and show a large X effect upon genetic analysis. Although there are few data bearing on this question, morphological anomalies among *Drosophila* hybrids do not appear to be limited to, or even more severe among, males (Weisbrot 1963; Coyne 1985b; Orr, unpublished observations). Third, there is dramatic evidence that "compensation control sequences" are conserved between even distantly related *Drosophila* species: the X-linked *D. pseudoobscura Hsp82* gene remains dosage compensated when it—and flanking DNA carrying a cis-acting compensation sequence—are transformed into various autosomal sites in *D. melanogaster* (Jaffe and Laird 1986). *D. melanogaster* signals therefore "recognize" compensation sequences from *D. pseudoobscura*. Working with more closely related species, Lakhotia et al. (1981) found apparently normal dosage compensation in sterile hybrids of *D. melanogaster* and *D. simulans*. Fourth, it is not clear how the dosage compensation hypothesis can account for *partial* hybrid sterility or inviability [e.g., only one of four *D. pseudoobscura-D. persimilis* hybrid F_1 females is sterile (Orr 1987)]: surely divergence of cis-acting or countersequences would disrupt compensation of most or all X-linked genes among hybrids. This condition would almost certainly be lethal to hybrid embryos. Thus, this hypothesis "works" only under rather restricted conditions: dosage compensation systems must diverge in such a way that most genes remain compensated in hybrids while a few do not. It is not at all obvious how this could occur.

Finally, it is difficult to envision how failures of dosage compensation could explain the *sterility* of hybrids that are viable and morphologically normal.

All of the hypotheses discussed above assume that X-linked and autosomal genes diverge at similar rates, but that the crossing schemes used—or the requirement of dosage compensation—favors detection of X effects. As we have seen, these explanations are not very satisfactory. This leaves us with the possibility that genetic divergence between species proceeds faster on sex chromosomes than autosomes.

Differences in evolutionary rate

Why would genes evolve more rapidly on sex chromosomes than on autosomes? Any explanation must depend on the facts that sex chromosomes are hemizygous in one sex and are only three-quarters as numerous in populations as any autosome. Noting the first fact, Haldane (1924) suggested that advantageous *recessive* alleles accumulate far more rapidly on sex chromosomes than on autosomes. Such alleles will not be selected on autosomes until they attain high enough frequencies to allow the appearance of homozygotes. Completely recessive alleles will then be effectively neutral when they first appear, and will usually be lost from populations. On sex chromosomes, however, such alleles will be subject to immediate positive selection in the heterogametic sex, which enjoys the beneficial effect with only a single copy of the allele.

Charlesworth et al. (1987) mathematically modeled this process, allowing advantageous alleles with varying degrees of dominance to arise on sex chromosomes and autosomes and calculating the relative rate of fixation (mutation rates are also lower on sex chromosomes because they are less numerous). The relative fixation rates on sex chromosomes versus autosomes turn out to depend only on the coefficient of dominance, h, the degree to which the selective advantage of homozygotes for an allele is expressed in heterozygotes. (The value of h ranges from 0, when the mutant allele is fully recessive, to 1, when it is completely dominant; $h = 0.5$ corresponds to additivity.) For example, when an allele is equally advantageous in both sexes and there is dosage compensation, its rate of evolution on an X chromosome relative to that on an autosome is

$$(2h + 1)/4h$$

When the allele is completely recessive ($h = 0$), this ratio is infinity, reflecting the low probability of survival of a recessive autosomal allele. When completely dominant ($h = 1$), the ratio is ¾. If the favorable allele is partially recessive ($h < 0.5$), the probability of fixation is higher for sex chromosomes than for autosomes. In other models, such as those assuming unequal selection on the sexes or no dosage compensation, this threshold changes, but in all cases except for selection affecting only the homogametic sex, increased recessivity causes faster evolution of the sex chromosomes. Other evolutionary models, such as those assuming fixation of slightly deleterious alleles by drift or the stochastic turnover of alleles under stabilizing selection, lead to rates of sex chromosome evolution that are either equal to or slightly lower than rates of autosomal evolution.

These considerations provide a possible explanation for the large effects of the sex chromosomes on postzygotic isolation. Alleles that ultimately cause sterility or inviability will tend to accumulate on the sex chromosomes if they are usually recessive and have favorable effects on their normal genetic background. (We are not, of course, suggesting that these alleles cause sterility or inviability when they first appear. These phenomena appear only among hybrids containing long-diverged genomes.)

This model can also explain Haldane's rule if an additional qualification is made. It seems reasonable to assume that the mutations ultimately causing hybrid male sterility or inviability are of two types—those whose original advantageous effects were limited to males, and those whose advantageous effects were expressed in both sexes. Similarly, the original, advantageous effects of female "sterility" or "inviability" genes were either limited to females or expressed in both sexes. (We presume that genes originally expressed in one sex but ultimately causing sterility/inviability of only the other sex are rare.) If this assumption is added to the assumption of Charlesworth et al. (1987) that most favorable mutations are partially recessive ($h < ½$), it can be shown that genes causing hybrid male sterility or inviability will accumulate faster than those causing hybrid female effects as long as some of the advantageous effects were limited to one sex (see Appendix). Hybrid males will therefore be affected more frequently and more severely than hybrid females, explaining Haldane's rule. It can also be shown (see Appendix) that *both* male and female sterility/inviability alleles accumulate faster on the X chromosome than on the autosomes. The theory therefore accounts for the observation that the X chromosome plays a large role in postzygotic isolation of both sexes.

It is important to note that this theory was constructed as a post facto explanation of Haldane's rule and X chromosome effects, and must therefore be tested with independent data not deriving from traditional backcross analysis. The theory also requires us to assume that the original, advantageous effects of mutations causing postzygotic isolation are usually fully or partially recessive, an assumption for which there is no evidence.

The theories of Charlesworth et al. (1987) and that outlined in our Appendix make at least two testable predictions. First, a pair of species goes through an intermediate stage of evolution in which sterility or inviability is limited to the heterogametic sex in *both* reciprocal crosses. Sterile or inviable homogametic hybrids should evolve later—*much* later if a large fraction of the mutations capable of causing postzygotic isolation is partially recessive. This prediction is met by our data correlating postzygotic isolation with genetic distance: we find a large temporal gap between the evolution of male and female postzygotic isolation in *Drosophila*. We can calculate the average genetic distance between pairs of species in each of our five sterility/inviability classes. As we have seen, classes 0, 0.25, and 0.5 (corresponding to none, one, or two of the four "sexes" inviable or sterile in reciprocal crosses) almost always include only males, with female effects almost completely limited to classes 0.75 and 1.00. Table 4 shows that classes 0, 0.25, and 0.5 contain species separated by similar and low genetic distances, whereas there is a large gap in genetic distance between members of classes 0.5 and 0.75. Sterility and inviability therefore arise fairly quickly in hybrid males, but much more slowly in females, as predicted.

The theory's second—and stronger—prediction derives from its assumption that the favorable effects of alleles ultimately causing postzygotic isolation are usually recessive. Because such alleles accumulate much faster on sex chromosomes than autosomes, it should be found that postzygotic isolation maps largely to the X chromosome in hybrids between recently diverged species, but to both X chromosomes and autosomes in hybrids between older pairs. Although not extensive, the data do support this prediction. In hybrids between the two subspecies *Drosophila pseudoobscura pseudoobscura* and *D. pseudoobscura bogotana* [Nei's $D = 0.194$ (Ayala and Dobzhansky 1974)], the X chromosome has a large effect on male sterility whereas the autosomes have a much smaller effect (Orr 1988). On the other hand, in hybrids between *D. pseudoobscura* and its more distant relative *D. persimilis* [$D = 0.3$–0.4 (Ayala and Dobzhansky 1974; Lakovaara et al. 1976)], the autosomes have a greater effect than in the younger pair of taxa (Orr 1987). An identical result appears in studies of male sterility in the *D. virilis* group (Orr and Coyne, 1989). It is obviously desirable to further test this prediction in other groups with well-known phylogenies.

It is important to note that any correlation between the age of a pair of taxa and the effect of the autosomes relative to the X would contradict the notion that X effects are merely artifacts of hemizygosity in backcross males. After all, the "hemizygosity" hypothesis posits that large X effects do not result from a time-dependent evolutionary process, and therefore cannot explain why the relative effects of autosomes increase with divergence time.

In conclusion, we have a nested set of facts and theories about postzygotic reproductive isolation. The first is Haldane's rule, the ubiquitous sterility and/or inviability of heterogametic hybrids early in speciation. Underlying this phenotypic rule is a genetic one: postzygotic isolation is caused largely by genetic changes on the X chromosome. Finally, the

TABLE 4. Mean (± 1 SE) genetic distance at which a given level of postzygotic isolation occurs in *Drosophila*.[a]

Isolation index	Mean genetic distance ± SE (*N*)
0.00	0.138 ± 0.058 (8)
0.25	0.251 ± 0.083 (5)
0.50	0.249 ± 0.032 (16)
0.75	0.722 ± 0.198 (5)
1.00	0.991 ± 0.127 (8)

[a]Data are phylogenetically corrected (see Coyne and Orr 1989), and isolation indices are rounded down to nearest 0.00, 0.25, 0.50, 0.75, 1.00.

evolutionary explanation of Charlesworth et al. (1987) and its extension in our Appendix appear to account for both the genetic and phenotypic rules. This explanation survives two empirical tests. Although neither test is conclusive, no other theory so fully explains the regularities of postzygotic isolation.

Nevertheless, this evolutionary explanation remains somewhat unsatisfactory because of its ad hoc assumption that the alleles causing postzygotic isolation were originally advantageous alleles that were partially recessive. This theory postulates that such alleles differ fundamentally from those causing morphological or behavioral differences between species, because genetic analyses show that the latter usually act additively in hybrids (see summary, Charlesworth et al. 1987). This difference between "morphological" and "sterility/inviability" genes contradicts the widely held view that these genes are really identical. After all, it is a tenet of the modern synthesis that reproductive isolation arises simply as a pleiotropic byproduct of the divergence of "ordinary" genes via natural selection or genetic drift. When the strength of the data is contrasted with the weakness of the assumption required to explain them, we sometimes get the feeling that we are missing a key insight that will make everything clear.

Finally, we need more research to test both the generality of our observations and the theories proposed to explain them. It is still possible that a breakdown in dosage compensation could explain both Haldane's rule and the large role of the X chromosome. This question can be resolved by genetic analysis of postzygotic isolation in hybridizations among species lacking dosage compensation, and by direct examination of dosage compensation in hybrids obeying Haldane's rule. Because our explanation of Haldane's rule and X effects depends on the hemizygosity, and hence heteromorphy, of entire sex chromosomes (or at least of large sex chromosome segments), our two rules of speciation should not be observed in taxa with single-locus or environmental sex determination. If they are, these two rules would require a radically different explanation. Many species of plants have heteromorphic sex chromosomes and hence should also obey these rules. Yet we are aware of no surveys of the pattern of postzygotic isolation or its genetic basis in closely related plant species of identical ploidy. Finally, we need more tests of the prediction of Charlesworth et al. (1987) that the relative role of the autosomes in producing postzygotic isolation should increase with divergence time between taxa. Such tests are fairly simple in species such as *Drosophila* that have well-established phylogenies and are easily crossed in the laboratory. Confirmation of this prediction would strengthen the evidence that Haldane's rule and the large effect of the X chromosome result from an evolutionary process and are not mere artifacts of genetic analysis.

NOTE ADDED IN PROOF

C. Langley (pers. comm.) recently pointed out that mutations producing favorable homozygotes but inferior heterozygotes will, like advantageous recessives, accumulate much faster on X chromosomes than on autosomes. These mutations are underdominant, and are immediately selected against if they occur on an autosome. If X linked, however, they can be advantageous in hemizygous males; and if this selection is strong enough, it can overcome the underdominance in females and lead to fixation.

We can easily incorporate underdominance into the "pleiotropy theory" discussed below. X-linked underdominant mutations with a greater effect in males than in females will be fixed more often than mutations having the opposite effect. If there is a correlation between the sex an underdominant mutation originally affects and the hybrid sex it later afflicts, it is clear that in species crosses hybrid males will be more afflicted than females, explaining Haldane's rule (see Equations 1a and 1b in the Appendix). Underdominant mutations with female-limited effects will be fixed far less often than advantageous recessive mutations with female-limited effects. Because of this, underdominance may be a much more powerful explanation of Haldane's rule than the "recessivity" theory we consider below. We must still assume, however, that postzygotic isolation is based on genes whose action is different from those causing morphological differences among species.

APPENDIX

We assume that the mutations ultimately causing hybrid sterility and inviability were selectively favored on their original genetic background (see Charlesworth et al. 1987). We further assume that there is some correlation between the sex in which a mutation was originally favored and the hybrid sex that is later afflicted by that mutation. In particular, we assume that the mutations capable of causing sterility or inviability of a particular hybrid sex are of two types: those selectively favored in that sex only and those selectively favored in both sexes. We presume that mutations originally favored in only one sex but that cause the sterility or inviability of the other hybrid sex only are negligibly rare. Thus, of the favorable mutations affecting hybrid male fertility and viability, a proportion p_m was favored in males only, and $1 - p_m$ was favored in both sexes. Similarly, p_f of the favorable mutations affecting hybrid female fertility or viability were advantageous in females only, and $1 - p_f$ in both sexes. [Obviously if $p_m = p_f = p$, then $pv + (1 - p)(v) = v$, where v is the per locus rate of mutation to favorable alleles].

We will consider (1) the conditions under which sterility and inviability genes afflicting male hybrids accumulate faster than those afflicting female hybrids (explaining the observed temporal lag between male and female hybrid effects and thus Haldane's rule), and (2) the conditions under which the X chromosome accumulates more male *and* female sterility/inviability genes than an autosome (explaining the large effect of the X on both male and female hybrid sterility and inviability).

Haldane's rule

For convenience, we consider males the heterogametic sex; the results, of course, hold when females are heterogametic. We assume that X-linked loci are dosage compensated (as they are in *Drosophila* and most mammals). We now derive a formula for the ratio, R, of the rates at which male versus female sterility and inviability genes accumulate. Haldane's rule (more extreme postzygotic isolation in males) is expected when this ratio exceeds one. By definition, $R = K_{male}/K_{female}$, where K_{male} is the rate at which alleles ultimately causing hybrid male sterility and inviability are substituted and K_{female} is the substitution rate for female sterility and inviability genes. K_{male} and K_{female} each includes two components: the substitution rate for mutations occurring on the X chromosome and the substitution rate for the autosomes (we ignore Y-linked mutations). A proportion x of all favorable mutations are X linked, and $1 - x$ are autosomal.

(If there are equal per locus mutation rates to favorable alleles on the X and the autosomes, x is the proportion of the genome that is X linked.) Of those favorable X-linked mutations, some are favored in one sex only and some are favored in both sexes; similarly, there are two types of favorable autosomal mutations. Thus

$$K_{\text{male}} = x[p_m X_m + (1 - p_m)X_b] + (1 - x)[p_m A_m + (1 - p_m)A_b] \tag{1a}$$

and

$$K_{\text{female}} = x[p_f X_f + (1 - p_f)X_b] + (1 - x)[p_f A_f + (1 - p_f)A_b] \tag{1b}$$

where X_m, X_f, and X_b are the substitution rates of X-linked sterility and inviability genes favored in males only, females only, or both sexes, respectively [e.g., $X_m = 1.5Nvu$, where N is the population size and u is the fixation probability for X-linked mutants favored in males only (see Charlesworth et al. 1987)]. Similarly, A_m, A_f, and A_b are the substitution rate of autosomal sterility and inviability genes favored in males only, females only, or both sexes, respectively. These substitution rates are given by Charlesworth et al. (1987) (substitution rates in that paper are expressed in time units of $(Nv)^{-1}$ generations and assume that all mutations are unique and that selection is weak). Substituting these rates into Eqs. (1a) and (1b) and rearranging, we find that

$$R = \frac{K_{\text{male}}}{K_{\text{female}}} = \frac{x(1 - 2h) + 2h(2 - p_m)}{x(1 - 2h)(1 - p_f) + 2h(2 - p_f)} \tag{2}$$

where h is the coefficient of dominance of the favorable mutations ($h = 0$ corresponds to complete recessivity, $h = ½$ corresponds to additivity, and $h = 1$ to complete dominance). We assume that the average selection coefficient is the same in males and females [in the nomenclature of Charlesworth et al. (1987) $s_1 = s_2 = s$]. Assuming for simplicity that equal proportions of hybrid male and female sterility/inviability mutations are originally favored in only one sex (i.e., $p_m = p_f = p$), we see that $R > 1$ whenever $h < ½$, and $x \neq 0$ and $p \neq 0$. In other words, alleles causing postzygotic isolation in hybrid males accumulate faster than those affecting females as long as some advantageous mutations occur on the X chromosome, some mutations are favored in only one sex, and the advantageous mutations are partially recessive.

Equation (2) has several interesting properties. First, R is independent of the strength of selection (although the quantity s appears in K_{male} and K_{female}, it does not appear in their ratio). Second, although the quantity x can affect the extent to which R differs from one, the value of x does *not* affect whether $R > 1$ as long as $x \neq 0$. Thus, if the other conditions hold ($h < ½$

and $p \neq 0$), hybrid male effects will accumulate at least slightly faster than female effects in all taxa, regardless of the proportion of the genome that is X linked.

Because it is difficult to visualize how R varies as a function of three variables, and because we are most interested in the effects of varying h and p, we ignore below variation in x. Assuming for simplicity that $x = \frac{1}{4}$, Eq. (2) becomes

$$R = \frac{1 + 14h - 8hp}{1 + 14h - p(1 + 6h)} \tag{3}$$

Again, $R > 1$ when $h < \frac{1}{2}$ and $p \neq 0$. Figure 5 shows how R in Eq. (3) varies with h and p. R is obviously greatest when h is small and p is large. Indeed, R approaches infinity as h approaches 0 and p simultaneously approaches 1.

Although Eq. (3) assumes that h takes some point value, we can generalize it by considering some frequency distribution of dominance for favorable mutations. In particular, if h is normally distributed with mean h' (and very small variance, allowing integration between $-\infty$ and ∞), then R becomes

$$R = \frac{1 + 14h' - 8h'p}{1 + 14h' - p(1 + 6h')} \tag{4}$$

Thus, alleles causing hybrid male sterility and inviability will accumulate faster than such female alleles if $p \neq 0$ and favorable mutations are on average partially recessive ($h' < \frac{1}{2}$).

It should be noted that the distribution of h' entering Eq.(4) is the frequency distribution of dominance coefficients for *newly* arising favorable mutations, not for those favorable mutations that are ultimately fixed. These distributions are probably quite different because it is easier to fix mutations that are more dominant (Wright 1929, 1977:511) (except when considering X-linked mutations with male-limited selection). The above model requires only that the mean dominance of new mutations be less than one-half. Although it is possible that this quantity is much less than one-half, little information is available (see Charlesworth et al. 1987). This uncertainty is unfortunate because, as Figure 5 shows, R in Eq.(4) is fairly sensitive to h'. If, for example, $h' = 0.1$, male sterility/inviability alleles will accumulate 25% faster than female alleles (assuming $p = \frac{1}{2}$). If h' increases to 0.25, however, alleles in males accumulate only 8% faster than those in females. We have no idea whether an 8% difference between male and female substitution rates is sufficient to explain Haldane's rule. In general, however, it appears that the present model requires that h' is fairly small (less than $\frac{1}{4}$).

Realistically, the present model also requires that p is not very small (probably greater than $\frac{1}{4}$). Fortunately, some information on the extent to

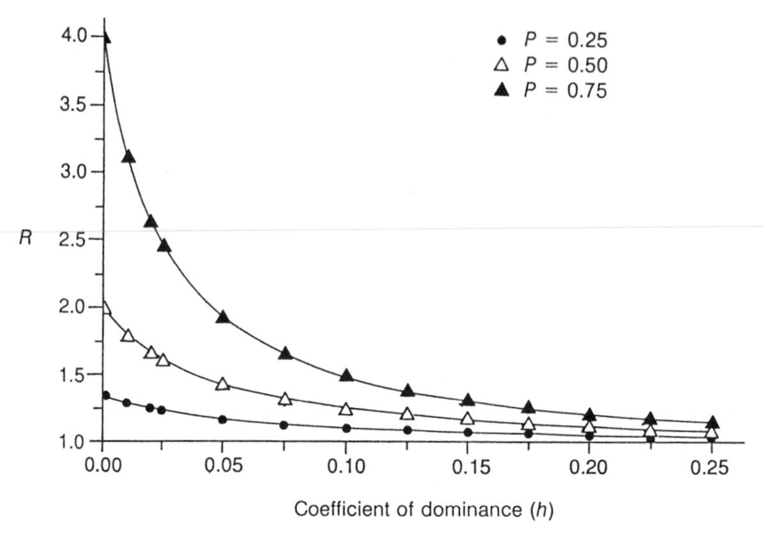

FIGURE 5. R (the ratio of male/female substitution rates for alleles ultimately causing postzygotic reproductive isolation) as a function of the coefficient of dominance (h) of the advantageous effects of these alleles. Three cases are shown for various values of p, the proportion of mutations favored in only one sex. Alleles causing postzygotic isolation in hybrid males are substituted much faster than those affecting females when h is small and p is large.

which fertility and viability mutations affect both sexes is available. However, although p represents the proportion of *new* mutations affecting one sex only, most data on the correlations of mutant effects between the sexes come from studies of standing variation. These studies demonstrate that most alleles affecting fertility do *not* affect the sexes equally. Temin (1966), for instance, found a correlation of fertility effects between the sexes of $r = 0.35$ for *D. melanogaster* second chromosome heterozygotes and $r = 0.15$ for homozygotes; Watanabe and Oshima (1973), who also studied *D. melanogaster* second chromosomes, found that $r = 0.10$ for heterozygotes and $r = 0.18$ for homozygotes. These low correlations show that most fertility mutations have sex-limited effects (or, more accurately, that the effect is smaller in one sex). Mitchell (1977) has calculated the analogous correlation for viability mutations. In this case *new* mutations arising on the *D. melanogaster* X chromosome were studied (females were heterozygous and males were hemizygous). Mitchell (1977) found that $r = 0.58 \pm 0.05$. Although this correlation is higher than for fertility (as might be expected biologically), it is still not close to one. Many viability mutations must have unequal effects on the sexes. Although some caution is required when interpreting these results (because, for instance, most of the mutations studied are deleterious, not favorable), it seems likely that many mutations affecting fertility and viability are sex limited or at least have quite unequal effects on the sexes. Thus the proportion p is probably fairly large, as the present theory requires.

It is important to realize that the present explanation of Haldane's rule is *not* the same as Muller's "hemizygosity hypothesis." According to Muller, Haldane's rule has nothing to do with the more rapid evolution of X-linked versus autosomal loci or the faster accumulation of hybrid male versus female sterility/inviability genes. Instead, Muller argued that Haldane's rule results simply from the expression of recessive X-linked alleles among hybrid males but not among hybrid females. To Muller, hybrid females are fertile and viable because they are heterozygous at all X-linked loci; if, like males, hybrid females expressed recessive X-linked alleles, they too would be sterile or inviable. In their explication of Haldane's rule, Charlesworth et al. (1987) also relied upon Muller's argument. As previously noted, Muller's hypothesis has been falsified experimentally: the fertility of F_1 hybrid females is *not* due to the masking of recessive X-linked alleles in heterozygous females since females carrying an F_1 male-like genotype (homozygous for one species' X) remain fertile. Instead, it is clear that the alleles causing hybrid male sterility simply do not affect females whether in heterozygous or homozygous state. In addition, in hybridizations that do produce sterile or inviable females, the X chromosome usually has a larger effect than any autosome, in spite of the fact that the X is not hemizygous in females (Orr 1987; Orr and Coyne, 1989). Finally, these sterile or inviable females are usually *heterozygous* and not homozygous for the X chromosome. This is inexplicable by Muller's hypothesis.

In our model, Haldane's rule does *not* result from the expression of recessive X-linked genes in hybrid F_1 males but not in females. Rather, it reflects the more rapid *substitution* of mutations causing male rather than female postzygotic isolation. This higher "male substitution rate" results simply from the rapid accumulation of partially recessive X-linked mutations that are favored only in males. Indeed, on our "pleiotropy model," more frequent F_1 male than female effects are expected even if all sterility and inviability genes behave as *dominants* on a hybrid genetic background. This is obviously not true for Muller's argument.

Finally, we point out that although Muller's hemizygosity hypothesis cannot completely account for the known data, hemizygosity among hybrid males may well play some role in Haldane's rule. Our pleiotropy model and the hemizygosity model are not mutually exclusive.

The large X effect

We now show that the pleiotropy model outlined can also account for th large role of the X chromosome in hybrid male and female sterility and inviability. To demonstrate this, we calculate the ratios, X/A, of the accumulation rates for X-linked versus autosomal mutations that cause hybrid sterility or inviability.

Males. Considering the two types of male sterility and inviability mutations (those originally affecting males or both sexes), the relevant ratio is

$$X/A_{\text{male}} = \frac{p_{\text{m}}X_{\text{m}} + (1 - p_{\text{m}})X_{\text{b}}}{p_{\text{m}}A_{\text{m}} + (1 - p_{\text{m}})A_{\text{b}}} \tag{5}$$

where X_{m}, X_{b}, A_{m}, and A_{b} are as defined above. Using the formulas provided by Charlesworth et al. (1987), it is easily shown that

$$X/A_{\text{male}} = \frac{2h(1 - p_{\text{m}}) + 1}{2h(2 - p_{\text{m}})} \tag{6}$$

This ratio exceeds 1 whenever $h < ½$, regardless of the value of p_{m}. Thus the X will accumulate hybrid male sterility and inviability genes faster than an equivalent-sized autosome whenever the favored mutations are partially recessive, no matter what proportion of mutations is originally favored in males only. The X would therefore have the largest effect of any chromosome on hybrid male sterility or inviability. This result is essentially identical to that of Charlesworth et al. (1987).

Females. An analogous equation can be derived for the ratio of the rates at which X-linked versus autosomal *female* sterility and inviability genes accumulate. This ratio is

$$X/A_{\text{female}} = \frac{2h - p_{\text{f}} + 1}{2h(2 - p_{\text{f}})} \tag{7}$$

This ratio is greater than 1 whenever $h < ½$ and $p \neq 1$. Thus the X will accumulate more female sterility and inviability genes than an equivalent-sized autosome as long as favorable mutations are usually partially recessive and some mutations are selectively favored in both sexes.

It is interesting to note that when $h < ½$, $X/A_{\text{male}} > X/A_{\text{female}} > 1$ (if $p_{\text{m}} = p_{\text{f}} = p$, and $p \neq 1$). This model thus predicts that in crosses producing sterile or inviable hybrid males *and* females, the effect of the X relative to the autosomes will be greater among backcross males than among females, but the effect of the X will be greater than any autosome in both sexes. Preliminary data from the *virilis* group supports this prediction (Orr and Coyne, 1989). It is important to note that Muller's explanation of the large X effect and Haldane's rule does not make this prediction; it instead predicts that the X will show a larger effect than an equivalent-sized autosome when considering male sterility or inviability, but that the X and an autosome will have equal effects on female postzygotic isolation, since the X is not hemizygous in females. Thus, the Muller hypothesis predicts that $X/A_{\text{male}} > X/A_{\text{female}}$, but $X/A_{\text{female}} = 1$.

Summary

In conclusion, our pleiotropy model simultaneously explains the time lag between the evolution of postzygotic isolation in male and female hybrids (explaining Haldane's rule) and the observed large effect of the X chromosome on both male and female sterility and inviability. The conditions that must be met are (1) favorable mutations are, on average, partially recessive ($h' < ½$); (2) some of these favorable mutation are X linked ($x \neq 0$); and (3) some of these mutations are favored in one sex only whereas others are favored in both sexes ($0 < p < 1$). Under these conditions more alleles causing male than female hybrid sterility and inviability will accumulate per unit time *and* these genes will cluster on the X faster than on the autosomes.

ACKNOWLEDGMENTS

This work was supported by grant GM 38462 from the National Institutes of Health to J.A.C. and genetics training grant GM 07197 from the National Institutes of Health to the University of Chicago. We thank Brian Charlesworth, Russ Lande, and John Maynard Smith for discussion and criticism.

LITERATURE CITED

Ayala, F. J., and T. Dobzhansky. 1974. A new subspecies of *Drosophila pseudoobscura* (Diptera:Drosophilidae). Pan-Pac. Entomol. 50:211–219.

Baker, B. S., and J. M. Belote. 1983. Sex determination and dosage compensation in *Drosophila melanogaster*. Annu. Rev. Genet. 17:345–394.

Baverstock, P. R., M. Adams, R. W. Polkinghorne, and M. Gelder. 1982. A sex-linked enzyme in birds: Z-chromosome conservation but no dosage compensation. Nature (London) 296:763–767.

Bigelow, R. S. 1960. Interspecific hybridization and speciation in the genus *Acheta* (Orthoptera, Gryllidae). Can. J. Zool. 38:509–524.

Bock, I. R. 1984. Interspecific hybridization in the genus *Drosophila*. Evol. Biol. 18:41–70.

Butlin, R. K., and G. M. Hewitt. 1988. Genetics of behavioural and morphological differences between parapatric subspecies of *Chorthippus parallelus* (Orthoptera:Acrididae). Biol. J. Linnaean Soc. 33:233–248.

Charlesworth, B., J. A. Coyne, and N. H. Barton. 1987. The relative rates of evolution of sex chromosomes and autosomes. Am. Natur. 130:113–146.

Crow, J. F. 1942. Cross fertility and isolating mechanisms in the *Drosophila mulleri* group. Univ. Texas Pub. 4228:53–67.

Coyne, J. A. 1983. Genetic basis of differences in gential morphology among three sibling species of *Drosophila*. Evolution 37:1101–1118.

Coyne, J. A. 1984. Genetic basis of male sterility in hybrids between two closely related species of *Drosophila*. Proc. Natl. Acad. Sci. U.S.A. 81:4444–4447.

Coyne, J. A. 1985a. The genetic basis of Haldane's rule. Nature (London) 314:736–738.

Coyne, J. A. 1985b. Genetic studies of three sibling species of *Drosophila* with relationship to theories of speciation. Genet. Res. 46:169–192.

Coyne, J. A., and B. Charlesworth. 1986. Location of an X-linked factor causing sterility in male hybrids of *Drosophila simulans* and *D. mauritiana*. Heredity 57:243–246.

Coyne, J. A., and B. Charlesworth. 1988. Genetic analysis of X-linked sterility in hybrids between three sibling species of *Drosophila*. Heredity, in press.

Coyne, J. A., and M. Kreitman. 1986. Evolutionary genetics of two sibling species of *Drosophila, D. simulans* and *D. mauritiana*. Evolution 40:673–691.
Coyne, J. A., and H. A. Orr. 1989. Patterns of speciation in *Drosophila*. Evolution 43:362–381.
Crow, J. F. 1942. Cross fertility and isolating mechanisms in the *Drosophila mulleri* group. Univ. Texas Pub. 4228:53–67.
Curtis, C. F. 1982. The mechanism of hybrid male sterility from crosses in the *Anopheles gambiae* and *Glossina morsitans* complexes. In: W. M. F. Steiner (ed.), *Recent Developments of Insect Disease Vectors*. Stipes Publishing Co., New York.
Davidson, G. 1974. *Genetic Control of Insect Pests*. Academic Press, London.
Detlefsen, J. A. 1914. Genetic studies on a cavy species cross. Carnegie Inst. Wash. Pub. 205.
Dobzhansky, Th. 1936. Studies on hybrid sterility. II. Localization of sterility factors in *Drosophila pseudoobscura* hybrids. Genetics 21:113–135.
Dobzhansky, Th. 1937. *Genetics and the Origin of Species*. Columbia University Press, New York.
Dobzhansky, Th. 1974. Genetic analysis of hybrid sterility within the species *Drosophila pseudoobscura*. Hereditas 77:81–88.
Gray, A. P. 1954. *Mammalian Hybrids*. Commonwealth Agricultural Bureaux, Farnham Royal, England.
Gray, A. P. 1958. *Bird Hybrids*. Commonwealth Agricultural Bureaux, Farnham Royal, England.
Grula, J. W., and O. R. Taylor. 1980. Some characteristics of hybrids derived from the sulfur butterflies *C. eurytheme* and *C. philodice*. Evolution 34:673–687.
Haldane, J. B. S. 1922. Sex-ratio and unisexual sterility in hybrid animals. J. Genet. 12:101–109.
Haldane, J. B. S. 1924. A mathematical theory of natural and artificial selection. Part I. Trans. Cambridge Philos. Soc. 23:19–41.
Haldane, J. B. S. 1932. *The Causes of Evolution*. Longmans, Green & Co., London.
Jaffe, E., and C. Laird. 1986. Dosage compensation in *Drosophila*. Trends Genet. December:316–321.
Johnson, M. S., and J. R. G. Turner. 1979. Absence of dosage compensation for a sex-linked enzyme in butterflies. Heredity 43:71–77.
Kacser, H., and J. A. Burns. 1981. The molecular basis of dominance. Genetics 97: 639–666.
Lakhotia, S. C., A. Mishra, and P. Sinha. 1981. Dosage compensation of X-chromosome activity in interspecific hybrids of *Drosophila melanogaster* and *D. simulans*. Chromosoma 82:229–236.
Lakovaara, S., A. Saura, P. Lankinen, L. Pohjola, and J. Lokki. 1976. The use of isoenzymes in tracing evolution and in classifying Drosophilidae. Zool. Scr. 5:173–179.
Lancefield, D. E. 1929. The genetic study of crosses of two races or physiological species of *Drosophila obscura*. Z. Induktiv. Abstamm. Vererbung. 52:287–317.
Lemeunier, F., and M. Ashburner. 1976. Relationships within the *melanogaster* subgroup of the genus *Drosophila* (Sophophora). II. Phylogenetic relationships between six species based upon polytene chromosome banding sequences. Proc. R. Soc. London Ser. B 193:275–294.
Lindsley, D. L., and E. Lifschytz. 1972. The genetic control of spermatogenesis in *Drosophila*. Pp. 203–221 in: R. A. Beatty and S. Gluecksohn-Waelsch (eds.), *Proceedings of the International Symposium on the Genetics of the Spermatozoon*. Bogtrykkeriet Forum, Copenhagen.
Lindsley, D. L., L. Sandler, B. A. David, R. C. Carpenter, R. E. Denell, J. C. Hall, H. Nozawa, D. M. Perry, and M. Gould-Somero. 1972. Segmental aneuploidy and the genetic gross structure of the *Drosophila* genome. Genetics 71:157–184.
Lucchesi, J. C. 1978. Gene dosage compensation and the evolution of sex chromosomes. Science 202:711–716.
Mayr, E. 1940. Speciation phenomena in birds. Am. Natur. 74:249–278.
Mayr, E. 1942. *Systematics and the Origin of Species*. Columbia University Press, New York.

Mitchell, J. A. 1977. Fitness effects of EMS-induced mutations on the X chromosome of *Drosophila melanogaster*. I. Viability effects and heterozygous fitness effects. Genetics 87:763–774.

Muller, H. J. 1940. Bearing of the *Drosophila* work on systematics. Pp. 185–268 in: J. S. Huxley (ed.), *The New Systematics*. Clarendon Press, Oxford.

Muller, H. J. 1950. Evidence of the precision of genetic adaptation. Harvey Lect. 43:165–229.

Naviera, H., and A. Fontdevila. 1986. The evolutionary history of *Drosophila bussatii*. XII. The genetic basis of sterility in hybrids between *D. buzzatii* and its sibling *D. serido* from Argentina. Genetics 114:841–857.

Nei, M. 1972. Genetic distance between populations. Am. Natur. 106:282–292.

Ohmachi, M., and S. Masaki. 1964. Interspecific crossing and the development of hybrids between the Japanese species of *Teleogryllus* (Orthoptera:Gryllidae). Evolution 18: 405–416.

Orr, H. A. 1987. Genetics of male and female sterility in hybrids of *Drosophila pseudoobscura* and *D. persimilis*. Genetics 116:555–563.

Orr, H. A. 1989. Genetics of sterility in hybrids between two subspecies of *Drosophila*. Evolution 43:180–189.

Orr, H. A., and J. A. Coyne. 1988. Exceptions to Haldane's rule in *Drosophila*. Drosophila Inform. Serv. 66:111.

Orr, H. A., and J. A. Coyne. 1989. The genetics of postzygotic isolation in the *Drosophila virilis* group. Genetics, in press.

Patterson, J. T., and R. K. Griffen. 1944. The genetic mechanism underlying species isolation. Univ. Texas Pub. 4445:212–223.

Shaw, D. D., P. Wilkinson, and C. Moran. 1979. A comparison of chromosomal and allozymal variation across a narrow hybrid zone in the grasshopper *Caledia captiva*. Chromosoma 75:333–351.

Sturtevant, A. H., and E. Novitski. 1941. Sterility in crosses of geographical races of *Drosophila micromelanica*. Proc. Natl. Acad. Sci. U.S.A. 27:392–394.

Tan, C. C. 1935. Salivary gland chromosomes in the two races of *Drosophila pseudoobscura*. Genetics 20:392–402.

Temin, R. G. 1966. Homozygous viability and fertility loads in *Drosophila melanogaster*. Genetics 53:27–46.

Throckmorton, L. H. 1982. The *virilis* species group. Pp. 227–296 in: M. Ashburner, H. L. Carson, and J. N. Thompson, Jr. (eds.), *The Genetics and Biology of Drosophila*, Vol. 3b. Academic Press, London.

Tucker, P. K., R. D. Sage, A. C. Wilson, and E. M. Eicher. 1988. The distribution of sex chromosomes in a hybrid zone between two species of European house mouse (*Mus domesticus* and *Mus musculus*). Genome 30 (suppl. 1):387.

Vigneault, G., and E. Zouros. 1986. The genetics of asymmetrical male sterility in *Drosophila mohavensis* and *Drosophila arizonensis* hybrids: Interactions between the Y-chromosome and autosomes. Evolution 40:1160–1170.

Watanabe, T., and Oshima, C. 1973. Fertility genes in natural populations of *Drosophila melanogaster*. II. Correlation between productivity and viability. Jpn. J. Genet. 48: 337–347.

Weisbrot, D. 1963. Studies on differences in the genetic architecture of related species of *Drosophila*. Genetics 48:1131–1139.

White, M. J. D. 1973. *Animal Cytology and Evolution*. Cambridge University Press, Cambridge.

Wright, S. 1929. Fisher's theory of dominance. Am. Natur. 63:274–279.

Wright, S. 1977. *Evolution and the Genetics of Populations*, Vol. 3. University of Chicago Press, Chicago.

Wu, C.-I., and A. T. Beckenbach. 1983. Evidence for extensive genetic differentiation between the sex-ratio and the standard arrangement of *Drosophila pseudoobscura* and *D. persimilis* and identification of hybrid sterility factors. Genetics 105:71–86.

Zouros, E., K. Lofdahl, and P. A. Martin. 1988. Male hybrid sterility in *Drosophila*: Interactions between autosomes and sex chromosomes in crosses of *D. mohavensis* and *D. arizonensis*. Evolution 42:1321–1331.

CHAPTER 9

CHROMOSOMAL DIVERGENCE AND REPRODUCTIVE ISOLATION IN DIK-DIKS

Oliver A. Ryder, Arlene T. Kumamoto, Barbara S. Durrant, and Kurt Benirschke

INTRODUCTION

The scientific discipline of systematics and the conservation of the Earth's biota are increasingly linked in the face of ever-increasing demands from humankind (Wilson 1985). Similarly, in the context of conservation issues, definitions of species and the evolutionary significance of populations definable below the level of the biological (isolation) concept of species are of more than theoretical interest as management decisions with potentially irreversible consequences affecting habitats and biological populations are made (Lovejoy 1980; Benirschke 1985; Ryder 1986a, 1986b, 1987).

DIK-DIKS

The small and graceful antelopes of the genus *Madoqua* occur in eastern and northeastern Africa, a single exception being *M. kirki damarensis* that occurs in southwest Africa. Somewhat larger than hares, the males and females alike have crests of erectile hair on their foreheads. Only males have small, pointed horns. Females average 20% larger and heavier than males. The name dik-dik derives from one of their vocalizations ("zick-zick;" often an alarm call) (Haltenorth and Diller 1980; Ansell 1971). Dik-diks are territorial and deposit secretions from their prominent preorbital glands to mark their territorial boundaries. A breeding pair is the unit of reproduction (Hendricks 1975).

Five species of *Madoqua* are generally recognized. *M. kirki* and *M. guentheri* constitute the subgenus *Rhynchotragus* because of their elongated probosces and accompanying strong curvature of the premaxillae (Ansell 1971). The ranges of these two species are presented in Figure 1.

Because of their diminutive size, appealing facial features, monogamous life-style, and other interesting life history attributes, dik-diks are desirable for exhibition in zoos. Of the five dik-dik species, only the two species of the subgenus *Rhynchotragus* have recently been exhibited in North American zoos. However, in spite of births within zoo collections, the establishment of a self-sustaining population has not been achieved. In addition to genetic factors such as inbreeding depression (Ballou and Ralls 1982) and the outbreeding depression described here, environmental factors such as husbandry and nutrition most likely have also detracted from efforts to establish self-sustaining captive populations.

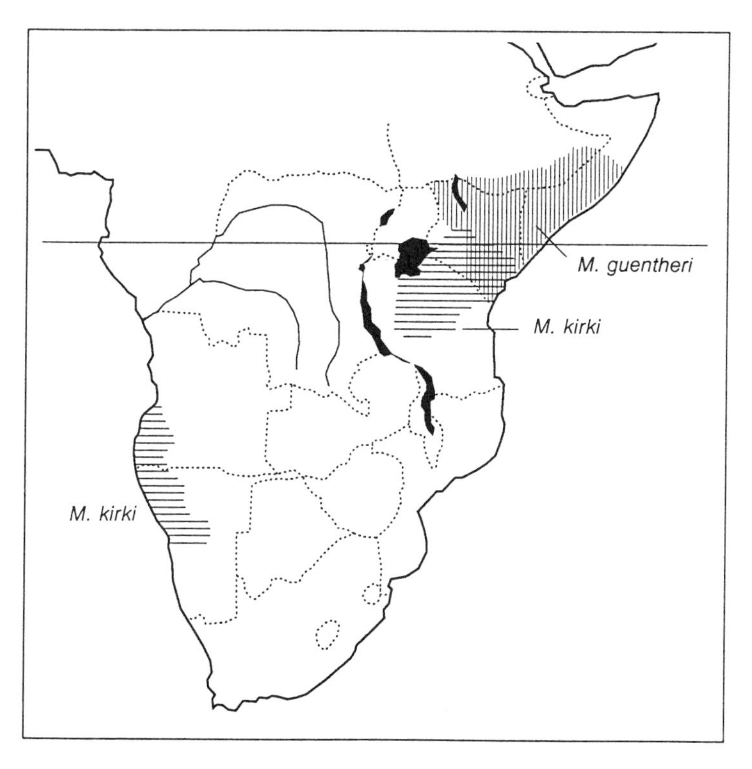

FIGURE 1. African distribution of *Madoqua kirki* and *M. guentheri*. After Kingdon (1982) for East Africa and Haltenorth and Diller (1980) for southwest Africa.

Previous cytogenetic studies of the subgenus *Rhynchotragus* demonstrated the chromosomal complexity of the long-snouted dik-diks. Chandra et al. (1967) studied a presumed *M. kirki* specimen from East Africa at the Philadelphia Zoo with findings of $2n = 46$ chromosomes, including one large pair of acrocentric chromosomes. A female Kirk's dik-dik held at the Cincinnati Zoo was studied by Platt and Soukup (1977) with findings of $2n = 48$ chromosomes, including two pairs of metacentrics. Benirschke and Kumamoto (1987) first definitively identified the presence of two karyotypic classes in *M. kirki*, one with a compound X chromosome. They also first described the karyotype of *M. guentheri* and the Robertsonian rearrangement involving MGU5 and MGU17, as well as the existence in captivity of cytological hybrids between the described karyotype classes. Unresolved by previous studies have been the reproductive consequences of the chromosomal diversity in dik-diks of the subgenus *Rhynchotragus* and the delineation of the specific chromosomes involved in the karyotypic heteromorphisms.

We describe here the results of a cytogenetic survey of 7 founder animals and 51 individuals bred in captivity with findings of remarkable karyotypic diversity and accompanying infertility. Cytotype hybrid animals possessing specific chromosomal rearrangements producing meiotic breakdown and thereby resulting in postmating reproductive isolation in the subgenus *Rhynchotragus* are identified.

MATERIALS AND METHODS

Primary tissue culture and subsequent harvesting of mitotic cells for cytogenetic analysis followed methods described by Kumamoto and Bogart (1984).

Photographs of many of the karyotyped individuals have been taken. Relatively few of the karyotyped animals have died and, consequently, voucher specimens have yet to be deposited. Testicular and epididymal tissue was obtained from necropsy following the accidental death of the *M. kirki* cytotype *b* male (ID181019; Table 1). Testicular and epididymal biopsies were performed surgically on *M. kirki* cytohybrid male (ID25433; Table 1) and a male *M. kirki* × *M. guentheri* hybrid (ID27969; Table 1).

Electroejaculation and surgical procedures were performed on anesthetized animals in accordance with NIH animal care guidelines. The Zoological Society of Philadelphia granted permission for nonterminal investigative studies of a male interspecies hybrid under their ownership.

RESULTS

Pedigrees depicting the 58 individuals studied are presented in Figure 2. To our knowledge, all the existing dik-diks in North American zoos could

TABLE 1. Dik-dik electroejaculation.

ID	Karyotype	Born	Weight (lbs)	Testicular measure (w/o epididymis)	Date	Volume (ml)	Concentration ($\times 10^6$)	Motility	SOP	% Ab
103534	Kirk's *b*	8/78	8.50	R: 24.7 × 11.1	3/10/87	0.10	85	95	5	39
	$2n = 47$			L: 26.6 × 11.9						
181019	Kirk's *b*	5/83	8.00	R: 20.7 × 14.8	3/27/87	0.01	10	10	2	75[a]
	$2n = 47$			L: 23.0 × 13.6	5/01/87	0.01	10	0	—	80[a]
25433	Kirk's *ab*	8/84	12.00	R: 22.5 × 16.0	3/27/87	0.01	0	—	—	[b]
	$2n = 47$			L: 25.0 × 16.0	5/01/87	0.01	0	—	—	—
27969	Guenther's × Kirk's *b*	1/82	8.75	R: 16.6 × 09.5	3/11/87	0.10	0	—	—	[b]
	$2n = 48$			L: 19.4 × 11.3	4/16/87	0.10	0	—	—	—
27970	Guenther's × Kirk's *b*	10/82	8.50	R: 17.0 × 10.1	3/11/87	0.10	0	—	—	[b]
	$2n = 48$			L: 17.3 × 10.0	4/16/87	0.01	0	—	—	—

[a] A few Sab spermatids observed in ejaculate.
[b] Sab spermatids only.

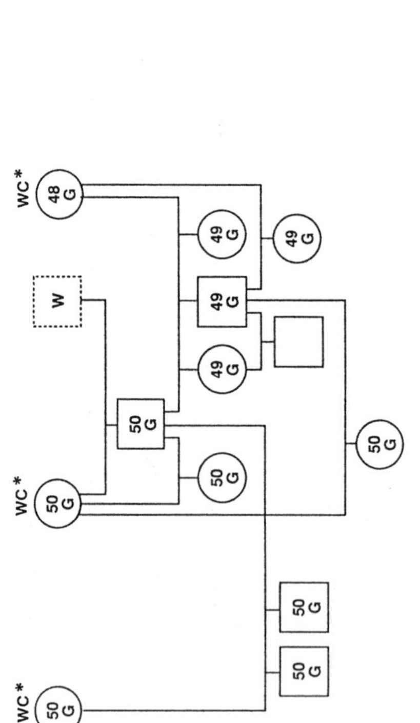

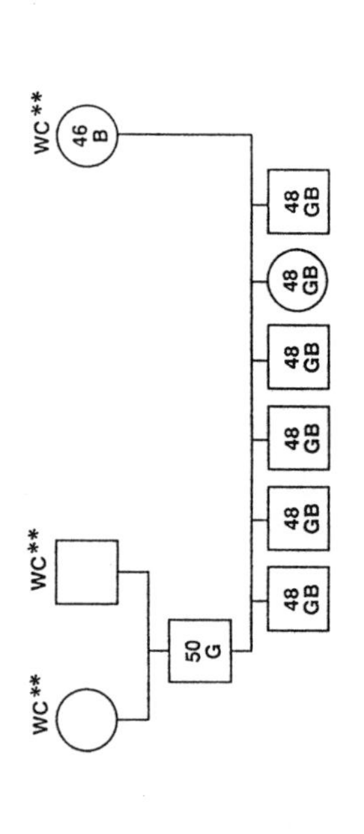

FIGURE 2. (A) Pedigree of *M. guentheri.* *1984 importation from Lake Baringo, Kenya. One of the females imported in 1984, pregnant by an unidentified male, subsequently gave birth to a $2n = 50$ male. ** 1978 importation from Mbalambala, Kenya. Note that one of the females in this importation was actually a *M. kirki* cytotype *a* ($2n = 46$). Subsequent mating of this female to a $2n = 50$ male *M. guentheri* resulted in interspecies hybrids. WC denotes wild-caught individuals. (B) Partial pedigree of *M. kirki.* *1977 importation from Garissa, Kenya. **1972 importation from Kenya. ***1962 importation, probably from Kenya. This individual is an F_1. The two females from the 1972 importation were pregnant on arrival in captivity, presumably having bred in the wild. Both females produced male offspring while in quarantine and parentage assignments were confused, as indicated by the broken lines and question marks. WC denoted wild-caught individuals.

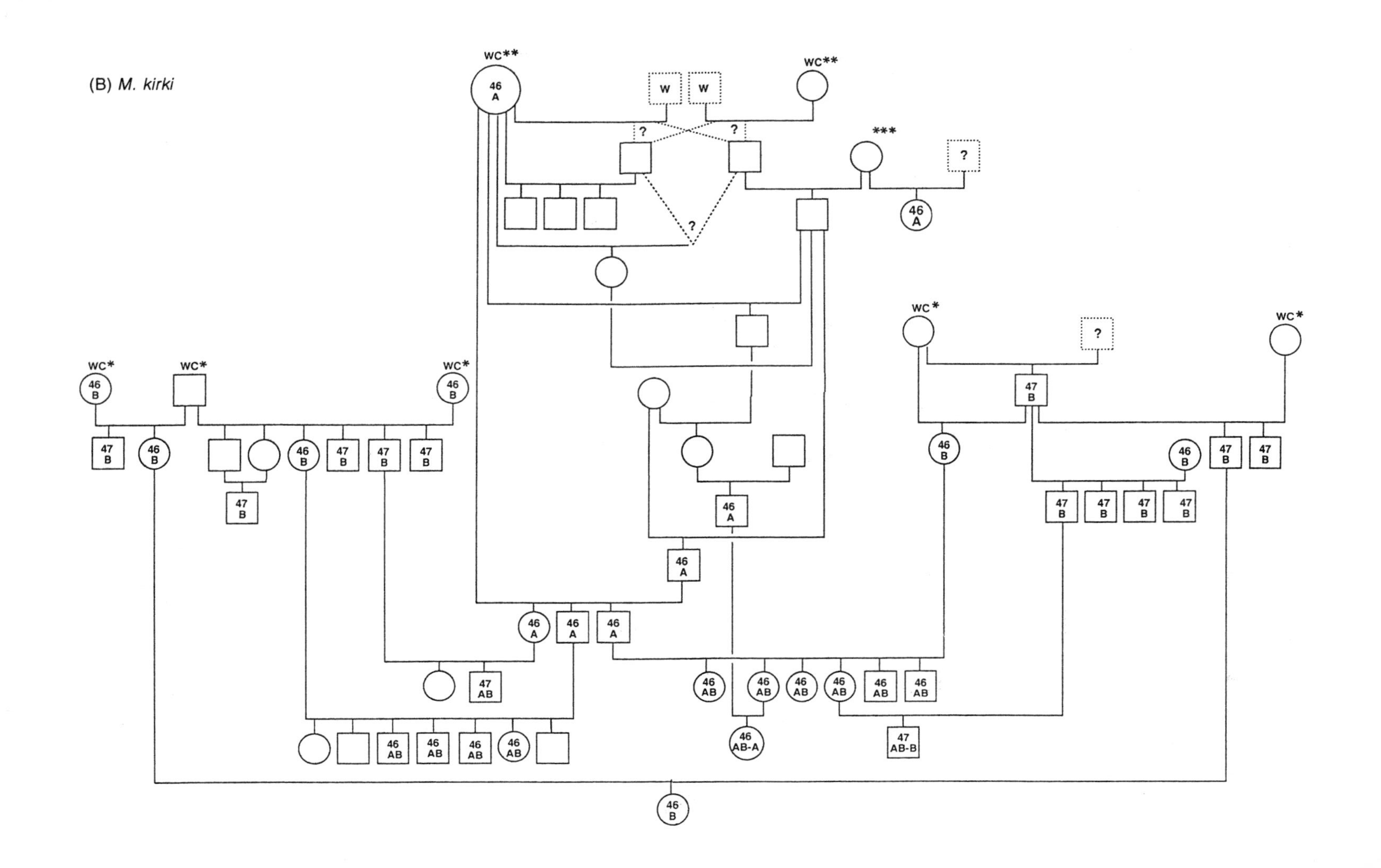

wc**
46
A
W
W
wc**
?
?
***
?
46
A
?
wc*
?
wc*
wc*
46
B
wc*
wc*
46
B
47
B
46
B
46
B
47
B
47
B
47
B
46
B
46
B
47
B
47
B
47
B
46
A
47
B
47
B
47
B
47
B
46
A
46
A
46
A
46
A
47
AB
46
AB
46
AB
46
AB
46
AB
46
AB
46
AB
46
AB
46
AB
46
AB
46
AB
46
AB-A
47
AB-B
46
B

(B) *M. kirki*

be derived from as few as 17 animals or as many as 19 individuals. Of the 14 animals imported to North America since the late 1960s, seven have been karyotyped. Chromosomal data from first-generation offspring infer karyotypic status of an additional six imported individuals.

Extensive chromosomal variation characterizes the collected set of karyotypes of the imported and first-generation captively-bred dik-diks. Comparison of the assembled karyotype data suggests that three generally distinct chromosomal races exist among the imported dik-dik specimens.

The karyotype of the *Madoqua guentheri* (Günter's dik-dik) individuals is quite distinct from all the other dik-diks studied. Of three imported animals studied (all females), two had diploid numbers of 50 and one had $2n = 48$ chromosomes, the karyotypes differing by a chromosome fission/fusion of the classical Robertsonian type (Figure 3). Offspring from the

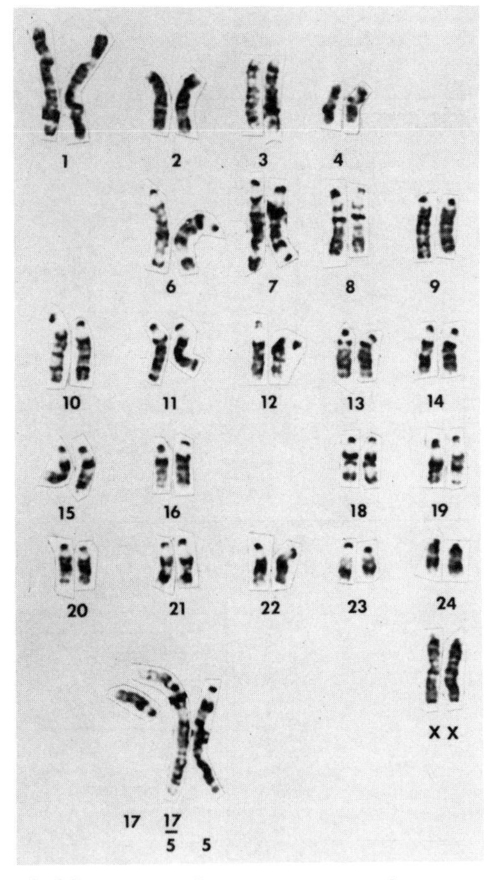

FIGURE 3. G-banded karyotype of $2n = 49$ *M. guentheri*. Note the single fusion product of chromosomes 17 and 5 with corresponding homologous elements adjacent.

imported $2n = 48$ female and a $2n = 50$ male possessed the expected 49 chromosomes (Benirschke and Kumamoto 1987). Both male and female *M. guentheri* with $2n = 49$ karyotypes have successfully reproduced in captivity, as depicted in the pedigree presented in Figure 2A.

The karyotypes of the examined specimens of Kirk's dik-dik, *M. kirki*, present greater intraspecific variability than that observed among *M. guentheri* specimens. Two distinct but nonhomogeneous cytotype classes are currently designated. Cytotype *a* individuals possess $2n = 46$ chromosomes (Figure 4A). In cytotype *b* individuals (Figure 4B) a translocation

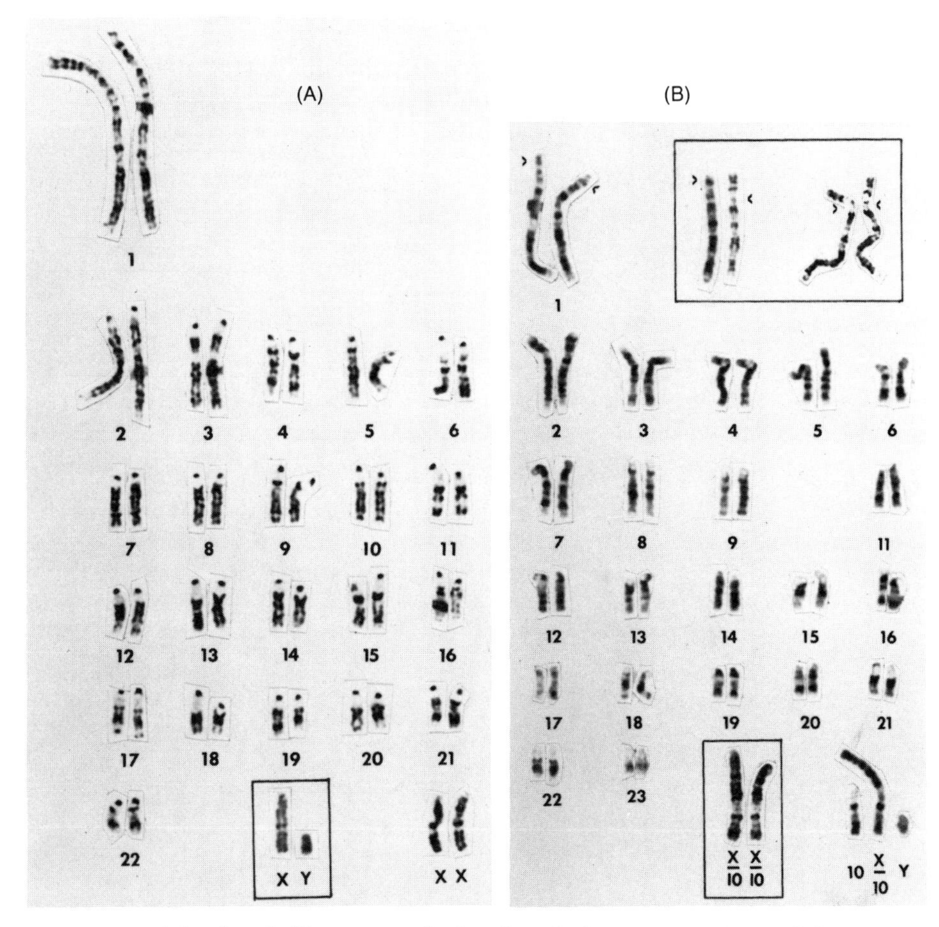

FIGURE 4. (A) G-banded karyotype of a female *M. kirki* cytotype *a*, $2n = 46$. Inset box shows banded sex chromosomes of a male. (B) G-banded karyotype of a male *M. kirki* cytotype *b*, $2n = 47$. Note translocation involving chromosomes 10 and X. Lower box inset shows banded sex chromosomes from a female. Note both Xs with translocated chromosome 10. Upper inset box depicts polymorphic condition of chromosome 1. Centromere positions are indicated by arrow heads.

involving the X chromosome and chromosome 10 has occurred such that males possess one more chromosome ($2n$ = 47), corresponding to the autosomal portion of the compound X chromosome, than do females of cytotype *b* ($2n$ = 46). Unlike the situation present in our sample of *M. guentheri*, there is no simple centric fusion/fission differentiating the two forms. Multiple rearrangements, including inversions and tandem fusions, distinguish the two *M. kirki* cytotypes (Figure 5). *M. kirki* cytotype *a* can be traced back to a 1972 importation from Kenya of two females, both pregnant with male offspring, whereas cytotype *b* can be traced to a 1977 importation from Garissa, Kenya, of one male and four females, one of which was pregnant (Figure 2).

Analysis of the pedigree data and population cytogenetics data for *M. kirki* (Figure 2B) reveals that only female offspring of matings between cytotype *a* and cytotype *b* parents have themselves produced offspring,

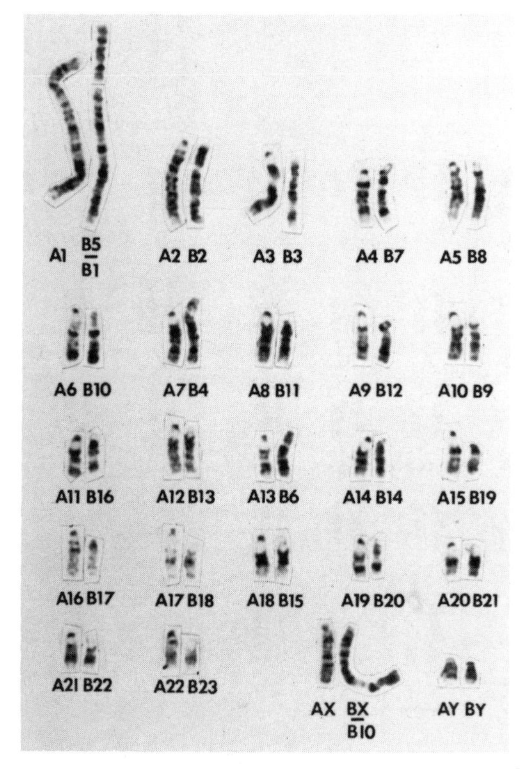

FIGURE 5. Composite karyotype consisting of G-banded haploid karyotypes of *M. kirki* cytotype *a* and cytotype *b*. Homologous chromosomes are adjacent. **A** denotes cytotype *a*; **B** denotes cytotype *b*. Numbers indicate assigned chromosome numbers as presented in Figure 4A and B.

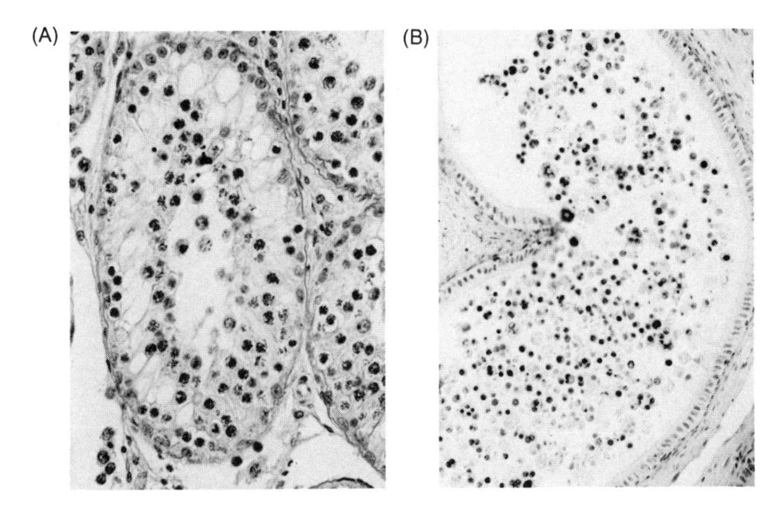

FIGURE 6. (A) Histology of testis in a male Kirk's dik-dik, the offspring of a male cytotype *b* and female cytotype *a*. (B) Histology of epididymis in the same male Kirk's dik-dik cytohybrid.

although individuals of both sexes have had opportunity to breed. Comparison of the haploid complements of *M. kirki* cytotype *a* and cytotype *b* (Figure 5) raises the possibility that, due to karyotypic divergence, individuals in populations of the two cytotypes are partially or completely reproductively isolated.

An assessment of the reproductive potential of male *M. kirki* cytotype *a* × *b* hybrids (cytohybrids) has been made from analysis of ejaculates obtained following mild rectal electrical stimulation [electroejaculation (Martin 1978)] (Table 1) and from histological analysis of testicular and epididymal sections (Figure 6A and B). Results of similar investigations on a male *M. kirki* cytotype *b* individual are included in Table 1 and in Figure 7. The *M. kirki* cytohybrid male was 3 years old and sexually mature. However, in spite of the presence of extensive meiotic activity, gametic maturation was completely arrested. Arrested meiotic products, including spermatocytes and cells with apparently haploid nuclei, could rarely be seen, but no spermatozoa were observed in either testicular tubules (Figure 6A) or in epididymal sections (Figure 6B). The results obtained following electroejaculation (Table 1) were consistent with the histological findings; no spermatozoa were obtained from the *M. kirki* cytohybrid male although some spermatids were noted.

The chromosomal differences between *M. guentheri* and either *M. kirki* cytotype are also quite dramatic and likely involve a larger number of rearrangements than for the two *M. kirki* cytotypes. A composite karotype

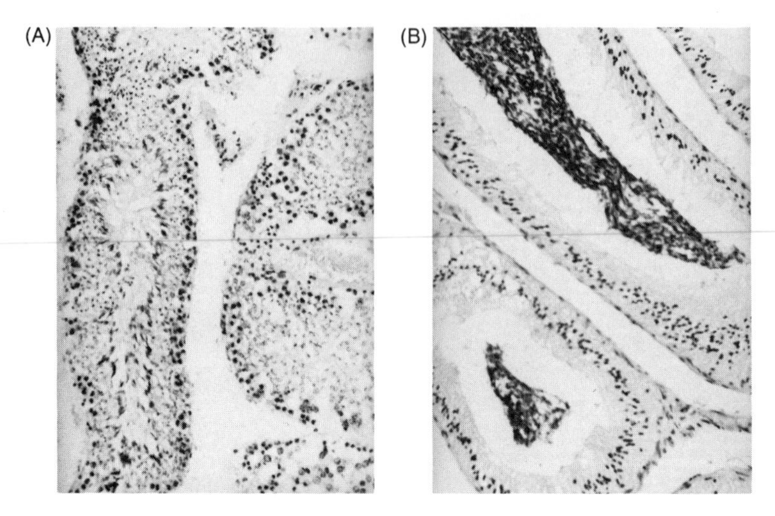

FIGURE 7. (A) Histology of testis in a male Kirk's dik-dik cytotype *b*. (B) Histology of epididymis from the same male.

of *M. kirki* cytotype *b* and *M. guentheri* allows comparison of the chromosomes of the two species (Figure 8). In the captive pedigree analyzed here, presumed interspecific hybrids have been produced involving only a female *M. kirki* cytotype *b* and a male *M. guentheri*. None of the six offspring (five males, one female) of this cross has reproduced.

Reproductive assessment of two *M. kirki* × *M. guentheri* male hybrids based on semen analysis (Table 1) or testicular histology (Figure 9) presents a picture of complete male sterility in both sexually mature animals. In four procedures no spermatozoa were obtained from electroejaculation. Testis size in the two hybrid males was notably smaller than in *M. kirki* cytotype *b* or cytohybrid males (Table 1). Histologically, a complete arrest of meiosis is noted, the testicular tubules are completely sterile, and the meiotic disturbance is considerably greater than that noted in the *M. kirki* cytohybrid male (Figure 9).

DISCUSSION

Zoo curators have recognized two species-level taxa among the imported dik-diks based on pelage and other external morphological differences: *M. kirki* and *M. guentheri*. The imported dik-diks and their descendants comprise three distinct chromosomal classes, two of which are themselves polymorphic. Relatively few of the imported specimens identified as *M. guentheri* or as cytotype *b M. kirki* have died. As a consequence voucher

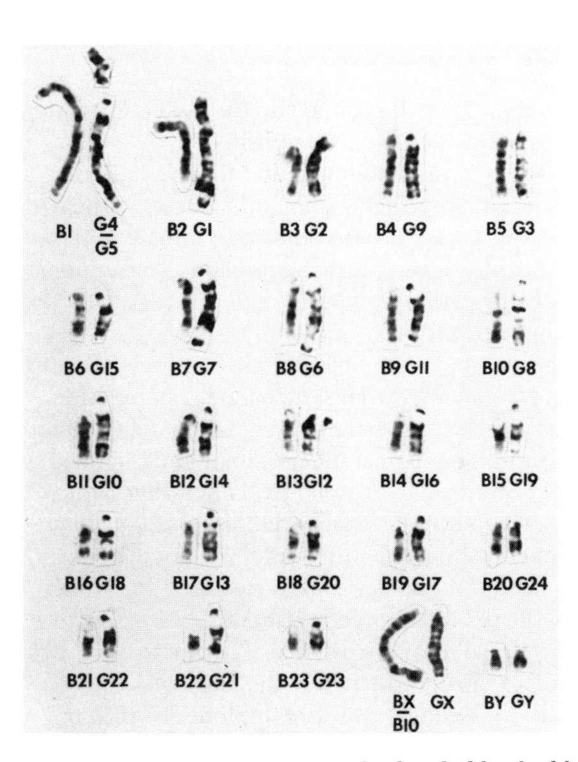

FIGURE 8. Composite comparison consisting of G-banded haploid karyotypes of *M. guentheri* and *M. kirki* cytotype *b*. **B** denotes *M. kirki* cytotype *b*; **G** denotes *M. guentheri*. Numbers indicate assigned chromosome numbers presented in Figures 4B and 3, respectively.

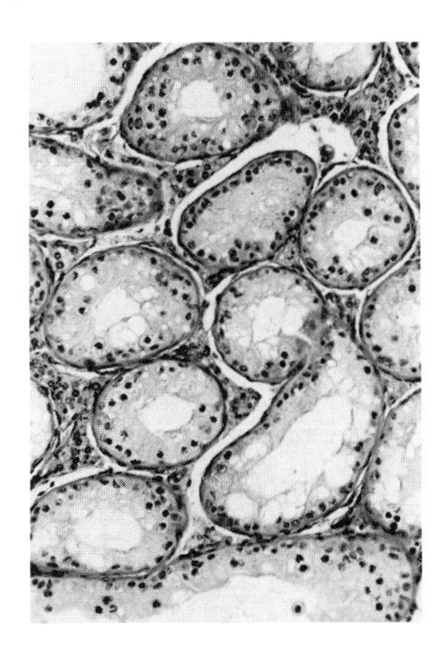

FIGURE 9. Histology of testis in a male hybrid offspring of a *M. guentheri* male and a *M. kirki* cytotype *b* female.

specimens have yet to be deposited and the collection of dental and cranial measurements remains to be accomplished.

Five apparently independent importations of dik-diks are depicted in Figure 2. Two of the importations include all the wild-caught *M. guentheri* that have been karyotyped (as well as one female *M. kirki* cytotype *b* individual). Three dik-dik importations provide the founder stock for the zoo populations of *M. kirki* cytotypes *a* and *b* as well as their descendant cytohybrid population.

One of the distinct chromosomal classes includes all the individuals identified as *M. guentheri*. This cytological population, although polymorphic for a Robertsonian arrangement involving chromosomes 5 and 17, may also constitute a biological species. Four geographical subspecies of *M. guentheri* are recognized by Ansell (1971). However, the issue of partitioning of intraspecific chromosomal variation in *M. guentheri* is obscured by lack of information about the ranges of the named *M. guentheri* subspecies and capture locale for the specimens destined for North American zoos. Nonetheless, the dik-diks discussed here as *M. guentheri* are clearly reproductively isolated from the dik-diks we refer to as *M. kirki* cytotype *b*, as evidenced by lack of reproduction in the *M. guentheri* × *M. kirki* cytotype *b* hybrids of either sex and by the complete sterility of hybrid males as demonstrated by semen evaluation and histological examination of testis and epididymis. Considering that the chromosomal rearrangements distinguishing the two groups include an X autosome translocation (MKI X; MKI 10) and a tandem fusion (MGU 4; MGU 5) in addition to additions, deletions, and, possibly, translocations of chromosomal material (Figure 8), this result is not particularly surprising. It is notable that, as there have been no matings between *M. guentheri* and *M. kirki* cytotype *a* individuals, no similar test of fertility exists for presumed interspecies hybrid individuals lacking an X autosome translocation.

Two of the discrete chromosomal classes are present in what zoo curators have considered a single species, *M. kirki*. Analysis of pedigrees (Figure 2) indicates that, most probably, cytotype *a* individuals comprised two of the importations (1972 and an earlier importation). The 1977 importation from Garissa, Kenya, was the source of the cytotype *b* individuals. (A single $2n = 46$ cytotype *b* individual was included in the 1978 importation from Mbalambala, Kenya that was also a source of *M. guentheri*.) Seven geographically defined subspecies of *M. kirki* are recognized by Ansell (1971) who followed Hollister (1924) with respect to the East African dik-diks. Kingdon (1982), noting the distribution of *M. guentheri* in east Africa and *M. kirki* in both east and southwest Africa, has suggested that the existence of morphological intergrades between the two species is evidence of secondary contact following isolation. The extensive geographic distribution of specimens displaying intermediate or mixed characters (Ansell 1968; cited in Kingdon 1982) would now seem to be less likely due to backcrossing adjacent to zones of hybridization (because of the sterility of F_1 hybrids) than to other causes.

Based upon inferences of geographic origin, one zoo curator identified individuals of the 1972 importation as *M. kirki hindei* and individuals of the 1977 importation as *M. kirki minor* (Dolan, personal communication, 1987). Whether or not the trinomial designations can be applied with confidence, the existence of two chromosomally distinct groups within *M. kirki* has been demonstrated. Furthermore, it is likely that the two *M. kirki* chromosomal races correspond to separate populations in nature and may be referred to as cytotypes. The infertility of the cytohybrid male (Table 1 and Figure 6) provides evidence for reproductive barriers between the source populations and suggests that the cytotype populations may be behaving like biological species (although cryptic morphologically). Only field studies could establish whether the cytotypes are distributed allopatrically or sympatrically.

As antelopes, dik-diks are remarkable for their monogamy and site fidelity that extends through the lifetimes of individuals (Hendricks 1975; Tilson and Tilson 1986). Offspring of territorial pairs are forcibly dispersed from their natal territory at 6–9 months of age (Hendricks 1975). Although some speciation theories emphasize chromosomal evolution through inbreeding in small demes (see Sites and Moritz 1987 for a review), insufficient behavioral data exist to establish to what extent inbreeding is favored by dispersal patterns in dik-diks. Nonetheless, if life history traits and social structure favor relatively rapid rates of chromosomal evolution and speciation in *Madoqua*, allopatric populations, geographic subspecies, and/or sympatric species may be chromosomally divergent. The existence of additional chromosomally divergent populations stands as an investigative challenge.

Additional cytogenetic studies of animals with known capture locales are now necessary to evaluate the systematics, zoogeography, and modes of speciation in dik-diks. Investigation of the chromosomes of the Damara dik-diks, *M. kirki damarensis*, in conjunction with electrophoretic and other molecular genetic estimates of genetic distance, should provide additional insights into dik-dik speciation and systematics.

Although the largest study of its kind, the presently available data on chromosomal evolution in dik-diks lack sufficient corroborative data to discriminate which of the available descriptive models of chromosomal evolution best applies to this group of species (see Sites and Moritz 1987 for a recent review). Representatives of the genus *Madoqua* appear in the fossil record in the Pliocene (possibly in the late Miocene) (Gentry 1987; Savage and Russell 1983), providing an upper age limit for the speciation events resulting in the extant forms within, approximately, the last 5 million years.

However, the morphological similarity of dik-diks that are reproductively isolated through chromosomal divergence might imply that the speciation event(s) that have resulted in reproductive isolation have occurred relatively recently, possibly not longer ago than the Pleistocene (see Kingdon 1982).

To date, relatively few data have been published relating the analysis of semen obtained by electroejaculation with reproductive performance in endangered species, with the notable exception of the cheetah (Wildt et al. 1983). Although we describe here the special case of semen analysis as an indicator of meiotic disruption in consequence of chromosomal incompatibilities, the concordance between the results of the electroejaculation procedures and the subsequent testicular and epididymal histology is striking and recommends further application of such an approach for assessment of infertility when chromosomal imbalances are known, inferred, or suspected.

Diagnosis of reproductive failure in zoological collections as a result of chromosomal incompatibility has been observed in other species, notably *Aotus trivirgatus* (Cicmanec and Campbell 1977; deBoer 1982) and *Gazella soemmeringi* (Benirschke et al. 1984). As knowledge of comparative cytogenetics advances, chromosomal analysis is indicated for increasing numbers of taxa in order that captively bred populations accurately reflect the genetic constitution of wild noncaptive populations (Ryder 1986b). Certainly, any newly imported dik-diks should have their chromosomes examined to ensure appropriate pairing for reproduction.

PROSPECTS FOR FUTURE RESEARCH

Chromosomal studies of dik-diks provide another example of the gaps that exist in our current knowledge of mammalian systematics. Lack of knowledge of the systematics and evolutionary genetics of numerous threatened taxa is increasingly recognized as an obstacle to conservation planning and management efforts.

The identification of morphologically cryptic taxa possessing sufficient chromosomal divergence to provide postmating reproductive isolation identifies dik-diks as a notable mammalian speciation model meriting additional investigation. The collection of additional information on the population ecology and genetics of dik-diks, including information on life history traits, dispersal, cytogenetics of populations of known geographic origin, mitochondrial DNA variation within and between populations, and intra- and interpopulation protein electrophoretic estimates of genetic distance will enable a clearer understanding of the systematics and the processes of speciation in *Madoqua*. Preliminary investigation of protein genetic distances between individuals of the three chromosomal races of dik-diks could provide insights into the temporal aspects of the remarkable chromosomal divergence observed for dik-diks in the subgenus *Rhynchotragus*.

SUMMARY

Five separate importations into U.S. zoos of dik-diks from East Africa, *Madoqua* sp., have occurred recently. Chromosomal studies of the wild-born, imported, individuals and/or their descendants have identified remarkably different karyotypes. To date, 58 individual dik-diks have been karyotyped as part of this study and three major cytotypes have been identified by G-banding analyses. All the wild-born dik-dik specimens were imported as *Madoqua kirki*; subsequently, all but one of the specimens imported in 1978 from Mbalambala, Kenya, have been reclassified as *Madoqua guentheri*.

The karyotype of *M. guentheri* is quite distinct from that of *M. kirki*. The diploid number of *M. guentheri* ranges for $2n = 48$ to $2n = 50$, due to a Robertsonian translocation. There are two quite distinct cytotypes of *M. kirki* that we designate *a* and *b*. *M. kirki* cytotype *a* has $2n = 46$, whereas *M. kirki* cytotype *b* has $2n = 47$ in males and $2n = 46$ in females. Unlike the situation present in our sample of *M. guentheri*, there is no simple centric fusion/fission differentiating the two forms. Multiple rearrangements, including inversions, and X autosome translocation, and tandem fusions, distinguish the two *M. kirki* cytotypes. *M. kirki* cytotype *a* can be traced back to a 1972 Kenyan importation whereas cytotype *b* can be traced to a 1977 importation from Garissa, Kenya. Although seven subspecies of *M. kirki* have been recently recognized, due to uncertainties in the distribution of the named subspecies and in the absence of definitive cranial data, we choose not to assign existing subspecific names to correspond to observed cytotypes.

Analysis of pedigrees in conjuction with the karyotypic information reveals that although offspring of matings between $2n = 48$ and $2n = 50$ *M. guentheri* produce fertile offspring of both sexes, the offspring of matings between $2n = 50$ *M. guentheri* and $2n = 46$ *M. kirki* have failed to reproduce, having had opportunity to do so. Histological examination of testis and epididymal sections obtained from a *M. guentheri* × *M. kirki* F_1 hybrid revealed a complete lack of spermatozoa. Furthermore, although the female offspring of mating between *M. kirki* cytotype *a* and cytotype *b* individuals have themselves produced offspring, none of the male offspring of such crosses is recorded as having been successful sires. Histological examination of testis sections revealed severe disruption of spermatogenesis and a complete lack of mature spermatozoa in a cytotype-hybrid male.

The degree of karyotypic differentiation between the species *M. kirki* and *M. guentheri* is itself remarkable. Perhaps of greater interest is the finding of sufficient karyological differentiation between individual dik-diks from two separate capture operations (and presumably belonging to two allopatric subspecies) to produce male sterility in cytotype-hybrid offspring. We conclude that the two populations of *M. kirki* are behaving like biological species, similar to Apennine *Mus* populations. Collection of cytogenetic data from dik-diks with known capture locale will enable a clearer understanding of systematics and speciation within *Madoqua*.

ACKNOWLEDGMENTS

Supported by the Zoological Society of San Diego, NIH Grant GM23073, a Conservation Project Grant from the Institute of Museum Services, and the Caeser Kleberg Foundation. We thank the many zoo curators and veterinarians who kindly provided information and samples for analysis. P. T. Robinson, DVM performed anesthesia and surgical procedures. Technical assistance was provided by Suellen Charter, Mike Graves, Marlys L. Houck, Patricia Sarver, Tally L. Wright, and Joan Yamada. Katherine Ralls and two anonymous reviewers provided helpful suggestions to an earlier version of the manuscript

LITERATURE CITED

Ansell, W. F. H. 1968. In: J. Meester and H. W. Setzer (eds.), *A Preliminary Guide to the Mammals of Africa*. Smithsonian Institution. Washington, D.C.

Ansell, W. F. H. 1971. Artiodactyla. In: J. Meester and H. W. Setzer (eds.), *The Mammals of Africa*. Smithsonian Institution. Washington, D.C.

Ballou, J., and Ralls, K. 1982. Inbreeding and juvenile mortality in small populations of ungulates: A detailed analysis. Biol. Conserv. 24:239–272.

Benirschke, K. 1985. the genetic management of exotic animals. In: J. P. Hearn and J. K. Hodges (eds.), *Advances in Animal Conservation*. Symp. Zool. Soc. London 54:71–88.

Benirschke, K., and Kumamoto, A. T. 1987. Challenges of Artiodactyl cytogenetics. La Chromosoma II-45, in press.

Benirschke, K., Kumamoto, A. T., Olsen, J. H., Williams, M. M., and Oosterhuis, J. 1984. On the chromosomes of *Gazella soemmeringi*, Cretzschmar, 1926 Z. Saügetierk. 49: 368–373.

Chandra, H. S., Hungerford, D. A., and Wagner, J. 1967. Chromosomes of five artiodactyl mammals. Chromosoma (Berlin) 21:211–220.

Cicmanec, J. C., and Campbell, A. K. 1977. Breeding owl monkeys (*Aotus trivirgatus*) in a laboratory environment. Lab. Anim. Sci. 27:512–517.

de Boer, L. E. M. 1982. Karyological problems in breeding owl monkeys. Int. Zoo Yb. 22:119–124.

Gentry, A. W. 1987. Artiodactyla. In: V. J. Maglio and H. B. S. Cooke (eds.), *Evolution of African Mammals*. Harvard University Press, Cambridge, MA.

Haltenorth, T., and Diller, H. 1980. *A Field Guide to the Mammals of Africa Including Madagascar*. Collins, London.

Hendricks, H. 1975. Change in a population of dikdik, *Madoqua (Rhynchotragus) kirki* (Günter, 1880). Z. Tierpsychol. 38:55–69.

Hollister, N. 1924. East African mammals in the U.S. National Museum. Bull. Am. Mus. Nat. Hist. 99(3):1–151.

Kingdon, J. 1982. *East African Mammals*. Vol. III, Part C: *Bovids*. Academic Press, London.
Kumamoto, A. T., and Bogart, M. H. 1984. The chromosomes of Cuvier's gazelle. In: O. A. Ryder and M. L. Byrd (eds), *One Medicine*. Springer-Verlag, New York.
Lovejoy, T. E. 1980. Tomorrow's ark: By invitation only. Int. Zoo Yb. 20:181–183.
Martin, I. C. A. 1978. The principles and practice of electroejaculation of mammals. Symp. Zool. Soc. London 43:127–152.
Platt, R. N., and Soukup, S. 1977. Banding studies in for artiodactyl and one primate species. Mammal. Chromos. Newsl. 18(4):122–123.
Ryder, O. A. 1986a. Species conservation and systematics: The dilemma of subspecies. Trends Ecol. Evol. 1:9–10.
Ryder, O. A. 1986b. Genetics investigations: Tools for supporting breeding programme goals. Int. Zoo Yb. 24/25:157–152.
Ryder, O. A. 1987. Conservation action for gazelles: An urgent need. Trends Ecol. Evol. 2:143–144.
Savage, D. E., and Russell, D. E. 1983. *Mammalian Paleofaunas of the World*. Addison-Wesley, Reading, MA.
Sites, J. W., Jr., and Mortiz, C. 1987. Chromosomal evolution and speciation revisited. Syst. Zool. 36:153–174.
Tilson, R. L., and Tilson, J. W. 1986. Population turnover in a monogamous antelope (*Madoqua kirki*) in Namibia. J. Mammal. 67:610–613.
Wildt, D. E., Bush, M., Howard, J. G., O'Brien, S. J., Meltyer, D., VanDyke, A., Ebedes, H., and Brand, D. J. 1983. Unique seminal quality in the South African cheetah and a comparative evaluation in the domestic cat. Biol. Reprod. 26:1019–1025.
Wilson, E. O. 1985. The biological diversity crisis. BioScience 35:700–706.

PART THREE

EFFECTS OF POPULATION STRUCTURE ON SPECIATION

CHAPTER 10

FOUNDER EFFECT SPECIATION

N. H. Barton

INTRODUCTION

There are many definitions of "species." Perhaps the most widely accepted is that of "biological species": a group of populations possessing inherited differences that prevent gene exchange with other such groups (Dobzhansky 1937; Mayr 1942). In this view, speciation consists of the permanent separation of lineages. Alternatively, a species can be seen as being at a distinct equilibrium, which is stable toward introgression of foreign genes (see Carson 1985). Speciation is now the branching of a population onto different "adaptive peaks." Though these two definitions are different, they overlap: most mechanisms of reproductive isolation lead to multiple stable equilibria, and, conversely, gene exchange between populations at different "adaptive peaks" will usually be impeded.

In either view, speciation is harder to explain than straightforward adaptation. By definition, movements away from the current stable equilibrium will be opposed, and novel genotypes that are partially reproductively isolated from the bulk of the species will leave fewer offspring, and so will be eliminated. The basic difficulty is that changes that might lead to reproductive isolation or to peak shifts will be hard to establish. Other reasons why species should constitute coherent units, resistant to change, have been put forward. Gene flow might swamp changes that evolve in limited regions, extensive coadaptation might demand that changes in one gene or character could occur only if accompanied by simultaneous change elsewhere, and the limited range of possible developmental pathways, perhaps enhanced by the evolution of homeostasis, might make the phenotype insensitive to genetic perturbations. If species were as inert as these arguments suggest, then adaptation as well as speciation would be impeded.

In reaction to these apparent difficulties, various models of "founder effect speciation" have been proposed (Mayr 1954, 1982; Carson 1968,

1975, 1982; Templeton 1980; Kaneshiro 1980). A few individuals found a population that is geographically isolated from the ancestral species and that rapidly expands to fill a new area. Random drift and changes in selection cause a rapid shift to a new, coadapted combination of alleles, and hence cause reproductive isolation. The stress is on the genetic, rather than the ecological, consequences of founder events: selection pressures are seen as changing because of a general reduction in heterozygosity (Mayr 1954), deviations from Hardy–Weinberg proportions (Templeton 1980), or rapid population growth (Carson 1968, 1975), rather than because the founding population is exposed to a different physical or biotic environment.

Whether or not founder events are a significant cause of speciation depends on a comparison of the relative plausibility of a variety of mechanisms. I will argue that the problems that have been raised are not in fact major obstacles to divergence, and that reproductive isolation can evolve in many ways. In particular, the classic view that reproductive isolation evolves as a side-effect of divergent selection pressures and the accumulation of different mutations remains the most economical explanation of speciation. Given the variety of possible speciation mechanisms, the problem may be to explain why speciation is so slow (cf. Felsenstein 1981).

Even having accepted the diversity of possible mechanisms, we must still determine whether founder effects have been significant in particular cases (for example, in the dramatic radiation of the Hawaiian *Drosophila*). There is no doubt that drastic founder events can cause divergence. However, there are strong theoretical arguments that make other mechanisms seem more plausible. Evidence from nature and from the laboratory is inconclusive, but does suggest that species differences are likely to have evolved in ways that would not be greatly aided by population bottlenecks.

MECHANISMS OF SPECIATION

How might a population come to rest on a different adaptive peak from its cousins? This question can be answered only by examining genetic models. Given such model, it is necessary to find the rate of shifts between peaks, and the amount of reproductive isolation contributed by each shift. To determine whether founder events are likely to make a significant contribution to isolating barriers, it is necessary to compare their effect with that of other population structures and other diversifying pressures.

These comparisons are difficult: we have only a rough idea of the long-term structure of populations, and still less understanding of the genetic basis of reproductive isolation. Some isolating factors are relatively simple and well understood—for example, chromosome rearrangements (White

1978; Lande 1979), infectious sterility in *Drosophila* (Ehrman and Parsons 1980), and major gene incompatibilities in plants (Gottlieb 1984; Christie and MacNair 1987). However, species differences are more often polygenic (Dobzhansky 1936; Coyne 1985a, 1986; Tanksley et al. 1982; Maynard Smith 1983; Barton and Charlesworth 1984; Coyne and Lande 1985; Barton and Hewitt 1985). Because reproductive isolation has a complex, and largely unknown, genetic basis, it is unwise to rely heavily on particular models. Fortunately, general results do emerge, which are independent of genetic details. I will illustrate these with a sequence that begins with a rather abstract model of a continuous character, but then incorporates progressively more genetic structure. These models will describe the genetic effects of sampling drift; changes in the selective environment may be more important, but are harder to quantify.

A POLYGENIC CHARACTER UNDER DISRUPTIVE SELECTION

Suppose that there is a bimodal relation between individual fitness (W) and the value of some quantitative character (z) (Figure 1A). This disruptive selection might arise from a bimodal distribution of some resource, combined with differences in resource use between genotypes. The resource is assumed to be unlimited, so that individual fitnesses do not depend on the current distribution of phenotypes. [Frequency dependence could maintain a stable polymorphism, which might lead to sympatric speciation without the aid of bottlenecks (Wilson and Turelli 1986; Wilson, this volume).] To begin, take the genetic and environmental variances of

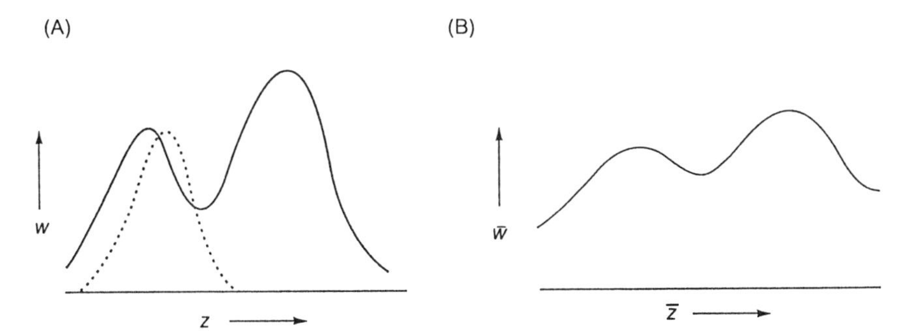

FIGURE 1. A polygenic character under disruptive selection. (A) The relation between individual fitness and phenotype [$W(z)$]. (B) The adaptive landscape: the relation between mean fitness and the population mean [$\overline{W}(\bar{z})$]. The dotted curve shows the distribution of the phenotype in the population, which links these two relations.

the character to be fixed. This can be seen as an ad hoc assumption, but is a good approximation if variation is due to many loosely linked genes with additive effects (Bulmer 1980).

The response to disruptive selection does not depend directly on the individual fitness $W(z)$, but rather on the relation between mean fitness ($\overline{W}$) and the mean of the character ($\bar{z}$) (Figure 1B, Kirkpatrick 1982a): the population will evolve so that the mean fitness increases, and so will move uphill on the "adaptive landscape" $\overline{W}(\bar{z})$. This adaptive landscape is a smoothed version of the individual fitness function. Provided that the variance is not too large, it will also be bimodal; the population will climb uphill to one or another of the two peaks.

DRIFT IN A STABLE POPULATION

First, consider a population of fixed size, N. There is some chance that a series of sampling accidents will knock the population from one peak to another. Such a shift is most likely to occur abruptly, and will take about the same time as would be needed for a response to selection alone (Newman et al. 1985). This is because the disruptive selection makes it unlikely that the population will dither for long between the two peaks, while it is also unlikely that sampling drift will consistently move the population in the direction needed for a very rapid shift. The expected time T before a shift depends primarily on the depth of the valley separating the peaks, relative to the height of the initial peak: $T \approx (\overline{W}_p/\overline{W}_v)^{2N}$ (Wright 1942; Barton and Rouhani 1987). Surprisingly, the expected time depends little on the distance between the peaks: a shift of many standard deviations may occur, provided that the intervening valley is shallow.

If peak shifts are to lead to speciation, they must cause reproductive isolation. The most straightforward possibility is that each shift causes some degree of isolation, and that a series of shifts eventually leads to full separation. If successful shifts involve independent sets of genes, then the net isolation will be the product of the effects of each shift, and we can measure the rate of accumulation of reproductive isolation by the isolation produced by each shift divided by the expected time between shifts. This approach will be used to compare the rate of speciation expected from different processes. However, an alternative view that appears in some texts on founder effect speciation is that although a peak shift may not itself cause much isolation, it triggers further divergence (e.g., Mayr 1954, 1982). [This is reminiscent of White's (1978) view of the role of chromosomes in speciation.] However, even if this is so, it still seems reasonable to judge the effectiveness of founder events by the rate at which they generate direct isolation. If further divergence is triggered by a partial barrier to gene flow, then it is fair to compare founder events with other processes by the strength of the barriers that they produce. If divergence is triggered by some kind of epistasis (such that alleles experience different marginal selection pressures at different adaptive peaks), then this epistasis can be incorporated into the models, as is done below. To find the consequences of peak shifts, we must therefore find how much the flow of genes between populations at different peaks is impeded.

If two populations are maintained at different stable equilibria by disruptive selection, then differences at selected loci can be preserved in spite of interchange of individuals. To measure the barrier to gene exchange between the populations, we must find the reduction in gene flow at marker loci not subject to disruptive selection. This will depend on the ratio between the rate at which introgressing genomes are eliminated by selection and the rate at which the marker genes recombine onto the new genetic background. This argument can be quantified in the case in which the populations meet in a continuous cline: there, the barrier to gene exchange is proportional to $(\overline{W}_p/\overline{W}_v)^{1/r}$, where r is the harmonic average rate of recombination between introgressing markers and the selected genes (Barton 1986a). This relation applies not only to disruptive selection on a quantitative character (described previously), but also to any model with fixed genotypic fitnesses. The only restrictive assumption is that the barrier should not be extremely strong. Remarkably, the barrier to gene exchange depends on the same ratio of mean fitnesses that determines the rate of peak shifts. This allows us to reach conclusions that do not depend on specific models.

These two relations imply that shifts that establish strong reproductive isolation must greatly reduce mean fitness, and so are necessarily unlikely to be established by sampling drift in a population of constant size. Thus, a given level of reproductive isolation is more likely to be built up in a series of small steps than in a few drastic "genetic revolutions." This argument was presented for chromosome rearrangements by Walsh (1982), and extended to any models that can be described by an "adaptive landscape" by Barton and Charlesworth (1984).

It is hard to make this argument more precise without knowing the typical shape of the "adaptive landscape" (cf. Kauffman and Levin 1987). But suppose, for the sake of argument, that a variety of peak shifts are available to a population: some involve only shallow valleys, and so give only weak isolation, while some involve deep valleys, and give strong isolation. We can measure the strength of these shifts by $(\overline{W}_p/\overline{W}_v) = e^{-s}$, and write their distribution as $\psi(s)$. A shift of strength s will cause a barrier to gene flow that depends on the ratio s/r. Since the chance of a peak shift is $\approx \exp(-2Ns)$ per unit time, the rate of evolution of reproductive isolation is $\approx \int \psi(s)e^{-2Ns}(s/r)\,ds$. If weak shifts are available in about the same numbers as strong ones, then $\psi \approx$ constant, and the net rate is $\approx \psi/[r(2N)^2]$.

This rough calculation has two implications. First, shifts will usually be weak ($s \approx 1/2N$). If (as seems likely) more weak shifts are available than strong, then ψ will decrease with s, and this tendency will be exaggerated. Second, though the rate of divergence does decrease as population size increases, the decrease is not drastic (N^{-2} as opposed to, say, e^{-2Ns}). Since the relevant N is the size of local demes, rather than of the whole population (Lande 1979), drastic bottlenecks do not seem necessary for drift to lead gradually to isolation.

THE GENETIC CONSEQUENCES OF FOUNDER EVENTS

This general argument shows that in a population of fixed size, drift is most likely to lead to the gradual accumulation of isolation by a series of weak peak shifts. How do the effects of occasional founder events compare? Before looking at the specific model of disruptive selection, consider the general genetic consequences of bottlenecks. Suppose that a few individuals found a new population. If extinction is to be avoided, it must expand within a few generations. The burst of sampling drift caused by the bottleneck has two inseparable effects: it will reduce genetic variance and heterozygosity (H) by a factor $H_\infty/H_0 = (1 - 1/2N_0)(1 - 1/2N_1)\ldots$, and it will produce a variance among bottlenecked populations in allele frequency and in the means of quantitative characters that is proportional to $1/2N_0 + 1/2N_1 (1 - 1/2N_0) \ldots = 1 - H_\infty/H_0$. (Note that $H = 1 - \Sigma p_i^2$ is used here as a measure of allelic diversity rather than of heterozygosity.)

It is important to realize that these two effects on the gene pool are both consequences of genetic drift. Founder events may also cause slight deviations from Hardy–Weinberg proportions and random fluctuations in linkage disequilibria, but these effects will be transient. Templeton (1980) has argued that different genetic effects of a bottleneck can be separated, so that under favorable circumstances genetic variability can remain high enough for the population to respond to new selective conditions, even though "inbreeding" effects are large enough to cause a "genetic transilience." The "transilence" is seen as being caused not by the straightforward drift of allele frequencies, but rather by deviations from Hardy–Weinberg proportions, and by the fact that "a given allele appears against a narrower spectrum of genetic backgrounds" (Templeton 1980:1015). Now, the first effect is small and transient: although the proportion of heterozygotes, averaged across an ensemble of populations, can be greatly reduced by a bottleneck, the effect *within* a population is small. If selfing is forbidden, there is a slight ($\approx 1/N$) excess of heterozygotes relative to Hardy–Weinberg expectations for that population. The second effect is a consequence of a general loss of variability, and cannot be separated from it. It seems inevitable, therefore, that a founder event severe enough to significantly perturb the gene pool must necessarily cause a severe reduction in variability.

So far in this section we have ignored the effects of selection during the bottleneck. Balancing selection, stabilizing selection, and associative overdominance will resist the loss of heterozygosity caused by drift, and will impede peak shifts. However, these complications can be ignored if these forces are not extremely strong relative to the rate of increase of the population (Rouhani and Barton 1987b). This is simply because the bottleneck is not likely to last long enough for selection to have a significant effect. Though neglect of selection is an approximation, it is accurate for quite strong selection (Rouhani and Barton 1987b). To the extent that selection *is* strong enough to have a significant effect, the approximation will lead us to overestimate the importance of founder events. Much attention has been given to possible changes in selection pressures during founder events. For example, Carson (1975) has suggested that selection pressures might be relaxed during the rapid increase of a founder population, and that this could allow new combinations of alleles to be established. However, the previous argument suggests that any changes during the brief flush would have little effect on the outcome (see also Charlesworth and Smith 1982).

FOUNDER EVENTS AND QUANTITATIVE CHARACTERS

Given that selection can be ignored, the chance of a peak shift is simply the chance that a neutral character will drift into the domain of attraction of the new peak. Suppose that the initial genetic variance, v_0, is entirely additive. This variance will decrease by a factor $(1 - 1/2N_i)$ in each generation, and the eventual loss will be $\Delta v = v_0/2N_0 + v_1/2N_1 + \ldots$. [In terms of the loss of heterozygosity, $\Delta v = v_0\,(1 - H_\infty/H_0)$.] At the same time, the mean will be subject to random perturbations with variance v_i/N_i in each generation. The net variance in the mean caused by the whole bottleneck is thus $\Sigma\, v_i/N_i = 2\Delta v$ (Rouhani and Barton 1987b). This can be no greater than twice the initial additive genetic variance, whatever the sequence of population increase (N_i) (Wright 1942; Bulmer 1980). Moreover, data on the loss of heterozygosity at neutral loci can be used to estimate the effect of the bottleneck on morphological characters. We can see that the chance of a peak shift depends primarily on the distance between the domain of attraction of the new peak and the initial position, relative to the initial genetic variance (cf. Lande 1980). A shift of more than $2\sqrt{2v_0}$ is unlikely, even in the most drastic founder event. If the initial heritability were 1/2, this corresponds to 2 phenotypic standard deviations for a founder event that removed all variation. In a less extreme case, in which 10% of heterozygosity is lost, a shift of more than 0.6 phenotypic standard deviations is unlikely.

How much reproductive isolation is likely to be produced by such a founder event? Three constraints are relevant. First, shifts become unlikely when they involve selection pressures that can have significant effects even during the bottleneck: shifts with extreme effects are therefore ruled out. Second, the distribution of breeding values must be wide, relative to the distance between peak and valley, if shifts are to be likely: the initial population must therefore suffer a high load, since an appreciable proportion of individuals will have intermediate phenotypes. Third, if the variance is greater than some threshold, the two adaptive peaks will merge into one (Figures 1 and 2). These limits combine to give a relation between the maximum reproductive isolation and the initial load on the population.

For example, suppose that individual fitness is $W = \exp[+sz^2(2 - z^2)]$; this function has two symmetric peaks at $z = \pm 1$. Suppose that the heritability is initially 1/2, and that the founder event eliminates half this genetic variation. Then, a population with phenotypic variance V is unlikely to shift by more than $2\sqrt{2(V/2)(1/2)} = \sqrt{2V}$. With this fitness function, the distance between the other peak and the central valley is $\sqrt{1 - 3V}$. (Selection is assumed to be fairly weak; if selection on the ancestral population is strong, the original fitness peak moves away from the cen-

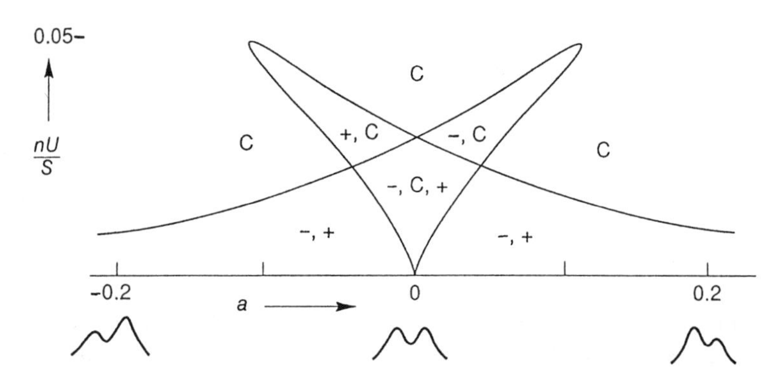

FIGURE 2. Disruptive selection on a polygenic character. This is a simplified version of Kirkpatrick's (1982a) model: $W = \exp\{s[z^2(2 - z^2) + az(1 - z^2/3)]\}$. The distribution of allelic effects at n equivalent loci is assumed to be Gaussian (Lande 1975), and linkage disequilibrium is ignored. The pattern of equilibria is sensitive to the rate of mutation, relative to selection (nU/s) and to the degree of asymmetry of the two fitness peaks (a). Six different configurations of stable equilibria are possible, and slight parameter changes can cause sudden shifts between configurations.

tral valley, reducing the chance of a shift (Charlesworth and Rouhani 1988). Thus, we have $\sqrt{2V} > \sqrt{1 - 3V}$; V must be less than 1/3 if the peaks are to remain distinct, but must be greater than 1/5 if a shift is to be likely. Now, consider the relation between the initial load and the likely amount of isolation. The load may be measured by the ratio between the maximum fitness and the mean fitness at equilibrium: $\Lambda = \exp[-s(4V + V^2)] + O(s^2)$. The amount of reproductive isolation can be measured by the ratio between the fitness of populations with variance V at peak and valley: $R = (\overline{W}_p/\overline{W}_v) = \exp[-s(1 - 3V)^2] + O(s^2)$.

These calculations show that the amount of isolation is likely to be less than the load associated with disruptive selection of this form. [In this model, $V > 1/5$, so $R \leqslant (2/3)L$.] Of course the exact relation will vary with the form of the fitness function. However, since the genetic variance must be delicately balanced if shifts between distinct peaks are to be likely, and since the load associated with disruptive selection is only a small fraction of the total load, it seems unlikely that founder events could generate much isolation.

THE GENETIC BASIS OF POLYGENIC VARIATION

The evolution of quantitative characters is most easily understood when variation is based on large numbers of additive, loosely linked loci. But in fact, a limited number of genes is involved, and dominance, epistasis, and linkage disequilibrium introduce important interactions.

Linkage disequilibrium can be generated by selection and by drift (Bulmer 1980). For a given level of variation at individual loci (maintained, for example, by mutation), both effects reduce the chances of peak shifts in a founder event (Charlesworth and Rouhani 1988). This is because stabilizing selection generates associations between alleles with opposite effects (+ −, − +). These associations reduce the net additive genetic variance, and so reduce the response of the mean to sampling drift. Random fluctuations in linkage disequilibrium cause random fluctuations in the genetic variance; on average, these also reduce the response to drift (Charlesworth and Rouhani 1988).

The chance that a peak shift will occur as a result of drift depends on the way the genetic variance changes through the bottleneck, which in turn depends on the distribution of effects on the character of individual genes. If genes are highly polymorphic, so that the distribution of effects is roughly Gaussian (Lande 1975), then the general arguments that have been given still apply. However, it is more likely that variation is caused by rare alleles, giving a highly leptokurtic distribution of effects (Turelli 1984). If alleles have strictly additive effects, the arguments are not much altered (Barton and Charlesworth 1984, Figure 1; Barton and Turelli

1987). However, Bryant et al. (1986a and 1986b) have pointed out that if rare alleles tend to be recessive, random drift can greatly increase the expressed genetic variance. This is because the phenotypic variance introduced by those alleles that increase in frequency outweighs the loss in variance associated with those alleles that decrease in frequency (Robertson 1952; Lande 1980; Barton and Charlesworth 1984). The increase in genetic variance caused by a bottleneck may facilitate peak shifts in several ways. It will make the population more responsive to changing selection, it may itself trigger a shift if individuals with phenotypes near a new, higher fitness peak are produced in sufficient numbers (Figure 2; Kirkpatrick 1982a), and most obviously, it will increase the response to sampling drift, and so make a purely stochastic shift more likely (Bryant et al. 1986a).

The arguments developed for a purely additive character can be extended to find the likely effect of dominance, in a rather general way. Note that if the bottleneck is brief, selection on the character can be neglected. Further, the distribution of allele frequencies, and hence of the character mean, depends only on a single parameter, which we can take to be the net loss of heterozygosity. This is because (provided that the diffusion approximation is accurate) the effects of drift depend only on $\Pi(1 - 1/2N_i)$, and not on the pattern of population increase. Consider a character affected by n loci; the recessive allele has frequency p, and the effect of a substitution is 2α. It is reasonable to assume that the direction of the effect varies between loci, so that *on average* drift does not affect the mean. Then, the character mean is $z = \Sigma \pm \alpha(2p^2 - 1)$, and the genetic variance in the base population is $v = 4n\alpha^2p^2(1 - p^2)$. The variance of the mean across an ensemble of bottlenecked populations is $\mathrm{var}(z) = 4n\alpha^2 \mathrm{var}(p^2)$. This variance can be calculated from the movements of the distribution of allele frequency (Crow and Kimura 1970), in terms of the net loss of heterozygosity ($F \equiv 1 - H_\infty/H_0$). When recessives are very rare ($p \ll 1$), $\mathrm{var}(z)/v = F^3(3 - 3F + 6F^2/5 - F^3/5)/p$. [Bryant et al. (1986b) have made similar calculations; their results differ slightly in that they calculate the effects of a single round of binomial sampling, and express the results relative to the additive genetic variance in the base population, rather than the total genetic variance (v).]

When the bottleneck is severe ($F \approx 1$), and recessives are rare, the mean can shift substantially from its original value. If the initial broad-sense heritability is 1/2, a founder event could cause a shift of up to $2\sqrt{2p}$ phenotypic standard deviations. However, the effect of dominance is much less important in less extreme founder events: $\mathrm{var}(z)/v \simeq 3F^3/p$ when $F \approx 0$. For the case $p = 10\%$, additive variation actually causes *more* divergence when $F < 20\%$ (Figure 3: see also Bryant et al. 1986b, Figure 1). We have assumed that selection acts only on the character of interest, so that once the mean shifts to the vicinity of the new peak, a permanent

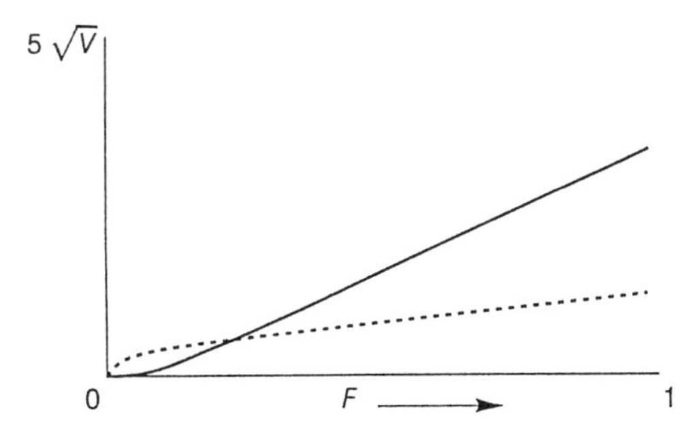

FIGURE 3. The largest shift in mean that is likely to be produced by a founder event (defined as 2 standard deviations). A broad-sense heritability of 1/2 is assumed; the size of the shift, relative to the phenotypic standard deviation, is plotted against the severity of the bottleneck, expressed as F, the factor by which heterozygosity is reduced. Dotted line: additive model. Solid curve: recessive model. The frequency of the rare recessives is $p = 10\%$.

shift will occur. However, recessives are likely to be rare because they are intrinsically deleterious (Bryant et al. 1986a). Thus, even if the mean is knocked across the fitness trough, it may return to its original position as selection eliminates harmful recessives. It is still possible that permanent isolation might result if populations recovered from the bottleneck in different ways: these possibilities are, of course, open to experimental test.

DIVERGENCE UNDER STABILIZING SELECTION

If polygenic variation is based on rare alleles maintained in a mutation–selection balance, then divergence can occur even under stabilizing selection: many different combinations of alleles can give the same optimal phenotype, and so the population can occupy many alternative stable equilibria (Wright 1935). One must therefore ask whether isolation is likely to evolve faster through shifts among different combinations of alleles within the same phenotype, or through shifts between different phenotypes. The question is then whether founder events are likely to contribute significantly to either process.

Consider an additive model in which mutation at a rate μ maintains variation at n biallelic loci, in spite of stabilizing selection s. Each allele has effect α. Then, the equilibrium genetic variance is $v = 4n\mu/s$. Any combination of plus and minus alleles that allows the mean to come close to the optimum is stable (Barton 1986b). Though calculation of the rate of shifts is complicated (Kimura 1981; Burger 1988; Foley 1987; Barton 1989), the general picture is clear. Provided the product of population size and selection on each locus is small ($Ns\alpha^2 < 10$), shifts may occur at an appreciable rate: if selection on each locus is weak ($s\alpha^2 \approx 10^{-3}$), then divergence is possible even in large populations ($N < 10^4$). As the product decreases to $Ns\alpha^2 < 1$, the rate of divergence approaches the neutral rate, μ. Thus, isolation will evolve slowly, over a timescale of $1/\mu$ or longer. The maximum possible level of reproductive isolation occurs when half the loci in two populations differ in state: then the reduction in mean fitness of a hybrid population is $R = -ns\alpha^2/4$. This can be compared with the equilibrium mutation load, $L = 2n\mu$: $R/L = s\alpha^2/8\mu = 1/4p$, where p is the frequency of the rarer allele. Thus, substantial isolation would eventually develop if the alleles responsible for variation are sufficiently rare.

Founder events will cause relatively little isolation under this model. This is because each gene is close to fixation, and because drift is effective even in quite large populations ($N < 10^4$). In the most severe bottleneck, one set of alleles from the base population will be fixed. The chance that the rarer allele will be fixed is then equal to its frequency, p. On average, a proportion p of loci will shift, and so the isolation produced will be a fraction $2p$ of the maximum; relative to the mutation load, $R < L/2$. An extreme bottleneck is approximately equivalent to $(p/\mu) = (2/s\alpha^2)$ generations of weaker drift in a large population. This could be large if variant alleles are fairly common; however, the maximum isolation that can be reached will then be weak. The main conclusion is that drift leads only to gradual divergence, whether it acts steadily or in bursts.

It is hard to see how this model could be extended to include the effects of dominance. If rare alleles were recessive, then they would segregate at higher frequency than if they were partly dominant. Stochastic shifts would therefore be facilitated. However, such shifts would cause dominant alleles to become rare, after which divergence would proceed more slowly than in the additive model. If recessives are in fact generally rare, this is most likely to be because they are deleterious: as previously discussed, such extraneous selection would greatly impede divergence.

OTHER MODELS

How far do these points apply to other models? The general conclusion is that in a population of fixed size, the chance of a peak shift depends on the depth of the valley in the adaptive landscape, whereas in a founder event it depends on the distance between the peaks. Many schemes fit easily into the same framework and give similar results.

Much attention has been paid to the establishment of underdominant chromosomal rearrangements (Wright 1941; White 1978; Lande 1979; Hedrick 1981). As with quantitative characters, new karyotypes can be established if the product of population size and selection pressure is not too large ($Ns < 10$). This implies that isolation is likely to evolve gradually

(Walsh 1982; see previous discussion). Founder events can allow the establishment of strongly selected incompatibilities; however, a suitable chromosomal mutation must occur in the small founding population. The probability of karyotypic change must therefore be less than the chromosomal mutation rate (see Hedrick and Levin 1984). If founder events are to be important in chromosomal evolution, it is likely to be within a network of demes rather than in isolated lineages (see White 1978; Lande 1985).

Any model in which variation is based on new mutations or rare alleles suffers from similar restrictions. If founder events are to lead to significant isolation, then the ancestral population must contain substantial polymorphic variation, which is subject to strong epistatic selection (Wright 1932; Templeton 1980). Thus, the most favorable model discussed so far is where a variable polygenic character is under disruptive selection. For the same reason, Carson (1975) has proposed that isolation develops when a set of alleles, kept polymorphic by heterozygote advantage, shifts from one coadapted combination to another. The difficulty with such schemes is that because they require high polymorphism in the base population, they impose a high load on the population (Charlesworth and Smith 1982, and previous discussion). This limits the amount of reproductive isolation that can be produced. Some genetic structures avoid this difficulty, at least in part: as previously discussed, rare recessives can mask a high level of genetic polymorphism. Some forms of epistasis may have similar effects: Goodnight (1987) has shown that nth-order additive × additive × . . . effects can be inflated by a factor of up to 2^n. However, the amount of isolation that may evolve is still limited by the sustainable load, and it is not clear that the actual heritable variance in fitness includes significant dominance and epistatic components.

Templeton (1980, 1981, 1982a, 1982b) has suggested that one or a few major genes, coupled with epistatically interacting modifiers, may be particularly sensitive to founder events. The basic argument is that "if a polygenic trait is controlled by a large number of loci, each with small effects, often very little change or reduction in variability in the polygenic trait would be expected. But, if there are a few major genes, the stochastic effects of a founder event cannot be ignored" (Templeton 1980:1015). Though intuitively appealing, this argument is incorrect: a bottleneck reduces within-population variance and increases between-population variance by the same factor (relative to the initial additive genetic variance), regardless of the number of loci involved. As the models previously discussed show, founder events can (if sufficiently intense) cause large shifts in polygenic traits, which may lead to significant isolation. Unfortunately, there is no simple relation between genetic architecture and the likely mode of speciation (Barton and Charlesworth 1984).

To take a simple example of Templeton's idea, consider a major gene with two alleles (A,a); when a predominates, alleles $m_1, m_2, \ldots, m_n$ are favored at n modifier loci. Conversely, when A predominates, alleles $M_1, M_2, \ldots, M_n$ are favored. The mean fitness of the population can be written $\overline{W} = 1 + s(p_A - p_a)\,\Sigma(p_{Mi} - p_{mi})$. To complete the model, we must include some mechanism that will maintain polymorphism in the ancestral population. The simplest possibility is symmetric recurrent mutation at a rate μ between each pair of alleles ($\mu \ll s$). At equilibrium, either $p_a = \mu/ns$, $p_m = \mu/s$, or $p_A = \mu/ns$, $p_M = \mu/s$. Now, consider the chance that a brief bottleneck will cause a shift from (a,m) to (A,M). This may happen in two ways. If A becomes fixed, the modifier alleles will slowly increase in response, over a time $\approx 1/s$. However, if A is not quite fixed, it will fall back before the modifiers increase enough to favor A rather than a. In an extreme bottleneck, the chance of fixation of A is μ/ns. Alternatively, the modifiers may rise to a high enough frequency so that A is favored. If n is large, and all loci become fixed after the bottleneck, then $\Sigma(p_M - p_m)$ is normally distributed, with expectation $-n(1 - 2\mu/s)$ and variance $4n\mu/s$. The chance that this variate will exceed zero is always smaller than μ/ns; so, for this particular model, the major gene is most likely to shift first. The simple conclusion is that there is a chance μ/ns of a shift that leads to reproductive isolation $R \approx ns$. Relative to the initial mutation load ($L = 2\mu$), the expected isolation is $E(R) = L/2$: this ratio is of the same order as for models of major genes alone, or for purely polygenic systems.

A wide range of other models can be imagined. If the modifiers are treated as one or more additive polygenic characters (cf. Lande 1983), results are essentially the same. Introducing epistasis between the modifiers, as well as with the major gene, might allow shifts between different combinations of modifiers. However, the major gene would then be irrelevant. Templeton (1982b, and personal communication) has suggested (by analogy with the *abnormal abdomen* polymorphism in *Drosophila mercatorum*) that fluctuating selection pressures might keep the major gene polymorphic. If fluctuations in selection are slow, then divergence will occur without the need for sampling drift. If selection fluctuates rapidly enough, and heterozygotes have the highest geometric mean fitness, a stable polymorphism will be maintained (Haldane and Jayakar 1963). Then it is not clear how sampling drift at the major locus would cause a peak shift or "transilience."

DIVERGENCE DUE TO CHANGING SELECTION PRESSURES

We have been considering divergence caused by random drift. Peak shifts may also occur in a purely deterministic way, through temporary changes in selection coefficients. These can occur for many reasons; models of founder effect speciation have stressed the importance of changes caused by inbreeding at individual loci (Templeton 1980, 1982a) or in the general genetic background (Mayr 1954, 1982), and by relaxed selection during population flushes (Carson 1968, 1975). However, these genetic and demographic factors do not seem likely to be important relative to the changes in physical and biological environment that must occur when a new habitat is colonized, or, for that matter, relative to the continual effects of coevolution and climate. It is hard to make any generalization about the likely patterns of selective change. It can be noted, though, that when selection on polymorphic systems allows multiple equilibria, small parameter changes can cause large shifts in the pattern of equilibria (Kirkpatrick 1982a, Figure 2). Thus, the systems that are most likely to change through drift during founder events are also most likely to change for other reasons.

BEYOND THE "ADAPTIVE LANDSCAPE"

So far, speciation has been seen as consisting of shifts between different peaks on an adaptive landscape. However, this picture may be misleading. There are difficulties in defining the "adaptive landscape" (Provine 1986), and there are evolutionary processes (notably recombination and frequency-dependent selection) that cannot be described by gradient models. But more important, our imagination is limited by viewing populations as being at equilibrium, and by neglecting the complicated geometry that channels evolution. If, as Fisher (1930) believed, populations are steadily accumulating new, favorable mutations, then genetic divergence is an inevitable side-effect of adaptation. This is because at any one time, the range of individual mutations is limited by the current genotype (or set of genotypes) and by finite population size. For example, a protein coded by 1000 base pairs can mutate only to $\approx$3000 alternative sequences. This is a tiny fraction of the $4^{1000} = 10^{602}$ possible sequences (Gillespie 1984). The contrast is accentuated when the whole genome, rather than a single gene, is included. So, although many sequences may be favorable at any one time, only one of them is likely to be available even in a very large population. Two populations will therefore diverge, even if they begin in identical states and are exposed to identical selection pressures. Once they contain different sets of genotypes, different mutations become available, and divergence soon becomes irreversible. In a sense, such divergence is driven by sampling drift, since in an infinite population, all possible sequences would be available simultaneously. However, it will occur in any species that could live on our planet.

Though reproductive isolation is not an inevitable consequence of adaptive divergence, it is highly likely: hybrid combinations will never have been tested by selection. This mode of speciation has been suggested many times (e.g., Fisher 1930; Muller 1939; Lewontin 1966; Templeton 1981). However, it has been rather neglected, perhaps because it is so simple, and because it is hard to make any robust generalizations about its likely rate. Kaufmann and Levin (1987) have discussed approaches to this problem, which may give predictions about the degree of isolation that is likely to develop populations that have accumulated different sets of favorable alleles. Within the model of stabilizing selection on a polygenic character, previously discussed, the process will occur as a result of fluctuations in the selective optimum. For example, if the optimum increases sufficiently, a shift of one of the minus loci to plus will be favored. However, which locus shifts is a matter of chance. Here, the rate is limited only by the strength of the fluctuations in selection, and it could be fast. More generally, a limit on the net rate is placed by the substitution load: if, as Haldane (1957) calculated, one favorable allele could be fixed every 30 generations or so, then isolation could build up very quickly. Founder events, of course, will in no way accelerate this process: if anything, they will impede it by eliminating variation.

Carson (1982) has suggested that runaway sexual selection could be triggered by a founder event. But, at least in the models of Lande and Kirkpatrick, divergence through sexual selection may occur readily, for a variety of reasons. Because there is no obvious selective barrier to be overcome, the image of "triggering" is misleading. Genetic drift during founder events does not seem a particularly likely cause of divergence here: there is no obstacle to change in large populations, and, in any case, these "quasineutral" processes cannot easily give permanent isolation. Kaneshiro's (1980) suggestion that elements of courtship could be lost during population bottlenecks, thus leading to prezygotic isolation, does not suffer from this objection. If elements of courtship are recovered in different ways in different populations, we return to a process rather similar to that previously discussed.

SOME GENERALIZATIONS

The general points that emerge from this survey of the theory can now be summarized:

1. Reproductive isolation can evolve through either random drift or changing selection. Founder events may involve both processes; however, the standard models of founder effect speciation stress genetic and demographic, rather than ecological, factors.
2. In speciation theory as a whole, most attention has been paid to random drift. This is reasonable, in that (in principle) predictions can be made based on measurable features such as population size. However, this

emphasis may distort our view of the relative importance of random "Wrightian" versus deterministic "Fisherian" speciation.

3. In a population of fixed size, the probability of a peak shift ($\approx \overline{W}^{2N}$) decreases directly with the amount of isolation produced by that shift ($\approx \overline{W}^{1/r}$). Therefore, isolation will tend to evolve gradually, through a series of weakly selected steps ($s \approx 1/2N$).
4. In contrast, the probability that a bottleneck will cause a peak shift is more or less independent of the isolation produced. It depends instead on the distance between the original equilibrium and the domain of attraction of the new state, relative to the original genetic variance.
5. High levels of selected polymorphisms must be present if a founder event is to generate strong isolation. The amount of isolation that is likely to be produced is therefore limited by the net genetic load on the ancestral population. This limitation applies to a wide range of models, including those involving polygenic traits, epistasis between polymorphic loci, and major genes interacting with a set of modifiers.
6. This limitation can be partly avoided if phenotypic variation is based on rare recessive alleles. Then, higher levels of genetic variation can be maintained for a given phenotypic variance. However, variation based on recessive alleles gives substantially greater divergence than additive variation only if the founder event is drastic.
7. There is no reason to suppose that coadaption and homeostasis bind a species into a coherent unit, resistant to change. Divergence can occur even under strong stabilizing selection, as a population shifts between different combinations of genes giving the same phenotype. The steady accumulation of favorable mutations is also likely to lead to reproductive isolation. Neither process is much accelerated by founder events.

THE EVIDENCE

How can we judge the importance of founder effects in speciation, relative to other processes? Here, the most important theoretical conclusion is that the genetic effects of a brief population bottleneck are given by the net reduction in variability: even allowing for the effects of dominance and epistasis, a series of events drastic enough to give significant isolation must necessarily cause a severe loss of genetic variance. We therefore expect that the prevalence of population bottlenecks will be revealed by the pattern of biochemical variation.

Sampling drift in natural populations

Drastic bottlenecks certainly occur, and can certainly cause substantial loss of variation. Where introductions of only a few colonists are known to have occurred, substantial loss of enzyme heterozygosity is seen. For example, the common mynah was introduced at the turn of the century to South Africa and to several Pacific regions; the average reduction in enzyme heterozygosity relative to the ancestral Indian population is $F = 0.23$ (Baker and Moeed 1987). The most obvious examples are the many human isolates, in which deleterious recessives, rare elsewhere, may drift to high frequency (Vogel and Motulsky 1986, Chap. 6). On a larger scale, the similarity between mitochondrial and β-globin DNA sequences between non-African populations, combined with their differentiation from African populations, suggests a sharp bottleneck $\approx 10^5$ years ago (Johnson et al 1983, Wainscoat et al. 1986; Cann et al. 1987; Jones and Rouhani 1986). In a broader sense, sampling drift can cause substantial variation between populations. The concordance of estimates of population structure made using different enzyme loci suggests that much geographic differentiation is due to sampling drift (Slatkin 1981, 1987). Divergence may be caused by accumulated drift in small, stable isolates—perhaps the most striking example is in the salamanders (Yanev and Wake 1981; Wake et al., this volume; Larson, this volume)—or through a series of colonizations, as a species expands into fresh habitat (for example, during the spread of lodgepole pine into Canada after the last glaciation; Cwynar and MacDonald 1987).

Laboratory experiments

These examples show that sharp bottlenecks, or sustained sampling drift, can have drastic effects on genetic variation. However, evidence that either bottlenecks, or drift in general, cause significant reproductive isolation is much weaker.

Comparisons between laboratory lines of *Drosophila* and *Musca* show that, given sufficient time, both pre- and postmating isolation can evolve (Ringo et al. 1985; Cohan and Hoffman 1986). These lines have inevitably been subject to both selection and drift: the problem is to resolve the relative contribution of each process. Powell (1978), Dodd and Powell (1987), and Ringo et al. (1985) have compared lines that were passed through repeated single-pair bottlenecks with controls. Powell found premating isolation in three of eight bottlenecked lines, compared with none involving the controls. Ringo et al. found erratic premating isolation, and also a significant decrease in F_1 female productivity, for the bottlenecked lines, but no consistent change for the controls. This postmating isolation gave a fitness loss of $\approx 4\%$ per bottleneck. Though small, this figure is greater than might be expected, given that the loss of variability caused by each founder event was only $F \approx 0.2$.

There are several problems in interpreting these results. The most striking consequence of drift in the experiment of Ringo et al. was a strong reduction of male mating propensity and female productivity (Ringo et al.

1986). This may be confounded with the much weaker isolation between lines. Perhaps most important, both Powell and Ringo et al. began with a mixture of flies from different populations. Thus, the initial allelic diversity at selected loci may have been created. Given the argument that the amount of reproductive isolation that can follow from a founder event is limited by the initial variation in fitness, this initial mixing could allow much stronger isolation to develop than would be common in nature. It can also be argued that drift merely accelerates the inevitable sifting out of new gene combinations from the heterogeneous base population. Ringo et al. (1985:655) argue that such effects of heterogeneity would not explain the stronger isolation found between the drift lines and the base population than among the drift lines. However, it is not at all clear how this pattern is to be explained under any model.

These laboratory experiments show that (weak) isolation can evolve, and suggest that, under favorable circumstances, drift can encourage this divergence. More work is needed to determine whether this is also true when founders are drawn from a single natural population, to find the genetic basis of changes in quantitative variation, and, most important, to relate experimental results to theoretical predictions.

Island radiations

The central issue of how reproductive isolation actually evolves in nature can be addressed only by work on natural populations. Here, as with most arguments over speciation, evidence is indirect and circumstantial.

The radiation of the Hawaiian *Drosophila,* which has stimulated so much of the argument for founder effect speciation, is dramatic not only in the number of species involved, but also for the striking ecological and sexual adaptations that have evolved (Carson and Kaneshiro 1976; Carson and Templeton 1984; Kaneshiro and Boake 1987). (We should bear in mind, though, that the rate of evolution of premating isolation, as measured in laboratory preference tests, is not so different from that of the continental *Drosophila;* Coyne and Orr, this volume.) It is clear that founder events have occurred in the colonization of the archipelago, of its constituent islands, and of patches of habitat ("kipukas") within islands. What is at issue is whether these founder events have played a causal role in speciation and, if so, whether through their ecological or their genetic effects. The main difficulty in accepting a major genetic effect is that there is no evidence that founder events have caused much loss of genetic variation: enzyme heterozygosity is similar to that in continental species (Ayala 1975; Craddock and Johnson 1979; Sene and Carson 1977; De Salle and Hunt 1987).

Templeton (1980, 1981) has argued that founder events are most likely to lead to speciation when there is little loss of variability, and yet a high level of "inbreeding." However, as previously discussed, if reproductive isolation is to evolve during a population bottleneck, it will be as a result of random changes in allele frequency: a loss of variability therefore inevitably accompanies any event drastic enough to cause significant drift. Of course, it may be that enzyme polymorphisms are maintained by balancing selection, so that variability would be recovered after a bottleneck. Such selection cannot be very strong: where bottlenecks are known to have occurred, enzyme heterozygosity is reduced (e.g., Baker and Moeed 1987; Cwynar and MacDonald 1986; Janson 1987). Also, strong balancing selection would act as a powerful buffer against the effects of drift at linked loci. It seems unlikely that, even if overall heterozygosity were restored by balancing selection, the same alleles would be recovered. The small genetic distances between related Hawaiian *Drosophila* species therefore provide strong evidence that founder events with drastic genetic effects have not occurred (Sene and Carson 1977; De Salle and Hunt 1987).

Since noncoding DNA variation is more likely to be strictly neutral than enzyme polymorphisms, these should be more reliable tracers of founder events. Most population data are from mitochondrial sequences; these are much more sensitive to drift and founder events, because a single inseminated female carries only one mitochondrial genome, compared with at least four nuclear genomes. If she is multiply inseminated, the discrepancy may be large: De Salle et al. (1987) estimate that in *D. mercatorum*, the ratio is 6 to 8. Interpopulation variation for mitochondrial DNA is thus much greater than for isozymes. For example, in *D. heteroneura* and *D. silvestris*, Fst is ≈ 0.6 for mitochondrial DNA (De Salle et al. 1986), whereas enzyme polymorphisms show much less spatial differentiation (Nei's D is ≈ 0.05; Sene and Carson 1977). However, there is still considerable variation within populations and within species (De Salle et al. 1986). Founder events involving single females must therefore be relatively infrequent. A rough estimate can be made of the time since such an event from the present level of polymorphism and the rate of sequence divergence. Though the rate of mitochondrial sequence divergence is faster than that of nuclear DNA, it is still slow [$\approx 2 \times 10^{-8}$ versus 5×10^{-9} per year (De Salle and Hunt 1987, Figure 4)]. Thus, the proportion of polymorphic sites, $p \approx 0.04$ (De Salle et al. 1986), suggests that there was at most one single female founder event at the colonization of Hawaii. The lack of homogeneity of even mitochondrial sequences is strong evidence that founder events are unlikely to have had a drastic effect on the nuclear sequences that are likely to contribute to reproductive isolation. Moreover, a population structure that fails to perturb weakly selected or neutral variation is unlikely to perturb the more strongly selected differences that might lead to reproductive isolation.

It has frequently been argued that the high frequency of single island endemism in the Hawaiian *Drosophila* implies that interisland colonization "is overwhelmingly conducive to speciation" (Carson and Templeton 1984:108; Carson and Kaneshiro 1976). Now, a high proportion of endemic species shows only that speciation is fast relative to the rate of interisland colonisation. Moreover, if the correlation were to be explained by speciation caused by colonization, then the probability of speciation in a single founder event would need to be very high. The genetic models and laboratory experiments previously discussed show that strong isolation can arise only from a long series of bottlenecks. A stronger argument is that sister species are often found on different islands: Carson and Templeton (1984) state that "of 26 [picture-winged] species on the Big Island (i.e. Hawaii), there are 19 lineages . . . related to a separate ancestral population on an older island." But again, this pattern might be better explained by ecology rather than by genetics: if colonization and speciation occur in a relatively short period after the formation of a new island, then most new species may come from the older islands. Of course, this pattern is incompatible with speciation through colonization of kipukas within islands.

Moving beyond the Hawaiian *Drosophila*, the many other examples of island radiations do argue, in a very general way, for some role of colonization in speciation. However, there is again little evidence implicating the genetic effects of bottlenecks: enzyme heterozygosities are often high (Johnson et al. 1986; Patton 1984). It seems more likely that "faunal drift," in which a random sample of species reaches an empty habitat, causes adaptive divergence into a variety of new niches. Indeed, workers on speciose groups usually invoke such explanations, rather than founder events (Johnson 1982; Grant 1986; Grant and Grant, this volume; Otte, this volume).

The genetic structure of natural populations

If population bottlenecks are to be important in speciation, there must be abundant polymorphism for genes that affect fitness in such a way that reproductive isolation may result. Carson's and Templeton's models postulate that particular kinds of variation are particularly favorable. Carson (1975) suggests that selection keeps certain *combinations* of alleles polymorphic (that is, epistasis maintains polymorphisms involving strong linkage disequilibrium). Templeton (1980, 1982b) invokes strongly selected major gene polymorphisms that interact with a large number of modifiers; there is epistasis between major genes and modifiers, as well as among the modifiers. Are such highly epistatic systems common in nature?

It is important to distinguish two kinds of epistasis; the distinction can be obscured by use of the blanket term "coadapted gene complex." First, there may be epistatic interactions between genes fixed for different alleles in different populations or species. Such interactions are widespread, and, indeed, are essential if reproductive isolation is to be based on anything more than heterozygote disadvantage at single loci. Second, there may be epistatic interactions between alleles segregating within one population. It is this type of polymorphic epistasis that is required by Carson's and Templeton's models. The existence of this form of "coadaptation" is suggested by *Drosophila* inversion polymorphisms, which protect distinct combinations of alleles from recombination. Similar tightly linked clusters are involved in Batesian mimicry and in incompatibility systems in plants. However, if reproductive isolation is to be produced, the interacting alleles must be broken up to produce unfit recombinants. The problem now is that there will be an excessive variation in fitness in the base population, which will make it hard to maintain a stable polymorphism (Turner 1977; Charlesworth and Smith 1982).

Some such cases are known; for example, in *Simulium erythrocephalum* there is strong linkage disequilibrium between an inversion that affects development rate and the sex-determining locus (Post 1985). But overall, such highly epistatic polymorphisms must be rare. In *Drosophila*, most variation in fitness can be accounted for by a simple mutation/selection balance (Mukai 1977; Charlesworth 1987). The component associated with epistasis and the accompanying linkage disequilibrium is at most a few percent (Mukai 1977; Charlesworth and Charlesworth 1976; Charlesworth 1987). Clegg et al. (1980) have found that though selection on the whole genome can have significant effects on embedded marker loci, these effects tend to accelerate the breakup of linkage disequilibria and to buffer random fluctuations. Clegg et al. (1980) argue that the net level of epistasis, relative to recombination, is too low to produce the highly structured polymorphisms postulated by Lewontin (1974) and Carson (1975). Certainly, there is no evidence that "a buildup of ever more complex polygenic balances" causes a "mature population to become saturated and . . . unable to respond to new conditions" (Carson and Templeton 1984:99).

Other types of fitness variation may more readily produce reproductive isolation after a founder event. For example, if a polygenic character is under disruptive selection, all that is required is that the initial genetic variance is high enough, relative to the distance to the new domain of attraction. However, though most quantitative characters are highly heritable, there is little evidence that they are under disruptive selection (for a possible exception, see Grant 1986, Grant and Grant, this volume, and Schluter, 1988). Thus, only a small fraction of the variance in fitness associated with quantitative characters is likely to contribute to isolation after a founder event.

Broad correlations

Information about the general nature of speciation mechanisms can come from comparisons across large numbers of species. Several such studies have found a strong correlation between the rate of speciation and the rate of chromosomal evolution: speciose groups tend also to be chromosomally diverse (Wilson et al. 1975; Bush et al. 1977). Though White (1978) has argued that chromosome rearrangements cause, or at least trigger, speciation, the fact that most of the rearrangements that have been established cause little reproductive isolation makes it more likely that the correlation reflects some common factor that increases the rate of both speciation and karyotypic change. The most obvious possibility is that drift in small populations can establish underdominant rearrangements (Wright 1942; Lande 1979) and also can promote speciation. This argument for the importance of drift in speciation is weakened, however, because chromosomal rearrangements may be established by selective or meiotic advantages that outweigh any underdominance caused by nondisjunction (see King 1987).

A second striking pattern is that if one sex of an interspecific cross is sterile or inviable, it is usually the heterogametic sex (Haldane 1922). A related pattern is that hybrid sterility or inviability involves the sex chromosomes much more often than would be expected. Both these patterns are found across a wide range of animal groups (Charlesworth et al. 1987). In a survey of different speciation mechanisms (including most of those discussed in this chapter), Charlesworth et al. (1987) found that models in which random drift knocks a population past some selective barrier would not account for these patterns. The only known process that would explain Haldane's rule and the excess involvement of the sex chromosomes is the accumulation of incompatible combinations of favorable recessive alleles. Such recessive alleles would accumulate much faster on the sex chromosomes, because they would then be unmasked in hemizygotes. This model successfully predicts the delay between the appearance of sterility or inviability in one sex of a cross and the completion of speciation, which is seen in comparisons of sibling pairs in *Drosophila* (Coyne and Orr 1988, and this volume).

CONCLUSIONS

This survey of speciation mechanisms shows that reproductive isolation may evolve in many ways. These mechanisms fall into (roughly) two classes. A population may overcome some selective barrier, and so passes from one "adaptive peak" to another. Alternatively, there may be a steady accumulation of favorable alleles; if different alleles accumulate in different populations, then isolation may result. These two processes correspond to the views of evolution held by Wright and by Fisher; natural selection and sampling drift play a role in both.

It is hard to distinguish between different mechanisms of speciation. There are strong theoretical arguments that the particular genetic models of founder effect speciation put forward by Mayr, Carson, and Templeton ar unlikely to be effective. However, the more general questions of how drift and selection interact to give new species, and whether founder events, through either their genetic or their ecological effects, promote speciation, remain open.

ACKNOWLEDGMENTS

Thanks are due to J. A. Coyne, J. Endler, G. Hewitt, S. Rouhani, M. Turelli, and D. S. Wilson for their helpful comments on the manuscript. This work was supported by grants from the SERC (GR/E/08507 and GR/C/91529).

LITERATURE CITED

Ayala, F. J. 1975. Genetic differentiation during the speciation process. Evol. Biol. 8:1–78.

Baker, A. J., and A. Moeed. 1987. Rapid genetic differentiation and founder effect in colonizing populations of common mynahs (*Acridotheras tristus*). Evolution 41:525–538.

Barton, N. H. 1986a. The effects of linkage and density-dependent regulation on gene flow. Heredity 57:415–426.

Barton, N. H. 1986b. The maintenance of polygenic variation through a balance between mutation and stabilising selection. Genet. Res. 47:209–216.

Barton, N. H. 1989. The rate of divergence under stabilising selection, mutation and drift. Genet. Res., in press.

Barton, N. H., and B. Charlesworth. 1984. Genetic revolutions, founder effects, and speciation. Annu. Rev. Ecol. Syst. 15:133–164.

Barton, N. H., and G. M. Hewitt. 1985. Analysis of hybrid zones. Annu. Rev. Ecol. System. 16:113–148.

Barton, N. H., and S. Rouhani. 1987. The frequency of shifts between alternative equilibria. J. Theor. Biol. 125:397–418.

Barton, N. H., and M. Turelli. 1987. Adaptive landscapes, genetic distance, and the evolution of quantitative characters. Genet. Res. 49:157–174.

Bryant, E. H., S. A. McCommas, and L. M. Combs. 1986a. Morphometric differentiation among experimental lines of the housefly in relation to a bottleneck. Genetics 114:1213–1223.

Bryant, E. H., S. A. McCommas, and L. M. Combs. 1986b. The effect of an experimental bottleneck upon additive genetic variation in the housefly. Genetics 114:1191–1211.

Bulmer, M. G. 1980. *The Mathematical Theory of Quantitative Genetics.* Oxford University Press, Oxford.

Burger, R. 1988. The mutation-selection balance and continuum of alleles models. Math. Bio. Sci., in press.

Bush, G. L., S. M. Case, A. C. Wilson, and J. L. Patton. 1977. Rapid speciation and chromosomal evolution in mammals. Proc. Nat. Acad. Sci. U.S.A. 74:3942–3946.

Cann, R. L., M. Stoneking, and A. C. Wilson. 1987. Mitochondrial DNA and human evolution. Nature (London) 325:31–35.

Carson, H. L. 1968. The population flush and its genetic consequences. Pp. 123–137 in: R. C. Lewontin (ed.), *Population Biology and Evolution.* Syracuse University Press, Syracuse, NY.
Carson, H. L. 1975. The genetics of speciation at the diploid level. Am. Natur. 109:73–92.
Carson, H. L. 1982. Speciation as a major reorganization of polygenic balances. Pp. 411–433 in: C. Barigozzi (ed.), *Mechanisms of Speciation.* Liss, New York.
Carson, H. L. 1985. Unification of speciation theory in plants and animals. Syst. Bot. 10:380–390.
Carson, H. L., and K. Y. Kaneshiro. 1976. *Drosophila* of Hawaii: Systematics and ecological genetics. Annu. Rev. Ecol. System. 7:311–346.
Carson, H. L., and A. R. Templeton. 1984. Genetic revolutions in relation to speciation phenomena: The founding of new populations. Annu. Rev. Ecol. System. 15:97–131.
Charlesworth, B. 1987. The heritability of fitness. In: J. Bradbury and M. Anderson (eds.), *Sexual Selection: Testing the Alternatives.* Dahlem Conference, J. Wiley, London.
Charlesworth, B., D. Charlesworth. 1976. An experiment on recombination load in *Drosophila melanogaster.* Genet. Res. 25:267.
Charlesworth, B., J. A. Coyne, and N. H. Barton. 1987. The relative rates of evolution of sex chromosomes and autosomes. Am. Natur. 129:113–146.
Charlesworth, B., and D. B. Smith. 1982. A computer model of speciation by founder effects. Genet. Res. 39:227–236.
Charlesworth, B. and S. Rouhani. 1988. The probability of peak shifts in a founder population. II. An additive polygenic trait. Evolution 42:1129–1145.
Christie, P. and M. R. MacNair. 1987. The distribution of postmating isolating genes in populations of the yellow monkey flower, *Mimulus guttatus.* Evolution 41:571–578.
Clegg, M. T., J. F. Kidwell, and C. R. Horch. 1980. Dynamics of correlated systems. V. Rates of decay of linkage disequilibrium in experimental populations of *Drosophila melanogaster.* Genetics 94:217–234.
Cohan, F. M., and A. A. Hoffmann. 1986. Genetic divergence under uniform selection. II. Different responses to selection for knockdown resistance to ethanol among *Drosophila melanogaster* populations and their replicates. Genetics 114:145–163.
Coyne, J. A. 1985a. Genetic studies of three sibling species of *Drosophila* with relationship to theories of speciation. Genet. Res. 46:169–192.
Coyne, J. A. 1985b. The genetic basis of Haldane's rule. Nature (London) 314:736–738.
Coyne, J. A., and R. Lande. 1985. The genetic basis of species differences in plants. Am. Natur. 126:141–145.
Coyne, J. A., and H. A. Orr. 1988. Patterns of speciation in *Drosophila.* Evolution, in press.
Craddock, E. M., and W. E. Johnson. 1979. Genetic variation in Hawaiian *Drosophila.* V. Chromosomal and allozymic diversity in *Drosophila silvestris* and its homosequential species. Evolution 33:137–155.
Crow, J. F. and M. Kimura. 1970. *An Introduction to Population Genetics Theory.* Harper & Row, New York.
Cwynar, L. C., and G. M. MacDonald. 1987. Geographical variation of lodgepole pine in relation to population history. Am. Natur. 129:463–469.
DeSalle, R., L. V. Giddings, and A. R. Templeton. 1986. Mitochondrial DNA variability in natural populations of Hawaiian *Drosophila.* I. Methods and levels of variability in *D. silvesetris* and *D. heteroneura.* Heredity 56:75–86.
DeSalle, R., and J. A. Hunt. 1987. Molecular evolution in Hawaiian Drosophilids. Trends Ecol. Evol. 2:212–216.
DeSalle, R., A. R. Templeton, I. Mori, S. Pletscher, and J. S. Johnston. 1987. Temporal and spatial heterogeneity of mtDNA polymorphism in natural populations of *Drosophila mercatorum.* Genetics 116:215–223.
Dobzhansky, Th. 1936. Studies on hybrid sterility. II. Localization of sterility factors in *Drosophila pseudoobscura* hybrids. Genetics 21:113–135.
Dobzhansky, Th. 1937. *Genetics of the Evolutionary Process.* Columbia University Press, New York.
Dodd, D. M. B., and J. R. Powell. 1987. Founder flush speciation: An update of experimental results with *Drosophila.* Evolution 39:1388–1393.

Ehrman, L. and Parsons, P. A. 1980. Sexual isolation among widely distributed populations of *Drosophila immigrans*. Behav. Genet. 10:401–407.

Felsenstein, J. 1981. Skepticism towards Santa Rosalia, or why are there so few kinds of animals? Evolution 35:124–138.

Fisher, R. A. 1930. *The Genetical Theory of Natural Selection*. Oxford University Press, Oxford.

Foley, P. 1987. Molecular clock rates at loci under stabilising selection. Proc. Natl. Acad. Sci. U.S.A. 84:7996–8000.

Gillespie, J. H. 1984. Molecular evolution over the mutational landscape. Evolution 38:1116–1129.

Goodnight, C. J. 1987. On the effect of founder events on epistatic genetic variance. Evolution 41:80–91.

Gottlieb, L. D. 1984. Genetics and morphological evolution in plants. Am. Natur. 123:681–709.

Grant, P. R. 1986. *The Evolution and Ecology of Darwin's Finches*. Princeton University Press, Princeton.

Haldane, J. B. S. 1922. Sex ratio and unisexual sterility in hybrid animals. J. Genet. 12:101.

Haldane, J. B. S. 1957. The cost of natural selection. J. Genet. 55:511–524.

Haldane, J. B. S., and S. D. Jayakar. 1963. Polymorphism due to selection of varying direction. J. Genet. 58:237–242.

Hedrick, P. W. 1981. The establishment of chromosomal variants. Evolution 35:322–332.

Hedrick, P. W. and D. A. Levin. 1984. Kin founding and the establishment of chromosomal variants. Am. Natur. 124:789–797.

Janson, K. 1987. Genetic drift in small and recently founded populations of the marine snail *Littorina saxatalis*. Heredity 58:31–38.

Johnson, M. J., D. C. Wallace, S. D. Ferris, M. G. Rattazzi, and L. L. Cavalli-Sforza. 1983. Radiation of human mitochondrial DNA types analyzed by restriction endonuclease cleavage patterns. J. Mol. Evol. 19:255–271.

Johnson, M. S. 1982. Polymorphism for direction of coil in *Partula suturalis:* Behavioral isolation and positive frequency dependent selection. Heredity 49:145–151.

Johnson, M. S., J. Murray and B. Clarke. 1986. Allozymic similarities among species of *Partula* on Moorea. Heredity 56:319–328.

Jones, J. S., and S. Rouhani. 1986. How small was the bottleneck? Nature (London) 319:449–450.

Kaneshiro, K. Y. 1980. Sexual isolation, speciation and the direction of evolution. Evolution 34:437–444.

Kaneshiro, K. Y., and C. R. B. Boake. 1987. Sexual selection and speciation: Issues raised by Hawaiian Drosophilids. Trends Ecol. Evol. 2:207–212.

Kauffman, S., and S. A. Levin. 1987. Towards a general theory of adaptive walks on a rugged landscape. J. Theoret. Biol. 128:11–46.

Kimura, M. 1981. Possibility of extensive neutral evolution under stabilising selection with special reference to non-random usage of synonymous codons. Proc. Natl. Acad. Sci. U.S.A. 78:5773–5777.

King, M. 1987. Chromosomal rearrangements, speciation, and the theoretical approach. Heredity 59:1–6.

Kirkpatrick, M. 1982a. Quantum evolution and punctuated equilibrium in continuous genetic characters. Am. Natur. 119:833–848.

Kirkpatrick, M. 1982b. Sexual selection and the evolution of female choice. Evolution 36:1–12.

Lande, R. 1975. The maintenance of genetic variability by mutation in a polygenic character with linked loci. Genet. Res. 26:221–236.

Lande, R. 1979. Effective deme sizes during long-term evolution estimated from rates of chromosomal rearrangement. Evolution 33:234–251.

Lande, R. 1980. Genetic variation and phenotypic evolution during allopatric speciation. Am. Natur. 116:463–479.

Lande, R. 1983. The response to selection on major and minor mutations affecting a metrical trait. Heredity 50:47–65.

Lande, R. 1985. The fixation of chromosomal rearrangements in a subdivided population with local extinction and recolonisation. Heredity 54:323-332.

Lewontin, R. C. 1966. Is nature probable or capricious? Bioscience 16:25-27.

Lewontin, R. C. 1974. *The Genetic Basis of Evolutionary Change.* Columbia University Press, New York.

Maynard Smith, J. 1983. The genetics of punctuation and stasis. Annu. Rev. Genet. 17:11-25.

Mayr, E. 1942. Systematics and the origin of species. Columbia University Press, New York.

Mayr, E. 1954. Change of genetic environment and evolution. Pp. 156-180 in: J. S. Huxley, A. C. Hardy, and E. B. Ford (eds.), *Evolution as a Process.* Allen & Unwin, London.

Mayr, E. 1982. Processes of speciation in animals. Pp. 1-19 in: C. Barigozzi (ed.), *Mechanisms of Speciation.* Liss, New York.

Mukai, T. 1977. Genetic variance for viability and linkage disequilibrium in natural populations of *Drosophila melanogaster.* Pp. 97-112 in: F. B. Christiansen and T. M. Fenchel (eds.), *Measuring Selection in Natural Populations.* Springer-Verlag, Berlin.

Muller, H. J. 1939. Reversibility in evolution considered from the standpoint of genetics. Biol. Rev. Cambridge Philos. Soc. 14:261-280.

Newman, C. M., J. E. Cohen, and C. Kipnis. 1985. Neo- Darwinian evolution implies punctuated equilibria. Nature (London) 315:400-402.

Patton, J. L. 1984. Genetical processes in the Galapagos. Biol. J. Linnean Soc. 21:97-113.

Post, R. J. 1985. Sex chromosome evolution in *Simulium erythrocephalum* (Diptera: Simuliidae). Heredity 54:149-158.

Powell, J. R. 1978. The founder-flush speciation theory—an experimental approach. Evolution 32:465-474.

Provine, W. 1986. *Sewall Wright and Evolutionary Biology.* University of Chicago Press, Chicago.

Ringo, J., D. Wood, R. Rockwell, and H. Dowse. 1985. An experiment testing two hypotheses of speciation. Am. Natur. 126:642-661.

Ringo, J., K. Barton, and H. Dowse. 1986. The effect of genetic drift on mating propensity, courtship behaviour, and postmating fitness in *Drosophila simulans.* Behaviour 97:226-233.

Robertson, A. 1952. The effect of inbreeding on the variation due to recessive genes. Genetics 37:189-207.

Rouhani, S., and N. H. Barton. 1987a. Speciation and the "shifting balance" in a continuous population. Theoret. Pop. Biol. 31:465-492.

Rouhani, S., and N. H. Barton. 1987b. The probability of peak shifts in a founder population. J. Theoret. Biol. 126:51-62.

Schluter, D. 1988. Estimating the form of natural selection on a quantitative trait. Evolution 42:849-861.

Sene, F. M., and H. L. Carson. 1977. Genetic variation in Hawaiian *Drosophila.* IV. Allozymic similarity between *D. silvestris* and *D. heteroneura* from the island of Hawaii. Genetics. 86:187-198.

Slatkin, M. 1981. Estimating levels of gene flow in natural populations. Genetics 99:323-335.

Slatkin, M. 1987. Gene flow and the geographic structure of natural populations. Science 236:787-792.

Tanksley, S. D., H. Medina-Filho, and C. M. Rick. 1982. Use of naturally occurring enzyme variation to detect and map genes controlling quantitative traits in an interspecific backcross of tomato. Heredity 49:11-26.

Templeton, A. R. 1980. The theory of speciation via the founder principle. Genetics 94:1011-1038.

Templeton, A. R. 1981. Mechanisms of speciation—a population genetic approach. Annu. Rev. Ecol. System. 12:23-48.

Templeton, A. R. 1982a. Adaptation and integration of evolutionary forces. Pp. 15-31 in: R. Milkman (ed.) *Perspectives on Evolution.* Sinauer, Sunderland, Mass.

Templeton, A. R. 1982b. Genetic architectures of speciation. Pp. 105–121 in: C. Barigozzi (ed.), *Mechanisms of Speciation.* Liss, New York.

Turelli, M. 1984. Heritable genetic variation via mutation-selection balance: Lerch's zeta meets the abdominal bristle. Theoret. Pop. Biol. 25:138–193.

Turner, J. R. G. 1977. Butterfly mimicry—the genetical evolution of an adaptation. Evol. Biol. 10:163–206.

Vogel, F., and A. G. Motulsky. 1986. *Human Genetics: Problems and Approaches.* Springer-Verlag, Berlin.

Wainscoat, J. S., A. V. S. Hill, A. L. Boyce, J. Flint, M. Hernandez, S. L. Thein, J. M. Old, J. R. Lynch, A. G. Falusi, D. J. Weatherall, and J. B. Clegg. 1986. Evolutionary relationships of human populations from an analysis of nuclear DNA polymorphisms. Nature (London) 319:491–493.

Walsh, J. B. 1982. Rate of accumulation of reproductive isolation by chromosome rearrangements. Am. Natur. 120:510–532.

White, M. J. D. 1978. *Modes of Speciation.* W. H. Freeman, San Francisco.

Wilson, A. C., G. L. Bush, S. M. Case, and M. C. King. 1975. Social structuring of mammalian populations and rate of chromosomal evolution. Proc. Natl. Acad. Sci. U.S.A. 72:5061–5065.

Wilson, D. S., and M. Turelli. 1986. Stable underdominance and the invasion of empty niches. Am. Natur. 127:835–850.

Wright, S. 1932. The roles of mutation, inbreeding, crossbreeding and selection in evolution. Proc. Sixth Int. Cong. Genet. 1:356–366.

Wright, S. 1935. Evolution in populations in approximate equilibrium. J. Genet. 30:257–266.

Wright, S. 1941. On the probability of fixation of reciprocal translocations. Am. Natur. 74:513–522.

Wright, S. 1978a. *Evolution and the Genetics of Populations.* IV. *Variability within and among Populations.* University of Chicago Press, Chicago.

Wright, S. 1978b. Review of "Modes of Speciation" by M. J. D. White Paleobiology 4:373–379.

Yanev, K. P., and D. B. Wake. 1981. Genic differentiation in a relict desert salamander, *Batroceps campi.* Herpetologica 37:16–28.

CHAPTER 11

MATING SYSTEM EVOLUTION AND SPECIATION IN HETEROSTYLOUS PLANTS

Spencer C. H. Barrett

INTRODUCTION

A rekindling of interest in the processes that result in reproductive isolation and speciation is evident in the recent evolutionary literature (White 1978; Barigozzi 1982; Carson and Templeton 1984; Barton and Charlesworth 1984; Brown 1985; Iwatsuki et al. 1986). Two major stimuli are largely responsible for renewed discussion of the mechanisms responsible for species formation. Electrophoretic surveys of enzyme polymorphisms have enabled comparative studies of the magnitude of genetic divergence among populations, races, and species. The results have highlighted our ignorance of the kinds and amount of genetic change that are required for species formation. Second, the controversy over the relative importance of punctuated versus gradual change during evolutionary diversification (e.g., Gould 1980; Charlesworth et al. 1982) has focused interest on the tempo of species formation and the role of founder events in fostering rapid speciation. In particular, the recent literature has addressed the question of whether reproductive isolation evolves slowly, as an incidental by-product of the adaptive divergence of gene pools, or whether new species are formed in rare events distinct from the normal processes of phyletic evolution.

A notable feature of much of the recent literature on the genetics and evolution of species formation is the paucity of information from plants. Where these are considered it is not uncommon that only diploid cross-fertilizing groups are discussed (e.g., Carson 1985). Yet these features are hardly representative of the diversity of genetic and reproductive systems that occur, particularly in flowering plants and ferns. The limited coverage of plant data in controversies such as the significance of sympatric speciation, or the role of small, isolated populations in the speciation process, is unfortunate since information from plants may serve to temper the polarization of viewpoints that often accompanies any healthy debate. By broadening the discussion to include organisms with diverse ecological and genetic characteristics, a more realistic *biological* viewpoint of what is undoubtedly a varied and complex process is likely to prevail. For it seems that the specific details of the speciation process will, in large part, be governed by the biological peculiarities of the organisms themselves. Among flowering plants, for example, annual species usually develop strong postzygotic isolating mechanisms, whereas in many long-lived perennials these are absent (Grant 1981). With such divergent patterns within a single group of organisms, it may be a futile exercise to attempt a truly unified theory of speciation for plants and animals, although Carson (1985) has recently suggested this is possible and a worthwhile goal.

This review focuses on just one aspect of the speciation process in plants: the role of mating system evolution in promoting character divergence and reproductive isolation in herbaceous plants. The approach taken is largely empirical and involves a review of experimental work on contemporary plant populations using the tools of population and evolutionary genetics. Since the work I discuss largely concerns the evolution of heterostylous reproductive systems, I begin by briefly summarizing some salient features of plant genetic systems relevant to the main themes of the chapter.

Plant reproductive systems

As originally recognized by Darwin (1876, 1877), flowering plants display a considerable diversity of reproductive systems. This variation has important implications for the modes of speciation that can occur (Baker 1959; Grant 1981). Although three modal reproductive classes are traditionally recognized (outbreeders, inbreeders, and apomicts), each of these groupings contains a variety of different mechanisms associated with pollination biology and sexual systems of individual taxa (see Richards 1986, for a recent review). A notable feature of many plant groups is the occurrence among closely related taxa of contrasting reproductive modes. This suggests that the evolution of floral syndromes, and their influence on mating patterns, is intimately associated with the development of reproductive isolation and speciation (Grant 1949; Baker 1961; Stebbins 1970).

One of the most prominent evolutionary pathways leading to species formation in herbaceous plants involves the replacement of outcrossing by a mating system based on predominant self-pollination (Stebbins 1957; Jain 1976). Numerous biosystematic studies have documented this shift in mating system among related species in genera such as *Amsinckia* (Ray and Chisaki 1957), *Arenaria* (Wyatt 1984), *Armeria* (Baker 1966), *Clarkia* (Lewis 1973), *Gaura* (Raven and Gregory 1972), *Gilia* (Grant and Grant 1965), *Lasthenia* (Ornduff 1966), *Leavenworthia* (Lloyd 1965), *Limnanthes* (McNeill and Jain 1983), *Lycopersicon* (Rick et al. 1977), *Melochia* (Martin 1967), *Oxalis* (Ornduff 1972), *Plectritis* (Carey and Ganders 1986), *Petrorhagia* (Thomas and Murray 1981), and *Stephanomeria* (Gottlieb 1973). The evolution of self-fertilization favors establishment of chromosomal rearrangements and new homozygous genotypes and contributes toward the reproductive isolation of the selfing derivative.

A frequent observation in genera with related outcrossers and selfers is the tendency for selfing taxa to occur at the geographic margins of the range of their outcrossing progenitors or in ecologically marginal sites within the ancestral species range (Jain 1976; Lloyd 1980). This pattern is in accord with the view that self-fertilizing individuals are at a selective advantage at low density and has led to the suggestion that they are favored in pioneer environments or under conditions in which population bottlenecks frequently occur.

It is evident from the literature on the evolution of selfing that many workers envision that genetic processes occurring rapidly in small populations play a significant role in the development of reproductive isolation and speciation in herbaceous plants. This viewpoint is of relevance to broader issues associated with the significance of founder effect speciation. Both Templeton (1981) and McNeill and Jain (1983) have suggested that although the vast majority of founder events do not lead to rapid speciation via genetic transilience, this mode of species formation may occur commonly in association with the evolution of selfing in plants. Unfortunately students of plant speciation have rarely examined the microevolutionary forces operating within populations in sufficient detail to provide evidence in support of this claim.

Polyploidy

A second important feature of evolutionary diversification in plants is the frequent occurrence of polyploidy and its role as an isolating mechanism. Since current estimates suggest that 70–80% of angiosperm species are of polyploid origin (Goldblatt 1980; Lewis 1980), it could be argued that this form of sympatric speciation is among the commonest modes of speciation in flowering plants. This is sometimes neglected from debates concerned with the likely occurrence of sympatric speciation in nature. Unfortunately, in spite of the widespread nature of polyploidy in flowering plants, we are still woefully ignorant of the population processes that lead to the origin and establishment of polyploid species. The wider adaptability and evolutionary success of many polyploids are believed to arise from extensive gene duplication and subsequent diversification, fixed heterozygosity, and the reduced effects of inbreeding depression (Roose and Gottlieb 1976; Stebbins 1980; Grant 1981; Levin 1983; Barrett and Shore 1989). Although heterosis may be involved in the success of many polyploids of hybrid origin, other changes that accompany polyploidization may also be involved. In particular, alterations in breeding system are frequently associated with changes in ploidal level so that among related taxa, diploids are often primarily outcrossing, whereas polyploids tend to be selfers (Grant 1981). The kind of polyploidy can also vary with breeding system. Selfing is commonly associated with allopolyploidy, whereas autopolyploidy is almost exlusively found in outcrossing species (Bingham 1980). These patterns indicate that to fully understand polyploid speciation in flowering plants all aspects of the evolution of their genetic systems, including both chromosomal and reproductive factors, need to be considered.

Heterostyly

Beginning with Darwin's original work (Darwin 1877), heteromorphic incompatibility systems (distyly and tristyly) have provided a rich source of experimental material for researchers interested in the evolution of plant genetic systems (see Ganders 1979; Barrett 1988b). These outbreeding floral polymorphisms, which have evolved independently in 24 angiosperm families, are simply inherited and particularly susceptible to genetic modifications that influence the mating systems of populations. In almost every distylous or tristylous genus there are selfing homostylous taxa that have originated as a result of the breakdown of heterostyly. In most cases the evolution of homostyly is associated with dramatic reductions in flower size, acquisition of the self-pollinating habit, and the evolution of reproductive isolation. One consequence of the differences between closely related heterostylous–homostylous pairs is that the homostylous derivatives are usually given separate taxonomic status (Ornduff 1969).

In this chapter I review two studies of mating system evolution in heterostylous groups currently being conducted in our laboratory, and discuss their overall relevance to speciation theory. In the first I examine the relationship between polyploidy and breeding system diversification in the *Turnera ulmifolia* complex and evaluate several hypotheses to account for the evolution of selfing taxa from distylous members of the complex. The second study examines the breakdown of tristyly to homostyly in *Eichhor-*

nia and assesses the potential role of population bottlenecks in initiating mating system change. A common theme in studies of both groups is that relatively simple genetic alterations in floral traits can have profound effects on mating patterns and that these changes have important ecological and evolutionary consequences.

POLYPLOID SPECIATION IN THE *TURNERA ULMIFOLIA* COMPLEX

Turnera ulmifolia L. (Turneraceae) is a polymorphic polyploid complex of herbaceous, perennial weeds native throughout much of the New World tropics and adventive in parts of Africa, India, and southeast Asia. Populations are conspicuous on roadsides and open waste ground but rarely colonize arable land (Barrett 1978a). The only monograph of the genus, by Urban (1883), recognized 12 intergrading taxonomic varieties within the *T. ulmifolia* complex. Later regional treatments have elevated some varieties to specific rank, whereas others are of dubious taxonomic status. In our studies we have followed Urban's treatment with the knowledge that most of the varieties we have studied are good biological species. Details of the morphological variation and crossability of selected varieties are presented in Shore and Barrett (1985a), and Arbo (1985) and Arbo and Fernández (1983) provide recent taxonomic treatments of some members of the complex.

Breeding systems and polyploidy

Surveys of the *T. ulmifolia* complex throughout much of its New World range have revealed an association between the breeding systems of populations and their chromosome numbers (Table 1). Diploid and tetraploid populations are uniformly distylous with strong self-incompatibility and a

TABLE 1. Relationships between ploidal level and breeding system in the *Turnera ulmifolia* complex.[a]

Ploidal level	Varieties	Populations	Compatibility	Stamen-style polymorphism
Diploid	3	15	SI	Distylous
Tetraploid	3	33	SI	Distylous
Hexaploid	3	25	SC	Homostylous

[a]SI, self-incompatible; SC, self-compatible; $x = 5$. After Barrett and Shore (1987).

1:1 ratio of the floral morphs, whereas hexaploid populations are homostylous and self-compatible. The three homostylous varieties that we have studied are differentiated for morphological traits and allozymes, occur at different margins of the geographical range of the species complex, and are intersterile. This indicates that distyly has broken down to homostyly on at least three separate occasions in the complex, each time in association with the hexaploid condition (Shore and Barrett 1985b). Cytological and isozyme studies indicate that although tetraploid varieties form quadrivalents and exhibit tetrasomic inheritance, hexaploids form bivalents and display considerable fixed heterozygosity at isozyme loci. These findings suggest that tetraploid populations are autopolyploids, whereas hexaploid populations may be allopolyploids, or at least the respective polyploids lie toward different ends of the auto–allopolyploid continuum (Barrett and Shore 1987).

Although it is unusual to find autopolyploidy and allopolyploidy within the same species complex, their association with outcrossing and selfing breeding systems in *T. ulmifolia* parallels a more general pattern found among other flowering plant groups. Stebbins (1980) has suggested that to be successful an autopolyploid is dependent on heterozygosity and that the biochemical and physiological advantages conferred by heteroallelism are important components of autopolyploid vigor. Evidently the polysomic condition cannot tolerate the homozygosity associated with self-pollination. It is significant that among the many polyploid crop cultivars there are no examples of successful polysomic polyploid species that are self-pollinated (Bingham 1980). In contrast, heterozygosity in allopolyploids results from gene multiplication rather than through allelic variation maintained by outcrossing. The simultaneous contribution of high biochemical diversity, as a result of fixed heterozygosity, and assured reproduction through selfing may be the key components of the genetic system responsible for the success of many allopolyploid weeds (Barrett and Shore 1989). In *T. ulmifolia* the three homostylous varieties occur at the geographic margins of the range, suggesting that the ability of homostyles to reproduce under conditions of low pollinator availability may have been important in their initial establishment and spread.

Origin and evolution of homostyly

What factors account for the association between the breakdown of distyly to homostyly and the hexaploid condition in the *T. ulmifolia* complex? Formal genetic analysis of distyly and homostyly has demonstrated that distyly is governed by a single gene "locus" with L plants of genotypes *ss* and *ssss* and S plants of genotypes *Ss* and *Ssss* in diploids and tetraploids, respectively. Homostyly arises by a crossover in the supergene that controls the polymorphism (Shore and Barrett 1985b). The shift from outcrossing to selfing is therefore a relatively simple genetic change that occurs in one mutational step as a result of recombination. There is no evidence that polyploidy per se promotes increased recombination in heterostylous plants as suggested by Dowrick (1956). In *T. ulmifolia* hexaploids, synthesized from triploids by the use of colchicine, remain distylous and self-incompatible. This indicates that at its inception hexaploidy does not stimulate homostyle formation, at least in the experimental material with which we have worked (Shore and Barrett 1986).

One explanation for the puzzling association between hexaploidy and homostyle formation in the *T. ulmifolia* complex concerns the hybrid origin of homostylous varieties. It is possible that the spread of recombinant homostyles is favored in allohexaploid populations because of reduced inbreeding depression in comparison with diploid and tetraploid populations. Several authors have modeled the spread of homostyles in distylous populations (Crosby 1949; Bodmer 1960; Charlesworth and Charlesworth 1979a) and it is clear that inbreeding depression is one of the most important factors restricting the spread of selfing phenotypes in outcrossing populations (Lloyd 1979; Lande and Schemske 1985). Although theoretical studies indicate that the magnitude of inbreeding depression resulting from deleterious recessive genes is reduced by population bottlenecks, a similar effect can also arise from polyploidy (Lande and Schemske 1985). This has been termed a "hybridity bottleneck" by Hedrick (1987), who suggests that genes that increase the rate of selfing are likely to increase in frequency in newly established allopolyploids as a result of their decreased genetic loads. Similar arguments may apply to *T. ulmifolia*, and it would be of interest to investigate the genetic load of populations with contrasting ploidal levels to examine whether inbreeding depression declines with increased ploidal level. Unfortunately, since homostyles in the complex are well-established species and may be of considerable age, an analysis of this type may not be particularly informative for inferring the past history of selection associated with the origin and establishment of homostyly.

Mating system evolution in homostyles

The evolution of homostyly in *T. ulmifolia* has not resulted in the adoption of selfing as the principal reproductive mode (Figure 1). Although some populations in the complex are small-flowered and highly autogamous, others have large, showy flowers and well-developed herkogamy (spatial separation of anthers and stigmas) and are outcrossing. Experimental studies on the range of homostylous floral phenotypes in the *T. ulmifolia* complex suggest that outcrossing phenotypes are derived from selfing phenotypes (Barrett and Shore 1987). This mating system change is particularly well illustrated in *T. ulmifolia* var. *augustifolia* in the Caribbean region. On small islands (e.g., Bahamas) small-flowered selfing phenotypes occur, whereas on the larger, more ecologically complex islands (e.g., Greater Antilles, Jamaica) large-flowered, outcrossing homostyles predominate (Figure 2). These patterns of floral variation suggest that frequent episodes of extinction and colonization favor selfing homostyles on small islands, whereas on larger islands selection pressures for outcrossing, associated with adaptive radiation, occur following the initial establishment of selfing colonists. The autochthonous development of outcrossing mechanisms in island plants is well known and has been reviewed by Baker (1967), Carlquist (1974), and Ehrendorfer (1979).

The reestablishment of outcrossing, following homostyle formation in *T. ulmifolia,* points to the evolutionary lability of the mating system and demonstrates its ability to respond to local selection pressures, if sufficient genetic variation for traits influencing mating behavior are maintained in populations. The allohexaploid status of homostyles must be important in this regard. What role processes such as intergenomic recombination, gene silencing, and regulatory divergence have played in releasing variation locked up within individuals as fixed heterozygosity is not known. It is notable that the spectacular adaptive radiation of *Bidens* species on the Hawaiian islands (Sun and Ganders 1988) also involves allohexaploid taxa. Interestingly, *Bidens* has also evolved alternative outcrossing systems (e.g., gynodioecism) following island colonization (Sun and Ganders 1986).

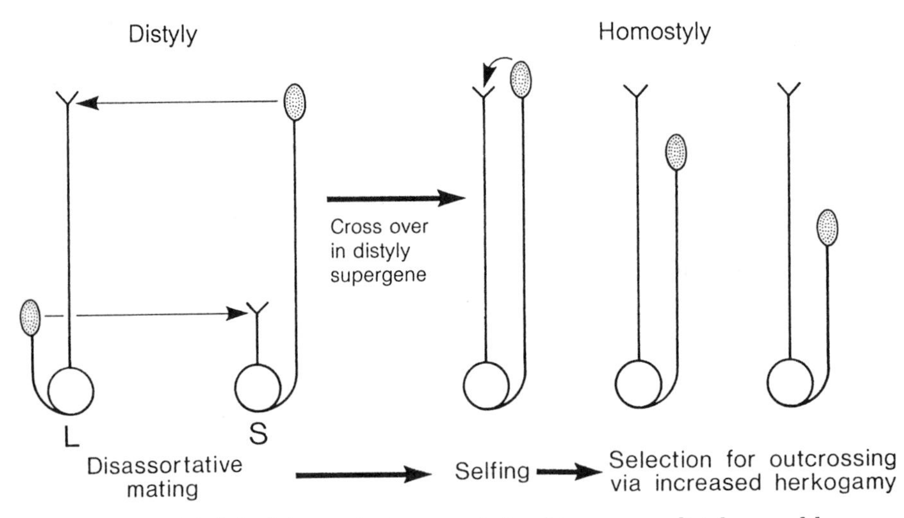

FIGURE 1. Model of the evolutionary relationships among distylous and homostylous forms within the *Turnera ulmifolia* complex.

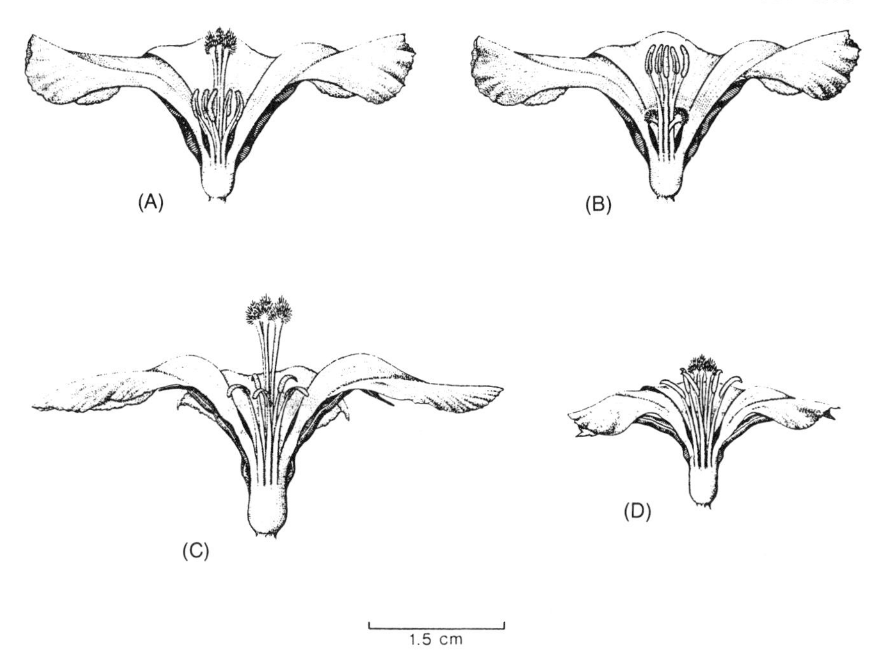

FIGURE 2. Floral variation in the *Turnera ulmifolia* complex. (A, B) Long- and short-styled morphs of distylous var. *intermedia* (4*x*) showing the reciprocal arrangement of stamens and styles; (C) var. *angustifolia* (6*x*) large-flowered outcrossing homostyle from Jamaica; (D) var. *augustifolia* (6*x*) small-flowered selfing homostyle from the Bahamas.

POPULATION BOTTLENECKS AND MATING SYSTEM SHIFTS IN *EICHHORNIA*

As discussed earlier, the evolution of self-fertilization in flowering plants is frequently associated with colonization of geographically or ecologically marginal environments. In several of the studies cited, population bottlenecks arising through drought or founder events have been implicated in the establishment of self-fertilizing variants. Although the role of founder events in the speciation process remains a contentious issue in most animal groups (see Barton, this volume), they play an important role in mating system shifts in many herbaceous plant groups. Barton and Charlesworth (1984, p. 144) state "there are no grounds for expecting a strong association between change in mating system and founder events, except in self-compatible hermaphrodites." Many cases of the shift from outcrossing to selfing in plants involve this condition. Our own studies of self-compatible tristylous species of *Eichhornia* (Pontederiaceae) indicate that genetic and ecological processes operating in small populations can be of major importance for initiating evolutionary changes in the mating system. The remainder of this chapter documents these changes, examines the selective forces responsible for the evolution of selfing, and discusses the genetic and developmental basis of changes in floral phenotype.

Interspecific patterns

Of the eight species of *Eichhornia,* three possess large, showy, entomophilous flowers and are primarily tristylous, and the remaining five are small-flowered, largely monomorphic for floral traits, and predominantly self-pollinating. Two sources of evidence indicate that the major pathway of breeding system evolution in the genus involves the repeated breakdown of tristyly to give rise to selfing, semihomostylous (with one set of stamens adjacent to the stigma) species (Barrett 1985a, 1988b). The first is the occurrence of semihomostylous variants in each of the three tristylous species, *E. azurea* (Barrett 1978b), *E. crassipes* (Barrett 1979), and *E. paniculata* (Barrett 1985b). This indicates that the microevolutionary forces responsible for the breakdown of tristyly occur under the present ecological conditions in which *Eichhornia* species are found. The second line of evidence is the presence of residual genetic polymorphisms associated with the tristylous syndrome in several autogamous semihomostylous species (e.g., *E. diversifolia, E. heterosperma, E. paradoxa*). The occurrence of these floral polymorphisms indicates that the semihomostylous condition is derived from tristyly, unless of course we are prepared to accept that tristyly is in *statu nascendi* within each semihomostylous species. Given the complexity of the polymorphism and its extreme rarity in the angiosperms as a whole, this seems unlikely (Eckenwalder and Barrett 1986).

Microevolutionary studies of the breakdown process

Our early studies on the evolutionary breakdown of tristyly in *E. azurea* and *E. crassipes* were restricted because of their long-lived clonal nature, the absence of sexual reproduction in many populations, and the rarity of semihomostyle formation. These features hinder ecogenetic studies and reduce the likelihood of detecting microevolutionary changes in the mating system. More recently we have focused our attention on the remaining tristylous species, *E. paniculata,* since it possesses the full range of mating systems represented in the genus and is a short-lived perennial or annual that is easily crossed and cultured under glasshouse conditions.

Eichhornia paniculata is an emergent aquatic that grows in seasonal pools, ditches, marshes, and rice fields. The species is distributed primarily in northeast Brazil and the Caribbean islands of Cuba and Jamaica, with single isolated collections reported from west Brazil, Ecuador, and Nicaragua. Two aspects of *E. paniculata* ecology make it particularly susceptible to population size fluctuations. The species' major center of distribution, northeast Brazil, is an arid region with one of the most unpredictable rainfall regimes in the world (Nimer 1972). Variation in rainfall has a major influence on the availability of aquatic habitats inhabited by *E. paniculata* and fluctuations in population size and frequent local extinctions are an integral feature of the population biology of the species. In addition, *E. paniculata* produces large numbers of tiny seeds and, like many aquatic plants, the species is particularly prone to both short- and long-distance dispersal by water birds. The founding of isolated populations by long-distance bird dispersal probably accounts for the scattered distribution of the species in the New World tropics. Colonizing events of this type are frequently associated with population bottlenecks, and, in *E. paniculata,* these have a major disruptive effect on the maintenance of tristyly.

Genetic studies of the inheritance of tristyly in *Eichhornia* indicate that the polymorphism is under the control of two diallelic loci (*S*, *M*) with *S* epistatic to *M* (S. C. H. Barrett, unpublished data). With this inheritance pattern and legitimate mating between the floral morphs, an isoplethic equilibrium (1:1:1) is the only possible condition in large populations if the morphs are of equal fitness (Heuch 1979). This expectation provides a logical starting point for studies concerned with the maintenance and breakdown of heterostyly and surveys of style morph frequency in natural populations can provide valuable clues to the mechanisms responsible for anisoplethy (Charlesworth 1979; Barrett et al. 1983; Morgan and Barrett 1988). Our surveys of morph frequencies in populations of *E. paniculata* from northeast Brazil and Jamaica indicate major deviations from isoplethy (Barrett 1985a; Barrett et al. 1989). Of 84 populations sampled in northeast Brazil, 58 were trimorphic, 21 were dimorphic with only the L and M morphs represented, and 5 were monomorphic containing only the M morph. On the island of Jamaica a total of 26 populations were sampled, of which 19 were monomorphic with M plants only and 7 were dimorphic with L and M plants. The S morph is absent from Jamaica, presumably as a result of founder events (see below). Similar patterns of floral morph distribution are evident in *E. crassipes* where the S morph is restricted to lowland tropical South America and is absent from the remainder of the species' New and Old World ranges, whereas the M morph predominates in most areas (Barrett 1977; Barrett and Forno 1982).

By plotting the frequency of floral morphs in Brazilian populations of *E. paniculata* it is possible to discern several patterns in population structure that aid in formulating a model of the breakdown process (Figure 3). Large areas of the triangle in Figure 3A are not occupied by populations, and trimorphic populations tend to be concentrated on the left side of the triangle as a result of a reduced frequency of the S morph in many pop-

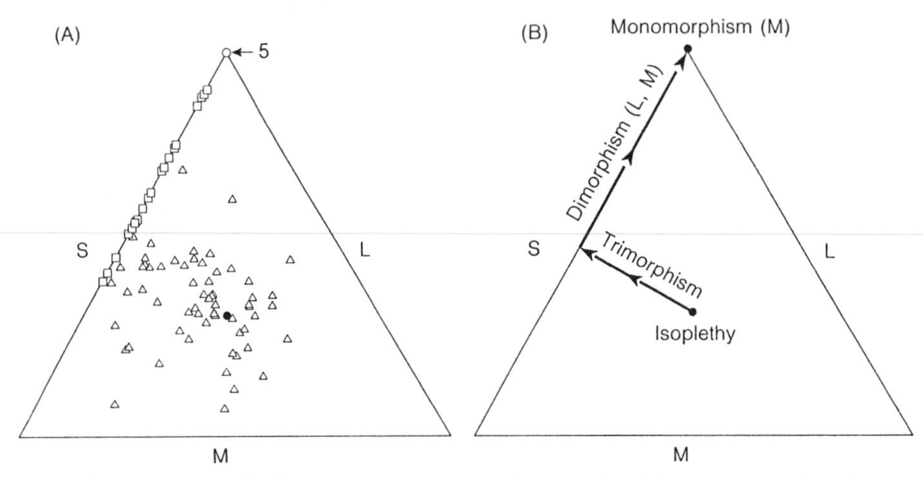

FIGURE 3. Morph frequencies in 84 populations of *Eichhornia paniculata* from northeast Brazil. (A) Small triangles are trimorphic populations, squares are dimorphic populations, and open circles are monomorphic populations. Isoplethy (filled circle) is equidistant from all axes and the distance of a point from each axis is proportional to the frequency of the morph in the population, e.g., points close to the S axis have a low frequency of the S morph. (B) Pathway from floral trimorphism through dimorphism to monomorphism associated with the breakdown of tristyly to semihomostyly in *Eichhornia paniculata.*

ulations. Among the sample of dimorphic populations the L morph is usually underrepresented in comparison with a 1:1 expectation, with the M morph predominating in most populations. These patterns indicate that the breakdown of tristyly to floral monomorphism is associated with two key stages: (1) loss of the S allele and hence the S morph from populations, and (2) loss of the *m* allele and thus the L morph. These stages are depicted in Figure 3B with arrows indicating the pathway of evolutionary change from floral trimorphism through dimorphism to monomorphism. The breakdown of tristyly may not always follow the trajectory illustrated in Figure 3B, for a variety of reasons associated with founder events and local conditions. However, the available data indicate that the most frequent pathway of mating system change in *E. paniculata* involves the stages outlined above. This pathway and associated changes in floral phenotype and mating system are illustrated schematically in Figure 4 and discussed in detail below.

Evolution of dimorphism from trimorphism

What factors account for the loss of the S morph from populations of *E. paniculata* and its absence from Jamaica? The most likely explanation involves stochastic influences on population size and the influence of founder events. Since *E. paniculata* is highly self-fertile, polymorphic populations can arise from selfing and segregation in genotypes heterozygous at the *S* and *M* loci. The *m* allele governing the L phenotype can be carried by all three style morphs, and the *M* allele, controlling the M phenotype, by the S and M morphs. In contrast, the dominant *S* allele is only carried by the S morph and separate introductions of this morph are therefore necessary for it to become established in populations. Presumably the S morph is absent from Jamaica because it was not involved in early dispersal events to the island. Computer simulation studies on the effects of random fluctuations in population size in tristylous systems verify that the S morph is most often lost from populations (Heuch 1980; Barrett et al. 1989).

Although similar processes on a neighborhood scale may contribute to the reduced frequency of the S morph in Brazilian populations of *E. paniculata,* selective factors also appear to play an important role. For tristyly to function effectively, populations should provide sufficient pollen and nectar rewards to attract specialized long-tongued pollinators, usually bees in the genera *Ancyloscelis* and *Florilegus* (Barrett 1985b). Where long-tongued pollinators are absent, either because of local ecological conditions or demographic factors associated with small population size, the maternal fitness of the S morph may suffer disproportionately, in comparison with L and M morphs, because of its concealed female reproductive parts. This effect has been implicated as the cause of low seed set in *E. crassipes* (Barrett 1977), and comparisons of fruit set in floral morphs from

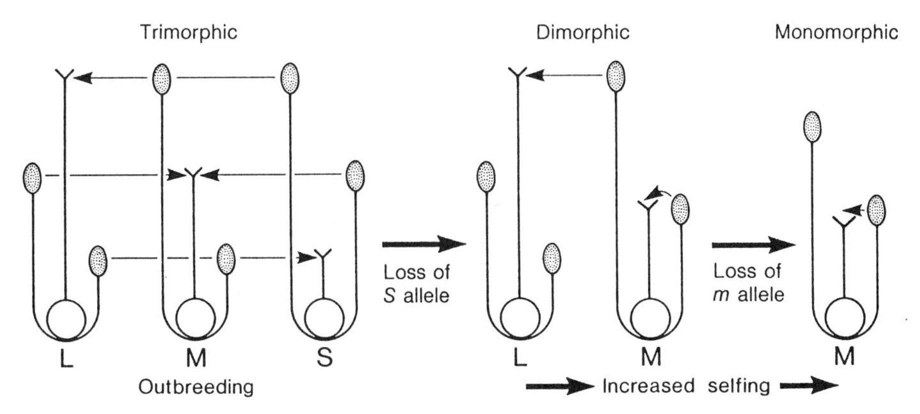

FIGURE 4. Model of the breakdown of tristyly to semihomostyly in *Eichhornia paniculata.* Arrows indicate the predominant matings that occur within populations. Note the modification in short-stamen position of the M morph in dimorphic and monomorphic populations and the reduction in dimensions of floral organs in the monomorphic population.

populations of *E. paniculata,* with and without the services of long-tongued bees, have demonstrated the potential fitness penalties to the S morph associated with the absence of specialized pollinators (Figure 5). Simulation studies confirm that reductions in the female fertility of the S morph will result in its loss from trimorphic populations. The decline in frequency of the S morph occurs more rapidly when mating patterns change from disassortative to random mating. This effect is likely in small populations serviced by generalist pollinators (Barrett et al. 1989).

Several lines of evidence suggest that nontrimorphic populations are more capable of persisting in marginal environments. Censuses of popula-

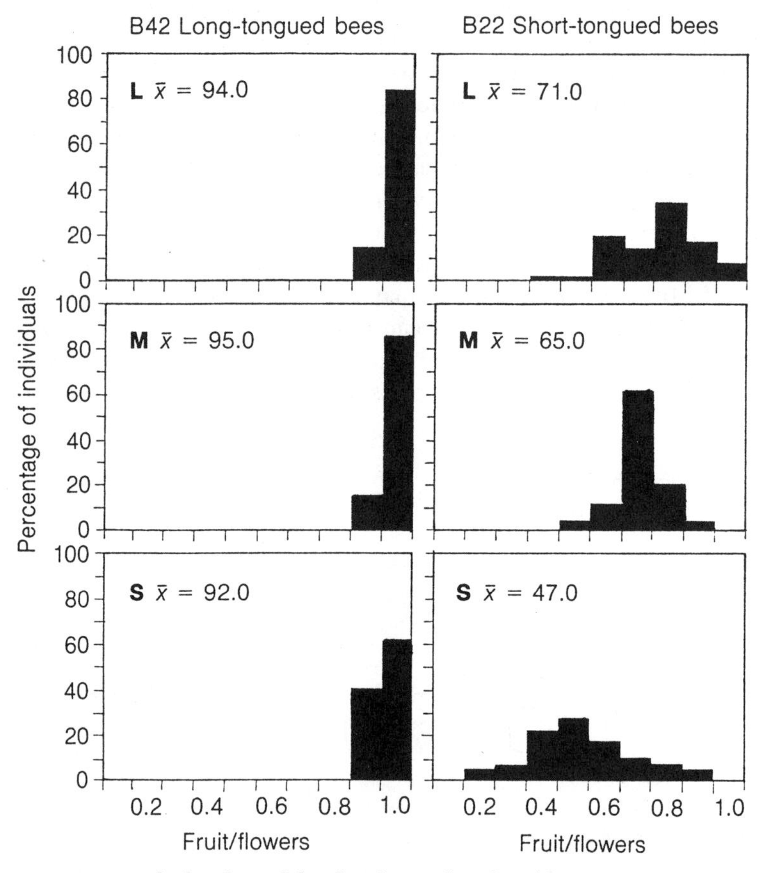

FIGURE 5. Female fertility of the floral morphs of *Eichhornia paniculata* in populations from northeast Brazil serviced by contrasting pollinators. The distributions are of percentage fruit set of the L, M, and S morphs in two populations. B42 was visited by long-tongued specialist bees (*Florilegus* and *Ancyloscelis*) and B22 was visited by short-tongued generalist bees (*Trigona* and *Apis mellifera*).

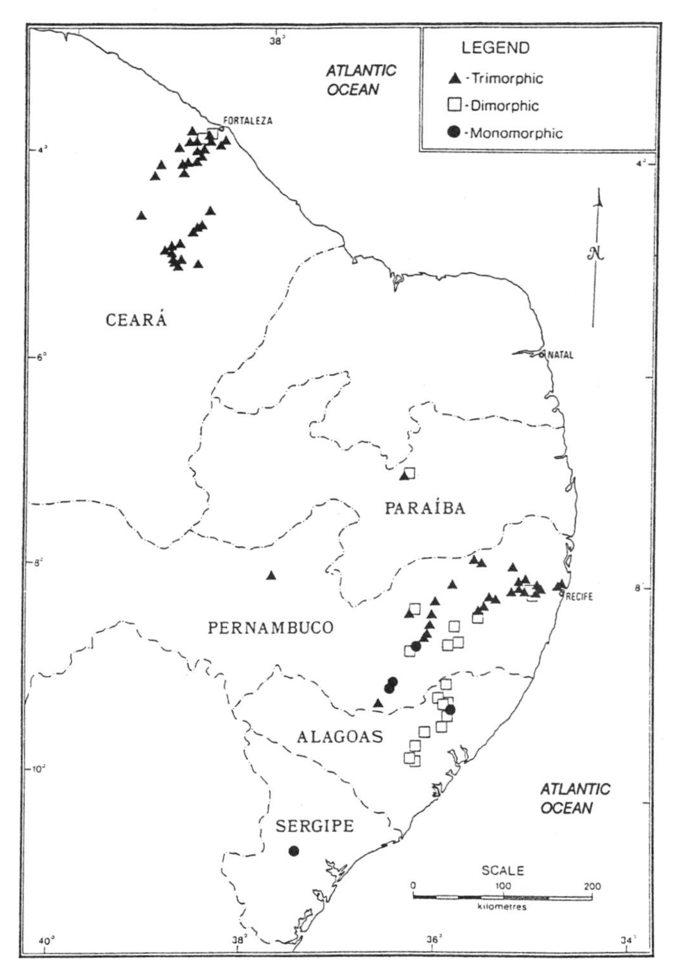

FIGURE 6. Geographical distribution of *Eichhornia paniculata* in northeast Brazil. Triangles are trimorphic populations, squares are dimorphic populations, and circles are monomorphic populations.

tion size in northeast Brazil indicate that although trimorphic populations range from large to small, nearly all monomorphic and dimorphic populations are composed of less than 100 individuals. In addition, whereas trimorphic populations are concentrated in two main areas of northeast Brazil, dimorphic and monomorphic populations are more commonly found at the southern margins of the region surveyed (Figure 6). The habitats occupied by many nontrimorphic populations are distinct from those in which trimorphic populations are found and, in addition, the density of plants in nontrimorphic populations is significantly lower than in trimorphic populations (Barrett et al. 1989). The association between selfing and low-density conditions is a recurrent theme in the mating system literature and data from *E. paniculata* provide support for the hypothesis that reproductive assurance is an important selective force in the evolution of selfing.

Studies of floral biology and estimates of outcrossing rate in *E. paniculata* populations have demonstrated major differences between trimorphic and nontrimorphic populations in their mating systems (Barrett 1985a; Glover and Barrett 1986; Barrett et al. 1987). Virtually all monomorphic and dimorphic populations that have been examined are composed of selfing variants of the M morph. The variants possess modified short-level stamens in which one, or more rarely two or three, stamens are elongated into the mid-level position adjacent to the stigma of the mid style (Figure 4). Because of the close proximity of their stigmas and anthers, the variants are highly self-pollinating. Floral modifications that increase selfing rates occur most commonly in the M morph and are found only rarely in the L and S morphs. Although they are frequent in dimorphic populations and usually fixed in monomorphic populations, the variants are uncommon in trimorphic populations. Figure 7 provides measurements of several floral traits from the L and M morph in two trimorphic populations and a dimorphic population in which selfing variants appear to have recently originated. As can be seen, alterations in the short-stamen level of the M morph result in a marked reduction in stigma–anther separation in the dimorphic population. In spite of this change the averge values of the remaining floral traits are largely unaffected, although an increased variance is evident, perhaps because of inbreeding effects (see below). The occurrence of dimorphic populations of *E. paniculata* with different frequencies of modified M plants provides a rare opportunity to investigate the initial stages of the evolution of selfing and, as we shall see, alterations in floral phenotype that promote self-pollination involve relatively minor genetic and developmental changes.

Evolution of monomorphism from dimorphism

Once selfing variants of the M morph have become established in dimorphic populations they are more likely than unmodified plants to found new populations because of their capacity to set seed in the absence of pollinators. The observation that all monomorphic populations in northeast Brazil and Jamaica are composed exclusively of selfing variants of the M morph provides strong evidence of the selective advantage of this morph during colonizing episodes. But what are the microevolutionary processes that operate *within* dimorphic populations? Can these populations retain an outcrossed mating system? Is a stable equilibrium of outcrossing and

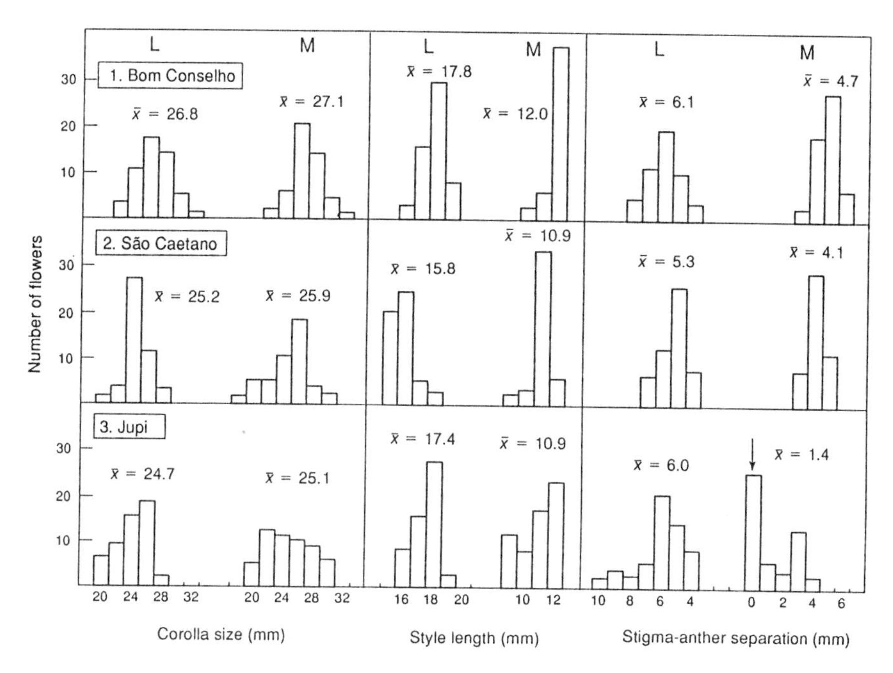

FIGURE 7. Distributions of corolla size, style length, and stigma–anther separation (mm) in the long- and mid-styled morphs (L and M, respectively) from two trimorphic (Bom Conselho, São Caetano) and one dimorphic (Jupi) population of *Eichhornia paniculata* from northeast Brazil. The arrow indicates the breakdown of herkogamy in the M morph. Flowers in the O class are capable of autonomous self-pollination. Note the wider distributions of floral traits in the dimorphic population. This may result from reduced canalization associated with inbreeding (see text).

selfing morphs likely, or will the selfing variants spread to fixation, giving rise to additional monomorphic populations? Although we do not have all of the answers to these questions, several lines of evidence suggest that the maintenance of a dimorphic mixed mating system can occur only under rather restrictive conditions and that the spread and ultimate fixation of the M morph are more likely outcomes.

Once selfing variants of the M morph have become established in dimorphic populations they enjoy several important fitness advantages over the L and unmodified M morphs. The most obvious of these, reproductive assurance in the absence of pollinators, has already been considered in the context of colonization. However, even where pollinator service is available, modified M plants may also experience an asymmetrical mating advantage. This is because genes that promote self-fertilization bias their own transmission through the mating cycle and thus tend to be automatically selected (Fisher 1941). As illustrated in Figure 4, selfing variants are potentially capable of transmitting genes via pollen and ovules through selfing as well as through outcrossing. In contrast, the L morph suffers under this model, since it acts principally as a female in matings with the selfing variant because its pollen is incapable of effectively outcompeting with the selfed pollen of the M variant.

Pollen discounting (Holsinger et al. 1984) and inbreeding depression (Lande and Schemske 1985) can potentially prevent the spread of selfing variants in outbreeding populations. However, these factors appear to be of minor significance in dimorphic populations of *E. paniculata.* Long-level anthers of selfing variants are unaffected by genetic changes in the position of "short-level" stamens and hence continue to contribute genes to the outcrossed pollen pool. Inbreeding depression in dimorphic populations appears to be mild enough that it presents no major obstacle to the spread of selfing variants (P. Toppings and S. C. H. Barrett, unpublished data). It is possible that the population bottlenecks that contribute to the loss of the S morph reduce genetic load sufficiently that if selfing variants do arise in populations they are almost always automatically selected unless, of course, they are lost by genetic drift when at low frequency.

Mating systems and floral morphology

Electrophoretic studies of the mating systems of floral morphs in populations of *E. paniculata* from northeast Brazil and Jamaica have demonstrated the important influence of floral morphology on outcrossing and selfing rates (Glover and Barrett 1986; B. C. Husband and S. C. H. Barrett, unpublished data). In trimorphic populations outcrossing rates are high, with no significant differences among the floral morphs. In dimorphic populations, however, the M morph experiences a high level of self-fertilization, whereas the L morph remains largely outcrossed (Figure 8). There are difficulties in estimating the outcrossing rates of monomorphic populations since they are often devoid of polymorphism at isozyme loci. In populations with some polymorphism high levels of self-fertilization have been observed, although it is likely that higher selfing rates occur in populations with no electrophoretic variation. Among 11 populations surveyed by Glover and Barrett (1987), outcrossing rates were significantly correlated with the number of polymorphic loci, alleles per locus, and observed heterozygosity of populations.

The floral phenotypes of selfing M variants in monomorphic populations usually differ from those that are present in dimorphic populations. Although only minor modifications in floral morphology are evident in most dimorphic populations (Figure 7), selfing phenotypes in monomorphic populations frequently display a suite of floral traits that normally distinguishes selfing species of *Eichhornia* from their outcrossing

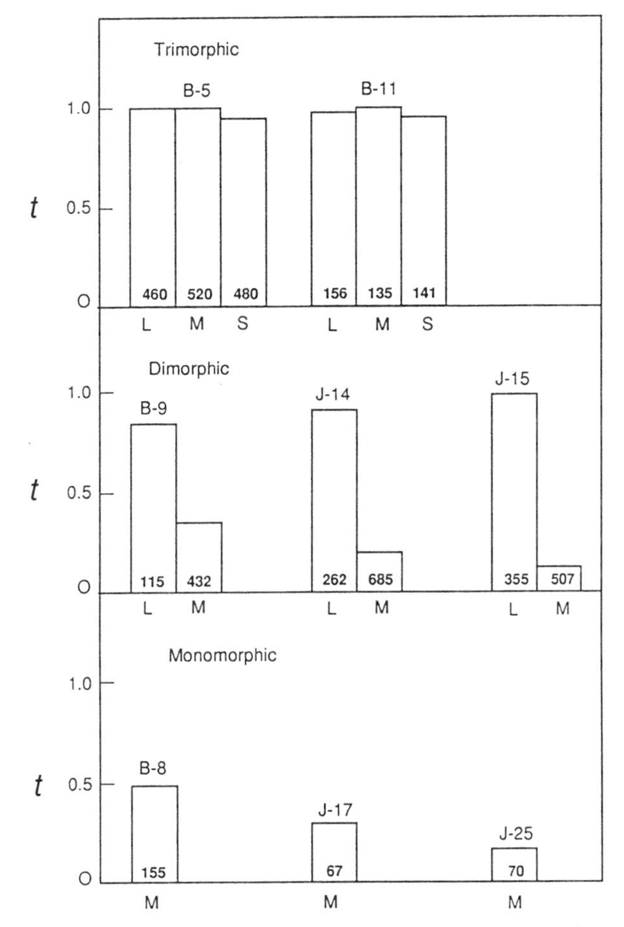

FIGURE 8. Estimates of the outcrossing rate (t) of floral morphs in trimorphic, dimorphic, and monomorphic populations of *Eichhornia paniculata* from northeast Brazil (B) and Jamaica (J). The number of progeny assayed per morph are indicated. Outcrossing rates were estimated using isozyme loci. Within trimorphic populations there were no significant differences between the outcrossing rate of floral morphs whereas in dimorphic populations the morphs differed significantly ($p \ll 0.001$) in outcrossing rate.

progenitors. These include smaller, less showy flowers, fewer flowers per inflorescence, lower pollen–ovule ratios, lower fruit and seed abortion rates, and reduced allocation of resources to male function (Barrett 1985, 1988a; Morgan and Barrett 1989). Populations that possess these attributes have all three "short-level" stamens adjacent to the mid stigma and are usually referred to as semihomostyles (Ornduff 1972). In spite of the major differences in reproductive biology between tristylous and semihomostylous flowers of *E. paniculata,* populations possessing these floral syndromes are fully interfertile. This indicates that not all changes in mating systems are necessarily accompanied by the evolution of reproductive isolation.

Genetic and developmental basis of floral modification

An important question in our studies of the evolution of self-fertilization in *E. paniculata* has concerned the processes responsible for the origin of variation in floral phenotype. What genetic and developmental mechanisms are involved and how does the variation initially arise in populations? Trimorphic populations usually exhibit a highly canalized floral phenotype with relatively minor phenotypic variation in stamen and style length (see Figure 7 for example). Constancy of floral traits is typical of many animal-pollinated species (Berg 1959) and is particularly likely in heterostylous species in which the precise positioning of reproductive parts is required to effect legitimate pollination between the floral morphs (Barrett and Glover 1985). A different pattern is encountered in nontrimorphic populations of *E. paniculata* that frequently display considerable developmental instability for a range of floral traits (Barrett 1985b; Seburn et al. 1989). Instability may arise from inbreeding, in normally outcrossing species, as a result of the exposure of deleterious recessive genes that affect developmental pathways (Lerner 1954). Genes of this type are more likely to be exposed in small inbred populations following population bottlenecks (Levin 1970).

Studies of the inheritance of stamen elongation in selfing variants of the M morph indicate recessive gene control (S. C. H. Barrett, unpublished data). This suggests that the association between developmental instability, the occurrence of floral modifications, and nontrimorphic population structure may result from biparental inbreeding in small populations. According to this hypothesis recessive genes with effects on floral morphology and mating systems are part of the segregational load of tristylous populations. The genes, however, are rarely exposed to selection because of the predominantly outcrossed mating system of populations. However, bottlenecks and subsequent inbreeding may expose mating system modifier genes to selection, and because of reduced genetic load and their effects on mating (previously discussed), the genes experience an automatic advantage. Since selfing variants of the M morph appear to have originated independently in different parts of the range of *E. paniculata,* it will be of particular interest to determine whether the same modifier genes are involved. Crosses between the variants and tests of allelism are currently underway to determine the genetic basis of mating system modification.

Simple genetic changes in floral morphology and reproductive physiology can often have important effects on the mating systems of plants

(Bachmann 1983; Hilu 1983; Gottlieb 1984). Frequently these changes are manifested relatively late in development and, as a result, they may have only minimal effects on other facets of floral phenotype when they first appear in populations. Developmental studies of the range of floral phenotypes in *E. paniculata* (Richards and Barrett 1985 and unpublished data) indicate that the most common phenotypic alteration that gives rise to selfing results from rapid changes in filament length that occur primarily in the 24 hours prior to anthesis (Figure 9). This modification involves only one stamen of the six that occur within an *E. paniculata* flower. The

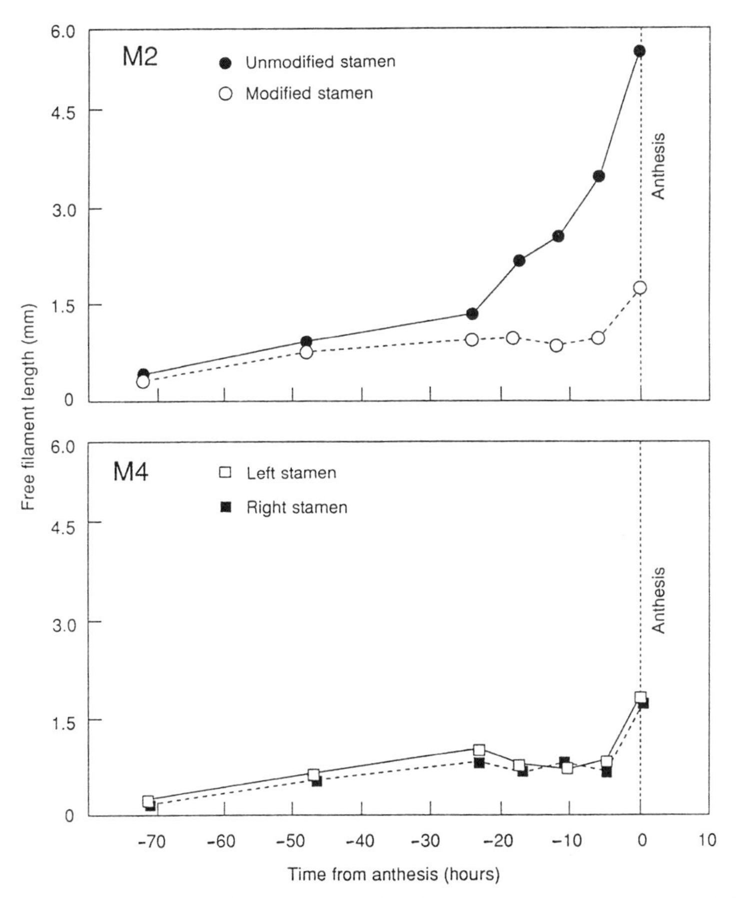

FIGURE 9. Developmental changes in the free filament length of short-level anthers in two genotypes of the mid-styled morph of *Eichhornia paniculata*. Genotype M2 possesses modified stamens that cause autonomous self-pollination of mid-level stigmas. Genotype M4 is unmodified and incapable of autonomous self-pollination. Only one stamen within each flower of genotype M2 is modified; the stamen can be either the left or right stamen of the short-level.

remaining stamens are unaffected by its change in position. The single stamen produces sufficient pollen, released at the commencement of anthesis, to self-fertilize the hundred or so ovules within a flower. It seems likely that the expansion of filament length in modified stamens of *E. paniculata* is regulated by hormones, such as gibberellic acid. These have been implicated in the regulation of reproductive organ size and position in a range of other flowering plants (Greyson and Tepfer 1967; Pharis and King 1985).

Although inheritance studies of the modified genotype illustrated in Figure 9 are consistent with single recessive gene control, many additional genes with small effects are likely to govern the syndrome of floral traits that has accompanied the evolution of semihomostyly in *E. paniculata.* However, the important point is that the initial change in floral phenotype that precedes the evolution of the selfing syndrome in *E. paniculata* can be under simple genetic control.

CONCLUSIONS

Our studies of mating system evolution in heterostylous plants indicate that changes in the reproductive behavior of flowering plants can often be rapid and result from relatively simple genetic and developmental modifications in floral plan. In the *Turnera ulmifolia* complex recombination in the distyly supergene, in association with polyploidy, leads to a quantum change in mating system with reproductive isolation arising in a single step. Unfortunately, the low frequency of recombination events in heterostylous plants (see Charlesworth and Charlesworth 1979b; Shore and Barrett 1985b) has prevented us from conducting population studies on the selection dynamics of homostyles in distylous populations of *T. ulmifolia.* However, our studies of homostyle evolution do provide empirical support for Charlesworth and Charlesworth's (1979a) theoretical model of the breakdown of distyly. They demonstrated that when the "allele" that determines short styles is dominant, as in *T. ulmifolia,* long homostyles (phenotypes with long styles and long-level stamens) are likely to spread to fixation with greater probability than other recombinant phenotypes. Each breakdown event in the *T. ulmifolia* complex has given rise to the long homostylous phenotype, notwithstanding the secondary evolutionary modifications in floral structure that have taken place in each taxon subsequent to its origin. It is remarkable that a simple genetic constraint, involving the dominance relationships at the distyly locus, can determine the particular pathway of floral evolution that is followed in heterostylous species.

In contrast to *T. ulmifolia,* the breakdown of heterostyly in *Eichhornia* involves several key steps with the mating system changing in stages from outcrossing to selfing. However, the initial modifications in floral morphology of the M morph appear, in some cases at least, to be simply inherited and to originate quite frequently in local populations. Semihomostylous species of *Eichhornia* are largely composed of populations that possess modified M phenotypes and selfing variants in the remaining tristylous species usually possess this phenotype (Barrett 1988b). This suggests that in common with *T. ulmifolia,* genetic constraints associated with floral development may play an important role in guiding the specific reproductive modifications that can occur.

Although the acquisition of predominant self-fertilization does not necessarily result in speciation, the effective reproductive isolation that results is likely to enhance character divergence, particularly in floral traits, and the gradual build-up of postzygotic isolating mechanisms. In many genera of herbaceous plants outcrossing species have given rise to clusters of selfing microspecies, many of which are notoriously difficult to classify (Davis and Heywood 1963; Grant 1981). In *Eichhornia* two putative progenitor-derivative species pairs (*E. azurea* tristylous, *E. heterosperma* homostylous, and *E. paniculata* tristylous, *E. meyeri* homostylous) involve outcrossing and selfing taxa. The derived homostyles are almost indistinguishable from their putative outcrossing progenitors in vegetative traits and as a result have not been recognized as distinct from them by most taxonomists. However, the homostyles cannot be crossed with the outcrossing taxa, perhaps because floral traits associated with the selfing syndrome impair reproductive function.

In selfers, the reduced dimensions of reproductive structures, low pollen production, and small pollen size must restrict opportunities for mating with outcrossers irrespective of whether postzygotic isolating mechanisms occur. Supplementing these intrinsic barriers to mating that result from reproductive character divergence, ecological and geographical isolation may arise because of the ability of selfing variants to invade environments not occupied by their outcrossing ancestors. In both *Turnera* and *Eichhornia,* for example, selfing variants have colonized areas at the geographic margins of the range, presumably because of the capacity of single individuals to establish colonies following long-distance dispersal (Baker 1955). Once established, the geographic isolation of selfing populations, exposure to novel selection pressures, and genetic drift in small populations are all likely to enhance genetic changes in other aspects of plant phenotype, leading to the evolution of reproductive isolation.

With the exception of changes in genetic system associated with polyploidy and chromosomal rearrangements, the evolution of reproductive isolation in plants appears to evolve most commonly in allopatric populations as a result of the normal processes of adaptive divergence. A growing body of evidence from population studies of herbaceous plants does, however, suggest that mating system shifts in small populations may play a more significant role in species formation than is generally recognized, by initiating processes that lead to reproductive isolation. Population sizes in many herbaceous plants are small, gene flow is often highly restricted, and local differentiation is widely observed (Levin 1978). These microevolutionary characteristics of plants, in association with the simple genetic basis of many morphological and physiological traits associated with the mating system, foster reproductive character divergence and speciation. Nowhere is this more evident than in association with the evolution of selfing from outcrossing in herbaceous flowering plants.

ACKNOWLEDGMENTS

I thank Nick Barton, Doug Futuyma, Les Gottlieb, and Alan Templeton for valuable discussion, Brian Husband, Martin Morgan, Jennifer Richards, and Peter Toppings for assistance and permission to cite unpublished studies, Elizabeth Campolin for drawing the figures, and the Natural Sciences and Engineering Research Coucil of Canada for financial support.

LITERATURE CITED

Arbo, M. M. 1985. Notas taxonómicas sobre Turneráceas sudamericanas. Candollea 40:175–191.

Arbo, M. M., and A. Fernández. 1983. Posición taxonómica, citología y palinología de tres niveles de ploidía de *Turnera subulata* smith. Bonplandia 5:211–226.

Bachmann, K. 1983. Evolutionary genetics and the genetic control of morphogenesis in flowering plants. Evol. Biol. 16:157–208.

Baker, H. G. 1955. Self-compatibility and establishment after "long-distance" dispersal. Evolution 9:347–348.

Baker, H. G. 1959. Reproductive methods as factors in speciation in flowering plants. Cold Spring Harbor Symp. Quant. Biol. 24:177–191.

Baker, H. G. 1961. Rapid speciation in relation to changes in the breeding systems of plants. Pp. 881–885 in: *Recent Advances in Botany.* University of Toronto Press, Toronto.

Baker, H. G. 1966. The evolution, functioning and breakdown of heteromorphic incompatibility systems. Evolution 20:349–368.

Baker, H. G. 1967. Support for Baker's Law—as a rule. Evolution 21:853–856.

Barigozzi, C. 1982. Mechanisms of speciation. *Progress in Clinical and Biological Research,* Vol. 96. Alan R. Liss, New York.

Barrett, S. C. H. 1977. Tristyly in *Eichhornia crassipes* (Mart.) Solms. (Water Hyacinth). Biotropica 9:230–238.

Barrett, S. C. H. 1978a. Heterostyly in a tropical weed: The reproductive biology of the *Turnera ulmifolia* complex (Turneraceae). Can. J. Bot. 56:1713–1725.

Barrett, S. C. H. 1978b. The floral biology of *Eichhornia azurea* (Swartz) Kunth (Pontederiaceae). Aquat. Bot. 5:217–228.

Barrett, S. C. H. 1979. The evolutionary breakdown of tristyly in *Eichhornia crassipes* (Mart.) Solms. (Water Hyacinth). Evolution 33:499–510.

Barrett, S. C. H. 1985a. Ecological genetics of breakdown in tristyly. Pp. 267–275 in: J. Haeck and J. W. Woldendorp (eds.), *Structure and Functioning of Plant Populations. II: Phenotypic and Genotypic Variation in Plant Populations.* North-Holland, Amsterdam, The Netherlands.

Barrett, S. C. H. 1985b. Floral trimorphism and monomorphism in continental and island populations of *Eichhornia paniculata* (Spreng.) Solms (Pontederiaceae). Biol. J. Linnean Soc. 25:41–60.

Barrett, S. C. H. 1988a. The evolution, maintenance and loss of self-incompatibility systems. Pp. 98–124 in J. and L. Lovett Doust (eds.), *Reproductive Ecology of Plants: Patterns and Strategies.* Oxford University Press, Oxford.

Barrett, S. C. H. 1988b. Evolution of breeding systems in *Eichhornia* (Pontederiaceae): A review. Ann. Missouri Bot. Garden 75:741–760.

Barrett, S. C. H., and I. W. Forno. 1982. Style morph distribution in New World populations of *Eichhornia crassipes* (Mart.) Solms-Laubach (Water Hyacinth). Aquat. Bot. 13:299–306.

Barrett, S. C. H., and D. E. Glover. 1985. On the Darwinian hypothesis of the adaptive significance of tristyly. Evolution 39:766–774.

Barrett, S. C. H., and J. S. Shore. 1987. Variation and evolution of breeding systems in the *Turnera ulmifolia* L. complex (Turneraceae). Evolution 41:340–354.

Barrett, S. C. H., and J. S. Shore. 1989. Isozyme variation in colonizing plants. In: D. and P. Soltis (eds.), *Isozymes in Plant Biology.* Dioscorides Press, in press.

Barrett, S. C. H., A. H. D. Brown, and J. S. Shore. 1987. Disassortative mating in tristylous *Eichhornia paniculata* (Pontederiaceae). Heredity 58:49–55.

Barrett, S. C. H., M. T. Morgan, and B. C. Husband. 1989. Dissolution of a complex genetic polymorphism: The evolution of self-fertilization in tristylous *Eichhornia paniculata.* Evolution, in press.

Barrett, S. C. H., S. D. Price, and J. S. Shore. 1983. Male fertility and anisoplethic population structure in *Pontederia cordata* (Pontederiaceae). Evolution 37:745–759.

Barton, N. H., and B. Charlesworth. 1984. Genetic revolutions, founder effects and speciation. Annu. Rev. Ecol. System. 15:133–164.

Berg, R. L. 1959. A general evolutionary principle underlying the origin of developmental homeostasis. Am. Natur. 93:103–105.

Bingham, E. T. 1980. Maximizing heterozygosity in autoploids. Pp. 471–490 in: W. H. Lewis (ed.), *Polyploidy: Biological Relevance.* Plenum Press, New York.

Bodmer, W. F. 1960. The genetics of homostyly in populations of *Primula vulgaris.* Philos. Trans. R. Soc. London 242:517–549.

Brown, G. K. 1985. Modes and mechanisms of plant speciation: Introduction. System. Bot. 10:379.

Carlquist, S. 1974. *Island Biology.* Columbia University Press, New York.

Carey, K., and F. R. Ganders. 1987. Patterns of isoenzyme variation in *Plectritis* (Valerianaceae). System. Bot. 12:125–132.

Carson, H. L. 1985. Unification of speciation theory in plants and animals. System. Bot. 10:380–390.

Carson, H. L., and A. R. Templeton. 1984. Genetic revolutions in relation to speciation phenomena: The founding of new populations. Annu. Rev. Ecol. System. 15: 97–132.

Charlesworth, B., and D. Charlesworth. 1979a. The maintenance and breakdown of distyly. Am. Natur. 114:499–513.

Charlesworth, B., R. Lande, and M. Slatkin. 1982. A neo-Darwinian commentary on macroevolution. Evolution 36:474–498.

Charlesworth, D. 1979. The evolution and breakdown of tristyly. Evolution 33:489–498.

Charlesworth, D., and B. Charlesworth. 1979b. A model for the evolution of heterostyly. Am. Natur. 114:467–498.

Crosby, J. 1949. Selection of an unfavorable gene-complex. Evolution 3:212–230.

Darwin, C. 1876. *The Effects of Cross- and Self-Fertilization in the Vegetable Kingdom.* John Murray, London.

Darwin, C. 1877. *The Different Forms of Flowers on Plants of the Same Species.* John Murray, London.

Davis, P. H., and V. H. Heywood. 1963. *Principles of Angiosperm Taxonomy.* Oliver and Boyd, Edinburgh.

Dowrick, V. P. J. 1956. Heterostyly and homostyly in *Primula obconica.* Heredity 10:219–236.

Eckenwalder, J. E., and S. C. H. Barrett. 1986. Phylogenetic systematics of Pontederiaceae. System. Bot. 11:373–391.

Ehrendorfer, F. 1979. Reproductive biology in island plants. Pp. 293–306 in: D. Bramwell (ed.), *Plants and Islands.* Academic Press, London.

Fisher, R. A. 1941. Average excess and average effect of a gene substitution. Ann. Eugen. 11:53–63.

Ganders, F. R. 1979. The biology of heterostyly. N. Z. J. Bot. 17:607–635.

Glover, D. E., and S. C. H. Barrett. 1986. Variation in the mating system of *Eichhornia paniculata* (Spreng.) Solms (Pontederiaceae). Evolution 40:1122–1131.

Glover, D. E., and S. C. H. Barrett. 1987. Genetic variation in continental and island populations of *Eichhornia paniculata* (Pontederiaceae). Heredity 59:7–17.

Goldblatt, P. 1980. Polyploidy in angiosperms: Monocotyledons. Pp. 219–240 in: W. Lewis (ed.), *Polyploidy: Biological Relevance.* Plenum Press, New York.

Gottlieb, L. D. 1973. Genetic differentiation, sympatric speciation, and the origin of a diploid species of *Stephanomeria.* Am. J. Bot. 60:545–553.

Gottlieb, L. D. 1984. Genetics and morphological evolution in plants. Am. Natur. 123:681–709.

Gould, S. J. 1980. Is a new and general theory of evolution emerging? Paleobiology 6:119–130.

Grant, V. 1949. Pollination systems as isolating mechanisms in flowering plants. Evolution 3:82–97.

Grant, V. 1981. *Plant Speciation,* 2nd ed. Columbia University Press, New York.

Grant, V., and K. A. Grant. 1965. *Flower Pollination in the Phlox Family.* Columbia University Press, New York.

Greyson, R. I., and S. S. Tepfer. 1967. Emasculation effects on the stamen filament of *Nigella hispanica* and their partial reversal by gibberellic acid. Am. J. Bot. 54:971–976.

Hedrick, P. W. 1987. Genetic load and the mating system in homosporous ferns. Evolution 41:1282–1289.

Heuch, I. 1979. Equilibrium populations of heterostylous plants. Theoret. Pop. Biol. 15: 43–57.

Heuch, I. 1980. Loss of incompatibility types in finite populations of the heterostylous plant *Lythrum salicaria.* Hereditas 92:53–57.

Holsinger, K. E., M. W. Feldman, and F. B. Christiansen. 1984. The evolution of self-fertilization in plants: A population genetic model. Am. Natur. 124:446–453.

Hilu, K. W. 1983. The role of single-gene mutations in the evolution of flowering plants. Evol. Biol. 16:97–128.

Iwatsuki, K., P. H. Raven, and W. J. Bock. 1986. *Modern Aspects of Species.* University of Tokyo Press, Tokyo.

Jain, S. K. 1976. The evolution of inbreeding in plants. Annu. Rev. Ecol. System. 10:173–200.

Lande, R., and D. W. Schemske. 1985. The evolution of self-fertilization and inbreeding depression in plants. I. Genetic models. Evolution 39:24–40.

Lerner, I. M. 1954. *Genetic Homeostasis.* Oliver and Boyd, London.

Levin, D. A. 1970. Developmental instability and evolution in peripheral isolates. Am. Natur. 104:343–353.

Levin, D. A. 1978. Some genetic consequences of being a plant. Pp. 189–212 in: P. Brussard (ed.), *Ecological Genetics: The Interface.* Springer-Verlag, New York.

Levin, D. A. 1983. Polyploidy and novelty in flowering plants. Am. Natur. 122:1–25.

Lewis, H. 1973. The origin of diploid neospecies in *Clarkia.* Am. Natur. 107:161–170.

Lewis, W. 1980. Polyploidy in angiosperms: Dicotyledons. Pp. 241–268 in: W. Lewis (ed.), *Polyploidy: Biological Relevance.* Plenum Press, New York.

Lloyd, D. G. 1965. Evolution of self-compatibility and racial differentiation in *Leavenworthia* (Cruciferae). Cont. Gray Herbarium, Harvard Univ. 195:3–134.

Lloyd, D. G. 1979. Some reproductive factors affecting the selection of self-fertilization in plants. Am. Natur. 113:67–79.

Lloyd, D. G. 1980. Demographic factors and mating patterns in Angiosperms. Pp. 67–88 in: O. T. Solbrig (ed.), *Demography and Evolution of Plant Populations.* Blackwell, Oxford.

Martin, F. W. 1967. Distyly, self-incompatibility, and evolution in *Melochia*. Evolution 21:493–499.
McNeill, C. I., and S. K. Jain. 1983. Genetic differentiation studies and phylogenetic inference in the plant genus *Limnanthes* (section Inflexae). Theoret. Appl. Genet. 66:257–269.
Morgan, M. T., and S. C. H. Barrett. 1988. Historical factors and anisoplethic population structure in tristylous *Pontederia cordata:* A reassessment. Evolution, 42:496–504.
Morgan, M. T., and S. C. H. Barrett. 1989. Reproductive correlates of mating system variation in *Eichhornia paniculata* (Pontederiaceae). J. Evol. Biol., in press.
Nimer, E. 1972. Climatologia da regiaõ Nordeste do Brasil. Rev. Bras. Geogr. 34:3–51.
Ornduff, R. 1966. A biosystematic survey of the goldfield genus *Lasthenia* (Compositae: Helenieae). Univ. Calif. Pub. Bot. 40:1–92.
Ornduff, R. 1969. Reproductive biology in relation to systematics. Taxon 18:121–133.
Ornduff, R. 1972. The breakdown of trimorphic incompatibility in *Oxalis* section Corniculatae. Evolution 26:52–65.
Pharis, R. P., and R. W. King. 1985. Gibberellins and reproductive development in seed plants. Annu. Rev. Plant Physiol. 36:517–568.
Raven, P. H., and D. P. Gregory. 1972. A revision of the genus *Gaura* (Onagraceae). Memoirs Torrey Bot. Club 23:1–96.
Ray, P. M., and H. F. Chisaki. 1957. Studies on *Amsinckia*. I. A synopsis of the genus, with a study of heterostyly in it. Am. J. Bot. 44:529–536.
Richards, A. J. 1986. *Plant Breeding Systems.* Allen & Unwin, London.
Richards, J. H., and S. C. H. Barrett. 1984. The developmental basis of tristyly in *Eichhornia paniculata* (Pontederiaceae). Am. J. Bot. 71:1347–1363.
Rick, C. M., J. F. Fobes, and M. Holle. 1977. Genetic variation in *Lycopersicon pimpinellifolium:* Evidence of evolutionary changes in mating systems. Plant Syst. Evol. 127:139–170.
Roose, M. L., and L. D. Gottlieb. 1976. Genetic and biochemical consequences of polyploidy in *Tragopogon.* Evolution 30:818–830.
Seburn, C. N., T. D. Dickinson, and S. C. H. Barrett. 1989. Floral instability in *Eichhornia paniculata* (Spreng.) Solms (Pontederiaceae). I. Stamen variation in genotypes from northeast Brazil. Submitted.
Shore, J. S., and S. C. H. Barrett. 1985a. Morphological differentiation and crossability among populations of the *Turnera ulmifolia* L. complex (Turneraceae). System. Bot. 10:308–321.
Shore, J. S., and S. C. H. Barrett. 1985b. The genetics of distyly and homostyly in *Turnera ulmifolia* L. (Turneraceae). Heredity 55:167–174.
Shore, J. S., and S. C. H. Barrett. 1986. Genetic modifications of dimorphic incompatibility in the *Turnera ulmifolia* L. complex (Turneraceae). Can. J. Genet. Cytol. 28:796–807.
Stebbins, G. L. 1957. Self-fertilization and population variability in the higher plants. Am. Natur. 41:337–354.
Stebbins, G. L. 1970. Adaptive radiation of reproductive characteristics in angiosperms. I. Pollination mechanisms. Annu. Rev. Ecol. System. 1:307–326.
Stebbins, G. L. 1980. Polyploidy in plants: Unsolved problems and prospects. Pp. 495–520 in: W. Lewis (ed.), *Polyploidy: Biological Relevance.* Plenum Press, New York.
Sun, M., and F. R. Ganders. 1986. Female frequencies in gynodioecious populations correlated with selfing rates in hermaphrodites. Am. J. Bot. 73:1634–1648.
Sun, M., and F. R. Ganders. 1988. Mixed mating systems in Hawaiian *Bidens* (Asteraceae). Evolution 42:516–527.
Templeton, A. R. 1981. Mechanisms of speciation—a population genetic approach. Annu. Rev. Ecol. System. 12:23–48.
Thomas, S. M., and B. G. Murray. 1981. Breeding systems and hybridization in *Petrorhagia* sect. *Kohlrauschia* (Caryophyllaceae). Plant System. Evol. 139:77–94.
White, M. J. D. 1978. *Modes of Speciation.* W. H. Freeman, San Francisco.
Wyatt, R. 1984. The evolution of self-pollination in granite outcrop species of *Arenaria*. I. Morphological correlates. Evolution 38:804–816.

CHAPTER 12

POPULATION STRUCTURE AND THE GENETIC AND MORPHOLOGIC DIVERGENCE AMONG POCKET GOPHER SPECIES (GENUS *THOMOMYS*)

James L. Patton and Margaret F. Smith

INTRODUCTION

Speciation is usually viewed in an allopatric context (see Bush 1975; Mayr 1970, for reviews), although alternatives are advocated for some cases (e.g., Bush 1975; Endler 1977; White 1978). In spite of a general acceptance of the allopatric mode, arguments prevail as to the explicit conditions favoring species divergence, whether divergence is most likely in isolated peripheral populations, whether gene flow serves solely to constrain differentiation, and whether genetic or morphologic change accompanies speciation.

Much current focus has been placed on the relationship between population structure, particularly that generating small, relatively isolated units, and both divergence potential and the degree of substantive morphologic and/or genetic change. Thus, Mayr's (1954) founder principle (or peripatric speciation; Mayr 1982) has received considerable attention by students of empirical and theoretical aspects of speciation. This model has direct implications for Wright's (1956) "interdemic selection," in which the most effective way to spread new adaptations is through the establishment of new local populations. Thus, if extinctions and recolonizations are frequent, group selection can lead to fixation of genes in populations even if those genes are opposed by natural selection (Wade 1978; Wade and McCauley 1980). Or, rapid genetic evolution can result from the combination of genetic drift and the new selection regime offered to the new populations, what Mayr called a "genetic revolution," a process similar to Carson's (1982) "founder-flush" and Templeton's (1980) "genetic transilience" theories (reviewed by Carson and Templeton 1984). Although doubt has been expressed as to whether any of these processes account for the formation of any species (Barton and Charlesworth 1984), the demographic instability of a widespread species would facilitate their operation (Slatkin 1987).

Over the past 15 years, we have focused on an analysis of the genetic structure of pocket gopher populations of the *Thomomys bottae* group, from both a demographic and a geographic perspective. This focus has been placed on pocket gophers because they are among the most genetically and phenotypically variable mammalian taxa, in spite of a general morphology and life history constrained by their fossorial habits (Nevo 1979). In the present chapter, we examine the relationship between population genetic structure, as viewed in a geographic context, and speciation potential and then ask questions concerning the amount of genetic and morphologic divergence that has accompanied species formation in pocket gophers. Only by such a set of questions can the role of genetic struc ture in forming new species and in developing evolutionary novelties be understood.

GENETIC STRUCTURE OF POCKET GOPHER POPULATIONS: A GEOGRAPHIC PERSPECTIVE

Pocket gophers are charaterized by extensive amounts of morphologic variation largely distributed among populations or geographic units. The approximately 215 geographic races currently recognized by taxonomists for members of the *bottae* group of *Thomomys* (Hall 1981) exemplify this statement. The degree of differentiation among populations is a reflection of several interacting factors, namely, low dispersal powers resulting from morphologic specializations for subterranean existence, a very wide range in the habitats occupied resulting in large among-habitat selection differentials, and patchy distributions resulting from discontinuities in available soils. However, this view of geographic population structure derived from current morphologic taxonomy is only limitedly supported by available genetic data.

Differentiation among local populations

Detailed analyses of the genetic demography of pocket gopher populations have been achieved directly for only one area representing the single species *Thomomys bottae* (Patton and Feder 1981; Daly and Patton, in preparation). The key elements from these studies relevant to the discussion are as follows: (1) population density is relatively low for a rodent, not exceeding 80 aduts per hectare; (2) there is a high variance in male reproductive success, as determined by paternity exclusion analysis; (3) a skew in adult sex ratio favoring females is typical; (4) adults rarely, if ever, shift their territories once established; (5) turnover in the adult population is high, with 55–75% of the breeding population each year recruited from the prior reproductive period; and (6) juvenile dispersal is extensive in terms of both proportion and distance. The result of these combined features is a short-term relatively small effective population size such that drift can produce substantial shifts in allele frequency among populations in adjacent fields, but much larger effective sizes over the long-term resulting from the combination of high turnover and high dispersal rates. Differentiation among local populations of pocket gophers, therefore, seems to be mediated by the interactions of features 1 through 4 counterbalan-

TABLE 1. Genetic demographic parameters for two populations of *Thomomys bottae* pocket gophers: Hastings Reservation in the central coast and Death Valley in the eastern desert of California.

Parameter	Population	
	Hastings Reservation	Death Valley
Population density (adults/hectare)	60	<6
Dispersion pattern of adult territories	Even	Highly clumped
Adult sex ratio (m:f)	1:2.9[a]	1:1[b]
Within-sex variance in reproductive success		
Male	High	Low (?)
Female	Low	Low
Gene flow rate	0.19 km/year[c]	(?) Lower

[a] $p < 0.001$.
[b] $p > 0.05$.
[c] Minimal estimate based on frequency distribution of dispersal distances between natal and breeding sites; see Daly and Patton (in preparation).

ced by 5 and 6. As a result, local populations (i.e., populations in adjacent grassland fields) can build up amounts of differentiation of about 7% [Wright's (1965) F_{ST} = 0.07, on average; Patton and Feder 1981; Patton and Smith, in preparation], although estimates of gene flow from both direct and indirect methods are quite high (Daly and Patton, in preparation).

Nevertheless, there is substantial geographic variation in these basic demographic parameters influencing local population genetic structure. The area from which the above picture is available (coastal central California) is an area in which gophers are extraordinarily abundant across geography, with little or no discontinuities in distribution over large areas (see Patton and Yang 1977; Patton 1985). This is in contrast to populations occupying patches of friable soils in the desert regions of California, for example. Here, limited demographic data from populations in the vicinity of Death Valley indicate much lower population densities, seasonally shifting adult territories, and even adult sex ratios (Table 1). The likelihood is, therefore, that geographic differences in demography will generate differences in the underlying genetic structure as well.

Hierarchical geographic genetic differentiation

Pocket gopher species typically exhibit very high levels of genetic differentiation among populations (Patton and Yang 1977; Zimmerman and Gayden 1981). Interpopulation differentiation within *Thomomys bottae*, for example, can reach 30% or more of the genome [Rogers' (1972) genetic distance based on allozyme analyses]. However, there is a distinct pattern to this geographic differentiation when it is viewed in the context of hierarchical F_{ST} measures. For example, the 131 population samples available that represent the total species' range were divided into four geographic units that, in turn, were divided into regions of approximately equal size. F_{ST} is then calculated as a function of variation among populations within region, among regions relative to geographic units, and among the latter relative to the entire species range (Table 2). As is evident, the amount of genetic diversity observed among populations within a regional area is equivalent to that observed among these areas within a regional area is equivalent to that observed among these areas within any single geographic unit, or between geographic units relative to the entire species range. Thus, genetic differentiation in this species increases rather abruptly among populations up to a certain geographic distance (about 100 km) but, beyond that threshold, differentiation is unrelated to distance per se.

TABLE 2. Hierarchical estimates of genetic differentiation for populations of *Thomomys bottae* sampled from throughout the species range based on Wright's (1965) standardized variance in allele frequency (F_{XY}).

Hierarchical comparison		
X	Y	F_{XY}
Fields	Local population	0.066[a]
Populations	Regions	0.311
Regions	Geographic units	0.314
Geographic units	Total range	0.280

[a]From Patton and Feder (1981).

Differential gene flow and variation in geographic population structure

The geographic structure to pocket gopher populations is best represented by a stepping-stone model, and the degree of structure is mediated by the amount of "connectedness," or gene flow, among populations (Patton and Yang 1977; Patton 1985). To test the generality of this view, we employed Slatkin's (1981, 1985) indirect methods to estimate the degree of gene flow among populations. For simplicity, analyses presented here are limited to comparisons between populations of *T. bottae* from California that were sampled over equivalent geographic distances but assignable to the central and coastal versus the interior desert regions of the state (i.e., these are areas of equivalent size but ones separated by the crest of the Sierra Nevada). The former represents an area of high "connectedness," that is, an area in which populations are rather uniformly and continuously distributed. The latter represents an area in which populations are physically isolated due to the patchiness of habitat suitable to support permanent populations of gophers.

Figure 1 presents curves of Slatkin's (1981) conditional allele frequency for these two regions. The quantitative method described by Larson et al. (1984) was used to determine the best fit of each empirical curve to the theoretical ones derived from a one-dimensional stepping-stone and an island model (Slatkin 1981:Figures 2a and b). Under both models, the central California area exhibits an estimated gene flow level that is an order of magnitude higher than that in the eastern deserts (2.50 versus 0.25). This difference is confirmed by application of Slatkin's (1985) quantitative estimation of Nm, the product of effective population size and gene flow rate, based on the average frequency of "private" alleles, those found in only one population. These values range over a factor of four, from 0.79 for the desert samples to 2.94 for those from central California. Since genetic drift will result in substantial local differentiation if *Nm* is less than 1, but not if it is greater than unity (Slatkin 1987), these differences suggest that the relative importance of evolutionary forces can be quite different between different geographic areas within the range of a single species.

The conclusions from these analyses are that some geographic areas within the range of *T. bottae*, for example, are characterized by estimated levels of gene flow several factors higher than those estimated in other areas. Typical of the latter are population samples largely isolated in enclosed desert basins or montane meadows. Both the evolutionary potential and mode of divergence are thus likely to vary considerably across the geographic range of the species. In particular, areas characterized by small and relatively isolated populations have the highest potential to be those from which evolutionary novelty will stem. These are the ones that combine the conditions conducive to drift and to differential selection that might lead to Mayr's "genetic revolution" or to Templeton's "genetic transilience" (see review by Carson and Templeton 1984), although Barton and Charlesworth (1984) challenge such widely held assumptions.

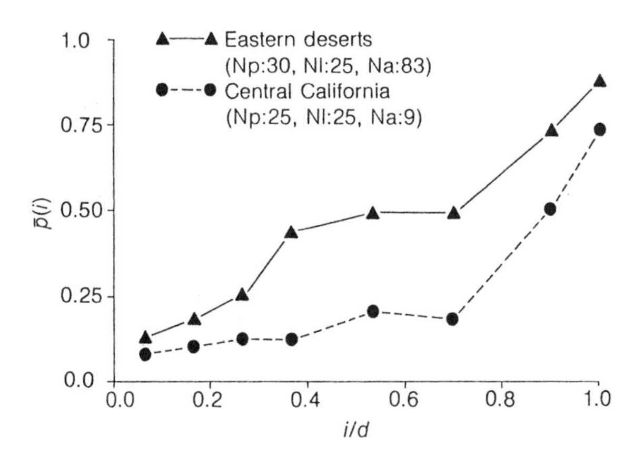

FIGURE 1. Slatkin's (1981) conditional allele frequencies [$\bar{p}(i)$] for two geographic subsets of population samples of *Thomomys bottae* pocket gophers from California: central and coastal (solid circles) and eastern desert (solid triangles) regions. The curve have been "smoothed" by averaging successive triplets of values of *i*/*d*, the proportion of the total number of populations (*d*) for which each *i* value is present. N_p, number of populations sampled; N_l, number of loci examined; N_a, number of alleles detected.

OPERATIONAL DEFINITION OF SPECIES

Pocket gopher taxa exhibit contiguously allopatric distribution patterns, be they recognized species or even genera (Miller 1964; Thaeler 1968a; Nevo 1979). As a consequence, determination of the species status of definable population units must rest with the examination of the genetic interactions, or lack thereof, between units in which they are in contact. For our purposes, we have followed a biological definition for species (sensu Mayr 1963), but one that emphasizes genetic rather than purely reproductive isolation. Our operational definition is as follows (Patton et al. 1979; Patton 1981):

1. Contact populations that do not form hybrids and that, therefore, are both reproductively and genetically isolated are considered separate species.
2. Contact populations that hybridize, but for which hybrid class individuals are limited to the F_1 generation, are genetically, if not reproductively, isolated, and are considered separate species.
3. Contact populations that hybridize and produce a full array of hybrid class individuals in expectation of random mating within the zone of contact are considered conspecific.

Our studies of contact zone dynamics between genetically and morphologically definable geographic units of the *Thomomys bottae* group of taxa have revealed interactions only of the second and third types defined. Thus, *T. bottae* and *T. umbrinus*, where they meet in southern Arizona (Patton 1973; Patton et al. 1972), or *T. bottae* and *T. townsendii*, where they meet in northeastern California (Thaeler 1968b; Patton et al. 1984), are genetically isolated in spite of F_1 hybrid production in both cases. On the other hand, various geographic units, equally definable on genetic or morphologic terms as these two sets of species pairs, are all referrable to the single species *T. bottae* because of complete intergradation at points of contact (see Hafner et al. 1983; Patton et al. 1979; Smith and Patton 1980; Smith et al. 1983).

The point that species definition rests on a biological construct is seminal to the discussion beyond regarding the ultimate significance of speciation events in pocket gophers.

HABITAT VARIABILITY AND THE EXPRESSION OF MORPHOLOGIC DIVERSITY IN POCKET GOPHER POPULATIONS

Pocket gophers of the *T. bottae* complex occupy virtually every available habitat type from desertscrub below sea level to alpine communities above timberline, spanning an elevational range of over 13,000 feet, often over very short horizontal distances. Throughout this habitat range, variability in body proportion and pelage coloration may be extreme, leading to the taxonomic recognition of a large number of local races. As mentioned previously, some 215 such races are currently listed for the group (Hall 1981), more than 40 within the state of California alone.

In our analyses of the factors leading to racial differentiation, we have focused on defining the amount of variation present in the cranium, for example, the descriptive basis for that divergence (be it primarily size related or an expression of shape change), and an understanding of the causal basis for the variation observed among populations. Much of our focus has been on comparisons between populations of the same "genetic stock" (as defined by allozyme similarity patterns) that span a very wide range in habitat type and quality over short geographic distances. Specifically, these have involved comparisons between natural desertscrub populations and derivatives that have recently invaded monocultures of alfalfa in the eastern California deserts. Such comparisons provide us with the opportunity to examine directly the impact of nutritional regime on the expression of body size, and the relationship of size changes to accompanying cranial shape parameters. In such a way, environmental components of variation can be determined and considered directly in the comparison of geographic units of any single species, or of different species, in questions relating to patterns and processes of evolutionary divergence in the group as a whole.

Patton and Brylski (1987) have shown, for example, that body mass increases nearly twofold in individuals inhabiting alfalfa, and that this size increase can occur in just a few generations. They have also shown that the size increase is largely due to differences in nutritional quality during early growth. Young, born and raised in the laboratory to females collected while pregnant from natural vegetation sites, grow at the same rate as do those in alfalfa fields and reach a significantly larger size at equivalent ages in comparison to natural vegetation animals. The body mass increase causes about a 25% increase in cranial size, but it does not affect individual character allometries and thus overall cranial shape; each character has the same allometric coefficient for both natural vegetation and alfalfa samples. However, populations representative of different genetic stocks (that is, major geographic electromorphic units that differ from one another by an average Rogers' genetic distance of 25–30%) can be distinguished by different character allometries in spite of similar ranges in overall size and similar pattern responses to habitat variability. Thus, cranial size appears to be largely a function of variation in environmental quality whereas cranial shape parameters are concordant with geographically definable genetic entities. We consider, therefore, that such shape changes reflect and identify underlying historical and long-term adaptive geographic units.

To further examine this relationship, Smith and Patton (1988) examined variation in cranial components for all population samples of *T. bottae* from the eastern California deserts, making additional comparisons between those from natural vegetation and alfalfa monocultures, and looking more broadly at the relationship between size and shape (as reflected in differential character allometries) among these units. All alfalfa field populations exhibit a substantial increase in size without any difference in cranial shape when compared to their natural vegetation relatives, and each geographic genetic unit is characterized by different sets of character allometric coefficients, indicating significantly different underlying cranial shapes.

The comparisons between samples taken from alfalfa monocultures and nearby natural vegetation communities suggest that habitat is a major component in the size variation observed, i.e., that size is an ecophenotypic variable. To test this hypothesis, we categorized each available individual sample as to habitat type, considered variation within as opposed to among population samples, and examined the relationship of character variation and habitat variation in the following way: The initial step was an ANOVA on log-transformed cranial variables, with population as the main effect. Most of the variation in the data is explained by variation among populations, which combines that across habitats, currently recognized subspecies, and geographic genetic units. The remaining, or residual, variation is that within populations. A principal components analysis (PCA) on the average within-population covariance matrix derived from these residuals thus produces eigenvectors that can be multiplied by the original mensural characters to obtain scores for each individual on new, within-group PCA axes. These scores can then be treated as variables in a second ANOVA with habitat, genetic group, and their interaction as sources of variation.

The paired alfalfa-natural vegetation samples demonstrate that PC-1 is a size axis, and the prediction stemming from these comparisons is that size is directly related to habitat quality. Figure 2 plots least-squares mean scores for each of the two geographic genetic units defined for the eastern California deserts for the first three PC axes as a function of habitat. The prediction that size (PC-1) scales positively with an increase in habitat quality is clearly confirmed. Both genetic stocks show the same general trend, although the details of size change as a function of habitat are different between them. The second and third PC axes represent size-independent changes; in general, there is no habitat effect on these axes, although both genetic groups exhibit distinct cranial shape changes. The conclusion from these analyses is that size scales allometrically across habitats within a given genetic group, although there is a different slope to this relationship between them. On the other hand, substantive changes in cranial shape do not result from habitat variation per se (see Smith and Patton 1988, for further details).

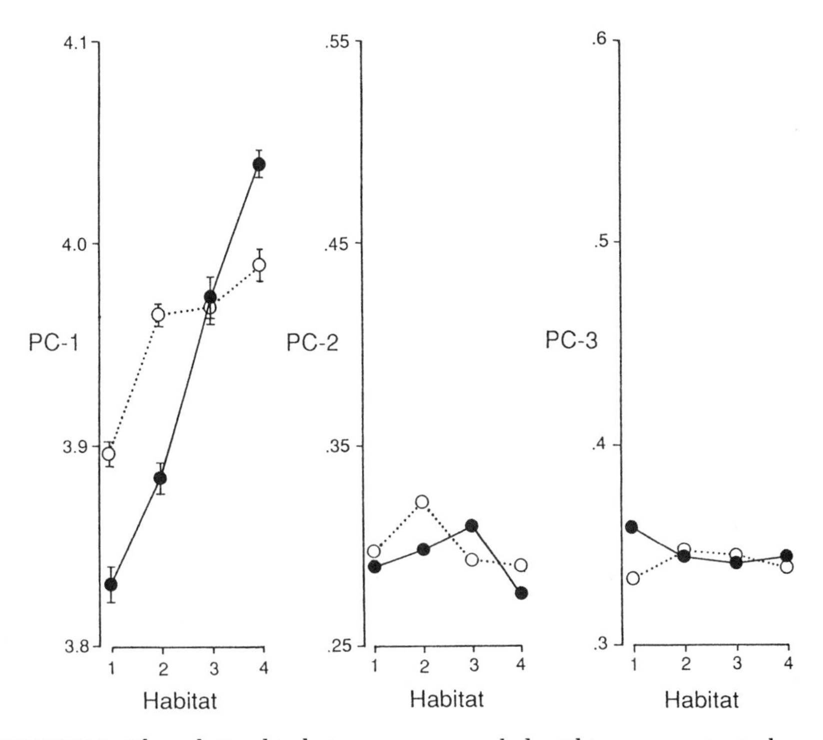

FIGURE 2. The relationship between mean pooled, within-group principal component axes for log-transformed cranial mensural variables and habitat scale for populations of *Thomomys bottae* from the eastern California desert region (open and closed symbols represent samples allocated to different geographic genetic stocks that occur in the area; see Smith and Patton 1988). Habitats of individual localities were categorized into four classes, in an implied gradient of ecological quality, as follows: (1) natural desertscrub communities, soils shallow and rocky; (2) lake-shoe or riparian salt grass natural communities with deep, friable soils; (3) non-cultivated, but human-produced pastureland with friable soils; and (4) alfalfa monocultures with deep, friable soils. Vertical bars indicate 95% confidence limits for the PC-1 means. See text for further explanation.

Although size variation among populations of pocket gophers may have a direct environmental determination, it may also exert an influence on the genetic structure of local populations. Patton and Brylski (1987) have shown that the body size variation observed in alfalfa versus natural vegetation populations accompanies a fundamental shift in their life history parameters as well. Included here are changes in total population size, in reproductive success among individual males, and in reproductive output per female (Table 3). These, when taken together, provide a substantially different potential for genetic change among the populations concerned,

TABLE 3. Correlative variation in life history parameters for populations of *Thomomys bottae* pocket gophers inhabiting natural vegetation communities and alfalfa monocultures in the eastern California deserts.

Parameter	Vegetation type	
	Natural	Alfalfa
Adult density	< 5/hectare	80/hectare
Litter size	4.2	5.6
Females breeding in season of birth (%)	0	46
Length of breeding season	< 3 months	> 6 months
Adult sex ratio (m:f)	1:1.12[a]	1:3.63[b]
Adult sexual dimorphism (male size/female size)	1.056	1.171

[a] $P > 0.05$.
[b] $P < 0.001$.

minimally, but not exclusively, by shifts in their resulting respective short-term effective population sizes. Hence, differences in life history of these types, directly related to habitat quality, may be reflected in the general geographic genetic differences observed throughout the species' ranges (see Patton and Yang 1977; Patton 1981, 1985; and discussion above).

PATTERN AND PROCESS OF SPECIES DIVERGENCE: THE CASE OF *THOMOMYS TOWNSENDII*

Genetic relationships

Thomomys townsendii is a large-bodied species of the *bottae* group inhabiting lacustrine and fluviatile soils of the enclosed basins and river valleys of the northern Great Basin (Davis 1937; Figure 3). That it is a species distinct from *T. bottae* is confirmed by analyses of contact zones in Honey Lake Valley of northeastern California (Thaeler 1968b; Patton et al. 1984). Here, hybridization is limited to F_1 production at very narrow zones of contact. As a consequence, the two forms are genetically isolated, if not reproductively so, as introgression of alleles from one parent to the other does not occur.

Phylogenetic analyses (Patton and Smith 1981) have shown, however, that *T. townsendii* is a derivative of one particular geographic genetic stock

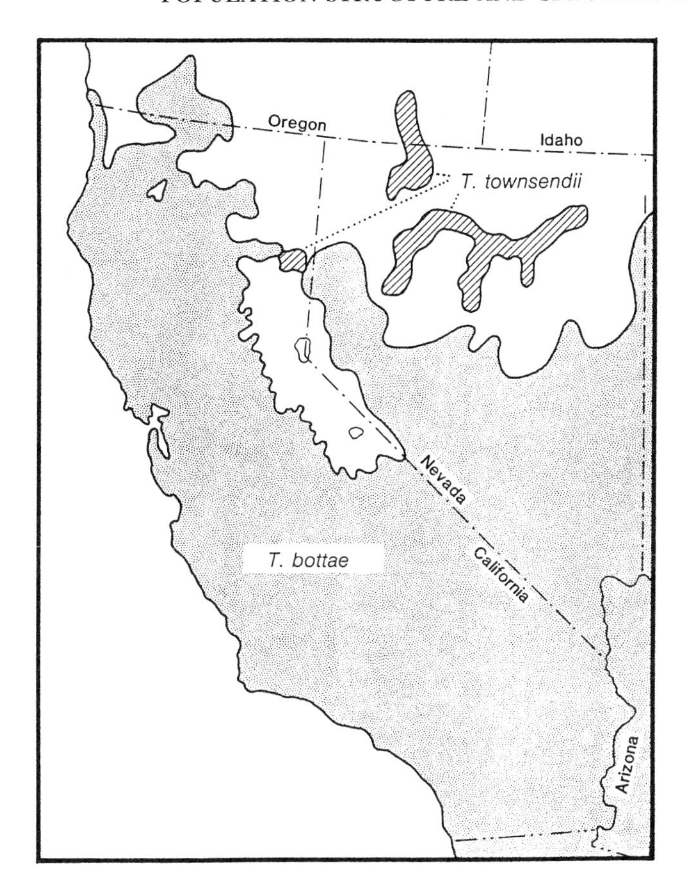

FIGURE 3. Map of the present geographic distribution of *Thomomys bottae* and its derivative, *Thomomys townsendii*, in California and Nevada. Populations of *T. townsendii* also occur in the Snake River basin of central Idaho (see Davis 1937).

of *T. bottae*. It contains no alleles that are unique to it, and it shares as much as 90% Rogers' genetic similarity with many *T. bottae* populations occurring in central California. This is only one-third the degree of differentiation found among the major geographic genetic components of *T. bottae* itself, and it is nearly equivalent to that observed among local populations less than a mile distant from one another (Patton and Feder 1981; Patton and Smith, in preparation). Essentially, *T. townsendii* exhibits no substantive degree of genetic differentiation, as measured at least by electromorphic characters, from *T. bottae*. Moreover, this example points to the fact that genetic isolation among pocket gophers is generally unrelated to the overall degree of genetic differentiation; geographic genetic units of *T.*

bottae that are substantively more differentiated exhibit no reproductive or genetic isolation where they contact one another (Hafner et al. 1983; Patton et al. 1979; Smith and Patton 1980; Smith et al. 1983).

Not only is *T. townsendii* indistinguishable genetically from *T. bottae*, it shares a particularly close relationship with specific geographic populations of the latter species. The map in Figure 4 illustrates the genetic relationships among 120 populations of pocket gophers of California and adjacent Nevada based on Rogers' genetic distance. As is evident, *T.*

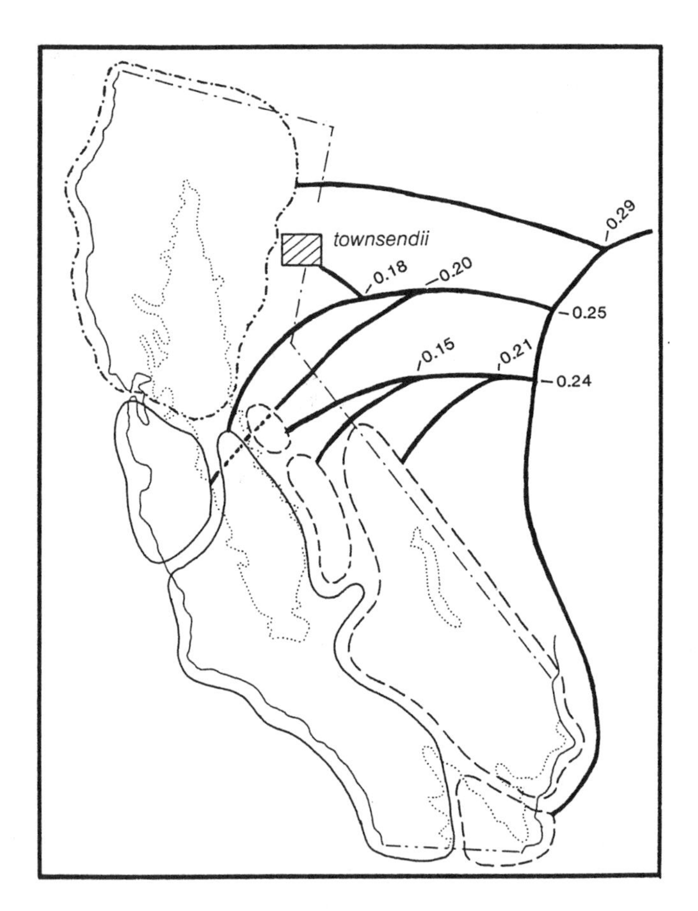

FIGURE 4. Phenetic relationships among the geographic components of *Thomomys bottae* and of *Thomomys townsendii* in California, based on a UPGMA tree of Rogers' genetic distances among 121 population samples. Genetic distances marking each internal node are indicated. Dotted lines identify the geographic position of the Central Valley, Death Valley, and the Colorado River basin of California.

townsendii is phenetically very similar to *T. bottae* from the central and southern parts of the state, much more so than are the geographic components of *T. bottae* among themselves. *Thomomys townsendii* is most probably a recent derivative of *T. bottae* stock; we see no other way to explain the pattern of shared genetic characters or overall degree of similarities (see Patton and Smith 1981).

Interestingly, *T. townsendii* is most genetically dissimilar to that *bottae* geographic unit with which it hybridizes but from which it is genetically isolated (Figure 4). Our operational definition of species of pocket gophers is based on interactions between differentiated populations that meet in nature. Therefore, would the degree of genetic incompatability of *townsendii* relative to *bottae* be less, or even nonexistent, were *townsendii* to meet and interact with a geographic unit of *bottae* that is closer to it genetically? Unfortunately, a simple test of this possibility is not feasible as gophers do not breed readily in the laboratory.

Morphologic relationships

As mentioned, *T. townsendii* is generally characterized by its large size and by its occupation of deep and friable soils in areas otherwise characterized by shallow and indurate soils of basin bajadas. The question, therefore, is whether the size change that characterizes *T. townsendii* has been accompanied by substantive shifts in character allometric coefficients, indicative of structural reorganization of cranial shape, or is merely of the type of change observed in among-habitat comparisons of *T. bottae*—that is, largely an ecophenotypic expression.

To examine this question, we performed a series of PC analyses, using two procedures to adjust for size. Following previous studies, PC-1 from analyses involving log-transformed variables is interpreted as a size axis; confirmation of this interpretation stems from independent analyses involving the a priori removal of size by Burnaby's (1966) technique (see Rohlf and Bookstein 1987; Smith and Patton 1988). Figure 5A illustrates the relationships along the first two PC axes for 155 population samples of *T. bottae* from California, *T. townsendii*, and *T. bulbivorous*, a member of the *bottae* group from central Oregon but a sister taxon to the clade representing *bottae* and *townsendii* (Patton and Smith 1981). As is evident, samples are broadly arrayed along PC-1, and their position on this axis is positively and significantly related to their overall size, as, for example, indexed by a measure of cranial length ($r = 0.9784, p < 0.0001$). In this view, therefore, *T. townsendii* is quite different from all populations of *T. bottae* within California, although not quite as divergent as is *T. bulbivorous*. However, if size is removed from the analysis, either by excluding PC-1 from any bivariate plot of PC axes, or by use of Burnaby's (1966) technique,

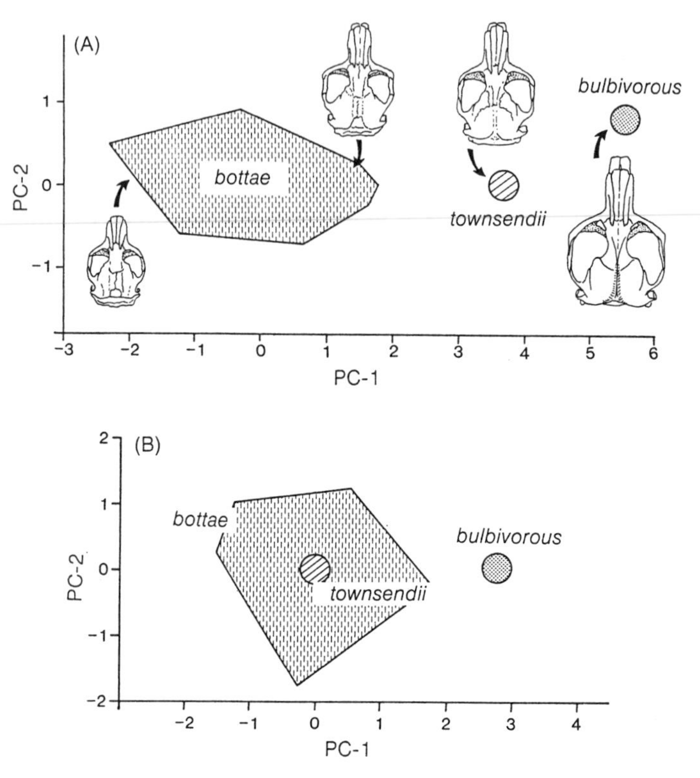

FIGURE 5. (A) Bivariate plot of the first two principal component axes derived from 13 log-transformed cranial measurements for 155 population samples of female *Thomomys bottae* from California, its derivative species *T. townsendii*, and the sister species, *T. bulbivorous* from central Oregon. These two axes account for 84% of the total variation in the data set. PC-1 is a general size axis, as is clearly indicated by the drawings of representative crania of each taxon. (B) Bivariate plot of the first two principal component axes from the same 13 log-transformed cranial mensural variables as in (A), but for which size adjustment was made by Burnaby's (1966) method prior to performing the PCA. Because of program limitations, only 58 of the 155 samples of *T. bottae* used in analysis (A) are included here. Note that *T. townsendii* is no longer distinguishable from *T. bottae*, as in (A), whereas *T. bulbivorous* remains distinct. See text for further explanation.

T. townsendii becomes indistinguishable from *T. bottae*, although *T. bulbivorous* remains outside to the *bottae* cluster (Figure 5B). In the simplest sense, *T. townsendii* does not exhibit any measurable cranial shape difference relative to *T. bottae* populations whereas *T. bulbivorous* does.

To take this analysis one step further, since populations of *T. bottae* within California do exhibit geographically uniform cranial shape differences that are concordant with geographic genetic groupings (Smith and Patton 1988), we examined the explicit cranial relationship of *T. townsendii* to the various sampled populations of *bottae* from California. If the PCA scores are used as variables in a clustering routine, *townsendii* falls outside of the entire cluster of *bottae* samples when PC-1 is included in the analysis, but clusters tightly with that geographic segment of *bottae* occupying the San Joaquin Valley when PC-1, the size variable, is excluded from the analysis (Figure 6). Importantly, populations from the San Joaquin Valley are part of the geographic unit of central and southern

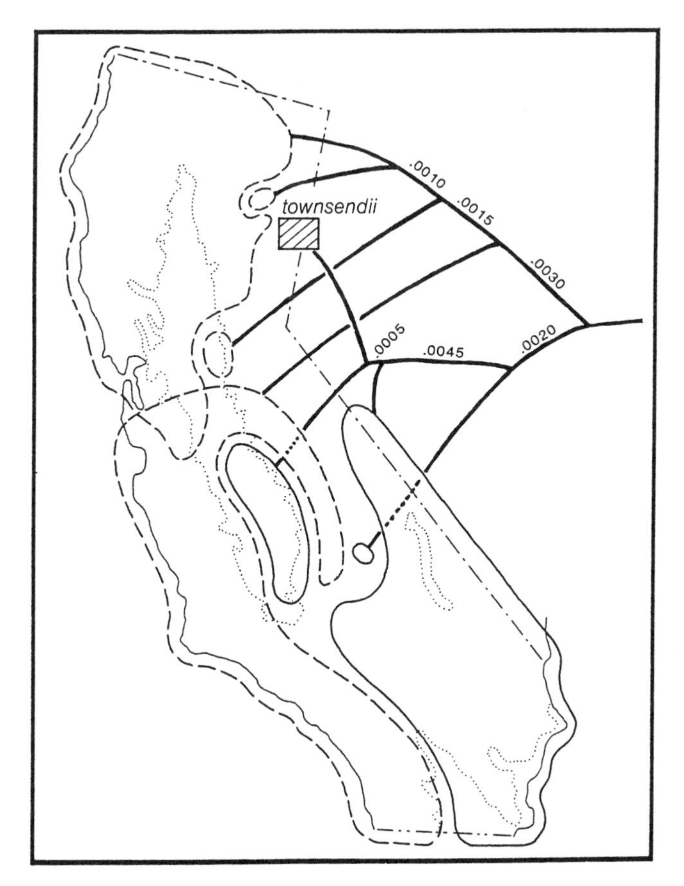

FIGURE 6. Phenetic relationships among geographic components of *Thomomys bottae* and *Thomomys townsendii* in California based on a UPGMA phenogram derived from a correlation matrix from cranial mensural variables adjusted for size by Burnaby's (1966) technique. Branch lengths are indicated. Note the phenetic similarity of *T. townsendii* with samples of *T. bottae* from the San Joaquin Valley.

California to which *townsendii* exhibits a genetic relationship (compare Figures 4 and 6). Although the phenograms in these two figures are not totally concordant, their respective matrices do exhibit significant positive covariation by the Mantel test (Student's $t = 6.098, p < 0.001$ with df = infinity; matrix correlation coefficient = 0.349; see Mantel 1967).

As a consequence, *T. townsendii*, although genetically isolated from the population of *T. bottae* with which it is in contact, exhibits very little genetic differentiation relative to *bottae* as a whole and its morphologic divergence involves size alone, unaccompanied by shifts in underlying cranial shape. The fact that size within *T. bottae* is largely an ecophenotypically plastic feature directly dependent upon environmental quality suggests that *T. townsendii* owes its size differentiation, at least initially, to the same phenomenon. Hence, in the evolution of *T. townsendii* as a species distinct from *T. bottae*, there may be a relationship between the ancestral invasion and isolation in deep soils bordering Pleistocene lake beds (that undoubtedly suppported rich vegetation communities relative to the surrounding desert), a resultant body size increase and correlated demographic attributes affecting genetic structure, and the eventual realization of genetic isolation. Whatever the basis of this isolation, however, it has not been accompanied by measurable electromorphic or morphologic divergence from the ancestral species.

CONSEQUENCES OF SPECIATION IN POCKET GOPHERS

Species formation is generally unrelated to the degree of genetic differentiation among populations of pocket gophers, as measured by electrophoretic variables. Rather, genetic isolation is often, but not always, associated with chromosomal rearrangements that lead to sterility in F_1 hybrids due to meiotic imbalances (Patton 1981; Hafner et al. 1987). Since the degree of gene flow among populations may be sufficient to prevent the fixation of negatively heterotic rearrangements due to drift alone, new species are unlikely to evolve by this means except under conditions in which habitat patchiness leads to small, isolated populations. Regardless of the mechanism of genetic isolation, however, gene flow between established populations is usually considered to serve a constraining role in evolutionary divergence (Slatkin 1987). Geographic variability in genetic structure characterizes widespread species of pocket gophers, such as *T. bottae*, as is indicated by variation in gene flow estimates and genetic demographic parameters. Those segments of current species that comprise series of small, isolated units thus would have a higher evolutionary potential, even though a majority of such populations may experience extinction. At least in terms of the nature of the ancestral population, pocket gophers seemingly fit Mayr's peripatric model. Several consequences result from this model, however, that need to be emphasized.

For one, speciation may be extremely rapid, resulting from the chance fixation of new genetic combinations not occurring in the ancestral species' range. If the derivation of *T. townsendii* from *T. bottae* is typical of pocket gophers, and we have no reason to believe otherwise for the moment, this new combination may not result in demonstrable divergence at either the genic (as detected by allozyme differentiation) or morphologic levels, yet would be sufficient to ensure genetic isolation of the derivative relative to the ancestral species. Unfortunately, at the moment we do not know the basis for the genetic isolation between *bottae* and *townsendii* where they meet in northeastern California. It does not appear to have a cytogenetic basis (Patton et al. 1984), as is true for other examples in *Thomomys* (Patton 1981).

A second result of such a pattern and mode of species formation is an ancestral species that is paraphyletic, in a cladistic sense, to the derivative one (see Patton and Smith 1981; Hafner et al. 1987). Such a result is a direct consequence of viewing the formation of species in a peripatric context while maintaining a definition of species status based on biological criteria (i.e., a genetic isolation species concept). In fact, all mechanisms of speciation that are currently advocated by evolutionary biologists (e.g., Bush 1975) will result in paraphyletic taxa as long as reproductive isolation forms the basis for species definition.

A third implication of these analyses is that morphologic divergence in pocket gophers is not necessarily coincidental with speciation events (e.g., Futuyma 1987), in spite of a population structure that might be expected to promote a punctuational event (following Eldredge and Gould 1972; Gould and Eldredge 1986). In part, this result may stem from the general constraints placed on morphologic variation and differentiation by the subterranean habitus.

SUMMARY

Speciation is generally considered in an allopatric context, with the "genetic cohesion" of large central populations, in which gene flow leads to genetic and phenotypic similarity, counterbalanced by small, peripheral populations in which founder effects reduce genetic variability and may initiate a genetic restructuring precipitating reproductive isolation and phenotypic novelty. Species formation in pocket gophers matches the major components of this model, although there are fundamental differences in the outcome. The genetic structure of local populations is conducive to differentiation by drift overcoming substantial gene flow potential, but the extreme patchy distribution of pocket gophers in peripheral areas ensures the long-term presence of isolated populations of small size. Most differentiation, whether measured at the phenotypic or genetic level, is accentuated in such populations. However, phenotypic divergence among populations has a strong environmental component, with the nutritional quality of the occupied habitat of fundamental importance in body size variation. Species divergence, where species are defined by genetic incompatibility at points of natural contact, may be unaccompanied by either demonstrable genetic or morphologic change. Rather, differentiation at the level of the external phenotype may involve only body size changes of the same type as that observed among populations of the same species existing under radically different nutritional regimes. Hence, real novelties do not necessarily stem from peripheral isolates in spite of conditions favoring genetic transilience. An explicit statement of diversification in pocket gophers is given by the comparison of *Thomomys bottae* and its derivative *Thomomys townsendii* in northeastern California.

ACKNOWLEDGMENTS

We are grateful to Carol Patton, Patty Garvey-Darda, and Phil Brylski for aid in the field, to Monica Frelow for laboratory analyses, and to Rich Strauss, Bruce Riska, and Enrique Lessa for analytical advice and for conceptual input. Karen Klitz drew Figure 2. Appreciation is extended to the game and fish departments of California, Nevada, and Arizona for necessary collecting permits. Financial support was provided by National Science Foundation grants to J.L.P.

LITERATURE CITED

Barton, N. H., and B. Charlesworth. 1984. Genetic revolutions, founder effects, and speciation. Annu. Rev. Ecol. System. 15:133–164.

Burnaby, T. P. 1966. Growth-invariant discriminant functions and generalized distances. Biometrics 22:96–110.

Bush, G. L. 1975. Modes of animal speciation. Annu. Rev. Ecol. System. 6:339–364.

Carson, H. L. 1982. Speciation as a major reorganization of polygenic balances. Pp. 411–433 in: C. Barigozzi (ed.), *Mechanisms of Speciation*. A. R. Liss, New York.

Carson, H. L., and A. R. Templeton. 1984. Genetic revolutions in relation to speciation phenomena: The founding of new populations. Annu. Rev. Ecol. System. 15:97–131.

Davis, W. B. 1937. Variations in Townsend pocket gophers. J. Mamal. 18:145–158.

Eldredge, N., and S. J. Gould. 1972. Punctuated equilibria: An alternative to phyletic gradualism. Pp. 82–115, in: T. J. M. Schopf (ed.), *Models in Paleobiology*. Freeman, Cooper and Co., San Francisco.

Endler, J. A. 1977. *Geographic Variation, Speciation, and Clines*. Princeton University Press, Princeton, New Jersey.

Futuyma, D. J. 1987. On the role of species in anagenesis. Am. Natur. 130:465–473.

Gould, S. J., and N. Eldredge. 1986. Punctuated equilibrium at the third state. System. Zool. 35:143–148.

Hafner, J. C., D. J. Hafner, J. L. Patton, and M. F. Smith. 1983. Contact zones and the genetics of speciation in the pocket gopher *Thomomys bottae* (Rodentia:Geomyidae). System. Zool. 32:1–20.

Hafner, M. S., J. C. Hafner, J. L. Patton, and M. F. Smith. 1987. Macrogeographic patterns of genetic differentiation in the pocket gopher *Thomomys umbrinus*. System. Zool. 36:18–34.

Hall, E. R. 1981. *The Mammals of North America*, Vol. 1. John Wiley, New York.
Larson, A., D. B. Wake, and K. P. Yanev. 1984. Measuring gene flow among populations having high levels of genetic fragmentation. Genetics 106:293–308.
Mantel, N. 1967. The detection of disease clustering and a generalized regression approach. Cancer Res. 27:209–220.
Mayr, E. 1954. Change of genetic environment and evolution. Pp. 157–180 in: J. Huxley, A. C. Hardy, and E. B. Ford (eds.), *Evolution as a Process*. Allen & Unwin, London.
Mayr, E. 1963. *Animal Species and Evolution*. Harvard University Press, Cambridge.
Mayr, E. 1982. Processes of speciation in animals. Pp. 1–19 in: C. Barigozzi (ed.), *Mechanisms of Speciation*. A. R. Liss, New York.
Miller, R. S. 1964. Ecology and distribution of pocket gophers (Geomyidae) in Colorado. Ecology 45:256–272.
Nevo, E. 1979. Adaptive convergence and divergence of subterranean mammals. Annu. Rev. Ecol. System. 10:269–308.
Patton, J. L. 1973. An analysis of natural hybridization between the pocket gophers *Thomomys bottae* and *Thomomys umbrinus* in Arizona. J. Mammal. 54:561–584.
Patton, J. L. 1981. Chromosomal and genic divergence, population structure, and speciation potential in *Thomomys bottae* pocket gophers. Pp. 255–295 in: O. A. Reig (ed.), *Ecologia y Genetica de la Especiacion Animal*. Equinoccio, Caracas, Venezuela.
Patton, J. L. 1985. Population structure and the genetics of speciation in pocket gophers, genus *Thomomys*. Acta Zool. Fenn. 170:109–114.
Patton, J. L., and P. V. Brylski. 1987. Pocket gophers in alfalfa fields: Causes and consequences of habitat-related body size variation. Am. Natur. 130:493–506.
Patton, J. L., and J. H. Feder. 1981. Microspatial genetic heterogeneity in pocket gophers: Non-random breeding and drift. Evolution 35:912–920.
Patton, J. L., J. C. Hafner, M. S. Hafner, and M. F. Smith. 1979. Hybrid zones in pocket gophers: Genetic, phenetic, and ecologic concordance patterns. Evolution 33:860–876.
Patton, J. L., R. K. Selander, and M. H. Smith. 1972. Genetic variation in hybridizing populations of gophers (genus *Thomomys*). System. Zool. 21:263–270.
Patton, J. L., M. F. Smith, R. D. Price, and R. A. Hellenthal. 1984. Genetics of hybridization between the pocket gophers *Thomomys bottae* and *Thomomys townsendii* in northeastern California. Great Basin Natur. 44:431–440.
Patton, J. L., and M. F. Smith. 1981. Molecular evolution in *Thomomomys*: Phyletic systematics, paraphyly, and rates of evolution. J. Mammal. 62:493–500.
Patton, J. L., and S. Y. Yang. 1977. Genetic variation in *Thomomys bottae* pocket gophers: Macrogeographic patterns. Evolution 31:697–720.
Rogers, J. S. 1972. Measures of genetic similarity and genetic distance. Studies Genet., Univ. Texas Pub. 7213:145–153.
Rohlf, F. J., and F. L. Bookstein. 1987. A comment on shearing as a method for "size correction." System. Zool. 36:356–367.
Slatkin, M. 1981. Estimating levels of gene flow in natural populations. Genetics 99: 323–335.
Slatkin, M. 1985. Rare alleles as indicators of gene flow. Evolution 39:53–65.
Slatkin, M. 1987. Gene flow and the geographic structure of natural populations. Science 236:787–792.
Smith, M. F., and J. L. Patton. 1980. Relationships of pocket gopher (*Thomomys bottae*) populations of the lower Colorado River. J. Mammal. 61:681–696.
Smith, M. R., and J. L. Patton. 1988. Subspecies of pocket gophers: Causal bases for geographic differentiation in *Thomomys bottae*. System. Zool. 37:163–178.
Smith, M. R., J. L. Patton, J. C. Hafner, and D. J. Hafner, 1983. *Thomomys bottae* pocket gophers of the central Rio grande Valley, New Mexico: Local differentiation, gene flow, and historical biogeography. Occasional Papers Mus. Southwestern Biol., Univ. New Mexico 2:1–16.
Templeton, A. R. 1980. The theory of speciation via the founder principle. Genetics 94:1011–1038.
Thaeler, C. S., Jr. 1968a. An analysis of the distribution of pocket gopher species in northeastern California (genus *Thomomys*). Univ. Calif. Pub. Zool. 86:1–46.

Thaeler, C. S., Jr. 1968b. An analysis of three hybrid populations of pocket gophers (genus *Thomomys*). Evolution 22:543–555.
Wade, M. J. 1978. A critical view of the models of group selection. Quart. Rev. Biol. 53:101–114.
Wade, M. J., and D. E. McCauley. 1980. Group selection: The phenotypic and genotypic differentiation of small populations. Evolution 34:799–812.
White, M. J. D. 1978. *Modes of Speciation.* W. H. Freeman, San Francisco.
Wright, S. 1956. Gene and organism. Am. Natur. 87:5–18.
Wright, S. 1965. The interpretation of population structure by F-statistics with special regard to systems of mating. Evolution 19:395–420.
Zimmerman, E. G., and N. A. Gayden. 1981. Analysis of genic heterogeneity among local populations of the pocket gopher, *Geomys bursarius*. Pp. 272–287 in: M. H. Smith and J. Joule (eds.), *Mammalian Population Genetics*. University of Georgia Press, Athens.

PART FOUR

BIOGEOGRAPHY AND ECOLOGY OF SPECIATION: BACKGROUND AND THEORY

CHAPTER 13

SYMPATRIC SPECIATION IN INSECTS: PERCEPTION AND PERSPECTIVE

Catherine A. Tauber and Maurice J. Tauber

> *I can by no means agree with this naturalist that migration and isolation are necessary elements for the formation of new species.* (Darwin 1859, 6th ed.; in reference to biologist J. T. Gulick; see Carson 1987a).

Ever since Darwin's time, sympatric speciation has remained controversial. Some biologists question its occurrence (e.g., Mayr 1963, 1970; Futuyma and Mayer 1980; Paterson 1981), and some accept the possibility but emphasize the general applicability of what they believe are "better-established alternatives" (e.g., Carson 1987b; Coyne and Barton 1988). Others conclude that sympatric speciation (along with allopatric and parapatric speciation) is a significant mode of evolution, particularly among certain groups of organisms (e.g., Scudder 1974; Bush 1975; White 1978; Kondrashov and Mina 1986). This is our position.

Although our chapter focuses on sympatric speciation, the issue requires a broad setting. The fundamental questions are not whether sympatric speciation occurs or how important it is, but what are the conditions that promote speciation, and what processes bring it about? Within this context, it is vital (as George Barlow stated during a lively discussion that followed Ernst Mayr's recent lecture at Berkeley) to consistently apply equally rigorous standards in the analysis of all proposed modes of speciation. It is also crucial to bring a well-balanced, wide ranging approach to speciation studies—an approach that integrates biological realism into theoretical studies and concomitantly promotes well-focused experimental and comparative studies that are strongly founded in theoretical advances.

Why does sympatric speciation remain controversial, and what is the source of the contention? First, because evolutionary processes are complex and because evolutionary time extends over many human life spans, the study of evolution does not lend itself readily to the establishment of discrete, falsifiable hypotheses (see Loehle 1987). Thus, observations and experimental results emanating from evolutionary studies often elicit broadly differing interpretations.

Historically, such an atmosphere creates a disproportionately large role for personalities and their interplay in the acceptance of ideas. In the case of speciation, Ernst Mayr (undoubtedly among the greatest evolutionary theorists of our time) has had a central role. Deference to his forceful opposition to the concept of sympatric speciation (e.g., Mayr 1963, 1970, 1976) has pervaded evolutionary biology for many years. Commendably, Mayr's challenges have stimulated rigorous thought and much research on the subject; nevertheless, the strength of his influence has strongly biased the application of theories of speciation to other areas of evolutionary biology, e.g., systematics, cladistics, biogeography, and macroevolution. In our view, this bias has constrained the theoretical development of these sciences. (However, for refreshing, independent overviews, see Endler 1982a; Vrba 1987.)

A second reason for the continuing controversy arises from the seemingly counterintuitive nature of sympatric speciation. Speciation involves the division of a species on an adaptive peak, so that each part moves onto a new adaptive peak (in the sense of Wright 1932), without either one going against the upward force of natural selection. This process is readily envisioned if a species becomes subdivided by a physical barrier and each part experiences different mutations, population fluctuations, and selective forces. Thus, we achieve the popular wisdom embodied in the allopatric model of speciation. In contrast, conceiving the process of sympatric speciation (the division of a single population and radiation onto separate peaks without geographical isolation) is intuitively more difficult and has generally depended on the development of appropriate population genetics models. Such models have predetermined assumptions, and they are restricted in the numbers and types of variables and processes they include. Therefore, the generality of their predictions is open to interpretation, challenge, and reinterpretation.

A third reason originates from the common association of the process of speciation with present patterns of species distributions. Systematists, in particular, are often very comfortable with the idea of allopatric speciation; it is compatible with all patterns of geographic distribution—including fully sympatric distributions, which are readily ascribed to secondary invasion. Less obvious, but equally reasonable, is the concept that sympatric speciation is also compatible with the diverse patterns of geographic distributions found in animals (e.g., Scudder 1974; Endler 1977, 1982b; Glazier 1987; also see Rosenzweig and Taylor 1980; Barton and Hewitt 1985; Levinton 1988).

Fourth, divergence among biologists over the occurrence and importance of sympatric speciation is reinforced by the taxonomically skewed distribution of proposed cases. Among animals, a very high proportion of examples originate from insects (e.g., Bush 1975; Kondrashov and Mina 1986; and see below) and from fish and mollusks (see McKaye 1980; Davis 1982; numerous authors in Echelle and Kornfield 1984; Gittenberger 1988; Meyer 1989; Wilson, this volume). Few come from birds or mammals, and these are frequently associated with chromosomal rearrangements (e.g., Bush 1981)—a notable exception being the large cactus finches of Isla Genovesa in the Galapagos (Grant and Grant 1979; Grant and Grant, this volume; also see Thorpe 1945). Such disparity may originate from substantive biological differences among taxa or from differences in how biologists define "species." Either situation generates diverse approaches to experimentation and analysis, as well as divergent views of speciation.

In spite of the continuing controversy, there now exist substantial indications that sympatric speciation may have a significant role in the evolution of insects. This chapter reviews some of the data; in doing so, it is not our purpose to refute past arguments against sympatric speciation (e.g., Mayr 1963, 1970; Futuyma and Mayer 1980; Paterson 1981). Recent reviews and papers provide cogent evidence that geographic isolation is not a requirement for speciation (Bush 1975; Endler 1977; Templeton 1980; Barton and Charlesworth 1984; Bush and Howard 1986; Kondrashov and Mina 1986; Feder et al. 1988; McPheron et al. 1988). Instead, this chapter focuses on studies indicating that the theoretical restrictions on sympatric speciation are much less severe than previously thought and that many groups of insects possess combinations of traits that fit within the current understanding of sympatric speciation. It is our conviction that the study of speciation is at a crucial stage in which a greater awareness and appreciation of the progress in each of several approaches to speciation would lead to significant advances through the synergistic interaction of theory and experiment.

DEFINITIONS AND BACKGROUND

Before beginning our discussion of genetic models and case studies, we set forth our terminology and perspective, largely to avoid misunderstandings that tend to occur in the literature. We follow the definition of a species that combines both evolutionary and biological concepts: a *species* consists of a group of populations whose evolutionary pathway is separate from that of other groups; this distinct evolutionary pathway is achieved by the group's reproductive isolation from other groups (slight modification of definition in Futuyma and Mayer 1980).

From this definition it follows that the process of speciation has at least two components (see Maynard Smith 1966). First is the diversification of populations, or, as Barton and Charlesworth (1984) describe it, the splitting of an ancestral population into populations located at different equilibria under selection. Second is the evolution of reproductive isolation, either through the direct or indirect effects of natural selection, sexual selection, genetic drift, or nonadaptive molecular or other changes.

Although speciation can be classified according to the many underlying processes (mutation, recombination, selection, gene flow, and stochastic processes) (e.g., Templeton 1980), it is inseparable from a geographic perspective. Many of the events and forces involved in speciation are influenced by the degree to which movement between populations is interrupted by physical barriers. Thus, there may be considerable differences in the mechanisms comprising each mode of speciation (allopatric, parapatric, and sympatric; see definitions by Futuyma and Mayer 1980).

First, according to the allopatric model of speciation, an ancestral species becomes subdivided by a physical barrier that prevents gene flow. Directional selection (natural and/or sexual) in each area produces adaptation along different lines, and the subdivided populations diverge polygenically. Allopatric speciation rarely involves a shift in host or habitat, but stochastic events may play an important role (see diverse views of Mayr 1982; Barton and Charlesworth 1984; Carson and Templeton 1984). Reproductive isolation is achieved either as a by-product of geographic (adaptive or nonadaptive) diversification at numerous loci or as the result of direct (natural or sexual) selection if divergent, partially incompatible populations come into secondary contact (for a variety of views, see Kaneshiro 1976; Endler 1977; Lande 1981; Sawyer and Hartl 1981; Kirkpatrick 1982; Nei et al. 1983; West-Eberhard 1983; Littlejohn and Watson 1985; Paterson 1985; Spencer et al. 1986; Butlin 1987a; excellent summary by Hedrick and Louis 1984; Coyne and Orr, this volume; Otte, this volume).

The second mode of speciation, parapatric speciation, involves the divergence of populations that are spatially separated, but that share a common border, so that a small proportion of individuals in each encounters the other. Geographically widespread species usually exhibit smooth or stepped clinal variation in morphological, behavioral, physiological, and life-history traits. The steepness and structure of the clines are functions of gradients in the abiotic and biotic environment, the characteristics of the variable traits, and the ease with which individuals move along the gradients. For example, variation in patterns of voltinism (or other major life-history traits) along a cline can influence a large suite of interrelated characteristics (e.g., for a discussion of crickets see Masaki and Walker 1987; Harrison and Rand, this volume). Therefore both differences in (natural and sexual) selection pressures along the cline and restrictions on gene flow (e.g., low vagility, linkage) contribute to parapatric diversification (Endler 1977; Barton and Hewitt 1981). Reproductive isolation between adjacent populations may evolve as an indirect result of genetic diversification through natural or sexual selection or as the direct result of natural selection for premating reproductive isolating mechanisms. Presumably, several processes accelerate its evolution: strong, disruptive selection in adjacent areas, reduction in gene flow along the cline, strong natural and/or sexual selection on premating traits in the zones of overlap, interaction among pleiotropic characters, and the subdivision of hybrid zones into mosaic populations (Endler 1977; Barton and Hewitt 1981; Lande 1982; Slatkin 1982; West-Eberhard 1983; Littlejohn and Watson 1985; Harrison 1986).

The third type of speciation occurs in sympatry. Its primary driving force is disruptive, frequency- or density-dependent natural selection on the manner in which a population uses resources. Thus, it involves a shift in host or habitat association or the partitioning of an essential, but limiting, resource. As in the other modes of speciation, the evolution of reproductive isolation between sympatric populations does not require direct selection for premating reproductive isolating mechanisms, although it is compatible with such a situation. Reproductive isolation may evolve as the indirect result of adaptation to diverse habitats or hosts (e.g. through pleiotropy or selection for modifier genes), through direct selection for assortative mating genes, or as a by-product of sexual selection (see Maynard Smith 1966; Endler 1977; Thornhill and Alcock 1983; West-Eberhard 1983). The evolution of reproductive isolation between sympatric subpopulations is promoted by the tendency to mate on the host or after habitat selection (Diehl and Bush, this volume). Simple genetic control of host or habitat selection and of phenological traits, as well as conditioning to the host or habitat, may also facilitate the process.

Distinctions among the modes of speciation are not sharp; the three different types form a continuum in both the degree of spatial separation and in the mechanisms underlying diversification. Also, processes involved in the evolution of reproductive isolation can be very similar in all three types. We recognize that sympatric divergence generally leads to some degree of spatial separation or other external restriction on gene flow among genotypes. Thus, we propose that the essential characteristic of sympatric speciation is that the *initiating* force in diversification is disruptive selection on a panmictic population for diversification in the use of resources or in the escape from natural enemies. Only after diversification has begun does the physical separation of subpopulations come into play. Therefore, sympatric speciation always involves the adoption of new resources (food, habitat) or the spatial or temporal partitioning of underexploited resources—i.e., the evolution of specialization.

With this discussion as background, the remainder of the chapter examines some theoretical and empirical studies that impinge upon sympatric speciation in insects. First, we discuss some implications derived from selected genetic models of speciation. Then, we discuss a series of experimental and comparative studies to illustrate that insects possess a variety of traits that are consistent with the sympatric mode of speciation. We conclude that progress in the study of speciation depends on close, synergistic interactions among the diverse approaches to the topic of speciation.

GENETIC MODELS OF SPECIATION

Population genetics models have great utility in the analysis of speciation because the process not only encompasses numerous, diverse interacting factors, but it also occurs too slowly to observe directly. Genetic models provide the means of evaluating particular parameters, and they allow for the precise development of predictions, whether intuitive or counterintuitive. Thus, they can function as powerful tools in formulating hypotheses and predicting the likelihood of events when various parameters within the process are modified. Nevertheless, genetic models of speciation suffer some severe limitations. To date, they have been able to address, in detail, only aspects of the process, and they are restricted in the numbers and kinds of interacting variables they can reasonably include. Perhaps the most severe, but readily corrected, problem with models of speciation is that they have been developed on a largely theoretical basis. Only recently have they been based on the ecological and genetic characteristics of specific taxa that are considered likely candidates for sympatric speciation. We regard this as a progressive change because, in general, a model's applicability and power reside in how closely its assumptions and predictions reflect conditions in nature.

It is not our purpose in this section to develop or evaluate the mathematical logic of various models. Rather, we illustrate a historical trend: as models increasingly reflect the life-styles of specific insect taxa, the conditions under which they predict that sympatric speciation (or events favorable to sympatric speciation) can occur have tended to become less restrictive.

In the beginning, Ernst Mayr, in his masterly synthesis of evolutionary biology (Mayr 1963), presented a commendable challenge to biologists—to devise realistic genetic models for the process of sympatric speciation. Soon afterward, Maynard Smith (1966) took up this challenge and proposed a genetic model, itself based on an earlier model by Levene (1953).

Maynard Smith's model, which considers genetic changes at only two loci, consists of two steps: first, the establishment of a stable polymorphism through disruptive selection in a two-niche situation, and, second, the evolution of reproductive isolation between the two morphs. For this model to operate, two conditions must be met: (1) the density-dependent factors that regulate population size must be independent in each niche, and (2) the selective differences between the morphs must be large.

The conditions for sympatric speciation demanded by Maynard Smith's model are quite severe (e.g., see later work by Maynard Smith 1970; Maynard Smith and Hoekstra 1980; Hoekstra et al. 1985), and the assumption of simple genetic control of niche association is probably not met by many species. Nevertheless, the model has great value in that (1) it subdivided the complex problem of speciation into clearly envisioned components, (2) it presented a general, but modifiable approach to the problem, and (3) it provided significant stimulus and a theoretical framework for the development of subsequent models.

An underlying premise of Maynard Smith's (1966) model, and many others that followed, is that sympatric speciation involves the evolution of specialization to novel or underused resources (hosts, habitats)—the invasion of "empty niches" (see Wilson and Turelli 1986). Numerous theoretical studies have addressed aspects of this process (see below). Some describe conditions for the maintenance of genetic variability or balanced polymorphisms for resource utilization. Others consider the evolution of host or habitat races under disruptive selection, the interaction between gene flow and disruptive selection in heterogeneous environments, the maintenance of underdominance under frequency-dependent selection, and the evolution of reproductive isolation between host- and habitat-adapted polymorphs. The various models differ greatly in the conditions for which they predict diversification and/or reproductive isolation.

The models also differ in the degree to which they reflect biological reality. For example, several recent models, which could have been broadly applicable to the divergence of host and/or habitat utilization in insects (Felsenstein 1981; Hoekstra et al. 1985; Futuyma 1986; Fialkowski 1988), unfortunately do not take into account the association between host or habitat preference and mating site (Bush and Diehl 1982; Rice 1984; Jaenike 1988; Diehl and Bush, this volume). For many insect species, this constitutes a major omission because mating occurs on the host or after dispersal to the habitat. Therefore, disruptive selection on genes that determine habitat or host selection leads pleiotropically to assortative mating. As a result, the very severe restrictions these models place on diversification are inappropriate for most proposed cases of sympatric speciation in insects.

In addition to this, many models of diversification in resource utilization do not take into account other pleiotropic effects of genes for host or habitat selection, selection for "modifier genes" (sensu Maynard Smith 1966), genetic covariance, or genotype–environment interactions (for some exceptions, see Endler 1977; Mani 1981; Smith-Gill 1983; Via and Lande 1985; Meyer 1987; Thompson 1987; Smith 1988). Through their influence on phenology, behavior, or physiology, these factors can readily increase differences in fitness across habitats. Moreover, they may contribute to the development of assortative mating (e.g., see Wood 1980; Wood and Guttman 1982, 1983). Thus, models of diversification and speciation should have a holistic approach—one that recognizes that genetic changes can have direct and indirect selective effects and that such changes can have far-reaching effects on interactions at several trophic levels. Exclusion of these factors can result in restricted (and perhaps erroneous) interpretations of data from studies of genetic variation in host or habitat association (e.g., aspects of Futuyma 1986 and Butlin 1987b).

Another crucial, but often neglected, consideration for sympatric speciation is the effect of intraspecific competition for limited resources on population structure ("competitive speciation" of Rosenzweig 1978). This concept, which is based on the dynamic nature of fitness and which is expressed as density- or frequency-dependent selection, was explicitly incorporated into several genetic models of adaptation to novel resources (e.g., Pimm 1979; Udovic 1980; Seger 1985; Wilson and Turelli 1986; Ginzburg et al. 1988; also see references therein). As a result, the conditions under which sympatric speciation is thought to occur were greatly expanded: (1) the environment need not be heterogeneous for sympatric speciation to occur; slight differences in resource utilization, under intraspecific competition, can lead to divergence, (2) during the initiation of adaptation to a new resource, phenotypes that are poorly adapted might have relatively high fitness if the resource is abundant and underexploited, and (3) stable polymorphisms involving at least two loci might be maintained indefinitely in a heterogeneous environment—thus providing sufficient time for the evolution of dominance, linkage, and reproductive isolation.

The latter point—stable polymorphisms involving more than one locus—is particularly significant for sympatric speciation. Most early models assume that sympatric speciation is restricted to cases in which host or habitat shifts result from very simple genetic changes. This situation generally requires that fitness and/or assortative mating be genetically correlated with habitat or host selection. The probability of sympatric diversification was considered to be very low if the gene(s) determining host selection behavior are independent of those that determine host-related fitness. However, in most insects that have been examined, changes in host or habitat association require allelic substitutions at several loci (e.g., Futuyma and Peterson 1985; Jaenike 1986a; but see Tauber and Tauber 1977a, 1977b), and few studies have demonstrated a genetic correlation between host selection and developmental or reproductive success (e.g., Via 1986).

Several recent models that are derived from studies of natural biological systems help to reconcile the above contradiction. Rausher's (1984a) model demonstrates that the maintenance of genetic variation in both preference for a new host and for increased fitness on that host does not require unduly restrictive conditions. The main factor in his model is independent, frequency-dependent selection in heterogeneous or unpredictable environments (see also Wilson and Turelli 1986). Diehl and Bush's model (this volume) demonstrates that *if* mating takes place within the habitat (a common characteristic of many host- and habitat-specific insects), sympatric divergence of polygenically controlled habitat association (one locus for habitat preference, two for fitness) occurs even with relatively weak habitat preference and moderate differences in fitness. Jaenike's (1988) model illustrates that early adult experience (e.g., contact with larval food) can lead to multilocus differentiation of a polyphagous population. In this model, experience augments genetic subdivision to a greater degree when both host preference and survival are genetically variable than when only preference has a genetic basis. In either case, mating on the host (or in the habitat) increases the degree of differentiation.

In considering a different, but related problem, Rice's (1984) model deals with the polygenic control of habitat preference. In his scenario, habitat preference is controlled by two loci and is genetically correlated with fitness; mating occurs after dispersal to the habitat. Relatively moderate levels of disruptive selection on habitat preference lead to positive assortative mating. Rice's model does not focus on the situation in which preference and fitness are controlled by independent genes, but a later paper (Rice 1987) illustrates the applicability of the model to habitat-specific organisms. Taken together, the models of Wilson and Turelli, Rausher, Rice, Jaenike, and Diehl and Bush illustrate that polygenic control of various aspects of host or habitat association is not an impediment to sympatric speciation.

Kondrashov's (1983a, 1983b, 1986) models lead to similar relaxation of restrictions on sympatric speciation. His theoretical work indicates that sympatric speciation is possible even if the evolution of reproductive isolation requires allelic changes in up to eight loci. Therefore, the presence of complex reproductive isolating mechanisms cannot be invoked as a reason for rejecting the concept of sympatric speciation.

From genetic and behavioral perspectives, Lande (1981, 1982) and

West-Eberhard (1983, 1984), respectively, examined the importance of sexual selection to speciation. They showed that sexual selection can cause very rapid divergence of both sympatric and allopatric populations, even with moderate levels of gene flow. Their ideas are of particular interest because sympatric species of insects often have divergent mating behavior, secondary sexual characters, and sex pheromones (e.g., Roelofs and Comeau 1969; Matthews and Matthews 1978; Blum and Blum 1979; Thornhill and Alcock 1983; Eberhard 1985; Phelan and Baker 1987). Changes in any of these traits could be involved in the evolution of reproductive isolation between sympatric populations, but as Kaneshiro and Boake (1987) point out, there is a need for theoreticians to develop speciation models that include assumptions concerning the functions and costs of sexually selected behavior patterns. To ensure biological realism, we add: these models should be based on the traits, life-styles, and population characteristics of diversifying or speciose taxa known to be under sexual selection.

What do recent theoretical studies reveal regarding sympatric speciation? They illustrate that under frequency-dependent selection, genetic variation and polymorphisms for resource utilization can be maintained in heterogeneous environments and that under some circumstances, sympatric speciation may occur with moderate levels of disruptive selection. Moreover, they show that under certain biologically realistic conditions, polygenically controlled ecological traits can diversify in sympatry and that sympatrically derived reproductive isolating mechanisms need not be based on only one or a few genes. Nor does the evolution of reproductive isolation in sympatry require the process of reinforcement. Additionally, recent models illustrate that sexual selection may result in very rapid diversification, even in the presence of moderate levels of gene flow. In theory, the requirements for sympatric speciation are more robust than previously thought.

Where do we go from here? Theoretical studies are now at the threshold of biological realism; therefore the assumptions and predictions of speciation models could be tested vis-á-vis natural systems of insects. As the following discussion illustrates, there is substantial background from experimental and comparative studies for the development and refinement of precise speciation models based on the variable traits, the innate life-styles, and the population characteristics of specific taxa in nature. A variety of insect groups provide excellent systems against which to develop and test such biologically meaningful genetic models of speciation.

CASE STUDIES

Within this section, we discuss a variety of insects for which sympatric speciation has been proposed. In doing so we do not focus on the strength of the claims for sympatric speciation. Rather, our objective is first, to illustrate that insects have numerous variable traits, that when subjected to realistic selective regimens can subserve sympatric speciation, and second, to show that these traits exist within the context of life-styles that could promote sympatric speciation.

Each case involves two or more sympatric populations or species that differ in their utilization of some essential resource (i.e., either they use different hosts or habitats or they partition the utilization of a single host or habitat). Thus, they conform to the spirit of Maynard Smith's (1966) model of sympatric speciation in a two-niche situation and/or Rosenzweig's (1978) model of competitive speciation. In most of the cases, mating occurs in association with the host and/or habitat. And, in each one, reproductive isolation between the host- or habitat-associated entities involves at least one of the four mechanisms that Maynard Smith (1966) proposed for the evolution of reproductive isolation: (1) habitat selection, (2) pleiotropic effects, (3) modifying genes, and (4) assortative mating genes.

Previously, Bush and Diehl (1982) and Zwölfer and Bush (1984) emphasized the importance of the host serving as the mating site, and Diehl and Bush (this volume) demonstrate theoretically and empirically how this mechanism promotes reproductive isolation along host or habitat lines. Therefore, our consideration generally focuses on other mechanisms related to sympatric speciation. We illustrate that insects have evolved a variety of reproductive isolating mechanisms that may not have required direct selection for assortative mating genes—a situation that facilitates sympatric speciation.

For convenience, we organized our examples into five categories: (1) host-associated herbivores, (2) host-associated parasitoids, (3) habitat-associated *Drosophila,* (4) host-associated predators, and (5) habitat-associated predators.

Host-associated herbivores

A number of systematic studies of herbivorous insects have invoked sympatric speciation as an explanation for the occurrence of host races and host-associated sibling species. A fine example is E. Gorton Linsley's monumental revision of the North American Cerambycidae (e.g., Linsley 1961; Linsley and Chemsak 1984). Several genera in the family (e.g., *Callidium, Xylotrechus, Neoclytus, Saperda*) contain many host-associated races and sibling species, and as early as the mid 1800s Benjamin D. Walsh (1864, 1865) recognized the relationship between host shifts and taxonomic diversification in the Cerambycidae. Linsley (1959, 1961) reemphasized that such taxa call for the consideration of sympatric speciation because (1) they show considerable evidence of rapid host changes; (2) as a general rule, the closely related sibling species that are sympatric usually use different host plants, whereas the closely related allopatric species have identical or similar host plants; (3) in many species, mating occurs in association with the host plant (i.e., either immediately after emergence or at the oviposition site after dispersal); and (4) conditioning may be involved in host selection.

Sympatric speciation has also received serious consideration in ecological literature dealing with herbivore–plant associations. In discussing the evolution of community patterns, Strong et al. (1984) concluded that host specificity, colonization of introduced hosts, and speciation via host association are significant determinants of change in phytophagous insect communities over time. They and others (e.g., Zwölfer 1987; Ginzburg et al. 1988) noted evolutionary trends toward greater variety in resources and in the ways insects exploit these resources. Concomitantly, they also noted trends toward greater species richness over evolutionary time. Strong et al. (1984) examined both allopatric and sympatric speciation via host–race formation to explain patterns of host use and colonization of new host plants. They discussed numerous cases of herbivorous insect groups that show host shifts, and they compiled examples of possible sympatric host–race formation (Strong et al. 1984:106–107). Studies of other herbivorous insects have led, or readily lead to similar conclusions. To the list by Strong et al. (1984) we add several other cases (Table 1).

Almost all the examples from Strong et al. (1984) and from Table 1 fit the criteria that subserve sympatric speciation via host–race formation as proposed by Bush (1974), Bush and Diehl (1982), and Diehl and Bush (this volume). First, mating occurs on the host plant—and therefore host selection promotes assortative mating. Second, all except the leafhoppers (Ross 1957; McClure and Price 1976) and the hummingbird flower mites (Colwell 1986a, 1986b) exhibit host-related differences in traits related to fitness. Third, the females determine the larval host and discriminate among hosts for oviposition. And, fourth, in almost all cases the phenology of the herbivore is related to the phenology of the host—a situation that can enhance assortative mating along host lines.

In examining the last trait—phenology—we note that temporal isolation in insects involves a variety of mechanisms. Most commonly, it involves genetically based differences among host races (Tauber et al. 1986). In these examples, the genes that confer host-related phenological differences function as modifying genes because they enhance fidelity and adaptation to the host (sensu Maynard Smith 1966; see, e.g. Smith 1988). Thus, any contribution they make to the evolution of reproductive isolation may occur without direct selection for assortative mating or reinforcement. Such genetically controlled phenological differences can be based on either polygenic or simple Mendelian inheritance (see Tauber et al. 1986).

Both of these mechanisms are fully compatible with the process of sympatric speciation (see Kondrashov 1983a, 1983b; Wilson and Turelli 1986; Diehl and Bush, this volume).

Direct genetic control is not the only way to attain seasonal isolation; phenological differences can also be host-induced and thus pleiotropic with genes that determine host selection. Although in his initial discussion, Maynard Smith (1966) thought that it was "unlikely" for pleitropy to have a large role in reproductive isolation among animals, recent evidence from phenological studies suggests that it may not be uncommon—particularly among homopterous insects that rely on cues from the host plant to initiate development after dormancy.

Treehoppers in the *Enchenopa binotata* species complex present a fine example in which host-induced phenological differences cause substantial assortative mating along host lines (Wood 1980; Wood and Guttman 1982, 1983; Wood, Olmstead, and Guttman, in preparation; Wood and Keese, in preparation). This species complex consists of at least six genetically differentiated species that utilize different but sympatric host plants (Guttman et al. 1981). Mating by these univoltine insects occurs during a very narrow period of the life cycle. Subsequently, females oviposit into the twigs of their host plants; eggs overwinter and hatch in the spring when fluids begin to flow through the twigs. The timing of fluid flow and the concomitant timing of hatch differ among the host plants. As a result, the life cycles of *Enchenopa* spp. on the various host plants are asynchronous. *Enchenopa's* combination of traits—univoltinism, brief periods of mating, and host-determined adult emergence times—results in allochronic life cycles, and therefore assortative mating, along host lines. Because adult emergence time is largely determined by the host, assortative mating is a pleiotropic effect of the genes that control host selection.

Conditioning to the host plant constitutes another mechanism that could promote sympatric speciation in herbivores; it can contribute to the divergence and/or the reproductive isolation of host-associated races (Jaenike 1988). First, it may facilitate population growth on a new host after accidental or occasional oviposition. Second, it may help maintain fidelity to the new host during genetic diversification and the evolution of other, genetically based reproductive isolating mechanisms.

It has been proposed that conditioning would be particularly important in initiating and maintaining new host associations if larval experience were to influence adult preference, but the occurrence of this phenomenon is problematical (see Corbet 1985; Jaenike 1988). However, conditioning of the adult can play a significant role. And, there exist several clear examples of conditioning in adult herbivores: *Rhagoletis* (apple maggot) and other fruit flies (Prokopy et al. 1982; Cooley et al. 1986; Prokopy and Papaj 1988), *Battus, Pieris,* and *Colias* butterflies (Stanton 1984;

TABLE 1. Additional examples of possible sympatric host–race formation or sympatric speciation via host–race formation.[a]

Reference	Insect	Notes
Linsley (1959, 1961)	Various Cerambycid genera (e.g., *Callidium, Xylotrechus, Neolytus, Saperda*)	Closely related sibling species that are sympatric use very different host plants, whereas those that are allopatric have identical or similar host plants. In many instances, mating occurs in association with the host plant
Müller (1958); Blackman (1979)	Various biotypes of aphids	Biotypes exhibit genetically controlled differences in host selection, as well as prezygotic and postzygotic reproductive isolating mechanisms even between closely related, host-alternating species, whose sexuals meet each year on a mutual primary host. Changes in host preference may occur during the parthenogenetic phase of the seasonal cycle
Katakura (1981)	Phytophagous ladybird beetles in the *Henosepilachna vigintioctomaculata* complex	Closely related forms differ in host preference and phenology. Although the author claims allopatric diversification is the rule, numerous traits of the beetles suggest that reexamination of the data is appropriate
Linsley and MacSwain (1958)	Anthophorid bees in the genus *Diadasia*	Sympatric forms differ in pollen plants and both seasonal and diurnal timing of flight. Mating of some species occurs in association with flowers, and, under stress, some species exhibit oligolecty
Thorp (1969)	Andrenid bees in the subgenus *Diandrena*	Results indicate phenotypic plasticity in utilization of pollen plants by oligolectic females under stress

Papaj 1986; Traynier 1986), and *Deloyala* (tortoise shell beetles) (Rausher 1983). Conditioning also occurs in the hummingbird flower mites, *Proctolaelaps* and *Rhinoseius,* but the stage at which it occurs is not known (Colwell 1986b).

These examples illustrate that conditioning in herbivorous insects that are thought to undergo sympatric speciation, particularly those that emerge as adults from their host material—e.g., cerambycid beetles (Craighead 1921; Linsley 1961), wild bees (Thorp 1969)—should be examined experimentally. Variability in this trait (see Papaj 1986), and its

TABLE 1. *(Continued)*

Reference	Insect	Notes
Ramírez (1970)	Fig wasps in the chalcidoid family Agaonidae (*Blastophaga* and *Tetrapus*)	Each species of *Ficus* has its own unique species of agaonid pollinator; the agaonids play a role in *Ficus* speciation; conversely the roles that adaptation to the host and dispersal play in diversification and speciation of the agaonids should be examined. Many agaonid traits are compatible with sympatric speciation
Seitz and Komma (1984)	Host-specific populations of the gall-former, *Tephritis conura*	The pattern of electrophoretic and phenological differences between host-associated populations of *T. conura* suggests divergence based on differences in flowering times of the hosts (*Cirsium oleraceum* and *C. heterophyllum*). Mating occurs on the host plant (Zwölfer 1974)
Goeden (1987, 1988)	Fruit flies in the genera *Tephritis* and *Trupanea*	Facultative gall formation by normally capitulum-infesting fruit flies could provide an avenue of escape from natural enemies
Sturgeon and Mitton (1986)	Mountain pine beetle *Dendroctonus ponderosae*	Sympatric populations from three hosts are genetically and morphologically differentiated. The traits that promote sympatric diversification of host races include host-related phenological differences, host-related pheromonal differences including host-dependent production of pheromones and responses to them, possible larval conditioning of adults to host trees, and periodic epizootics

[a]See Strong et al. (1984:106–107) for the initial list.

role in the adaptation and reproductive isolation of herbivores, needs detailed investigation (see Jaenike 1988).

In addition to pleiotropy and conditioning, sexual selection represents a factor that could profoundly affect the diversification of host-specific herbivorous insects. A prime example comes from the hummingbird flower mites (*Proctolaelaps* and *Rhinoseius*) (Colwell 1986a, 1986b), in which sexual selection appears to be the primary selective force in the evolution of host fidelity. Hummingbird flower mites spend most of their lives in flowers where they feed on pollen and nectar and where they mate and produce young. They move between flowers by stowing away in the nostrils of hummingbirds—hence their name. In some parts of the tropics as many as 20 species may coexist in one area—each in its own species of flower. Among the species there is very little overlap in hosts.

Unlike the herbivores discussed above, flower mites within *Proctolaelaps* and *Rhinoseius* do not form typical host races with host-related differences in survival or success. Although mating occurs on the host, and females discriminate among hosts (as in Bush's model), Colwell (1986a, 1986b) considers host fidelity in the flower mites to be largely related to the ability to find mates. As a result, sexual rather than natural selection may have the primary role in the evolution of host differentiation and speciation among the sympatric populations of the mites.

Given the previous example of how sexual selection may enhance the evolution of host-specific herbivores, it is timely to reconsider how the evolution of secondary sexual characters could function in the divergence and reproductive isolation of other host-specific herbivorous insects. For example, consideration of reproductive behavior and sexual selection might help elucidate the enigmatic problem of speciation in the *Erythroneura* leafhoppers on sycamore and *Nephotettix* leafhoppers on rice (see Ross 1957; McClure and Price 1976; Inoue 1982).

Learned, or host-induced, responses to the chemical stimuli involved in reproduction may also influence the diversification of herbivorous insects (e.g., Roitberg and Prokopy 1981; Cardé and Baker 1984). The most compelling cases involve the production of sex pheromones by moths and beetles (e.g., Roelofs and Comeau 1969; Lanier and Burkholder 1974; Borden et al. 1980; Cardé and Baker 1984; Phelan and Baker 1987; see Cardé 1986 for a balanced and thorough approach to the role of pheromones in reproductive isolation and speciation). These cases invite the analysis of intraspecific variation in pheromone production and reception. A particularly intriguing question is whether the chemistry of the larval host can alter the structure or mixture of sex pheromones produced by adults.

We conclude our discussion of herbivorous insects with a word of caution. As Futuyma (1986) correctly pointed out, explanations of how host specificity evolves (whether in sympatry or otherwise) are commonly based on the assumption that there are trade-offs in fitness associated with shifts in host plants. Many fine developmental and genetic studies are aimed at detecting such trade-offs—e.g., direct relationships between host selection and performance on the host (e.g., Smiley 1978; Futuyma and Wasserman 1981; Rausher 1984b; Via 1986; Thompson 1988). These studies produced dissimilar results, and from this it is frequently concluded that host specificity is most likely to evolve in physically separated populations (e.g., Futuyma 1986; Butlin 1987b). This conclusion is flawed on several counts—the main one being that it overlooks the indirect effects of host association—i.e., effects of adopting a new host on interactions with organisms other than the host, e.g., escape from parasitoids, predators, pathogens, and/or competitors (e.g., Doutt 1960; Price 1981, 1986; Jaenike 1985; Lawton 1986; several authors in Boethel and Eikenbery 1986; James et al. 1988), or on the ability of the herbivore to obtain mates or cope with high densities of conspecifics (Alstad and Edmunds 1983; Colwell 1986a, 1986b). Relatively few studies of host selection in herbivores examine such interactions (e.g., Alstad and Edmunds 1983; Rausher and Papaj 1983; Colwell 1986a, 1986b; Damman and Feeny 1988; Elizabeth Bernays, unpublished; Thomas Wood, unpublished), but the results illustrate that the effects may be considerable. Analysis of the evolutionary diversification of host-specific herbivores requires a broader conceptual framework and a broader experimental approach than it has generally received in the past.

Host-associated parasitoids

In spite of the wealth of opportunities that parasitoids offer for investigating the mechanisms of speciation, these insects are very poorly known. Parasitoids display many characteristics that make them prime candidates for nonallopatric modes of speciation (e.g., Bush 1975; Matthews 1975; Gordh 1977; Khasimuddin and DeBach 1976; Price 1980; Waage and Greathead 1986). Among these are traits that are shared by host-associated herbivores—traits that reflect specialization on small, patchy resources. These include (1) mating on or near the host (e.g., Matthews 1975; van den Assem et al. 1980), (2) selection of larval host by adult females (e.g. Slobodchikoff 1973), (3) relatively high degrees of host specificity among many groups, and (4) phenological differences among host races and sibling species (e.g., van den Bosch et al. 1979; Tauber et al. 1983; Murakami 1988). Thus, we can expect models of host-race formation (e.g., Bush 1969, 1974; Diehl and Bush, this volume) to apply to many groups of parasitoids. In addition, the haplo-diploid method of sex determination, the occurrence of sexual selection on courtship behavior and secondary sexual characteristics, the prevalence of sib-matings, and the infrequency of repeated inseminations among parasitoids (see Matthews 1975; Gordh and DeBach 1978; van den Assem 1986) can contribute to rapid divergence along host lines (e.g., Askew 1968; Rao and DeBach 1969; Matthews 1975).

As in herbivores, a variety of biochemical, physiological, behavioral, and ecological traits determine host associations in parasitoids. These function at various stages in the parasitoid's interaction with its host—from seasonal synchrony to habitat selection, host habitat finding, and successful development in specific hosts (e.g., Doutt et al. 1976; Vinson 1981; Weseloh 1981; Arthur 1981; van Lenteren 1981; Beckage 1985; Strand 1986). But, there has been very little work on the genetic variability in these factors (e.g., Slobodchikoff 1973; Messenger et al. 1976; Vinson et al. 1977; Arthur 1981; Price 1981; Vet and van Opzeeland 1984; Caltagirone 1985; Bouletreau 1986; Ruberson et al. 1989).

One trait that has received special attention is conditioning of the adult parasitoid for habitat or host selection (Arthur 1981; Vinson et al. 1977; van Alphen and Vet 1986; Wardle and Borden 1986). Particularly significant is the recent finding that adult parasitoids can become conditioned to the odor emitted by the feces of their hosts without coming into direct contact with the host (Lewis and Tumlinson 1988). Such behavior facilitates the evolution of host fidelity because newly emerged adults can become conditioned to their larval host before dispersing and seeking sites for mating and oviposition.

Among the few relatively well-studied cases of host-associated parasitoids are the braconids *Asobara tabida* and *Asobara rufescens* that parasitize two species of drosophilid larvae in different microhabitats (Vet and Janse 1984). These parasitoid species show characteristics that fall within Bush's model of speciation via host–race formation. (1) Mating occurs in the microhabitat immediately after emergence. (2) Although the two parasitoids are hybridizable under laboratory conditions, they show species-specific differences in fitness across habitats. (3) Habitat finding and host location are based on genetically determined odor preferences that are modified by adult experience (Vet and van Opzeeland 1984). This system offers excellent opportunities for theoreticians and experimentalists to investigate in detail the dynamic behavioral, physiological, and population phenomena involved in speciation.

The ichneumonid genus *Megarhyssa* provides another interesting but controversial case of proposed sympatic speciation in parasitoids. Gibbons' (1979) analysis of published data (largely from Heatwole et al. 1963, 1964; Heatwole and Davis 1965) led him to conclude that this group fits Rosenzweig's (1978) model of competitive speciation. According to the published data, diversification involves at least partial partitioning of a single limited resource. Three very closely related, sympatric species of *Megarhyssa* parasitize *Tremex* larvae inside of rotting logs, but each of the species tends to use a different part of the *Tremex* population as hosts. Only the species with the longest ovipositor can attack larvae that are deep in the logs; the species with the shortest ovipositor is restricted to larvae that are close to the surface, etc. Although there is overlap among the three *Megarhyssa* species in their host utilization (R. W. Matthews, personal communication), partial partitioning of larval resources may reduce the levels of interspecific competition relative to intraspecific competition. If so, Rosenzweig's (1978) model of competitive speciation may apply.

In *Megarhyssa*, intraspecific competition apparently is not restricted to competition among females for larval hosts. There also may be strong competition among males for mates. This situation is reflected in premating and mating behavior; males tend to (1) form aggregations on the bark of infested trees, (2) jostle each other for position, and (3) remain very faithful to particular trees (Matthews et al. 1979). Moreover, in all three species, females mate very early in their adult lives; over 80% mate before emerging through the bark and the remainder mate shortly after emergence (Crankshaw and Matthews 1981). Female *Megarhyssa* have very long abdomens—much longer than those of males, and until Crankshaw and Matthews' study it was generally believed that preemergence mating was unlikely. Their study very nicely showed that males fully insert their abdomens and part of their thoraces into the female emergence tunnel, and that transfer of sperm to the uniquely placed female genital opening is accomplished in the tunnel.

Megarhyssa males display interesting behavior patterns in that they (1) form mixed-species aggregations, (2) insert their abdomens into emergence tunnels without regard to the species (or sex) of the emergent, and (3) occasionally mate interspecifically with postemergent females. This suggests a degree of nonselectivity among males for choice of mate, and leads us to pose an intriguing question for future research: Since females of the three species differ in size, does the habit of preemergence mating result in positive assortative mating along the lines of size? If this is so, and if ovipositor length is correlated with body size, the partitioning of resources, assortative mating, and the diversification of the three species may result from frequency-dependent selection on a single trait—size—in both males and females.

Parasitoids in the pteromalid genus *Muscidifurax* provide another case that warrants further study. The work of Legner (1969, 1987), Kogan and Legner (1970), and van den Assem and Povel (1973) demonstrates the occurrence of several cryptic sibling species in what was previously thought to be a monotypic genus. Some of the taxa are sympatric. All attack the housefly; however, they are distinguished morphometrically and behaviorally (by differences in courtship). Divergence of some strains apparently occurred after the housefly invaded America, and thus could have been very recent (see van den Assem and Povel 1973; Legner 1987). The implications of current findings for sympatric speciation invite further comparative analyses of the distribution, ecology, behavior, and genetics of these interesting parasitoids.

In summary, insect parasitoids encompass a vast variety of life-styles and variable traits that will be of immense value to consider vis-á-vis the diverse models of speciation. As a group, they offer enormous, unexplored opportunities for studying the mechanisms involved in speciation.

Habitat-associated *Drosophila*

Like the parasitoids, the genus *Drosophila* shows many traits that, under certain circumstances, could lead to rapid diversification and speciation without geographic isolation. Among the most important are the tendency to mate after dispersal to the habitat, a high degree of host–habitat specialization, the ability of the adults to become conditioned to their habitats, and sexual selection (Heed 1971; Craddock 1975; Kaneshiro 1976; Parsons 1981a; Jaenike 1982, 1988; Rice 1984; Kaneshiro and Boake 1987; Hoffman 1988). Parsons (1981a) and Richardson (1982) illustrate the association between taxonomic divergence and diversification in resource utilization within the genus, and both authors make very strong cases for examining the feasibility of sympatric speciation in the group. Even among the diverse and speciose *Drosophila* of the Hawaiian Islands, which are largely associated with allopatric speciation, the occurrence of pairs of sympatric sibling species suggests sympatric speciation and invites further study (e.g., Kaneshiro et al. 1973). Here, we deal with the continental species first, and then the Hawaiian fauna.

After reading Gibbons' (1979) study with the ichneumonid wasps, Parsons (1981b) suggested that competitive speciation (through the partitioning of a single resource) might also have occurred in *Drosophila.* In Parsons' example, three closely related species of *Drosophila* differ in their use of a single food resource—fermented fruit. The substrate constitutes a heterogeneous resource in that its ethanol content varies considerably. The three species of *Drosophila* differ in their abilities to tolerate ethanol; therefore ethanol levels determine the flies' utilization of the resource as larval food, as a mating place, and as an oviposition site.

Apparently, competition for resources is very strong within each of the three *Drosophila* species, but weaker between species. Thus, Parsons considers this group to be candidates for competitive speciation, and he calls for greater emphasis on detailed comparative studies of resource utilization in *Drosophila.*

Food specificity in *Drosophila* species can involve both behavioral and physiological mechanisms, i.e., the nutritive quality and/or the attractants and repellants of host plants (e.g., reviews by Kircher and Heed 1970; Mueller 1985). Although many species feed on a variety of fermenting plant materials, females are selective in their choice of oviposition sites. Relative competitive abilities also have a role in determining food utilization (e.g., Fellows and Heed 1972). Thus, an understanding of host and habitat utilization in *Drosophila* requires study of the genetic variability and selective value of a variety of traits (e.g., Ehrman and Parsons 1976; Parsons 1978, 1983; Jaenike and Grimaldi 1983; Jaenike 1986b; Sokolowski et al. 1986; various authors in Huettel 1986).

Such studies would be of special value to the analysis of speciation in the *Drosophila* of the Hawaiian Islands. Radiation of this fauna is notably associated with diversification of elaborate secondary sexual traits and courtship displays, as well as the evolution of lek behavior. And it is clear that sexual selection has played a major role in the speciation of Hawaiian *Drosophila* (see Spieth 1974; Kaneshiro 1976; Kaneshiro and Boake 1987). Unfortunately, because of difficulties in studying this group in the field (Hardy 1974), comparative studies of host and habitat utilization are not common (see e.g., Heed 1968, 1971, Carson 1971).

Although most speciation events in the Hawaiian taxa are attributable to geographic isolation coupled with differentiation of sexual traits (see summaries by Carson 1974, 1982; Spieth 1974; Kaneshiro 1976; Kaneshiro and Boake 1987; see also Ehrman and Wasserman 1987), some may involve sympatric diversification via host or habitat diversification (e.g., Richardson 1974; Craddock 1975). For example, some of the sexually isolated species also use different food or habitats (Heed 1968, 1971; Carson 1971; Kaneshiro et al. 1973; Richardson 1974). An excellent example is found in the fully sympatric sibling species on the Island of Hawaii, *Drosophila silvarentis* and *Drosophila heedi*. After careful study in the field, it was found that these two species feed exclusively on the flux of a single species of tree, but in different places—one on the trunk well above the ground and the other in the soil (Kaneshiro et al. 1973).

The question of whether the primary impetus for divergence of behaviorally isolated *Drosophila* is geographic or ecological remains open for study. Current findings with sympatric species of Hawaiian *Drosophila* invite further comparative investigation of resource utilization and host selection behavior, as well as sexual behavior. Such studies would also provide incentive for investigating the factors that promote speciation in other groups of insects (e.g., several genera of microlepidopterans, coleopterans, and neuropterans) that have radiated on the Hawaiian Islands.

Over the years, disruptive selection experiments with *Drosophila* have constituted an important avenue for investigating sympatric speciation on an experimental level (e.g., Thoday and Gibson 1962, 1970; other references in Halliburton and Gall 1981; Rice 1985). There have been two general types of selection experiments. First are those that select directly for positive assortative mating. Generally, these experiments produced significant, but incomplete, reproductive isolation. The second type of laboratory experiment selects indirectly for reproductive isolation. After disruptive selection is applied to a particular trait, the selected subpopulations are assessed for mating preference. Under this type of selection, the development of reproductive isolation has been variable (see Halliburton and Gall 1981; Markow 1981; Rice 1985 and references therein; also Spiess and Wilke 1984; Bird and Semeonoff 1986).

In striving to increase the behavioral complexity and biological realism of selective regimes in laboratory experiments with *Drosophila*, Rice (1985) merged aspects of the two general approaches. His experimental approach involved a relatively complex arena that consistently required the flies to search for, and find, specific "habitats." Thus, it provided for direct selection on habitat choice, with pleiotropic effects on assortative mating. Because mating generally occurs after *Drosophila* have undergone postemergence dispersal, Rice's selective regime simulated the natural sequence of behavior. His experiment resulted in substantial reproductive isolation between the subpopulations utilizing the various "habitats."

Disruptive selection in Rice's experiment was very strong—probably much stronger than would occur in nature. Therefore, this study, like its predecessors, does not confirm that disruptive selection on traits such as habitat preference can result in reproductive isolation under natural conditions. But it demonstrates two important points: (1) under laboratory conditions, disruptive selection on habitat preference can result in significant assortative mating, and (2) selection experiments should simulate or reflect aspects of natural conditions whenever possible.

In summary, the genus *Drosophila* offers abundant opportunities for the experimental analysis of mechanisms involved in genetic divergence and reproductive isolation. In addition to selection on genetically controlled ecological and sexual traits, mechanisms such as nongenetic sterility factors (e.g., Ehrman et al. 1987; Thompson 1987), transposable elements, repetitive elements, and conditioning to hosts or habitats should receive more emphasis, in relation to speciation, than they have in the past (e.g., see Hedrick and Louis 1984).

Prey-associated predators

This category emphasizes the significant point that host fidelity is not restricted to herbivorous insects and that the role of food shifts in speciation is a valid consideration for predaceous insects, as well as for herbivores and parasitoids. Insect predators are generally difficult to study experimentally and their food habits are very poorly known. Nevertheless, the literature provides clear evidence that predaceous insects encompass the full range of prey associations from broadly generalized to narrowly specialized (see Thompson 1951; Hodek 1966, 1973; van den Bosch and Messenger 1973; DeBach 1974; Coppel and Mertins 1977; Henry et al. 1986; Tauber and Tauber 1987b).

Moreover, among groups of insect predators that have been examined closely, the mechanisms underlying prey specificity range from simple to highly complex. For example, seasonal synchrony with a single type of abundant, acceptable prey may provide a high degree of specificity (Begon and Mortimer 1981; Evans 1982). Or, specificity may stem from an intricate series of physiological and behavioral traits (e.g., Hagen et al. 1976; Thornhill 1976, 1979; Greany and Hagen 1981; Vinson 1981; Tauber and Tauber 1987b; Yeargan 1987). As with herbivores and parasitoids, the adult females of predaceous insects may greatly influence larval food specificity through selection of an oviposition site.

Studies of variability in prey association have largely concentrated on nongenetic mechanisms. They have included theoretical and experimental analyses of "switching," conditioning, and changes in searching behavior in response to habitat quality, predator satiation, or age (e.g., Murdoch 1969; Lawton et al. 1974; Murdoch and Sih 1978; Morse 1980; Sih 1980; Carter and Dixon 1982; Inoue and Matsura 1983; Luck 1984; Dicke 1988).

We are aware only of one study that emphasized genetic variability in prey specificity in relation to the speciation of predaceous insects, and this investigation, which compares a pair of closely related (hybridizable) neuropteran species, has particular relevance for sympatric speciation (Tauber and Tauber 1987b). One of the species, *Chrysopa quadripunctata*, is a general predator; it feeds on a variety of soft-bodied arthropods that occur on oak, maple, fruit trees, roses, etc. The other species, *Chrysopa slossonae*, feeds only on a single species of prey—the woolly alder aphid *Prociphilus tesselatus*—and only when the aphids are on their secondary host, alder (during late summer). Although the woolly alder aphid colonies are protected by ants, *C. slossonae* larvae circumvent these ants and attack the aphids by camouflaging themselves with the waxy secretions produced by the aphids (Eisner et al. 1978).

A number of genetically controlled traits differentiate the two species and adapt them to their different food resources. The specialist has much larger eggs than the general feeder, and its larval behavior is highly specialized. Although each species may take the prey of the other, development is prolonged when they feed on the unnatural prey. Unlike the generalist, which is multivoltine, the specialist has a univoltine life cycle that is synchronized with the peak occurrence of its prey. Furthermore, the specialist has specific prey requirements for reproduction, and mating appears to be associated with the specific prey. This pattern of interspecific variation suggests that reproductive isolation between the two species evolved mainly as a by-product of adaptation to different hosts.

Our field and laboratory studies show that the biology and behavior of these two predaceous species are more consistent with a model of sympatric speciation than with allopatric speciation. First, all of the traits known to differentiate the two species appear to be of adaptive significance in relation to the predators' feeding habits; thus, a shift in prey association appears to be the primary focus of diversification. Second, there is no evidence that the two species were geographically separated in the past. In fact, the pattern of host alternation by the prey of the specialized predator provides evidence to the contrary (see Tauber and Tauber 1987b). Aspects of Diehl and Bush's (this volume) model of host–race formation, Rosenzweig's (1978) model of competitive speciation, and Wilson and Turelli's (1986) model for the diversification of traits under frequency-dependent selection appear to be applicable to this case.

Habitat-associated predators

This category adds another dimension to the consideration of sympatric speciation because, unlike most of the previous examples, it involves specialization to the habitat, not to food. The important ingredients in this category appear to be strong selective pressure on adaptation to different habitats, phenological diversification, simple genetic control of key ecological traits, and perhaps sexual selection.

Our primary example involves the *Chrysoperla carnea* species complex that currently consists of two described species: *Chrysoperla carnea* and *Chrysoperla downesi* (Tauber and Tauber 1977a, 1977b, 1982, 1987a). In eastern North America these two species are reproductively isolated in nature but fully hybridizable in the laboratory. Many of the major differences between the species are related to differences in their habitats. One, *C. carnea,* occurs in fields and meadows during the summer and migrates to deciduous trees in autumn. The other species, *C. downesi,* finds its habitat in coniferous trees. Adults of each species are cryptically colored only in their own habitat: *C. carnea* is bright green in spring and summer and reddish brown in autumn and winter; *C. downesi* is dark green all year. The high degree of crypsis in the two species, year around, suggests strong selection by vertebrate predators in both the deciduous and evergreen habitats.

The larvae of both species prey on a variety of soft-bodied arthropods; the adults eat honeydew and pollen. *C. carnea,* which occurs in fields in which prey are abundant all summer, is multivoltine and has several generations per year. *C. downesi,* which inhabits coniferous trees in which prey appears to be more seasonal, is restricted to a single generation in the spring. In addition, males and females tend to overwinter in conspecific aggregations in which mating may occur before dispersal (Sheldon and MacLeod 1974). Thus, reproductive isolation is largely the result of differential habitat association and asynchronous breeding seasons. Extensive hybridization studies have demonstrated that pivotal aspects of each of these traits are based on relatively simple genetic systems (Tauber et al. 1977; Tauber and Tauber 1977a, 1977b, 1987a).

Diversification in the *C. carnea* species complex also involves changes in the unique, substrate-borne songs that males and females produce during courtship. The species-specific differences in courtship songs are polygenically based (Henry 1985a). In the laboratory, interspecific matings may sometimes be impeded by the differences in courtship songs (Henry 1985a, 1985b); however, our experience has been that males and females from the various species and geographic populations in the *C. carnea* species complex readily hybridize in the laboratory (Tauber and Tauber 1977a, 1977b, 1986). In the field, the songs may not function to attract the sexes to each other; rather they are produced after males and females have come into close range and have initiated courtship. Thus, it is not known to what degree the songs function as barriers to hybridization in the field. Specifically, it is unknown whether males and females of the different species will be attracted to each other in the field and initiate courtship to the point where songs are produced.

The *C. carnea* species complex of western North America and Europe exhibits considerable geographic variation in habitat association, phenology, and courtship songs (Tauber and Tauber 1982, 1986, 1987a; Henry 1985a, 1985b). In general, geographic variation for habitat association and phenology has a variety of expressions. Some areas contain one (allopatric) or more (sympatric) monomorphic populations; populations from other areas are polymorphic for the traits (Tauber and Tauber 1986). Geographic variation in courtship calls apparently involves only pure allopatric or sympatric populations of one or two song types—no polymorphisms (Henry 1985a, 1985b). Within North America, the variation in all the traits appears to be based on allelic substitutions at homologous loci, and the traits appear to vary independently of each other. Thus, a geographic mosaic of combinations occurs.

The diversity in the *C. carnea* species complex has generated two models of speciation. One is an allopatric model based on divergence in courtship songs; the other is a sympatric model based on habitat diversification and seasonal isolation. The allopatric model proposes that divergence in courtship songs is the "driving force" of speciation for the *C. carnea* species complex (Henry 1985a). This model implies that the courtship songs diversified through genetic drift and sexual selection in geographically isolated populations; the sympatric occurrence of populations with different songs is attributable to secondary invasion. Such a scenario predicts the occurrence of song-differentiation populations that do not have other ecologically based barriers to interbreeding (i.e., populations that occupy the same microhabitat and have diurnally and seasonally overlapping periods of mating). It also predicts heterospecific courtship in sympatric, song-differentiated populations (this courtship should proceed through singing—until the heterospecific partners recognize their mistake and terminate courtship).

Although this model of speciation via song divergence is untestable for the habitat-diversified populations of eastern North America, it could apply to some populations in western North America. But, to date, the evidence is sparse. The habitat associations and phenological traits of song-differentiated populations from the west have not been identified; nor has the degree of reproductive isolation or genetic divergence been established for these populations. Thus, although this type of speciation certainly is possible for the *C. carnea* species complex, the question remains as to whether or not it is likely to occur.

The other model proposed for the *C. carnea* species complex depicts habitat diversification as the primary impetus to speciation (Tauber and Tauber 1977a, 1977b). In this group of lacewings, cryptic adult coloration is highly correlated with habitat association, and disruptive selection on genetic variation in color could provide the initial step in diversification along habitat lines. This scenario involves the establishment of a stable polymorphism (underdominance) for habitat association. Subsequently, modifying genes that enhance adaptation to the habitat (e.g., genes that control phenology, habitat selection, and diurnal patterns of activity) are selected; these changes result in the development of reproductive isolation as a by-product of adaptation to the habitat. In this model, differentiation of courtship songs may occur through genetic drift and/or different regimes of sexual selection on the habitat-associated predators.

This model of speciation, based on habitat divergence, is readily applicable to populations in eastern North America and parts of northwestern North America in which clear, species-specific differences in habitat association are known. The model predicts the occurrence of populations that are polymorphic for traits underlying habitat association and phenology—a situation that occurs in many parts of western North America (Tauber and Tauber 1986). Whether habitat diversification has a role in the evolution of song-differentiated populations in the mountains of western North America is not known. This question requires analysis of their habitat association and phenology.

In summary, it is apparent that a variety of diverse factors are important in diversification and speciation of the *C. carnea* species complex. However, some populations clearly have combinations of traits that strongly suggest the occurrence of sympatric speciation. Primary among these are selective (habitat-associated) differences in adult coloration, seasonal barriers to interbreeding, and the simple genetic control of key aspects of habitat association and phenology.

CONCLUSION

Theoretically, a wide variety of biologically realistic conditions can promote the sympatric diversification of reproductively isolated specialists. These conditions may apply to a wide variety of insects. Given the high degree of specialization in many insect groups, it is timely to ask whether the vast diversity and large numbers of species in the Insecta are explainable, in part, by the phenomenon of sympatric speciation.

Whether or not biologists agree with this reasoning is not crucial. Furthermore, whether or not sympatric speciation occurs is not the fundamental issue. A much more significant and vital question is *how* does speciation occur—what *processes* are involved? If this is so, evolutionary biologists should examine the mechanisms involved in all modes of speciation. For example, investigations should ask: What is the genetic basis for ecologically significant differences among species? What are the broad-ranging pleiotropic effects of genetic changes in key traits that determine host or habitat association? What selective pressures act on the various traits? How does habitat selection affect gene flow in the field? How does partial seasonal asynchrony affect gene flow in the field? Under what circumstances can sexual selection lead to reproductive isolation of sympatric populations; is host or habitat differentiation a necessary prerequisite? How rigorous are the models that describe the various modes of speciation? How do the models stand up against natural situations?

Answers to these questions require the integration of theoretical, experimental, and comparative studies on a variety of specific taxa as they function in nature. Such a broad ranging approach will allow evolutionary biologists to apply equally rigorous standards to all proposed modes of speciation, and, in doing so, will concomitantly illuminate mechanisms responsible for sympatric and other modes speciation.

SUMMARY

Theoretical, comparative, and experimental studies indicate that sympatric speciation may be an important component in the evolutionary diversification of a large number of different insect groups. Recent genetic models, dealing with the role of frequency-dependent selection in heterogeneous environments, the diversification of polygenic and pleiotropic traits, and the evolution of reproductive isolation, suggest that the restrictions on sympatric speciation are much less severe than earlier models predicted. Case studies from a variety of insects with very different life-styles—host-associated herbivores, host-associated parasitoids, habitat-associated *Drosophila,* host-associated predators, and habitat-associated predators—indicate that insects exhibit variability in numerous behavioral, physiological, ecological, and genetic traits that could subserve sympatric speciation. These include mating in association with the host or after habitat selection, genes that simultaneously improve adaptation to the host or habitat and promote assortative mating, simple genetic control of variation in host or habitat fidelity and phenology, host- or habitat-induced responses that cause assortative mating as a pleiotropic effect of host or habitat selection, conditioning to hosts or habitats, intraspecific competition for resources, and sexual selection. Moreover, these traits occur within the context of insect life-styles that could promote sympatric speciation. We conclude that advances in the study of speciation require the focusing of three approaches (theoretical, experimental, and comparative) on a number of individual taxa, with emphasis on their variable traits, specific life-styles, and selective pressures in nature.

ACKNOWLEDGMENTS

We dedicate this chapter to Professor E. Gorton Linsley, Dean Emeritus, University of California, Berkeley.

We gratefully acknowledge L. E. Ehler (University of California, Davis) for hospitality during a sabbatical leave (M.J.T.). We thank E. G. Linsley (University of California, Berkeley), R. W. Thorp (University of California, Davis), T. K. Wood (University of Delaware), and R. W. Matthews (University of Georgia, Athens) for discussing their studies on speciation, and an anonymous reviewer for some constructive comments on a very early draft of the manuscript. We also thank D. J. Futuyma (State University of New York, Stony Brook), J. W. Neal, Jr. (USDA/ARS, Beltsville), and J. R. Ruberson (Cornell University) for bringing literature to our attention. We acknowledge the National Science Foundation for continuing support.

LITERATURE CITED

Alstad, D. N., and G. F. Edmunds, Jr. 1983. Adaptation, host specificity, and gene flow in the black pineleaf scale. Pp. 413–426 in: R. F. Denno and M. S. McClure (eds.), *Variable Plants and Herbivores in Natural and Managed Systems.* Academic Press, New York.

Arthur, A. P. 1981. Host acceptance by parasitoids. Pp. 97–120 in: D. A. Nordlund, R. L. Jones, and W. J. Lewis (eds.), *Semiochemicals.* John Wiley, New York.

Askew, R. R. 1968. Considerations on speciation in Chalcidoidea (Hymenoptera). Evolution 22:642–645.

Barton, N. H., and B. Charlesworth. 1984. Genetic revolutions, founder effects, and speciation. Annu. Rev. Ecol. System. 15:133–164.

Barton, N. H., and G. M. Hewitt. 1981. Hybrid zones and speciation. Pp. 109–145 in: W. R. Atchley and D. S. Woodruff (eds.), *Evolution and Speciation.* Cambridge University Press, Cambridge.

Barton, N. H., and G. M. Hewitt. 1985. Analysis of hybrid zones. Annu. Rev. Ecol. System. 16:113–148.

Beckage, N. E. 1985. Endocrine interactions between endoparasitic insects and their hosts. Annu. Rev. Entomol. 30:371–413.
Begon, M., and M. Mortimer. 1981. *Population Ecology.* Blackwell Scientific Publications, Oxford.
Bird, S. R., and R. Semeonoff. 1986. Selection for oviposition preference in *Drosophila melanogaster.* Genet. Res., Cambridge 48:151–160.
Blackman, R. L. 1979. Stability and variation in aphid clonal lineages. Biol. J. Linnean Soc. 11:259–277.
Blum, M. S., and N. A. Blum (eds.). 1979. *Sexual Selection and Reproductive Competition in Insects.* Academic Press, New York.
Boethel, D. J., and R. D. Eikenbary (eds.) 1986. *Interactions of Plant Resistance and Parasitoids and Predators of Insects.* Ellis Horwood, Ltd., Chichester.
Borden, J. H., J. R. Handley, J. A. McLean, R. M. Silverstein, L. Chong, K. N. Slessor, B. D. Johnston, and H. R. Schuler. 1980. Enantiomer-based specificity in pheromone communication by two sympatric *Gnathotrichus* species (Coleoptera: Scolytidae). J. Chem. Ecol. 6:445–456.
Bouletreau, M. 1986. The genetic and coevolutionary interactions between parasitoids and their hosts. Pp. 169–200 in: J. Waage and D. Greathead (eds.), *Insect Parasitoids.* Academic Press, London.
Bush, G. L. 1969. Sympatric host race formation and speciation in frugivorous flies of the genus *Rhagoletis* (Diptera, Tephritidae). Evolution 23:237–251.
Bush, G. L. 1974. The mechanism of sympatric host race formation in the true fruit flies. Pp. 3–23 in: M. J. D. White (ed.), *Genetic Mechanisms of Speciation in Insects.* D. Reidel, Boston.
Bush, G. L. 1975. Modes of animal speciation. Annu. Rev. Ecol. System. 6:339–364.
Bush, G. L. 1981. Stasipatric speciation and rapid evolution in animals. Pp. 201–218 in: W. R. Atchley and D. S. Woodruff (eds.), *Evolution and Speciation.* Cambridge University Press, Cambridge.
Bush, G. L., and S. R. Diehl. 1982. Host shifts, genetic models of sympatric speciation and the origin of parasitic insect species. Pp. 297–305 in: J. H. Visser and A. K. Minks (eds.), *Insect and Host Plant.* Pudoc, Wageningen.
Bush, G. L., and D. J. Howard. 1986. Allopatric and non-allopatric speciation; assumptions and evidence. Pp. 411–438 in: S. Karlin and E. Nevo (eds.), *Evolutionary Processes and Theory.* Academic Press, Orlando, FL.
Butlin, R. 1987a. Speciation by reinforcement. Trends Ecol. Evol. 2:8–13.
Butlin, R. 1987b. A new approach to sympatric speciation. Trends Ecol. Evol. 2:310–311.
Caltagirone, L. E. 1985. Identifying and discriminating among biotypes of parasites and predators. Pp. 189–200 in: M. A. Hoy and D. C. Herzog (eds.), *Biological Control in Agricultural IPM Systems.* Academic Press, Orlando, FL.
Cardé, R. T. 1986. The role of pheromones in reproductive isolation and speciation of insects. Pp. 303–317 in: M. D. Huettel (ed.), *Evolutionary Genetics of Invertebrate Behavior.* Plenum Press, New York.
Cardé, R. T., and T. C. Baker. 1984. Sexual communication and pheromones. Pp. 355–383 in: W. J. Bell and R. T. Cardé (eds.), *Chemical Ecology of Insects.* Sinauer Associates, Sunderland, MA.
Carson, H. L. 1971. The ecology of *Drosophila* breeding sites. Univ. Hawaii, Harold L. Lyon Arboretum Lect. 2:1–27.
Carson, H. L. 1974. Patterns of speciation in Hawaiian *Drosophila* inferred from ancient chromosomal patterns. Pp. 81–93 in: M. J. D. White (ed.), *Genetic Mechanisms of Speciation in Insects.* Australia and New Zealand Book Co, Sydney.
Carson, H. L. 1987a. The process whereby species originate. BioScience 37:715–720.
Carson, H. L. 1987b. The genetic system, the deme, and the origin of species. Annu. Rev. Genet. 21:405–423.
Carson, H. L. 1982. Evolution of *Drosophila* on the newer Hawaiian volcanoes. Heredity 48:3–25.
Carson, H. L., and A. R. Templeton. 1984. Genetic revolutions in relation to speciation phenomena: The founding of new populations. Annu. Rev. Ecol. System. 15:97–131.

Carter, M. C., and A. F. G. Dixon. 1982. Habitat quality and the foraging behaviour of coccinellid larvae. J. Anim. Ecol. 51:865–878.

Colwell, R. K. 1986a. Population structure and sexual selection for host fidelity in the speciation of hummingbird flower mites. Pp. 475–495 in: S. Karlin and E. Nevo (eds.), *Evolutionary Processes and Theory.* Academic Press, Orlando, FL.

Colwell, R. K. 1986b. Community biology and sexual selection: Lessons from hummingbird flower mites. Pp. 406–424 in: J. Diamond and T. J. Case, *Community Ecology.* Harper & Row, New York.

Cooley, S. S., R. J. Prokopy, P. T. McDonald, and T. T. Y. Wong. 1986. Learning in oviposition site selection by *Ceratitis capitata* flies. Entomol. Exp. Appl. 40:47–51.

Coppel, H. C., and J. W. Mertins. 1977. *Biological Insect Pest Suppression.* Springer-Verlag, Berlin.

Corbet, S. A. 1985. Insect chemosensory responses: A chemical legacy hypothesis. Ecol. Entomol. 10:143–153.

Coyne, J. A., and N. H. Barton. 1988. What do we know about speciation? Nature (London) 331:485–486.

Craddock, E. M. 1975. Reproductive relationships between homosequential species of Hawaiian *Drosophila.* Evolution 28:593–606.

Craighead, F. C. 1921. Hopkins host-selection principle as related to certain cerambycid beetles. J. Agricul. Res. 22:189–220.

Crankshaw, O. S., and R. W. Matthews. 1981. Sexual behavior among parasitic *Megarhyssa* wasps (Hymenoptera: Ichneumonidae). Behav. Ecol. Sociobiol. 9:1–7.

Damman, H., and P. Feeny. 1988. Mechanisms and consequences of selective oviposition by the zebra swallowtail butterfly. Anim. Behav. 36:563–573.

Darwin, C. 1959. *The Origin of Species* (6th ed., 1872). Oxford University Press, London.

Davis, Jr., G. F. 1982. Historical and ecological factors in the evolution, adaptive radiation, and biogeography of freshwater mollusks. Am. Zool. 22:375–395.

DeBach, P. 1974. *Biological Control by Natural Enemies.* Cambridge University Press, Cambridge.

Dicke, M. 1988. Prey preference of the phytoseiid mite *Typhlodromus pyri.* 1. Response to volatile kairomones. Exp. Appl. Acarol. 4:1–13.

Doutt, R. L. 1960. Natural enemies and insect speciation. Pan-Pacific Entomol. 36:1–14.

Doutt, R. L., D. P. Annecke, and E. Tremblay. 1976. Biology and host relationships of parasitoids. Pp. 143–168 in: C. B. Huffaker and P. S. Messenger (eds.), *Theory and Practice of Biological Control.* Academic Press, New York.

Eberhard, W. G. 1985. *Sexual Selection and Animal Genitalia.* Harvard University Press, Cambridge.

Echelle, A. A., and I. Kornfield (eds.). 1984. *Evolution of Fish Species Flocks.* University of Maine at Orono Press.

Ehrman, L., and P. A. Parsons. 1976. *The Genetics of Behavior.* Sinauer Associates, Sunderland, MA.

Ehrman, L., and M. Wasserman. 1987. The significance of asymmetrical sexual isolation. Pp. 1–20 in: M. K. Hecht, B. Wallace, and G. T. Prance (eds.), *Evolutionary Biology,* Vol. 12. Plenum Press, New York.

Ehrman, L., N. L. Somerson, and F. J. Gottlieb. 1987. Reproductive isolation in a Neotropical insect: Behavior and microbiology. Pp. 97–108 in M. D. Huettel (ed.), *Evolutionary Genetics of Invertebrate Behavior.* Plenum Press, New York.

Eisner, T., K. Hicks, M. Eisner, and D. S. Robson. 1978. "Wolf-in-sheep's clothing" strategy of a predaceous insect larva. Science 199:790–794.

Endler, J. A. 1977. *Geographic Variation, Speciation, and Clines.* Princeton University Press, Princeton.

Endler, J. A. 1982a. Alternative hypotheses in biogeography (Symposium). Am. Zool. 22:347–471.

Endler, J. A. 1982b. Problems in distinguishing historical from ecological factors in biogeography. Am. Zool. 22:441–452.

Evans, E. W. 1982. Timing of reproduction by predatory stinkbugs (Hemiptera: Pentatomidae): Patterns and consequences for a generalist and a specialist. Ecology 63:147–158.

Feder, J. L., C. A. Chilcote, and G. L. Bush. 1988. Genetic differentiation between sympatric host races of the apple maggot fly *Rhagoletis pomonella*. Nature 336:61–64.

Fellows, D. P., and W. B. Heed. 1972. Factors affecting host plant selection in desert-adapted cactiphilic *Drosophila*. Ecology 53:850–858.

Felsenstein, J. 1981. Skepticism towards Santa Rosalia, or why are there so few kinds of animals? Evolution 35:124–138.

Fialkowski, K. R. 1988. Lottery of sympatric speciation—a computer model. J. Theoret. Biol. 130: 379–390.

Futuyma, D. J. 1986. The role of behavior in host-associated divergence in herbivorous insects. Pp. 295–302 in: M. D. Huettel (ed.), *Evolutionary Genetics of Invertebrate Behavior*. Plenum Press, New York.

Futuyma, D. J., and G. C. Mayer. 1980. Non-allopatric speciation in animals. System. Zool. 29:254–271.

Futuyma, D. J., and S. C. Peterson. 1985. Genetic variation in the use of resources by insects. Annu. Rev. Entomol. 30:217–238.

Futuyma, D. J., and S. S. Wasserman. 1981. Food plant specialization and feeding efficiency in the tent caterpillars *Malacosoma disstria* and *M. americanum*. Entomol. Exp. Appl. 30:106–110.

Gibbons, J. R. H. 1979. A model for sympatric speciation in *Megarhyssa* (Hymenoptera: Ichneumonidae): Competitive speciation. Am. Natur. 114:719–741.

Ginzburg, L. R., H. R. Akcakaya, and J. Kim. 1988. Evolution of community structure: Competition. J. Theor. Biol. 133:513–523.

Gittenberger, E. 1988. Sympatric speciation in snails: A largely neglected model. Evolution 42: 826–828.

Glazier, D. S. 1987. Toward a predictive theory of speciation: The ecology of isolate selection. J. Theoret. Biol. 126:323–333.

Goeden, R. D. 1987. Life history of *Trupanea conjuncta* (Adams) on *Trixus californica* Kellog in southern California (Diptera: Tephritidae). Pan-Pacific Entomol. 63:284–291.

Goeden, R. D. 1988. Gall formation by the capitulum-infesting fruit fly, *Tephritis stigmatica* (Diptera: Tephritidae). Proc. Entomol. Soc. Wash. 90:37–43.

Gordh, G. 1977. Biosystematics of natural enemies. Pp. 125–148 in: R. L. Ridgway and S. B. Vinson (eds.), *Biological Control by Augmentation of Natural Enemies*. Plenum Press, New York.

Gordh, G., and P. DeBach. 1978. Courtship behavior in the *Aphytis lingnanenis* group, its potential usefulness in taxonomy, and a review of sexual behavior in the parasitic Hymenoptera (Chalcidoidea: Aphelinidae). Hilgardia 46:37–75.

Grant, B. R., and P. R. Grant. 1979. Darwin's finches: Population variation and sympatric speciation. Proc. Natl. Acad. Sci. U.S.A. 76:2359–2363.

Greany, P. D., and K. S. Hagen. 1981. Prey selection. Pp. 121–135 in: D. A. Nordlund, R. L. Jones, and W. J. Lewis (eds.), *Semiochemicals*. John Wiley, New York.

Guttman, S. I., T. K. Wood, and A. A. Karlin. 1981. Genetic differentiation along host plant lines in the sympatric *Enchenopa binotata* Say complex (Homoptera: Membracidae). Evolution 35:205–217.

Hagen, K. S., S. Bombosch, and J. A. McMurtry. 1976. The biology and impact of predators. Pp. 93–142 in: C. B. Huffaker and P. S. Messenger (eds.), *Theory and Practice of Biological Control*. Academic Press, New York.

Halliburton, R., and G. A. E. Gall. 1981. Disruptive selection and assortative mating in *Tribolium castaneum*. Evolution 35: 829–843.

Hardy, D. E. 1974. Introduction and background information. Pp. 71–80 in M. J. D. White (ed.), *Genetic Mechanisms of Speciation in Insects*. Australia and New Zealand Book Co., Sydney.

Harrison, R. G. 1986. Pattern and process in a narrow hybrid zone. Heredity 56:337–349.

Heatwole, H., and D. M. Davis. 1965. Ecology of three sympatric species of parasitic insects of the genus *Megarhyssa* (Hymenoptera: Ichneumonidae). Ecology 46:140–150.

Heatwole, H., D. M. Davis, and A. M. Wenner. 1963. The behaviour of *Megarhyssa*, a genus of parasitic Hymenopterans (Ichneumonidae: Ephialtinae). Z. Tierpsychol. 19:652–664.

Heatwole, H., D. M. Davis, and A. M. Wenner. 1964. Detection of mates and hosts by parasitic insects of the genus *Megarhyssa* (Hymenoptera: Ichneumonidae). Am. Midland Natur. 71:374–381.

Hedrick, P. W., and E. J. Louis. 1984. Speciation: A population genetics perpective. Pp. 251–262 in: B. C. Joshi, R. P. Sharma, and H. C. Bansal (eds.), *Proceedings of the XV International Congress of Genetics*, Vol. 4. Oxford & IBH Publ. Co., New Delhi.

Heed, W. B. 1968. Ecology of the Hawaiian Drosophilidae. Pp. 387–419 in: M. R. Wheeler (ed.), *Studies in Genetics. IV. Research Reports.* University of Texas Publication 6818, Austin.

Heed, W. B. 1971. Host plant specificity and speciation in Hawaiian *Drosophila.* Taxon 20:115–121.

Henry, C. S. 1985a. Sibling species, call differences, and speciation in green lacewings (Neuroptera: Chrysopidae: *Chrysoperla*). Evolution 39:965–984.

Henry, C. S. 1985b. The proliferation of cryptic species in *Chrysoperla* green lacewings through song divergence. Florida Entomol. 68:18–38.

Henry, T. J., J. W. Neal, Jr., and K. M. Gott. 1986. *Stethoconus japonicus* (Heteroptera: Miridae): A predator of *Stephanitis* lace bugs newly discovered in the United States, promising in the biocontrol of azalea lace bug (Heteroptera: Tingidae). Proc. Entomol. Soc. Washington 88:722–730.

Hodek, I. (ed.) 1966. *Ecology of Aphidophagous Insects.* Junk N. V., The Hague.

Hodek, I. 1973. *Biology of Coccinellidae.* Junk N. V., The Hague.

Hoekstra, R. F., R. Bijlsma, and A. J. Dolman. 1985. Polymorphism for environmental heterogeneity: Models are only robust if the heterozygote is close in fitness to the favoured homozygote in each environment. Genet. Res., Cambridge 45:299–314.

Hoffmann, A. A. 1988. Early adult experience in *Drosophila melanogaster.* J. Insect Physiol. 34:197–204.

Huettel, M. D. (ed.) 1986. *Evolutionary Genetics of Invertebrate Behavior.* Plenum Press, New York.

Inoue, H. 1982. Species-specific calling sounds as a reproductive isolating mechanism in *Nephotettix* spp. (Hemiptera: Cicadellidae). Appl. Entomol. Zool. 17:253–262.

Inoue, T., and T. Matsura. 1983. Foraging strategy of a mantid, *Paratenodera angustipennis* S.: Mechanisms of switching tactics between ambush and active search. Oecologia 56:264–271.

Jaenike, J. 1982. Environmental modification of oviposition behavior in *Drosophila.* Am. Natur. 119:784–802.

Jaenike, J. 1985. Parasite pressure and the evolution of amanitin tolerance in *Drosophila.* Evolution 39:1295–1301.

Jaenike, J. 1986a. Genetic complexity of host-selection behavior in *Drosophila.* Proc. Natl. Acad. Sci. U.S.A. 83:2148–2151.

Jaenike, J. 1986b. Intraspecific variation for resource use in *Drosophila.* Biol. J. Linnean Soc. 27:47–56.

Jaenike, J. 1988. Effects of early adult experience on host selection in insects: Some experimental and theoretical results. J. Insect Behav. 1:3–15.

Jaenike, J., and D. Grimaldi. 1983. Genetic variation for host preference within and among populations of *Drosophila tripunctata.* Evolution 37:1023–1033.

James, A. C., J. Jakubczak, M. P. Riley, and J. Jaenike. 1988. On the causes of monophagy in *Drosophila quinaria.* Evolution 42:626–630.

Kaneshiro, K. Y. 1976. Ethological isolation and phylogeny in the *plantibia* subgroup of Hawaiian *Drosophila.* Evolution 30:740–745.

Kaneshiro, K. Y., and C. R. B. Boake. 1987. Sexual selection and speciation: Issues raised by Hawaiian drosophilids. Trends Ecol. Evol. 2:207–212.

Kaneshiro, K. Y., H. L. Carson, F. E. Clayton, and W. B. Heed. 1973. Niche separation in a pair of homosequential *Drosophila* species from the island of Hawaii. Am. Natur. 107:766–774.

Katakura, H. 1981. Classification and evolution of the phytophagous ladybirds belonging to *Henosepilachna vigintioctomacultata* complex (Coleoptera, Coccinellidae). J. Faculty Sci., Hokkaido Univ., Ser. 6, Zool. 22:301–378.

Khasimuddin, S., and P. DeBach. 1976. Biosystematic and evolutionary statuses of two sympatric populations of *Aphytis mytilaspidis* (Hym.: Aphelinidae). Entomophaga 21:113–122.

Kircher, H. W., and W. B. Heed. 1970. Phytochemistry and host plant specificity in *Drosophila*. Pp. 191–209 in: C. Steelink and V. C. Runeckles (eds.), *Recent Advances in Phytochemistry*, Vol. 3. Appleton-Century-Crofts, New York.

Kirkpatrick, M. 1982. Sexual selection and the evolution of female choice. Evolution 36:1–12.

Kogan, M., and E. F. Legner. 1970. A biosystematic revision of the genus *Muscidifurax* (Hymenoptera: Pteromalidae) with descriptions of four new species. Can. Entomol. 102:1268–2190.

Kondrashov, A. S. 1983a. Multilocus model of sympatric speciation. I. One character. Theoret. Pop. Biol. 24:121–135.

Kondrashov, A. S. 1983b. Multilocus model of sympatric speciation. II. Two characters. Theoret. Pop. Biol. 24:136–144.

Kondrashov, A. S. 1986. Multilocus model of sympatric speciation. III. Computer simulations. Theoret. Pop. Biol. 29:1–15.

Kondrashov, A. S., and M. V. Mina. 1986 Sympatric speciation: When is it possible? Biol. J. Linnean Soc. 27:201–223.

Lande, R. 1981. Models of speciation by sexual selection on polygenic traits. Proc. Natl. Acad. Sci. U.S.A. 78:3721–3725.

Lande, R. 1982. Rapid origin of sexual isolation and character divergence in a cline. Evolution 36:213–223.

Lanier, G. N., and W. E. Burkholder. 1974. Pheromones in speciation of Coleoptera. Pp. 161–189 in: M. C. Birch (ed.), *Pheromones*. Elsevier, New York.

Lawton, J. H. 1986. The effects of parasitoids on phytophagous insect communities. Pp. 265–287 in: J. Waage and D. Greathead (eds.), *Insect Parasitoids*. Academic Press, London.

Lawton, J. H., J. R. Beddington, and R. Bonser. 1974. Switching in invertebrate predators. Pp. 141–158 in: M. B. Usher and M. H. Williamson (eds.), *Ecological Stability*. Chapman & Hall, London.

Legner, E. F. 1969. Reproductive isolation and size variation in the *Muscidifurax raptor* complex. Ann. Entomol. Soc. Am. 62:382–385.

Legner, E. F. 1987. Transfer of thelytoky to arrhenotokous *Muscidifurax raptor* Girault and Sanders (Hymenoptera: Pteromalidae). Can. Entomol. 119:265–271.

Levene, H. 1953. Genetic equilibrium when more than one ecological niche is available. Am. Natur. 87:331–333.

Levinton, J. 1988. *Genetics, Paleontology, and Macroevolution*. Cambridge University Press, Cambridge.

Lewis, W. J., and J. H. Tumlinson. 1988. Host detection by chemically mediated associative learning in a parasitic wasp. Nature (London) 331:257–259.

Linsley, E. G. 1959. Ecology of Cerambycidae. Annu. Rev. Entomol. 4:99–138.

Linsley, E. G. 1961. The Cerambycidae of North America. Part I. Introduction. Univ. Calif. Pub. Entomol. 18:1–135.

Linsley, E. G., and J. A. Chemsak. 1984. The Cerambycidae of North America, Part VII, No. 1: Taxonomy and classification of the subfamily Lamiinae, Tribes Parmenini through Acanthoderini. Univ. Calif. Pub. Entomol. 102:1–258.

Linsley, E. G., and MacSwain, J. W. 1958. The significance of floral constancy among bees of the genus *Diadasia* (Hymenoptera, Anthophoridae). Evolution 12:219–223.

Littlejohn, M. J., and G. F. Watson. 1985. Hybrid zones and homogamy in Australian frogs. Annu. Rev. Ecol. System. 16:85–112.

Loehle, C. 1987. Hypothesis testing in ecology: Psychological aspects and the importance of theory maturation. Quart. Rev. Biol. 62:397–409.

Luck, R. F. 1984. Principles of arthropod predation. Pp. 497–529 in: C. B. Huffaker and R. L. Rabb (eds.), *Ecological Entomology*. John Wiley, New York.

Mani, G. S. 1981. Conditions for balanced polymorphism in the presence of differential delay in developmental time. Theoret. Pop. Biol. 20:363–393.

Markow, T. A. 1981. Mating preference is not predictive of the direction of evolution in experimental populations of *Drosophila*. Science 213:1405–1407.

Masaki, S., and T. J. Walker. 1987. Cricket life cycles. Pp. 349–423 in: M. K. Hecht, B. Wallace, and G. T. Prance (eds.), *Evolutionary Biology*, Vol. 12. Plenum Press, New York.

Matthews, R. W. 1975. Courtship in parasitic wasps. Pp. 66–86 in: P. W. Price (ed.), *Evolutionary Strategies of Parasitic Insects and Mites.* Plenum Press, New York.

Matthews, R. W., and J. R. Matthews. 1978. *Insect Behavior.* John Wiley, New York.

Matthews, R. W., J. R. Matthews, and O. Crankshaw. 1979. Aggregation in male parasitic wasps of the genus *Megarhyssa:* I. Sexual discrimination, tergal stroking behavior, and description of associated anal structures. Florida Entomol. 62:3–8.

Maynard Smith, J. 1966. Sympatric speciation. Am. Natur. 100:637–650.

Maynard Smith, J. 1970. Genetic polymorphism in a varied environment. Am. Natur. 104:487–490.

Maynard Smith, J., and R. Hoekstra. 1980. Polymorphism in a varied environment: How robust are the models? Genet. Res., Cambridge 35:45–57.

Mayr, E. 1963. *Animal Species and Evolution.* Belknap Press of Harvard University Press, Cambridge, MA.

Mayr, E. 1970. *Populations, Species, and Evolution.* Belknap Press, Cambridge, MA.

Mayr, E. 1976. Sympatric speciation. Pp. 144–175 in: E. Mayr, *Evolution and the Diversity of Life, Selected Essays.* Belknap Press, Cambridge.

Mayr, E. 1982. Adaptation and selection. Biologisches Zentrablatt 101:161–174.

McClure, M. S., and P. W. Price. 1976. Ecotype characteristics of coexisting *Erythroneura* leafhoppers (Homoptera: Cicadellidae) on sycamore. Ecology 57:928–940.

McKaye, K. R. 1980. Seasonality in habitat selection by the gold color morph of *Cichlasoma citrinellum* and its relevance to sympatric speciation in the family Cichlidae. Environ. Biol. Fish 5:75–78.

McPheron, B. A., D. C. Smith, and S. H. Berlocher. 1988. Genetic differences between host races of *Rhagoletis pomonella.* Nature 336:64–66.

Messenger, P. S., F. Wilson, and M. J. Whitten. 1976. Variation, fitness, and adaptability of natural enemies. Pp. 209–231 in: C. B. Huffaker and P. S. Messenger (eds.), *Theory and Practice of Biological Control.* Academic Press, New York.

Meyer, A. 1987. Phenotypic plasticity and heterochrony in *Cichlasoma managuense* (Pisces, Cichlidae) and their implications for speciation in cichlid fishes. Evolution 41:1357–1369.

Meyer, A. 1989. Trophic polymorphisms in cichlid fishes: Do they represent intermediate steps during sympatric speciation and explain their rapid adaptive radiation? In press in J.-H. Schroder (ed.), *New Trends in Ichthyology,* Paul Parey, Berlin.

Morse, D. H. 1980. *Behavioral Mechanisms in Ecology.* Harvard University Press, Cambridge, MA.

Mueller, L. D. 1985. The evolutionary ecology of *Drosophila.* Pp. 37–98 in: M. K. Hecht, B. Wallace, and G. T. Prance (eds.), *Evolutionary Biology,* Vol. 9. Plenum Press, New York.

Müller, F. P. 1985. Biotype formation and sympatric speciation in aphids (Homoptera: Aphidinea). Entomol. Gen. 10:161–181.

Murakami, Y. 1988. Ecotypes of *Torymus (Syntopmaspis) beneficus* Yasumatsu et Kamijo (Hymenoptera: Torymidae) with different seasonal prevalences of adult emergence. Appl. Entomol. Zool. 23:81–87.

Murdoch, W. W. 1969. Switching in general predators: Experiments on predator specificity and stability of prey populations. Ecol. Monog. 39:335–354.

Murdoch, W. W., and A. Sih. 1978. Age-dependent interference in a predatory insect. J. Anim. Ecol. 47:581–592.

Nei, M., T. Maruyama, and C.-I. Wu. 1983. Modes of evolution of reproductive isolation. Genetics 103:557–579.

Papaj, D. R. 1986. Interpopulation differences in host preference and the evolution of learning in the butterfly *Battus philenor.* Evolution 40:518–530.

Parsons, P. A. 1978. Habitat selection and evolutionary strategies in *Drosophila:* An invited address. Behav. Genet. 8:511–526.

Parsons, P. A. 1981a. Habitat selection and speciation in *Drosophila.* Pp. 219–240 in: W. R. Atchley and D. S. Woodruff (eds.), *Evolution and Speciation.* Cambridge University Press, Cambridge.

Parsons, P. A. 1981b. Sympatric speciation in *Drosophila?*: Ethanol threshold metrics and habitat subdivision. Am. Natur. 117:1023–1026.
Parsons, P. A. 1983. Ecobehavioral genetics: Habitats and colonists. Annu. Rev. Ecol. System. 14:35–55.
Paterson, H. E. H. 1981. The continuing search for the unknown and unknowable: A critique of contemporary ideas on speciation. African J. Sci. 77:113–119.
Paterson, H. E. H. 1985. The recognition concept of species. Pp. 21–29 in: E. S. Vrba (ed.), *Species and Speciation.* Transvaal Museum Monograph No. 4, Transvaal Museum, Pretoria.
Phelan, P. L., and T. C. Baker. 1987. Evolution of male pheromones in moths: Reproductive isolation through sexual selection? Science 235:205–207.
Pimm, S. L. 1979. Sympatric speciation: A simulation model. Biol. J. Linnean Soc. 11:131–139.
Price, P. W. 1980. *Evolutionary Biology of Parasites.* Princeton University Press, Princeton.
Price, P. W. 1981. Semiochemicals in evolutionary time. Pp. 251–279 in: D. A. Nordlund, R. L. Jones, and W. J. Lewis (eds.), *Semiochemicals.* Wiley, New York.
Price, P. W. 1986. Ecological aspects of host plant resistance and biological control: Interactions among three trophic levels. Pp. 11–30 in: D. J. Boethel and R. D. Eikenbary (eds.), *Interactions of Plant Resistance and Parasitoids and Predators of Insects.* Ellis Horwood, Ltd., Chichester.
Prokopy, R. J., A. L. Averill, S. S. Cooley, and C. A. Roitberg. 1982. Associative learning in egglaying site selection by apple maggot flies. Science 218:76–77.
Prokopy, R. J., and D. R. Papaj. 1988. Learning of apple fruit biotypes by apple maggot flies. J. Insect Behav. 1:67–74.
Ramírez, B. W. 1970. Host specificity of fig wasp (Agaonidae). Evolution 24:680–691.
Rao, S. V., and P. DeBach. 1969. Experimental studies on hybridization and sexual isolation between some *Aphytis* species (Hymenoptera: Aphelinidae). III. The significance of reproductive isolation between interspecific hybrids and parental species. Evolution 23:525–533.
Rausher, M. D. 1983. Conditioning and genetic variation as causes of individual variation in oviposition behavior of the tortoise beetle, *Deloyala guttata.* Anim. Behav. 31:743–747.
Rausher, M. D. 1984a. The evolution of habitat preference in subdivided populations. Evolution 38:596–608.
Rausher, M. D. 1984b. Tradeoffs in performance on different hosts: Evidence from within- and between-site variation in the beetle *Deloyala guttata.* Evolution 38:582–595.
Rausher, M. D., and D. J. Papaj. 1983. Demographic consequences of discrimination among conspecific host plants by *Battus philenor* butterflies. Ecology 64:1402–1410.
Rice, W. R. 1984. Disruptive selection on habitat preference and the evolution of reproductive isolation: A simulation study. Evolution 38:1251–1260.
Rice, W. R. 1985. Disruptive selection on habitat preference and the evolution of reproductive isolation: An exploratory experiment. Evolution 39:645–656.
Rice, W. R. 1987. Speciation via habitat specialization: The evolution of reproductive isolation as correlated character. Evol. Ecol. 1:301–314.
Richardson, R. H. 1974. Effects of dispersal, habitat selection and competition on a speciation pattern of *Drosophila* endemic to Hawaii. Pp. 140–164 in: M. J. D. White (ed.), *Genetic Mechanisms of Speciation in Insects.* Australia and New Zealand Book Co., Sydney.
Richardson, R. H. 1982. Phyletic species packing and the formation of sibling (crptic) species clusters. Pp. 107–123 in: J. S. F. Barker and W. T. Starmer (eds.), *Ecological Genetics and Evolution.* Academic Press, Sydney.
Roelofs, W. L., and A. Comeau. 1969. Sex pheromone specificity: Taxonomic and evolutionary aspects in Lepidoptera. Science 165:398–400.
Roitberg, B. D., and R. J. Prokopy. 1981. Experience required for pheromone recognition by the apple maggot fly. Nature (London) 292:540–541.
Rosenzweig, M. L. 1978. Competitive speciation. Biol. J. Linnean Soc. 10:275–289.
Rosenzweig, M. L., and J. A. Taylor. 1980. Speciation and diversity in Ordovician invertebrates: Filling niches quickly and carefully. Oikos 35:236–243.
Ross, H. H. 1957. Principles of natural coexistence indicated by leafhopper populations. Evolution 11:113–129.

Ruberson, J. R., M. J. Tauber, and C. A. Tauber. 1989. Intraspecific variability in hymenopteran parasitoids: Comparative studies of two biotypes of the egg parasitoid *Edovum puttleri* (Hymenoptera: Eulophidae). J. Entomol. Soc. Kansas, in press.

Sawyer, S., and D. Hartl. 1981. On the evolution of behavioral reproductive isolation: The Wallace effect. Theoret. Pop. Biol. 19:261–273.

Scudder, G. G. E. 1974. Species concepts and speciation. Can. J. Zool. 52:1121–1134.

Seger, J. 1985. Intraspecific resource competition as a cause of sympatric speciation. Pp. 43–53 in: P. J. Greenwood, P. H. Harvey, and M. Slatkin (eds.), *Evolution*. Cambridge University Press, Cambridge.

Seitz, A., and M. Komma. 1984. Genetic polymorphism and its ecological background in Tephritid populations (Diptera: Tephritidae). Pp. 141–158 in: K. Wöhrmann and V. Loeschcke (eds.), *Population Biology and Evolution*. Springer-Verlag, Berlin.

Sheldon, J. K., and E. G. MacLeod. 1974. Studies on the biology of the Chrysopidae IV. A field and laboratory study of the seasonal cycle of *Chrysopa carnea* Stephens in central Illinois (Neuroptera: Chrysopidae). Trans. Am. Entomol. Soc. 100:437–512.

Sih, A. 1980. Optimal behavior: Can foragers balance two conflicting demands? Science 210:1041–1043.

Slatkin, M. 1982. Pleitrophy and parapatric speciation. Evolution 36:263–270.

Slobodchikoff, C. N. 1973. Behavioral studies of three morphotypes of *Therion circumflexum* (Hymenoptera: Ichneumonidae). Pan-Pacific Entomol. 49:197–206.

Smiley, J. 1978. Plant chemistry and the evolution of host specificity: New evidence from *Heliconius* and *Passiflora*. Science 201:745–747.

Smith, D. C. 1988. Heritable divergence of *Rhagoletis pomonella* host races by seasonal asynchrony. Nature 336:66–67.

Smith-Gill, S. J. 1983. Developmental plasticity: Developmental conversion *versus* phenotypic modulation. Amer. Zool. 23:47–55.

Sokolowski, M. B., S. J. Bauer, V. Wai-ping, L. Rodriguez, J. L. Wong, and C. Kent. 1986. Ecological genetics and behavior of *Drosophila melanogaster* larvae in nature. Anim. Behav. 34:403–408.

Spencer, H. G., B. H. McArdle, and D. M. Lambert. 1986. A theoretical investigation of speciation by reinforcement. Am. Natur. 128:241–262.

Spiess, E. B., and C. M. Wilke. 1984. Still another attempt to achieve assortative mating by disruptive selection in *Drosophila*. Evolution 38:505–515.

Spieth, H. T. 1974. Mating behavior and evolution of the Hawaiian *Drosophila*. Pp. 94–101 in: M. J. D. White (ed.), *Genetic Mechanisms of Speciation in Insects*. Australia and New Zealand Book Co., Sydney.

Stanton, M. L. 1984. Short-term learning and the searching accuracy of egg-laying butterflies. Anim. Behav. 32:33–40.

Strand, M. R. 1986. The physiological interactions of parasitoids with their hosts and their influence on reproductive strategies. Pp. 97–136 in: J. Waage and D. Greathead (eds.), *Insect Parasitoids*. Academic Press, London.

Strong, D. R., J. H. Lawton, and T. R. E. Southwood. 1984. *Insects on Plants*. Harvard University Press, Cambridge, MA.

Sturgeon, K. B., and J. B. Mitton. 1986. Allozyme and morphological differentiation of mountain pine beetles *Dendroctonus ponderosae* Hopkins (Coleoptera: Scolytidae) associated with host tree. Evolution 40:290–302.

Tauber, C. A., and M. J. Tauber. 1977a. A genetic model for sympatric speciation through habitat diversification and seasonal isolation. Nature (London) 268:702–705.

Tauber, C. A., and M. J. Tauber. 1977b. Sympatric speciation based on allelic changes at three loci: Evidence from natural populations in two habitats. Science 197:1298–1299.

Tauber, C. A., and M. J. Tauber. 1982. Evolution of seasonal adaptations and life history traits in *Chrysopa*: Response to diverse selective pressures. Pp. 51–72 in: H. Dingle and J. P. Hegmann (eds.), *Evolution and Genetics of Life Histories*. Springer-Verlag, New York.

Tauber, C. A., and M. J. Tauber. 1986. Ecophysiological responses in life-history evolution: Evidence for their importance in a geographically widespread insect species complex. Can. J. Zool. 64:875–884.

Tauber, C. A., and M. J. Tauber. 1987a. Inheritance of seasonal cycles in *Chrysoperla* (Insecta: Neuroptera). Genet. Res., Cambridge 49:215–223.

Tauber, C. A., and M. J. Tauber. 1987b. Food specificity in predacious insects: A comparative ecophysiological and genetic study. Evol. Ecol. 1:175–186.

Tauber, C. A., M. J. Tauber, and J. R. Nechols. 1977. Two genes control seasonal isolation in sibling species. Science 197:592–593.

Tauber, M. J., C. A. Tauber, J. R. Nechols, and J. J. Obrycki. 1983. Seasonal activity of parasitoids: Control by external, internal, and genetic factors. Pp. 87–108 in: V. K. Brown and I. Hodek (eds.), *Diapause and Life Cycle Strategies in Insects*. Junk, N. V. The Hague.

Tauber, M. J., C. A. Tauber, and S. Masaki. 1986. *Seasonal Adaptations of Insects*. Oxford University Press, New York.

Templeton, A. R. 1980. Modes of speciation and inferences based on genetic distances. Evolution 34:719–729.

Thoday, J. M., and J. B. Gibson. 1962. Isolation by disruptive selection. Nature (London) 193:1164–1166.

Thoday, J. M., and J. B. Gibson. 1970. The probability of isolation by disruptive selection. Am. Natur. 104:219–230.

Thompson, J. N. 1987. Symbiont-induced speciation. Biol. J. Linnean Soc. 32:385–393.

Thompson, J. N. 1988. Evolutionary ecology of the relationship between oviposition preference and performance of offspring in phytophagous insects. Entomol. Exp. Appl. 47:3–14.

Thompson, W. R. 1951. The specificity of host relations in predacious insects. Can. Entomol. 83:262–269.

Thornhill, R. 1976. Sexual selection and paternal investment in insects. Am. Natur. 110:153–163.

Thornhill, R. 1979. Adaptive female-mimicking behavior in a scorpionfly. Science 205:412–414.

Thornhill, R., and J. Alcock. 1983. *The Evolution of Insect Mating Systems*. Harvard University Press, Cambridge.

Thorp, R. W. 1969. Systematics and ecology of bees of the subgenus *Diandrena* (Hymenoptera: Andrenidae). Univ. Calif. Pub. Entomol. 52:1–146.

Thorpe, W. H. 1945. The evolutionary significance of habitat selection. J. Anim. Ecol. 14:67–70.

Traynier, R. M. M. 1986. Visual learning in assays of sinigrin solution as an oviposition releaser for the cabbage butterfly, *Pieris rapae*. Entomol. Exp. Appl. 40:25–33.

Udovic, D. 1980. Frequency-dependent selection, disruptive selection, and the evolution of reproductive isolation. Am. Natur. 116:621–641.

van Alphen, J. J. M., and L. E. M. Vet. 1986. An evolutionary approach to host finding and selection. Pp. 23–61 in: J. Waage and D. Greathead (eds.), *Insect Parasitoids*. Academic Press, London.

van den Assem, J. 1986. Mating behaviour in parasitic wasps. Pp. 137–167 in: J. Waage and D. Greathead (eds.), *Insect Parasitoids*. Academic Press, London.

van den Assem, J., and G. D. E. Povel. 1973. Courtship behaviour of some *Muscidifurax* species (Hym., Pteromalidae): A possible example of a recently evolved ethological isolating mechanism. Neth. J. Zool. 23:465–487.

van den Assem, J., M. J. Gijswijt, and B. K. Nübel. 1980. Observations on courtship and mating strategies in a few species of parasitic wasps (Chalcidoidea). Neth. J. Zool. 30:208–227.

van den Bosch, R., and P. S. Messenger. 1973. *Biological Control*. Intext Educational Publishers, New York.

van den Bosch, R., R. Hom, P. Matteson, B. D. Frazer, P. S. Messenger, and C. S. Davis. 1979. Biological control of the walnut aphid in California: Impact of the parasite, *Trioxys pallidus*. Hilgardia 47:1–13.

van Lenteren, J. C. 1981. Host discrimination by parasitoids. Pp. 153–179 in: D. A. Nordlund, R. L. Jones, and W. J. Lewis (eds.), *Semiochemicals*. John Wiley, New York.

Vet, L. E. M., and C. J. Janse. 1984. Fitness of two sibling species of *Asobara* (Braconidae: Alysiinae), larval parasitoids of Drosophilidae in different microhabitats. Ecol. Entomol. 9:345–354.

Vet, L. E. M. and K. van Opzeeland. 1984. The influence of conditioning on olfactory microhabitat and host location in *Asobara tabida* (Nees) and *A. rufescens* (Foerster) (Braconidae: Alysiinae) larval parasitoids of Drosophilidae. Oecologia 63:171–177.

Via, S. 1986. Genetic covariance between oviposition preference and larval performance in an insect herbivore. Evolution 40:778–785.

Via, S., and R. Lande. 1985. Genotype-environment interaction and the environment of phenotypic plasticity. Evolution 39:505–522.

Vinson, S. B. 1981. Habitat location. Pp. 51–77 in: D. A. Nordlund, R. L. Jones, and W. J. Lewis (eds.), *Semiochemicals*. John Wiley, New York.

Vinson, S. B., C. S. Barfield, and R. D. Henson. 1977. Oviposition behaviour of *Bracon mellitor*, a parasitoid of the boll weevil (*Anthonomus grandis*). II Associative learning. Physiol. Entomol. 2:157–164.

Vrba, E. S. 1987. Ecology in relation to speciation rates: Some case histories of Miocene-Recent mammal clades. Evol. Ecol. 1:283–300.

Waage, J., and D. Greathead (eds.). 1986. *Insect Parasitoids*. Academic Press, London.

Walsh, B. D. 1864. On phytophagic varieties and phytophagic species. Proc. Entomol. Soc. Philadelphia 3:403–430.

Walsh, B. D. 1865. Graduation from "individual peculiarities" to species in insects. Ann. Mag. Nat. Hist., Ser. 3, 16:383–384.

Wardle, A. R., and J. H. Borden. 1986. Detrimental effect of prior conditioning on host habitat location by *Exeristes roborator*. Naturwissenschaften 73:559–560.

Weseloh, R. M. 1981. Host location by parasitoids. Pp. 79–95 in: D. A. Nordlund, R. L. Jones, and W. J. Lewis (eds.), *Semiochemicals*. John Wiley, New York.

West-Eberhard, M. J. 1983. Sexual selection, social competition, and speciation. Quart. Rev. Biol. 58:155–183.

West-Eberhard, M. J. 1984. Sexual selection, competitive communication and species-specific signals in insects. Pp. 284–324 in: T. Lewis (ed.), *Insect Communication*. Academic Press, London.

White, M. J. D. 1978. *Modes of Speciation*. W. H. Freeman & Co., San Francisco.

Wilson, D. S., and M. Turelli. 1986. Stable underdominance and the evolutionary invasion of empty niches. Am. Natur. 127:835–850.

Wood, T. K. 1980. Divergence in the *Enchenopa binotata* Say complex (Homoptera: Membracidae) effected by host plant adaptation. Evolution 34:147–160.

Wood, T. K., and S. I. Guttman. 1982. Ecological and behavioral basis for reproductive isolation in the sympatric *Enchenopa binotata* complex (Homoptera: Membracidae). Evolution 36:233–242.

Wood, T. K., and S. I. Guttman. 1983. *Enchenopa binotata* complex: Sympatric speciation? Science 220:310–312.

Wright, S. 1932. The roles of mutation, inbreeding, crossbreeding and selection in evolution. Proc. Sixth Int. Congr. Genet. (Ithaca, NY) 1:356–366.

Yeargan, K. V. 1987. Ecology of a bolas spider, *Mastophora hutchinsoni;* Phenology, hunting tactics, and evidence for aggressive chemical mimicry. Oecologia 74:524–530.

Zwölfer, H. 1974. Das Treffpunkt-Prinzip als Kommunikationsstrategie und Isolationsmechanism bei Bohrfliegen (Diptera: Trypetidae). Entomol. German. 1:11–20.

Zwölfer, H. 1987. Species richness, species packing, and evolution in insect-plant systems. Pp. 301–319 in: E. -D. Schulze and H. Zwölfer (eds.), *Ecological Studies*, Vol. 61. Springer-Verlag, Berlin.

Zwölfer, H., and G. L. Bush. 1984. Sympatrische und parapatrische Artbildung. Z. Zool. System. Evol. 22:211–233.

CHAPTER 14

THE ROLE OF HABITAT PREFERENCE IN ADAPTATION AND SPECIATION

Scott R. Diehl and Guy L. Bush

INTRODUCTION

Speciation of sexually reproducing animals and most plants involves the development of *intrinsic* barriers to gene flow that act even when populations reside in the same place (Walsh 1864; Dobzhansky 1937; Mayr 1963). *Intrinsic* barriers to gene exchange are inherent in the organisms themselves, and include both postmating reproductive incompatibilities and premating barriers due to behavioral or ecological characteristics that limit gene flow between populations that would otherwise interbreed freely. Allopatric isolation requires *extrinsic* barriers that are external to the organism involved, such as isolation by distance or some physical barrier. Of course, the strength of any particular *extrinsic* barrier varies widely among different taxa confronting the barrier, and depends on inherent mobility. In the absence of *extrinsic* barriers, we regard populations utilizing different habitats to be sympatric when all individuals can readily move between habitats within the lifetime of an individual. Although habitats are almost always spatially distinct on a local scale, we argue that the use of the term "microallopatry" to describe such a situation is misleading because *extrinsic* barriers to gene exchange do not exist. It should be clear that allopatric and sympatric isolation represent extremes of a continuum, with parapatric isolation falling in between.

A commonly held view is that allopatric isolation is the only way (with rare exceptions) gene flow can be reduced adequately so that conspecific populations might evolve *intrinsic* isolation (Mayr 1963, 1978; Futuyma and Mayer 1980; Futuyma 1986; Paterson 1981). Others have argued that nonallopatric speciation may occur in certain organisms in which gene flow is reduced only by *intrinsic* barriers, without the requirement of allopatric isolation (Bush 1975a; White 1978; Diehl and Bush 1984; Bush and Howard 1986).

Several authors have developed theoretical models to explore the range of biological conditions under which nonallopatric speciation might occur. Most are based on a model investigating the maintenance of polymorphism proposed by Levine (1953). Levine demonstrated that polymorphism could be maintained at loci affecting fitness in two different sympatric habitats if population sizes are regulated independently in the two habitats. Maynard Smith (1966), using an analytical approach, showed that this situation could lead to sympatric speciation at least under some circumstances. He considered primarily *intrinsic* isolation due to a positive assortative mating mechanism that acts independent of habitat choice or utilization, but also suggested that isolation could arise by nongenetic "imprinting" of habitat preferences. Felsenstein (1981) investigated the habitat-independent assortative mating model using computer simulation and explored a much wider range of genetic and ecological conditions. In Felsenstein's model, genotypes exhibit positive assortative mating within habitats based on a single locus, but their movement between habitats is random. His results indicate that progress toward speciation (development of reproductive isolation) is unlikely unless natural selection is unrealistically intense and assortative mating is nearly perfect.

Futuyma and Peterson (1985) and Futuyma (1986) have argued that Felsenstein's results support their view that nonallopatric speciation occurs very rarely, if at all. However, we argued that Felsenstein's model fails to appropriately represent the biology underlying situations in which *intrinsic* isolation results from divergence in habitat preference (Bush and Diehl 1982). The common observation in many animals that preferential mating occurs among individuals utilizing the same host or habitat (Price 1980; Diehl and Bush 1984) has motivated the development of habitat preference speciation models (Bush 1969, 1975b).

The principal goal of this chapter is to evaluate the likelihood of nonallopatric speciation when assortative mating is coupled with habitat preference. For reasons discussed later, we will assess progress toward speciation by comparing (1) the rate at which initially rare alleles at loci affecting reproductive isolation and fitness in different habitats increase to polymorphic frequencies and (2) the magnitude of disequilibria generated between isolating and fitness loci. We based our habitat preference speciation (HPS) model on Felsenstein's habitat-independent speciation (HIS) model, except for the mechanism of *intrinsic* isolation. In the HPS model, habitat preference is nonrandom and controlled by a single locus. Furthermore, all mating takes place within each of the two habitats in the HPS model, so there is positive assortative mating by habitat preference genotype. To study the adaptive role of habitat preference unassociated with the speciation process, we also consider a third model that incorporates genetically based habitat preference, but where mating does not take place within the habitat. Since in this model there is habitat preference with random mating, we designate it the HPRM model.

THE MODELS

Common features and assumptions

All models presented incorporate the following common assumptions. We consider situations in which two ecologically independent habitats are distributed either sympatrically or parapatrically. Individuals are haploid and have two unlinked loci, each with two alleles (B, b, and C, c) that influence fitness in the different habitats. Natural selection affects these loci by influencing fertility, viability, or both, according to the fitness scheme shown in Table 1. These models are deterministic and deal with populations of infinite size. Fitness is regulated independently in the two habitats, according to the model of Levine (1953), and this results in frequency-dependent selection as discussed later. The sizes of the habitats are assumed to be equal. We also assume that there is excess reproduction, to the extent that the populations utilizing each habitat are of equal relative size after natural selection, regardless of the frequency of migration between the habitats in the previous generation.

TABLE 1. Relative fitness in different habitats.[a]

	Fitness in habitat		
Genotype	I	II	Mean
BC	$1 + 2s + xs^2$	1	$1 + s + (xs^2)/2$
Bc	$1 + s$	$1 + s$	$1 + s$
bC	$1 + s$	$1 + s$	$1 + s$
bc	1	$1 + 2s + xs^2$	$1 + s + (xs^2)/2$

[a] Alleles B and C confer relatively high fitness in habitat I but low fitness in habitat II, and conversely for alleles a and b. The parameter x determines whether these fitness loci interact multiplicatively (when $x = 1$) or additively (when $x = 0$). When these loci interact multiplicatively, the coupling genotypes (BC and bc) experience a higher mean fitness (averaged over both habitats), whereas their additive interaction results in equal mean fitness for all four genotypes. Selection intensity is varied by the parameter s (following Felsenstein's treatment), and is related to the traditional selection (S) by $S = 1 - 1/(1 + 2s + xs^2)$.

Specific features

All three models include a third, unliked locus with two alleles (A, a). In Felsenstein's HIS model, identical genotypes at this locus mate with positive assortment. The parameter d controls penetrance of assortative mating, and ranges from 0.0 (random mating) to 1.0 (complete isolation) according to the following scheme, where p is the frequency of the A allele:

	A	a
A	$p^2(1-d)+pd$	$p(1-p)(1-d)$
a	$p(1-p)(1-d)$	$(1-p)^2(1-d)+(1-p)d$

Individuals in the HIS model do not exhibit habitat preference, since their movement between habitats is independent of genotypes at all three loci. The level of migration between habitats is influenced solely by extrinsic barriers in this model as follows:

		Move to habitat	
		I	II
Reside in habitat	I	$1-m$	m
	II	m	$1-m$

In sympatry (m = 0.5), migration results in a complete mixing of the subpopulations associated with the two habitats every generation. Parapatric situations occur when migration is limited due to an *extrinsic* barrier (m < 0.5). The life cycle of the HIS model considered here proceeds with natural selection acting on the B and C loci, followed by migration between habitats that is independent of genotypes, then with recombination and positive assortative mating within each habitat. Recombination in all models occurs within diploid individuals that temporarily form by the union of two haploids during reproduction.

The life cycle of the HPS model, in contrast, consists of natural selection acting on the B and C loci, followed by genetically based habitat preference (migration influenced by the A locus), then by recombination and random mating within each habitat. We assume that A genotypes move preferentially to habitat I and a to habitat II. The penetrance of habitat preference is specified by the parameter g, which ranges from 0.0 (random habitat choice and mating) to 1.0 (complete habitat fidelity and isolation). In sympatry (m = 0.5), the proportion of individuals moving to each habitat depends solely on their A locus genotype. Under parapatric conditions (m = 0.1), the movement of individuals depends both on the A locus genotype and their habitat prior to migration:

			Move to habitat	
			I	II
		Genotype		
Reside in habitat	I	A	$g + (1 - m)(1 - g)$	$m(1 - g)$
		a	$(1 - m)(1 - g)$	$g + m(1 - g)$
	II	A	$g + m(1 - g)$	$(1 - m)(1 - g)$
		a	$m(1 - g)$	$g + (1 - m)(1 - g)$

Note that in sympatry ($m = 0.5$) the equations influencing migration are equivalent, irrespective of whether individuals reside in habitat I or II prior to migration. The parameter g in the HPS model is analogous to the parameter d in the HIS model, in that the degree of reproductive isolation is a function of their magnitudes. However, d influences the degree of assortative mating between A locus genotypes within each habitat, but has no effect on migration between habitats. By contrast, g influences migration of A locus genotypes, but has no effect on assortative mating within habitats. Reproductive isolation between habitats is a consequence of the fact that habitat choice is nonrandom and because mating occurs within the habitat. Habitat choice could involve behavioral preferences for alternative habitats, differences in seasonal activity periods corresponding to temporal variation in habitat availabilities, and other factors (Diehl and Bush 1984). There are many examples in natural populations that generally conform to the biological assumptions inherent in the HPS model (Diehl and Bush 1984).

The third model that we consider here is intended to represent organisms that utilize two different habitats, but mate at random independent of the habitats, then exhibit nonrandom habitat preference in choosing the habitat that their progeny will utilize. Specifically, the life cycle in the HPRM model involves natural selection acting on the B and C loci, followed by genotypically random migration between habitats as specified for the HIS model above. Migration is then followed by recombination and random mating among individuals in the same location. If the two habitats are located sympatrically, mating is random with respect to the habitat from which individuals originated, and can be thought of as mating in some central location independent of either habitat. If the habitats are parapatric, the genotypically random migration preceding mating results in only a partial mixing of the populations associated with the two habitats. In this case, mating is nonrandom with respect to the habitat from which individuals originated. The life cycle of the HPRM model is completed by habitat choice influenced by the A locus as previously specified for the HPS model. This determines the habitat that progeny will utilize in the next generation. Examples of this model include organisms such as insects, where immature stages may be genetically adapted to utilize different host plants or habitats, but where mating takes place independent of the habitat, and postmating habitat preference occurs prior to egg laying.

Initial conditions and methods of analysis

We developed PASCAL programs derived from those provided by Felsenstein (1981) to represent these biological relationships. Our initial conditions reflect a situation in which a population has been utilizing a single habitat to which it is genetically adapted. We believe it unlikely that mutations conferring adaptation to a new habitat would occur simultaneously at two or more loci in the same individual. Therefore, we specified identical initial populations in each habitat, with frequencies of 0.01 for the *A*, *B*, and *C* alleles, and initial genotype frequencies were set at $f(abc) = 0.97, f(Abc) = f(aBc) = f(abC) = 0.01$. The initial pairwise disequilibrium between all three loci, therefore, was 1% of the maximum possible value for these allele frequencies (−0.0001) and the residual three locus disequilibrium was zero. We analyzed the behavior of the three models under sympatric ($m = 0.5$) and two parapatric conditions ($m = 0.1, 0.3$), for natural selection intensities (s) of 0.0, 0.1, 0.3, and 0.5, and for penetrance of habitat independent assortative mating (d) or habitat preference (g) ranging from 0.0 to 0.8, in increments of 0.2. We also considered values of d and g of 0.99, rather than 1.0, because a value of 1.0 results in an initial condition of complete reproductive isolation in the HIS and HPS models, and the subsequent outcome is obvious. We performed simulations for both multiplicative ($x = 1.0$) and additive ($x = 0.0$) interaction of the loci affecting fitness.

We ran the models for 1000 generations for each combination of parameters, and evaluated changes in allele frequencies and disequilibria. In most cases, genotype frequencies stabilized well before this time at frequencies of 0.5 for both alleles at all loci. Our criteria for assessing equilibrium were that allele frequencies at all three loci had reached at least 0.495, or that no change in allele frequencies (to five decimal places) occurred for 100 generations. For some parameters (listed below), it was necessary to run the simulations for up to 2725 generations before frequencies at the *B* and *C* loci stabilized. The frequency of the A allele did not increase by natural selection beyond its low initial value for most parameters tested in the HIS model. To consider how much disequilibrium would develop if allele frequencies reached high levels due to genetic drift in small founder populations, we performed additional simulations with initial allele frequencies of 0.5 at all loci.

The initial single "species" in all models is divided into subpopulations utilizing two habitats. Therefore, disequilibrium between pairs of loci within the species taken as a whole (D_{TOT}) is the combination of the Wahlund disequilibrium (D_{BET}) that results when allele frequencies at two loci differ in a consistent manner (co-vary) between the subpopulations in the two habitats (Smouse and Neel 1977) and disequilibrium within each of the subpopulations ($D_{W1} + D_{W2}$). It has been shown that $D_{TOT} = D_{W1}/2 + D_{W2} / 2 + D_{BET}$ (Nei and Li 1973). In analyzing the HIS model, Felsenstein (1981) reported only whether disequilibrium within each subpopulation remained at a value greater than zero. Here, we report levels of disequilibrium between the isolating locus A and either of the habitat-specific fitness loci B and C after stable equilibrium conditions are attained. We consider the disequilibrium within the total species consisting of both subpopulations pooled (D_{TOT}), as well as its partitioning into disequilibrium within each subpopulation considered separately (D_{W1} and D_{W2}), and disequilibrium due to covariance in allele frequencies between the subpopulations (D_{BET}). We also compare rates of approach to stable equilibrium under a variety of biological conditions.

RESULTS AND CONCLUSIONS

The results of our simulations demonstrate that progress toward nonallopatric speciation may occur under a broad range of biological conditions when habitat preference is coupled with mate choice as in the HPS model. This conclusion is supported by three aspects of our results. First, we found that alleles at a locus that influences habitat preference are readily maintained at high frequencies, even in the absence of genetically based differences in ability to utilize different habitats. Second, genetic variation for habitat preference substantially accelerates the increase of rare alleles at habitat-specific fitness loci. Third, we observed that disequilibrium is maintained between an unlinked habitat preference locus and loci that affect fitness in different habitats, even under completely sympatric conditions. In contrast, the HIS model (in which there is assortative mating but habitat choice is random) usually failed to demonstrate progress toward speciation when evaluated by these criteria.

Three other important conclusions can be drawn from our analyses. We found that increasing *extrinsic* isolation between habitats (i.e., parapatry) both accelerated rates of approach to equilibrium allele frequencies and increased the magnitude of disequilibrium between habitat preference and habitat-specific fitness loci. With one exception (noted below), there was only a small effect of having the two loci affecting fitness interact additively versus multiplicatively. Finally, even when genetic variation influencing habitat preference occurs within a randomly mating population (as in the HPRM model), significant disequilibrium is still maintained between unlinked loci affecting habitat choice and fitness in alternative habitats.

The number of generations required for genotype frequencies to reach equilibrium varied substantially. For all three models, it required less than 175 generations for genotype frequencies to stabilize when $s = 0.0$ (in this case, only the *A* locus undergoes any change), less than 350 generations when $s = 0.3$ and less than 150 generations when $s = 0.5$. When $s = 0.1$, it required up to 2725 generations to reach equilibrium under sympatric conditions, but substantially fewer generations when habitat preference was strong (e.g., HPS model, $g = 0.8$ required 375 generations). In contrast, the penetrance (d) of the assortative mating locus *A* in the HIS model had only a very slight effect on the rate of approach to equilibrium. All models

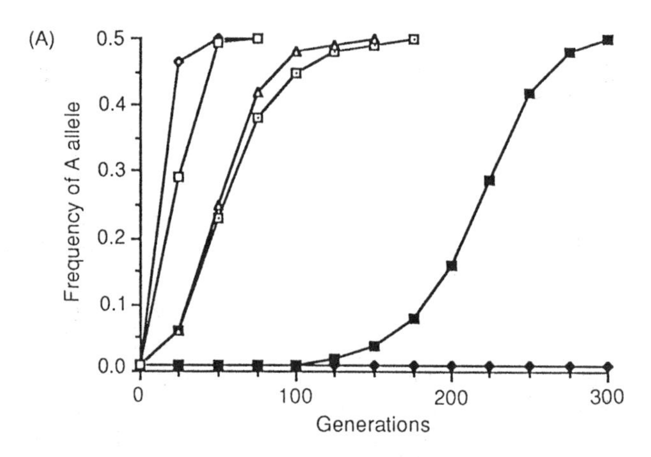

FIGURE 1. (A) Increase in frequency of the *A* allele. The HPS and HPRM models are graphically indistinguishable, and equilibrium at the habitat preference locus (*A*) is quickly attained under sympatric conditions ($m = 0.5$) even when $s = 0$, and is reached faster with higher penetrance of habitat preference (⊡, $g = 0.2$; ◇, $g = 0.4$). Increasing intensity of natural selection has only a trivial effect on the rate of increase of the *A* allele in the HPS and HPRM models (△, $m = 0.5, s = 0.5, g = 0.2$). The habitat preference locus increases to equilibrium faster under parapatric conditions when compared to its rate for identical conditions in sympatry (□, $m = 0.1$, $s = 0.5$, $g = 0.2$). In the HIS model, the *A* allele that controls assortative mating within habitats did not increase from its low initial frequency for most parameters tested (e.g., ♦, $m = 0.5, s = 0.5, d = 0.8$), except with very high penetrance of assortative mating and natural selection intensity (■, $m = 0.5, s = 0.5, d = 0.99$). (B) Increase in frequency of the *B* allele. Without assortative mating or habitat preference ($d = g = 0.0$), the HIS and HPS models are identical. The rate of increase of

approached equilibrium faster under parapatric conditions (e.g., HIS model, $s = 0.1$, $m = 0.1$ required 325 generations).

We will now examine in more detail the rate of approach to equilibrium genotype frequencies at the *A* locus (Figure 1A) and the *B* or *C* loci (Figure 1B). Figure 1 illustrates results obtained when the habitat-specific fitness loci interact multiplicatively (additive interaction of fitness loci, discussed below). For any genetically based isolating mechanism to limit gene flow between populations using different habitats, the initially rare alleles at the loci involved must reach high frequencies. First, we consider the rate of increase of the initially rare *A* allele in the absence of any natural selection acting on the *B* or *C* loci ($s = 0.0$). As shown in Figure 1A, the *A* allele increases very rapidly in the HPS and HPRM models under these conditions, and equilibrium is attained faster with higher penetrance of habitat preference. The *A* locus is stably maintained even when $s = 0.0$ due to frequency-

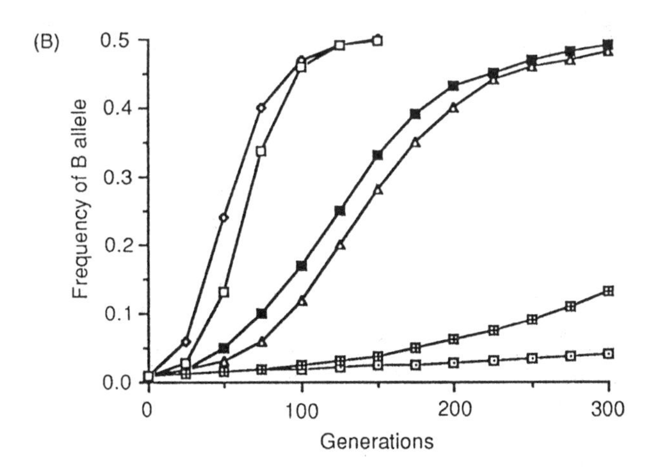

the initially rare habitat-specific fitness allelle (*B*) is strongly influenced by the intensity of natural selection (*e.g.*, in sympatry, $m = 0.5$, ⊡; $s = 0.1$; ◇, $s = 0.5$). In the HPS model, habitat preference considerably enhances the rate of increase of the *B* allele ($\triangle, m = 0.5, s = 0.1, g = 0.8$). Parapatric conditions also substantially increase the rate of approach to equilibrium at the *B* locus in the HPS model ($\square, m = 0.1, s = 0.1, g = 0.8$). In the HPRM model, the rate of increase of the *B* allele is enhanced by habitat preference ($\boxplus, m = 0.5, s = 0.1, g = 0.8$), but not as much as when habitat preference is coupled with mate choice as in the HPS model ($\triangle$). In the HIS model, assortative mating had no effect on the rate of increase of the *B* allele for most parameters (e.g., ■, $m = 0.5$, $s = 0.3$, $d = 0.0$–0.8). Even at very high penetrance of assortative mating ($d = 0.99$) and natural selection intensity ($s = 0.3$ or 0.5) in sympatry or parapatry ($m = 0.5$ or 0.1), the effect of assortative mating on the frequency of the *B* allele was trivial to the point of being graphically indistinguishable.

dependent selection. This arises from the fact that both habitats contribute equally to the total species' population each generation, irrespective of the frequency of genotypes moving into each habitat in the preceding generation. These results obtained for the HPRM model agree with Rausher's (1984) findings regarding the maintenance of polymorphism for habitat preference with random mating and without genetic variation affecting fitness in different habitats. These results are also consistent with analysis of genotype-dependent migration reported by Nagylaki and Moody (1980) and Moody (1981). In contrast, the initially rare *A* allele at the habitat-independent assortative mating locus in the HIS model never increased in frequency without natural selection acting on the *B* and *C* loci.

Alleles at the *B* and *C* loci always eventually reach an equilibrium frequency of 0.5 for any intensity of natural selection greater than zero. This is due to the fact that populations utilizing different habitats are regulated independently and migration and natural selection intensities are symmetric. However, their rate of approach to equilibrium is strongly influenced by the parameters of the model. First, we present results obtained in the absence of assortative mating or habitat preference (i.e., $d = g = 0.0$). Under these conditions, the *A* locus is neutral, and so its alleles do not change from their initial frequencies. The rate of approach to equilibrium at the *B* or *C* loci is much faster at higher intensities of natural selection (Figure 1B). These results are illustrated for the *B* locus only, as results for the *C* locus are identical.

It has been argued that one limitation on nonallopatric speciation is the fact that habitat utilization polymorphisms cannot be maintained without unrealistic assumptions of symmetry or severe constraints on the relative fitness of each allele in each habitat (reviewed by Futuyma and Peterson 1985). Wilson and Turelli (1986), on the other hand, have shown that conditions favoring the maintenance of polymorphism are greatly relaxed when fitness in different habitats is frequency dependent, even without any barriers to interbreeding between individuals using different habitats. Maintenance of polymorphism is also facilitated if organisms choose the habitat most favorable to their genotype (Taylor 1976; Jones 1980; Jones and Probert 1980). However, since we doubt that such nongenetic optimal habitat preference occurs frequently in nature, we have modeled habitat choice and habitat-specific fitness under the control of separate loci. It has been demonstrated elsewhere (Garcia-Dorado 1986) that a habitat preference locus broadens the range of conditions under which polymorphism can be maintained at a habitat-specific fitness locus.

Next, we compare rates of approach to equilibrium when there is both natural selection acting on the *B* and *C* loci and habitat preference or habitat-independent assortative mating influenced by the *A* locus. The intensity of natural selection has only a trivial effect on the rate of increase of the *A* allele in either the HPS and HPRM models (Figure 1A). However, the *A* allele did not increase from its low initial frequency for most parameters tested for the HIS model, except with very high penetrance of assortative mating and very intense natural selection (Figure 1A). Habitat preference considerably enhanced the rate of approach to equilibrium of the initially rare *B* allele in the HPS model (Figure 1B). Habitat preference has a similar effect in the HPRM model (Figure 1B), although not as pronounced as when habitat and mate choice are coupled, as in the HPS model. The level of penetrance of habitat-independent assortative mating in the HIS model had no effect on the approach to equilibrium at the *B* and *C* loci, except when very high penetrance of assortative mating was combined with very intense natural selection, and even then the effect was trivial (Figure 1B). All three models stabilized in their genotype frequencies substantially faster under parapatric versus sympatric conditions. This occurred at both the *A* locus that influences habitat preference and assortative mating (Figure 1A) and the *B* and *C* loci that affect fitness (Figure 1B).

We found that the effects of additive versus multiplicative interaction of the *B* and *C* fitness loci on the approach to equilibrium by the *A* locus were very minor for the HPS and HPRM models. In the HIS model, it required significantly more generations for the *A* locus to attain equilibrium under additive interaction of the fitness loci compared to when these loci interact mulitplicatively (e.g., for $m = 0.3$, $s = 0.3$, $d = 0.99$ requires 750 generations if additive, versus 350 generations if multiplicative). In addition, in two cases we observed that the *A* allele failed to increase at all if the fitness loci interact additively, whereas it did increase if these loci interact multiplicatively ($m = 0.5$, $s = 0.3$ or 0.5, $d = 0.99$). In nature, loci affecting fitness in different habitats could interact either additively or multiplicatively. Since the HPS and HPRM models are relatively robust to this assumption, divergence under the mechanisms embodied in these models should be more likely in nature than under the habitat-independent mechanism of assortative mating of the HIS model.

In contrast, the *B* and *C* loci attained their equilibrium frequencies slightly faster when interacting additively versus multiplicatively in all three models. This latter difference was most pronounced at low natural selection intensity and low penetrance of habitat preference in the HPS and HPRM models. As previously noted, the penetrance level of assortative mating in the HIS model had very little or no effect on the rate of increase of the *B* and *C* fitness loci. Thus, for all values of penetrance of habitat preference (g) greater than zero, the *B* and *C* loci increased toward equilibrium faster in the HPS and HPRM models than in the HIS model. We suggest that the fitness loci require a greater number of generations to attain equilibrium when they interact multiplicatively because of the lower mean fitness of repulsion genotypes compared to coupling genotypes under this fitness regime. While initially rare alleles at the two fitness loci are increasing in frequency, a large proportion of repulsion genotypes will be produced. Since the repulsion genotypes experience reduced fitness if the *B* and *C* loci interact multiplicatively, their rate of increase to equilibrium will be reduced.

We now consider the levels of disequilibrium generated between the assortative mating or habitat preference locus (*A*) and one of the habitat-specific fitness loci (*B*). These results are presented in Figure 2 for all three models, for two levels of natural selection intensity with additive interaction of the *B* and *C* loci, and five levels of penetrance of assortative mating (*d*) or habitat preference (*g*). We found that some disequilibrium was generated for all conditions tested for the HPS and HPRM models in both sympatry (Figure 2A) and parapatry (Figure 2B). The magnitude of the disequilibrium increases at higher penetrance of habitat preference and selection intensity. Less disequilibrium is generated in the HPRM model, where there is habitat choice but random mating, than in the HPS model, where habitat and mate choice are biologically coupled. By contrast, no disequilibrium is generated in the HIS model except when there is very high penetrance of assortative mating in parapatric conditions. The magnitude of disequilibrium between the *A* and *B* loci is greater in parapatry for all three models.

Even in sympatry much of the disequilibrium generated in the HPS model is distributed as between-population covariance in gene frequencies (D_{BET}), due to the inherent subdivision of the total population resulting from the coupling of habitat and mate preference. However, it is important to note that even though the within-population component of disequilibrium is very small in absolute magnitude, it may be actually quite near its theoretical maximum. The theoretical maximum disequilibrium is small because allele frequencies within populations deviate substantially from 0.5. In biological terms, this means that within each population even a habitat preference allele that confers a preference for the alternate habitat is usually associated with its appropriate habitat-specific fitness alleles (*ABC* and *abc*). Unlike the HPS model, the disequilibrium generated in sympatry in the HPRM model is distributed entirely within the populations (D_W), since random mating between these populations each generation eliminates all differences between populations. Similarly, when disequilibrium is generated in the HIS model in parapatry, it is distributed primarily within the populations, since there are only *extrinsic* barriers in this model.

The disequilibria generated in the HPS and HPRM models were slightly higher when the habitat-specific fitness loci interact multiplicatively. The HIS model failed to generate any disequilibrium in sympatry for all parameters tested (including $s = 0.5$, not shown) when there is

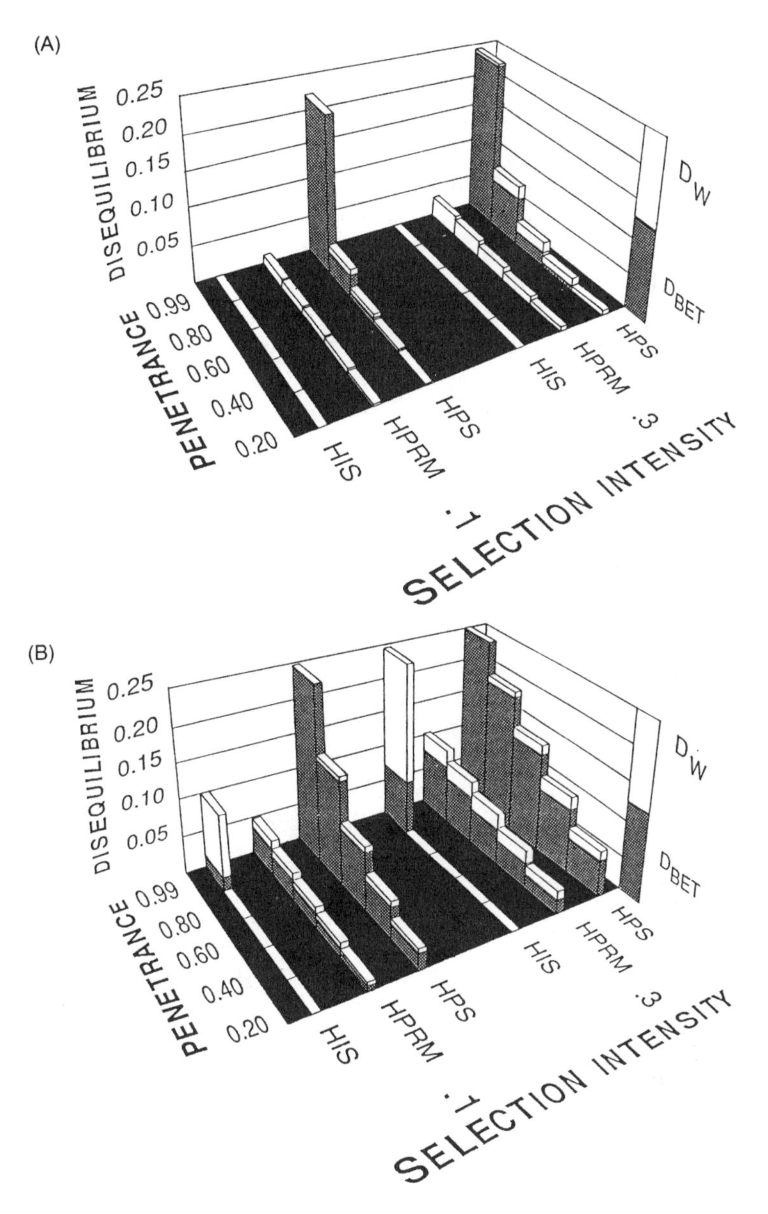

FIGURE 2. Disequilibrium between the *A* and *B* loci. (A) Sympatric conditions (m = 0.5) (B) Parapatric conditions (m = 0.1). The histograms represent the equilibrium states attained for two levels of natural selection intensity acting at the habitat-specific fitness loci (*B* and *C*) and five levels of penetrance of the *A* locus that influences habitat-independent assortative mating (d) in the HIS model or habitat preference (g) in the HPS or HPRM models. The shaded portion of each histogram represents the between-population component (D_{BET}) of the total disequilibrium and the unshaded portion represents the within-population component (D_W).

additive interaction of the habitat-specific fitness loci (Figure 2A). However, this model did generate disequilibrium in sympatry when these loci interact multiplicatively, if only for two extreme conditions tested (s = 0.3 or 0.5, d = 0.99). All of this disequilibrium was distributed within populations because of the lack of any barrier to gene flow between populations in the HIS model in sympatry. We carried out an additional simulation for the HIS model under sympatric conditions with additive interaction of the habitat-specific fitness loci with very extreme conditions of natural selection intensity and habitat-independent assortative mating (s = 10.0, d = 0.99). Even under these unrealistic conditions, the HIS model failed to exhibit any progress toward speciation in sympatry. The A locus did not increase from its low initial frequency, and disequilibrium did not develop. Thus, since the HPS and HPRM models appear to be less sensitive to the effects of multiplicative and additive interaction at loci affecting fitness than the HIS model, we conclude that the former models should be more widely applicable to natural populations.

Last, we briefly examined the possible effect of different starting conditions on the outcome for the three models. Although a rare allele at the habitat-independent assortative mating locus could drift to high frequency in small populations, it would not be selectively maintained, and might drift to fixation. Furthermore, we found that even when both alleles at the assortative mating locus were initialized at a frequency of 0.5, this had no effect on the magnitude of disequilibrium generated after stable conditions are attained. In fact, the levels of disequilibrium in all three models were unaffected by initial conditions.

DISCUSSION

We expect that speciation due to adaptation to different habitats would result in two new species, each adapted to optimally utilize the resources associated with its own habitat. Because loci that directly influence fitness in the different habitats are very unlikely to isolate the new species, an association must be established between the habitat-specific fitness loci and other loci that can isolate the populations in the different habitats. This is why we regard the disequilibria values between the isolating locus and one of the fitness loci presented in Figure 2 as a measure of progress toward speciation. We recognize that these associations do not complete the speciation process. Additional progress toward speciation could occur by the incorporation of new mutations at the loci modeled here or at other loci that increase isolation or habitat-specific fitness differences. In sympatry, the disequilibria that develop in the HPS model are distributed primarily as covariances in allele frequencies between populations utilizing different habitats. In contrast, all of the disequilibria developed under sympatric conditions in the HPRM and HIS models are entirely within populations. This is due to the fact that there is an *intrinsic* barrier to gene flow in the HPS model (genetically based habitat preference with mating limited to the habitat). It can be shown that when $s = 0.0$, migration between habitats in the HPS model in sympatry is reduced from 0.5 to $(1 - g^2)/2$. This *intrinsic* isolating mechanism is absent in the HPRM and HIS models, and the migration fraction between sympatric habitats is always 0.5, even when very intense natural selection and highly penetrant assortative mating produce large within-population disequilibrium.

The distribution of the disequilibrium generated in the HPS model as between-population covariances in gene frequencies accounts for its success in spite of the fact that this is a "two-allele" model in which different isolating alleles become fixed in the diverging populations. Because recombination reduces only the within-population component of disequilibrium, it has little effect on most of the disequilibrium (D_{BET}) generated in the HPS model, but a major effect on that generated in the HPRM and HIS models in sympatry. Felsenstein (1981) previously noted that "two-allele" models are generally less successful in generating progress toward speciation than "one-allele" models, in which the same isolating allele becomes fixed in both of the diverging populations. For example, O'Donald (1960) and Endler (1977) have explored "one-allele" models in which substitution of a new allele at one locus causes individuals to mate assortatively based on another locus that is differentially selected in two or more habitats. Alternatively, substitution of a new allele that causes a reduction in migration between sympatric or parapatric habitats, regardless of the habitat in which individuals originate, can also lead to isolation (Balkau and Feldman 1973). Imprinting of mate or habitat preference is a possible biological mechanism underlying "one-allele" isolation. Although these models appear theoretically sound, we believe that many situations involving isolation between natural populations are more appropriately represented by "two-allele" speciation models.

Rice (1984) has developed a "two-allele" model of sympatric speciation based on divergence in habitat preference and analyzed it by computer simulation. It differs from our HPS model in several important respects. For substantial divergence to develop in Rice's model, it must be assumed that assortative mating takes place between individuals with similar habitat preference as in our HPS model. However, it must also be assumed that genotypes with strong preference for either habitat experience superior fitness compared to genotypes with intermediate habitat preference (i.e., disruptive selection on habitat preference). There are two biological mechanisms we can envision that might generate this disruptive selection. Either habitat preference loci also affect fitness in different habitats (pleiotropy), or genotypes with intermediate habitat preference experience reduced fitness due to foraging inefficiency. The problem of recombination reducing disequilibrium between an isolating locus and habitat-specific fitness loci is circumvented by either of these mechanisms. Elsewhere, Rice (1987) has discussed the importance of genetic adaptation to utilize different habitats (i.e., the role played by our *B* and *C* loci). However, he does not seem to appreciate that these traits are very likely to be controlled by loci different from those influencing habitat preference, and the consequential requirement of the development of disequilibrium between habitat preference and habitat-specific fitness loci for progress to be made toward speciation. Although Rice's model of speciation may apply in some circumstances, we believe that it is more commonly the case that separate loci control habitat choice and fitness in different habitats, as in our HPS model.

One recurring criticism of nonallopatric speciation has been that the loci involved in speciation must be tightly linked for the process to occur under biologically realistic conditions because recombination would otherwise break down disequilibria among coadapted loci (Futuyma and Mayer 1980; Futuyma 1986; Felsenstein 1981). In the HPS model presented here, substantial progress toward speciation developed even though all loci were unlinked. This occurred because the *intrinsic* isolation due to the coupling of habitat and mate choice results in disequilibria that are primarily distributed as between-population covariances that are unaffected by recombination. In fact, although linkage may increase the level of divergence at equilibrium, it may initially impede the association of new mutations in the separate loci affecting habitat-specific fitness and habitat preference in the HPS model. It seems very unlikely that all of the loci necessary for adaptation to a new habitat would simultaneously mutate to new alleles in the same individual. Individuals with a mixture of alleles, with preference for one habitat but adapted to utilize the other habitat, would experience reduced fitness. Thus, if linkage between habitat preference loci and habitat-specific fitness loci is extremely tight, a finite population colonizing a new habitat might go extinct before the right combination of genotypes can develop by recombination.

Our results would appear to indicate that parapatric speciation occurs much more readily than sympatric speciation under the assumptions of both the HIS and HPS models. However, this conclusion may be due partly to the oversimplifications of natural situations inherent in our models. In nature, colonization of a new host or habitat probably requires many repeated attempts before the right combination of new alleles permits the efficient exploitation of the new resource. If the old and new habitats are distributed parapatrically, fewer migrants would contact the new habitat each generation than if the habitats are sympatric. Thus, it would probably require substantially longer for suitable genotypes to colonize the new habitat in parapatry than in sympatry. Our simple models do not take these aspects of the process into account. Populations in both habitats are of infinite size and can never go extinct, and we began each simulation with all of the genetic variation necessary for efficient use of the new habitat already present. In a finite population model, rare mutants might not migrate to the other habitat if parapatric, and thus rare genotypes adapted to the new habitat would experience low fitness in the original habitat and quickly go extinct. Another possibility not incorporated into our simple model is that, in sympatry, "imprinting" or "learning" could cause individuals to prefer a host or habitat which they have utilized previously (Papaj and Prokopy 1986; Mayr 1963:568). This would have precisely the same effect as an *extrinsic* barrier between parapatrically distributed habitats and might thus increase the likelihood of sympatric speciation.

Another argument frequently voiced against nonallopatric speciation is that reinforcement of postmating isolation by the evolution of premating isolation is unlikely to occur (Paterson 1981). This conclusion is partially based on the very high intensity of natural selection and penetrance of assortative mating required for divergence in Felsenstein's (1981) HIS model (Butlin 1987, and this volume; Coyne and Barton 1988). The traditional view of reinforcement involves a situation in which genetic incompatibilities between two "races" that have diverged in allopatry result in postmating hybrid inferiority (Dobzhansky 1940). It is proposed that this postmating barrier to interbreeding can be "reinforced" by the evolution of premating barriers that further reduce gene flow. The premating barriers are selected for because they reduce the production of inferior hybrids.

We argue that Felsenstein's model does not accurately represent the biological conditions proposed for the classic reinforcement model. First, the hybrid inferiority in the HIS model is due solely to the multiplicative interaction of the B and C loci that confer adaptation to different habitats. Without this multiplicative interaction, there is no hybrid inferiority. Second, the level of hybrid inferiority is relatively very small ($s^2/2$) even with multiplicative interaction of these loci, compared to the selection coefficients with respect to adaptation to the different habitats $(1 + s)^2$. Because of this relationship, unrealistically large selection coefficients within populations are required to generate even very low levels of hybrid inferiority. We suggest that a more appropriate model of reinforcement would not necessarily couple hybrid inferiority with a scheme of multiplicative interaction of loci adapting genotypes to different habitats. We predict that such a model would substantially relax the conditions necessary for reinforcement to occur.

Finally, we wish to consider the relevance of our HPS model to the issue of reinforcement. To begin with, if postmating hybrid inferiority exists, we predict that its reinforcement would be much more likely to occur by the evolution of premating isolation if it was due to habitat preference coupled with mating in the habitat than some habitat-independent mechanism. This is suggested by the fact that the HPS model readily produces premating isolation when multiplicative interaction of the habitat-specific fitness loci results in hybrid inviability. However, the HPS model generates nearly as much premating isolation in the absence of hybrid inviability when the habitat-specific fitness loci interact additively. This indicates that the isolation that evolves in the HPS model is not primarily the outcome of selection to reduce the production of inferior hybrids. Instead, isolation is a by-product of adaptation to different habitats. This is further supported by the fact that a moderately strong association develops between habitat preference and habitat-specific fitness loci in the HPRM model, in a situation in which isolation can never evolve. In this respect, the HPS model somewhat resembles Muller's (1940) view of the origin of isolating mechanisms as a by-product of adaptation to the environment, with the major distinction that Muller envisioned the process as occurring only in allopatry, whereas the HPS model does not require allopatric isolation.

We nevertheless speculate that reinforcement of hybrid inviability unrelated to adaptation to different habitats would still be much more likely to occur if premating isolation is based on habitat preference coupled with mating within the habitat. First, premating isolation by habitat preference leads to a subdivision of the population. This causes the disequilibrium between the isolating locus and the locus causing hybrid inviability to be distributed primarily as covariances in gene frequencies between populations. As previously discussed, this form of disequilibrium is not affected by recombination, so we expect it to be maintained at higher levels than disequilibrium evolving with a habitat-independent assortative mating mechanism. Second, unlike other models of reinforcement that have been proposed, isolation due to divergence in habitat preference would result in the ecological separation of the two populations. If the densities of the two populations were independently regulated due to their utilization of different habitats, they would be more likely to coexist.

Although there are many different species in which mate and habitat choice are coupled, the number of such species is not unlimited. Previously, it has been argued that genetic constraints prohibit nonallopatric speciation. However, our results indicate that sympatric and parapatric divergence can readily occur under genetically realistic conditions, if given sufficient time. We suggest that the overall frequency of nonallopatric speciation, as well as allopatric speciation, may more often be limited by the stability of distinct hosts or habitats over time. Divergence required hundreds of generations for some of the biological conditions we modeled here, and in many cases either one or both habitats may not persist in abundance long enough for this to occur. Furthermore, many hosts or habitats are already utilized by well-adapted species, and their presence would competitively exclude new colonists. We also suggest that the lack of suitable genetic variation within populations that are otherwise in a position to undergo allopatric or nonallopatric speciation probably also limits the overall frequency of species formation. In fact, knowledge of the actual genes controlling premating and postmating isolation, as well as the distribution of habitats in time and space, is generally lacking. This unfortunate gap between theoretical and empirical studies of speciation makes it currently impossible to definitively evaluate the relative likelihood of alternative models of speciation.

SUMMARY

We modeled speciation by divergence in habitat preference for a sympatric distribution of two independent habitats. Two unlinked loci control fitness in these habitats. A third unlinked locus influences preference for one or the other habitats each generation prior to mating and reproduction within the habitat. We determined equilibrium states by computer simulation for a range of natural selection intensities and habitat preference penetrances. Appreciable divergence, measured by linkage disequilibrium and differences in gene frequencies, developed between populations utilizing different habitats. This occurred even with moderate fitness differences and weak habitat preferences. We also further analyzed a previously developed sympatric speciation model with positive assortative mating within habitats but random movement between habitats. Our results demonstrate that considerable progress toward nonallopatric speciation occurs much more readily when mate choice is closely coupled with habitat preference, as commonly occurs in many different taxa.

ACKNOWLEDGMENTS

We thank P. Smouse for assistance regarding the partitioning of linkage disequilibrium and careful criticism of the manuscript. J. Felsenstein kindly provided annotated listings of his speciation computer programs. P. Hong and R. Tarquini assisted with programming. S. Berlocher, R. Bush, J. Endler, D. Futuyma, D. Howard, B. Milligan, A. Weis, S. Williams, and D. Wool also offered useful suggestions. S.R.D. acknowledges support from the computation centers of the Universities of Massachusetts and Michigan, and NSF (DEB 81-09381) and USDA (8200184) grants (R. Prokopy, P.I.). G.L.B. received support from Michigan State University and from NSF (DEB 82-11155) and USDA (86-CR-1-1398).

LITERATURE CITED

Balkau, B., and M. W. Feldman. 1973. Selection for migration modification. Genetics 74:171–174.

Bush, G. L. 1969. Sympatric host race formation and speciation in frugivorous flies of the genus *Rhagoletis* (Diptera, Tephritidae). Evolution 23:237–251.

Bush, G. L. 1975a. Modes of animal speciation. Annu. Rev. Ecol. Syst. 6:339–364.

Bush, G. L. 1975b. Pp. 187–206 in: *Evolutionary Strategies of Parasitic Insects and Mites*. Plenum, New York.

Bush, G. L., and S. R. Diehl. 1982. Host shifts, genetic models of sympatric speciation and the origin of parasitic insect species. Proc. 5th Int. Symp. Insect-Plant Relat. (Pudoc, Wageningen, Netherlands), 297–306.

Bush, G. L., and D. J. Howard. 1986. Allopatric and non-allopatric speciation; assumptions and evidence. Pp. 411–438 in: S. Karlin and E. Nevo (eds.), *Evolutionary Processes and Theory*. Academic Press, Orlando, FL.

Butlin, R. 1987. Speciation by reinforcement. Trends Ecol. Evol. 2:8–13.

Coyne, J. A., and N. H. Barton. 1988. What do we know about speciation? Nature (London) 131:485–486.

Darwin, C. 1859. *On the Origin of Species*. John Murray, London.

Diehl, S. R., and G. L. Bush. 1984. An evolutionary and applied perspective of insect biotypes. Annu. Rev. Entomol. 29:471–504.

Dobzhansky, Th. 1937. *Genetics and the Origin of Species*. Columbia University Press, New York.

Dobzhansky, Th. 1940. Speciation as a stage in evolutionary divergence. Am. Natur. 74: 312–321.

Endler, J. A. 1977. *Geographic Variation, Speciation and Clines*. Princeton University Press, Princeton, NJ.

Felsenstein, J. 1981. Skepticism towards Santa Rosalia, or why are there so few kinds of animals. Evolution 35:124–138.

Futuyma, D. J. 1986. *Evolutionary Biology*, 2nd ed. Sinauer Associates, Sunderland, MA.

Futuyma, D., and G. C. Mayer. 1980. Non-allopatric speciation in animals. Syst. Zool. 29:254–271.

Futuyma, D. J., and S. C. Peterson. 1985. Genetic variation in the use of resources by insects. Annu. Rev. Entomol. 30:217–238.

Garcia-Dorado, A. 1986. The effect of niche preference on polymorphism protection in a heterogeneous environment. Evolution 40:936–945.

Jones, J. S. 1980. Can genes chose habitats? Nature (London) 286:757–758.

Jones, J. S., and R. F. Probert. 1980. Habitat selection on a deleterious allele in a heterogeneous environment. Nature (London) 287:632–633.

Levine, H. 1953. Genetic equilibrium when more than ecological niche is available. Am. Natur. 87:331–333.

Maynard Smith, J. 1966. Sympatric speciation. Am. Natur. 100:637–650.

Mayr, E. 1963. *Animal Species and Evolution*. Harvard University Press, Cambridge, MA.

Mayr, E. 1978. Review of *Modes of Speciation* by M. J. D. White. Syst. Zool. 27:478–482.

Moody, M. 1981. Polymorphism with selection and genotype-dependent migration. J. Math. Biol. 11:245–267.

Muller, H. J. 1940. Bearings of the *Drosophila* work on systematics. Pp. 185–268 in: J. S. Huxley (ed.), *The New Systematics*. Clarendon Press, Oxford.

Nagylaki, T., and M. Moody. 1980. Diffusion model for genotype-dependent migration. Proc. Natl. Acad. Sci. U.S.A. 77:4842–4846.

Nei, M., and W.-H. Li. 1973. Linkage disequilibrium in subdivided populations. Genetics 75:213–219.

O'Donald, P. 1960. Assortative mating in a population in which 2 alleles are segregating. Heredity 15:389–396.

Papaj, D. R., and R. J. Prokopy. 1986. Phytochemical basis of learning in *Rhagoletis pomonella* and other herbivorous insects. J. Chem. Ecol. 12:1125–1143.

Paterson, H. E. H. 1981. The continuing search for the unknown and the unknowable: A critique of contemporary ideas on speciation. South Africa J. Sci. 77:133–119.
Price, P. W. 1980. *Evolutionary Biology of Parasites.* Princeton University Press, Princeton, NJ.
Rausher, M. D. 1984. The evolution of habitat selection in subdivided populations. Evolution 38:596–608.
Rice, W. R. 1984. Disruptive selection on habitat preference and the evolution of reproductive isolation: A simulation study. Evolution 38:1251–1260.
Rice, W. R. 1987. Speciation via habitat specialization: The evolution of reproductive isolation as a correlated character. Evol. Ecol. 1:301–314.
Smouse, P. E., and J. V. Neel. 1977. Multivariate analysis of gametic disequilibrium in the Yanomama. Genetics 85:733–752.
Taylor, C. E. 1976. Genetic variation in heterogeneous environments. Genetics 83: 887–894.
Walsh, B. 1864. On phytophagic varieties and phytophagic species. Proc. Entomol. Soc. Philadelphia 3:403–430.
White, M. J. D. 1978. *Modes of Speciation.* Freeman, San Francisco.
Wilson, D. S., and M. Turelli. 1986. Stable underdominance and the evolutionary invasion of empty niches. Am. Natur. 127:835–850.

CHAPTER 15

THE DIVERSIFICATION OF SINGLE GENE POOLS BY DENSITY- AND FREQUENCY-DEPENDENT SELECTION

David Sloan Wilson

Species in nature frequently coexist by doing different things: feeding on different resources, occupying different habitats, avoiding predators in different ways, and so on. The idea that natural environments afford multiple niches, allowing for the coexistence of functionally diverse species, is so obvious that few ecologists would seek to question it.

Less obvious is the idea that the environment of a single species can provide multiple niches, leading to the evolution and coexistence of functionally diverse forms within a single gene pool. An example will help make this idea clear. In some lakes the Cladoceran zooplankton species *Daphnia pulex* exists as two genetically distinct forms (Goulden et al., in preparation). One form has a combination of traits that is adaptive in the presence of intense fish predation: individuals are clear in appearance, migrate diurnally, and have a life history geared toward early reproduction at a small size. The second form possesses hemoglobin that colors the animal red (increasing vulnerability to predation), but allows it to descend into deeper deoxygenated waters that are food poor and predator safe. This form never migrates into the surface waters and has a life history geared toward late reproduction at a large size. Thus, it appears that some lakes provide *D. pulex* with not one but two niches, each requiring a different coordinated suite of physiology, life history, and behavior.

Although species are often conceived as homogeneous units occupying a single "niche," it is likely that many populations are faced with more than one possible strategy for survival and reproduction. If so, then the evolution of functional diversity in multiple-niche environments becomes an important theoretical question. How easy is it for disruptive selection to produce two or more specialized forms in a population, each with its own unique configuration of traits?

As with most basic questions, this one has a long and complex history that differs for the various subdisciplines of ecology and evolution. In some ways it is a very traditional question, but attempts to answer it frequently encounter difficulties. After all, if the splitting of a population into separate forms were a straightforward process, sympatric speciation would not be the controversial subject that it is. At least four problems can be identified.

1. *Crossing maladaptive valleys*. If we begin with a population adapted to one niche, and if adaptation to a second niche requires a different suite of traits, how can the entire suite arise by microevolution (Frazzetta 1975)? It seems unreasonable to expect the entire suite to appear as a single mutation, and yet when the elements arise separately they are maladaptive against the background of the original form. In *Daphnia pulex*, for example, hemoglobin by itself is disadvantageous unless coupled with an altered pattern of vertical migration. Using the metaphor of adaptive landscapes, we may envision the well-adapted forms as peaks of high fitness surrounded by valleys of unfit combinations of traits, which prevent the splitting of single gene pools into specialized forms. Wright's shifting balance theory was invented, in part, to solve this problem (Provine 1986).
2. *Mating between forms*. Even if we begin with two specialized forms occupying separate adaptive peaks, they are members of the same species and will mate with each other, producing progeny that are intermediate in form and therefore unfit. If mating is at random, and if one specialized form is less common than the other, its progeny will consist mostly of unfit intermediates and it ultimately will go extinct. This problem is identical to the unstable polymorphism produced in a single locus model with constant fitness values and heterozygote disadvantage. It is Paterson's (1978, 1980) primary argument against the evolution of reproductive isolation among incipient species in sympatry.
3. *The importance of population structure*. Virtually all attempts to model the evolution of functional diversity invoke population structure (e.g., Maynard Smith 1966; Bush 1975; reviewed by Templeton 1981; Futuyma and Mayer 1980; Rice 1987). In other words, not only are multiple niches required, but the niches must be separated from each other in space or in time. The particular role of population structure varies from model to model, but usually it serves to disrupt gene flow, ameliorating

the two problems previously outlined. In the extreme case the populations within each niche exist as isolated gene pools, and the evolution of functional diversity is straightforward. These models are worthwhile because multiple niches often do have an associated population structure in nature (e.g., the *D. pulex* example described). Nevertheless, for reasons that will become apparent, the concept of multiple niches should not be linked too closely with the concept of population structure.

4. *Genetic and developmental constraints.* If correlations among characters are fixed by genetic and developmental constraints (Raff and Kaufman 1983; Alberch 1982), then the evolution of multiple forms, each with separate correlations among characters, will be correspondingly difficult. If meaningful evolution is confined to speciation events (Eldredge and Gould 1972; Gould 1982; Stanley et al. 1983), then disruptive selection on large populations should not have an effect. In general, any factors that constrain natural selection should doubly constrain the evolution of functional diversity.

DENSITY- AND FREQUENCY-DEPENDENT SELECTION

Among the threads that run through the history of this subject is an idea developed first by Ludwig (1950) and more recently by Pimm (1978) and Rosenzweig (1978). The idea also exists as a verbal scenario familiar to most evolutionists. The scenario begins with a single population adapted to one niche, and a second niche that is unexploited. A mutant form appears that is maladaptive in niche 1, but crudely adapted to niche 2. The mutant is poorly adapted to both niches, but may still have a high fitness because niche 1 is "full" whereas niche 2 is "empty." The mutant form may therefore increase in frequency until its own numbers "fill" niche 2, leading to a stable polymorphism.

The words "full" and "empty" imply that the fitness of a form changes with the density and genetic composition of the evolving population. In general, attempts to model the evolution of functional diversity pay scant attention to density and frequency dependence (Udovic 1980), although both appear implicitly as "soft selection" in subdivided population models (Levene 1953; Maynard Smith 1966: Christiansen 1975; Maynard Smith and Hoekstra 1986; Hedrick 1986; Diehl and Bush, this volume). More recently, Wilson and Turelli (1986) attempted to model the verbal scenario in a way that makes density and frequency dependence explicit, and removes the confounding influence of population structure. An abbreviated version of the model is presented here.

We begin with two resources (the "niches") and one randomly mating consumer population that is polymorphic at a single locus. The three genotypes differ in their ability to consume the two resources, but otherwise are identical. The equilibrium density of resource i is

$$\hat{R}_i = \frac{K_i}{N(p^2 c_{i,AA} + 2pq c_{i,Aa} + q^2 c_{i,aa})} \tag{1}$$

where K_i is the density of resource i in the absence of consumption, N is the density of the consumer population, p is the frequency of the A allele, and $c_{i,AA}$, $c_{i,Aa}$, $c_{i,aa}$ are the consumption rates of the three genotypes on resource i. This equation is a simple way to model resources that decline with the density and harvesting rate of their consumers. Formally, it assumes that resource population dynamics are fast relative to consumer population dynamics, so that resource densities equilibrate to consumer density, even when the latter is not in equilibrium (MacArthur 1972).

The absolute fitnesses of three genotypes are

$$W_{AA} = 1 + ec_{1,AA}R_1 + ec_{2,AA}R_2 - T \tag{2}$$

$$W_{Aa} = 1 + ec_{1,Aa}R_1 + ec_{2,Aa}R_2 - T \tag{3}$$

$$W_{aa} = 1 + ec_{1,aa}R_1 + ec_{2,aa}R_2 - T. \tag{4}$$

Notice that consumer fitness increases with harvesting rate and resource density. These are identical to MacArthur's (1972) standard equations for consumer population dynamics, in which birth is a product of resource abundance and consumption rate (times a factor e that converts units of resource into units of consumer), and death is represented by a constant number T. When birth rate balances death rate, the absolute fitness of a genotype is one. These equations are explicitly density and frequency dependent, because the fitness of genotypes depends on resource abundance, and resource abundance depends on the density and genetic composition of the consumer population.

Next we must assign values for consumption rates (Figure 1). The verbal scenario assumes that the two niches require different adaptations, so that a single form cannot successfully utilize both. This is represented by the concave line in Figure 1, which gives the boundary of possible phenotypes. The AA genotype is well adapted to resource 1 but poorly adapted to resource 2, in keeping with the verbal scenario. The mutant Aa genotype is slightly better at consuming resource 2, but substantially worse at consuming resource 1. The aa genotype, which hardly ever appears in the population when the a allele is at mutation frequency, is still better on resource 2 and worse on resource 1.

Two additional aspects of Figure 1 deserve comment. First, notice that the Aa and aa phenotypes lie below the concave line, signifying that they are not the best that evolution can produce. This is intended to represent the fact that even beneficial mutations frequently do not produce the best

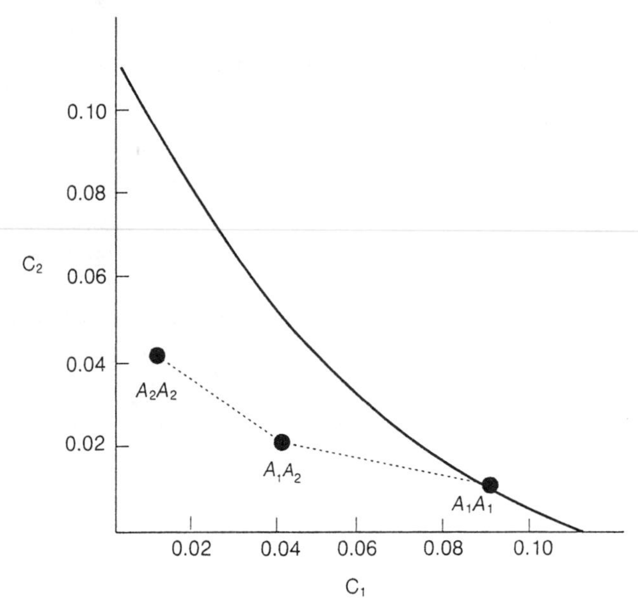

FIGURE 1. Consumption rates on resource 1 (C_1) and resource 2 (C_2). The solid line is the boundary of the phenotype set, representing the best that natural selection can produce. The three points represent the consumption rates of the *AA*, *Aa*, and *aa*, genotypes. In keeping with the verbal scenario, the *Aa* and *aa* genotypes are crudely adapted to resource 2 relative to the *AA* genotype, but at the expense of being poorly adapted to resource 1. See text for additional explanation. (From Wilson and Turelli 1986.)

possible phenotype. The endpoint of natural selection may lie on the boundary of the phenotype set (by definition), but the intermediate stages must often pass through the interior region. Second, notice that the three genotypes follow the contour of the boundary, so that the dashed line connecting them is similarly concave. This is important for the results of the model, and is discussed in more detail below.

Because the results of the model will be presented in the form of adaptive landscapes, it is important to be explicit about this common but poorly understood concept. As documented by Provine (1986:307–317), Wright actually developed two notions of adaptive landscapes. In the first, a point on the landscape refers to the mean fitness of an entire population. Natural selection is supposed to maximize mean fitness, which causes populations to evolve to the top of adaptive peaks. Unfortunately, this generalization breaks down when fitness is frequency and density dependent, which severely limits the utility of the concept. In the second notion, a point on the landscape refers to the fitness of a single genotype. A slope on the landscape signifies directional selection favoring a given genotype, whereas two peaks separated by a valley signifies disruptive selection favoring extreme genotypes at the expense of intermediates. This is the type of landscape employed below.

The effect of density- and frequency- dependent selection on the evolution of functional diversity can be demonstrated by following a single simulation run of the model. As shown in Figure 2A, we begin with the *A* allele close to fixation ($p = 1$) and the consumer population at very low density. With few consumers, both resources are close to their carrying capacities, and the fitness of the three genotypes is governed mostly by their consumption rates. The *AA* genotype is the best adapted consumer, and the adaptive landscape takes the form of a negative slope, representing directional selection for the *A* allele. We expect the consumer population to grow, and the frequency of the *A* allele to remain close to 1.

In Figure 2B, the consumer population has reached its carrying capacity. Both resources have been reduced to low levels, but since the consumer population consists mostly of *AA*, resource 1 is depressed more than resource 2. Now the low consumption rates of *Aa* and *aa* are compensated by more resources, and the adaptive landscape has shifted to a positive slope, representing directional selection for the *a* allele. We expect the consumer population to remain at high numbers, and the frequency of the *A* allele to decline.

In Figure 2C, the consumer population has reached an evolutionary equilibrium of $p = 0.67$, with corresponding shifts in resource abundance. A large number of *AA* genotypes still exist, which effectively graze resource 1. An appreciable number of *aa* genotypes also exist, which less effectively graze resource 2. With its intermediate consumption rates, the *Aa* genotype is a relatively inefficient generalist among specialists. The adaptive landscape has become two peaks separated by a valley of low fitness, representing selection against the intermediate form. Pimm (1978) reached a similar conclusion using a model based on the Lotka–Volterra equations.

When fitnesses are constant, polymorphisms with heterozygote disadvantage are unstable, leading to the fixation of the most common allele (see problem 2 previously outlined). We can test the stability of this polymorphism by making p close to 0, as shown in Figure 2D. Now the consumer population consists mostly of *aa* genotypes, creating an abundance of resource 1. The adaptive landscape shifts once again to a negative slope, favoring the *A* allele and returning the polymorphism to it equilibrium value of $p = 0.67$.

So far our single-locus model fails to address the problem of how adaptive combinations of traits evolve, whose elements are maladaptive in other

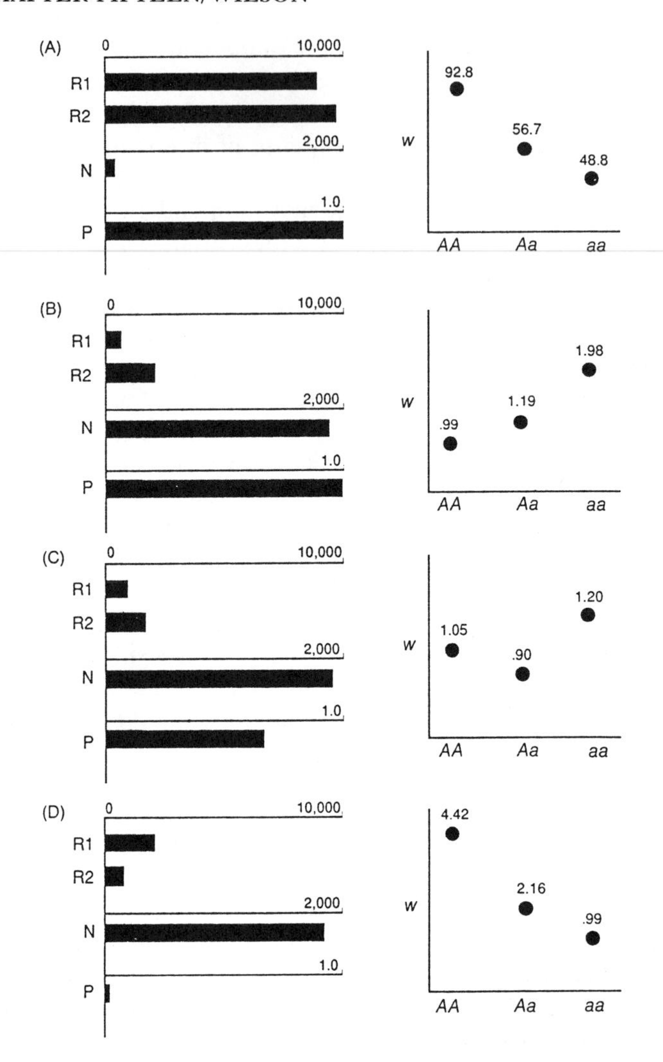

FIGURE 2. Four stages of a single simulation run in which $K_1 = K_2 = 10{,}000$, $e = 0.1$, $T = 1$, and consumption rates are as shown in Figure 1. Each stage shows equilibrium resource densities (top two histograms), consumer population density (third histogram), frequency (p) of the A allele (fourth histogram), and the absolute fitness of the three genotypes. The simulation begins with the consumer population at a low density and $p \approx 1$ (A). The next three stages are (B) consumer population at carrying capacity with $p \approx 1$, (C) consumer population at carrying capacity with $p = 0.67$ (evolutionary equilibrium), and (D) consumer population at carrying capacity with $p \approx 0$. Notice that the adaptive landscapes change from a slope favoring the A allele, to a slope favoring the a allele, to two peaks separated by a valley. The result is a stable polymorphism with heterozygote disadvantage.

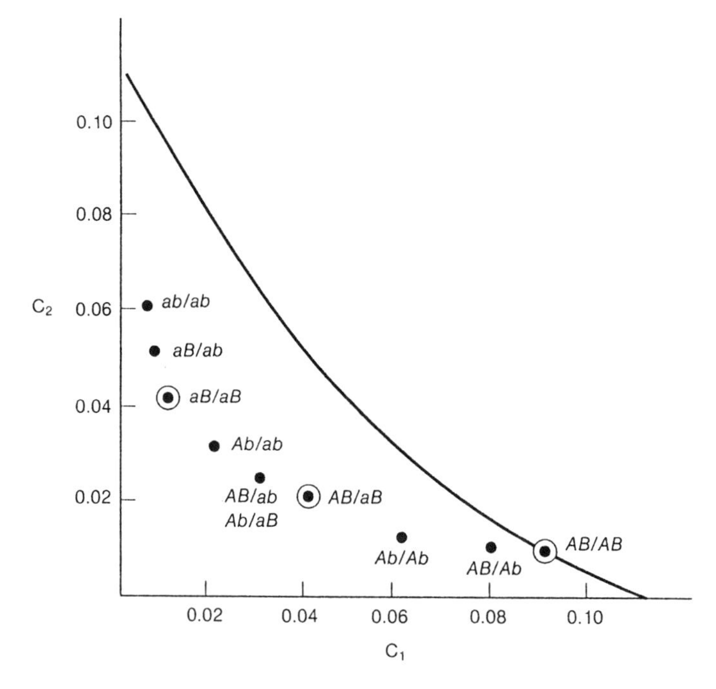

FIGURE 3. A two-locus version of the model, in which a polymorphic *B* locus is added to the *A* locus of Figure 1. *AB/AB* genotypes are best adapted to resource 1, whereas *ab/ab* genotypes are best adapted to resource 2. (From Wilson and Turelli 1986.)

combinations. We can begin to address this issue by considering a mutation at a second freely recombining locus, as shown in Figure 3. The *Bb* genotype is maladaptive in combination with *AA*, because it decreases consumption of resource 1 without increasing consumption of resource 2. In combination with *Aa* or *aa*, however, it increases specialization on resource 2. Figure 4 shows the resulting three-dimensional adaptive landscapes. As for the one-locus model, the topography shifts from a slope favoring the AB/AB homozygote, to a slope favoring the ab/ab homozygote, to two peaks separated by a valley of low fitness, depending on the allele frequencies at the two loci. The result is a stable two-locus polymorphism.

To summarize, ecologically plausible forms of density- and frequency-dependent selection can lead to stable polymorphisms with intermediate phenotypes relatively unfit. It is important to emphasize, however, that this by itself does not produce a bimodal distribution of phenotypes. Natural selection repeatedly acts against the intermediates, but sexual reproduction repeatedly recreates them (Felsenstein 1979). In the two-locus model

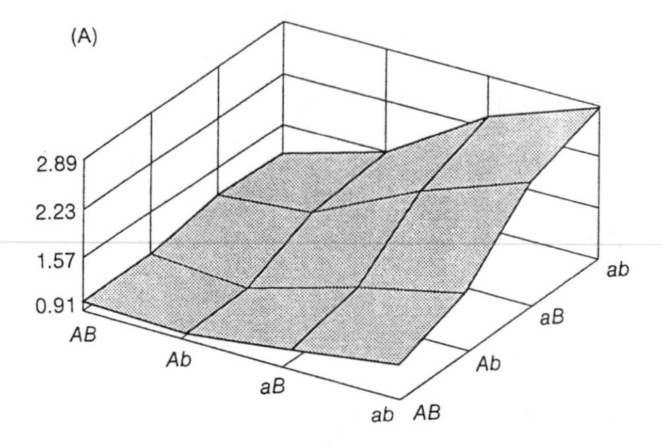
(A)
2.89
2.23
1.57
0.91
AB
Ab
aB
ab
AB
Ab
aB
ab

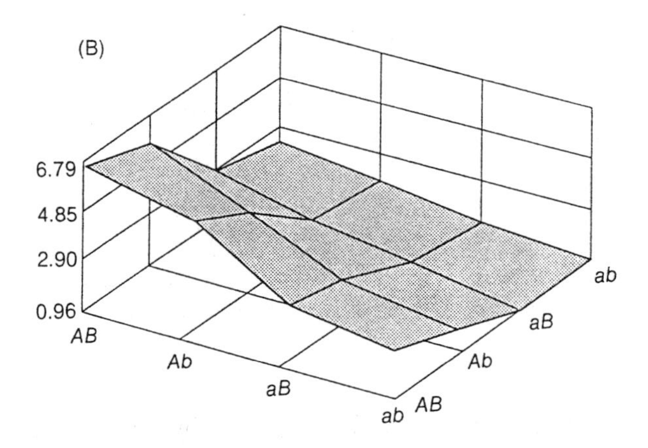
(B)
6.79
4.85
2.90
0.96
AB
Ab
aB
ab
AB
Ab
aB
ab

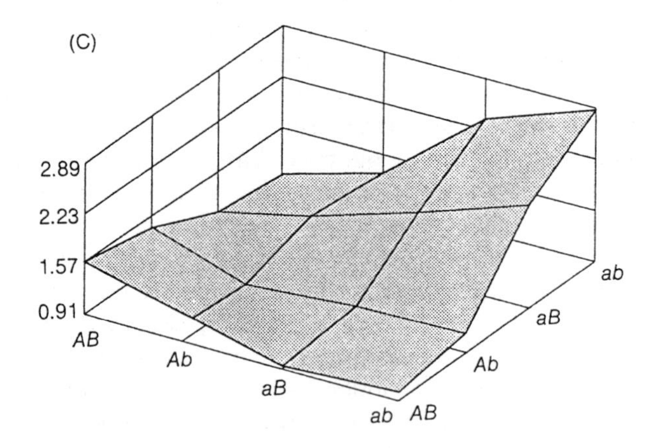
(C)
2.89
2.23
1.57
0.91
AB
Ab
aB
ab
AB
Ab
aB
ab

described, for example, 86% of the population are unfit intermediates at evolutionary equilibrium. Nevertheless, disruptive selection does set the stage for the evolution of additional traits that eliminate the intermediate phenotypes.

Intermediate phenotypes can be eliminated in three ways. The first is by the evolution of closer linkage, which by combining the two loci into a single "supergene" can reduce the number of genotypes from 10 to 3 (Hedrick et al. 1978). The second is by the evolution of dominance and other epistasis modifiers that cause heterozygotes to resemble homozygotes phenotypically. The third is by the evolution of assortative mating, which in the extreme becomes a speciation event. These three ways to eliminate intermediate forms lead to very different outcomes, but they should be regarded as alternative adaptive solutions to the same problem. Understanding the relative likelihood of the three solutions will be a rewarding area of future research.

Stable polymorphisms with heterozygote disadvantage are a common feature of the Wilson–Turelli model, whenever the consumption rates of the three genotypes describe a concave surface as in Figure 1. All genotypes have the same fitness at equilibrium when their consumption rates form a straight line, and heterozygotes are most fit when the consumption rates describe a convex surface. All three conditions yield a stable polymorphism, but only with heterozygote disadvantage do we expect subsequent evolution to eliminate the intermediate forms.

In evaluating the likelihood of stable polymorphisms with heterozygote disadvantage, it is important to distinguish between a trait (such as a morphological structure or a behavior) and the consumption rates produced by the trait. Among centrarchid fish, for example, plankivores differ from piscivores in mouth size, how protusible the mouth is, the spacing between the gill rakers, and a variety of other morphological traits. A fish that is morphologically halfway between a planktivore and a piscivore may not be half as good at catching plankton and fish as the respective specialists (Werner 1977). Thus, an intermediate morphology that lies on the line connecting those of two homozygotes may often yield consumption rates that lie below the line connecting the consumption rates of the two

◄ **FIGURE 4.** Three-dimensional adaptive landscapes derived from Figure 3, showing that (a) directional selection favors the *ab/ab* genotype when p_A and p_B are ≈ 1, (b) directional selection favors the *AB/AB* genotype when p_A and p_B are ≈ 0, and (c) the adaptive landscape becomes two peaks separated by a valley at evolutionary equilibrium. This two-locus polymorphism is stable, in spite of the fact that random mating causes the majority of individuals in the population to be relatively unfit intermediate genotypes.

homozygotes. In general, it is reasonable to expect the three genotypes to describe a concave surface for the same reasons that it is reasonable to expect the boundary of the fitness set to describe a concave surface.

The model presented here only begins to explore the implications of density- and frequency-dependent selection for the evolution of functional diversity (see Udovic 1980; Wilson and Turelli 1986 for more formal treatments). Nevertheless, it conveys an image of adaptive landscapes that fluctuates with the density and genetic composition of the evolving populations, in ways that may solve some of the problems encountered by other models. Crossing adaptive peaks is not necessarily a problem, because the peaks and valleys do not exist until after the polymorphism evolves. Mating between forms does not necessarily disrupt the stability of the polymorphism. Population structure may well facilitate the evolution of functional diversity, but it is not required. Finally, we can say nothing about genetic and developmental constraints, which were not built into our model.

FUNCTIONAL DIVERSITY IN THE MORPHOLOGY AND BEHAVIOR OF BLUEGILL SUNFISH (*LEPOMIS MACHROCHIRUS*)

The model that has been described suggests that single species in multiple-niche environments should sometimes differentiate into specialized forms, even in single panmictic gene pools. T. J. Ehlinger and I are studying a possible example in the bluegill sunfish (*Lepomis macrochirus*; Ehlinger and Wilson, in press). In this section we will briefly summarize our evidence and review examples of functional diversification in other organisms.

Bluegill are often the most common fish species inhabiting the lakes and ponds of eastern North America, and they have been studied extensively from the standpoint of foraging ecology (reviewed by Werner 1984), functional morphology (Webb 1982, 1984; Lauder and Lanyon 1980; Lauder 1980; Wainwright and Lauder 1986), social behavior (Brown and Colgan 1985, 1986), mating behavior (Avila 1976; Dominey 1980; Gross and Charnov 1980; Gross 1982; Colman et al. 1985), and genetic diversity (Avise and Felley 1979; Felley and Avise 1980; Avise et al. 1984; Pasdar et al. 1984; Whitmore 1986; Dawley 1987). Bluegill already have been shown to be functionally diverse in male mating behavior (Gross 1982; Dominey 1980). A "parental " form matures at a large size and defends territories, whereas a "cuckolding" form matures at a small size and sexually parasitizes the nests of parental males.

The life history of bluegill includes three distinct niche shifts (Werner and Gilliam 1984). Shortly after hatching, larvae leave the nests of parental males and move into the open water to feed on zooplankton. The open water is also a relatively safe habitat for the growing larvae since their small size, inconspicuous coloration and rapid escape responses make detection and capture by fish predators difficult. As the larvae grow, however, they become increasingly susceptible to fish predation and eventually move into the littoral zone and the safety of aquatic vegetation. As juveniles, bluegill are confined to the vegetation by predators (primarily the large-mouth bass, *Micropterus salmoides*), remaining there even when the open water habitat offers more food (Mittelbach 1981; Werner et al. 1983a, 1983b). Larger bluegill (> 80 mm standard length) are relatively immune to predation, and move freely between habitats as food abundance changes.

In terms of their foraging ecology, bluegill are usually regarded as generalist small-mouthed predators, capable of foraging in a wide range of habitats on a wide variety of prey types. They are a favorite species for optimal foraging studies, and have been shown to select both habitats and prey within habitats in ways that (roughly) maximize energy gain (Werner et al. 1983a). They also can modify the details of their search and capture techniques to effectively capture specific prey types (Ehlinger 1986, in press a, b).

Previous studies of bluegill foraging implicitly assume that the populations of single lakes are monomorphic. All fish of a given size are expected to be found in either the vegetation or the open water habitat, depending on the size-specific foraging rates and risks of predation (Werner et al. 1983a, 1983b). Our own studies, however, suggest that the populations of single lakes are functionally diversified into vegetation and open water specialists (Ehlinger and Wilson 1988; Ehlinger in press a, b). Our evidence can be summarized in six statements:

1. Open water and vegetation habitats require different techniques for successful foraging. Bluegill search for prey while hovering motionless, using their pectoral fins to maintain their position in the water (Ehlinger 1986, in press a), If no prey item is detected, they move to another location and again hover. The open water is an unstructured habitat with dispersed prey that are relatively conspicuous and easy to capture once found. Foraging rate is maximized by short hover durations and quick movement between hovers. The vegetation is a structured habitat with cryptic prey that often hide if they sense the approach of a predator (Charnov et al. 1976). Foraging rate is maximized by long hover durations and slow, stealthy movements between hovers (Ehlinger and Wilson 1988; Ehlinger in press a, b).
2. Laboratory experiments show that individual bluegill adjust their foraging tactics to increase their feeding rates within both habitats (Mittelbach 1981; Ehlinger 1986, in press a). This learning process can require a period of days and a fish that moves from one habitat to another

does not automatically adopt the habitat-specific foraging tactics. Therefore, in a new habitat it begins with a low foraging rate that increases gradually over a number of feeding trials (Ehlinger 1986, in press a, b). The same process of gradual behavioral adjustment to new habitats has been inferred from field gut-content data (Werner et al. 1981).

3. In spite of their ability to learn, bluegill within single lakes exhibit inflexible differences in their prey searching behavior. These differences are not unimodally distributed, but tend to cluster into "short-hover" and "long-hover" types, which perform best in the open water and vegetation, respectively (Figure 5). This pattern has been documented twice in separate cohorts from the same lake, and behavioral differences between individual fish remain stable for at least 20 weeks.
4. Bluegill captured in the open water and vegetation of a given lake are morphologically different. Fish from the vegetation are deeper bodied,

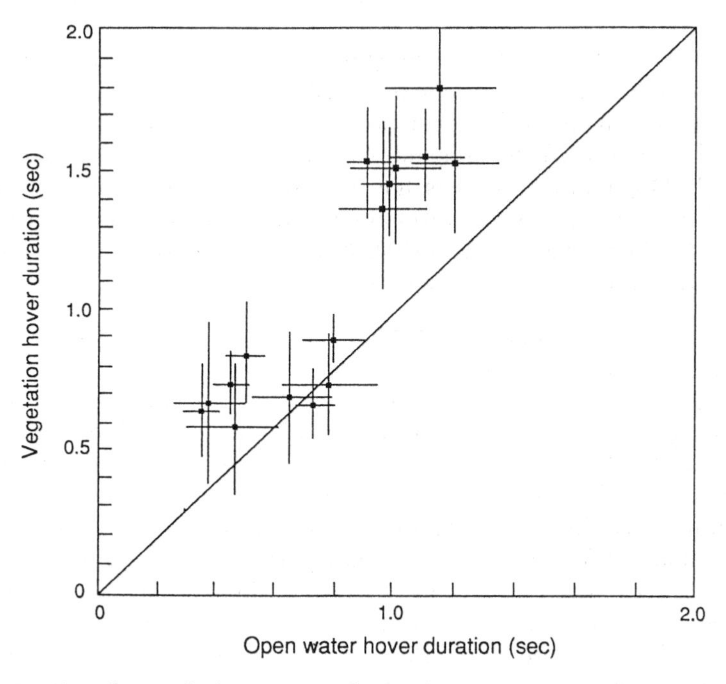

FIGURE 5. Behavioral phenotypic polymorphism in the bluegill sunfish. Hover duration of individual fish were measured in both open water (x axis) and vegetation (y axis) habitat. The fact that most points lie above the diagonal line means that individual fish hover longer in the vegetation habitat. The fact that the points form two clusters means that inflexible differences exist between fish, with "long hover" and "short hover" forms. (From Ehlinger and Wilson 1988.)

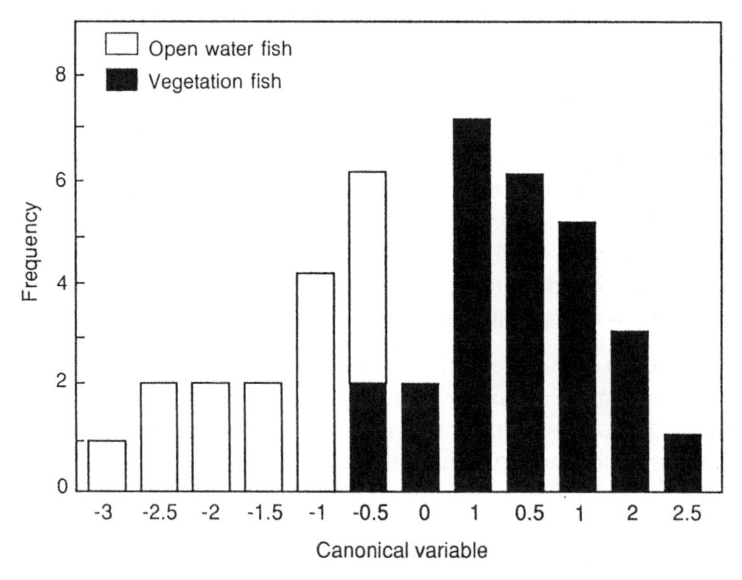

FIGURE 6. Frequency distribution of the morphological canonical variable for bluegill sunfish collected from the open water (open bars) and vegetation (shaded bars) habitats of a single lake. The standardized canonical coefficients are 0.41 × body depth, 1.02 × pectoral fin length, 0.59 × horizontal pectoral fin position, 1.17 × pelvic fin length, and −0.60 × vertical pectoral fin position. Fish from the two habitats are morphologically different from each other ($p < 0.0001$), and the entire population appears to be bimodally distributed in morphological space. The bimodal distribution exists only for the canonical variable; single measurements (e.g., pectoral fin length, body depth) appear to be unimodally distributed. (From Ehlinger and Wilson 1988.)

have longer pectoral and pelvic fins, and the pectoral fins are inserted further back on the body. Fish from the open water are more fusiform. These differences can be interpreted as morphological adaptations to swimming in structured versus unstructured habitats (Webb 1984). The differences exist for both males and females, and for the full size range that exists in both habitats. When fish from both habitats are considered together, single measurements (such as body depth relative to standard length) are unimodally distributed, but discriminant functions and principal components that combine several measurements into single indices are often bimodally distributed (Figure 6). At present we have no data on the heritability of these morphological forms.

5. Morphology, behavior, habitat use, and foraging rate are intercorrelated. Bluegill with deep bodies, long pectoral fins, etc. use long hovers, have higher foraging rates in the vegetation, and are found mostly in the vegetation habitat of lakes.

6. Bluegill caught from the open water and vegetation carry different parasites (unpublished data in collaboration with P. Muzall). Open water fish are infested primarily by a cestode whose intermediate host is a copepod, whereas vegetation fish are infested primarily by a digenetic trematode, whose intermediate host is a snail. The large difference in parasites signifies that the trophic "forms" spend substantial amounts of time in their respective habitats, and may also be a factor in habitat choice along with food and predators. It should be emphasized that all bluegill breed in the littoral zone, so that habitat segregation while foraging does not result in segregation while mating.

To summarize, the bluegill populations of single lakes are not monomorphic, but instead are behaviorally and morphologically diverse. A continuum of phenotypes exists, with a suggestion of correlated "clusters" of behavior and morphology adapted for foraging in the open water and vegetation habitats. Rather than a population of generalists, bluegill appear to be a population of multiple specialists. As a result, bluegill of the same size are not all found in one habitat or the other, but rather segregate into the habitat to which they are best adapted.

Bluegill are by no means the only example of functional diversity in single populations. Many species of fish (e.g., Liem 1980; Kornfield et al. 1982) and other organisms (e.g., Lively 1986; Collins and Cheek 1983; West-Eberhard 1983) are differentiated into ecologically distinct forms, which sometimes are so different that they were originally classified as separate species (Kornfield and Koehn 1975). Often the phenotypic diversity does not have a genetic basis, but rather is caused entirely by environmental induction. In other cases, a genetic component is indicated (e.g., Grant 1986; Turner et al. 1984; Gross and Philipp 1988; Goulden et al. 1988). Either way, the evolution and maintenance of phenotypic diversity in single gene pools may possibly serve as a starting point for speciation events (Pimm 1978; Rosenzweig 1978; West-Eberhard 1983, 1986; Wilson and Turelli 1986).

DISCUSSION

In his preface to the second edition of *Darwin's Finches*, Lack (1961) informed his readers of a shift in evolutionary thought. Adaptive explanations of differences between subspecies and closely related species, which were unacceptable when he wrote the book in 1947, had become acceptable! The new emphasis on natural selection was a hallmark of the modern synthesis (Gould and Lewontin 1979). Nevertheless, Lack's statement also implies that differences between populations within subspecies, and differences between individuals within populations, should perhaps not be regarded as adaptive. This is consistent with Mayr's (1963) vision of species as integrated genetic systems united by gene flow, which prevents adaptation to local circumstances and requires small isolated populations for significant evolutionary change. Thus, the modern synthesis invoked natural selection primarily to explain differences between species, as opposed to differences between populations and individuals within species.

It has taken decades since then to appreciate that differences between local populations, sometimes only a few meters apart from each other, can also have adaptive explanations (e.g., Endler 1983; Schemske 1984). Natural selection often is far more intense (Endler 1986) and gene flow often is far more limited (Ehrlich and Raven 1969) than envisioned by the architects of the modern synthesis. In many ways the concept of local adaptation represents a shift in evolutionary thought that is even more profound than the shift that occurred between 1947 and 1961.

Still, it is likely that this long-term historical trend has not yet reached its endpoint. In particular, the nature of individual differences within single populations has not received the attention accorded to differences between local populations and between species (Clark and Ehlinger 1986). In most cases, we do not know if differences represent the product of natural selection (an adaptive explanation), as opposed to the raw material on which natural selection acts (a nonadaptive explanation). Attitudes on this subject are often based more on theoretical models than on empirical data, and therefore vary from one subdiscipline to another in evolution and ecology. It is interesting to categorize these subdisciplines according to their emphasis on frequency- and density-dependent selection. For example, Van Valen's (1965) "niche variation hypothesis," Maynard Smith's (1982) concept of "evolutionarily stable strategies," and the subject of mimicry (Gilbert 1983) all explicitly assume that the fitness of genotypes is not a constant value, but depends on other genotypes in the population. All three subdisciplines predict the existence of phenotypic diversity in single populations. On the other hand, the field of population genetics has historically been dominated by constant-fitness models. As late as 1958 Lewontin's exploration of frequency- and density-dependent selection could be regarded as pathbreaking, and as late as 1975 Charlesworth and Charlesworth could write "equilibria with heterozygote disadvantage are, perhaps, unexpected . . . their plausibility in biological systems has so far seemed dubious" (p. 301). Against this background, the evolution of functional diversity within populations appears problematical.

Theories of speciation, both verbal and mathematical, are particularly dominated by constant-fitness thinking. Again and again, incipient species and their hybrid progeny are represented as either "fit" or "unfit" in a given habitat, without any reference to the presence of other genotypes in the population. It is therefore not surprising that the differentiation of single gene pools, ultimately leading to speciation, is regarded not only as unlikely but positively heretical (e.g., Paterson 1978, 1980)!

In spite of this grand mosaic of contrasting views, it is likely that ecologically plausible forms of density- and frequency-dependent selection can cause a randomly mating population to bifurcate into ecologically distinct forms whose intermediates are relatively unfit—two adaptive peaks separated by a valley. Furthermore, the existence of open water and vegetation forms in a species as well studied as the bluegill sunfish suggests that functional diversity must be specifically looked for to be found, and may be far more common in nature than currently perceived. Nevertheless, this does not mean that truly sympatric speciation has suddenly become a straightforward process. Sympatric speciation is a two-stage event, in which the evolution of separate forms must be followed by the evolution of reproductive isolation. While frequency- and density-dependent selection may solve many of the problems associated with the first stage, on which arguments against sympatric speciation traditionally have been based, numerous problems may still be associated with the second stage (Felsenstein 1981; Wilson and Hedrick 1982; Spencer et al. 1986; but see also Rice 1987; Diehl and Bush, this volume). If so, then the long-standing attitudes against sympatric speciation will be right for the wrong reasons. Clearly, the whole problem of functional diversification and its consequences deserves a fresh look, both theoretically and empirically, unburdened by assumptions from the past.

ACKNOWLEDGMENTS

This chapter reviews theoretical work done in collaboration with M. Turelli and empirical work done in collaboration with T. J. Ehlinger. I also thank A. B. Clark, J. Brown, and the ecology group at the Kellogg Biological Station for helpful comments and discussion. Supported by NSF Grant BSR-8320457.

LITERATURE CITED

Alberch, P. 1982. Developmental constraints in evolutionary process. Pp. 313–332 in: J. T. Bonner (ed.), *Evolution and Development*. Dahlem Konferenzen. Springer-Verlag, New York.

Avila, V. L. 1976. A field study of nesting behavior of male bluegill sunfish (*Lepomis macrochirus* Rafinesque). Am. Midl. Natur. 96:125–206.

Avise, J. C., and J. Felley. 1979. Population structure of freshwater fishes. I. Genetic variation of bluegill (*Lepomis macrochirus*) populations in man-made reservoirs. Evolution 33:15–26.

Avise, J. C., E. Bermingham, L. G. Kessler, and N. C. Saunders. 1984. Characterization of mitochondrial DNA variability in a hybrid swarm between subspecies of bluegill sunfish (*Lepomis macrochirus*). Evolution 38:931–941.

Brown, J. A., and P. W. Colgan. 1985. The ontogeny of social behavior in four species of centrarchid fish. Behavior 92:254–276.

Brown, J. A., and P. W. Colgan. 1986. Individual and species recognition in centrarchid fishes: Evidence and hypotheses. Behav. Ecol. Sociobiol. 19:373–379.

Bush, G. L. 1975. Modes of animal speciation. Annu. Rev. Ecol. System. 6:339–364.

Charlesworth, D., and B. Charlesworth. 1975. Theoretical genetics of Batesian mimicry. I. Single-locus models. J. Theoret. Biol. 55:283–303.

Charnov, E. L., G. H. Orians, and K. Hyatt. 1976. Ecological implications of resource depression. Am. Natur. 110:247–259.

Christiansen, F. B. 1975. Hard and soft selection in a subdivided population. Am. Natur. 126:418–249.

Clark, A. B., and T. J. Ehlinger. 1987. Pattern and adaptation in individual differences. Pp. 1–47 in: P. P. G. Bateson and P. Klopfer (eds.), *Perspectives in Ethology*, Vol. 7. Plenum Press, New York.

Coleman, R. M., M. R. Gross, and R. C. Sargent. 1985. Parental decision rules: A test in bluegill sunfish, *Lepomis macrochirus*. Behav. Ecol. Sociobiol. 18:59–66.

Collins, J. P., and J. E. Cheek. 1983. Effect of food and density on development of typical and cannibalistic salamander larvae in *Ambystoma tigrinum nebulosum*. Am. Zool. 23:77–84.

Dawley, R. M. 1987. Hybridization and polyploidy in a community of three sunfish species (Pisces:Centrarachidae). Copeia 2:326–335.

Dominey, W. J. 1980. Female mimicry in male bluegill sunfish—a genetic polymorphism? Nature (London) 284:546–548.

Ehlinger, T. J. 1986. Learning, sampling and the role of individual variability in the foraging behavior of bluegill sunfish. Ph.D. Thesis, Michigan State University.

Ehlinger, T. J. In press, a. Learning and individual variation in bluegill foraging: Habitat specific techniques. Anim. Behav.

Ehlinger, T. J. In press, b. Learning and individual variation in bluegill foraging: Habitat switching and individual specializations. Am. Natur.

Ehlinger, T. J., and D. S. Wilson. 1988. Complex foraging polymorphism in bluegill sunfish. Proc. Natl. Acad. Sci. U.S.A., 85:1878–1882.

Ehrlich, P. R., and P. H. Raven. 1969. Differentiation of populations. Science 165: 1228–1232.

Eldredge, N., and S. J. Gould. 1972. Punctuated equilibria: An alternative to phyletic gradualism. Pp. 82–115 in: T. J. M. Schopf (ed.), *Models in Paleobiology*. Freeman, Cooper and Co., San Francisco.

Endler, J. A. 1983. Natural selection on color patterns in poeciliid fishes. Environ. Biol. Fishes 9:173–190.

Endler, J. A. 1986. *Natural Selection in the Wild*. Princeton University Press, Princeton, NJ.

Felley, J. D., and J. C. Avise. 1980. Genetic and morphological variation of bluegill populations in Florida lakes. Trans. Am. Fish. Soc. 109:108–115.

Felsenstein, J. 1979. Excursions along the interface between disruptive and stabilizing selection. Genetics 93:773–795.

Felsenstein, J. 1981. Skepticism towards Santa Rosalia, or why are there so few kinds of animals? Evolution 35:124–138.

Frazzetta, T. H. 1975. *Complex Adaptations in Evolving Populations*. Sinauer Associates, Sunderland, MA.

Futuyma, D. J., and G. C. Mayer, 1980. Non-allopatric speciation in animals. System. Zool. 29:254–271.

Gilbert, L. E. 1983. Coevolution and mimicry. Pp. 263–281 in: D. J. Futuyma and M. Slatkin (eds.), *Coevolution*. Sinauer Associates, Sunderland, MA.

Gould, S. J. 1982. The meaning of punctuated equilibrium and its role in validating a hierarchical approach to macroevolution. Pp. 83–104 in: R. Milkman (ed.), *Perspectives in Evolution*. Sinauer Associates, Sunderland, MA.

Gould, S. J., and R. C. Lewontin. 1979. The spandrels of San Marco and the panglossian paradigm: A critique of the adaptationist program. Proc. R. Soc. London B Ser. 205: 581–598.

Goulden, C. E., D. Walton, J. A. Bebak, and D. S. Wilson. In preparation. The ecological significance of vertical stratification of *Daphnia pulex* genotypes.
Grant, P. R. 1986. *Ecology and Evolution of Darwin's Finches*. Princeton University Press, Princeton, NJ.
Gross, M. R. 1982. Sneakers, satellites and parentals: Polymorphic mating strategies in North American sunfishes. Z. Tierpsychol. 60:1–26.
Gross, M. R., and E. L. Charnov. 1980. Alternative male life histories in bluegill sunfish. Proc. Natl. Acad. Sci. U.S.A. 77:6937–6940.
Gross, M.R., and D. P. Phillip. 1988. Genetic inheritance of alternative male life histories in fishes. In manuscript.
Hedrick, P. W. 1986. Genetic polymorphism in heterogenous environments: A decade later. Annu. Rev. Ecol. System. 17:535–567.
Hedrick, P. W., S. Jain, and L. Holden. 1978. Multilocus systems in evolution. Evol. Biol. 11:101–184.
Kornfield, I. I., and R. K. Koehn. 1975. Genetic variation and speciation in new world Cichlids. Evolution 29:427–437.
Kornfield, I. I., D. C. Smith, P. S. Gagnon, and J. N. Taylor. 1982. The cichlid fish of Cuarto Cienegas, Mexico: Direct evidence of conspecificity among distinct trophic morphs. Evolution 36:658–664.
Lack, D. 1961. *Darwin's Finches*, 2nd ed. Cambridge University Press, Cambridge.
Lauder, G. V. 1980. The suction feeding mechanism in sunfishes (*Lepomis*): An experimental analysis. J. Exp. Biol. 88:49–72.
Lauder, G. V., and L. E. Lanyon. 1980. Functional anatomy of feeding in the bluegill sunfish (*Lepomis macrochirus*): In Vivo measurement of bone strain. J. Exp. Biol. 84:33–55.
Levene, H. 1953. Genetic equilibrium when more than one ecological niche is available. Am. Natur. 87:331–333.
Lewontin, R. C. 1958. A general method for investigating the equilibrium of gene frequencies in a population. Genetics 43:419–434.
Liem, K. F. 1980. Adaptive significance of intra- and interspecific differences in the feeding repertoires of cichlid fishes. Am. Zool. 20:295–314.
Lively, C. M. 1986. Predator-induced shell dimorphism in the acorn barnacle. (*Chthamalus anisopoma*). Evolution 40:232–242.
Ludwig, W. 1950. Zur Theorie der Konkurrenz. Die Annidation (Einnischung) als funfter Evolutionsfaktor. Neue Erbeg. Probleme Zool, Klatt-Festschrift 1950, 516–537.
MacArthur, R. H. 1972. *Geographical Ecology.* Harper & Row, New York.
Maynard Smith, J. 1966. Sympatric speciation. Am. Natur. 104:487–490.
Maynard Smith, J. 1982. *Evolution and the Theory of Games*. Cambridge University Press, Cambridge.
Maynard Smith, J., and R. Hoekstra. 1980. Polymorphism in a varied environment: How robust are the models? Genet. Res. Cambridge 35:45–57.
Mayr, E. 1963. *Animal Species and Evolution.* Belknap Press, Cambridge, MA.
Mittelbach, G. G. 1981. Foraging efficiency and body size: A study of optimal diet and habitat use by bluegills. Ecology 62:1370–1386.
Pasdar, M., D. P Philipp, W. A. Mohammad, and G. S. Whitt. 1984. Differences in tissue expressions of enzyme activities in interspecific sunfish centrarchidae hybrids and their backcross progeny. Biochem. Genet. 22:931–956.
Paterson, H. E. H. 1978. More evidence against speciation by reinforcement. South Africa J. Sci. 74:369–371.
Paterson, H. E. H. 1980. A comment on "mate recognition systems." Evolution 34: 330–331.
Pimm, S. L. 1978. Sympatric speciation: A simulation model. Biol J. Linnean Soc. 11: 131–139.
Provine, W. B. 1986. *Sewall Wright and Evolutionary Biology.* University of Chicago Press, Chicago.
Raff, R. A., and T. C. Kauffman. 1983. *Embryos, Genes, and Evolution: The Developmental-Genetic Basis of Evolutionary Change*. Macmillan, New York.

Rice, W. R. 1987. Speciation via habitat specialization: The evolution of reproductive isolation as a correlated character. Evol. Ecol. 1:301–315.
Rosenzweig, M. L. 1978. Competitive speciation. Biol. J. Linnean Soc. 10:274–289.
Schemske, D. W. 1984. Population structure and local selection in *Impatiens pallida*, a selfing annual. Evolution 38:817–832.
Spencer, H. G., B. H. McArdle, and D. M. Lambert. 1986. A theoretical investigation of speciation by reinforcement. Am. Natur. 129:241–262.
Stanley, S. M., B. Van Valkenburgh, and R. S. Steneck. 1983. Coevolution and the fossil record. Pp. 328–349 in: D. J. Futuyma and M. Slatkin (eds.), *Coevolution*. Sinauer Associates, Sunderland, MA.
Templeton, A. R. 1981. Mechanisms of speciation—a population genetic approach. Annu. Rev. Ecol. System. 12:23–48.
Turner, B. J., T. A. Grudzien, K. P. Adkisson, and M. M. White. 1984. Evolutionary genetics of trophic differentiation in Goodied fishes of the genus *Ilyodon*. Environ. Biol. Fishes 9:159–172.
Udovic, D. 1980. Frequency-dependent selection, disruptive selection and the evolution of reproductive isolation. Am. Natur. 116:621–641.
Van Valen, L. 1965. Morphological variation and the width of the ecological niche. Am. Natur. 99:337–390.
Wainwright, P. C., and G. V. Lauder. 1986. Feeding biology of sunfishes: Patterns of variation in the feeding mechanism. Zool. J. Linnean Soc. 88:217–228.
Webb, P. W. 1982. Locomotor patterns in the evolution of Actinopterygian fishes. Am. Zool. 22:329–342.
Webb, P. W. 1984. Body form, locomotion and foraging in aquatic vertebrates. Am. Zool. 24:107–120.
Werner, E. E. 1977. Competition and habitat shift in two sunfishes (Centrarchidae). Ecology 58:869–876.
Werner, E. E. 1984. The mechanisms of species interactions and community organization in fish. Pp. 360–382 in: D. Strong et al. (eds.), *Ecological Communities: Conceptual Issues and Evidence*. Princeton University Press, Princeton, NJ.
Werner, E. E. and J. F. Gilliam. 1984. The ontogenetic niche and species interactions in size-structured populations. Annu. Rev. Ecol. System. 15:393–425.
Werner, E. E., G. G. Mittelbach, and D. J. Hall. 1981. The role of foraging profitability and experience in habitat use by bluegill sunfish. Ecology 62:116–125.
Werner, E. E., G. G. Mittelbach, D. J. Hall, and J. F. Gilliam. 1983a. Experimental tests of optimal habitat use in fish: The role of relative habitat profitability. Ecology 64: 1525–2539.
Werner, E. E., J. F. Gilliam, D. J. Hall, and G. G. Mittelbach. 1983b. An experimental test of the effects of predation risk on habitat use in fish. Ecology 64:140–148.
West-Eberhard, M. J. 1983. Sexual selection, social competition, and speciation. Quart. Rev. Biol. 58:155–183.
West-Eberhard, M. J. 1986. Alternative adaptations, speciation and phylogeny: A review. Proc. Natl. Acad. Sci. U.S.A. 83:1388–1392.
Whitmore, D. H. 1986. Identification of sunfish species by muscle protein isoelectric focusing. Comp. Biochem Physiol. 84:177–180.
Wilson, D. S., and A. Hedrick. 1982. Speciation and the economics of mate choice. Evol. Theory 6:15–24.
Wilson, D. S., and M. Turelli. 1986. Stable underdominance and the evolutionary invasion of empty niches. Am. Natur. 127:835–850.

CHAPTER 16

GENOTYPIC DIVERSITY AND COEXISTENCE AMONG SEXUAL AND CLONAL LINEAGES OF *POECILIOPSIS*

Robert C. Vrijenhoek

INTRODUCTION

We now have a large body of genetic and ecological theory that seeks to explain the overwhelming predominance of sexually reproducing species of plants and animals. The primary issue concerning sexuality in higher organisms is to explain how the benefits of genetic recombination can offset the twofold cost of producing males (Williams 1975; Maynard Smith 1978). For dioecious organisms, the production of a genotypically diverse brood, half being males, must double the fitness of a sexual parent to balance the immediate advantage gained from producing a similar number of all-female offspring, all else being equal. Since all-female cloning has arisen in broadly scattered taxa of higher plants and animals, we need to explain why clones have generally failed to replace their sexual ancestors.

Attempts to use the comparative method to draw broad contrasts between the life-styles of sexual versus clonal organisms have largely failed to discriminate among competing hypotheses concerning the benefits of sex (cf. Bell 1982; Bierzychudek 1987; Burt and Bell 1987; Charlesworth 1987). This failure is the result of many factors. First, there has been a lack of clearly defined alternative hypotheses leading to testable predictions. Second, severe constraints affect the opportunity for clonal lineages to arise in many taxa, confounding attempts to make broad phylogenetic comparisons. And third, our general lack of detailed knowledge about the life histories of most organisms prevents simple groupings of sexual versus asexual organisms for comparative purposes. Thus, a leading proponent of this approach recently concluded that "the extensive comparative work on sexuality should be complemented by increased effort to devise short-term experimental tests of the rival theories" (Bell 1987:134). Similar pleas for experiments are made by Bierzychudek (1987) and Bremerman (1987). If a few good experiments can show that the ecological benefits of genetic diversity can compensate for its costs, we have no real problem explaining the predominance of sexual species.

An excellent example of a comprehensive experimental program addressing these problems is found in studies of a perennial grass *Anthoxanthum odoratum* (Antonovics and Ellstrand 1984; Ellstrand and Antonovics 1985; Schmitt and Antonovics 1986; Kelley et al. 1988). Comparisons of survival and reproduction in sexually versus clonally derived sibships suggest that genotypic diversity provides a frequency-dependent advantage. On average, sexually produced progeny have net reproductive rates summed over 2 years that are 1.43 times that of cloned sibships. The sexual advantage apparently is the result of attacks by parasites and predators, particularly aphids. Circumstantial evidence suggests that minority genotypes in sexual sibships have a greater probability of surviving, because the parasites focus their attacks on common genotypes. Although the 1.43 advantage is not the twofold benefit required to maintain sex, these experiments represent an important beginning. Because only a portion of the life cycle was examined, it is possible that the sexual advantage is amplified in other life stages; and furthermore, individual clones might suffer cumulative negative effects over several generations as specific parasites increase in frequency (J. Antonovics, personal communication). The results with *Anthoxanthum* are consistent with the Red Queen model (Bell 1982). According to this model, genotypic diversity compensates for the cost of sex in organisms that are faced with frequency-dependent attacks by parasites, pathogens, and predators. It is clear that more theoretical and experimental work with this hypothesis is warranted (see for example, Rice 1983; Bremerman 1987; Bell 1987).

Another hypothesis for the immediate ecological benefits of genotypic diversity has been labeled the Tangled Bank model (Bell 1982) after a hypothesis first suggested by Ghiselin (1974). According to this model, genetic diversity facilitates exploitation of a heterogeneous environment, because density-dependent selection favors diverse sibships that can use a broad range of resource patches. Individual clones, on the other hand, are severely restricted in their ecological breadth and thus suffer intense sib competition within patches. Case and Taper (1986) found that as long as the niche width of the sexual population is broader than that of a combined group of asexual clones, sex will be maintained. Koella (1988) has recently extended these results. Sex may generally be advantageous in saturated environments.

During the past 20 years, I have been engaged in field and experimental studies that focus on the genetics and ecology of sexual and clonal lineages of fishes that live in seasonally and spatially heterogeneous environments. Broader reviews of the research program with *Poeciliopsis* (Atheriniformes: Poecliidae) can be found elsewhere (Schultz 1977; Moore 1984; Vrijenhoek 1984a; Wetherington et al. 1989a). I report here on some studies that focus specifically on the ecological consequences of genotypic diversity for sexual and clonal lineages that must live together and compete for limiting resources in desert streams. These studies provide strong support for several assumptions behind the Tangled Bank model.

DESERT STREAMS AND CLONAL FISHES

Annual dry and rainy seasons in Sonora, Mexico, cause massive fluctuations in the water levels and consequently affect the abundance and diversity of food and spatial resources in desert streams (Schenck and Vrijenhoek

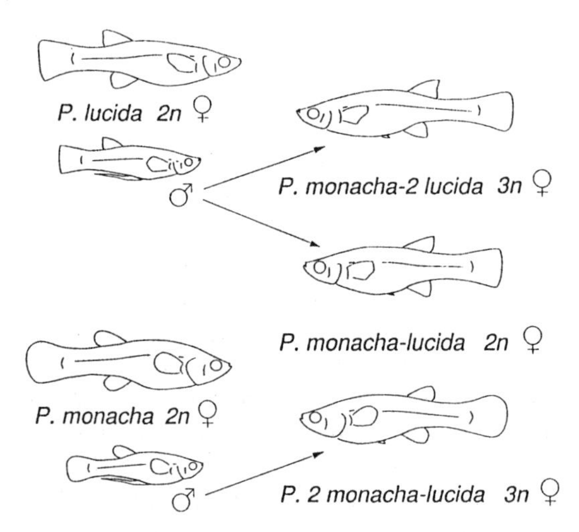

FIGURE 1. Sexual and clonal members of the *Poeciliopsis monacha* and *lucida* complexes. A complex contains a sexual species and one or more sperm-dependent unisexual biotypes (connected by arrows). Each unisexual biotype is given a name that reflects its genomic composition and ploidy level (Schultz 1969). For example, *P. 2 monacha-lucida (MML)* is an allotriploid having two haploid *monacha* genomes and one *lucida* genome.

1986, 1989). During severe droughts, fish are often crowded into residual pools and springs, and local extinctions are common. I have focused my studies on an assemblage of sexual and clonal fishes inhabiting headwater tributaries of Río Fuerte (Figure 1). *Poeciliopsis monacha* is a robust bodied, pool-dwelling species that inhabits mountain springs and arroyos. *P. lucida* is a streamlined species that inhabits more permanent streams. Both species reproduce sexually, having 1:1 primary sex ratios. In regions in which they overlap, hybridization has produced three all-female biotypes (Schultz 1969). Electrophoretic and tissue grafting studies revealed that each all-female biotype comprises several distinct clones (Vrijenhoek et al. 1977, 1978; Angus and Schultz 1979; Moore 1977). Each clone is designated with an acronym reflecting its biotype and Roman numerals reflecting its genotype (e.g., *MML/II* is *P. 2 monacha-lucida* electromorph clone II).

Two nonrecombinant modes of reproduction occur in the all-female biotypes (Figure 2). The allotriploid biotypes, *MML* and *MLL*, are gynogenetic (Schultz 1967). Each allotriploid form relies on males of the most closely similar sexual species for insemination, but the sperm contribute nothing genetically to the offspring. Gynogenetic inheritance is strictly clonal. However, the allodiploid biotype, *ML*, is hybridogenetic, a hemiclonal mode of reproduction (Schultz 1969). Only the *M* genome is transmitted to the eggs of *ML* females. The *L* genome is expelled prior to meiosis, thereby precluding synapsis and recombination. Haploid *M* ova are fertilized by sperm from *P. lucida* males, generating a new *ML′* hybrid genotype in each generation. Cytogenetic, electrophoretic, and tissue grafting studies have verified these reproductive modes (Cimino 1972a, 1972b; Moore 1977; Vrijenhoek et al. 1977; Angus and Schultz 1979).

Sperm dependence

Gynogenetic and hybridogenetic *Poeciliopsis* cannot escape their hosts to invade new habitats, nor can they competitively exclude their hosts, for in doing so, they lose their sperm source and ensure their own extinction. They are forced into a parasitic relationship with a closely related sexual species. Moore (1976) showed that dynamic equilibrium could exist between sexual host and clonal parasite. First, he assumed that sexual and clonal females are simply alternative phenotypes with equal fertility and survival (i.e., “primary fitness”) and with completely overlapping niches. If given a choice, males strongly prefer conspecific sexual females as mates. When the sexual species is rare, the solitary males mate preferentially with sexual females. But when the sexual species is abundant, males form dominance hierarchies, and subordinate males mate with clonal females (McKay 1971). Thus, the probability of clonal females getting inseminated is negatively correlated with their frequency in the population.

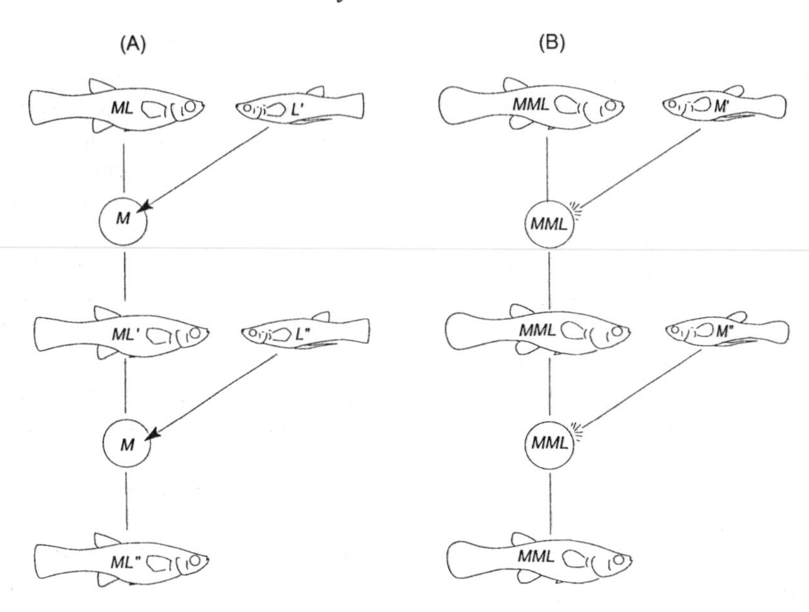

FIGURE 2. (A) Hybridogenetic and (B) gynogenetic modes of reproduction in *Poeciliopsis*. The letters *M* and *L* represent *monacha* and *lucida* genomes, respectively. The primes (′) represent different allelic markers associated with the paternal *lucida* genome. During hybridogenesis the paternal *L* genomes are substituted in each generation, but during gynogenesis the *M* genomes in sperm are not incorporated into the zygote. Other hybridogenetic biotypes, *P. monacha-occidentalis* and *P. monacha-latidens*, also transmit a hemiclonal *M* genome between generations, but they are inseminated by males of *P. occidentalis* and *P. latidens*, respectively.

Moore (1976) predicted the equilibrium frequency of clonal females, $\hat{Q}$, as a function of the ratio of primary fitnesses, $R = W_{\text{unisexual}}/W_{\text{bisexual}}$, of the two kinds of females. Schultz (1961) had previously shown that fecundity of the clonal females was roughly equivalent to that of the sexual females, but less is known about survival in nature. Assuming that primary fitnesses are equal, $\hat{Q}$ should be about 85% of the fish population. Raising the fitness of clonal females has little effect on $\hat{Q}$. According to Moore's model, the clonal females will constitute a majority of the fish population as long as their primary fitness is slightly better than one-half that of the sexual females.

Some natural populations of hybridogens, such as *P. monacha-occidentalis* in the Río Mayo and *P. monacha-lucida* in the Río Fuerte, fit this expectation for $\hat{Q}$, but populations inhabiting many sites in seven other river systems do not. In several rivers the hybridogenetic fish rarely exceed 10% of the fish (Moore et al. 1970). To explain the low frequency of hybridogens in these rivers, Moore (1976) suggested that their primary fitness must be limited by ecological factors that varied geographically. He suggested that the sexual ancestors of unisexual *Poeciliopsis* are adapted to different regions along broad ecological gradients. Because hybrids tend to be intermediate in morphology, and presumably in other ecologically relevant traits, the unisexual hybrids should be best suited for ecotonal regions in which both parental species are inferior competitors (Moore 1976, 1984). Moore's Intermediate Niche hypothesis was proposed before the discovery of abundant clonal diversity in roughly half of the rivers containing unisexual *Poeciliopsis* (Vrijenhoek 1979a). The hypothesis was not formulated to explain the ecology of individual clones, or, for that matter, the maintenance of clonal diversity (Moore 1984). The Intermediate Niche model may have some validity on broad geographic scales in which fishes are limited by their tolerance of environmental extremes, but as I describe below, it clearly does not explain the distribution and abundance of unisexual biotypes and individual clones on a local scale.

I proposed the Frozen Niche-Variation model to explain stable assemblages composed of multiple clones with one another and with their sexual ancestors (Vrijenhoek 1979a, 1984b). This model is really a syllogism based on two assumptions. First, different clones have multiple origins from sexual ancestors. Second, as new clones arise they "freeze" and faithfully replicate genotypes that affect their use of food and spatial resources. Each new clone expresses a distinct phenotype based on the combination of additive, dominance, and epistatic factors it froze from the sexual gene pool. Thus, interclonal selection should produce a structured assemblage of clones with reduced resource overlap and high efficiency within portions of the ancestral niche. Differential use of food and spatial resources should facilitate coexistence among clones, as well as facilitate coexistence with a broad-niched sexual ancestor.

Both assumptions of this model have been the focus of direct experimental attack. The first assumption, "multiple origins," is clearly supported through electrophoretic studies of natural populations (Vrijenhoek et al. 1977, 1978) and through experiments that created many new unisexual hybrids in the laboratory (Schultz 1973; Wetherington et al. 1987). The second assumption, "frozen genotypic variation," is the subject of ongoing research that has recently been reported elsewhere (Wetherington et al. 1987, 1989a, 1989b). Laboratory-synthesized hemiclones of *P. monacha-lucida* exhibit substantial differences in life history traits, including size at birth, growth rate, fertility, and survival, that would permit considerable scope for interclonal selection. We are presently examining morphological and physiological traits and behavior that might affect use of food and spatial resources in these synthetic hemiclones. Most significantly, these studies have revealed a tremendous wealth of ecologically relevant variability within the *P. monacha* ancestors of these hemiclonal *M* genomes.

Data supporting the conclusion to this syllogism started accumulating shortly after our discoveries of extensive clonal diversity in many *Poeciliopsis* populations. Field experiments involved members of the *monacha* complex, including *P. monacha*, and two gynogenetic clones *MML/I* and *MML/II*. Spatial segregation in the desert pools and streams is a primary factor leading to resource partitioning among sexual and clonal members of the *monacha* complex (Vrijenhoek 1978). *P. monacha* and *MML/II* tend to displace one another across an upstream–downstream gradient. The sexual species is most frequent in shady and less productive headwater pools, whereas *MML/II* is most frequent in sunny and more productive downstream habitats. Frequencies of *MML/I* do not vary much across this gradient. I first observed these broad-scale patterns in 1975, and except for the unusual situation discussed below, frequencies of these fishes have remained relatively stable during the past 10 years, roughly 30 *Poeciliopsis* generations (Schenck and Vrijenhoek 1986).

The presence or absence of water currents plays a primary role in determining the feeding modes and local distributions of these fish. I briefly summarize recent field and experimental studies below; for details of the experimental design see Schenck and Vrijenhoek (1986, 1989). Adult females of *MML/I* and *MML/II* both prefer to "drift feed" in areas with an influx of water, whereas *P. monacha* females tend to forage in deep still-water areas. Both clones increase the quantity of insects in their diet by drift feeding; however, *P. monacha*'s diet is more diverse and less affected by feeding mode. Demographic factors also affect local distributions. Juveniles of *P. monacha* and *MML/II* behave similarly to adults, favoring pools and currents, respectively. However, *MML/I* juveniles behave in a manner opposite to their adults, favoring still-water areas. More attention needs to be paid to demographic plasticity in ecological studies. Our tendency to focus only on adult organisms could lead us astray in circumstances such as these. Nevertheless, it is clear that members of the *monacha* complex exhibit substantial differences in feeding behaviors that are affected by small-scale spatial heterogeneity.

Water currents disappear in most desert pools during the end of the dry season. To simulate extreme dry season conditions, we eliminated water flow by shunting it around several natural pools that measured only a few meters across (Schenck and Vrijenhoek 1989). *P. monacha* females that had formerly been concentrated in deeper, still-water areas quickly spread into shallower areas that had current. Clone *MML/II* females did the opposite; they moved from shallower areas that had current into deeper still-water areas. Clone *MML/I* females did not significantly alter their use of space as a result of these manipulations. The broad-scale patterns of abundance that I had observed 10 years earlier across an upstream–downstream gradient (Vrijenhoek 1978) are clearly reflected in this small-scale pattern of displacement between *P. monacha* and *MML/II*, and the general stasis of *MML/I* within individual pools. Apparently, local interactions during the long dry season, when fish are crowded in residual pools, legislate the overall abundance of these three forms. Patterns of food and spatial partitioning and seasonal variation among clones have also been identified in other multiclonal assemblages of animals (Bell 1982; Vrijenhoek 1989a).

The ability of an asexual population to supplant a closely related sexual population should depend on clonal diversity. Multiple clones, each of which is somewhat specialized, could displace a generalized sexual lineage from much of the available resources, but a single clone could not. Bell (1982:131) made this prediction based on the Tangled Bank model, but he apparently was unaware that I demonstrated this result 3 years earlier (Vrijenhoek 1979a). In a sample of 15 localities containing multiple hemiclones of hybridogenetic fish, the all-female forms comprised 66.3% (21.6 SD) of the total *Poeciliopsis* females. However, in a sample of 33 localities containing a single hemiclone, hybridogens comprised only 7.2% (9.3 SD). Genotypically diverse populations of hybridogens dominated both current and still-water habitats within rivers, but monoclonal populations were generally restricted to peripheral habitats (Vrijenhoek 1984b).

The ecological success of hybridogenetic populations depends on the opportunity for freezing new hemiclonal genotypes from the sexual gene pool. Ecologically diverse, multiclonal assemblages occur in rivers where recurrent hybridization events can generate new clonal genotypes. Monoclonal populations occur in rivers lacking *P. monacha*, where migration and mutation are the sole sources of hemiclonal diversity. Apparently mutation does not generate sufficient phenotypic variation for ecological diversification among clonal lineages of *Poeciliopsis* (but see Lynch and Gabriel 1983, for a different view). Frozen variation resulting from multiple origins of clones provides the opportunity for interclonal selection, and consequently for the production of local assemblages of highly fit genotypes. Recombinational diversity that is frozen from the sexual gene pool is also good for clonal populations.

FOUNDER EVENTS AND SEXUAL POPULATION DYNAMICS

We have seen that genotypic diversity contributes to the ecological breadth of unisexual populations, but to understand the interaction between sexual and asexual lineages, it is not sufficient to study clonal diversity alone. Genetic variability in sexual lineages might also affect their ability to contend with spatially and seasonally heterogeneous environments. A long-term study of gene diversity and population dynamics among members of the *monacha* complex in the Arroyo de Jaguari and its tributaries has allowed me to examine this matter (Vrijenhoek and Lerman 1982; Vrijenhoek 1985). A local extinction during a severe drought in 1976 and subsequent recolonization had essentially eliminated gene diversity in a partially isolated, headwater population of *P. monacha* (site PL-1; Vrijenhoek 1978). Average gene diversity in the founder population inhabiting the upper portion of the Arroyo de los Platanos was low compared with populations that survived in downstream refugia. Clone *MML/I* and subsequently *MML/II* recolonized the upper Platanos, but their heterozygosity is fixed at a high level by virtue of clonal reproduction.

Loss of heterozygosity in the sexual founders clearly altered their ability to coexist with the clonal population (Figure 3). Before the local extinction event in 1975, *P. monacha* constituted the majority of the fish in the uppermost pools of the Arroyo de los Platanos. After the extinction–recolonization event in 1976, the situation had reversed. Now the clonal population constituted from 90 to 95% of the fish in these pools from 1978 to 1983 (about 15 *Poeciliopsis* generations). The decrease in the proportion of sexual fish coincided with a loss of gene diversity. In 1983, I removed 30 *P. monacha* females from a founder pool in the upper Platanos and replaced them with 30 similar females taken from a genetically diverse mainstream population (site JA; Vrijenhoek 1979b). Within 2 years (1985), gene diversity in the upper Platanos population of *P. monacha* rose to a level that was about equal to that of the Jaguari population. Concomitantly, *P. monacha* rebounded to its former frequency, about 80% of the fish population. The recovered gene diversity and corresponding population dynamic of the sexual species have remained relatively unchanged in samples taken through 1986.

Apparently, ecological interactions between the sexual and clonal fishes are controlled in part by gene diversity in the sexual population. During the same 10-year period, no comparable shifts in gene diversity or population dynamics occurred in parallel samples taken from the mainstream Jaguari pool, or in nearby sites containing the *monacha* complex (sites NA and TA in Vrijenhoek 1979b). A stable level of gene diversity in the sexual population appears to be associated with a stable frequency of unisexuals. Several factors might have compromised the ecological abilities of the homozygous *P. monacha* population. First, the sexual founders exhibited excessive fluctuating asymmetry relative to *MML/I* individuals at the same site (Vrijenhoek and Lerman 1982). Such effects are often attributable to a loss of developmental stability associated with inbreeding depression, and thus survival and fertility are likely to exhibit correlated effects (Lerner 1954). Second, the founder event also decreased genotypic diversity among sexual individuals, perhaps limiting

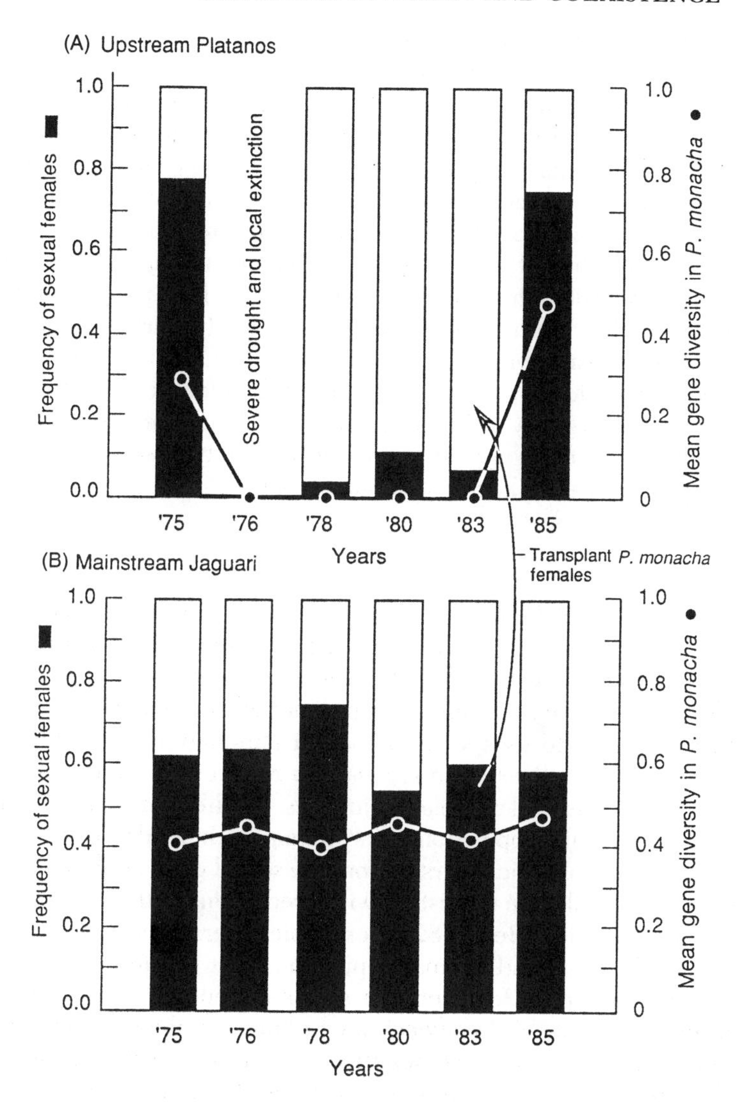

FIGURE 3. Gene diversity in *P. monacha* and the proportion of the fish complex composed of sexual individuals in (A) the upper Platanos population and (B) the Arroyo de Jaguari population. The gray area of the histogram represents the frequency of *P. monacha*, and the white area represents the combined frequency of the triploid clones *MML/I* and *MML/II*. The solid line represents the mean gene diversity based on four polymorphic loci (*Ldh-1*, *Idh-2*, *Pgd*, and *Mp-3*). The gene loci and allelic frequencies in this area have previously been described (Vrijenhoek 1979b).

their ability to exploit a broad range of food and spatial resources. A reduction in niche breadth would render the sexual individuals more susceptible to competitive displacement by the clones, especially during the dry season, when spatial overlap is highest. Third, I have recently completed a series of experiments that suggest genotypic diversity in *P. monacha* contributes to survival during acute hot, cold, and hypoxic stresses. These stresses occur naturally across seasons and across short distances in streams that are often punctuated by natural hot springs. Preliminary analysis of the field experiments revealed that allozyme genotypes at four polymorphic loci differentially affect survival. These allozymes clearly mark seasonally balanced polymorphisms, but more research needs to be undertaken before the actual foci of this pattern of selection can be identified. Thus, overall, the loss of heterozygosity within sexual individuals and the loss of genotypic diversity among them appear to govern the ability of sexual lineages to compete with closely related clonal genotypes.

CONCLUSIONS

Maynard Smith (1971) asked, "what use is sex?" Not surprisingly, the answer appears to be the production of genetic diversity. The function of sex would appear to explain its maintenance in nature, at least for *P. monacha*. Fortunately, having clones as experimental alternatives to sexual individuals allows us to escape the apparent tautology of this conclusion. Eliminating genetic diversity in *P. monacha* reduces its ability to live and compete with closely related clonal lineages. Furthermore, the ecological success of all-female populations of *Poeciliopsis* depends on the opportunity to freeze genotypic diversity from the sexual gene pool (Vrijenhoek 1979a). These results are consistent with predictions of the Tangled Bank model (Ghiselin 1974; Bell 1982). Genotypic diversity contributes to the ability of both sexual and asexual populations that live in heterogeneous natural environments. However, the sperm-dependent relationship between sexual and clonal *Poeciliopsis* might limit our ability to draw broad conclusions about the adaptive benefits of sex in these fish. It is tempting therefore to dismiss our results as a special case (cf. Lynch 1984). Clearly, sperm-dependent all-female clones cannot completely replace their sexually reproducing hosts. But as we have seen, sperm dependence does not provide the whole explanation for the population dynamics of sexual and clonal forms of *Poeciliopsis*. Individual genotypes are important. We are currently extending our studies by examining density-dependent selection and genotypic diversity in laboratory populations. In the laboratory, we can eliminate the constraint of sperm dependence by artificially inseminating the fish. We also need to explore the possibility that frequency-dependent factors might favor genotypic diversity in *Poeciliopsis*. Experiments that test the ability of certain fish pathogens to infect genotypically diverse versus uniform sibships are possible and should be undertaken.

To determine the generality of these results, I hope to encourage additional long-term and multidisciplinary studies of other mixed reproductive complexes. If additional field and laboratory experiments such as those with *Anthoxanthum* and *Poeciliopsis* reveal that genotypic diversity can compensate for the costs of sex, then explaining the near monopoly of sexual species will cease to be a problem. It has historically been argued that the lack of genetic plasticity dooms clonal lineages to an increased probability of extinction (Fisher 1930; Muller 1932). Furthermore, cladogenetic events and sexual recombination may be closely intertwined, decreasing the "speciation" rate of asexual lineages (Stanley 1975). Of course, the biological species concept is difficult to apply to clonal organisms, because its major criteria include the potential for interbreeding among individuals within a species and reproductive isolation from other species (Mayr 1963). Each time a new clone arises, it severs genetic continuity with its sexual ancestors. Although mutation plays a role in generating additional clonal diversity, many of the asexual "species" of animals that are currently recognized contain multiple clones that arose polyphyletically (White 1978; Vrijenhoek 1984a; Suomalainen et al. 1987; Hebert et al. 1989). It would be useful to know if the same is true for strictly asexual plants.

White (1978) estimated that about 0.1% of animal species are strictly thelytokous (i.e., have all-female, clonal reproduction). He excluded from this estimate many lower invertebrates that exhibit vegetative reproduction, because in most cases they also reproduce sexually at some stage of their life cycles. Bell (1982) provided an extensive compilation of asexuality in over 200 genera and 80 families of insects, solidifying White's earlier estimate of about 0.1%. A recent compilation of clonally reproducing vertebrates identified 51 biotypes among 20 genera and 12 families of fishes, amphibia, and reptiles (Vrijenhoek et al. 1989). Again, this number represents roughly 0.1% of the known vertebrate species. In spite of the taxonomic ambiguities associated with clonal organisms, the consistency of this estimate for insects and vertebrates suggests an underlying process that keeps the numbers of distinct, nonrecombinant lineages low relative to sexual lineages. Processes controlling the "standing crop" of clonal lineages in insects and vertebrates are probably analogous to a mutation/selection equilibrium: new clones arise in rare "mutation-like" events that disrupt normal gametogenesis, but compared with their sexual ancestors, most new clonal genotypes are deleterious (Templeton 1982; Vrijenhoek 1989a). A low rate of clonal origin might simply be balanced by a high rate of extinction.

ACKNOWLEDGMENTS

Research was supported by grants from the National Science Foundation (INT84-16427 and BSR86-00661).

LITERATURE CITED

Angus, R. A., and R. J. Schultz. 1979. Clonal diversity in the unisexual fish *Poeciliopsis monacha-lucida*: A tissue graft analysis. Evolution 33:27–40.

Antonovics, J., and N. C. Ellstrand. 1984. Experimental studies on the evolutionary significance of sexual reproduction. I. A test of the frequency-dependent selection hypothesis. Evolution 38:103–115.

Bell, G. 1982. *The Masterpiece of Nature: The Evolution and Genetics of Sexuality*. University of California Press, Berkeley.

Bell, G. 1987. Two theories of sex and variation. Pp. 117–134 in: S. C. Stearns (ed.), *The Evolution of Sex and Its Consequences*. Birkhäuser Verlag, Basel.

Bierzychudek, P. 1987. Resolving the paradox of sexual reproduction: A review of experimental tests. Pp. 163–174 in: S. C. Stearns (ed.), *The Evolution of Sex and Its Consequences*. Birkhäuser Verlag, Basel.

Bremerman, H. J. 1987. The adaptive significance of sexuality. Pp. 135–162 in: S. C. Stearns (ed.), *The Evolution of Sex and Its Consequences*. Birkhäuser Verlag, Basel.

Burt, A., and G. Bell. 1987. Red Queen versus Tangled Bank models. Nature (London) 330:118.

Case, M. L., and T. J. Taper. 1986. On the coexistence and coevolution of asexual and sexual competitors. Evolution 40:366–387.

Charlesworth, B. 1987. Red Queen versus Tangled Bank models. Nature (London) 330: 116–117.

Cimino, M. C. 1972a. Meiosis in triploid all-female fish (*Poeciliopsis*, Poeciliidae). Science 175:1484–1486.

Cimino, M. C. 1972b. Egg production, polyploidization and evolution in a diploid all-female fish of the genus *Poeciliopsis*. Evolution 26:294–306.

Ellstrand, N. C., and J. Antonovics. 1985. Experimental studies of the evolutionary significance of sexual reproduction. II. A test of the density-dependent selection hypothesis. Evolution 39:657–666.

Fisher, R. A. 1930. *The Genetical Theory of Natural Selection*. Oxford University Press, Oxford.

Ghiselin, M. T. 1974. *The Economy of Nature and the Evolution of Sex*. University of California Press, Berkeley.

Hebert, P. D. N., M. J. Beaton, and S. S. Schwartz. 1989. Polyphyletic origins of asexuality in *Daphnia pulex*. I. Breeding system variation and levels of clonal diversity. Evolution 43:(in press).

Kelley, S. E., J. Antonovics, and J. Schmitt. 1988. A test of the short-term advantage of sexual reproduction. Nature (London) 331:714–716.

Koella, J. C. 1988. The tangled bank: The maintenance of sexual reproduction through competitive interactions. J. Evol. Biol. 1:95–116.

Lerner, I. M. 1954. *Genetic Homeostasis*. Oliver and Boyd, Edinburgh.

Lynch, M., and W. Gabriel. 1983. Phenotypic evolution and parthenogenesis. Am. Natur. 122:745–764.

Maynard Smith, J. 1971. What use is sex? J. Theoret. Biol. 30:319–335.

Maynard Smith, J. 1978. *The Evolution of Sex*. Cambridge University Press, Cambridge.

Mayr, E. 1963. *Animal Species and Evolution*. Harvard, Belknap Press, Cambridge, MA.

McKay, F. E. 1971. Behavioral aspects of population dynamics in unisexual-bisexual *Poeciliopsis* (Pisces:Poeciliidae). Ecology 52:778–790.

Moore, W. S. 1976. Components of fitness in the unisexual fish *Poeciliopsis monacha-occidentalis*. Evolution 30:564–578.

Moore, W. S. 1977. A histocompatibility analysis of inheritance in the unisexual fish *Poeciliopsis 2 monacha-lucida*. Copeia 1977:213–223.

Moore, W. S. 1984. Evolutionary ecology of unisexual fishes. Pp. 329–398 in: B. J. Turner (ed.), *Evolutionary Genetics of Fishes*. Plenum Press, New York.

Moore, W. S., R. R. Miller, and R. J. Schultz. 1970. Distribution, adaptation, and probable origin of an all-female form of *Poeciliopsis* (Pisces: Poeciliidae) in northwestern Mexico. Evolution 24:806–812.

Muller, H. J. 1932. Some genetic aspects of sex. Am. Natur. 66:118–138.

Rice, W. R. 1983. Parent-offspring pathogen transmission: A selective agent promoting sexual reproduction. Am. Natur. 121:1317–1320.

Schenck, R. A., and R. C. Vrijenhoek. 1986. Spatial and temporal factors affecting coexistence among sexual and clonal of *Poeciliopsis*. Evolution 40:1060–1070.

Schenck, R. A., and R. C. Vrijenhoek. 1989. Habitat selection and feeding behavior of sexual and clonal *Poeciliopsis*. In: R. Dawley and J. Bogart (eds.), *Unisexual Vertebrates*. State University of New York Press, Albany.

Schmitt, J., and J. Antonovics. 1986. Experimental studies of the evolutionary significance of sexual reproduction. IV. Effect of neighbor relatedness and aphid infestation on seedling performance. Evolution 40:830–836.

Schultz, R. J. 1961. Reproductive mechanism of unisexual and bisexual strains of the viviparous fish *Poeciliopsis*. Evolution 25:302–325.

Schultz, R. J. 1967. Gynogenesis and triploidy in the viviparous fish *Poeciliopsis*. Science 157:1564–1567.

Schultz, R. J. 1969. Hybridization, unisexuality, and polyploidy in the teleost *Poeciliopsis* (Poeciliidae) and other vertebrates. Am. Natur. 103:605–619.

Schultz, R. J. 1973. Unisexual fish: Laboratory synthesis of a species. Science 179: 180–181.

Schultz, R. J. 1977. Evolution and ecology of unisexual fishes. Evol. Biol. 10:277–331.

Stanley, S. M. 1975. Clades versus clones in evolution: Why we have sex. Science 190:382–383.

Suomalainen, E., A. Saura, and J. Lokki. 1987. *Cytology and Evolution in Parthenogenesis*. CRC Press, Boca Raton, Florida.

Templeton, A. R. 1982. The prophecies of parthenogenesis. Pp. 75–101 in: H. Dingle and J. P. Hegmann (eds.), *Evolution and Genetics of Life Histories*. Springer-Verlag, Berlin.

Vrijenhoek, R. C. 1978. Coexistence of clones in a heterogeneous environment. Science 199:549–552.

Vrijenhoek, R. C. 1979a. Factors affecting clonal diversity and coexistence. Am. Zool. 19:787–797.

Vrijenhoek, R. C. 1979b. Genetics of a sexually reproducing fish in a highly fluctuating environment. Am. Natur. 113:17–29.

Vrinjenhoek, R. C. 1984a. The evolution of clonal diversity in *Poeciliopsis*. Pp. 399–429 in: B. J. Turner (ed.), *Evolutionary Genetics of Fishes*. Plenum Press, New York.

Vrijenhoek, R. C. 1984b. Ecological differentiation among clones: The frozen niche variation model. Pp. 217–231 in: K. Wöhrmann and V. Loeschcke (eds.), *Population Biology and Evolution*. Springer-Verlag, Berlin.

Vrijenhoek, R. C. 1985. Animal population genetics and disturbance: The effects of local extinctions and recolonizations on heterozygosity and fitness. Pp. 265–285 in: S. T. A. Pickett and P. White (eds.), *The Ecology of Natural Disturbance and Patch Dynamics*. Academic Press, New York.

Vrijenhoek, R. C. 1989a. Genetic diversity and the ecology of asexual populations. In: K. Wohrmann and S. Jain (eds.), *Population Biology and Evolution*. Springer-Verlag, Berlin.

Vrijenhoek, R. C. 1989b. Genetic and ecological constraints on the origins and establishment of unisexual vertebrates. In: R. M. Dawley and J. P. Bogart (eds.), *Evolution and Ecology of Unisexual Vertebrates*. State University of New York Press, Albany.

Vrijenhoek, R. C., and S. Lerman. 1982. Heterozygosity and developmental stability under sexual and asexual breeding systems. Evolution 36:768–776.

Vrijenhoek, R. C., R. A. Angus, and R. J. Schultz. 1977. Variation and heterozygosity in sexually vs. clonally reproducing populations of *Poeciliopsis*. Evolution 31:767–781.
Vrijenhoek, R. C., R. A. Angus, and R. J. Schultz. 1978. Variation and clonal structure in a unisexual fish. Am. Natur. 112:41–55.
Vrijenhoek, R. C., R. M. Dawley, C. J. Cole, and J. P. Bogart. 1989. A list of known unisexual vertebrates. In: R. M. Dawley and J. P. Bogart (eds.), *Evolution and Ecology of Unisexual Vertebrates*. State University of New York Press, Albany.
Wetherington, J. D., K. E. Kotora, and R. C. Vrijenhoek. 1987. A test of the spontaneous heterosis hypothesis for unisexual vertebrates. Evolution 41:721–731.
Wetherington, J. D., R. E. Schenck, and R. C. Vrijenhoek. 1989a. The origins and ecological success of unisexual *Poeciliopsis*: The Frozen Niche-Variation model. In: G. K. Meffe and F. F. Snelson, Jr. (eds), *The Ecology and Evolution of Unisexual Fishes (Poeciliidae)*. Prentice Hall, New York.
Wetherington, J. D., S. C. Weeks, K. E. Kotora, and R. C. Vrijenhoek. 1989b. Genotypic and environmental components of variation in growth and reproduction of fish hemiclones (*Poeciliopsis*:Poeciliidae) Evolution (in press).
White, M. J. D. 1978. *Modes of Speciation*. Freeman, San Francisco.
Williams, G. C. 1975. *Sex and Evolution*. Princeton University Press, Princeton, NJ.

PART FIVE

BIOGEOGRAPHY AND ECOLOGY OF SPECIATION: EMPIRICAL STUDIES

CHAPTER 17

TAXON CYCLE AMONG *ANOLIS* LIZARD POPULATIONS: Review of Evidence

Jonathan Roughgarden and Stephen Pacala

INTRODUCTION

What makes a fauna? Is the fauna of a region simply the set of all the species that happened to arrive there, or is it a subset consisting of those species that fit together particularly well? These age-old issues in biology have, for many years, been explored through research on islands. The small and isolated faunas on islands might be hoped to provide a glimpse at early stages in the development of the large continental faunas with which most people are more familiar.

This chapter offers a summary of how faunas appear to have developed on a set of small islands in the eastern Caribbean. The organisms involved are lizards from the genus *Anolis*. These organisms are significant zoologically and ecologically. Anoles comprise about 300 species, which is about 5–10% of the world's present-day lizard fauna. And on Caribbean islands, anoles replace ground-feeding insectivorous birds such as robins and jays. Their poikilothermic habit leads to an abundance usually between 0.1 and 1.0 lizards per square meter, and to a total abundance on a typical 400-km^2 island of order 10^8.

THIS DISCOVERY OF *ANOLIS*

Early research by Ernest Williams from the Museum of Comparative Zoology, Albert Schwartz from Miami-Dade Community College, and P. Hummelinck from the Caribbean Marine Biological Institute in Curacao, together with their associates, led to the collection and identification of the anoles in the eastern Caribbean. Next, species groups were clarified by R. Etheridge, G. Gorman, and H. Dessauer with osteological, karyotypic, electrophoretic, and immunogenetic data. The result of four decades of field collecting and systematic research can be summarized, for the eastern Caribbean, in the phylogenetic tree of Figure 1, in which the species have been arranged, so far as possible, in clockwise order beginning with Puerto Rico and ending with Curacao in the Netherlands Antilles. Refer also to Figure 2 for a map of the eastern Caribbean. (Both figures are from Roughgarden 1989.)

The northern Lesser Antilles are populated by the Bimaculatus group, a sister lineage to the Cristatellus group of Puerto Rico. The Bimaculatus group itself contains the Wattsi series, all of which are small brown lizards that perch within a few feet of the ground, and the Bimaculatus series, which are large or medium-sized usually green or gray-green lizards with white, blue, or red accent markings, that are relatively arboreal. On the northern island banks with two species (Anguilla, Antigua, and St. Kitts), the smaller species is always a member of the Wattsi series, and the larger is always a member of the Bimaculatus series. Early workers envisaged the Wattsi and Bimaculatus series as representing separate invasions from Puerto Rico, with the Wattsi series being the more recent.

The southern Lesser Antilles are populated by the Roquet group, a lineage found only on islands in a triangular region of the southeastern Caribbean extending from Martinique in the Lesser Antilles to Bonaire in the Netherlands Antilles.

Finally, the species on the central islands of the Lesser Antilles have spectacular geographic variation in body color, body size, squamation, and secondary sexual characteristics such as head and tail crests. This geographic variation was recognized by systematists through the naming of 12

FIGURE 1. Phylogenetic tree for *Anolis* in the eastern Caribbean,. The tree is ► derived from data on squamation, karyotypes, the electrophoresis of proteins, and the immunogenetic assay of albumins. The taxa being classified are the populations on present-day island banks, irrespective of their current nomenclatural status as "species" or "subspecies." Populations showing geographical variation that resulted in named subspecies (geographic races) are indicated as vertical lines with horizontal tick marks to denote each subspecific name instead of single dots for monotypic species. Notice only the central islands in the arc have such internal subspecific differentiation and also that the Bimaculatus group is sister to the Cristatellus group of Puerto Rico, and Roquet group is quite unrelated to the Cristatellus-Bimaculatus clade. Intersecting lines with numbers refer to character state changes (see Roughgarden 1989).

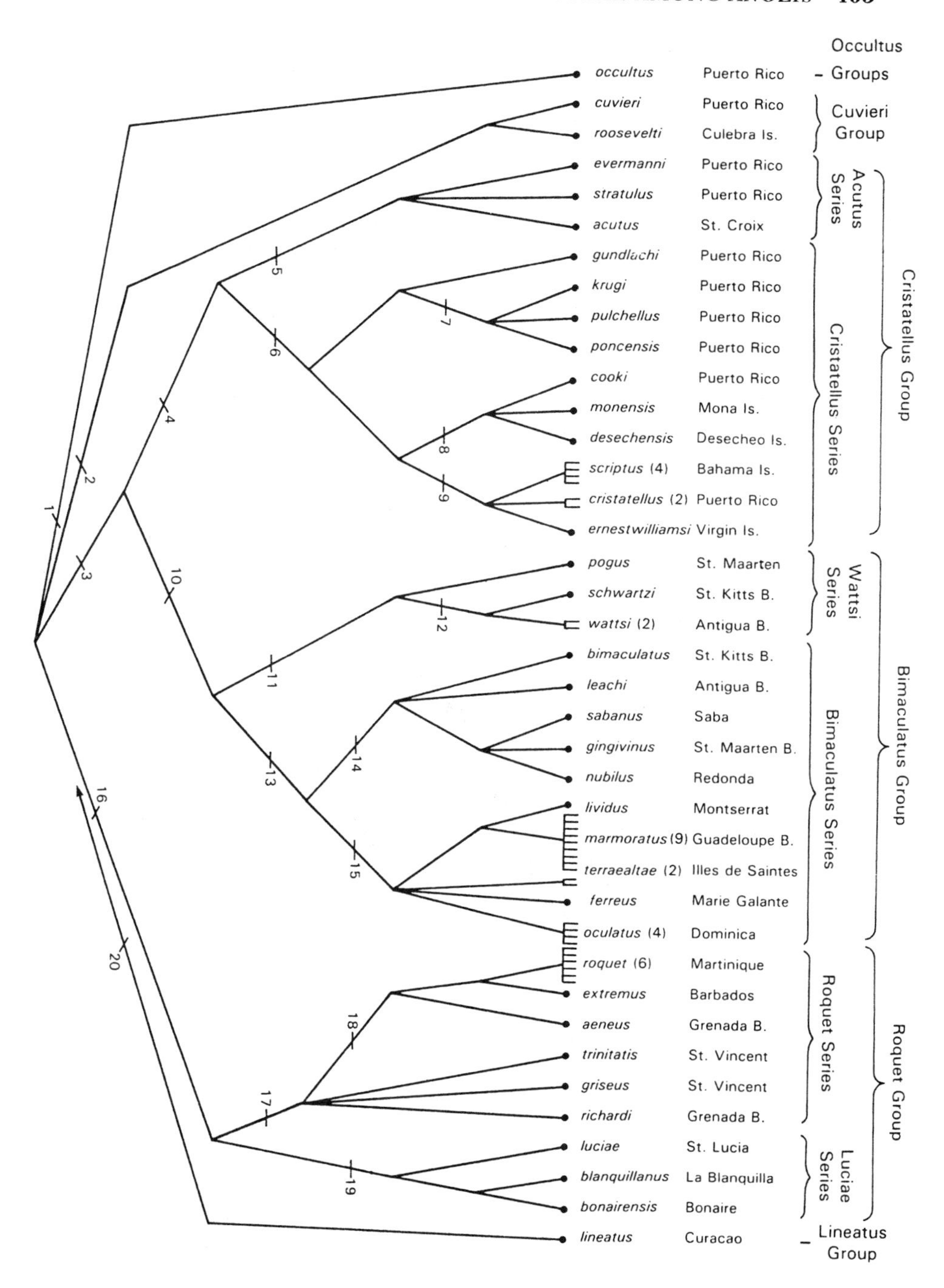
occultus
Puerto Rico
Occultus Groups
cuvieri
Puerto Rico
roosevelti
Culebra Is.
Cuvieri Group
evermanni
Puerto Rico
stratulus
Puerto Rico
acutus
St. Croix
Acutus Series
gundlachi
Puerto Rico
krugi
Puerto Rico
pulchellus
Puerto Rico
poncensis
Puerto Rico
cooki
Puerto Rico
monensis
Mona Is.
desechensis
Desecheo Is.
scriptus (4)
Bahama Is.
cristatellus (2)
Puerto Rico
ernestwilliamsi
Virgin Is.
Cristatellus Series
Cristatellus Group
pogus
St. Maarten
schwartzi
St. Kitts B.
wattsi (2)
Antigua B.
Wattsi Series
bimaculatus
St. Kitts B.
leachi
Antigua B.
sabanus
Saba
gingivinus
St. Maarten B.
nubilus
Redonda
lividus
Montserrat
marmoratus (9)
Guadeloupe B.
terraealtae (2)
Illes de Saintes
ferreus
Marie Galante
oculatus (4)
Dominica
Bimaculatus Series
Bimaculatus Group
roquet (6)
Martinique
extremus
Barbados
aeneus
Grenada B.
trinitatis
St. Vincent
griseus
St. Vincent
richardi
Grenada B.
Roquet Series
luciae
St. Lucia
blanquillanus
La Blanquilla
bonairensis
Bonaire
Luciae Series
Roquet Group
lineatus
Curacao
Lineatus Group
1
2
3
4
5
6
7
8
9
10
11
12
13
14
15
16
17
18
19
20

"subspecies" within the Guadeloupe archipelago, four in Dominica, and six in Martinique. The species of other islands are geographically homogeneous, except for slight variation in background hue from place to place.

The eastern Caribbean differs from the Bahamas, which have been receiving important study by Schoener (1968), Schoener and Schoener (1978, 1982, 1983a, 1983b, 1983c), and Schoener and Toft (1983). Unlike the Lesser Antilles, which has a native fauna of its own, the anoles of the Bahamas are not taxonomically distinct from populations on Cuba and Hispaniola.

BODY SIZE: A BIOGEOGRAPHIC REGULARITY

Early research established that each island bank (group of islands separated from others by deep water) possesses only one or only two species of anoles in natural habitat (i.e., not counting small enclaves of introduced anoles living near houses). As the phylogenetic tree in Figure 1 documents, the species on each island bank are endemic (native) to that bank. A biogeographic regularity involving the body sizes of these anoles was quickly identified, and summarized as a pair of "rules" by Williams (1972).

Rule 1. Species from islands in which only one anole is present have a characteristic body size (the "solitary size") intermediate between the extreme body sizes seen on islands in which two anole species coexist. As a statistic to describe body size, Williams chose the snout-vent length of the largest male specimen that had been collected (data from Schoener 1969; Lazell 1972). By this statistic, the size of a solitary species ranges from 65 to 80 mm. This rule is correct for 11 of 12 island banks having a solitary species and applies throughout the eastern Caribbean. Two kinds of exceptions were noted, however. First, specimens collected from forests growing on mountain slopes are often rather large, near 95 mm. This exception was

FIGURE 2. *Anolis* in the northern Lesser Antilles. Each island bank is drawn with the −50 m, present-day sea level, and +200 m contours. The circles present "vignettes" of the community ecology on each island. Within a vignette, the horizontal axis refers to body size, and the vertical axis to relative abundance. Also, squares represent whether the perch positions of two co-occurring species are displaced horizontally or vertically from each other. The islands all contain Bimaculatus group anoles, and on these islands all co-occurring species partition space vertically, and not horizontally as in the southern Lesser Antilles (not depicted here). ►

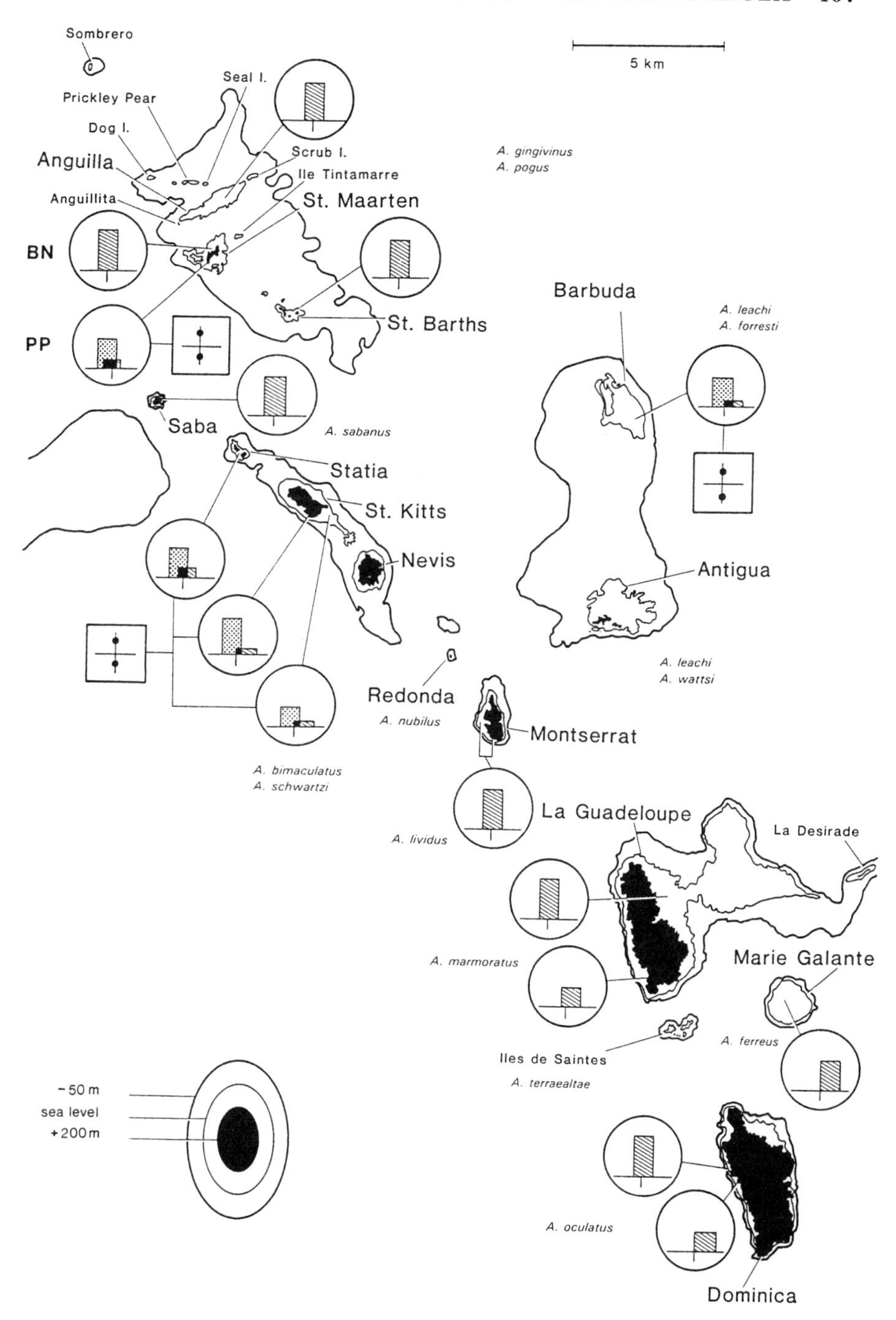

Sombrero
Seal I.
Prickley Pear
Dog I.
Anguilla
Scrub I.
Ile Tintamarre
Anguillita
St. Maarten
BN
PP
St. Barths
5 km
A. gingivinus
A. pogus
Barbuda
A. leachi
A. forresti
Saba
A. sabanus
Statia
St. Kitts
Nevis
Antigua
A. leachi
A. wattsi
Redonda
A. nubilus
Montserrat
A. bimaculatus
A. schwartzi
A. lividus
La Guadeloupe
La Desirade
A. marmoratus
Marie Galante
A. ferreus
Iles de Saintes
A. terraealtae
-50 m
sea level
+200 m
A. oculatus
Dominica

later explained as part of a general pattern of clinal variation in body size. In solitary species, the size of males, and, to a lesser extent, females, is an increasing function of the insect supply (Roughgarden and Fuentes 1977). More insects are collected at mesic locations along the slopes of mountains where the lizards are relatively large, than at xeric locations near sea level where they are smaller. Second, lizards from Marie Galante, near Guadeloupe, are exceptionally large, with males exceeding 100 mm.

Rule 2. Species on a two-species island differ in body length by a factor of 1.5 to 2.0. In particular, the body length of the larger species exceeds 100 mm, and the smaller is less than or equal to the minimum solitary size of 65 mm. This rule is correct for four of the five island banks that possess two *Anolis* species, and applies throughout the eastern Caribbean. Again an exception was noted: St. Maarten has two species of nearly the same body length; the ratio of the larger to the smaller is about 1.3. Also, the range of the smaller species on St. Maarten is confined to hills in the center of the island, whereas the larger species is found throughout the entire island; this smaller species is not restricted to small wooded pockets as stated by Lazell (1972). A range map for the St. Maarten anoles is found in Roughgarden et al. (1984).

A FIRST HYPOTHESIS: CHARACTER DISPLACEMENT

The biogeographic pattern in body size immediately suggested an hypothesis for the assembly of the eastern Caribbean anole fauna in terms of the popular idea of character displacement (Brown and Wilson 1956). Character displacement is the simultaneous evolution of two species away from the use of each other's resources, thereby lowering their interspecific competition. The divergence proceeds until the benefit of further reduction in interspecific competition balances the disadvantage of shifting further from the center of the resource distribution. This idea could be applied to *Anolis* because Schoener and Gorman (1968) had demonstrated a correlation between a lizard's body size and the average size of its prey. Two species of similar body size overlap more in their prey and, it was assumed, therefore suffer stronger interspecific competition than two species of greatly differing body size.

Williams (1972) suggested that the differing size of the species in the two-species communities (Rule 2) was the outcome of character displacement. He offered the scenario sketched in Figure 3. An island with a solitary-sized lizard is invaded by another solitary-sized species. Each then evolutionarily diverges away from the other yielding a stable two-species community. Moreover, it was further hypothesized that the original resident on an island was a member of the Bimaculatus series, and that the invading species is a member of the Wattsi series, in keeping with the belief

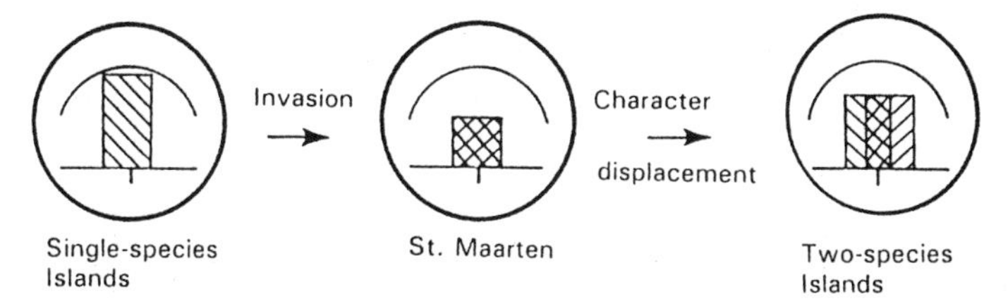

FIGURE 3. The character-displacement hypothesis for assembly of the two-species communities. The circles represent stages in the faunal assembly. Within a circle the horizontal axis refers to body size, the vertical axis to abundance, and the overlying curve represents the resource spectrum. An island with one solitary-sized species is invaded by another solitary-sized species and they then evolutionarily diverge in size yielding a stable two-species community.

that the Wattsi series is a more recent arrival to the Lesser Antilles than the Bimaculatus series.

The exceptions to the size rule would be explained under this hypothesis as follows: St. Maarten, with two species of nearly the same size, would be the initial (primitive) condition preceding the evolution of character displacement. The other two-species islands would be derived from this initial condition after character displacement had occurred. Marie Galante, a solitary island with an exceptionally large lizard, would be explained as some special case. One might postulate that a smaller species on it has very recently become extinct, perhaps as a result of human habitat disturbance. Actually, Williams (1972) hypothesized that invasions to Marie Galante from Guadeloupe were frequent enough to cause character displacement, even though no smaller species was permanently resident on Marie Galante.

LESSER ANTILLES: A MODEL FOR THE GREATER ANTILLES

In the Greater Antilles as many as seven species of anoles cooccur in natural habitat, and three species are typical. Rand (1964, 1967) described the typical perch positions and body sizes of anoles in these communities, and the idea emerged that the lowland anole communities were fashioned from the same architectural blueprint. For example, both Puerto Rico and Hispaniola have medium-sized green anoles perching in the canopy and high up on tree trunks, small gray-green anoles perching on tree trunks between the canopy and the base of the trunk, ending with medium-sized brown anoles perching at the base of the tree trunks. Rand (1964) termed these characteristic habitats "structural niches," and observed that the standard architecture of lowland communities consists of the co-occurrence of species representing three structural niches that were named the "trunk-crown," "trunk," and "trunk-ground" niches. Williams (1983) extended the approach by identifying more structural niches for other islands and habitats.

The presumably simple one- and two-species communities were regarded as the first stages in the assembly of a complex fauna. Indeed, as a thought experiment, Williams (1972) "constructed" the anole fauna of Puerto Rico beginning with an island containing a solitary anole, followed by invasion of another solitary form, followed by character displacement, followed by another invasion, and so forth, leading eventually to a large fauna. Thus, the one- and two-species stages in the formation of Puerto Rico's fauna were modeled after the one- and two-species islands in the Lesser Antilles today.

COMPARATIVE ECOLOGY OF TWO-SPECIES COMMUNITIES

Roughgarden et al. (1983) surveyed the community ecology of the two-species islands and discovered that the Bimaculatus communities differ qualitatively from the Roquet communities. In the north (Antigua and St. Kitts Banks) two species on an island differ in body size, as previously discussed, and in perch position. The large species perches above the smaller. Both species have the same body temperature, and the effective temperature of the microsites they perch in are the same. Furthermore, in all habitats ranging from xeric scrub at sea level to mesic forests on mountain slopes the smaller species is more abundant than the larger species. The ranges of both species completely overlap and cover the island.

In the south (St. Vincent and Grenada) the two species again differ in body size and perch position. But here the smaller species has a higher body temperature and perches in hotter microsites than the larger species. The species show no consistent separation in the height of their perching spots. Also, the relative abundance of the two species varies with habitat. The smaller species is more abundant than the larger species in xeric scrubby habitat near the coast and vice versa in the forests on mountain slopes.

Since the bimaculatus and roquet two-species communities are qualitatively different, even the simplest anole communities are not fashioned from one architectural blueprint. Instead, the relative geometry of competing species in niches space appears to reflect phylogenetic ancestry.

These qualitative differences in both the systematics and community ecology of the northern and southern Lesser Antillean islands imply that they ought to be discussed separately. Also, the northern and southern halves of the Lesser Antilles appear to have had separate geologic origins (Roughgarden et al. 1987). Thus, the remaining discussion will continue to focus almost exclusively on the northern islands.

A SECOND HYPOTHESIS: TAXON CYCLE

An alternative hypothesis for the assembly of the anole faunas on islands in the northeastern Caribbean that potentially accounts for both the exceptions and the regularities in the biogeographic pattern of body sizes was offered by Roughgarden et al. (1983) and developed further in Rummel and Roughgarden (1985a). By this hypothesis, an island exists with one medium-sized species awaiting the introduction of a larger species. After a larger species arrives both the original resident and the invader simultaneously evolve smaller body sizes. During this coevolution the range of the original resident gradually contracts until it becomes extinct. The invading species then converges to the medium size that the original resident had, with the net result being a species substitution. This cycle resembles the original "taxon cycle" proposed by Wilson (1961) for the melanesian ant fauna. As used here, the "taxon cycle" should not be viewed as a regular oscillation because, in theory, the time between consecutive cycles is not specified; it depends on the waiting time between suitably sized invaders.

The islands in the Bimaculatus group can be arranged as though each represents a stage in this cycle. Montserrat was suggested as an island that had never been invaded by a larger species (a "preinvasion solitary island") because the anole there has habits and an appearance similar to populations from the Wattsi series. The two-species communities in which the larger species exceeds 100 mm can be placed next in the cycle, followed by St. Eustatius where the larger species is about 90 mm, culminating with St. Maarten where the larger species is about 65 mm and the smaller species has a greatly restricted range. Anguilla, where the smaller species has recently become extinct, is a "postinvasion solitary island." Marie Galante was hypothesized to be an island where the smaller species had recently become extinct and where the larger species has not yet completely converged to the solitary size.

By this hypothesis the body size differences on the two-species islands express what was needed for a second species to invade an island successfully; the body size differences are not the result of in situ coevolutionary divergence between two species that were initially similar. Also, by this hypothesis the Bimaculatus series, and not the Wattsi series, is the more recent invader on the two-species islands because the Bimaculatus series anoles are larger than the Wattsi series anoles.

The theoretical rationale to the taxon cycle hypothesis is as follows: The only empty niche space open to an invader, given that a medium-sized anole is already there, is the space available to a large species. Competition with the resident prevents a species with the same size as the resident from invading, and there is also no niche space available to a species smaller than the resident because of asymmetrical competition (a large animal has a stronger effect on a small animal than vice versa). Next, establishment of the large species causes some decline in the abundance of the resident. Moreover, the resident becomes smaller as it evolves away from competition with the invader. These two developments open up niche space in the center of the resource axis that, in turn, leads to selection on the invader to become smaller than it was when it entered. Thus, both species become smaller after the invasion: the resident to move away from the invader, and the invader to enter the newly opened space at the center of the axis. On a small island (with a narrow resource spectrum) the resident will eventually become extinct, but on a large island it is conceivable that the resident will coexist as a small lizard with the now medium-sized invader. On small islands, such as those of the northern Lesser Antilles, extinction of the original resident returns the island to the state it had prior to the invasion, thus producing a cycle. In principle, another go-around would be triggered on the arrival of another large invading species.

The taxon cycle hypothesis per se does not address the question of where the invading species comes from, or when it comes. Instead, the hypothesis focuses on what happens on the small island that receives the invader. The invader may, of course, come from a nearby large island that harbors a complex fauna with some large species in it, or, as discussed later, from a nearby small island with special habitat (such as moist middle-elevation forest) in which even a solitary species evolves an unusually large size.

Both the taxon cycle and character displacement hypotheses assume that the niche width of a species stays rather constant whereas the niche position shifts evolutionarily. That is, suppose the niche axis is scaled in units of the logarithm of prey length. Then the variance of prey lengths consumed by a species of *Anolis* is assumed to remain rather constant even as the average prey length consumed by the species changes evolutionarily in accordance with evolutionary change in average body size. Alternatively, if the niche width of a solitary species is assumed to spread out quickly to use all of the available niche space, then invasion by a second species would be precluded. Thus, the hypotheses being considered assume a "species cohesion" for body size that is not easily broken down by the evolutionary effects of intraspecific competition (cf. Roughgarden 1987). If both these hypotheses are eventually falsified, then this assumption could be the culprit, whereas if one of these hypotheses is validated, then species cohesion would emerge as a constraint to be considered in theoretical discussions of a species' evolutionary plasticity.

EXPERIMENTAL EVIDENCE OF COMPETITION

The proliferation of hypotheses about competition created a serious need for experimental evidence of its reality. Roughgarden et al. (1984) introduced anoles to an offshore cay on the Anguilla bank. Pacala and Roughgarden (1985) and Rummel and Roughgarden (1985b) constructed 12 × 12 m experimental enclosures for anoles in natural habitat on St. Maarten and St. Eustatius. These experiments proved the existence of strong present-day competition between the two species on St. Maarten, as the earlier indirect evidence had indicated, and also that competition between the two species on St. Eustatius is weak and barely detectable. Specifically, on an offshore cay near Anguilla *A. gingivinus* greatly lowers the survival of introduced *A. pogus*. The addition of *A. pogus* to enclosures containing *A. gingivinus* lowers *A. gingivinus*' male and female growth rates, female fecundity, and the quantity of food in stomachs. The addition of *A. pogus* also causes *A. gingivinus* to perch higher in the vegetation than it does in the absence of *A. pogus*. Experiments with an identical design on St. Eustatius, where the two species differ in body size more than on St. Maarten, reveal much weaker defects.

Further experimental studies point to food, more than space, as a principal limiting resource. Removal of lizards from experimental enclosures leads to a doubling of insect abundance on the forest floor and to a 10- to 30 fold increase in the quantity of spiders in the vegetation (Pacala and Roughgarden 1984), thereby confirming experimentally a negative correlation between spiders and lizards discovered by Schoener and Toft (1983). These experiments have themselves been repeated by Schoener and Spiller (1987) with very similar results. Also, food augmentation increases lizard growth rates in nature (Licht 1974; Stamps 1977). In contrast, experimentally increasing the overlap in perch positions of the two species on St. Eustatius produces little effect (Rummel and Roughgarden 1985) and interspecific territoriality cannot be detected on St. Maarten except between animals of nearly identical body sizes (Bohlen 1983). Nonetheless, some cannibalism and cross-predation on juveniles imply that competition for food is not the only mechanism of competition.

EVIDENCE OF COMPETITION FROM ACCIDENTAL INTRODUCTIONS

The record of recently introduced anoles supplies more evidence of interspecific competition. To aid the biological control of the fruit fly,

Ceratitis capitata, in 1905 the Director of Agriculture of Bermuda liberated 71 individuals of *A. grahami* in public gardens (Wingate 1965). *A. grahami* is a medium-sized lizard collected from the Kingston area of Jamaica. By 1940 *A. grahami* was abundantly distributed throughout Bermuda. About 1940 the large anole from Antigua, *A. leachi*, first appeared. The species is now established in the center of Bermuda (Warwick area), primarily in woods. The spread of this large anole has taken place on territory already occupied by the medium-sized anole. Finally, the solitary-sized anole, *A. extremus*, from Barbados was noted in 1953 at the western end of Bermuda (Somerset and Ireland areas) where it still has a patchy distribution. Thus, *A. grahami* has preempted *A. extremus*.

Hispaniola provides two well-documented instances of "enclaves" of introduced anoles; these typically form when natural habitat is cleared and replaced with plantings (Williams 1977). The Cuban green anole, *A. porcatus*, has been known since 1970 from a few city blocks in Santo Domingo at the site of former trade fairs. Its range has remained static through 1977 and is surrounded by the native green anole, *A. chlorocyanus*. The Puerto Rican trunk-ground anole, *A. cristatellus*, has been known since 1956 to be abundant in gardens in the port city of La Romana in the Dominican Republic. The town is the site of a sugar mill constructed by a Puerto Rican-based company. The range of the anole has been static for over 20 years and is surrounded by the native trunk-ground anole, *A. cybotes*. Schoener and Schoener (1983a) also produced enclaves in the Bahamas with experimental introductions to tiny cays that were already inhabited by an anole. In contrast, introductions to empty cays large enough to support anoles invariably produced a population explosion of the introduced species.

Introduced small anoles evidently do not form enclaves when surrounded by larger anoles. A population of *A. wattsi* known from the botanical garden in St. Lucia has become extinct (Underwood 1959; Williams 1977). *A. wattsi* was introduced four times to a cay near Anguilla in which *A. gingivinus* was already living and *A. wattsi* eventually became extinct each time. The waiting time to extinction could be doubled simply by removing half of the *A. gingivinus* at the site prior to the introduction (Roughgarden et al. 1984).

Thus, the record of introductions suggests that a large anole can become established in the presence of a smaller anole, an anole introduced to newly opened habitat can form a virtually static enclave in the range of another anole with similar size and habits, and a small anole evidently cannot be introduced successfully in the presence of a larger anole.

LONG-TERM MONITORING ON ST. MAARTEN

Sustained monitoring of *Anolis* populations on St. Maarten reveals an exceptional speed and stability in the population dynamics of *Anolis* on a tropical island. Our laboratory has censused two sites on St. Maarten once or twice a year since 1977. Figure 4 shows the abundance plotted as a function of the month when the data were taken. Year after year the same cyclic pattern recurs even though two hurricanes passed the sites and the annual rainfall may vary by a factor of up to 2 in consecutive years. The abundance is lowest during the summer and doubles during the winter with the production of juveniles. The between-year variation in abundance compared at the same point in the cycle is less than the within-year variation, and the summer abundance has been essentially constant at the two sites for over 10 years. This census interval is long enough for two to three complete population turnovers. These long-term data show more between-year constancy and higher abundances than the records of anoles in Central America (Andrews and Rand 1982; Sexton 1985).

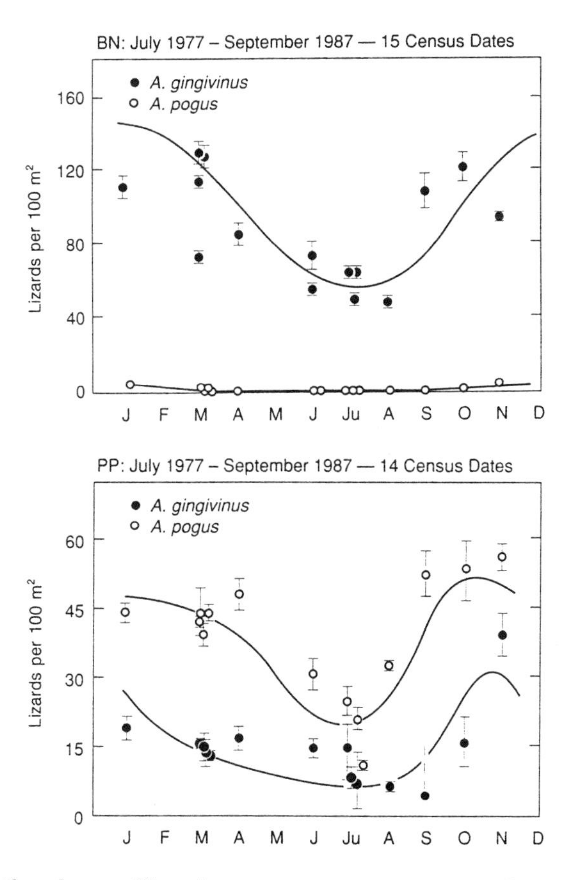

FIGURE 4. Abundance of lizards at two sites on St. Maarten from 1977 to 1987. The horizontal axis is the month when the census was taken and the vertical axis is the abundance of lizards per 100 m^2.

FOSSIL ANOLES

Skin fragments from a hatchling *Anolis* encased in late Oligocene to early Miocene amber (~ 25 my BP) have been collected from Chiapas, Mexico (Lazell 1965). In the West Indies, a complete fossil *Anolis* has been found encased in early Miocene amber (~ 20–23 my BP) from Cordillera Septentrional in the Dominican Republic (Rieppel 1980). This may be even older than originally reported (lower Oligocene or upper Eocene, 35–40 my BP, Poiner and Cannatella 1987). The fossil anole seems identical to the present-day green anoles of the Dominican Republic, *A. chlorocyanus* and *A. aliniger*. Thus, anoles have been established in the Caribbean theater even as the Caribbean itself has been forming geologically.

All other fossils of anoles are latest Pleistocene to Holocene. Cave deposits, presumably left by the extinct barn owl, *Tyto cavaticus*, have been examined from the Bahamas, Jamaica, Hispaniola, Puerto Rico, Barbuda, and Antigua (Etheridge 1964, 1965, 1966; Hecht 1951; Pregill 1981, 1982, 1984; Steadman et al. 1984). Also, new data from our excavations on Anguilla are reported here.

The cave deposits reveal the present-day herpetofauna of the northern Lesser Antilles to be nearly as diverse as the former fauna. Wholesale extinctions have not occurred since the Pleistocene, in contrast with insular birds (Olson and James 1982), except until the recently introduced mongoose (*Herpestes*) began endangering ground lizards (*Ameiva* and *Iguana*) and snakes (*Alsophis*). The only clear extinction since the Pleistocene consists of populations of *Leiocephalus*, a ground lizard estimated at 150–200 mm snout-vent length, that have become extinct on Puerto Rico, the Antigua bank, and, as our data show, the Anguilla bank. Its strongly tricuspid teeth are similar to the largely herbivorous lizard, *Dipsosaurus dorsalis* (Hotton 1955).

Contemporary research on fossils in the Lesser Antilles has yet to equal the spectacular finding of a bear-sized rodent, *Anblyrhiza inundata*, from the Anguilla bank (Cope 1883). This creature belongs to an endemic Antillean family, Heptaxodontidae, and its existence is a reminder that much happened before the late Pleistocene about which current fossil collections are silent.

Nonetheless, the Pleistocene cave deposits are unequivocal in showing that larger anoles have become smaller. On the Antigua bank, both Etheridge (1964) and Steadman et al. (1984), in separate excavations, show that *A. leachi*, the larger of the two anoles today on the Antigua bank, was even larger in the past—it has become smaller since the late Pleistocene. Also, the largest anoles from Puerto Rico have become smaller since the Pleistocene (Pregill 1981).

EXCAVATIONS ON ANGUILLA

Cave deposits are often found as isolated clumps of bones, discouraging the reconstruction of temporal sequence. Also, coarse screens have typically been used and information on the smaller species is scanty.

To obtain a quantitative record with improved stratigraphic resolution, we excavated a fissure in a cliff bordering Katouche Canyon on Anguilla (see map in Figure 5). Nests of the American kestrel (*Falco sparvarius*) are near the fissure, and these birds are the probable agents of fossil deposition.

A. pogus was last collected on Anguilla in 1922 (Lazell 1972); only *A. gingivinus* remains today. Figure 6 shows the lengths for all *Anolis* dentaries; the top layer presumably contains jaws only from *A. gingivinus* whereas the next layer contains jaws of *A. gingivinus* and *A. pogus* com-

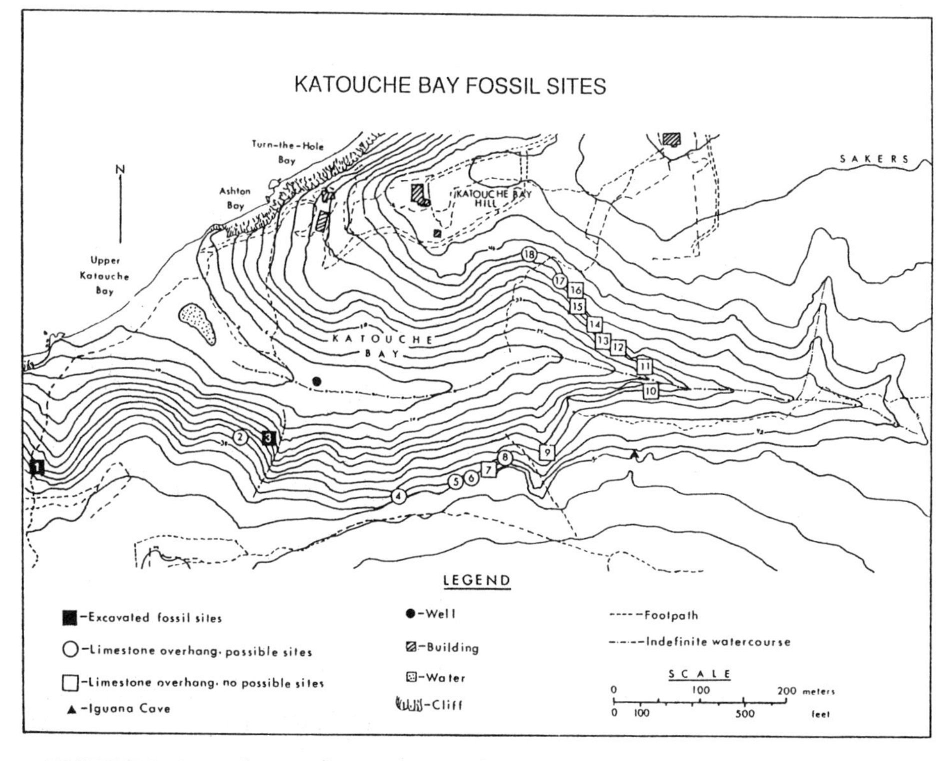

FIGURE 5. Map of Katouche Canyon on the north coast of Anguilla. Sites of fossil excavations are numbered. The productive site is #1 in the lower left corner of the map, where the exposure allowed little weathering.

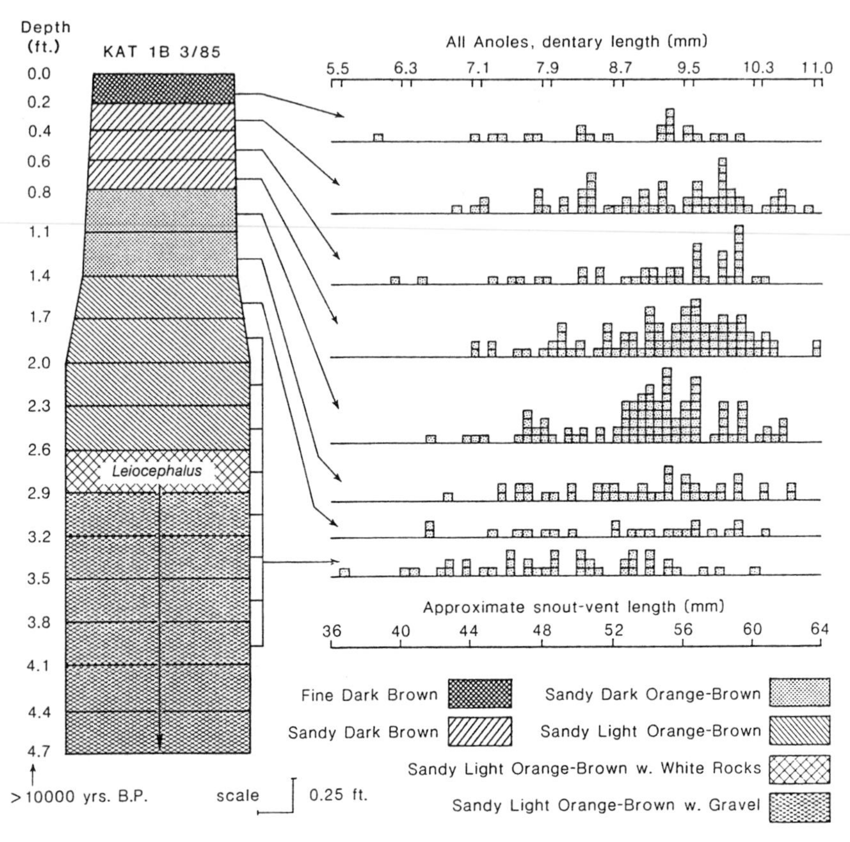

FIGURE 6. Results of an excavation for fossil *Anolis* on Anguilla. The material was excavated from a fissure in a cliff facing Katouche Canyon. The profile of strata is indicated at the left, and histograms depicting the distribution of jaw lengths at each stratum appear at the right. Notice the increase in jaw length between the 0.2- and 0.4-foot strata and the increase in the number of bones at about 1.4 feet. Charcoal fragments at the base of the excavation were dated as exceeding 10,000 years BP. Remains of *Leiocephalus* were found from 2.5 feet and below.

bined. The bottom scale is the snout-vent length corresponding to the dentary length from a regression using recent material. Table 1 presents the descriptive statistics for each layer, and for pooled data from 0.4 to 1.4 feet and 1.7 to 4.1 feet.

The excavation extended nearly 5 feet in depth. Charcoal fragments at the bottom were dated as greater than 10,000 years BP (Washington State University, Radiocarbon Dating Laboratory, Sample #3196, Reported on September 9, 1985).

The main features are (1) the jaws are significantly smaller in the top layer than in the next layer ($p < 0.02$), (2) numerous relatively large jaws extend from the second layer to about 1.5 feet, (3) below 1.5 feet the jaws are fewer and smaller ($p < 0.001$), and (4) fragments from the jaws of *Leiocephalus* begin at about 3 feet and extend to the bottom.

An interpretation of Figure 6 is (1) *A. gingivinus* has become about 10% smaller since the extinction of *A. pogus*. (2) *A. gingivinus* entered the island about 3000 to 4000 years ago. This interpretation is consistent with the appearance of both larger and more numerous jaws at the 1.5 foot level. (3) *Leiocephalus* became extinct about 5000 to 6000 years ago. (4) The original *A. pogus* was about 10 percent larger than it now is on St. Maarten.

The first appearance of anole fossils at 4 foot depth should not be interpreted as the entry of *A. pogus* because decomposition of fossils could explain the absence of anole jaws sufficiently deep in the profile; the number of all fossils declines toward the bottom of the excavation. In contrast, the absence of large jaws in the layers between 1.5 and 4 feet cannot be an ar-

TABLE 1. Anguilla excavation of *Anolis* jaw bones (dentaries).

Depth (feet)	Number	Average length (mm)	Standard error of mean (mm)	Standard deviation (mm)
0.2	22	85.5	2.3	10.8
0.4	63	91.0	1.2	9.9
0.6	42	90.3	1.6	10.3
0.8	92	91.8	0.9	8.7
1.1	96	89.6	0.9	8.7
1.4	43	89.0	1.5	9.7
1.7	23	86.9	2.3	11.1
2.0	7	85.3	2.0	5.2
2.3	6	78.3	5.3	12.9
2.6	12	78.5	3.4	11.8
2.9	5	77.8	5.1	11.5
3.2	4	82.8	7.8	15.5
3.5	3	83.3	2.3	4.0
3.8	1	71		
4.1	2	87.5	8.5	12.0
4.4	0			
4.7	0			
0.4–1.4	336	90.5	0.5	9.2
1.7–4.1	63	82.9	1.4	11.1

tifact of fossil decomposition because larger jaws that preserve better are absent, whereas smaller jaws are present. A caveat is that the largest jaws coincide with the largest sample sizes. The previous interpretation would be an artifact if some mechanism of deposition were more active during the time interval between the second layer and the 1.5-foot layer, resulting in the collection of exceptionally large specimens combined with the selective preservation of those large specimens.

EARLY MAN

Archaeologists have located remains of stone age culture in the northern islands (St. Kitts and Antigua) dating to approximately 3000 years BP (Rouse and Allaire 1978). Before this the Lesser Antilles may have been uninhabited. The arrival of Amerindians occurred about 2000 years BP with the spread of the Arawaks from the Orinoco delta. Thus, stone-age man could have introduced *A. gingivinus* to Anguilla from Saba (where its closest relative lives), because lithic man may have entered the Lesser Antilles at about the time the Anguilla excavation suggests that *A. gingivinus* appeared and, at least today, man brings about many more introductions of *Anolis* than would occur by natural means. The cleanly separated distribution of Bimaculatus and Roquet groups across the island arc, however, argues against a major role for human introductions in the zoogeography of *Anolis*.

The extinction of *Leiocephalus* appears to predate the arrival of man. The sea level 14,000 years BP was 40 m lower, by 8000 years BP it was 15 m lower, and by 6000 years BP had achieved its current height (Heatwole and MacKenzie 1967). Hence, the extinction of *Leiocephalus* approximately coincides with the culmination of the sea level increase that reduced the exposed area of the Anguilla bank to its present size.

THE HYPOTHESES TESTED

The first hypothesis, that the two-species communities are assembled from two medium-sized species that subsequently diverge in body size, is falsified by the data. Three empirical findings each directly contradict the character displacement hypothesis:

1. St. Maarten, where the two lizards have nearly the same body size, is the critical case. By the character displacement hypothesis, one or both of the species on St. Maarten must be a new arrival, and the island should represent the condition preceding the divergent evolution of body size. In fact, both species are visibly and biochemically differentiated from other *Anolis* in the region, and so neither is a very recent arrival from an

adjacent island. Moreover, the phylogenetic tree shows that both these species are rooted in the center of the Lesser Antilles, and *A. wattsi* is not a recent invader from Puerto Rico, as proposed in early literature.

2. By the character displacement hypothesis, medium-sized lizards become larger through evolutionary time. In fact, the fossil record shows larger lizards becoming smaller and converging toward a medium size; no fossil data show a medium-sized species becoming larger.
3. Finally, by the character displacement hypothesis, a single-species community begins to develop into a two-species community when the second species arrives that is the same size as the first species. This sets the stage for their subsequent divergence from one another. In fact, both in experimental studies of competition, and in accidental introductions, an invader very similar in size to an established resident does not "take." Thus, the initial condition of the character displacement hypothesis—that two medium-sized species coexist—is ecologically impossible.

Therefore, the character displacement hypothesis is contradicted by the phylogenetic status of the anoles on St. Maarten, the fossil record, and the ecology of colonization for *Anolis*.

The second hypothesis, that the one- and two-species communities are stages in a taxon cycle, is supported specifically for the islands of the Antigua, St. Kitts, and Anguilla banks. As presented earlier, the taxon cycle hypothesis is

> An island with a solitary-sized species is invaded by a larger species. Both then simultaneously evolve toward smaller sizes: the invader converges toward the solitary size; as the resident becomes smaller than the solitary size, its range contracts, and it eventually becomes extinct. Meanwhile, the invader completes its convergence to the solitary size, reestablishing the initial point of the cycle. The island remains in the solitary state until it is presented with another large invader.

This cyclic scenario is diagrammed in Figure 7. Twelve facts directly pertain to this hypothesis, as follows. This list does not include facts that are simply consistent with the hypothesis, but that are not logically or causally required by it. The facts are numbered in the list and in the figure, so that the relevant place in the scenario can be readily identified.

1. On the Antigua bank, separate excavations from Barbuda and from Antigua proper show that the larger anole there today, *A leachi,* was even larger in the past; it has become smaller since the later Pleistocene, directly supporting the taxon cycle hypothesis.
2. On the St. Kitts bank the larger anole, *A. bimaculatus,* is smaller on St. Eustatius than on St. Kitts proper, which can be interpreted as a derived condition. If so, this situation would represent one stage beyond that documented on the Antigua bank.

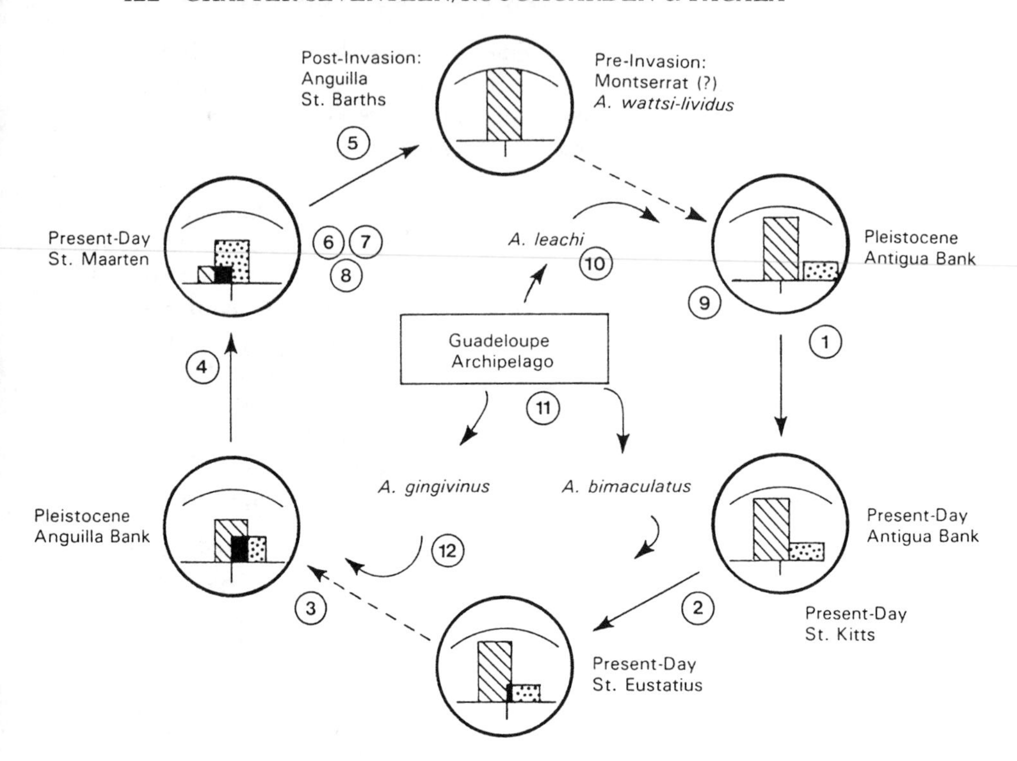

FIGURE 7. Taxon cycle in the northern Lesser Antilles. Large circles depict stages in the cycle. Within a circle the horizontal axis represents body size and the vertical axis represents relative abundance. The overlying curve represents the resource spectrum. The cycle begins when an island with a medium-sized species is invaded by a larger species from the Guadeloupe archipelago. Both then evolve a smaller size. As the invader's size approaches the medium size of the original resident, the range of the original resident contracts, followed by eventual extinction. The invader then completes its evolution toward the medium size that the original solitary species possessed, thus reestablishing the point at which the cycle began. Numbers depict specific places in the cycle for which evidence exists, as discussed in the text.

3. On Anguilla, the fossil excavations reported here show a large increase in the number of medium-sized lizards at approximately 4000 years BP. This increase in the number of fossils can be interpreted as reflecting the invasion of *A. gingivinus*. Moreover, the invader is larger than the species to have preceded it, *A. pogus*. Evidently, however, *A. gingivinus* entered the Anguilla bank at a relatively small size to begin with; that is, it was larger than *A. pogus* (and larger than it is today) but smaller than even the present-day sizes of the larger species of the Antigua and St. Kitts banks.

4. The size of *A. gingivinus* on Anguilla in the surface stratum is statistically significantly smaller than in the stratum immediately below the top 3-cm stratum. This finding can be interpreted as reflecting a reduction in body size of *A. gingivinus* that coincides with the loss of *A. pogus* on Anguilla, as mentioned next.
5. On Anguilla, the smaller species *A. pogus* has become extinct in historic times. Thus, the only known extinction of an anole in the Lesser Antilles consists of the smaller species on a two-species island, as expected by the taxon cycle hypothesis.
6. The smaller anole on St. Maarten, where there is strong present-day competition, has its range confined to the center of the island. In contrast, the slightly larger anole has a range encompassing the entire island. This condition can be interpreted as a reduction in the range of the smaller anole following an invasion by the larger anole, and to be a prelude to the eventual extinction of the smaller anole on St. Maarten, as has happened on Anguilla.
7. The size of *A. pogus* on St. Maarten is smaller than its relatives, *A. schwartzi* of the St. Kitts bank, and both *A. wattsi forresti* of Barbuda and *A. wattsi wattsi* of Antigua proper on the Antigua bank. This can be interpreted as the most extensive evolution toward a size below the solitary size on any of the two-species islands, as expected by the taxon cycle hypothesis of any original resident that has reached the stage immediately prior to its extinction.
8. Neither species on St. Maarten is a recent invader from a nearby island; both have the morphological and biochemical differentiation normal to distinct species. Thus, the closeness of the body sizes of the species on St. Maarten can be interpreted as a derived condition resulting from the evolution that has already occurred there, as required by the taxon cycle hypothesis.
9. Only the wattsi series has geographic variation within an island bank of the northern Lesser Antilles, the Antigua bank, that has led to a subspecific nomenclature. The Antigua bank harbors *A. wattsi wattsi* on Antigua proper and *A. wattsi forresti* on Barbuda. No bimaculatus-lineage anole has subspecific variation within an island bank north of Guadeloupe. This may be interpreted as indicating a longer presence of the smaller lizards than of the larger lizards on the northern Lesser Antillean islands, as required by the taxon cycle hypothesis.
10. Both experimental introductions and historical records of accidental introductions show that a large anole can successfully invade the habitat already occupied by a medium-sized species, provided the habitat is large enough. Conversely, an invader cannot "take" if it is the same size as an established resident. These ecological facts underly the taxon cycle hypothesis.

11. The putative source for the large bimaculatus-lineage anoles that have invaded the northern islands is Guadeloupe, where the geologically oldest habitat of the Lesser Antilles occurs (indeed as old as the oldest formations in Puerto Rico) and where a solitary anole exists with spectacular geographic differentiation into nine named subspecies. Moreover, the larger anole of the St. Kitts bank, *A. bimaculatus*, is extremely similar in appearance and habits to the subspecies on Guadeloupe, *A. marmoratus alliaceus*, living in middle-elevation rain forest on the Soufriere. Indeed, the similarity led to early taxonomic confusion between *A. bimaculatus* of St. Kitts and this type on Guadeloupe. This rain-forest habitat is typically where the largest anoles on a solitary island can be found, thus providing a source of large lizards as required by the taxon cycle hypothesis. [The scanty collections of *A. m. alliaceus* do not exhibit unusually large body sizes. The anoles from similar habitat on nearby Dominica do, however (Roughgarden and Fuentes 1977).] Furthermore, the prevailing currents lead north from Guadeloupe (Richardson and Walsh 1986), carrying vegetation (and people) from the Guadeloupe area to the northern Lesser Antilles. Thus, given that the anole on Guadeloupe was present from the beginning on the part of that island that was originally formed as a fragment of Puerto Rico, dispersal to the relatively new habitat being generated nearby between Guadeloupe and the Anagada Passage seems almost a certainty.
12. The apparent timing of the introduction of *A. gingivinus* to the Anguilla bank coincides with the dates of stone-age villages on the Antigua bank. These villages predate the better known and widely distributed Arawak Indian artifacts. The direction of human migration into the Lesser Antilles has been from South America, particularly from the Orinoco Delta. Today the largest impact of man on anoles is to effect introductions, not to cause extinctions, as is well-known in Florida, the Greater Antilles, and Bermuda. Thus, the taxon cycle on Anguilla may have been initiated by early man.

Figure 7 summarizes where these 12 points pertain to the taxon cycle hypothesis for the northern Lesser Antilles. Because each of the three island banks has been colonized separately by the second species on it, three instances of a cycle seem to be occurring—one per bank. But no single bank has gone around an entire cycle. The Antigua bank, for example, is still at an early two-species stage whereas St. Eustatius of the St. Kitts bank is further along. The Anguilla bank evidently was never invaded to begin with by a second species as large as that on the other two banks, and thus the early phases of the cycle were bypassed there.

DIFFICULTIES WITH THE TAXON CYCLE HYPOTHESIS

Although evidence supports the taxon cycle hypothesis for the anole communities in the northern Lesser Antilles, six important difficulties or qualifications remain.

The first, and most thorny, problem continues to be the exceptionally large body size of Marie Galante. We originally hypothesized Marie Galante to be a community at the stage on the taxon cycle just after extinction of the small species and prior to complete convergence to the solitary size. But this interpretation is not tenable seeing that the size of the solitary anole on Marie Galante is much larger than the size of the larger anole on St. Maarten. If St. Maarten really were prior to Marie Galante in the taxon cycle, then the larger anole on St. Maarten should be larger than that remaining on Marie Galante. Yet the reverse is true. Therefore, the Marie Galante anole must be considered apart from a taxon cycle involving the Antigua, St. Kitts, and Anguilla banks. Thus, a taxon cycle, if one exists, is indicated only for the islands north of Guadeloupe, between Guadeloupe and the Anagada Passage.

An independent argument is therefore needed to treat the case of Marie Galante. This island near Guadeloupe is at or very near the oldest habitat in the Lesser Antilles. It is near basement exposed on La Deserade that is reliably dated to the late Jurassic (Fink 1972). Although undoubtedly not above water until some time in the Tertiary, this area of the Lesser Antilles was probably the first habitat available for *Anolis*. It appears that much of the fauna of the Guadeloupe region entered the Lesser Antilles from Puerto Rico through a vicariance mechanism. That is, proto-Guadeloupe split tectonically from Puerto Rico carrying a Puerto Rican fauna on it that then evolved in isolation. Thereafter the Guadeloupe region served as the source pool for overwater colonists to newly forming habitat close by in the Lesser Antilles (Roughgarden et al. 1987).

Male *A. ferreus* of Marie Galante are very large, the females less strikingly so, leading to a very high sexual dimorphism. The maximum snout-vent length of a male (~120 mm) is about twice the maximum length of a female; this is the largest dimorphism in size of any northern Lesser Antillean anole (Lazell 1972). We conjecture therefore that the stock of anoles on Marie Galante has been present there long enough to evolve a wide enough niche to preclude invasion by a second species. This conjecture rests on the notion that breaking down species cohesion in body size requires a long time (cf. Roughgarden 1987). If so, this island has neither participated in any taxon cycle nor been subject to character displacement, because *A. ferreus* has so wide a niche that no second species of *any* size can invade.

Second, we offer no hypothesis to account for the one- and two-species states of the islands in the southern Lesser Antilles, where the Roquet group anoles are found. As previously mentioned, our studies have shown that the anole community structure on the two-species islands in the southern Lesser Antilles is qualitatively different from that on the northern two-species islands. In the south the two species of an island are associated with different habitats, even though both overlap in a great many areas. One species may be thought of as a montane form at home in the shade of forests, the other a creature of open xeric habitat. Thus, the "dimensionality" of niche space on the southern islands is perhaps higher than in the north, and no theory formulated so far may apply. Furthermore, the southern Lesser Antilles seem to have had a qualitatively different geological history from the north (Roughgarden et al. 1987) and the applicability of a colonization-type hypothesis would need to be justified relative to the geological history there.

Third, not only is the taxon cycle occurring in a limited area, but there is also reason to believe that such a cycle is rare, and may, in the instance of Anguilla, have been facilitated by early man. The distribution of anoles throughout the eastern Caribbean, taken together with a cladogram describing the phylogenetic relationships among the anoles, faithfully coincides with hypothesized events in the geologic origin of the Caribbean (Roughgarden et al. 1987). This circumstance allows anoles to be used much as though they were "biological strata" to aid in reconstructing the geological origin of the Caribbean. But if a taxon cycle were frequently occurring, then the distributions of lineages would be homogenized, and spotted with gaps representing extinctions, so that any congruence between the biogeographic pattern of distributions and geological events would have long been erased. Moreover, as discussed earlier, the timing of the arrival of the larger species on St. Maarten appears, on the limited evidence available, to coincide with the spread of lithic man into the northern Lesser Antilles.

Fourth, the presence of a wattsi series anole as the smaller anole of each two-species island is improbable (though possible) by independent random colonization from Guadeloupe. It would therefore appear that the Anguilla–St. Kitts–Antigua banks were themselves united at some time in the past when they were colonized once by a common ancestor to *A. marmoratus* of Guadeloupe. After splitting these banks have been independently colonized a second time from Guadeloupe.

A fifth difficulty may not be a difficulty at all. The evidence of reduction in size for the larger lizard on Anguilla, presented in Figure 6, is limited to the difference between the surface stratum and the one immediately below. This reduction in size would slightly predate, or coincide with, the extinction of the smaller anole on Anguilla during the 1920s. The problem is that there is no evidence of change in size for the larger anole during the preceding 3000 to 4000 years, extending back to the time when the larger lizard appears to have been introduced. This fact, if genuine, would suggest that a small anole (50 mm in length) can precariously coexist with a solitary-sized anole (65 mm in length). The small anole drops out only when the habitat is destroyed, which is the coup de grace given its strong competition with a larger anole. Still, the smaller anole does seem to have become somewhat smaller during that time interval (even though the larger anole did not change), because the smaller anole was larger at the time its competitor was introduced than it now is in its only remaining location on St. Maarten.

Finally, another possible concern is that the observed reduction in body size of the large species, as on the Antigua bank, is actually caused by some other factor not part of the taxon cycle hypothesis. Pregill (1986) has suggested that the size change of the large species on Barbuda since the Pleistocene reflects a response to human habitat modification, and to the introduction of animals associated with man. Human activities, particularly wood cutting, renders the habitat more open. Pregill imagines that the food supply is reduced because of desertification (resulting in slower growth rates for lizards) and that lizards receive more predation pressure (implying that they do not live long enough to obtain the size they did prior to human activity). An open habitat, however, can also lead to a longer daily activity period, and human activities can also reduce the number of lizard predators, both factors that would lead to a faster growth rate and a longer average life span. Any conjecture is possible, and we cannot say, a priori, what man's net effect is on the evolution of lizard body size.

Some correlative cases would therefore seem instructive. The islands in the vicinity of Madagascar have been studied by Arnold and his associates (Arnold 1976; Taylor et al. 1979; Arnold 1979, 1980; Bullock et al. 1985). Aldabra had crocodiles, a large iguanid, two or three geckoes, and two skinks that no longer occur there. Yet Arnold points out that human influence is not likely to be involved in these extinctions as there is no evidence of permanent settlement and close relatives of these extinct species coexist with man elsewhere. Arnold attributes these extinctions to a drastic reduction in the size of the atoll following the sea level rise at the end of the Pleistocene. In contrast, early colonists to Mauritius almost surely hunted the large turtles there to extinction, and a number of smaller reptiles that were known on Mauritius are now restricted to tiny shelf islands nearby, presumably as a result of human activities. The key point, it would seem, is that the effect of man depends on both the island and the man. No straightforward generalization has emerged for the effect of man on reptiles other than that very large reptiles are usually hunted to extinction. No humans are known to hunt small lizards to an extent that could produce an extinction.

Fewer extinctions of lizards are known in the Lesser Antilles than on the islands near Madagascar. As discussed earlier, only the *Leiocephalus* line has become extinct, one species from the Antigua bank, and one from the Anguilla bank. These extinctions appear to predate the arrival even of lithic man, and, like Aldabra, seem attributable to the last sea level change. The only other known extinction of a small lizard in the Lesser Antilles is *A. pogus*, a member of the wattsi series on Anguilla, as discussed extensively. Human activities would seem deleterious to *A. pogus*, and those activities, combined with the strong competition it received from *A. gingivinus*, have evidently caused the extinction. But the wattsi series populations on the other banks, where presumably an equal or greater degree of human disturbance has taken place, have not become extinct—the difference is that the wattsi form on Anguilla was receiving strong competition from another anole, whereas the wattsi forms on the other island banks are not.

From the standpoint of anoles, the focus on human-caused extinctions seems over emphasized. Similarly, Richman et al. (1988) concluded on the basis of an extensive review, combined with a detailed analysis of the lizards on the Baja California Islands and the South Australian Islands, that human disturbance simply exacerbates the processes of extinction that are already in place. It is worth emphasizing then that the most obvious effect of man on anoles is to introduce them to new locations. The great number of known introductions of anoles contrasts with the scarcity of extinctions of anoles.

Just as the effect of man on the extinction of lizards depends on circumstances, the effect of man on the evolution of body size is also likely to depend on circumstances. Yet Pregill (1987) reports that a reduction in body size of large insular lizards is a general phenomenon. It seems likely that the selective forces underlying the taxon cycle are experienced by species on all islands, and, if so, a universality of the selective pressure could account for the seeming generality of the observation that large lizards have been evolving smaller body sizes since the Pleistocene on many islands throughout the world.

CONCLUSION

If a taxon cycle exists in the northern Lesser Antilles as suggested, then small islands may not offer glimpses of the early stages in the assembly of a complex fauna, as long believed. The reason is that present-day large islands may never have passed through a small stage; their faunas may be assembled through the mixing of entire communities residing on geologic blocks that suture to form large land masses. Hispaniola, for example, is sutured from at least three blocks, Cuba from two, and the faunas on these large islands may never have passed through any stage comparable to what is happening on small islands.

Small islands can, however, reveal the consequences of species interactions with great clarity. When this information is spliced together with studies of other processes, including population responses to large-scale habitat complexity, then an ecological theory for complex faunas is possible. The fauna on a small island is like a tissue culture—it supplies invaluable information about the organism but is not a precursor to an organism, it is merely a piece of the organism seen clearly, yet out of context.

Research on the assembly of island faunas offers certain insights for the biology of continental areas. First, faunas are not stable over evolutionary time—evolution does not produce an integrated community whose members fit together in a harmonious way. Instead, after a new species adds to a community extinctions result, even over evolutionary time. Although people may be familiar with the idea that an introduced species can displace a previously established species, this is thought to be an ecological phenomenon only. What the taxon cycle suggests is that even when an invasion does not immediately displace a resident, that is, even when the invasion results in an augmented community in which the invader and resident coexist ecologically, the final result over evolutionary time may still be the elimination of one of the species. Second, community structure is strongly influenced by the identity of what arrives at the community's boundary. This imparts a strong historical and phylogenetic component to the explanation of community structure. The success of the plate tectonic theory in geology, however, makes the reconstruction of species movements more accessible to scientific investigation than might even have been hoped years ago. Finally, many island communities are special and highly derived extractions from complex communities, not early stages in the assembly of a complex fauna. But the processes that occur on islands are among those that occur in continental regions as well, and islands continue to serve as natural laboratories in which these processes are revealed with unusual clarity.

SUMMARY

A "taxon cycle" involving populations of *Anolis* lizards appears to be occurring on the Antigua, St. Kitts, and Anguilla banks in the northern Lesser Antilles. The cycle begins with an island that has a medium-sized species; it is invaded by a larger species that has come from the Guadeloupe archipelago. Both then evolve a smaller size. As the invader's size approaches the medium size of the original resident, the original resident's species range contracts, culminating in extinction. Meanwhile, the invader completes its evolution toward the medium size that the (now extinct) original resident once possessed. Thus, the condition at which the cycle began is eventually reestablished. Data on the ecology of colonization, phylogenetic relationships, biogeographic distribution, and the fossil record of anoles since the Pleistocene support this hypothesis, and falsify the alternative hypothesis of faunal buildup through invasion followed by character displacement. A taxon cycle does not appear to happen often, however; nor is there evidence of such a cycle anywhere else in the Lesser Antilles.

ACKNOWLEDGMENTS

We thank Walter Brown, John McLaughlin, Mary Anne Murray, and Kenneth Naganuma for assistance during the fossil excavations on Anguilla, and two anonymous reviewers for helpful comments. We also thank the National Science Foundation for the financial support that made these studies possible, most recently through Grant BSR-8719597.

LITERATURE CITED

Andrews, R. and S. Rand. 1982. Seasonal breeding and long-term population fluctuations in the lizard *Anolis limifrons*. Pp. 405–412 in: E. Leigh, S. Rand, and D. Windsor (eds.), *The Ecology of a Tropical Forest: Seasonal Rhythms and Long-term Changes*. Smithsonian Institution Press, Washington, D.C.

Arnold, E. N. 1976. Fossil reptiles from Aldabra atoll, Indian Ocean. Bull. Br. Mus. (Nat. Hist.) Zool. 29:83–116.

Arnold, E. N. 1979. Indian Ocean giant tortoises: Their systematics and island adaptations. Philos. Trans. R. Soc. Ser. B 286:127–145.

Arnold, E. N. 1980. Recently extinct populations from Mauritius and Reunion, Indian Ocean. J. Zool. London 191:33–47.

Bohlen, C. 1983. Competition for space in the anoles of the island of St. Maarten (Neth. Antilles). Unpublished Master's thesis, Stanford University, Stanford, CA.

Brown, W. L., Jr., and E. O. Wilson. 1956. Character displacement. System. Zool. 7:49–64.

Bullock, D. J., E. N. Arnold, and Q. Bloxam. 1985. A new endemic gecko (Reptilia: Gekkonidae) from Mauritius. J. Zool. London 206:591–599.

Cope, E. D. 1883. On the contents of a bone cave in the island of Anguilla (West Indies). Smithson. Cont. Knowl. 489:1–30.

Ethridge, R. 1964. Late Pleistocene lizards from Barbuda, British West Indies. Bull. Florida State Mus. 9:43–75.

Etheridge, R. 1965. Fossil lizards from the Dominican Republic. Quart. J. Florida Acad. Sci. 28:83–105.

Etheridge, R. 1966. An extinct lizard of the genus *Leiocephalus* from Jamaica. Quart. J. Florida Acad. Sci. 29:47–59.

Fink, L. K., Jr. 1972. Bathymetric and geologic studies of the Guadeloupe region, Lesser Antilles arc. Marine Geol. 12:267–288.

Heatwole, H., and F. MacKenzie. 1967. Herpetogeography of Puerto Rico. IV. Paleogeography, faunal similarity, and endemism. Evolution 21:429–438.

Hecht, M. K. 1951. Fossil lizards of the West Indian genus *Aristelliger* (Gekkonidae). Am. Mus. Novit. 1538:1–33.

Hotton, N. III. 1955. A survey of adaptive relationships of dentition to diet in North American Iguanids. Am. Mid. Natur. 53:88-114.

Lazell, J. D., Jr. 1965. An *Anolis* (Sauria, Iguanidae) in amber. J. Paleontol. 39:379-382.

Lazell, J. D., Jr. 1972. The Anoles (Sauria, Iguanidae) of the Lesser Antilles. Bull. Mus. Comp. Zool. 143:1-115.

Licht, P. 1974. Response of *Anolis* lizards to food supplementation in nature. Copeia 1974:215-221.

Olson, S. L., and H. F. James. 1982. Fossil birds from the Hawaiian Islands: Evidence for wholesale extinction by man before western contact. Science 717:633-635.

Pacala, S. W., and J. D. Roughgarden. 1985. Population experiments with the *Anolis* lizards of St. Maarten and St. Eustatius. Ecology 66:129-141.

Poinar, G. O., Jr., and D. C. Cannatella. 1987. An Upper Eocene frog from the Dominican Republic and its implications for Caribbean biogeography. Science 237:1215-1216.

Pregill, G. K. 1981. Late Pleistocene herpetofaunas from Puerto Rico. Misc. Publ. Univ. Kansas Mus. Nat. Hist. 71:1-72.

Pregill, G. K. 1982. Fossil amphibians and reptiles from New Providence, Bahamas. Smithson. Contrib. Paleobiol. 48:8-22.

Pregill, G. K. 1984. An extinct species of *Leiocephalus* from Haiti (Sauria: Iguanidae). Proc. Biol. Soc. Wash. 97:827-833.

Pregill, G. K. 1986. Body size of insular lizards: A pattern of Holocene dwarfism. Evolution 40:997-1008.

Rand, A. S. 1964. Ecological distribution of the anoline lizards of Puerto Rico. Ecology 45:745-752.

Rand, A. S. 1967. Ecological distribution of the anoline lizards around Kingston, Jamaica. Breviora 22:1-18.

Richardson, P. L., and D. Walsh. 1986. Mapping climatological seasonal variations of surface currents in the tropical Atlantic using ship drifts. J. Geophys. Res. 91: 10537-10550.

Richman, A. D., T. J. Case, and T. Schwaner. 1988. Natural and unnatural extinction rates of island reptiles. Am. Natur. 31:611-630.

Rieppel, O. 1980. Green anole in Dominican amber. Nature (London) 286:486-487.

Roughgarden, J. D. 1987. Community coevolution: A comment. Evolution 41:1130-1134.

Roughgarden, J. D. 1989. *The Anoles of the Eastern Caribbean: Competition, Coevolution and Plate Tectonics*. Cambridge University Press, Cambridge, in preparation.

Roughgarden, J. D., and E. R. Fuentes. 1977. The environmental determinants of size in solitary populations of West Indian *Anolis* lizards. Oikos 29:44-51.

Roughgarden, J. D., D. Heckel, and E. R. Fuentes. 1983. Coevolutionary theory and the biogeography and community structure of *Anolis*. Pp. 371-410 in: R. B. Huey, E. R. Pianka, and T. W. Schoener (eds.), *Lizard Ecology, Studies of a Model Organism*. Harvard University Press, Cambridge, MA.

Roughgarden, J. D. , S. W. Pacala, and J. D. Rummel. 1984. Strong present-day competition between the *Anolis* lizard populations of St. Maarten (Neth. Antilles). Pp. 203-220 in: B. Shorrocks (ed.), *Evolutionary Ecology*. Blackwell Scientific Publications, London.

Roughgarden, J. D., S. D. Gaines, and S. W. Pacala. 1987. Supply side ecology: The role of physical transport processes. Pp. 459-486 in: P. Giller, and J. Gee (eds.), *Organization of Communities: Past and Present*. Blackwell Scientific Publications, London.

Rouse, I., and L. Allaire. 1978. Caribbean. Pp. 431-483 in: R. E. Taylor and C. Meighan (eds.), *Chronologies in New World Archeology*. Academic Press, New York.

Rummel, J. D., and J. D. Roughgarden. 1985a. A theory of faunal buildup for competition communities. Evolution 39:1009-1033.

Rummel, J. D., and J. D. Roughgarden. 1985b. Effects of reduced perch-height separation on competition between two *Anolis* lizards. Ecology 66:430-444.

Schoener, T. W. 1968. The *Anolis* lizards of Bimini: Resource partitioning in a complex fauna. Ecology 49:704-726.

Schoener, T. W. 1969. Size patterns in West Indian *Anolis* lizards: I. Size and species diversity. System. Zool. 18:386-401.

Schoener, T. W., and G. C. Gorman. 1968. Some niche differences in three Lesser Antillean lizards of the genus *Anolis*. Ecology 49:819–830.

Schoener, T. W., and D. H. Janzen. 1968. Notes on environmental determinants of tropical versus temperate insect size patterns. Am. Natur. 102:207–225.

Schoener, T. W., and A. Schoener. 1978. Inverse relation of survival of lizards with island size and avifaunal richness. Nature (London) 274:685–687.

Schoener, T. W., and A. Schoener. 1982. The ecological correlates of survival in some Bahamian *Anolis* lizards. Oikos 39:1–16.

Schoener, T. W., and A. Schoener. 1983a. Distribution of vertebrates on some very small islands. I. Occurrence sequences of individual species. J. Anim. Ecol. 52:209–235.

Schoener, T. W., and A. Schoener. 1983b. Distribution of vertebrates on some very small islands. II. Patterns in species number. J. Anim. Ecol. 52:237–262.

Schoener, T. W., and A. Schoener. 1983c. The time to extinction of a colonizing propagule of lizards increases with island area. Nature (London) 302:332–334.

Schoener, T. W., and D. Spiller. 1987. Effect of lizards on spider populations: Manipulative reconstructions of a natural experiment. Science 236:949–952.

Schoener, T. W., and C. A. Toft. 1983. Spider populations: Extraordinarily high densities on islands without top predators. Science 219:1353–1355.

Sexton, O. 1985. Reproductive cycles of Panamanian reptiles. Pp. 95–110 in: W. G. D'Arcy and M. D. Correa A. (eds.), *The Botany and Natural History of Panama: La Botanica e Historia Natural de Panama*. Missouri Botanical Garden, St. Louis.

Stamps, J. A. 1977. Rainfall, moisture, and dry season growth rates in *Anolis aeneus*. Copeia 1977:415–419.

Steadman, D. W., G. K. Pregill, and S. L. Olson. 1984. Fossil vertebrates from Antigua, Lesser Antilles: Evidence for late Holocene human-caused extinctions in the West Indies. Proc. Natl. Acad. Sci. U.S.A. 81:4448–4451.

Taylor, J. D., C. J. R. Braithwaite, J. F. Peake, and E. N. Arnold. 1979. Terrestrial faunas and habitats of Aldabra during the late Pleistocene. Philos. Trans. R. Soc. London Serv. B 286:47–66.

Underwood, G. 1959. Revisionary Notes. The anoles of the Eastern Caribbean (Sauria, Iguanidae). Part III. Bull. Mus. Comp. Zool. 121:191–226.

Williams, E. E. 1972. The origin of faunas. Evolution of lizard congeners in a complex island fauna: A trial analysis. Evol. Biol. 6:47–88.

Williams, E. E. 1977. Anoles out of place: Introduced anoles. Pp. 110–118 in: E. E. Williams (ed.), *The Third Anolis Newsletter*. Harvard University, Cambridge, MA.

Williams, E. E. 1983. Ecomorphs, faunas, island size, and diverse end points in island radiations of *Anolis*. Pp. 326–370 in: R. B. Huey, E. R. Pianka, and T. W. Schoener (eds.), *Lizard Ecology, Studies of a Model Organism*. Harvard University Press, Cambridge, MA.

Wilson, E. O. 1961. The nature of the taxon cycle in the Melanesian ant fauna. Am. Natur. 95:169–193.

Wingate, D. 1965. Terrestrial herpetofauna of Bermuda. Herpetologica 21:202–218.

CHAPTER 18

SYMPATRIC SPECIATION AND DARWIN'S FINCHES

P. R. Grant and B. R. Grant

INTRODUCTION

The allopatric model of speciation is widely accepted as the most probable mode of speciation for vertebrates. Many naturalists are not convinced that it is sufficient to account for all the taxonomic diversity they see, principally because they would have to invoke an apparently improbable number of range splittings and rejoinings to explain the existence of many closely related species in local assemblages. They are led, therefore, to consider the alternatives of parapatric speciation and sympatric speciation (Bush 1975; Endler 1977).

Sympatric speciation can account for the evolutionary development of a correspondence between the large number of niches in a heterogeneous environment and the large number of species in one or more related taxa (cf. Hutchinson 1968). But how could it occur? How can a single population of interbreeding organisms be converted into two reproductively isolated segments in the absence of spatial barriers or hindrances to gene exchange? Through polyploidy, is one answer applicable to many plant taxa, many invertebrate taxa, and some vertebrates. But since polyploidy is not found in all taxa it does not provide a general mechanism for speciation. Up to 1966 it could be said that this lack of a general mechanism was the greatest weakness of the theory of sympatric speciation (Mayr 1963).

Maynard Smith (1966) provided the missing mechanism. He showed, with a simple one-locus two-allele model, how a stable polymorphism can exist (without heterozygous advantage) in a heterogeneous environment with two niches, even when the adults form a single randomly mating population. This is the first step toward sympatric speciation. The second is the subsequent evolution of reproductive isolation between the populations in the two niches. The conditions for this to happen are stringent. The least likely to be realized involve pleiotropic effects of the niche-adapting alleles, effects that directly govern mate choice. A more likely condition is habitat (or niche) selection, with individuals breeding only in the habitat, or in association with the niche (resource or host), to which they are adapted. Alternatively assortative mating could be under the control of alleles at another locus, with either modifying effects on the niche-adapting alleles or independent effects. Maynard Smith (1966:649) concluded "Whether this paper is regarded as an argument for or against sympatric speciation will depend on how likely such a polymorphism is thought to be, and this in turn depends on whether a single gene difference can produce selective coefficients large enough to satisfy the necessary conditions."

The selective coefficients can be made large in models (Dickinson and Antonovics 1973; Pimm 1979; Rice 1984), by assigning low fitness to heterozygotes at or near polymorphic equilibria (Rosenzweig 1978; Pimm 1979, Loeschcke and Christiansen 1984; Wilson and Turelli 1986). Models have been made increasingly realistic by explicit incorporation of density-dependent and frequency-dependent competition for resources among phenotypes (Roughgarden 1972; Matessi and Jayakar 1976), distributions of resources that are continuous (Bengtsson 1979; Loeschcke and Christiansen 1984; Seger 1985) or discrete (Rice 1984), and dynamics of resource consumption and renewal (Wilson and Turelli 1986). Thus, the first step in developing a stable polymorphism, which Maynard Smith (1966) regarded as crucial, poses no special theoretical difficulties.

The second step, the evolution of reproductive isolation, is equally crucial yet less well investigated theoretically (Hendrickson 1978; Futuyma and Mayer 1980; Kondrashov and Mina 1986). It is generally thought to require a genetic association between traits conferring adaptation to a niche and traits causing mating preferences. The association may be created by strong linkage disequilibrium, but this is opposed by reassortment of alleles at different loci through recombination (Maynard Smith 1966; Felsenstein 1981; Rice 1984). Alternatively pleiotropy may be involved. Either individuals breed only in the habitat to which they are adapted (Rice 1984, 1987), a passive form of reproductive segregation, or they actively choose to breed with similar types that are adapted to the same niche as themselves (Seger 1985). Another class of models was introduced by O'Donald (1960) to consider assortative mating of phenotypes (and genotypes), without explicit attention being given to ecological factors such as resources or habitats. Since a gene for assortative mating can increase to fixation sympatrically (Kalmus and Maynard Smith 1966; Seiger 1967; Endler 1977), speciation could be the outcome of sexual selection alone (Lande 1981; West-Eberhard 1983).

Given the sympatric evolution of two species from one in models, how likely is it to occur in nature? There is no clear answer. Experiments in the laboratory help to reduce the gap between theoretical investigation of what can happen and the empirical demonstration of what does happen in nature. For example in the laboratory it has been shown that disruptive selection in a two-niche environment can maintain a stable genetic polymorphism (Thoday and Gibson 1970; Hedrick et al. 1976; Halliburton and Gall 1981; Hedrick 1986). A large degree of reproductive isolation can also evolve under similar experimental conditions (Pimentel et al. 1967; Soans et al. 1974; Rice 1985).

In nature itself the evidence is ambiguous. On the one hand events leading to speciation in sympatry can be plausibly reconstructed by using ecological, behavioral, and genetic data from a few, unusually suitable, organisms, mainly insects (Bush 1975; Tauber and Tauber 1977a, 1977b, 1987; Tauber et al. 1977; Gibbons 1979; Bush and Diehl 1982; Feder et al. 1988; McPheron et al. 1988; Smith 1988). On the other hand, even these well-investigated examples can be explained alternatively in terms of the classical model of allopatric speciation (Hendrickson 1978; Futuyma and Mayer 1980), and it is nearly impossible to clearly reject one mode of speciation as an explanation in favor of the other. In response to this dilemma some biologists stress genetic rather than geographic criteria for classifying modes of speciation (Templeton 1981), although this does not eliminate the problem of identifying and understanding the causes of sympatric divergence.

Those causes are best investigated with polymorphic species that show current signs of splitting into two. We will now describe the results of a field study of one such species, the large cactus finch *Geospiza conirostris* on Isla Genovesa, Galápagos.

The background, briefly, is as follows; for a fuller statement see Grant and Grant (1989a). We began the study in 1978 with the intention of determining why beak size is exceptionally variable in this population. Sympatric speciation was not in our minds, but in the first field season we made several observations that were clearly relevant to theoretical ideas about it (Grant and Grant 1979). The population appeared to be ecologically subdivided into two groups, and there were hints of a reproductive subdivision too. Subdivision is incipient speciation, or at least it provides the potential for speciation. Subdivision can arise in two ways. The population could be in the secondary contact phase of allotropic speciation, in which the descendants of immigrants from a partly differentiated conspecific population are in the process of diverging under selection from residents, both ecologically and reproductively. Arguing against an immigration hypothesis, the nearest conspecific population is situated very far away, at the south of the archipelago on Isla Española and its satellite

Gardner (Figure 1). Moreover there is no evidence among recently discovered fossils that the species has been present on the intervening island of Santa Cruz in the last 2000 years (D. W. Steadman, personal communication). This forces us to confront the sympatric alternative: in situ origin of the differences, with a stage having been reached on the way to full sympatric speciation.

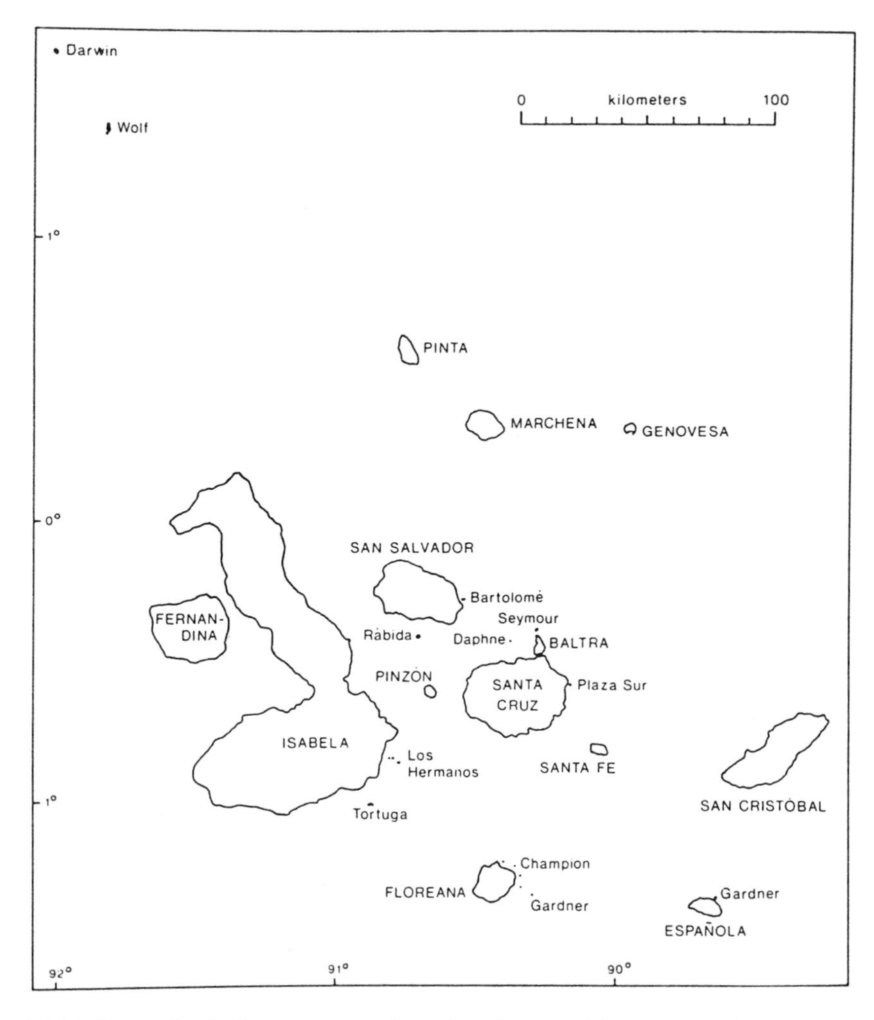

FIGURE 1. The Galápagos archipelago. Populations of *Geospiza conirostris* occur on I. Genovesa in the northeast and on I. Española and the neighboring I. Gardner in the south.

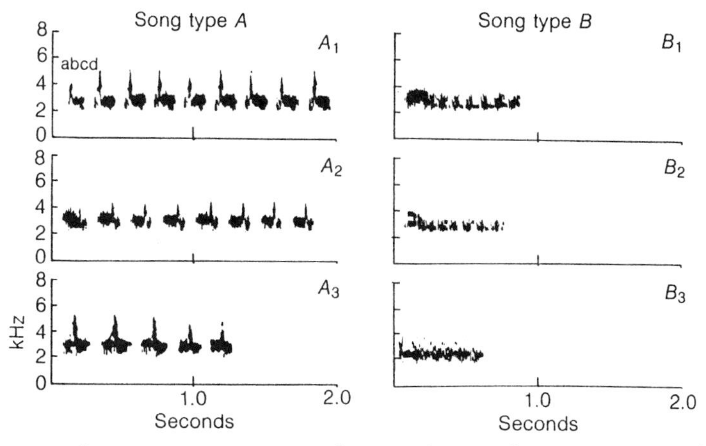

FIGURE 2. The two songs types in the population of I. Genovesa, and three variants of each of them. (From Grant 1984.)

In the remainder of this chapter we first present the details of our initial observations, then we give the results of a decade-long field study, and finally draw conclusions and consider the implications for sympatric speciation theory.

INITIAL OBSERVATIONS

Ecological isolation

Males repeatedly sing a single song type, thereby advertising their ownership of territories and their position to their mates, neighbors, and offspring. Two types of song are sung by members of the population, which we designate as *A* and *B* (Figure 2). To our surprise we found that males that sang the *A* song had significantly longer beaks than those that sang the *B* song, on average by 6.3% of the smaller mean. The beaks also tended to be narrower and shallower among the *A* singers, so beak shape differed between the groups (Figure 3).

A few observations made in the dry season of 1978 indicated a dietary difference between the groups. Five males, all song *A* type, were observed to feed from cactus fruits by drilling a hole with the beak, removing seeds, and eating the surrounding arils. Six males, all song *B* type, were observed ripping open cactus pads with their beaks to feed on the exposed insect larvae and pupae. It seems extraordinary that no song *A* males ripped open pads, and no song *B* males opened fruits, as the difference in bill dimensions, though statistically significant, is not profound. In contrast to this

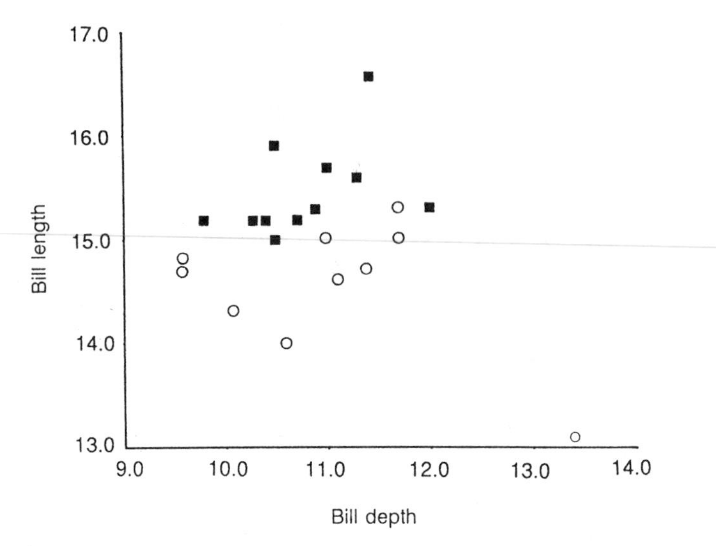

FIGURE 3. Beak characteristics of males in 1978 that sang the *A* song (○) or the *B* song (■). (From Grant 1983.)

feeding difference, eight males of the two song groups shared another cactus resource: pollen and nectar extracted from open flowers. Altogether the observations reject the hypothesis that the two groups of males used the three cactus-feeding activities with the same frequencies ($X_2^2 = 11.48$, $p < 0.005$).

The difference in beak shape between the two groups of males is functionally related to the associated foraging differences, because a long bill is advantageous in probing whereas a short and relatively stout bill is advantageous in crushing or tearing large and hard food items or surrounding plant tissue (Bowman 1961; Abbott et al. 1977; Smith et al. 1978). Each beak shape is well suited to the less demanding task of reaching into fully open flowers to obtain nectar and pollen (Grant and Grant 1981).

The dietary difference was pronounced in the dry season, when a diminishing food supply limited the population sizes (Grant and Grant 1980; Grant 1986), and less marked in the wet season at a time of plentiful food (Grant and Grant 1979, 1980).

Reproductive isolation

We did not witness pair formation in 1978, and had no direct evidence of a reproductive separation between the two song groups of males. However, an unexpected pattern in the territories suggested the possibility of some degree of reproductive isolation.

The surprise was the discovery that no two mated males of the same song type had adjacent territories (Figure 4). In contrast, the few unmated males showed no such regularity. Their territories were next to males that sang either the same song type as themselves, which we call homotypic, or the opposite one (heterotypic), or both. Among the mated males the ratio of 11 boundaries between heterotypic neighbors to 0 boundaries between homotypic neighbors cannot be attributed to chance (binomial test, $p < 0.002$). The territories of unmated males were randomly distributed with respect to song type of neighbors ($p > 0.1$).

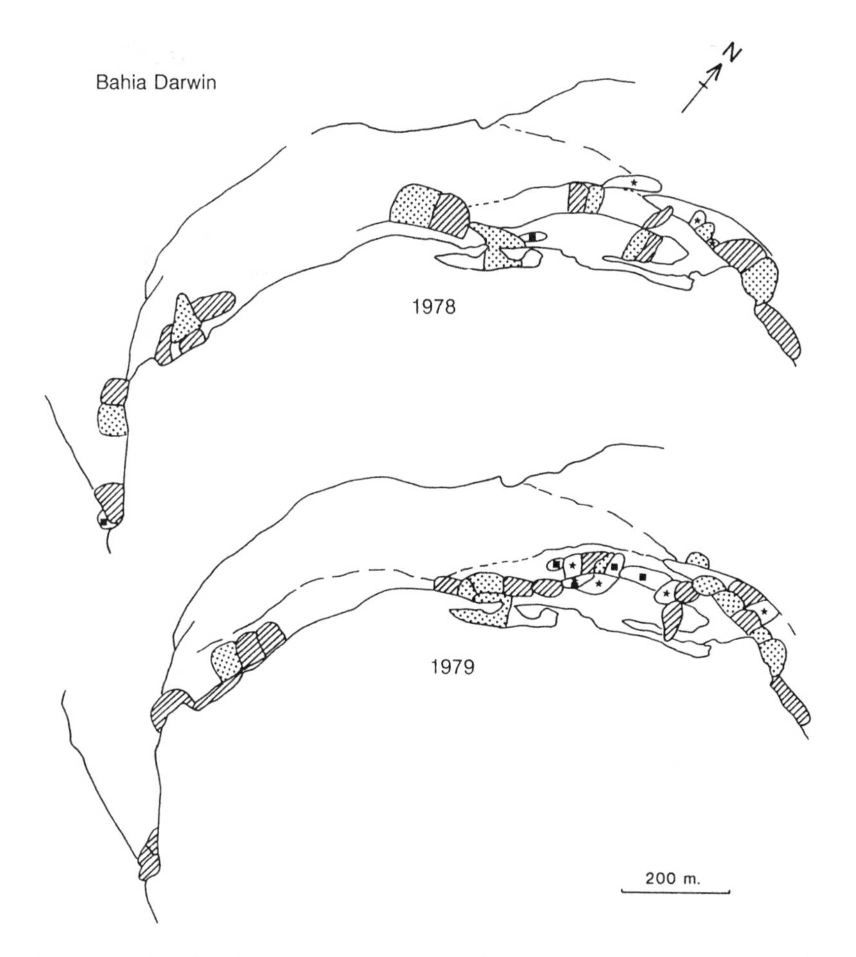

FIGURE 4. The distribution of territories in the main study area in 1978 and in 1979. Symbols: stippled signifies a mated *A* male, lines signify a mated *B* male, a star shows an unmated *A* male, and a square shows an unmated *B* male. (From Grant and Grant 1983.)

Most importantly, the difference between the territory patterns of the mated males and the unmated males is statistically significant (Fisher's exact test, $p = 0.02$). The difference could have arisen from female choice of particular males, as elsewhere in the archipelago our observations of pair formation in Darwin's Finches had indicated a large element of female choice of males (Ratcliffe and Grant 1983, 1985). Females could have actively chosen to pair with males who had heterotypic neighbors.

Thus, the spatial nonrandomness implies two things: females are responsive to the difference in song types, as well as to the particular song type of neighbors. We gained no insights into why females should pay attention to the neighbor of a potential mate, but an influence of song type on mate choice suggested the possibility of assortative mating with respect to paternal song type. And if assortative mating were taking place, the population was at least partly subdivided reproductively.

SUBSEQUENT OBSERVATIONS

Several questions raised by our initial observations could be answered only by following the fates of offspring of known parents. To take one example, to answer the question "Is mating assortative with respect to song type?" we had to find out if sons sing the same song as their fathers, and if daughters pair with males who sing the same song as their own fathers.

It was soon apparent that a long-term study would be needed to obtain the answers. Only 6 offspring of a total of 120 banded as nestlings in 1978 survived for 1 year, and only 4 of them bred in the study area. The study was extended for 10 years to build up the requisite sample sizes.

Song and mating

Some facts are consistent with the idea that the population is structured along the lines of song. Sons do sing their father's song type, and even reproduce much of the fine structure in their own songs (Grant 1984). Thirty sons of known fathers were heard to sing. Most of them were tape recorded. Twenty-six of them sang the same type as their father's. Such a consistent following of father's song type is not expected by chance (binomial test, $p < 0.001$). It probably arises through a copying process (imprinting) in early life (Bowman 1979, 1983; Grant 1986). Among the remainder, one male never sang in either of his two breeding seasons, one sang the song type opposite to his father's, and two sang both song types. Thus, there is a continuity between generations in the structure of the population determined by song, although the continuity is not perfect.

On the other hand, daughters do not initially pair assortatively with regard to paternal song type (Fisher's exact test, $p = 0.28$). Of the 30 females with an identified *A*- or *B*-singing father, 14 (47%) paired homotypically and 16 paired heterotypically. Three of the latter subsequently deserted their mates. Two of them then paired homotypically whereas the third paired heterotypically once again. No temporal pattern of regularity occurred either, such as the first few females all pairing homotypically, and the later ones all pairing heterotypically or randomly.

It is hard to escape the conclusion that choice of a male by a female is independent of any conditioning influence of her father's song type. Moreover there is no evidence of assortative mating by bill size. In none of the years was there a significant parametric correlation between mates in bill size or any other measured morphological trait (Grant 1983). In none of the years did the females mated to song *A* males differ significantly in any of the measured morphological traits from those mated to song *B* males. Hence there is no direct evidence of reproductive subdivision of the population.

A less likely possibility is that the offspring fit enough to survive and breed come predominantly from the assortative matings with respect to song type. If this did occur it would effect a reproductive subdivision of the population in the face of random mating. Our data indicate a trend, if anything, in the opposite direction. Out of eight recruits from known parents and grandparents, six were produced by mothers who had paired disassortatively with respect to paternal song type.

Morphology and ecology

The random mating of females implies that the difference in bill size between the two song groups should disappear in subsequent generations, the rate depending on the degree to which bill size is heritable. The capture and measurement of offspring of known and measured parents allowed us to determine that bill size is highly heritable (Table 1). And in line with the above reasoning, the difference in average bill length between the *A* males and the *B* males disappeared after 1978 as a result of recruitment to both groups of young males (Table 2). If their mothers had mated in the same way as the group of birds whose initial pairings were known, i.e., randomly, half of them would have had maternal grandfathers of the song type opposite to their fathers'.

With the breakdown of the morphological subdivision of the population, the ecological subdivision disappeared too. In the dry season of 1981, for example, the different methods of feeding on cactus resources still tended to be associated with different beak sizes and shapes in the same manner as observed in 1978, but the association of each of them with song types was no longer present. Finally, the alternating pattern of territories of mated males disappeared in 1979, to reappear weakly at the beginning of

TABLE 1. Heritabilities and repeatabilities of morphological traits.[a]

	Heritability (h^2)	Repeatability (r)
Mass (g)	0.82 ± 0.25** 31	0.75***
Wing length (mm)	0.68 ± 0.20** 33	0.82***
Tarsus length (mm)	0.88 ± 0.19*** 31	0.83***
Bill length (mm)	0.97 ± 0.19*** 34	0.75***
Bill depth (mm)	0.71 ± 0.10*** 34	0.95***
Bill width (mm)	0.71 ± 0.20** 32	0.96***

[a]Estimates are given with one standard error, and sample sizes are below them. Heritabilities were calculated by regressing mean offspring value on midparent value. Repeatabilities were estimated from 37 individuals each measured twice from less than 1 to 7 years apart (see Lessells and Boag 1987 for details of the method). Statistical significance is indicated by **($p < 0.01$) or ***($p < 0.001$).

the breeding seasons of 1980 and 1982 (Grant and Grant 1983). The regular pattern disappeared because new recruits established new territories next to previously established ones (Figure 4), and some females chose males with homotypic neighbors as mates. It did not disappear because males reestablished their territories in different places; once hav-

TABLE 2. Comparison of bill measurements between song *A* males and song *B* males in 1978 and 1980.[a]

Years	Males	*N*	Bill length	Bill depth	Bill width
1978	Song *A*	11	15.47 ± 0.14	10.79 ± 0.18	9.99 ± 0.15
	Song *B*	10	14.55 ± 0.20	11.02 ± 0.36	10.23 ± 0.23
1980	Song *A*	13	15.12 ± 0.25	10.82 ± 0.18	9.93 ± 0.13
	Song *B*	19	14.76 ± 0.17	10.79 ± 0.20	10.03 ± 0.14

[a]A significant difference in bill length was present in 1978 ($p < 0.002$) but not in 1980 ($p > 0.1$).

ing bred, males hold their territories for life. This fact, combined with the long life of these birds, was responsible for the brief reappearance of the nonrandomness in the territory boundaries of mated males.

Note from Figure 4 that territories are interchangeable between males of the two song types; the territory of an *A* male one year may be occupied by a *B* male in another year. There is no systematic association between males of a song type and habitat features, and therefore no spatial isolation of the two groups other than that provided by territoriality itself. This applies to the nonbreeding (dry) season as well as to the breeding (wet) season.

Therefore the signs of subdivision, both ecological and reproductive, that we had observed in 1978 disappeared thereafter.

HOW THE POPULATION BECAME SUBDIVIDED

The change in population structure gave us the same disadvantage experienced when attempting to reconstruct the events leading to speciation, namely the processes of interest had already occurred. In our case the subdivision of the population occurred prior to our study. In the year immediately preceding the study, 1977, very dry conditions prevailed (Grant and Grant, 1983). To judge from events on Isla Daphne Major at this time (Boag and Grant 1981, 1984), there was probably no (successful) breeding and substantial mortality (Grant and Grant 1983). There were no males in 1978 wearing the plumage of 1-year-olds. The stressful conditions of 1977 may have created the structure in 1978 in some way.

The important question to answer is how the beak size difference came to be associated with the song difference in males.

Random effects

One possibility is that it was simply a matter of chance. The original study area constitutes 10% of the island area that is occupied by breeding cactus finches (Grant and Grant 1980). Events in the study area may not be typical of the island as a whole, and chance is likely to play a greater role in small samples than in the total population. By chance, heavy mortality in 1977 may have been biased with respect to beak size in the two groups of males.

Two methodological checks failed to turn up evidence of bias of this type. In 1987 we did what we should have done in 1978; we walked around the island and censused birds to see if the proportions of *A* and *B* males in the study area differed from those outside. They did not (Table 3). This does not directly address the question of whether the beak size difference between the song groups in 1978 was representative of a population-wide

TABLE 3. Frequencies of males that sang song *A*, song *B*, or both *A* and *B* in the study area and outside in 1987.[a]

Males	Study area	Outside
Song *A*	0.32	0.33
Song *B*	0.66	0.64
Song *A* and *B*	0.02	0.03
Total individuals	38	124

[a]The study area constitutes 10% of the island area occupied by breeding cactus finches. The censused area outside, on the west, north, and east coasts and in the center, amounts to an additional 23%.

difference, but at least it gives no indication that the study area and the finches in it are atypical. In vegetation also the study area is representative of the whole island (Grant and Grant 1980; Hamann 1981).

The second check was on the beak difference itself. To ascertain that the difference was real and not the result of measurement error or of different regimes of growth and abrasion experienced by the two groups of males, we recaptured and remeasured as many as possible of the original 21 males in the years 1979 to 1981, and compared the original and second set of measurements. No differences were found (Grant 1983). The recaptured *A* males (4) and *B* males (4) differed significantly in bill length at second measurement ($p < 0.01$; Grant and Grant 1983), just as the same individuals did at first measurement ($p < 0.01$). Repeatability of measurements is fairly high (Table 1). Thus the bill differences were real and not an artifact of sampling.

Random effects cannot be dismissed; nevertheless we have sought an explanation for the association between song type and bill size in terms of the known characteristics of the birds.

Systematic effects, and the development of an hypothesis

In 1983 we reasoned that an association between bill size and song type is likely to arise in the breeding season because song is used principally at that time (Grant and Grant 1983). It could arise, for example, if females of a particular morphology choose to pair with males of a particular song type, for then the male offspring would inherit some of the alleles that govern that particular morphology from their mother, and acquire the song from their father. From this reasoning we developed the following hypothesis to account for the bill length–song type association observed in 1978.

Large females come into breeding condition earlier than small females, and are the first to choose mates. They pair to a disproportionate extent with males that have heterotypic neighbors. Males that sing the *A* song have heterotypic neighbors more frequently than do the *B* males simply as a result of their relative scarcity; hence not only do they have a frequency-dependent mating advantage, but they tend to pair with the largest females. The slight tendency for males of the two groups in the next generation to differ in size is enhanced by disruptive selection in the dry season. Disruptive selection occurs on beak size; the longest and the shortest billed males are favored over intermediates as a result of the feeding advantage each enjoys in the two cactus niches of fruit and pad exploitation. Long-billed birds come predominantly from the *A* males and short-billed birds come predominantly from the *B* males, since bill size is strongly correlated with body size (Grant 1981a, 1983). In a drought, such as occurred in 1977, disruptive selection is intense, with the result that the association between bill length and song type is exceptionally strong in the following year (e.g., 1978).

Consistent facts and a test

Retrospective hypotheses are satisfactory to the extent that they are consistent with the known facts. They are useful to the extent they predict events that can be tested against future observations. We will treat these two features in turn.

The three most important facts consistent with the hypothesis are as follows. First, among the new pairs formed in 1980 and 1981 large females paired significantly earlier than small females (Grant and Grant 1983). Nonsignificant tendencies in the same direction were observed in the other three breeding seasons for which we have sufficient data: 1982, 1986, and 1987. Second, males with heterotypic neighbors had a mating advantage over those without heterotypic neighbors in each of the years from 1980 to 1982 (Grant 1985). This may have come about largely through female preference for experienced males, but we cannot rule out the possibility of a direct preference for males with heterotypic neighbors (B. R. Grant and P. R. Grant 1987).

Third, there is evidence of disruptive selection, although it is indirect. If disruptive selection had occurred in 1977 the variance in bill size of the survivors in 1978 would have been greater than the variance of the total sample in 1976 before the drought. We lack information from 1976 to make the direct comparison, but if we assume that adults in 1982 had the same characteristics as their predecessors in 1976, then we can compare the 1978 adults with the 1982 adults. This was done (Grant 1985). The comparison showed that the 1978 survivors were indeed exceptionally variable in bill length (Levene's test, $p < 0.05$). When the total adult sample was partitioned by sex, the trend was evident in both sexes but the difference was statistically demonstrable only among the females. Sample sizes are small, however, and probably as a result there is no indication of bimodality in the bill length frequency distribution in 1978 (Figure 3), which would be expected in a large sample after strong disruptive selection had occurred.

Predictions from the hypothesis

Droughts occur in the Galápagos archipelago on average about once a decade (Grant and Boag 1980). We were fortunate to witness a complete drought in 1985; no rain whatever was recorded in the normal wet season months of January to May, and birds did not breed. This should have been the ideal circumstance for observing the events that previously had only been hypothesized to occur, because we knew the sizes and breeding histories of many individuals in the preceding year of 1984. The resulting characteristics of the population in 1986 should have been the same as those we encountered in 1978.

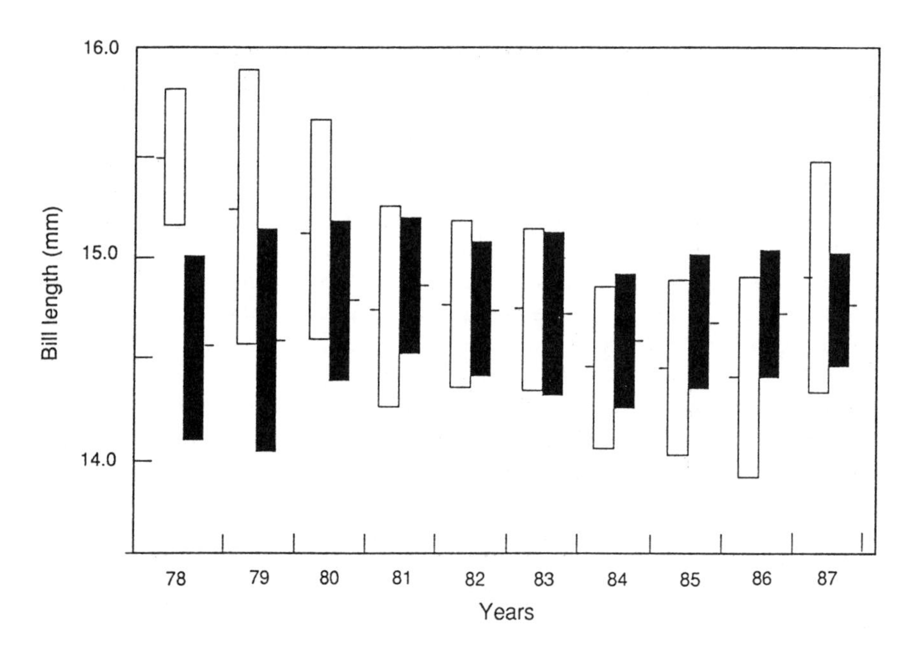

FIGURE 5. Beak lengths of *A* males (open bars) and *B* males (solid bars). Bars represent 95% confidence limits. Means are shown by short horizontal lines. See Table 4 for the sample sizes.

TABLE 4. The samples of males used in Figure 5.[a]

	Number of measured males[b]		
Year	**Type A**	**Type B**	**Proportion of total males**
1978	11	10	0.84
1979	10	9	0.58
1980	13	19	0.76
1981	17	29	0.79
1982	21	30	0.88
1983	22	29	0.56
1984	22	30	?
1985	18	24	?
1986	14	19	0.97
1987	12	17	0.76

[a]The combined sample of *A* males and *B* males is also expressed as a proportion of the total number of males that held territories on the study area each year. Proportions are not known in 2 years because few birds bred in 1984 and none did so in 1985.
[b]Numbers changed from year to year through gains (recruitment) and losses (mortality). Annual turnover was generally low, but although maximum longevity exceeds 10 years none of the males measured in 1978 was alive in 1987.

Disruptive selection did not occur in 1985. Variances did not change, and song *A* and song *B* males did not differ in bill length, or in any other measurement, in the following year (Figure 5 and Table 4).

The reason was apparent even before the drought began. In 1983 there occurred an El Niño event that has been described as the most severe oceanographic disturbance of the region this century (Cane 1983). More than 2400 mm of rain fell on Genovesa. This is to be contrasted with our previous wet season maximum of 172 mm (P. R. Grant and B. R. Grant 1987)! One of the consequences was an extreme growth of vines, principally *Ipomoea habeliana.* They smothered and killed many cactus bushes (see Plate 53 in Grant 1986). The poor condition of the surviving cactus continued through the dry year of 1984 (57 mm total rainfall) and into 1985. Very little flowering occurred, and to a rough approximation it may be said that one of the two cactus niches had disappeared. So although the other niche, in the form of insect larvae and pupae in rotting pads, was abundantly present, the conditions for disruptive selection to occur were not.

We took advantage of this initially disappointing situation to predict the occurrence of directional selection on beak size in a fundamentally altered

TABLE 5. Selection on morphological characters associated with survival during the dry years of 1984 and 1985.[a]

Character	Coefficients			
	1983–1984		1984–1985	
	β ± SE	s	β ± SE	s
Bill length	−0.22 ± 0.08**	−0.24**	−0.05 ± 0.18	0.14
Bill depth	−0.05 ± 0.08	−0.07	0.37 ± 0.19*	0.31**
Tarsus length	−0.02 ± 0.07	−0.10	0.14 ± 0.19	0.22*
Numbers (survival)	137 (0.54)		74 (0.39)	

[a]Standardized selection gradients (β ± standard error) measure the direct effect of selection on each character, and standardized selection differentials (s) measure the combined direct and indirect effects from correlated characters. The number of individuals present at the beginning of each interval (July) is followed by the proportion surviving in parentheses. Statistical significance of the coefficients is indicated by *($p < 0.05$) or **($p < 0.01$); see Lande and Arnold (1983), Grant (1985), and Endler (1986) for details of the analysis.

environment. In the near absence of cactus fruits and scarcity of flowers cactus finches could feed only on insects in cactus pads, on insects and termites in dead wood obtained by stripping off the bark, and occasionally on seeds. These foraging and feeding tasks require behavioral skills that are enhanced by a deep but short beak (Grant 1985). Therefore the prediction was for relatively long-billed birds to be at a selective disadvantage and relatively deep-billed birds to be at a selective advantage. Over the period from 1983 to 1985 this is exactly what we found (Table 5). Survivors had significantly shorter bills than nonsurvivors in the first year, and those surviving the second year had significantly deeper bills than those that did not survive (Grant and Grant 1989b).

CONCLUSIONS

The population was partly subdivided ecologically, for a brief period and only once in the decade. We understood how the subdivision collapsed through the combined effects of random mating with respect to song and the transmission to the offspring of parental features of song and beak morphology. We understand less well how the subdivision was created.

Thus, *Geospiza conirostris* is not currently undergoing a split into two species sympatrically on Genovesa. There are four interrelated reasons for this. First, in spite of an occasional subdivision of the population, the dry season niches are not different enough to support two very different feeding types. Second, extreme fluctuations of the environment prevent prolonged divergence of two feeding types. Third, there is neither a detectable deviation from random mating nor an identifiable factor that would favor it. Fourth, mate choice is not genetically correlated with the use of a particular niche.

DISCUSSION

The temporary subdivision of the Genovesa population of *conirostris* represents a stage, potentially, on the way toward speciation. The likelihood of the process of speciation ever going to completion sympatrically appears to be very low. We will first discuss this in the context of speciation of Darwin's Finches as a whole, and then consider what general remarks can be made about sympatric speciation of vertebrates in the light of our study.

All sympatric species of ground finches differ by at least 15% in at least one bill dimension (Grant 1981b, 1986). Since their niche differences are directly proportional to their bill differences (Abbott et al. 1977), there is a minimum niche difference between them too. These differences are believed to have evolved under directional selection, entirely in allopatry in some cases and partly in allopatry and partly in sympatry in others (Grant 1981b, 1986). Reproductive isolation evolved as a correlated effect of the ecological and morphological divergence (Grant 1986). Sympatric species do not normally interbreed because they discriminate between conspecific and heterospecific potential mates at least partly on the basis of their bill size and shape (Ratcliffe and Grant 1983).

The two groups of *G. conirostris* males differed at maximum in our study by only 6%. This is probably too small to foster any discrimination in a mating context. Admittedly in the secondary contact phase of allopatric speciation incipient species are likely to differ initially by less than the final amount of 15% or more in some case, but 6% falls so far short of the minimum apparently required for coexistence that something else would have to cause further divergence to make speciation possible. Furthermore one of the dry season food niches was observed to decline catastrophically, reminding us that a degree of temporal stability would be needed for further divergence to occur.

The "something else" required for further divergence would have to be another feeding niche not presently exploited, or seldom exploited, and differing markedly from the other two. Two candidates are easy to identify. One comprises the many types of small seeds produced by grasses and herbs, and the small quantities of nectar in the small but plentiful flowers of *Waltheria ovata*. The other comprises the large and hard seeds of *Opuntia* cactus and the even larger and harder seeds of *Cordia lutea*. Mature *Opun-*

tia seeds are exploited rarely by the cactus finch individuals with the largest bills, whereas *Cordia* seeds appear to be too hard for any of them to crack. The large-seed niche could be exploited efficiently by individuals with even larger beaks than are currently observed in the population. An obvious ecological reason why there are no small seed specialists or large seed specialists in the population now is that two other species, *G. difficilis* and *G. magnirostris,* occupy the niches (Grant and Grant 1980, 1982; Schluter and Grant 1982). Thus the absence of one or both of these species would appear to be an essential condition for sympatric speciation to occur.

The situation is interestingly different on the other two islands occupied by *G. conirostris,* I. Española and its satellite I. Gardner. The large-seed niche provided by *Opuntia* and *Cordia* is present, but the species to exploit it, *G. magnirostris,* is absent. The populations of *G. conirostris* vary even more in bill dimensions than on Genovesa (Grant et al. 1985), and individuals do exploit the large seeds (Grant and Grant 1982). Nevertheless there is no evidence that the population on each island is currently splitting into two species sympatrically. The evolutionary response to the absence of the competitor has taken the form instead of a large average beak size and a change in beak proportions (Lack 1947; Grant and Grant 1982). This may be equivalent to uniform dietary generalization among members of the population rather than separate individual specializations. Field studies on this island have been of short duration, and we know nothing about the possibilities of ecological subdivision of the population and nonrandom mating. We are left to speculate. We suppose the difference in dry season niches is not so great, discretely different, and stable as to penalize repeatedly the individuals with intermediate phenotypes. Therefore there is nothing to promote nonrandom mating. In other words, even under this apparently exceptionally favorable circumstance for sympatric divergence, the coherence given to the population by random mating is not broken by strong disruptive selection.

Another exceptionally favorable circumstance has been identified by Ford et al. (1973) on I. Santa Cruz. The frequency distribution of beak depths of a sample of male medium ground finches (*G. fortis*) at Bahía Academía displayed bimodality. Ford et al. (1973) suggested the population may be in the process of splitting into two under disruptive selection. This locality is certainly diverse floristically (Abbott et al. 1977). Other explanations are possible. The bimodality could have arisen through hybridization with the large ground finch, *G. magnirostris* (Bowman 1961; Snow 1966). Alternatively it might be the product of recent immigration of a partially differentiated form of *G. fortis* from another island such as I. San Cristóbal, and some degree of assortative mating (Grant et al. 1976). In other words the population could be either in the secondary contact phase of allopatric speciation, or at an advanced stage of sympatric speciation (Grant 1986).

In the light of this study, carried out under entirely natural conditions, what general remarks can be made about the conditions for sympatric speciation to occur and the likelihood of its occurrence?

The factor of greatest importance suggested by our study is the way in which the environment is heterogeneous. Only when the heterogeneity is strong and the difference between the extremes is large will disruptive selection be strong enough to produce the divergence that can lead to speciation (see also Maynard Smith and Hoekstra 1980; Wilson and Turelli 1986). This opinion derived from an empirical study echoes Maynard Smith's (1966: 649) emerging from a theoretical one: "[The] argument for or against sympatric speciation will depend on how likely such a polymorphism is thought to be, and this in turn depends on whether a single gene difference can produce selective coefficients large enough to satisfy the necessary conditions."

But just as important as heterogeneity is the factor of environmental constancy or lack of it. Theoretical models often assume constant population sizes, as a reflection of constant environments (e.g., Maynard Smith 1966; Dickinson and Antonovics 1973). Conclusions drawn from the models about the conditions under which sympatric speciation can occur could be crucially dependent on the absence of perturbations to the environment. For example, Wilson and Turelli's (1986) model suggests, as does Pimm's (1979), that ecologically differentiated incipient species can coexist for a long enough time for prereproductive isolation to evolve, resulting in speciation. Results of our study suggest the opposite, at least for the population we studied. Admittedly the environmental fluctuations we have described may be unusually large. Nevertheless all environments vary in time to some extent. Lack of agreement between model expectations and field observations reveals a need for greater realism in the models. There is a need to explore the effects of varying the amplitude, duration and predictability of the fluctuations on the evolution of sympatric divergence.

We doubt if favorable environmental conditions for sympatric speciation will often occur for vertebrates. In general their behavioral flexibility allows each individual to cope with a broad range of conditions. As a consequence they show relatively little (heritable) niche or habitat specialization. Specializations in habitat use are opportunistic (Nettleship 1972; Robertson 1972), governed by availability of space or resources and competition for them (Svärdson 1949; Fretwell and Lucas 1970). Phenotype-dependent specializations in resource use that are not related to sex or age are subtle and matters of degree (e.g., Price 1987). Although it is convenient to model phenotypes as if they take discretely different and equal elements from a continuous resource distribution (Seger 1985), in fact they take overlapping and unequal segments, and the overlap is far greater than the separation (Grant et al. 1976), although there are occasional exceptions (Price 1987; Werner and Sherry 1987). In the language of niche theory, the within-phenotype component of the total niche width of a population is far greater than the between-phenotype component (Roughgarden 1972; Taper and Case 1985).

Nevertheless there are two reasons for believing that sympatric speciation may occur among vertebrates, albeit extremely rarely. The first is that genetic drift may do what selection does not, or cannot, do. Chance association of alleles under sustained conditions of low population size (Wright 1980) may result in a coupling of niche adaptation and selective mating in a heterogeneous environment from which, under selection, speciation could proceed. The interaction of selection and drift has not been explored in models of sympatric speciation. Nor, for that matter, have other biologically important factors such as mating system, dispersal, population size fluctuations, and disease susceptibility.

The second reason is that peculiar variation exhibited by some vertebrate species appears to provide a basis for sympatric speciation to occur. Inasmuch as the phenotypic variation reflects underlying genetic variation it suggests there are some genetic systems that, more than others, make it possible for a population to split into two under the right environmental circumstances.

An outstanding example of unusual variation is provided by a species of fish, *Cichlasoma minckleyi*, in Mexico (Kornfield and Koehn 1975; Sage and Selander 1975; Kornfield et al. 1982; Liem and Kaufman 1984). A snail-eating type with molariform teeth is produced in the same brood as a detritus- or alga-eating type with papilliform teeth. The discrete polymorphism in dentition may arise from a sharp threshold of developmental reactivity to stimuli impinging on the pharyngeal plate, and could be controlled by a switch gene with two alleles at a single locus (Sage and Selander 1975). The important point is that a major difference in dentition and diet might be produced by a relatively simple (although unknown) genetic difference. Alternatively the morphological difference may be environmentally induced, and thus the result and not the cause of the dietary difference, as appears to be the case in the related species *C. managuense* (Meyer 1987). Of relevance to the issue of sympatric divergence and speciation, Sage and Selander (1975) suggest the morphs of *C. minckleyi* are unlikely to be segregated spatially. They do not mate assortatively (Kornfield et al. 1982). Ehlinger and Wilson (1988) and Wilson (this volume) have investigated a similar phenomenon of morphological, ecological, and behavioral differentiation in populations of bluegill sunfish (*Lepomis macrochirus*) in which once again the genetic basis of the differences needs to be shown.

A second example is provided by bird species of the genus *Pyrenestes* in

Africa (Chapin 1924, 1954). Within each sex of these finches an extreme bimodality is expressed in the frequency distribution of beak size. Small and large morphs appear to be produced from the same brood. In addition to nonrandom mating and disruptive selection, an unusual genetic system could be contributing to this highly unusual form of phenotypic variation (Smith 1987), perhaps along the same lines as the one operative in the *Cichlasoma* fish.

These two cases, though extreme, are not unique. Unusual variation is shown in other fish (McKaye 1980; Hinder and Jonsson 1982; Liem and Kaufman 1984; Smith and Todd 1984; Kondrashov and Mina 1986; Meyer 1987) and other birds (Ford et al. 1973; Smith and Temple 1982). These are enough to justify further search for sympatric speciation.

SUMMARY

Models of sympatric speciation, with varying degrees of complexity and realism, show the conditions under which a single interbreeding population can split sympatrically into two. They describe mechanisms for the evolution of niche differences between two segments of a population, and the subsequent evolution of reproductive isolation between them.

We studied a population of large cactus finches, *Geospiza conirostris*, on Isla Genovesa, Galápagos, for 10 years. In the first year the population appeared to be ecologically subdivided into two groups, and there were hints of a reproductive subdivision too. However, the ecological subdivision disappeared subsequently through the effects of random mating. The subdivision may have been created by disruptive selection under the stressful conditions of a drought in the year immediately preceding our study. A second drought occurred in the eighth year of study, but the hypothesis of disruptive selection could not be tested because the environment had changed; one of the two principal dry season cactus foods had been almost eliminated by the smothering effects of extensive vine growth in the exceptionally long and wet growing season 2 years earlier.

The major lesson we draw from this study is that environmental heterogeneity has to be of a special kind to foster sympatric divergence leading to speciation. The niches or habitats to which different members of the population adapt should be markedly different and display a long-term persistence, although not necessarily a constancy. Unusual phenotypic variation among some vertebrates, however, suggests that with this group of animals sympatric speciation may occur, albeit extremely rarely.

ACKNOWLEDGMENTS

Our research on I. Genovesa has been funded by grants from N.S.E.R.C. (Canada) and N.S.F., and supported by the Dirección General de Desarrollo Forestal, Quito, and the Charles Darwin Research Station on I. Santa Cruz, Galápagos. We thank T. K. Werner and T. W. Sherry for their hospitality while the first draft of this chapter was being prepared, and D. J. Futuyma, T. B. Smith, and an anonymous reviewer for comments on the manuscript.

LITERATURE CITED

Abbott, I., L. K. Abbott, and P. R. Grant. 1977. Comparative ecology of Galápagos Ground Finches (*Geospiza* Gould): Evaluation of the importance of floristic diversity and interspecific competition. Ecol. Monogr. 47:151–184.

Bengtsson, B. 1979. Theoretical models of speciation. Zool. Scripta 8:303–304.

Boag, P. T., and P. R. Grant. 1981. Intense natural selection in a population of Darwin's Finches (Geospizinae) in the Galápagos. Science 214:82–85.

Boag, P. T., and P. R. Grant. 1984. Darwin's Finches (*Geospiza*) on Isla Daphne Major, Galápagos: Breeding and feeding ecology in a climatically variable environment. Ecol. Monogr. 54:463–489.

Bowman, R. I. 1961. Morphological differentiation and adaptation in the Galápagos finches. Univ. Calif. Pub. Zool. 58:1–302.

Bowman, R. I. 1979. Adaptive morphology of song dialects in Darwin's Finches. J. Ornithol. 120:353–389.

Bowman, R. I. 1983. The evolution of song in Darwin's Finches. Pp. 237–537 in: R. I. Bowman, M. Berson, and A. E. Leviton (eds.), *Patterns of Evolution in Galápagos Organisms.* American Association for the Advancement of Science, Pacific Division, San Francisco, CA.

Bush, G. L. 1975. Modes of animal speciation. Annu. Rev. Ecol. System. 6:339–364.

Bush, G. L., and S. R. Diehl. 1982. Host shifts, genetic models of sympatric speciation and the origin of parasitic insect species. In: J. H. Visser and A. K. Minks (eds.), *5th International Symposium on Insect-Plant Relationships.* PUDOC, Wageningen.

Cane, M. A. 1983. Oceanographic events during El Niño. Science 222:1189–1195.

Chapin, J. P. 1924. Size-variation in *Pyrenestes,* a genus of Weaver-Finches. Bull. Am. Mus. Nat. Hist. 49:415–441.

Chapin, J. P. 1954. The birds of the Belgian Congo. Pt. IV. Bull. Am. Mus. Nat. Hist. 75B:1–846.

Dickinson, H., and J. Antonovics. 1973. Theoretical considerations of sympatric divergence. Am. Natur. 107:256–274.

Ehlinger, T. J., and D. S. Wilson. 1988. Complex foraging polymorphism in the Bluegill Sunfish. Proc. Natl. Acad. Sci. U.S.A. 85:1878–1882.

Endler, J. A. 1977. *Geographic Variation, Speciation, and Clines.* Princeton University Press, Princeton, NJ.

Endler, J. A. 1986. *Natural Selection in the Wild.* Princeton University Press, Princeton, NJ.

Feder, J. L., C. A. Chilcote, and G. L. Bush. 1988. Genetic differentiation between sympatric host races of the apple maggot fly *Rhagoletis pomonella.* Nature (London) 336:61–64.

Felsenstein, J. 1981. Skepticism towards Santa Rosalia, or why are there so few kinds of animals? Evolution 35:124–138.

Ford, H. A., D. T. Parkin, and A. W. Ewing. 1973. Divergence and evolution in Darwin's Finches. Biol. J. Linnean Soc. 5:289–295.

Fretwell, S. D., and H. L. Lucas, Jr. 1970. On territorial behavior and other factors influencing habitat distribution in birds. I. Theoretical development. Acta Biotheoret. 19:16–36.

Futuyma, D. J., and G. C. Mayer. 1980. Non-allopatric speciation in animals. System. Zool. 29:254–271.

Gibbons, J. R. H. 1979. A model for sympatric speciation in *Megarhyssa* (Hymenoptera: Ichneumonidae): Competitive speciation. Am. Natur. 114:719–741.

Grant, B. R. 1984. The significance of song variation in a population of Darwin's Finches. Behaviour 89:90–116.

Grant, B. R. 1985. Selection on bill characters in a population of Darwin's Finches: *Geospiza conirostris* on Isla Genovesa, Galápagos. Evolution 39:523–532.
Grant, B. R., and P. R. Grant. 1979. Darwin's Finches: Population variation and sympatric speciation. Proc. Natl. Acad. Sci. U.S.A. 76:2359–2363.
Grant, B. R., and P. R. Grant. 1981. Exploitation of *Opuntia* cactus by birds on the Galápagos. Oecologia 49:179–187.
Grant, B. R., and P. R. Grant. 1982. Niche shifts and competition in Darwin's Finches: *Geospiza conirostris* and congeners. Evolution 36:637–657.
Grant, B. R., and P. R. Grant. 1983. Fission and fusion in a population of Darwin's Finches: An example of the value of studying individuals in ecology. Oikos 41:530–547.
Grant, B. R., and P. R. Grant. 1987. Mate choice in Darwin's Finches. Biol. J. Linnean Soc. 32:247–270.
Grant, B. R., and P. R. Grant. 1989a. *Evolutionary Dynamics of a Natural Population: The Large Cactus Finch of the Galápagos.* University of Chicago Press, Chicago.
Grant, B. R., and P. R. Grant. 1989b. Natural selection in a population of Darwin's Finches. Am. Natur., in press.
Grant, P. R. 1981a. Patterns of growth in Darwin's Finches. Proc. R. Soc. London Ser. B 212:403–432.
Grant, P. R. 1981b. Speciation and the adaptive radiation of Darwin's Finches. Am. Sci. 69:653–663.
Grant, P. R. 1983. Inheritance of size and shape in a population of Darwin's Finches, *Geospiza conirostris.* Proc. R. Soc. London Ser. B 220:219–236.
Grant, P. R. 1986. *Ecology and Evolution of Darwin's Finches.* Princeton University Press, Princeton, NJ.
Grant, P. R., I. Abbott, D. Schluter, R. L. Curry, and L. K. Abbott. 1985. Variation in the size and shape of Darwin's Finches. Biol. J. Linnean Soc. 25:1–39.
Grant, P. R., and P. T. Boag. 1980. Rainfall on the Galápagos and the demography of Darwin's Finches. Auk 97:227–244.
Grant, P. R., and B. R. Grant. 1980. The breeding and feeding characteristics of Darwin's Finches on Isla Genovesa, Galápagos. Ecol. Monogr. 50:381–410.
Grant, P. R., and B. R. Grant. 1987. The extraordinary El Niño event of 1982–83: Effects on Darwin's Finches on Isla Genovesa, Galápagos. Oikos 49:55–66.
Grant, P. R., B. R. Grant, J. N. M. Smith, I. Abbott, and L. K. Abbott. 1976. Darwin's Finches: Population variation and natural selection. Proc. Natl. Acad. Sci. U.S.A. 73:257–261.
Halliburton, R., and G. A. E. Gall. 1981. Disruptive selection and assortative mating in *Tribolium castaneum.* Evolution 35:829–843.
Hamann, O. 1981. Plant communities of the Galápagos Islands. Dansk Bot. Arch. 34:1–163.
Hedrick, P. W. 1986. Genetic polymorphism in heterogeneous environments: A decade later. Annu. Rev. Ecol. System. 17:535–566.
Hedrick, P. W., M. E. Ginevan, and E. P. Ewing. 1976. Genetic polymorphism in heterogeneous environments. Annu. Rev. Ecol. System. 7:1–32.
Hendrickson, H. T. 1978. Sympatric speciation: Evidence? Science 200:345–346.
Hinder, K. K., and B. Jonsson. 1982. Habitat and food segregation of dwarf and normal arctic charr (*Salvelinus alpinus*) from Vangsvatnet Lake, western Norway. Can. J. Fish. Aquatic Sci. 39:1030–1045.
Hutchinson, G. E. 1968. When are species necessary? Pp. 177–186 in: R. C. Lewontin (ed.), *Population Biology and Evolution.* Syracuse University Press, Syracuse, NY.
Kalmus, H., and J. Maynard Smith. 1966. Some evolutionary consequences of pegmatypic mating systems (imprinting). Am. Natur. 100:619–635.
Kondrashov, A. S. and M. V. Mina. 1986. Sympatric speciation: When is it possible? Biol. J. Linnean Soc. 27:201–223.
Kornfield, I. L., and R. K. Koehn. 1975. Genetic variation and evolution in New World cichlids. Evolution 29:427–437.
Kornfield, I., D. C. Smith, P. S. Gagnon, and J. N. Taylor. 1982. The cichlid fish of Cuatro Ciénegas, Mexico: Direct evidence of conspecificity among distinct trophic morphs. Evolution 36:658–664.
Lack, D. 1947. *Darwin's Finches.* Cambridge University Press, Cambridge.

Lande, R. 1981. Models of speciation by sexual selection on polygenic traits. Proc. Natl. Acad. Sci. U.S.A. 78:3721–3725.
Lande, R., and S. J. Arnold. 1983. The measurement of selection on correlated characters. Evolution 37:1210–1226.
Lessells, C. M., and P. T. Boag. 1987. Unrepeatable repeatabilities: A common mistake. Auk 104:116–121.
Liem, K. F. and L. S. Kaufman. 1984. Intraspecific macroevolution: Functional biology of the polymorphic cichlid species *Cichlasoma minckleyi*. Pp. 203–215 in: A. A. Echelle and I. Kornfield (eds.), *Evolution of Fish Species Flocks*. University of Maine at Orono Press, Orono.
Loeschcke, V., and F. B. Christiansen. 1984. Evolution and intraspecific exploitative competition. II. A two-locus model for additive gene effects. Theoret. Pop. Biol. 26:228–264.
Matessi, C., and S. D. Jayakar. 1976. Models of density-and frequency-dependent selection for the exploitation of resources. I. Intraspecific competition. Pp. 707–721 in: S. Karlin and E. Nevo (eds.), *Population Genetics and Evolution*. Academic Press, New York.
Maynard Smith, J. 1966. Sympatric speciation. Am. Natur. 100:637–650.
Maynard Smith, J., and R. Hoekstra. 1980. Polymorphism in a varied environment: How robust are the models. Genet. Res. 35:45–57.
Mayr, E. 1963. *Animal Species and Evolution*. Harvard University Press, Cambridge, MA.
McKaye, K. R. 1980. Seasonality in habitat selection by the gold color morph of *Cichlasoma citrinellum* and its relevance to sympatric speciation in the family Cichlidae. Environ. Biol. Fish. 5:75–78.
McPheron, B. A., D. C. Smith, and S. H. Berlocher. 1988. Genetic differences between host races of *Rhagoletis pomonella*. Nature (London) 336:64–66.
Meyer, A. 1987. Phenotypic plasticity and heterochrony in *Cichlasoma managuense* (Pisces, Cichlidae) and their implications for speciation in cichlid fishes. Evolution 41:1357–1369.
Nettleship, D. N. 1972. Breeding success of the Common Puffin (*Fratercula arctica* L.) on different habitats at Great Island, Newfoundland. Ecol. Monogr. 42:239–268.
O'Donald, P. 1960. Assortative mating in a population in which two alleles are segregating. Heredity 15:389–396.
Pimentel, D., G. J. C. Smith, and J. Soans. 1967. A population model of sympatric speciation. Am. Natur. 101:493–504.
Pimm, S. L. 1979. Sympatric speciation: A simulation model. Biol. J. Linnean Soc. 11:131–139.
Price, T. D. 1987. Diet variation in a population of Darwin's Finches. Ecology 68:1015–1028.
Ratcliffe, L. M., and P. R. Grant. 1983. Species recognition in Darwin's Finches (*Geospiza*, Gould). I. Discrimination by morphological cues. Anim. Behav. 31:1139–1153.
Ratcliffe, L. M., and P. R. Grant. 1985. Species recognition in Darwin's Finches (*Geospiza*, Gould). III. Male responses to playback of different song types, dialects and heterospecific songs. Anim. Behav. 33:290–307.
Rice, W. R. 1984. Disruptive selection on habitat preference and the evolution of reproductive isolation: a simulation study. Evolution 38:1251–1260.
Rice, W. R. 1985. Disruptive selection on habitat preference and the evolution of reproductive isolation: An exploratory experiment. Evolution 39:645–656.
Rice, W. R. 1987. Speciation via habitat specialization: the evolution of reproductive isolation as a correlated character. Evol. Ecol. 1:301–314.
Robertson, R. J. 1972. Optimal niche space of the redwinged blackbird (*Agelaius phoeniceus*). I. Nesting success in marsh and upland habitat. Can. J. Zool. 50:247–263.
Rosenzweig, M. L. 1978. Competitive speciation. Biol. J. Linnean Soc. 10:275–289.
Roughgarden, J. 1972. Evolution of niche width. Am. Natur. 106:683–718.
Sage, R. D., and R. K. Selander. 1975. Trophic radiation through polymorphism in cichlid fishes. Proc. Natl. Acad. Sci. U.S.A. 72:4669–4673.
Schluter, D., and P. R. Grant. 1982. The distribution of *Geospiza difficilis* in relation to *G. fuliginosa* in the Galápagos islands: Tests of three hypotheses. Evolution 36:1213–1226.

Seger, J. 1985. Intraspecific resource competition as a cause of sympatric speciation. Pp. 43–53 in: P. J. Greenwood, P. H. Harvey, and M. Slatkin (eds.), *Evolution, Essays in honour of John Maynard Smith.* Cambridge University Press, Cambridge.

Seiger, M. B. 1967. A computer simulation study of the inflence of imprinting on population structure. Am. Natur. 101:47–57.

Smith, D. C. 1988. Heritable divergence of *Rhagoletis pomonella* host races by seasonal asynchrony. Nature (London) 336:66–67.

Smith, G. R., and T. N. Todd. 1984. Evolution of species flocks of fishes in north temperate lakes. Pp. 45–68 in: A. A. Echelle and I. Kornfield (eds.), *Evolution of Fish Species Flocks.* University of Maine at Orono Press, Orono.

Smith, J. N. M., P. R. Grant, B. R. Grant, I. Abbott, and L. K. Abbott. 1978. Seasonal variation in feeding habits of Darwin's Ground Finches. Ecology 59:1137–1150.

Smith, T. B. 1987. Bill size polymorphism and intraspecific niche utilization in an African finch. Nature (London) 329:717–719.

Smith, T. B., and S. A. Temple. 1982. Feeding habits and bill polymorphism in Hook-billed Kites. Auk 99:197–207.

Snow, D. 1966. Moult and the breeding cycle in Darwin's Finches. J. Ornithol. 107:283–291.

Soans, A. B., D. Pimentel, and J. S. Soans. 1974. Evolution of reproductive isolation in allopatric and sympatric populations. Am. Natur. 108:116–124.

Svärdson, G. 1949. Competition and habitat selection in birds. Oikos 1:156–174.

Taper, M. L., and T. J. Case. 1985. Quantitative genetic models for the coevolution of character displacement. Ecology 66:355–371.

Tauber, C. A., and M. J. Tauber. 1977a. Sympatric speciation based on allelic changes at three loci: Evidence from natural populations in two habitats. Science 197:1298–1299.

Tauber, C. A., and M. J. Tauber. 1977b. A genetic model for sympatric speciation through habitat diversification and seasonal isolation. Nature (London) 268:702–705.

Tauber, C. A., and M. J. Tauber. 1987. Inheritance of seasonal cycles in *Chrysoperla* (Insecta: Neuroptera). Genet. Res. 49:215–223.

Tauber, C. A., M. J. Tauber, and J. R. Nechols. 1977. Two genes control seasonal isolation in sibling species. Science 197: 592–593.

Templeton, A. R. 1981. Mechanisms of speciation—a population genetic approach. Annu. Rev. Ecol. System. 12:23–48.

Thoday, J. M., and J. B. Gibson. 1970. The probability of isolation by disruptive selection. Am. Natur. 104:219–230.

Werner, T. K., and T. W. Sherry. 1987. Behavioral feeding specialization in *Pinaroloxias inornata,* the "Darwin's Finch" of Cocos Island, Costa Rica. Proc. Natl. Acad. Sci. U.S.A. 84:5506–5510.

West-Eberhard, M. J. 1983. Sexual selection, social competition, and speciation. Quart. Rev. Biol. 58:155–183.

Wilson, D. S., and M. Turelli. 1986. Stable underdominance and the evolutionary invasion of empty niches. Am. Natur. 127:835–850.

Wright, S. 1980. Genic and organismic selection. Evolution 34:825–843.

CHAPTER 19

FRUITING FAILURE, POLLINATOR INEFFICIENCY, AND SPECIATION IN ORCHIDS

Douglas E. Gill

INTRODUCTION

Widely regarded as the most "speciose"[1] flowering plant family, the Orchidaceae is also the most species-rich angiosperm family. It alone comprises over 7% of all flowering plant species. The estimated number of known orchid species varies from the conservative 18,000 (Heywood 1985) to the seemingly liberal (but perhaps realistic) 30,000 (Garay 1964; Sanford 1974). Tropical America has the greatest species richness, with at least 8266 species in 306 genera; tropical Asia, including the island-studded Malaysian archipelago, ranks second with about 6800 species in 250 genera (Dressler 1981). The known number of natural orchid species is sure to grow as more groups become better studied. The closest competitor for greatest number of species is the family Asteraceae (the composites), which number about 22,000 (Heywood et al. 1978).

The orchids divide rather neatly into two ecophysiological groups of near equal size: the terrestrials and the epiphytes. Whereas terrestrials are distributed throughout the world and penetrate to latitudes and elevations

[1]I take pleasure in using the adjectives "speciose" and *species-rich* in the same sentence to call attention to the fact that they are not synonymous and that "speciose" is not even a proper word: it is not found in any standard dictionary. If it were, it would derive from the Latin *speciosus*, and would mean "*beautiful*," as it does in its common use as *speciosa* in many scientific names, and the etymologically identical *specious* (= deceptively beautiful). I urge that the misuse of "speciose" in the evolutionary biological literature cease.

as high as any flowering plant, epiphytes are restricted to tropical and subtropical areas. Epiphytic orchids are particularly amenable to artificial hybridization and horticultural development. There is no doubt that the terrestrial habit is primitive, and because several tribes have both terrestrial and epiphytic members, it can be inferred that epiphytism has evolved independently several times.

Because of the beauty of some genera (e.g., *Cymbidium*, *Cattleya*), orchids have become prized for elegant social displays and personal adornment. Their popularity has spawned more than 400 local orchid societies around the world, a lucrative industry of horticulture and competitive showings, and, unfortunately, serious exploitation of the wild (offensive to most conservation sensibilities). Thus, orchids fascinate the scientist, nature lover, and lay person.

Orchids present several intriguing problems to students of evolution and speciation. Foremost among the questions they pose is, "What factors promoted (and presumably still are promoting) speciation in this family to a greater extent than any other flowering plant family?" This question can be broken down into others of sharper focus, three of which I wish to discuss here:

1. Have the rates of speciation been unusually high in the Orchidaceae compared to other taxa?
2. Are the mechanisms of reproductive isolation of species of orchids any different from those in other flowering plants?
3. How do deceptive orchids, those that offer no reward to their pollinators, speciate or maintain their specific identify?

In this chapter I shall briefly review the opinions of others on the first two questions, and summarize some field ecological data of my own on *Cypripedium acaule* Ait. that bear on the third question. Rather than answering the major question of how so many species of orchids evolved, I raise new issues that make answering the question even more problematical.

ORIGIN AND RATES OF SPECIATION OF ORCHIDS

Unfortunately no positive or useful fossil record of orchids exists (Schmid and Schmid 1977). The Massolongo specimens of *Protorchis* and *Paleorchis* dating from the Eocene are the oldest known, but their orchidaceous identity is still controversial (Schmid and Schmid 1977). Inferences about orchid phylogeny and paleontological origins derive primarily from phytogeography, anatomy, cytogenetics, and biochemistry, and are frustratingly tentative and speculative.

Orchid evolution as interpreted from current phytogeography has been the subject of considerable discussion and debate (Garay 1964; Sanford

1974; Dressler 1981). The conclusion of Garay (1960, 1972) that orchids originated in Malaysia during the late Cretaceous is widely accepted (Withner 1974; Dressler 1981). Raven and Axelrod (1974) suggested the possibility of a Laurasian origin somewhat earlier. The fact that major taxa have radiated endemically in the three major tropical centers of Asia, Africa, and South America argues strongly that much of the great orchid diversification must have occurred after the last breakup of the continents during the Jurassic and early Cretaceous periods. Because most dicot and monocot plant families had evolved to recognizable distinction by the early Tertiary, and because orchids are almost certainly (see below) derived from the Liliaceae, which itself is an advanced monocotyledonous family, it is necessary to conclude that the Orchidaceae originated comparatively recently, maybe as late as the early Tertiary, and are phylogenetically younger than most angiosperms (Dodson 1974; Schultes 1960, 1966).

Many features of orchid phytogeography can be explained only by major transoceanic dispersals between the continents and oceanic islands during the Tertiary (Schmid and Schmid 1977). That orchids are capable of long-distance dispersal is best seen in the famous history of plant colonization of the island Krakatoa since its spectacular volcanic creation in 1883. Thirteen years later 3 species of orchids were found, and in 1933 17 epiphytic and 18 terrestrial species of orchids had successfully colonized the new island. These remarkably rapid rates of colonization are in part explained by the fact that Krakatoa is located equidistant (45 km) from Sumatra and Java in the Sunda Strait, and, therefore is rather near two rich source pools of vagile species. In general, the patterns of species diversity of orchids in the Malay archipelago is consistent with the principles of island biogeography that relate richness to island area and distance from potential sources (MacArthur and Wilson 1967; Sanford 1974).

The widespread belief that orchids are derived from liliaceous ancestors has its roots in Charles Darwin's own meticulous studies of their anatomy and functional morphology. Having synthesized and published his theory of evolution by natural section only 3 years earlier, Darwin (1862) chose to illustrate the principles of adaptive evolution in the most stunning way possible—the origins of the exquisite pollination systems of orchids. After careful dissection of the vascular traces in orchid flowers, Darwin (1877) proved such nonobvious facts as that the labellum is actually a complex organ derived from fusion of a petal and two staminodes (of the outer whorl), and that the parts of the complex column are developmental derivatives of the standard liliaceous gynoecium and androecium. Of course, Darwin's assertions were new and daring then because they provided a powerful evolutionary foundation to the bewildering revisions of the Linnean taxonomy of orchids that were sprouting in Europe at the the time.

The only serious challenge to Darwin's (1877) conclusion that orchids evolved from liliaceous ancestry comes from recent studies of monocot phytochemistry. Williams (1979) studied the flavonoids of 142 species of orchids, and concluded that they are more similar to those of the Commelinaceae, Iridaceae, and Bromeliaceae than those of the Liliaceae (Williams 1982). This observation raises the puzzling problem of whether evolutionary convergence occurred (on a broad scale) in a large class of plant compounds or in an important set of anatomical (floral) structures between major monocot taxa.

Even given the assumption that orchids are derived from liliaceous ancestors, it has not yet been determined which plants are the sister family of the Orchidaceae. Many authors nominate the pantropical, saprophytic Burmanniaceae. The point of greatest controversy, however, is whether to include the Apostasioideae in the Orchidaceae. Their half-lily half-orchid flowers certainly make them an attractive missing link, but Withner (1974) is convinced that they are primitive orchids. However, Dressler (1981) cautions against calling either of the two genera *Apostasia* and *Neuwiedia* (with their 16 species) primitive and ancestral to any of the other modern subfamilies of orchids, including the Cypripedioideae, because their character states may be so derived as to represent an old orchidaceous lineage all their own.

In summary, I conclude that (1) orchids apparently originated quite recently and must have diversified in less time than most other flowering plant families, and (2) the family is this most species-rich of angiospermous families. Therefore, comparatively speaking, the *rate of speciation* in the Orchidaceae must have been exceptionally high for angiosperms because the numerator of the ratio is very large and the denominator is one of the smallest. Very roughly, a new species of orchid evolved every 10,000 years since the end of the Cretaceous. Being of similar age but less species-rich (Stebbins 1950), the other two most advanced families (Asteraceae and Poaceae) have somewhat slower rates of speciation. This conclusion supports Stebbins' (1950) opinion that evolution in the Orchidaceae is exceptionally tachytelic.

MECHANISMS OF REPRODUCTIVE ISOLATION

Several mechanisms of reproductive isolation have been proposed as most important in orchid speciation: chromosome diversity, mycorrhizal associations and biochemical specializations, adaptation to specialized pollinators, and spatial isolation itself.

Chromosome diversity

Diploid chromosome numbers in the Orchidaceae range from 10 to ± 200, with high frequencies at $2n = 28$, $2n = 38\text{–}42$, and $2n = 56$ (Tanaka and

Kamemoto 1974). Sanford (1974) observes that orchids have a rather high incidence of polyploidy: about 60% in Africa, Ceylon, and Jamaica and 53% in Great Britain. Citing Hagerup's (1944) studies on the influence of polyploidy on the fertilization of the common Eurasian terrestrial *Orchis maculatus* L., Stebbins (1950) concluded that the differences in chromosome numbers apparently provide the most important isolating mechanism in some temperate genera of orchids. Yet interspecific and intergeneric hybrids, both natural and artificial, within the subtribe Oncidiinae (subfamily Vanoideae) involve parents of strikingly diverse chromosome numbers (Dressler 1981:265). A quarter of a century later, Stebbins (1974) changed his view and advocated that the most important feature of the adaptive radiation in orchids is the diverse adaptations to pollination by distinct groups of Hymenoptera.

Today, nobody is claiming that cytogenetic diversity is the primary driving force of orchid speciation. There is no doubt that some hybrids fail to succeed because of cytogenetic incompatibilities, but cytogenetics does not seem to provide a sufficient explanation of the profuse species barriers in the Orchidaceae.

Mycorrhizal associations and biochemical specializations

All orchids depend on intimate associations with mycorrhizal fungi for seed germination, and in most species ectotrophic mycorrhizae are required for adult survival (Withner 1974; Dressler 1981). For a long time there was a suspicion that the diversity of orchid species was species-specifically dependent on an equally great diversity of mycorrrhizae. Curtis (1939) was the first to disprove the wish for species specificity (Hadley 1982). Comparatively few species of mycorrhizae have been found with orchids; numerous species of orchids either share the same species of fungus in nature or can be grown vigorously on them in the laboratory (Curtis 1939; Hijner and Arditti 1973; Peschke and Volz 1978; Warcup 1971). Conversely, certain species of orchids have been found to be infected with several species of fungi. For example, as many as eight species of fungi, including several species of *Rhizoctonia*, are able to invade the roots and live symbiotically with *Dactylorhiza purpurella* (Harvais and Hadley 1967; Williamson and Hadley 1970; Withner 1974). To be sure, orchid growth and performance do improve with certain species of fungi, but the great species diversity of orchids cannot be explained by a complementary diversity of associated mycorrhizae.

A diversity of flavonoids and alkaloids exists in orchids (Williams 1979; Luening 1974; Slaytor 1977), but no important concordances with species diversity emerge, at least relative to other plant taxa. Gas chromatographic analysis of the constituents of extrafloral nectar reveals no detectable variation from floral nectar (Baskin and Bliss 1969), and thin-layer chromatography of floral nectar reveals no interesting correlations between the types of sugars and pollinators of select orchid species (Jeffrey et al. 1970).

Although the biochemistry of orchid vegetative tissues is not implicated in orchid diversification, floral chemistry certainly is. Most famous are the monoterpenes and simple aromatic compounds that attract male bees of the Euglossini in a species-specific way (Dressler 1968; Dodson et al. 1969; Williams and Dodson 1972; Janzen et al. 1982). These exclusively neotropical bees go through such stereotyped behaviors in the presence of orchid flowers, particularly collecting fragrant material from flower parts (Williams 1982), that it must be concluded that they are rewarded for visiting orchid flowers. Yet exactly how these floral fragrances function as rewards to the bees remains obscure (Williams 1982). Insightful experiments have proved that bees distinguish very similar orchid species, particularly in the genera *Cynoches* and *Catasetum*, on the basis of these fragrances alone (Dodson et al. 1969; Hills 1972; Williams and Dodson 1972). Thus, the diversity of floral fragrance compounds is basic to present-day species limits and, by implication, the profuse evolution in at least six tribes of the vandoid orchids, and perhaps other subfamilies as well (Williams 1982).

Negatively correlated evolutionary diversifications

Anatomically, the orchids offer textbook examples of *negatively correlated evolutionary diversifications*: repetitively dull features in one trait and riotously variable states in another character. One such contrast is evident in vegetative structures, such as pseudobulbs and leaves, versus floral displays. Although it is true that orchids vary in size from the minute *Platystele jungermannoides* and *Bulbophyllum* spp. to the gigantic *Grammatophyllum* spp. (Dressler 1981), Stebbins (1974) himself comments disparagingly about the lack of anatomical and physiological innovations of the vegetative structures in orchids compared to their flamboyant flowers. In this way orchids contrast with euphorbs, all of which have unattractive, severely reduced flowers (not including the colorful subtending bracts), but which have plant forms that range from obscure herbs to stately "cactuses" in one genus, *Euphorbia*. A second contrast pits the conservative fruit capsule of orchids against their flamboyant flowers. In this regard, the Orchidaceae is the opposite of the Rosaceae, in which the structural uniformity of the pretty flowers contrasts with the remarkable fruit diversity in the family. It is impossible not to be impressed by orchid flowers.

Adaptation to specialized pollinators

It is conventional wisdom that the integrity of existing orchid species, and indeed the origin of new species, is nearly entirely dependent on the fidelity of highly specialized pollinators (Dodson 1962, 1965; Dodson and Frymire 1961). Ever since Darwin (1877) argued that orchid flowers function as clever contrivances by which pollination by insects is accomplished, students of orchid biology (e.g., Faegri and van der Pijl 1979; Dodson and Frymire 1961; Dressler 1981) have been convinced of the adaptive significance of orchid floral specializations in the context of their pollination systems. Everyone says that it is the diversity of relative positions of the pollinia and receptive stigmas in the column, the effectiveness of the rostellum and adhesive nature of the viscidium, the placement of the pollinarium on a discrete part of the pollinating vector's body, and the actual movement of flower parts and caudicles of the pollinarium that effectively promote outcrossing and prevent interspecific hybridization in nearly all species of orchids.

Dodson (1962, 1965) has been a major advocate of Darwin's (1877) thesis. There are innumerable examples of morphologically similar orchids occupying similar habitats, but differing conspicuously in their floral visitors (van der Pijl and Dodson 1966). Dodson (1962) presented the classic case of "leap-frog speciation" in the genus *Stanhopea* in the New World tropics. The adaptative radiation in the genus concerns only flower size and structure; the 15 or so species are indistinguishable by vegetative characters alone. These bizarre orchids successively attracted new species of euglossine bees each time they invaded a new isolated valley in tropical America. In the case of *Stanhopea tigrina*, *S. saccata*, *S. florida*, and *S. anfracta*, speciation involved returning to the plesiomorphic pollinating genus *Euglossa* from the apomorphic genera *Eulaema* and *Eufriesea* (Dodson 1962; van der Pijl and Dodson 1966; Dressler 1981). This is a classic example of an evolutionary reversal in pollination biology.

Although many people think that the classiest examples of coevolution are found in the relationships of orchids and their pollinators, van der Pijl and Dodson (1966), Dressler (1968), and Williams (1982) stress that there is no convincing evidence that any insect has in fact undergone evolutionary modification because of its association with orchids; all the evolutionary momentum is arguably on the part of the plants responding to availability of pollinators. For example, in the famous cases of female-insect mimics (the European *Ophrys*, the Australian *Caladenia* and *Drakaea*, etc.) that deceive male insects into pollination during pseudocopulations, the modifications are entirely on the part of the orchids (i.e., mimetic morphology, female sex pheromones, etc.). No novel anatomical structures or behaviors exclusively associated with orchid visits are known in the male insects—they are simply responding to sexual stimuli in ordinary passionate fashion.

Two features of the breeding systems of orchids are particularly noteworthy in the context of the speciation problem: (1) most species are self-compatible but never naturally self-fertilize (Dressler 1981), and (2) interspecific and intergeneric viability and fertility are common in the laboratory, but hybridizations in nature are quite rare (Kallunki 1976).

Self-compatibility

In light of the morphological "locks" that can be opened only by the precision "keys" held in the stereotyped behaviors of (largely) insect pollinators, it may not be surprising that most orchids are self-compatible. It can be argued that orchids have neatly avoided the genetic problems of inbreeding by mechanically preventing autogamous pollinations through pollinator manipulation. Thus, energy is saved by not dragging along additional, perhaps costly mechanisms of self-incompatibility. It seems that self-incompatibility exists primarily in those genera, like *Oncidium*, and species that have large inflorescences and that experience geitonogamous pollinations regularly. van Steenis (1969) concludes that orchid species have evolved through the development of pollinator specificity rather than incompatibility mechanisms.

Interfertility

Perhaps the feature of orchid breeding systems that most startles students of speciation, particularly those from zoological backgrounds, is the rampant interbreeding that is possible between species, genera, and even subtribes! It has been estimated that over 300 new species of orchids are created by artificial intertaxonomic hybridization each month by countless orchid enthusiasts competing for big prizes (Heywood 1985). In illustrative polygonal diagrams, Dressler (1981) summarizes the known cases of hybridization, both artificial and natural, in several of the major subtribes of orchids. The length of the list, as well as the complicated rules of nomenclature imposed on the pedigreed hybrids, is nothing short of mind-boggling.

The astonishing lack of intrinsic (postmating) barriers to hybridization conflicts with one aspect of the (now) traditional theory of the origin of biological species (Mayr 1963). Simply stated, two sister populations will inevitably diverge genetically after a long period of geographic separation. As genetic differences accumulate, abnormal development in hybrids and reduced reproductive performance in the F_1 or F_2 generations become manifest. Then, prezygotic isolating mechanisms, particularly assortative mating behaviors, that function to avoid the unproductive consequences of cross-matings, are expected to evolve. Thus, reinforcement of intrinsic developmental incompatibilities by specific mate recognition systems (Paterson 1980, 1981) accelerates the attainment of species status (reproductive isolation), the strongest mode of speciation.

However, as has been discussed extensively in this volume, the hard evidence that premating mechanisms of reinforcement actually evolve *after* postmating mechanisms are already in place, or that they evolve more rapidly in the face of hybrid unfitness, is surprisingly lacking (Spencer et al. 1986, 1987). Premating isolating mechanisms often seem to appear simultaneously, or even before postmating isolating mechanisms as undirected, random by-products of genetic divergence in allopatric populations. Coyne and Orr (this volume) do present some new evidence for reinforcement in *Drosophila* speciation.

Orchids pose an extreme case: specific mate-recognition systems through pollination specialization abound in the absence of postmating barriers. It is not at all clear how these premating (prezygotic) isolating mechanisms could have evolved when there is little or no evidence of fitness deficiencies in orchid hybrids. Healthy, viable, propagatable hybrid orchids are an everyday product of commercial and amateur growers. The conventional explanation (van der Pijl and Dodson 1966; Dressler 1981) of the rarity of natural hybrids is that the integrity of orchid species is maintained by the specializations of the pollinator attraction and visitation rather than by intrinsic genetic barriers. The exact evolutionary steps by which these pollination specializations originated and are maintained in sympatry remain unclear.

I am unaware of any rigorous model of population genetics that accounts for the situation in the orchids. On the other hand, the conventional experiences of orchid enthusiasts seem to conflict with the predictions of the models by Lande and Schemske (1985; Schemske and Lande 1985). They predict that there must be severe inbreeding depression to account for the overwhelming evolutionary direction toward obligate outbreeding through the pollination system. Everyone agrees that obligate outbreeding is the hallmark of orchid pollination systems, yet evidence of inbreeding depression is not evident among hybrid orchids grown in the greenhouse. To be sure, measures of relative fitness, including estimates of inbreeding depression, are totally lacking in this family under natural conditions. But it is unlikely that such hybrid fitness deficiencies will appear.

In this regard, orchids vaguely resemble other plant populations that undergo vigorous morphological differentiation without accompanying mechanisms of reproductive isolation when evolving in contrasting environments. But the adaptive differentiation in orchids concerns flower morphology and pollination, the very characters promoting specific intransigence in these plants. Thus, the failure to link the morphological diversification with a precise mechanism of speciation is troublesome.

Spatial isolation

van der Pijl and Dodson (1966:172) argue that the single most important isolating factor is the simple spatial isolation of populations itself. When orchids invade a new habitat by an accidental long-distance dispersal of seed(s), maturing invaders face a foreign array of potential pollinators. Evolutionary modification of floral morphology, color, and scents that soon attract one of the potential pollinators in the new surroundings is supposed to occur rapidly. van der Pijl and Dodson (1966) argue that it is the spatial isolation itself that is enough to generate the divergence, and that genetic divergence with respect to postzygotic isolating mechanisms is simply not a necessity; they discuss the extraordinary *Stanhopea* case in exactly this light. Although the developmental genetics behind this evolutionary plasticity is not yet known, orchids do seem extraordinary in their ability to change floral features quickly in evolution.

The hypothesis of spatial isolation assumes that the task before the invading orchid is attracting any creature that will effectively pollinate it. Presumably this must entail seducing an insect (say) away from its usual nectar sources or other rewarding stations. In general, two polar groups of potential pollinators exist in any such habitat: (1) the set of unspecialized vagabond foragers, and (2) the set of specialized pollinators, perhaps already specific to a particular species of flower. Thus, the problem facing the invading orchid is successfully competing with other floral or resource systems that are already rewarding the target.

Attraction of new pollinators seems easy when the invading orchid provides an even more attractive reward than the current rewarding stations. There may be no problem for orchid species that provide copious nectar rewards or critical floral fragrance compounds (e.g., for euglossine bees). But there is a major problem for any plant species whose flowers offer no reward for the potentially pollinating animals. Although some bees might visit initially out of curiosity, early exploration (Ackerman 1983), or mimicry (Dafni 1984), further visits will quickly cease (Real and Caraco 1986). Thus, successful invasion is unlikely, and the seeds of an adaptive radiation through speciation aborted.

These problems are serious for the Orchidaceae because there are many species of orchids that offer no rewards to pollinators. Porsch (1909) counted at least 1000 species of orchids that offer no nectar rewards; van der Pijl and Dodson (1966) estimate that probably 8000, or about one-third of all orchid species, are fundamentally deceptive! All of the subfamilies and tribes have nonrewarding members. In spite of the fact that they are fragrant, none of the 115 or so species in the subfamily Cypripedioideae rewards their pollinators (Dressler 1981; but see Holman et al. 1981). A more common expectation of isolation is the evolution of cleistogamy and self-fertilization (Stebbins 1950). Darwin (1877) and Sundermann (1977) argued that isolation from its normal (obligate) pollinator, *Eucera tuberculata* (Dafni 1987), accounts for the origin of autogamy in the bee orchid *Ophrys apifera* in England. Thus, it seems difficult to explain how the spatial isolation model would apply to the many nonrewarding species of orchids.

I am not satisfied that the rich diversity of orchid species evolved by pollinator specialization without intrinsic genetic incompatibility between incipient species. Normally, the phrase "pollinator specialization" implies increased efficiency of pollination. As I will discuss, this notion is particularly difficult to reconcile with field data that show the pollination systems of deceptive orchids are woefully inefficient.

FRUITING FAILURE OF DECEPTIVE ORCHIDS

The concern that nonrewarding orchids do not attract visitors and, as a consequence, produce few fruits per flower is illustrated by new detailed results of long-term field studies on *Cypripedium acaule* Ait. by Gill and Mock (1989) and Gill (1989) and scattered evidence in the literature.

The pink lady's-slipper, *Cypripedium acaule* Ait. (Cypripedioideae)

My students and I began a detailed demography study of a discrete population of pink lady's slippers (or moccasin flowers) in the heavily forested mountains of northwestern Virginia in 1978. Draped across a ridge-top with a moist north-facing slope and a drier, pine-studded south-facing slope, and encompassing nearly all of the orchids in the stand, a 32 × 72-m main plot was gridded into 168 4 × 4-m quadrats. A 384-m^2 experimental extension plot was appended in 1982. Every orchid in the plots was tagged and notes on its size and reproductive status were taken. A thorough census of the main plot has been taken every year since 1977; the extension plot has had a census every year since its origin. Every previously tagged plant was relocated (with the aid of a map), and every newly discovered plant was tagged and a record initiated. After 10 years (1977–1986) we have ongoing records of nearly 2500 individual plants.

Trained as zoologists with skills in field ecology, we were unfamiliar with much of the esoteric literature (e.g., Dodson 1966; Stoutamire 1967; Curtis 1954) on the reproductive biology of orchids when we began our studies. Thus, we rediscovered for ourselves several features of lady's-slipper biology that were previously known to a few naturalists, and also made observations that contradict previously published accounts. The detailed results of the population dynamics over time, the stage-specific survivorship of year cohorts, individual histories of flowering, frequency of temporary dormancies after emergence, patterns of flowering, fruiting, and recruitment, and analysis of population projections from this long-term demographic study will be reported elsewhere (Gill and Mock 1989; Gill 1989). I summarize the main points here.

Population size. The number of pink lady's-slipper plants visible above ground in the plot each summer has stabilized around a geometric mean of

680.6 with only minor oscillations over the 10-year period. These results resemble those reported by Curtis (1954). The orchids have been spatially clumped in three high-density centers since the beginning of the study. Although as many as 17 different states (one to six leaves, zero to three flowers, fruiting or not) have been observed, two vegetative states (one leaf only, two leaves only) and one mature state (two leaves, one flower) dominate the stand.

Individual size and age. Most nonflowering plants increased in size, and flowered only when they were in the largest class size. Mature plants usually retained large statures and flowered irregularly for a sequence of years, but some individuals stopped flowering, shrank in size, and may or may not have regrown to full mature size again. Such transitions among states by single individuals required the use of stage-specific and size-specific models (Lefkovitch 1965; Kirkpatrick 1984) rather than the more familiar models based on age (Leslie 1945) in my demographic analyses (Gill 1989).

Vegetative growth. My pink lady's-slippers had no significant vegetative reproduction in the form of rhizomatous or stoloniferous cloning. A minority of plants produced multiple buds, and some older (larger) plants sometimes had one or two extra pairs of leaves and one to three flower stalks, but individuals (genets) were distinctive.

Flowering history. Mature individuals flowered erratically, and no synchrony among individuals was evident. Most individuals in my study plot have never flowered. I have had only one plant (out of nearly 2500 records) that has flowered every year for 10 years. All others showed patternless "on" and "off" histories.

Fruiting success. Only 20 (2%) of the 895 flowers that were present on the plot over the 10-year period produced capsules naturally (Table 1). Hand-pollination experiments confirmed that these orchids were self-compatible, and 100% of flowers hand-pollinated with either xenogamous or autogamous pollen produced turgid capsules packed with myriad seeds. Rates of pollinia removal were exceedingly low. These results proved that the failure to produce fruit under natural, unmanipulated conditions was the result of pollinators failing to visit the flowers effectively. Examination of the flowers has failed to reveal any nectaries. Natural floral fragrance was imperceptible to my nose, but was detectable by bottling flowers for a couple of hours.

Subterranean states. Many plants, whether flowering or simply vegetative, disappeared for 1–5 years, and then reappeared above ground. I call

TABLE 1. Natural fruit production in a stand of pink lady's slipper, *Cypripedium acaule,* in Rockingham County, Virginia.[a]

Year	Number of flowers	Number of fruits	Percentage
1977	95	0	0.0
1978	5	0	0.0
1979	60	0	0.0
1980	121	1	0.8
1981	61	3	4.9
1982	77	0	0.0
1983	149	1	0.7
1984	62	2	3.2
1985	113	8	7.1
1986	152	5	3.2

[a]Study plot is 2364 m^2 at 688.8 m elevation.

these "dormancies" for want of a better word. When they reappeared above ground, they were most often in a small, vegetative state, but I have records of plants reappearing in large-sized flowering states after 1–3 years' absence. These observations support Summerhayes' (1951) reports of similar behavior in British orchids, but directly contradict the claim by Case (1964) that no North American orchid has such dormancies. My dormant plants were indeed alive, in spite of being nonphotosynthetic and underground; I surmise that they were functional parasites on mycorrhizal fungi. Thus, short-term absenteeism did not imply mortality in these plants.

Seedling recruitment. Roughly 130 new seedlings emerged per year in spite of the very low rates of capsule production. Spatial patterning of the seedling recruits was tightly correlated with the clumps of adults. The process of capsule dehiscence resulted in the standard, strongly leptokurtic seed dispersal curve; long-distance seed dispersal was highly improbable. I have no direct evidence of the length of the protocormal stage before emergence, but I infer from precise knowledge of location and year of capsule production by the 19 capsule-producing plants that the underground protocormal stage may be substantially greater than the 4 years reported by Curtis (1943), Kano (1968), and Stoutamire (1974).

Annual survivorship. The rate of annual survival averaged 88.35 ± 9.58%

across all plant categories, but was an impressive 97.14 ± 2.98% for those (mature) plants that flowered at least once during the 10-year study. Overall median life expectancy is estimated to be 5.6 years, but the median mature plants in my study plot can expect to live 23.89 years. These plants are very long lived once they have gotten past the critical first few years of emergence.

DISCUSSION

Just as Delpino (1874), van der Pijl (1966), Dressler (1981), and Nilsson (1980, 1981) concluded in their studies, I surmise that the pollination system of these orchids in my study area is one of deceiving newly emerged, naive queen bumblebees of the genus *Bombus*. I presume that the bees are at a low-energy state after hibernation, and their activity of searching for and establishing new nest sites in the spring exhausts their low-energy reserves. The display of large, pink, showy flowers with nectar guides must be irresistible to such starved bees. But the trap mechanism and subsequent lack of reward must negatively reinforce repeated visiting behavior (Dafni 1987).

Because *Cypripedium acaule* is so widespread and in so many different habitats, I doubt that it is mimic of any particular rewarding model species in a manner such as the large, pink, acid bog orchids *Colopogon tuberosus* and *Pogonia ophioglossoides* (Thien and Marcks 1972), or the persuasive cases of mimicry by *Orchis caspia* (Dafni 1983), *O. collina* (Dafni and Ivri 1979), *O. israelitica* (Dafni and Ivri 1980, 1981; Dafni and Baumann 1982), and *O. galilea* (Bino et al. 1982) in Israel. In my study area, the large, showy, pink, very fragrant, nectar-producing azalea *Rhododendron roseum* blooms about the same time as *C. acaule* and is a possible, but remote candidate for the role of model. It is implausible as an evolutionary stimulus simply because its range overlaps very little of the range of *C. acaule*. Rather, I view the pink lady's-slipper as simply presenting a grand, bold, glamorous, and enticing flower, complete with fake nectar guides, that deceives bumblebees in their desperate search for high-energy nutrition (Dafni 1987:89). Thus, it fits the category of nonmodel mimicry suggested by Dafni (1986).

The rarity of successful pollination in these orchids and the inconsistency of individuals being successful suggest to me that individual reproductive success in this species is indistinguishable from pure luck. I am faced with the incredible fact that after 10 years no statistical correlation of any feature of the orchids with success is possible. This dilemma raises the disturbing third question of this chapter: How did (do) the intricate, finely-tuned, apparently adaptive features of deceptive orchid flowers evolve when reproductive success is a matter of chance?

The fact that the pink lady's-slippers in my study recruit entire cohorts through only a few (often one) individuals implies that local populations repeatedly go through very narrow genetic bottlenecks every generation. I predict that the amount of genetic variation in local *C. acaule* populations is exceedingly low because I envision the whole process of colonization and population establishment as a multiplicative sequence of improbables: the combined likelihood of any seed arriving at a new site, of encountering a useful mycorrhizal fungus, of successfully germinating and getting through the prolonged protocormal stage, of reaching maturity, of attracting any pollinator, and of having that same individual pollinator return to the same single plant again to self-pollinate effectively is infinitessimally small. If all this were to happen, then millions of full sib progeny would be found in the population. Hence the subsequent reproductive biology of the population would be subject to severe inbreeding and rapid elimination of genetic variation that might have existed in the founder plant.

More dilemmas

Because my population of pink lady's-slipper orchids has shown a persistent pattern of poor fruiting at least over the first 10 years of this study, I naturally asked four questions:

1. Is my particular population representative of the species *C. acaule*?
2. Do other nonrewarding species of orchids show similar low levels of capsule production?

The next two questions address more general evolutionary issues:

3. Are the current patterns of population-wide fruiting failure all evolutionary stable strategies (ESS) or can these species be invaded by a mutant orchid that does a better job of reproduction?
4. How did deceptive orchids evolve their states of nonreward and concomitant reproductive failure if the present state is not an ESS?

The answer to the first question is yes. I have made observations on two other stands. A large stand of *C. acaule* at the foot of the west slope of the Blue Ridge Mountains at Luray had no fruit capsules present after the 1987 flowering season (personal observation). On 26 July 1987 I surveyed a stand of 99 plants (all that could be located) on a small forested peninsula on the Maine coast near Medomak; 55 of the 99 plants had flowered several weeks earlier and none had produced a fruit capsule.

Correspondence with several individuals studying populations of *C. acaule* in various parts of the Untied States reveals similar low rates of fruit (capsule) production per flower. Dr. Ernst Mayr writes (personal communication) that a stand of pink lady's -slippers in New Hampshire that he has studied for a quarter of a century had only 2 fruit capsules and 327 flowers in 1987. Ms. Margaret E. Cochran (1989, personal communication) reports that in several years study of a large population (2729 individuals) of *C. acaule* on Clinch Mountain in eastern Tennessee, fruit set from natural pollinations varied from a low of 3.4% to a high of 23.0% with the usual level being about 8%. Kurfess (1965) reports 1 capsule out of 100 flowers in his stand. Davis (1986) reports only 9.4% (N = 278 in 1983) and 4.3% (N = 986 in 1984) success in natural pollinations in Massachusetts, with the cause being pollinator limitation. Dr. Richard Primack (personal communication) reports similar low rates of pollination success in a stand near Boston, Massachusetts.

I conclude that throughout the range of *C. acaule* in northeastern North America more than 90% of individuals fail to fruit every year. The consistent explanation is failure of pollinators to visit the flowers in spite of their large and showy appearance. The principal reason for the low rate of pollinator visitations is that there is no reward provided by the flowers. Recalling that it takes a minimum of two flower visits to effect pollination in this obligately outcrossing system, and reflecting on the speed at which bees, particularly bumblebees, learn to reject nonrewarding stations (Real and Caraco 1986), it is surprising that any pink lady's-slipper succeeds in setting seed.

There is no reason to suspect that the cause of these patterns of widespread fruiting failure is due to man-made interferences, such as the accidental elimination of required or potential pollinator populations. Even in the absence of aerial spraying, rates of pollinia removal from *C. acaule* in New Brunswick, Canada are only 3–6% (Plowright et al 1980). The answer seems unambiguous: nonrewarding pink lady's-slippers do not attract available pollinators.

Other orchid species

Is the fruiting failure of *Cypripedium acaule* typical of other nonrewarding orchid species? The answer is yes, but that data on fruiting success as a percentage of flowers are rather scarce in the literature. (I will be pleased to have my attention drawn to reports I have overlooked.) In the case of other species in the Cypripedioideae, there is only Ackerman's (1981) report on the pollination biology of *Calypso bulbosa* var. *occidentalis* in northern California. Ackerman (1981) reports pollination rates of 11–34% and confirms Stoutamire's (1971) report that it is nectarless and otherwise nonrewarding. Boyden (1982) supports Ackerman's (1981) conclusion that the pollination system of *Calypso bulbosa* is one of deception.

I believe Darwin himself (1877) was the first to report the visitation rates to and fruit production in several species of British terrestrials with which he was familiar. His tables show the same pattern: those that produce nectar have abundant fruits and those that produce no nectar have low visit rates and capsule formation. This dichotomy was true even when the two kinds of orchid grew together in common pastures or other habitats. Darwin (1877:49) was puzzled about why the fly orchid (*Ophrys muscifera*, now called *O. insectifera*) did not produce flowers more attractive to insects. Throughout his book, he could not get himself to accept the idea of deliberate deception, because he could not believe that insects could so readily be duped. Of course, today the concept of plants having floral morphologies and fragrances that deceive insects is widely accepted.

Several recent studies of north temperate *Orchis* species and their allies provide the best quantitative data on relative fruiting success of nectarless and nectariferous orchids. In the presence of at least two nectar-producing species, including the unusual nectariferous model *O. israelitica*, *O. caspia* succeeds in setting seed in 67.1–86.0% of its flowers, but alone capsule production drops to 11.3–16.0% (Dafni 1983). The nectarless, rarely visited *O. collina* produced capsules from only 25–32% of its flowers in the 3 years 1973–1975, whereas its fragrant, nectar-producing, attractive relative *O. coriophora* produced 78–95% in the same years and region (Dafni and Ivri 1979). In the absence of the rewarding and attractive liliaceous *Bellevalia flexuosa*, the mimetic and nonrewarding *O. israelitica* repeatedly produces capsules from only 2.8–4.7% of its 200+ flowers (Dafni and Ivri 1981). In spite of sporting a non-nectar-producing spur and widely thought of as nonrewarding, *Dactylorhiza fuchsii* was shown to attract honeybees, *Apis mellifera*, in England with a glucose-rich stigmatic exudate, and succeeded in producing capsules from 53% of its 25 flowers (Dafni and Woodell 1986).

Reports of fruit production in other subfamiles of the Orchidaceae also support the dichotomous pattern: in nectariferous orchids pollination percentage is high, often 100%; orchids that produce no nectar or other rewards have very low (often less than 10%) frequencies of capsule formation. Whigham and McWethy (1980) report high levels of natural fruit production in the nectar-producing crane-fly orchid *Tipularia discolor* in Maryland. Ackerman and Montero Oliver (1985) report that natural pollination rates of the nonrewarding *Oncidium variegatum* in Puerto Rico are less than 4%. Both pollinator limitation and resource limitation contributed to the low fruit set (6%) in the nonrewarding orchid *Ionopsis utricularioides* in Puerto Rico (Montalvo and Ackerman 1987). Ackerman (1983) argues that *Cochleanthes lipscombiae* (subfamily Epidendroideae, Tribe Vandeae, subtribe Zygopetalinae) is perhaps typical of the fragrant, nectarless orchids that attract and are pollinated specifically by male euglossine bees, and may mimic genuine food sources, Deceptive as it is, it also has a lower fruit set (15%) under natural conditions on Barro Colorado Island, Panama. Inoue (1985, 1986) reports variable pollination success in Japanese species of the terrestrial genus *Platanthera* depending on whether noctuid or sphingid moths were responsible; several instances of low capsule set as a result of pollinator limitation are reported.

I conclude that we now have ample documentation that fruiting success in orchids often is pollinator-limited, just as it is in many other plants (Bierzychudek 1981), and the reason often is that orchid flowers simply do not attract visitors, not because the pollinators are not there due to some other catastrophe. These cases do not fit the concept of "specialized pollination systems that have evolved because of pollination efficiency." If all (8000) deceptive orchid species prove to have pollination that is equally as poor as these case examples, then they can hardly be held up as premier examples of evolving quickly because of improved efficiency in their pollination.

Evolutionarily stable strategies

These results bring me to the third question: "Are the current pollination systems of orchids ESSs?" My answer is no. In the case of the pink lady's-slipper, I ask the same question that Darwin (1877) asked about the fly orchid. Both deceptive species seem very vulnerable to invasion by a mutant genotype that is more attractive to insect visitors.

Consider the case of the beautiful bee orchid (*Ophrys apifera*), which is congeneric to the fly orchid, and probably the most celebrated of the pseudocopulation syndromes. Darwin (1877) determined that the flower morphology and movement of the caudicles in England are based on a system of autogamous self-fertilization. In contrast to the allegedly obligate outcrossing pollination system of bee orchids on the continent of Europe, Darwin (1877) concluded that the absence of effective pollinators on the depauperate English island favored a modification of the famous flower in a way that permitted self-pollination. As mentioned earlier, Sundermann (1977) supported this conclusion.

By analogy, I can envision at least three simple modifications of the pollination system of *Cypripedium acaule* that, were they to occur in an individual pink lady's-slipper in any population in North America, would be many times more successful than the current barren state in which most moccasin flowers currently find themselves. For example, given the fact that they are self-compatible, a slight modification of the anatomy of the flower that permitted autogamous self-pollinations, just like the bee orchid, would result in an enormous increase in individual reproductive fitness, Not only would it be superior to nearly all other individuals in any given year, but it also would repeat its performance every year it flowered in its long 25-year life span. It seems obvious that the speed of local fixation, if not species-wide fixation, of a mutant genotype in *C. acaule* that selfed would be very rapid.

A second kind of mutant that could invade any of the current populations of *C. acaule* would be one that offers a nectar reward to visiting bumblebees. In this case I envision an immediate positive reinforcement, rather than the current negative reinforcement, to the visiting behavior of any individual bee that happens to visit the nectar-producing plant. Encouraged to visit at least one other moccasin flower, the individual bee will deliver the pollen onto the receptive stigma of the second plant, even if it is nonrewarding (which it is likely to be if the nectar-producing plant is a new rare mutant). Hence the rewarding mutant genotype has an immediate fitness gain through *male function*; it is not likely to be the single plant producing the seed, but it fathered the rare seed-setting plants in the population.

A third option is flowering later in the season, particularly in mid-summer when there is an ever-increasing abundance of naive worker bumblebees emerging from nests. The greater density of potential pollinators alone would increase the likelihood that a late-flowering lady-slipper would be visited by at least one deceived bee. It is possible that this mutant might not work because the shaded understory of mid-summer, eastern mixed deciduous forests would discourage foraging by heliophilic bumblebees, irrespective of their naivete.

The annoying, untestable, yet common answer often given these evolutionary speculations is that such mutations have not yet happened in any population of pink lady's-slippers. This historical constraint–accident hypothesis seems unlikely considering how abundant *C. acaule* is and how long (about 100 mya) the genus *Cypripedium* has been around. If such mutants have occurred, I cannot think of what barriers would prevent the immediate invasion and complete fixation of such mutants in the current populations of *C. acaule*. Similarly, I do not see why such mutants would not be instantaneously successful in any of the species of deceptive orchids.

These problems bring me to my fourth question: How did so many orchids evolve deceptive pollination systems when to do so involves reduced reproductive performance? If the plesiomorphic condition is considered to be nectar producing, as Davni (1987) concludes, then how is the evolutionary *loss* of rewarding systems by natural selection explained? How did 8000 deceptive species of orchids, many of which are apparently not part of a model–mimicry system, originate? At present, I have no idea.

SUMMARY

The Orchidaceae is the most species-rich family of angiosperms, and exhibits some of the most advanced modifications of flower morphology. The orchids have recently and rapidly diversified, and may have the highest rates of speciation of all flowering plants, The driving evolutionary factor of their speciation is said to be reproductive isolation through pollinator specialization; orchids are often cited as the epitome of coevolution between plant and pollinators. Nearly one-third of known orchid species offer no floral rewards (nectar, pollen, etc.) and depend on deception to attract their flower visitors. Data from the literature reveal that deceptive orchids are rarely visited by pollinators and have relatively poor reproduction compared to nectariferous species. My long-term field studies of pink lady's-slippers, *Cypripedium acaule* Ait., show one severe case of persistent reproductive failure. The problem of how deceptive, reproductively failing species of orchids evolved is discussed.

ACKNOWLEDGMENTS

Field assistance has been most generously been offered by B. A. Mock, K. A. Berven, T. G. Halverson, R. Harris, C. von Dohlen, and W. Walton. We thank Mr. George Blomstrom, District Ranger, for permission to carry out these studies in the George Washington National Forest. I thank Drs. E. Mayr, R. Primack, J. Ackerman, and M. Cochran for so courteously transmitting manuscripts, unpublished data, and their experiences with orchids. I thank Mr. T. Lindsay, Naturalist of the Shenandoah National Park, for assistance in locating stands of moccasin flowers in the SNP, and D. H. Morse for showing me the stand at Medomak, Maine. National Science Foundation Grants DEB 77-04817, DEB 78-10832, DEB 80-05080, and BSR 86-05197 helped support this work.

LITERATURE CITED

Ackerman, J. D. 1981. Pollination biology of *Calypso bulbosa* var. *occidentalis* (Orchidaceae): A food deception scheme. Madrono 28(3):101–110.

Ackerman, J. D. 1983. Euglossine bee pollination of the orchid, *Cochleanthes lipscombiae*: A food source mimic. Am. J. Bot. 70(6):830–834.

Ackerman, J.D., and J. C. Montero Oliver. 1985. Reproductive biology of *Oncidium variegatum*: Moon phases, pollination, and fruit set. Am. Orchid Soc. Bull. 54(3): 326–329.

Baskin, S., and C. Bliss. 1969. Sugar content in extrafloral exudates from orchids. Phytochemistry 8:1139–1145.

Bierzychudek, P. 1981. Pollinator limitation of plant reproductive effort. Am. Natur. 117:838–840.

Bino, J. R., A. Dafni, and A. D. J. Meeuse. 1982. The pollination ecology of *Orchis galilea* (Bornm. et Schulze) Schltr. (Orchidaceae). New Phytol. 90:315–319.

Boyden, T. C. 1982. The pollination biology of *Calypso bulbosa* var. *americana* (Orchidaceae): Initial deception of bumblebee visitors. Oecologia 55:178–184.

Case, F. 1964. *Orchids of the Western Great Lakes Region*. Bulletin 48, Cranbrook Institute of Science, Bloomfield Hills, MI.

Cochran, M. E. 1989. Consequences of pollination by chance in the pink lady's slipper, *Cypripedium acaule*. Ph.D. dissertation, University of Tennessee, Knoxville.

Curtis, J. T. 1939. The regulation of specificity of orchid mycorrhizal fungi to the problem of symbiosis. Am. J. Bot. 26:390–399.

Curtis, J. T. 1943. Germination and seedling development in five species of *Cypripedium* L. Am. J. Bot. 30:199–205.
Curtis, J. T. 1954. Annual fluctuation in rate of flower production in native cypripediums during two decades. Bull. Torrey Bot. Club 81:340–352.
Dafni, A. 1983. Pollination of *Orchid caspia*—a nectarless plant which deceives the pollinators of nectariferous species from other plant families. J. Ecol. 71:467–474.
Dafni, A. 1984. Mimicry and deception in pollination. Annu. Rev. Ecol. System. 15: 253–278.
Dafni, A. 1986. Floral mimicry-mutualism and undirectional [sic] exploitation of insects by plants. Pp 81–90 in: B. Juniper and Sir Richard Southwood (eds., *Insects and the Plant Surface*. Edward Arnold, London.
Dafni, A. 1987. Pollination in *Orchis* and related genera: Evolution from reward to deception. Pp. 80–104 in: J. Arditti (ed.), *Orchid Biology, Reviews and Perspectives. IV*. Cornell University Press, Ithaca, NY.
Dafni, A., and H. Baumann. 1982. Biometrical analysis in populations of *Orchis israelitica* Baumann and Dafni, *O. caspia* Trautv. and their hybrids. Plant System. Evol. 140: 87–94.
Dafni, A., and Y. Ivri. 1979. Pollination ecology of, and hybridization between, *Orchis coriophora* L. and *O. collina* Sol. ex Russ. (Orchidaceae) in Israel. New Phytol. 83: 181–187.
Dafni, A., and Y. Ivri. 1980. Deceptive pollination syndromes in some orchids in Israel. Acta Bot. Neel. 29:55 (Abstr.).
Dafni, A., and Y. Ivri. 1981. Floral mimicry between *Orchis israelitica* Baumann and Dafni (Orchidaceae) and *Bellevalia flexuosa* Boiss. (Liliaceae). Oecologia 49:229–232.
Dafni, A., and S. R. J. Woodell. 1986. Stigmatic exudate and the pollination of *Dactylorhiza fuchsii* (Druce) Soo. Flora 178:343–350.
Darwin, C. 1862. *On the Various Contrivances by Which British and Foreign Orchids Are Fertilised by Insects*, 1st ed. John Murray, London.
Darwin, C. 1877. *The Various Contrivances by Which Orchids Are Fertilised by Insects*, 2nd ed., rev. D. Appleton, NY.
Davis, R. W. 1986. The pollination biology of *Cypripedium acaule* (Orchidaceae). Rhodora 88:445–450.
Delpino, F. 1874. Ulteriori osservasione e considerazioni sulla dicogamia del regno vegetale. 2 (IV). Delle piante zoidifile. Atti Soc. Ital. Sci. Natur. 16:151–200.
Dodson, C. H. 1962. The importance of pollination in the evolution of the orchids of tropical America. Am. Orchid Soc. Bull. 31:525–534, 641–649, 731–735.
Dodson, C. H. 1965. Studies in orchid pollination: The genus *Coryanthes*. Am. Orchid Soc. Bull. 34:680–687.
Dodson, C. H. 1966. Studies in orchid pollination: *Cypripedium, Phragmopedium*, and allied genera. Am. Orchid Soc. Bull. 35:125–128.
Dodson, C. H. 1974. Orchidales. Encyclopaed. Britt. 13:648–656.
Dodson, C. H., R. L. Dressler, H. G. Hills, R. M. Adams, and N. H. Williams. 1969. Biologically active compounds in orchid fragrances. Science 164:1243–1249.
Dodson, C. H., and G. P. Frymire. 1961. Natural pollination of orchids. Missouri Bot. Garden Bull. 49:133–152.
Dressler, R. L. 1968. Pollination by euglossine bees. Evolution 22:202–210.
Dressler, R. L. 1981. *The Orchids: Natural History and Classification*. Harvard University Press, Cambridge, MA.
Faegri, K., and L. van der Pijl. 1979. *The Principles of Pollination Biology*, 3rd ed. Pergamon Press, NY.
Garay, L. A. 1960. On the origin of the Orchidaceae. Bot. Mus. Leaflets (Harvard Univ.) 19:57–96.
Garay, L. A. 1964. Evolutionary significance of geographical distribution of orchids. Pp. 170–187 in: *Proceedings of the Fourth World Orchid Conference*.
Garay, L. A. 1972. On the origin of the Orchidaceae. II. J. Arnold Arbor. 53:202–215.
Gill, D. E. 1989. Population biology of the pink lady's slipper, *Cypripedium acaule* Ait. (Orchidaceae). II. Transition analysis and population projection. Manuscript.

Gill, D. E., and B. A. Mock. 1989. Population biology of the pink lady's slipper, *Cypripedium acaule* Ait. (Orchidaceae). I. Long-term field studies and demography. Manuscript.
Hadley, G. 1982. Orchid mycorrhizae. Pp. 83–118 in:J. Arditti (ed.), *Orchid Biology. Reviews and Perspectives, II.* Cornell University Press, Ithaca, NY.
Hagerup, V. H. 1944. On fertilization, polyploidy, and haploidy in *Orchis maculatus* L. (*sens. lat.*). Dansk Bot. Arkiv. 11(5):1–25.
Harvais, G. and G. Hadley. 1967. The relationship between host and endophyte in orchid mycorrhiza. New Phytol. 66:205–216.
Heywood, V. H. 1985. *Flowering Plants of the World.* Equinox (Oxford), Oxford.
Heywood, V. H., J. B. Harborne, and B. L. Turner. 1978. *The Biology and Chemistry of the Compositae*, Vols. I and II. Academic Press, London.
Hijner, J. A. , and J. Arditti. 1973. Orchid mycorrhiza: Vitamin production and requirements by the symbionts. Am. J. Bot. 60:829–835.
Hills, H. G. 1972. Floral fragrances and isolating mechanisms in the genus *Catasetum* (Orchidaceae). Biotropica 4:61–76.
Holman, R. T., W. R. Cunningham, and E. S. Swanson. 1981. A closer look at the glandular hairs on the ovaries of cypripediums. Am. Orchid Soc. Bull. 50(6):683–687.
Inoue, K. 1985. Reproductive biology of two platantherans (Orchidaceae) in the island of Hachijo. Jpn. J. Ecol. 35:77–83.
Inoue, K. 1986. Different effects of sphingid and noctuid moths on the fecundity of *Platanthera metabifolia* (Orchidaceae) in Hokkaido. Ecol. Res. 1:25–36.
Janzen, D. H., P. J. DeVries, M. L. Higgins, and L. S. Kimsey. 1982. Seasonal and site variation in Costa Rican euglossine bees at chemical baits in lowland deciduous and evergreen forests. Ecology 63(1):66–74.
Jeffrey, D. J., Arditti, and H. Koopowitz. 1970. Sugar content in floral and extrafloral exudates of orchids: Pollination, myrmecology, and chemotaxonomy implications. New Phytol. 69:187–195.
Kallunki, J. A. 1976. Population studies of *Goodyera* (Orchidaceae) with emphasis on the hybrid origin of *G. tesselata*. Brittonia 28:53–75.
Kano, K. 1968. Acceleration of the germination of so-called "hard to germinate" orchid seeds. Am. Orchid Soc. Bull. 37:690–698.
Kurfess, J. F. 1965. Through the letter slot. Am. Orchid Soc. Bull. 34:914.
Kirkpatrick, M. 1984. Demographic models based on size, not age, for organisms with indeterminate growth. Ecology 65:1874–1884.
Lande, R., and D. W. Schemske. 1985. The evolution of self-fertilization and inbreeding depression in plants. I. Genetic models. Evolution 39(1):24–40.
Lefkovitch, L. P. 1965. The study of population growth in organisms grouped by stages. Biometrics 21:1–18.
Leslie, P. H. 1945. On the use of matrices in certain population mathematics. Biometrika 33:183–212.
Luening, B. 1974. Alkaloids on the Orchidaceae. Pp. 349–382 in: C. L. Withner (ed.), *The Orchids. Scientific Studies.* John Wiley, New York.
MacArthur, R. H., and E. O. Wilson. 1967. *The Theory of Island Biogeography.* Princeton University Press, Princeton, NJ.
Mayr, E. 1963. *Animal Species and Evolution.* Belknap Press, Harvard University Press, Cambridge, MA.
Montalvo, A. M., and J. D. Ackerman. 1987. Limitations to fruit production in *Ionopsis utricularioidides* (Orchidaceae). Biotropica 19(1):24–31.
Nilsson, L. A. 1980. The pollination biology of *Dactylorhiza sambucina* (Orchidaceae). Bot. Notes 133:367–385.
Nilsson, L. A. 1981. Pollination ecology and evolutionary processes in six species of orchids. Acta Univ. Uppsaliensis 593:1–40.
Paterson, H. E. H. 1980. A comment on "mate recognition systems." Evolution 34:330–331.
Paterson, H. E. H. 1981. The continuing search for the unknown and the unknowable: A critique of contemporary ideas on speciation. South African J. Sci. 77:113–119.
Peschke, H. C., and P. A. Volz. 1978. *Fusarium moniliforme* Sheld. association with species of orchids. Phytologia 40(4):347–356.

Plowright, R. C., J. D. Thomson, and G. R. Thaler. 1980. Pollen removal in *Cypripedium acaule* (Orchidaceae) in relation to aerial fenithrothion spraying in New Brunswick, Canada. Entomologst 112:765–770.
Porsch, O. 1909. Die Honigersatzmittel der Orchideenbluete. Bot. Wandt. 111:496–509.
Raven, P. H., and D. E. Axelrod. 1974. Angiosperm biogeography and past continental movements. Ann. Missouri Bot. Garden 61:539–673.
Real, L., and T. Caraco. 1986. Risk and foraging in stochastic environments. Annu. Rev. Ecol. System. 17:371–390.
Sanford, W. W. 1974. The ecology of orchids. Pp. 1–100 in: C. L. Withner (ed.), *The Orchids. Scientific Studies*. John Wiley, New York.
Schemske, D. W., and R. Lande. 1985. The evolution of self-fertilization and inbreeding depression in plants. II. Empirical observations. Evolution 39(1):41–52.
Schmid, R., and M. J. Schmid. 1977. Fossil history of the Orchidaceae. Pp. 25–45 in: J. Arditti (ed.), *Orchid Biology. Reviews and Perspectives, I.* Cornell University Press, Ithaca, NY.
Shultes, R. 1960. *Native Orchids of Trinidad and Tobago*. Pergamon Press, NY.
Shultes, R. E. 1966, Orchid. Encyclopaed. Britt. 16:1041–1043.
Slaytor, M. D. 1977. The distribution and chemistry of alkaloids in the Orchidaceae. Pp. 95–115 in: J. Arditti (ed.), *Orchid Biology. Reviews and Perspectives, I.* Cornell University Press, Ithaca, NY.
Spencer, H. G., B. H. McArdle, and D. M. Lambert. 1986. A theoretical investigation of speciation by reinforcement. Am. Natur. 128:241–262.
Spencer, H. G., D. M. Lambert, and B. H. McCardle. 1987. Reinforcement, species, and speciation: A reply to Butlin. Am. Natur. 130:958–962.
Stebbins, G. L. 1950. *Variation and Evolution in Plants*. Columbia University Press, New York.
Stebbins, G. L. 1974. Flowering Plants. *Evolution above the Species Level*. Belknap Press, Harvard University Press, Cambridge, MA.
Stoutamire, W. P. 1967. Flowering biology of the lady's-slippers (Orchidaceae: *Cypripedium*). Michigan Bot. 6:158–175.
Stoutamire, W. P. 1971. Pollination in temperate American orchids. Pp. 233–243 in: M. J. G. Corrigan (ed.), *Proceedings of the 6th World Orchid Conference*. Halsted Press, Sydney.
Stoutamire, W. 1974. Terrestrial orchid seedlings. Pp. 101–128 in: C. L. Withner (ed.), *The Orchids. Scientific Studies*. John Wiley, New York.
Summerhayes, V. S. 1951. *Wild Orchids of Britain with a Key to the Species*. Collins, London.
Sundermann, H. 1977. The genus *Ophrys*—an example of the importance of isolation for speciation. AM. Orchid Soc. Bull. 46:825–830.
Tanaka, R.,, and H. Kamemoto. 1974. List of chromosome numbers in species of the Orchidaceae. Pp. 411–483 in: C. L. Withner (ed.), *The Orchids. Scientific Studies*. John Wiley, New York.
Thien, L. B., and B. G Marcks. 1972. The floral biology of *Arethusa bulbosa, Calopogon tuberosus, and Pogonia ophioglossoides* (Orchidaceae). Can. J. Bot. 50:2319–2325.
van der Pijl, L. 1966. Pollination mechanisms in orchids. Pp. 61–75 in: J. G. Hawkes (ed.), *Reproductive Biology and Taxonomy of Higher Plants*. Pergamon Press, Oxford.
van der Pijl, L., and C. H. Dodson. 1966. *Orchid Flowers, Their Pollination and Evolution*. University of Miami Press, Coral Gables, FL.
van Steenis, C. G. G. J. 1969. Plant speciation in Malesia, with special reference to the theory of non-adaptive strategy evolution. Biol. J. Linnean Soc. 1:97–133.
Warcup. J. H. 1971. Specificity of mycorrhizal association in some Australian terrestrial orchids. New Phytol. 70:41–46.
Whigham D. F., and M. McWethy. 1980. Studies on the pollination ecology of *Tipularia discolor* (Orchidaceae). AM. J. Bot. 67(4):550–555.
Williams, C. A. 1979. The leaf flavonoids of the Orchidaceae. Phytochemistry 18:803–813.
Williams N. H. 1982. The biology of orchids and euglossine bees. Pp. 119–171 in: J. Arditti (ed.), *Orchid Biology. Reviews and Perspectives. II*. Cornell University Press, Ithaca, NY.

Williams, N. H., and C. H. Dodson. 1972. Selective attraction of male euglossine bees to orchid floral fragrances and its importance in long distance pollen flow. Evolution 26:84–95.

Williamson, B., and G. Hadley. 1970. Penetration and infection of orchid protocorms by *Thanatephorus cucumeris* and other *Rhizoctonia solani* isolates. Phytopathology 60:1092–1096.

Withner, C. L. 1974. Developments in orchid physiology. Pp. 129–168 in: C. L. Withner (ed.), *The Orchids. Scientific Studies*. John Wiley, New York.

CHAPTER 20

SPECIATION IN HAWAIIAN CRICKETS

Daniel Otte

Hawaii has at least twice as many cricket species[1] as the continental United States; in numbers per unit area it is 1200 to 2500 times as rich. What modes of origin are indicated by such an outburst of speciation? As with other Hawaiian plants and animals, this diversity can be explained only by some favorable balance between rates of speciation and extinction. This chapter attempts through the study of crickets to address several unanswered questions about speciation in Hawaii. Where possible, I make comparison with *Drosophila*. The arguments are drawn from data appearing in detail in a forthcoming book on the biogeography and systematics of Hawaiian crickets (Otte and Rice 1989).

Since the archipelago is isolated by expanses of water great enough to impose barriers to all but a minute number of lucky dispersalists, its biota has evolved more or less independently of those elsewhere on the globe. Interisland isolation is also great and allows for a degree of independent evolution of communities on each island; therefore, within the archipelago a set of natural experiments can test the effects of age, topography, island size, and other factors on speciation rate and the structure of communities. Because of the rarity of immigrants, local (within-island or within-volcano) communities could evolve their own unique characteristics without being disturbed by a continual inflow of migrants.

In this chapter I focus on questions that since Darwin have remained at

[1]I use Wright's (1978) population concept of species: "The term species has now come to refer to arrays of populations where there is sufficient continuity of interbreeding to give intergradations of all characteristics among local populations, but absence of such intergradation with populations that are considered to constitute other species. [The] occurrence in the same region of two distinguishable populations which do not interbreed is ordinarily taken to indicate that these belong to different species. [And because speciation is often still in progress] all sorts of mixtures and gradations are to be expected."

the forefront of discussions on the origins of species: When do the signal differences (premating barriers) arise in relation to geographic and ecological isolation? What are the principal creative forces that produce the differences? What kinds of extrinsic barriers or partial barriers to interbreeding lead to speciation? Finally I will address the question of why there are so many Hawaiian cricket species and whether Hawaiian speciation is unique in other than trivial ways. I believe Littlejohn (1988) is quite correct in his assessment that "most of the speciational situations with which we deal will already be stabilized, optimized states, with the appropriate evolutionary adjustments having already occurred, so that only the successful systems are available for study. Indeed, it may require a great deal of good fortune to encounter dynamic evolution of a biocommunicative system in natural populations."

BACKGROUND

Origins. The Hawaiian endemic crickets appear to be derived from four original colonizing species and from two probable sources: a tree cricket (Oecanthinae) and a sword-tail cricket (Trigonidiinae) from the Americas, and two ground crickets (Nemobiinae) from the western Pacific region (Figure 1). Three of the four original species radiated extensively, although patterns of multiplication along the archipelago have varied according to where each group originated and how it dispersed. Much later, and probably aided by man, eight additional crickets colonized the islands, six from Asia (via the western Pacific), one from Africa, and one from either Africa or Southeast Asia (Otte and Rice 1989). These introduced species are widely distributed along the island chain.

All original colonists probably belonged to flightless species that arrived in Hawaii as eggs deposited in floating vegetation[2]; two of them (tree crickets and sword-tail crickets) were probably forest-inhabiting species and two others inhabited shorelines (coral or rocky shores, or mangrove swamps). Tree and sword-tail crickets may have gained a foothold in lowland forests; from there they expanded into all wet lowland forests and finally into montane forests. Being flightless, the lateral movement of the early colonists within islands must have been relatively slow (in contrast to the more rapid expansion of recently introduced flying crickets), being impeded on younger volcanoes by lava fields, and on older islands by inhospitable ridges and cliffs. Dispersal may have been easiest at a time when there were few competing species to act as obstacles.

[2]The recent (October 1988) flight of a swarm of African desert locusts (*Schistocerca gregaria*) across the Atlantic in a tropical storm originating off the African coast holds open the possibility that crickets could similarly have been transported from the American coasts to Hawaii on storms, which occasionally follow that path.

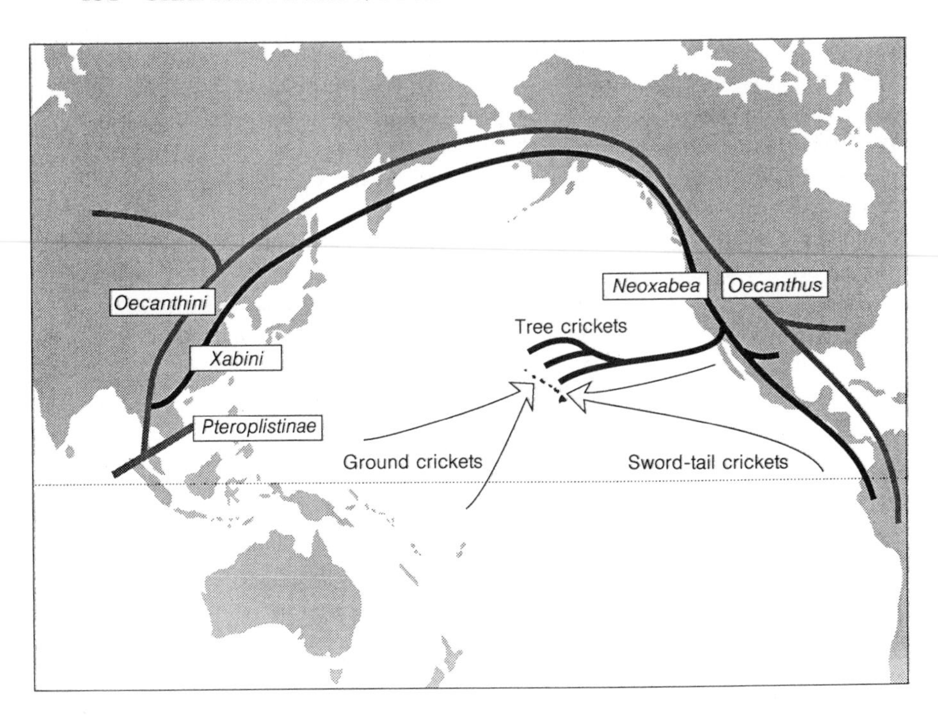

FIGURE 1. Probable origins of the endemic Hawaiian crickets. (From Otte and Rice 1989.)

Aside from the ground crickets, which are rock-adapted and live in coastal or interior lava, virtually the entire fauna is adapted to forests below 4000 feet elevation. This inability to colonize higher forests or to penetrate dry forests on leeward slopes remains puzzling, given the overall success of crickets on the islands. Did becoming adapted to a relatively stable and benign forest habitat preclude them from entering habitats that would call for coping with more extreme conditions of temperature and humidity? Vulnerability even within the zone of best adaptation may be seen in the fact that large tracts of forest are uninhabited, even though various lineages must have passed through them to achieve their present range.

Tree crickets (Oecanthinae). Tree crickets probably first colonized Kauai or Oahu (probably the latter), and within a span of 2.5 million years radiated into three genera (*Prognathogryllus*, *Leptogryllus*, and *Thaumatogryllus*), five subgenera, and 54 species, with the greatest diversification seen in the older islands (Figures 2A and B). Cladistics and biogeography suggest that nearly all speciation has taken place within islands.

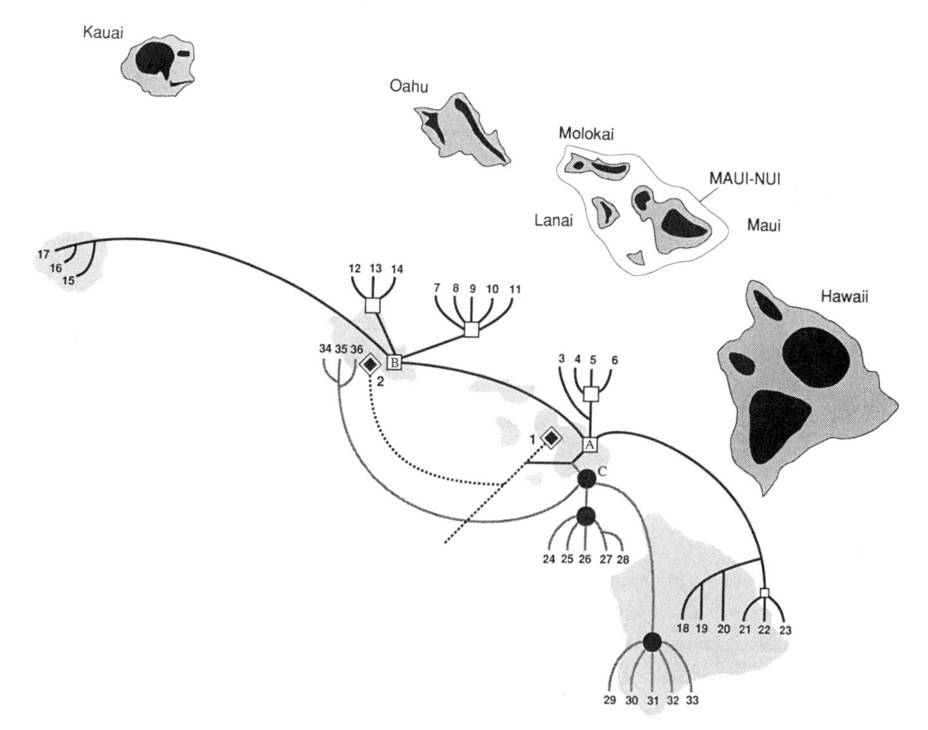

FIGURE 2. *Top*: Hawaiian archipelago showing the major highlands (black areas); Maui, Molokai, and Lanai were united into a single large island (Maui-Nui) during the Pleistocene. *Bottom*: Hypothetical phylogeny of *Prolaupala* (1,2) and *Laupala* (3-33).

This seems to agree with the speciation pattern see in some plant genera, such as *Cyrtandra* (with 129 species on Oahu alone), and in land snails, *Achatinella* (over 100 species on Oahu) (Carlquist 1980), but differs somewhat from that seen in Drosophilidae, in which a greater degree of between-island speciation may be common (Carson et al. 1970). Today 43% of the world's known tree cricket species are Hawaiian. Though this statistic can also be taken to reflect badly on the state of taxonomy elsewhere, the number of species is still impressive. Furthermore, diversity in form and ecology exceeds that of the remaining world's tree crickets; Hawaiian oecanthines have radiated into adaptive zones that elsewhere are occupied by the members of several other subfamilies. Some species have lost all traces of the sound-producing and -receiving apparatus, some have adapted to a life under bark, others hide in hollow twigs, some have become terrestrial, and still others are subterranean. From the standpoint of signaling the *Leptogryllus–Thaumatogryllus* group of genera are highly specialized members of the subfamily. The group has lost its sound-producing and -receiving structures completely and probably has adopted a largely olfactory mode of communication.

Sword-tail crickets (Trigonidiinae). Sword-tail crickets also appear to have originated on Oahu (or, less likely, on the Maui complex) between 1.5 and 2.5 million years ago and diversified into three genera (*Anaxipha*, *Prolaupala*, and *Laupala*), five subgenera, and more than 150 species. Although fewer adaptive zones were occupied as this group diversified (all species are at least partially arboreal, none has occupied caves, and no species has become mute), the speciation rate has been much higher. Unlike tree crickets, the diversity is higher in younger islands, with *Laupala* being much more diverse on the youngest island than on the oldest (Figure 2).

Ground crickets (Nemobiinae). *Caconemobius* is a genus of wingless and mute species derived from an oceanic group of crickets widespread through the Indian and Pacific Oceans and inhabiting seashores (coral and rocky outcrops and mangrove swamps) (Figure 1). In Hawaii it is represented by one species inhabiting the shores of all islands, and, probably derived from this shore species, an unknown number of inland species inhabiting lava crevices and lava tubes. Speciation in *Caconemobius* appears to be the result of the shore species repeatedly penetrating into island interiors along lava flows (either on the surface or below the surface, in crevices and perhaps in caves) and becoming isolated in different lava flows. The diversity of species seems to be inversely related to island age. The paucity of species on older islands may be due to habitat destruction—the disappearance of lava tubes and the extensive network of crevices, which become filled with erosional debris. A second genus, *Thetella*, is represented only by *T. tarnis*, a widely distributed species throughout the Pacific Ocean, from eastern Australia to Hawaii. The genus appears not to have speciated in Hawaii, perhaps due to its extraordinary vagility.

The introduced crickets. *Gryllus bimaculatus* (an Afro-Asian species) began its Hawaiian odyssey at the port in Lihue, Kauai. We know the year and even the month it must have entered. It was confined to the port area for about one year. Then, within months, it exploded to cover the entire island. It moved into all habitats except forest, on the wet as well as the dry side of the island, and now ranges from sea level to the high, often mist-covered central massif. Although it has since been reported in Oahu and Maui, it was not heard on these islands in August 1988. A second African species, *Acheta domesticus*, though it is widespread over the globe, is usually only locally distributed; in Hawaii it is known only from Honolulu and one small patch of desertlike habitat on the extreme west side of Kauai. *Teleogryllus oceanicus*, *Modicogryllus* sp., and *Trigonidomorpha sjostedti* are Asian species that are widespread in pastures, cane fields, and other disturbed habitats on all islands. Aside from *G. bimaculatus* we know little about when the last four species were introduced.

Distribution patterns. Taxonomic work shows some general patterns of distribution in numerous species, but reveals detailed patterns in only a few species (Otte and Rice 1989). Tree cricket distribution patterns vary considerably from island to island. On Kauai, *Prognathogryllus* is represented by several widespread species and a number of small peripheral isolates. On Oahu a greater proportion of the species are wide-ranging, resulting in considerable overlap among species. But the widest distributions of any *Prognathogryllus* species is still less than a large United States city, even though larger areas of suitable habitat appear to be available for colonization. The smallest distributions appear to be no larger than a few city blocks. By contrast, oecanthine species on the North American continent extend for many hundreds and even thousands of miles, and continental species introduced into Hawaii have colonized all of the Hawaiian islands. Distribution patterns of Hawaiian endemics suggest that some species originated within the overall range of the parental species and that virtually all speciation takes place within islands, much of it probably within volcanoes. Maui and Hawaii *Prognathogryllus* show how much wholly isolated cognate populations have differentiated. But what causes the species to break up into relatively small populations in an essentially continuous forest devoid of closely related (possibly competing) congeners?

THE CALLING SONG

In crickets it is possible to work directly with signals that we know the organisms themselves use to differentiate between their own and other species with whom they occur syntopically.[3] Readers may question the importance of song in the origin and maintenance of reproductive barriers among cricket species, particularly in cases in which song types are virtually indistinguishable morphologically. How do we know that song types correspond to what we normally think of as species? Differences in song within and between populations have, in all cases tested, been shown to be the result of genetic differences (Fulton 1933, 1937; Hörmann-Heck 1957; Bigelow 1960; Alexander 1957; Walker 1962). Sympatric (and syn-

[3]*Syntopic*: occupying the same microhabitat; *allotopic*: not occupying the same microhabitat even when sympatric (Rivas 1964).

topic) species virtually always have distinctive calling songs even when they are otherwise indistinguishable (Fulton 1952; Alexander 1962; Otte and Alexander 1983). No examples of within-population song varieties are known.

These songs are intense, rhythmic sounds produced by males to attract sexually responsive females (Regen 1913; Walker 1957; Alexander 1960). Experiments show that females of some species can discriminate even between slight differences among syntopic species (Walker 1957). Since songs promote homogamy, song differences have long been thought to be important either in speciation or postspeciational adjustment of populations (Alexander 1957, 1960, 1962, 1968; Walker 1957, 1962, 1974; Bigelow 1960, 1964). All existing evidence points to the conclusion that the calling song is the most important (even if not the only) premating barrier in species that communicate acoustically. The fact that songs are under both polygenic and multichromosomal control is of interest given the theory that speciation may occur faster when the number of loci involved in reproductive isolation is larger than when it is small (Nei 1976; Nei et al. 1983).

Research on cricket signaling has focused more on the sender than on the receiver. To better understand the role of the calling song in speciation it is necessary to know more about the coordination between sender and receiver and how this evolves. Given that each permanent signal change must be accompanied by a change in receivers, how do the large differences that often characterize syntopic species evolve? Small changes in male signals can probably be more easily tracked by females and could in time accumulate to yield big differences. But are large changes always the result of a gradual accumulation of small changes, or can changes be saltational? Barton and Charlesworth (1984) note that reproductive isolation usually involves many genes and that this tends to exclude macromutations as a mode of speciation and that isolation is built up in a series of small steps. Field observations show that saltational variation does occur, although in cricket songs it is possible that a large phenotypic change could result from a minor genetic one. Whether large song changes in signaling can spread and become fixed depends on whether the receiving mechanism is functionally linked to it. We note three examples of striking, within-population variation:

1. On the slopes of Puu Kukui (West Maui) we heard a male of *Prolaupala kukui* switch back and forth between his normal rate of singing and a rate half the normal speed. Only one male was heard making the switch. In the eastern United States *Oecanthus fultoni* and *O. quadripunctatus* do this too, particularly when starting up singing in the early evening (Alexander, personal communication).

2. In *Anaxipha* sp. on the east slopes of Mt. Waialialae on Kauai, we heard a number of males switching back and forth between the normal song and a song with half the pulse rate. Although the species was distributed continuously for hundreds of meters, the peculiar song variation was found in a stretch of fern about 50 m long.
3. On Hualalai volcano above Kailua-Kona we heard three males, separated by 50 m or so, singing an abnormal song. The normal song was a trill, and these three males dropped pulses to produce an erratic song.

The examples show that the pulse-timing mechanisms may be altered to fire at half the normal speed. In the latter two cases, a mutation causing the change may have made its appearance at least in the previous generation. If the neuromuscular mechanism of sending and receiving are part of a single system, sons and daughters of the affected individual can simultaneously respond to the new pulse rates. A linkage allows sender and receiver to evolve as a unit and simplifies the process of evolutionary change in a communication system. Alexander (1962) asked: "Is it possible that in some or many cases the genetic difference that causes the song difference—perhaps even the particular difference in the structure of the central nervous system itself—is exactly the same as the difference that causes the response difference?" Huber (1962) showed that some of the components necessary for production of the song pattern may also reside in the female's nervous system. More recently it has been postulated that single neurons or networks of neurons could be involved in both the generation of temporally patterned signals and recognition of these patterns, i.e., the central pattern generators and neural filters/sensory-motor templates could be one and the same (Hoy 1978; Doherty and Hoy 1985).

There is still little direct evidence for a "hard-wired" sender-receiver coupling, and the results of crossing experiments are equivocal. Crosses between *Teleogryllus* species show that male F_1 hybrids have songs with spectral and temporal characteristics that are intermediate between those of the parents *and* F_1 females respond best to hybrid songs, preferring the songs of siblings over those of reciprocal hybrid crosses (Hoy and Paul 1973; Hoy 1974; Hoy et al. 1977). But since signal generation and recognition are both under polygenic and multichromosomal control (Bentley and Hoy 1972; Hoy 1974; Elsner 1974; Elsner and Popov 1978) it is to be expected that hybridization will yield intermediate neural pattern generators and sensory templates in hybrids. Therefore, sender and receiver mechanisms may be under separate genetic control. Even if sender and receiver mechanisms are genetically independent, signaling could become coordinated through coevolution (Helversen and Helversen 1975a, 1975b;

Elsner and Popov 1978; Doherty and Hoy 1985) and sender and receiver could in time become genetically linked.

Synchronizing behavior in *Laupala* (two species from Mt. Tantalus) suggests a close linkage between sound reception and sound generation. In these two species males close to one another often synchronize perfectly and instantaneously. In *O. fultoni* there is an entrainment period during which one male gradually comes to synchronize with a neighbor.

ORIGIN OF SONG DIFFERENCES

When do species recognition differences arise? What is their role in the acquisition of species status? Do they ever cause speciation? Muller (1942), Mayr (1963), and later Nei (1976) held that all or most signal differentiation among related species is probably due to incidental differences that evolve in a period of isolation, while Dobzhansky (1937, 1970), Lack (1947, 1971), and others thought that much (though not all) differentiation is due to coevolution taking place in sympatry. Alexander (1962) surmised that new modes of singing in crickets developed in connection with keeping interspecific confusion to a minimum. It is difficult to resolve this controversy, or to decide in any given case when signal differences arose, because conclusive evidence is extremely difficult to obtain.

Divergence during isolation (independent evolution)

How much differentiation takes place while populations are isolated? It is a certainty that the signals of isolates will diverge in time due to various factors acting individually or jointly: (a) selection pressures from the physical environment (Littlejohn 1988; Ryan and Brenowitz 1985), or (b) from the bioacoustic environment (Littlejohn 1988; Ryan and Brenowitz 1985); (c) effects of natural selection on other characters (Muller 1942; Mayr 1963); (d) effects of mutation and random sampling (Nei 1976; Nei et al. 1983); and (e) effects of social selection (West-Eberhard 1983). But what is the contribution of each, and can the differences that separate sympatric species be accounted for by these factors? To determine what kinds of song differences arise outside the influences of other species is difficult, for it requires locating species that are isolated not only from cognates,[4] but from all species with signaling systems similar enough to cause interference. Hawaii provides few such cases, and no instances in which a species is isolated from all other cricket species. The nearest we can come to complete isolation is to examine species that live only with members of un-

[4]*Cognates:* two daughter species, or a mother and daughter species sharing an exclusive common ancestry.

related genera whose songs are so different that they probably interfere only minimally.

On both Maui and Hawaii a single lineage of *Prognathogryllus* has spread over each island and has formed isolated populations. The songs have diverged in some of the populations, in some cases only very slightly, in others sufficiently to suggest that species status has been reached. But the songs are not (with one exception) as different as among coexisting *Prognathogryllus*. The exception is the song of *P. waikemoi*, which is not only widely divergent, but is peculiar in having one of the most variable songs known in crickets. The possibility that the song evolved its high variability under conditions of relaxed selection is discussed in the following section.

Ecological release

Ecologists have considered the consequences of removing a species from competition with other species more than systematists have. Thus, Hutchinson (1958) speculated on how traits that act as intrinsic barriers to interbreeding might be affected by the number of related species: "When we have a great variety of allied sympatric species there must be an emphasis on very accurate mating barriers which is unnecessary where virtually no sympatric allies occur." Do organisms released from the competitive pressure of related species become less tightly adapted to a particular situation, and are elements of signals lost or do they become more variable? Lack (1971) notes cases in which fewer species on islands tend to have broader habitats and feeding stations than birds in richer mainland communities. On the basis of studies on island birds, Lack (1947) and P. Grant (1965, 1966) note that when species no long interact with related species, selection for species-identifying characteristics is relaxed and results in the loss of elaborate or distinctive morphology, evolution of much individual variation within species, and a return to simpler song types. Wilson and Bossert (1963) suggested that selection for species specificity of pheromones results in the synthesis of larger molecules, since the number of distinctive and unique forms that a molecule can take increases with its size. Available data on pheromones indicate that sex attractants, which demand high species specificity, are generally large molecules, whereas alarm and recruiting substances, which need not be highly specific, have lower molecular weights (Wilson 1971).

Among Hawaiian crickets the two species of *Prognathogryllus* that show the greatest amount of intrademic variability are isolated on Maui and Hawaii, respectively. *Prognathogryllus waikemoi* is so variable that after hearing the first three males of the species we believed that we had heard three different species. It is perhaps the most variable cricket song that has ever been recorded. In most *Prognathogryllus* (and crickets in general) temporal song rhythms are extremely constant and can easily characterize sympatric species. The song of this one species is as variable as the song of an entire genus. The only element of the song that remains constant is the pulse duration. Perhaps in this species individual pulses act as beacons of sound, while the interpulse intervals have become irrelevant.

Prognathogryllus mauka from the slopes of Hualalai and Mauna Loa on the Big Island is also highly variable; in this species, also, single pulses of sound may be produced.

It is difficult to argue from just two cases, but the complete absence of highly variable songs in multispecies communities supports the notion that song variability can be influenced by the presence of other species.

Character displacement and reinforcement (character coevolution)

To students of natural populations, divergence occurring between two groups in sympatry is inherently more interesting than divergence of geographically isolated groups. Brown and Wilson (1956) gave the name character displacement to the coevolutionary divergence occurring between populations living in the same area, which causes them to subdivide the resources of an environment on a smaller scale than they were subdivided previously. [Mayr (1963) used "character divergence" to refer to the same process, and V. Grant (1966, 1971) refers to it as the Wallace effect.] The term reinforcement has a similar meaning but was applied principally to coevolutionary enhancement of those signal differences that promote homogamy and at the same time discourage heterogamy. In spite of recent efforts to make the distinction between reproductive character displacement and reinforcement more explicit (Butlin 1985), the two concepts do not refer to two fundamentally different processes. Perhaps the terms are useful in defining when during speciation the coevolutionary adjustment occurs. If the groups involved are distinct species at the time they interact, their divergence may be called character displacement. If the groups are well on their way to becoming distinct species, the divergence should be called reinforcement.

Dobzhansky believed that character displacement commonly resulted from selection that enhanced sexual isolation—that the differences between coexisting species are "ad hoc contrivances" built by natural selection that guard against interbreeding. But the recent literature gives the impression that character displacement is rare, does not occur, or is unlikely (Paterson 1978, 1982a, 1982b, 1985; Butlin 1985; Carson 1986; West-Eberhardt 1983) and that differences are accidental by-products of divergence caused by other factors. Paterson states: "the isolating mechanism . . . which so much fuss has been made since 1935, comprise nothing more than an artificial, man-made class of 'effects.' They are, none of them, evolved adaptations to protect the 'integrity' of a species by preventing hybridization." My reading of Dobzhansky reveals no instance in which the character displacement (reinforcement) between cognates was not built upon a preexisting difference (evolved in allopatry)—a difference "visible" to, and improved by, selection.

Walker (1974) searched for examples of character displacement among United States acoustic insects (crickets and katydids) where two species in a zone of sympatry are more different than are allopatric parts of the same species. Not finding any convincing cases he asked: "Why should examples of character displacement be scarce in crickets and katydids?" He gives various possibilities:

1. Character displacement does not occur.
2. Character displacement occurs but is difficult to detect, perhaps because the process does not produce the pattern predicted by Brown and Wilson (1956). For example, a species might arise from a small peripheral population in which sympatric divergence could occur prior to the expansion of the new species into an allopatric range.
3. Character displacement occurs and produces the Brown and Wilson pattern but either not enough cases have been examined to properly assess how common it is, or the diagnostic pattern of character displacement is rapidly lost in the zone of sympatry as the trait rapidly spreads into the zone of allopatry.

Wallace (1968) and Hill et al. (1972) give an explanation of why few cases of character displacement have been discovered: a species could arise from another as a small border population; when this small population makes contact with the parent populations virtually all members of the daughter population interact directly with the parental population and are therefore under intense selection to acquire signal differences. The larger population is not under a similarly strong selection and may show no response to the presence of the smaller population. The expected result is that the small population will evolve in its entirety away from the larger population, while the larger population remains relatively unchanged. Subsequently the daughter species expands to become largely allopatric.

Cricket taxonomists and biologists (beginning with Fulton 1933) have long noticed that the songs of coexisting (synchronic and syntopic) species are always widely spaced, whereas those of related species separated by geography or season are more similar and may even be identical (Alexander 1962). The theory that coexisting species probably adjust their songs one to another in a fashion that reduces signal interference and confusion is given credence by experiments that show that acoustical signals of one insect species may interfere with those of another (Walker 1957; Hill et al. 1972; Loftus-Hills et al. 1971; Otte 1979a). Paterson (1978) argues that such differences are accidental and are not the result of one species adjusting its song to that of another.

To examine the question of whether the differences among sympatric species are accidental or not, the dispersion (or spacing) of songs in the sound space among coexisting sets of species can be compared with a random dispersion. If spacing is found to be hyperdispersed (as taxonomists have usually assumed it is), then the dispersion may be due to direct interaction among species. Calculating distances between songs in the hyperspace is complicated when songs differ in a number of parameters (pulse rate, pulse length, chirp rate, chirp length, etc.). We have attempted the procedure in Hawaiian *Laupala*, whose simple songs differ only in pulse rate (i.e., they are very nearly arranged along a single axis of the sound space). We began by plotting the pulse rate of the 46 taxa at 68 localities where two or more species coexist (Figure 3). This gives a picture of spacing among the taxa making up each community. Song differences among these coexisting taxa were then assembled into a frequency histogram. We then took pulse rates of the same 46 taxa and reassembled them at random, producing the same number of communities and community sizes as before, but not allowing any taxon to be grouped with itself. Two such random assemblages and frequencies of differences are shown in Figure 3. The results indicate, first, that observed communities of species have a greater average separation than would be expected by chance alone, and second, that observed communities show relatively few instances in which closely similar song types coexist in one community. Such hyperdispersion is probably due directly to an interaction among the species.

However, these results do not tell how communities came to be spaced in this fashion. The pattern could be due to character displacement (a coevolutionary response) or it could be due to trial and error mixing, with only species sufficiently different in pulse rate in the first place being permitted to coexist in stable communities. In the first hypothesis, incompatible species become compatible through evolutionary changes; in the second similar (incompatible) species fail to mix, whereas dissimilar (compatible) species are permitted to coexist.

Can the results of coevolution and ecological mixing be distinguished? Coevolutionary effects would have to be discounted (1) if there were no greater differences in sympatry than in allopatry in taxa that have partially overlapping ranges, (2) if clines in two or more coexisting species either did not diverge or did not follow parallel tracks, or (3) if taxa living alone

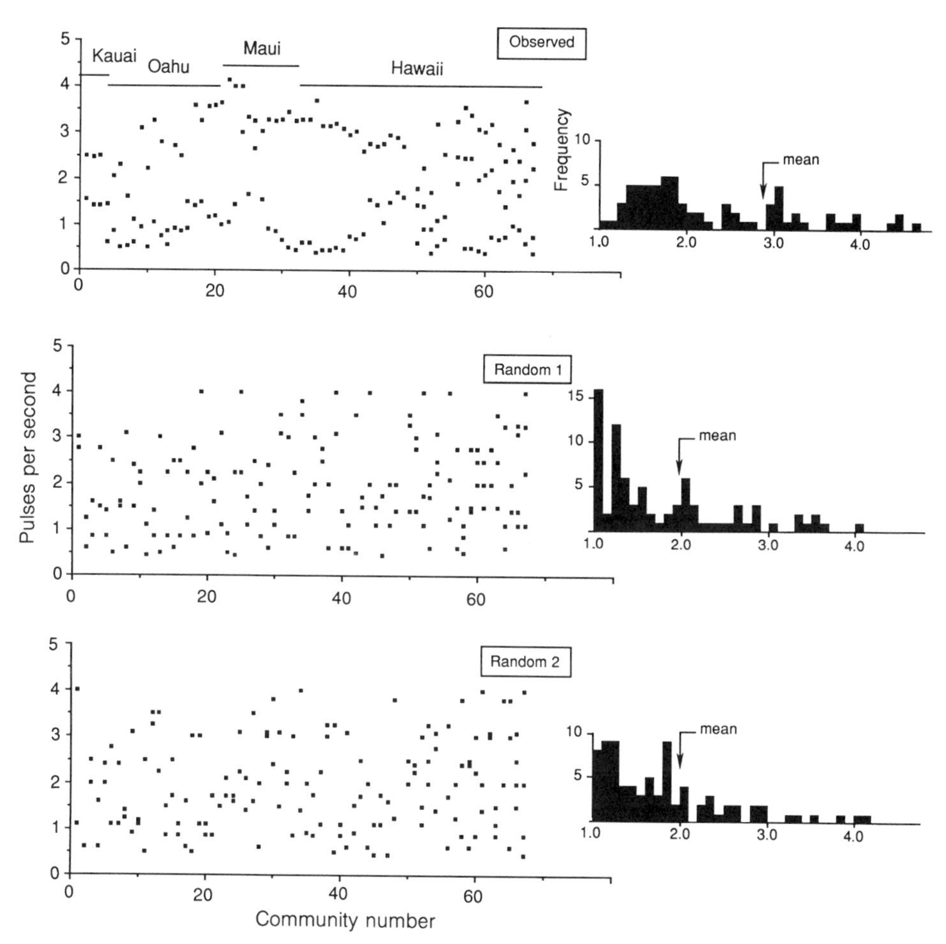

FIGURE 3. *Top*: Actual spacing of songs among syntopic *Laupala* species; each dot represents the mean pulse rate of each species at a locality. The histogram to the right plots the frequency of different degrees of spacing between neighboring song types in a community. Spacing on the (X axis) is calculated by dividing the pulse rate of a species by its next slowest neighbor (e.g., in the set of three species A = 4/second, B = 1.5/second, C = 1.4/second, spacing between A and B is 2.67, and that between B and C is 1.07). Spacing of 5.0 or greater is not shown. *Middle and bottom*: Random reassembly of song types from above into communities equal in size and in number to the above (but no song type is grouped with itself). Observed communities show a significantly lower than chance tendency for nearly identical song types to coexist.

showed the same or a lesser amount of phenotypic variation than those living in a multispecies community.

The distribution of *Laupala* species and the pattern of song within and between species is beginning to suggest that song differences are due to coevolutionary adjustments between interacting species.

Kaui. Song variation along a transect about 3 km in length (Kuilau Ridge Trail) (Figures 4 and 5) gives preliminary evidence of recent and perhaps incomplete speciation in *Laupala.* First results from a survey of songs along the transect come from the winter of 1977–78 (open histogram, Figure 4). The data indicate a single species at the upper end of the transect, and two well-differentiated species at the lower end. At two places along the transect the pulse rates of the common species were significantly slower (at zones D and E, Figure 4). Both slow-down zones were, at most, a few tens of meters wide. The presence of intermediate songs suggested that all songs were produced by one species. In my symposium presentation I suggested that the slowing down at these sites was perhaps due to the presence of an as yet undetected species. A survey conducted in the summer of 1988 revealed slower songs in precisely the same locales as in 1978 (areas 1 and 4, Figure 5). After correcting for effects of temperature (to 20°C), it became apparent that summer and winter songs were significantly different over much of the transect, especially near the upper end. The possible existence of two seasonally separated (allochronic) but syntopic species with slightly different song rates suggested an explanation for the slowing down of songs in certain places. A breakdown of the seasonal barrier in certain places could result in interbreeding and cause the slower, primarily winter species to adjust downwards relative to the slightly faster summer species. It was shown in Figure 3 that the degree of song difference displayed by winter and summer "species" does not exist among synchronic and syntopic *Laupala* species.

In December 1988 an additional 450+ males were tape recorded along the transect (Figures 4 and 5). These data confirm a seasonal difference in song, although the difference is less pronounced in areas 8 and 9 than elsewhere. Several differences between summer 1988 and winter 1988 are noteworthy (Figure 5).

(1) In area 1 the summer population of males has the appearance of a hybrid swarm (presumably between the summer and winter species). In area 4 song rates are clearly bimodal, suggesting the existence of two species, and the slow song type more or less replaced the faster type.
(2) The winter populations of males in area 1 are slower and less variable. In area 4 the absence of bimodality suggests that all songs are by a single species.

(3) There is an additional slow-down zone at area 6 that was not seen in the summer.
(4) At the lower end of the transect (area 10), two clearly defined synchronic and syntopic species are present both winter and summer in a thicket of hibiscus. But in the winter of 1988 the slow species was rare. These data also indicate a seasonality in the slow species in area 10.

A survey in December 1988 along the Powerline Road is shown in Figure 5. These songs have the same pulse rate as those across Keahua and Kawi streams. This survey was conducted to see if slower songs detected along this road in 1978 were present in 1988. Two areas with slower males were found at the same place as in 1978. The bimodal pattern suggests that there are two species here.

In summary, these preliminary data suggest the following hypothesis: there are three "species" along the Kuilau transect: a slow species at the lower end that is more common in the summer than in the winter; a summer species over the entire length of the transect (fastest song type); and a winter species (intermediate song type) completely syntopic with the summer species. Although summer and winter species are largely allochronic, separation of breeding seasons has broken down in several places where the winter form breeds all year round. Interbreeding with the summer species has caused considerable hybridization in area 1 and microgeographic parapatry in area 4. In both places the song of the rarer winter species has evolved to a slower rate (approaching the more normal separation between synchronic and syntopic *Laupala* species).

This interpretation hinges on the assumption that the differences in song are largely due to genetic differences, and the possibility that even tropical crickets can be highly seasonal. The long-term persistence of the slow down areas and the absence of geographic or ecological correlates of the slow-down areas suggests a genetic basis for the patterns.

On the matter of seasonality, we know of several *Laupala* populations on Kauai that are highly seasonal; in one of them males can be heard singing only in the summer. Since various species may be discontinuously distributed geographically in essentially continuous habitat, it seems reasonable to expect that temporal distribution may also be discontinuous and that seasonal separation of populations may also contribute to the extrinsic isolation of populations.

Regardless of what interpretation one adopts, the Kauai data give dramatic evidence of how sedentary and unchanging Hawaiian cricket populations are. In a flying species on a continent, small, locally adapted populations cannot become established since genetic novelties arising locally would be obliterated by the constant movement of individuals. It demonstrates the need to examine geographic variation in stable tropical areas on a very small scale.

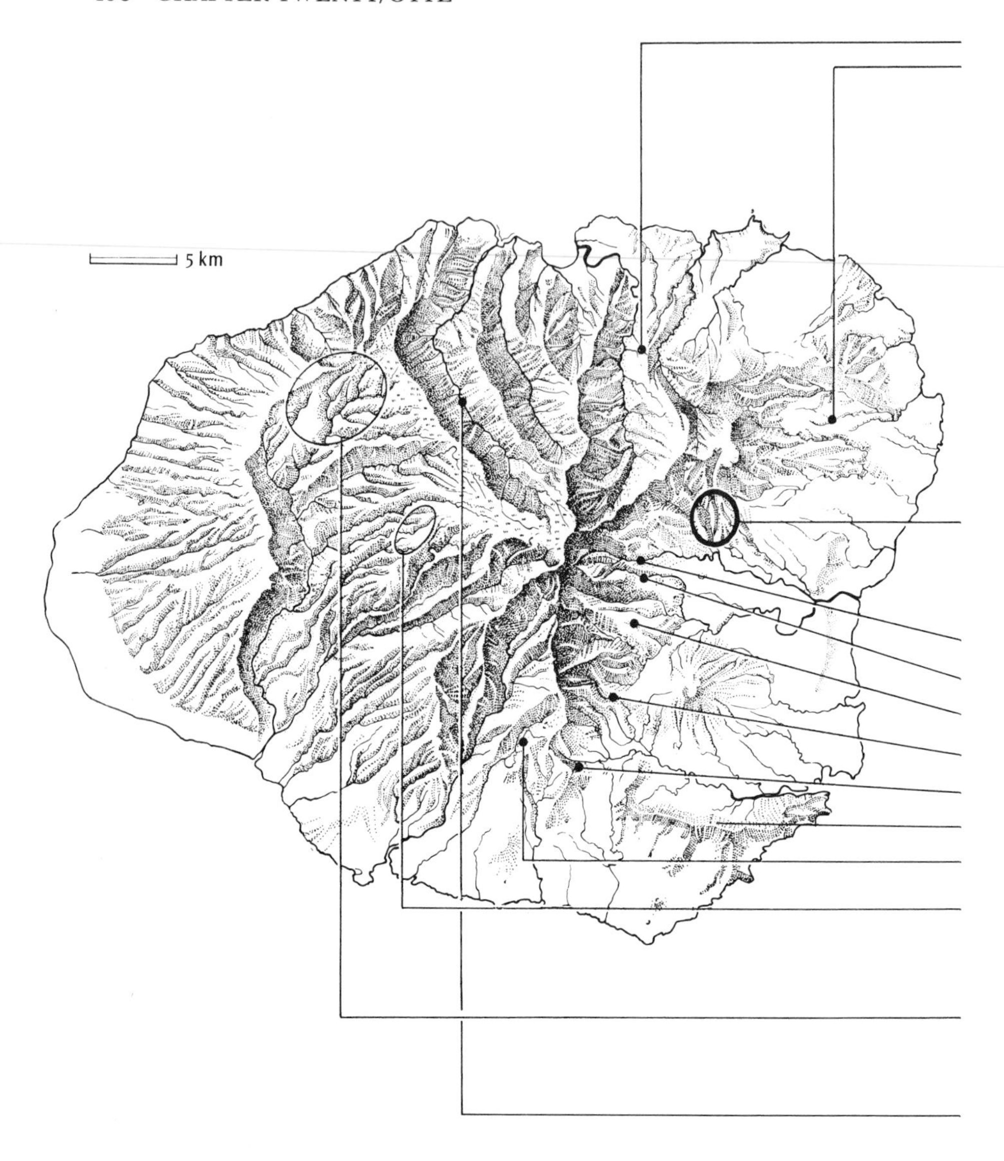

FIGURE 4. *Left*: map of Kauai. *Center*: Pulse rates among *Laupala* species corrected to 20°C. Open histograms: Winter data (Dec.–Jan. 1977–1978). Closed histograms: Summer data (Jun.–Aug.). In the transect B to C there are three "species": (1) a slow species at zone C; (2) a fast summer species (above dashed line); and (3) an intermediate primarily winter species (below dashed line). The last species has markedly slower songs at zones D and E, where it also breeds in the summer. At zone D (locality 51), intermediate songs suggest hybridization between summer

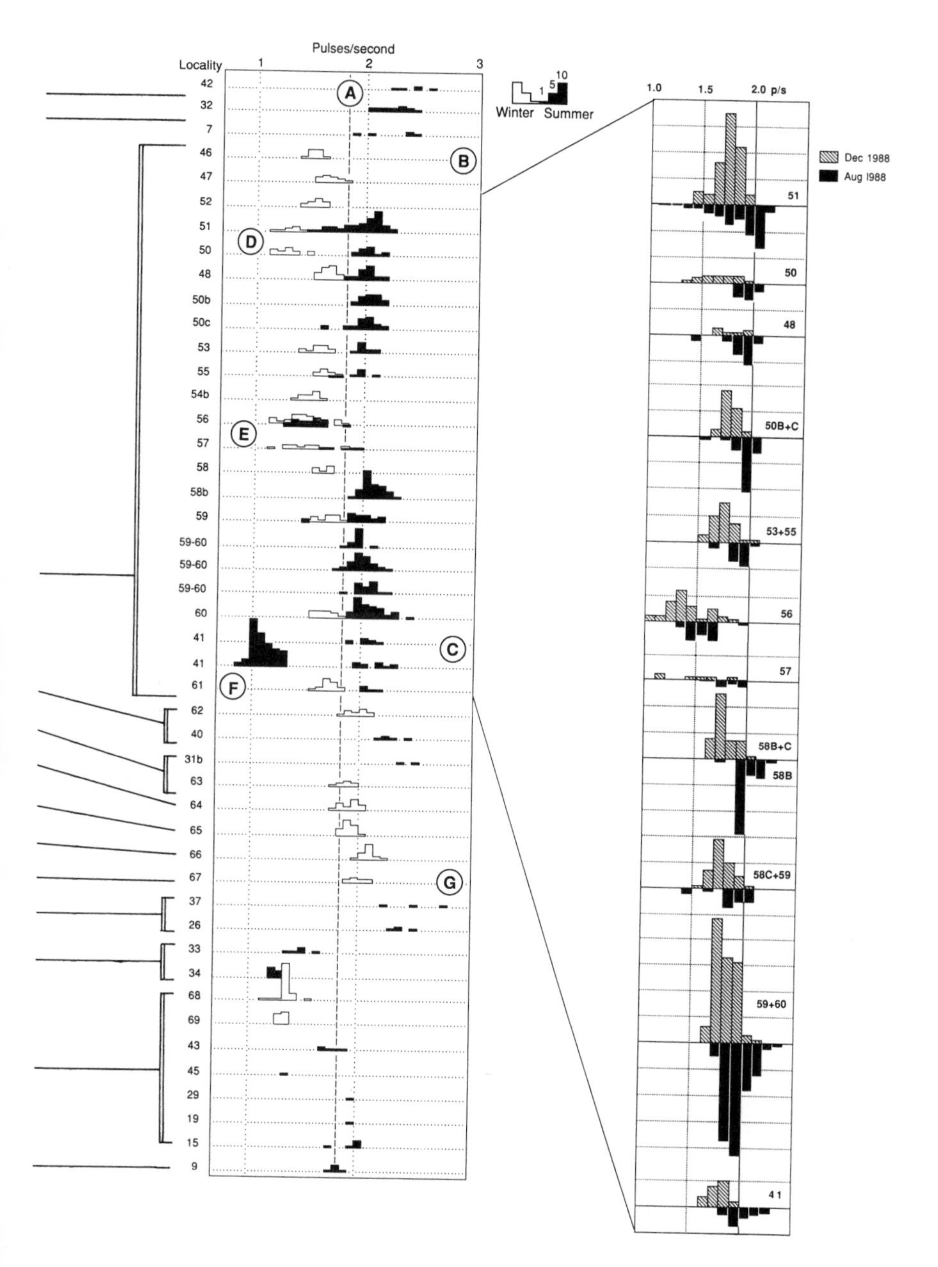

and winter species (better shown in Figure 5). At zone E the intermediate species replaces the faster species over a distance of some 60 m. The breakdown of allochrony between winter and summer species and consequent hybridization may have resulted in the evolution of slower songs in the winter species. *Far right:* Comparison of summer (Aug.) 1988 and winter (Dec.) 1988 data (corrected to 20°C) along the Kuilau transect confirms the summer–winter differences in song.

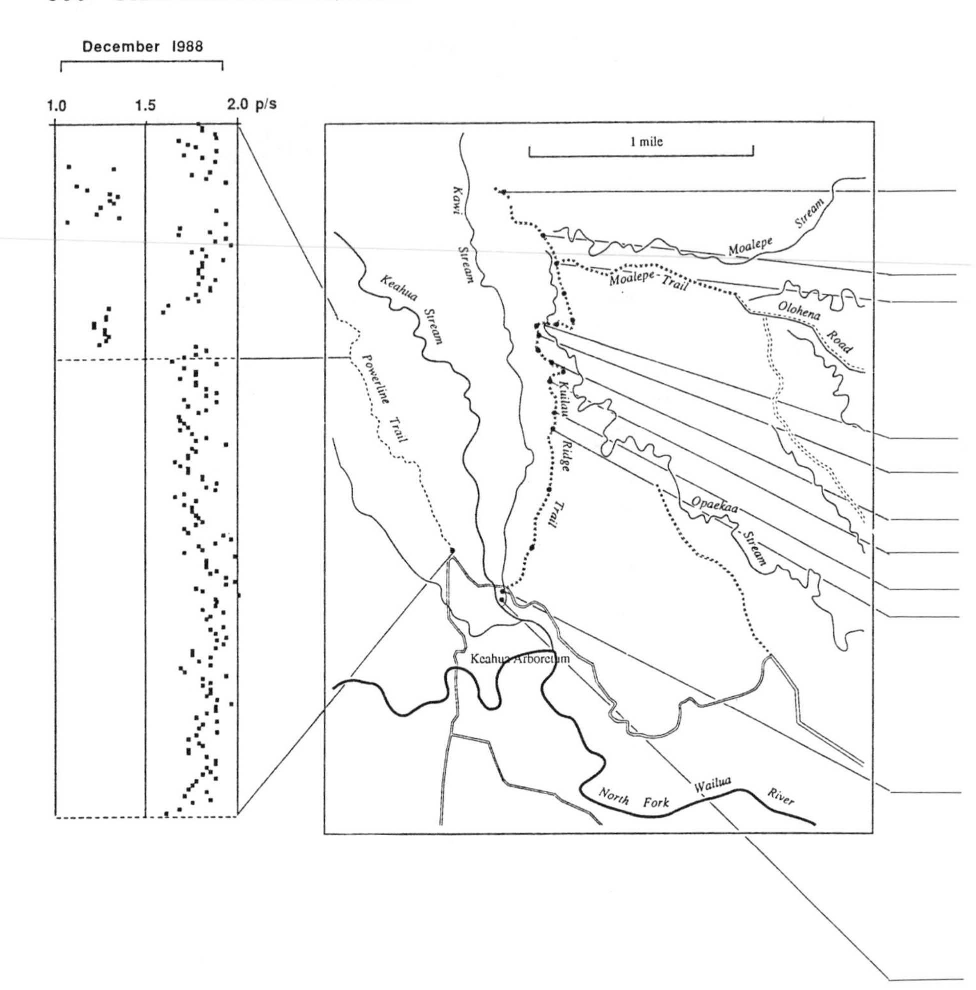

FIGURE 5. Comparison of pulse rates between summer (Aug.) 1988 and winter (Dec.) 1988 along the Kuilau Ridge and Powerline Trails. Each dot represents the pulse rate of a single male corrected to 20°C. These are arranged in the sequence in which they were recorded. The width of a recording area (1–10) is a function of the number of males recorded in that area. *Right hand scattergrams:* These diagrams show that the slow songs recorded in December coincide with those places where the winter species also breeds during the summer. It also suggests clinal variation in the winter species—for example from areas 3 to 4 and 7 to 6. There is also a greater tendency of the winter form to grade gradually into the slower song types; this may suggest a genetic continuity that is not as apparent between the common summer species and the slower song type. An exception is in area 1, where the summer species seems to grade into a hybrid population. *Far left scattergram:* Diagram shows a common species with pulse rates similar to the winter species along the Kuilau Ridge Trail and several small zones containing males

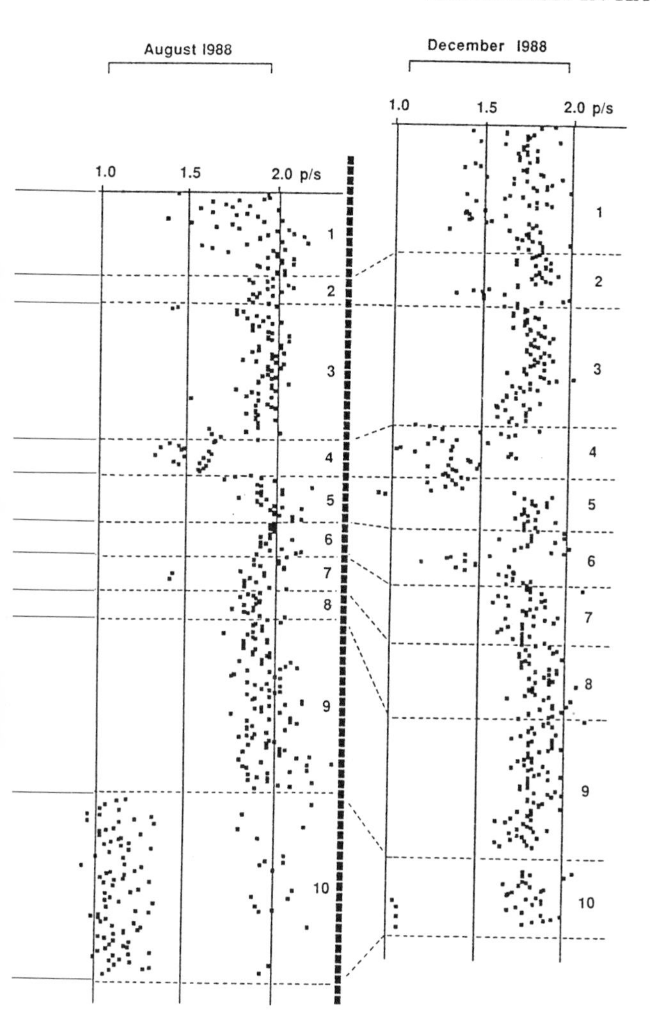

with slower songs. This pattern has also persisted for at least 10 years. The absence of intermediate songs suggests that speciation between faster and slower song types is complete. (See also Figure 4.)

Oahu. Transects on Oahu give a complicated picture in which putative species are geographically fractionated by ridges and valleys and in which the isolates show differentiation in song. Although we have tentatively grouped populations of singing males into "species" on the basis of their similarity, we do not know if isolated populations belong to the same species or not. Although detailed, meter-by-meter data on songs have not

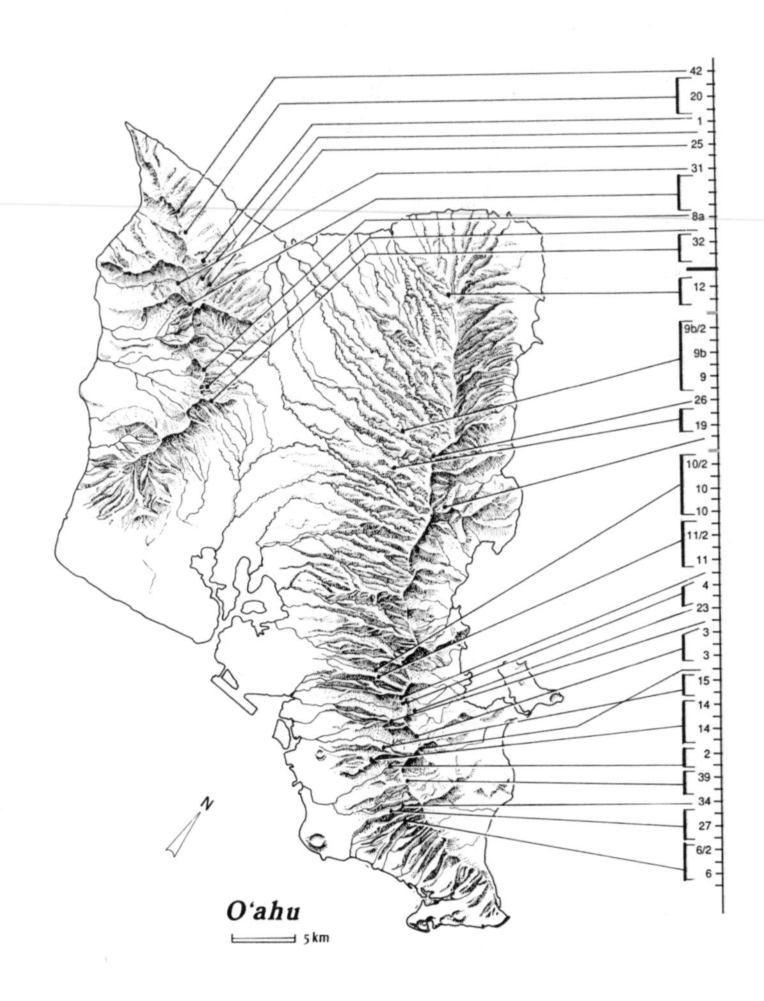

been collected on Oahu, the overall patterns suggest a few intriguing possibilities.

1. At the southern end of the Koolau Range (at A in Figure 6) the song of the species *L. palolo* becomes faster where it overlaps with *L. spisa*.
2. Also in the same region, *L. tantalis* is more rapid in the region where it overlaps with *L. nui* than farther north. However, *L. nui* does not change its pulse rate in a similar fashion.
3. In region B there appear to be parallel clines between *L. nui*(?) and *L. palolo*(?).

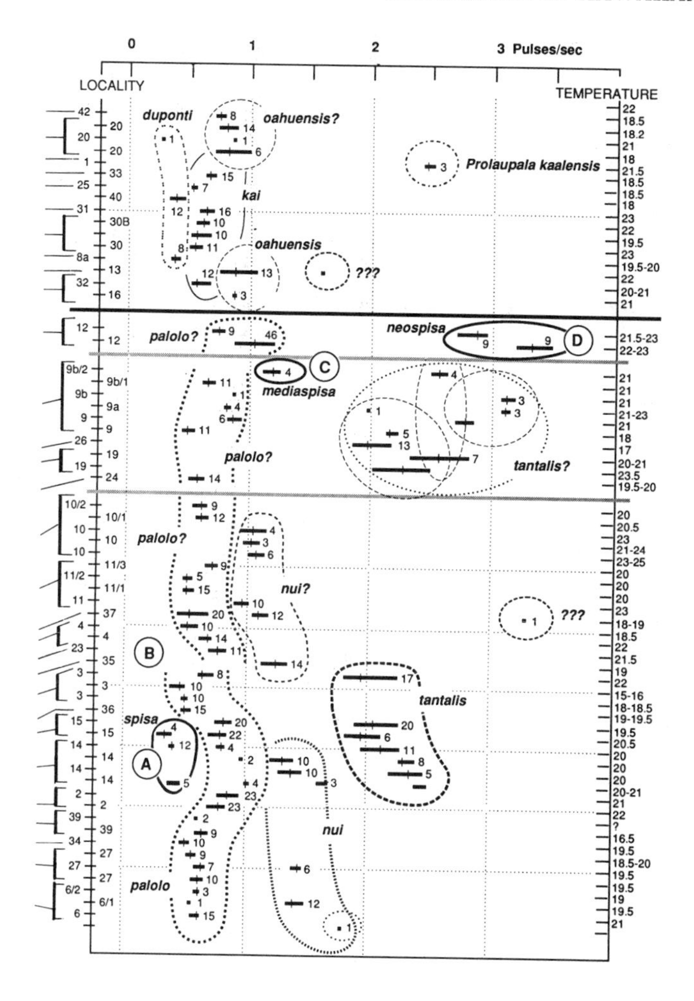

FIGURE 6. Song variation in *Laupala* species of Oahu (range, mean, and *N*). Song types are tentatively grouped into putative species on the basis of song similarity and, in some cases, morphology. Possible cases of character displacement are indicated by letters: In zone A the species *palolo* has a faster song where it is sympatric with the slower *spisa* than on either side where *spisa* does not occur. In the same region *tantalis* is faster where it is sympatric with the slower *nui*. Near zone B *palolo*(?) and *nui*(?) appear to have parallel clines. The Spisa group (*neospisa*, *mediaspisa*, and *spisa*—A, C, D) belong to a group which entered Oahu more recently than the Pacificum group (remaining species). If they are derived from a single introduction, then the group has fitted itself into three different parts of the acoustic environment.

4. The Spisa group, indicated by heavy lines at A, C, and D, is a group which probably came to Oahu from the Maui-Nui complex long after the Pacificum group (all other species) was established. The southern species (*spisa*) has the slowest song of any *Laupala* on the island, the northern species (*neospisa*) has the fastest song, and, in between, the third species (*mediaspisa*) has an intermediate song. If the three species in the group are indeed derived from a single ancestor, then the three lineages may have evolved in several different directions according to the mix of species each encountered. Ecologically, morphologically, and chromosomally these three species are indistinguishable. But the song differences are now so great that one wonders if the three song types would merge, even if no postmating barriers separated them. (But Littlejohn, 1988, argues against speciation operating in this fashion.)

Hawaii. Detailed transects on Big Island *Laupala* show the distribution of a number of putative species (Figure 7). One of the more interesting aspects of *Laupala* on Hawaii is the number of species along the windward slopes. At the north end there are two species. The number of species then increases in the central section (slopes of Mauna Kea) and decreases to one species on Kilauea and Mauna Loa. Coinciding with the greater number of species is a spreading out of the pulse rate such that the fastest and slowest species are to be found in the region of greatest diversity. A number of clines are also to be seen in this region. We can focus on some of the more interesting parts (Figure 7).

1. In sp. 1 there is a steady north to south cline in region A (localities 29X to 75) after which the pulse rate levels out. This may be due to the presence of sp. 5 in this part of the forest.
2. The songs of sp. 3 and sp. 4 are most similar to one another in region B, but that of sp. 4 appears to be displaced where it is syntopic with sp. 3. Farther north it becomes sympatric with sp. 4 and becomes faster still. At regions C and D there is evidence from songs that some hybridization occurs (Figure 8).
3. It is possible, and perhaps likely, that spp. 1-4 each represent several species. Only detailed, meter-by-meter surveys of song changes can give one a better information of numbers of species and interfaces between species. Also, it is likely that species separated by season will be discovered.

The overall impressions gained from Big Island data are, first, that the spreading out of the pulse rates is greatest where species are most numerous. Second, the divergence in signals seems to be greater between coexisting populations than between allopatric populations. Third, the example of two species with parallel clines suggests a coevolutionary response in two species.

Summary

Patterns of song variation in *Laupala* lead to these tentative conclusions:

1. Song evolution seems to be molded by the bioacoustic environment and particularly by the signals of more closely related species; character displacement is probably common in Hawaii and elsewhere, even if the principal clue—clinal character displacement in a zone of overlap—is rarely seen; if character displacement is common, then reinforcement must also be common (though less frequent than character displacement) for there must be situations in which speciation is incomplete. The phenomenon may seem to be rare only because it is difficult to see. Character displacement and reinforcement—similarly difficult to see—may initiate many of the key recognition differences we see among species. To see examples it is necessary to be in precisely the right circumstance, or arrange for it to occur.
2. The high variability seen in *Prognathogryllus* species that are isolated from all congeners tends to support the notion that the presence of other species causes songs of a species to fall within fairly narrow limits or to be fairly invariable; it is a difficult proposition to prove because there are species in Hawaii and elsewhere that have relatively invariable songs; it is noteworthy, however, that such variable songs are unknown in species that live in multispecies communities.
3. Some transects in *Laupala* show marked divergence occurring over short distances (tens to hundreds of meters). It seems likely, given the close proximity and absence of noticeable distribution gaps, that a genetic continuity exists.
4. Mayr favored an allopatric origin of behavioral reproductive barriers and yet saw an important role for ecological character displacement in speciation: "Those individuals of two overlapping species would be most favored by selection that have the least need for the resources jointly utilized by both species." He does not say why premating barriers would not be affected similarly. In *Laupala* species, the songs diverge while ecology remains unchanged.
5. Lande's (1980) notion that "In a population which is geographically isolated from the main range of a species, reproductive isolation can evolve only incidentally as a by-product of genetic divergence occurring for other reasons," is not justified insofar as the "isolated" deme may be influenced by related species other than cognates; therefore reproductive isolation evolves under the influences of the same kinds of pressures.

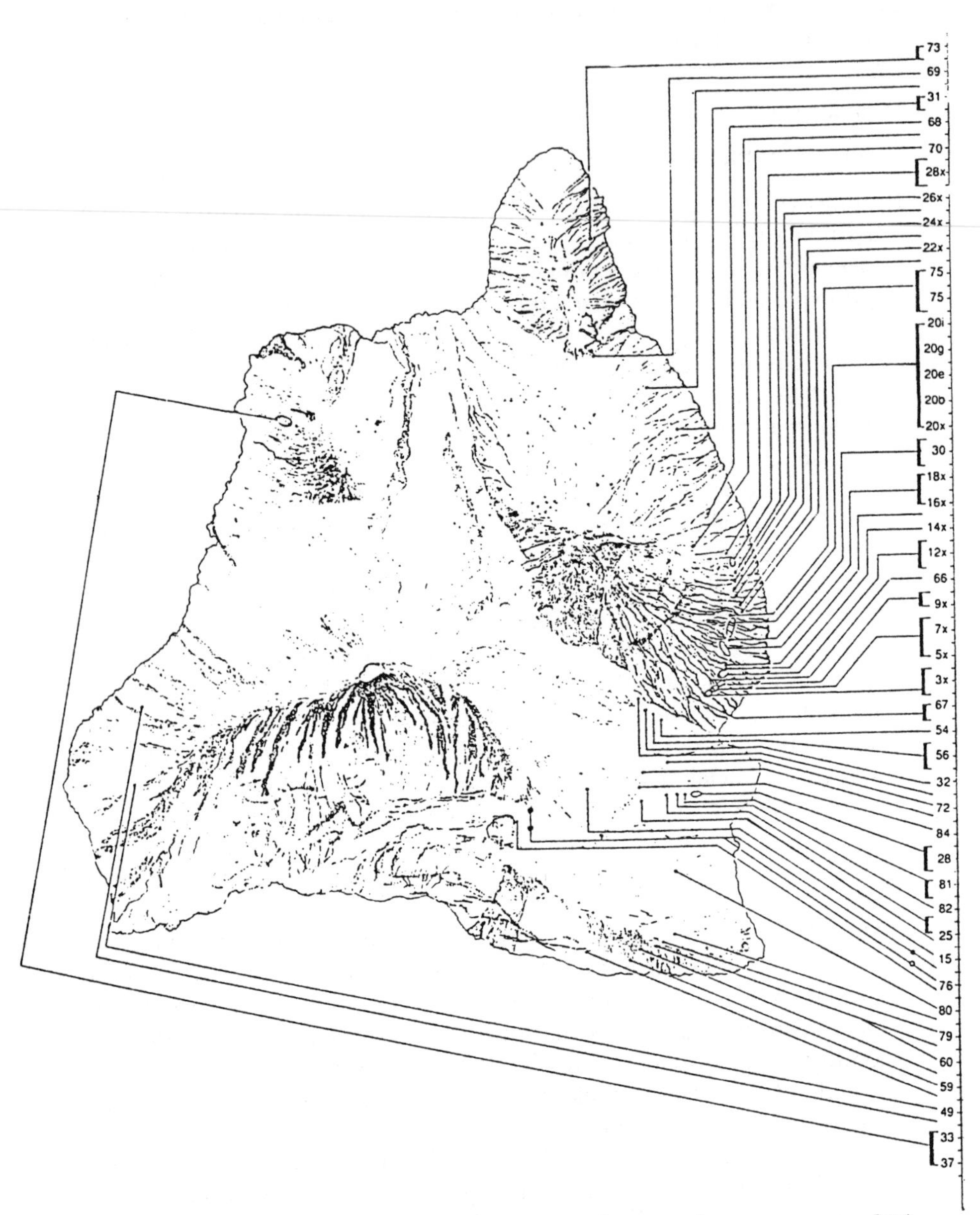

FIGURE 7. Song variation in *Laupala* species of Hawaii (range, mean, and *N*). Songs are tentatively grouped into putative species on the basis of song similarity, geographic continuity, and morphology. Sp. 1 shows a steady cline in pulse rate in zone A, perhaps due to the presence of a species with an intermediate song. The songs of spp. 3 and 4 are more similar to one another where they are allotopic than

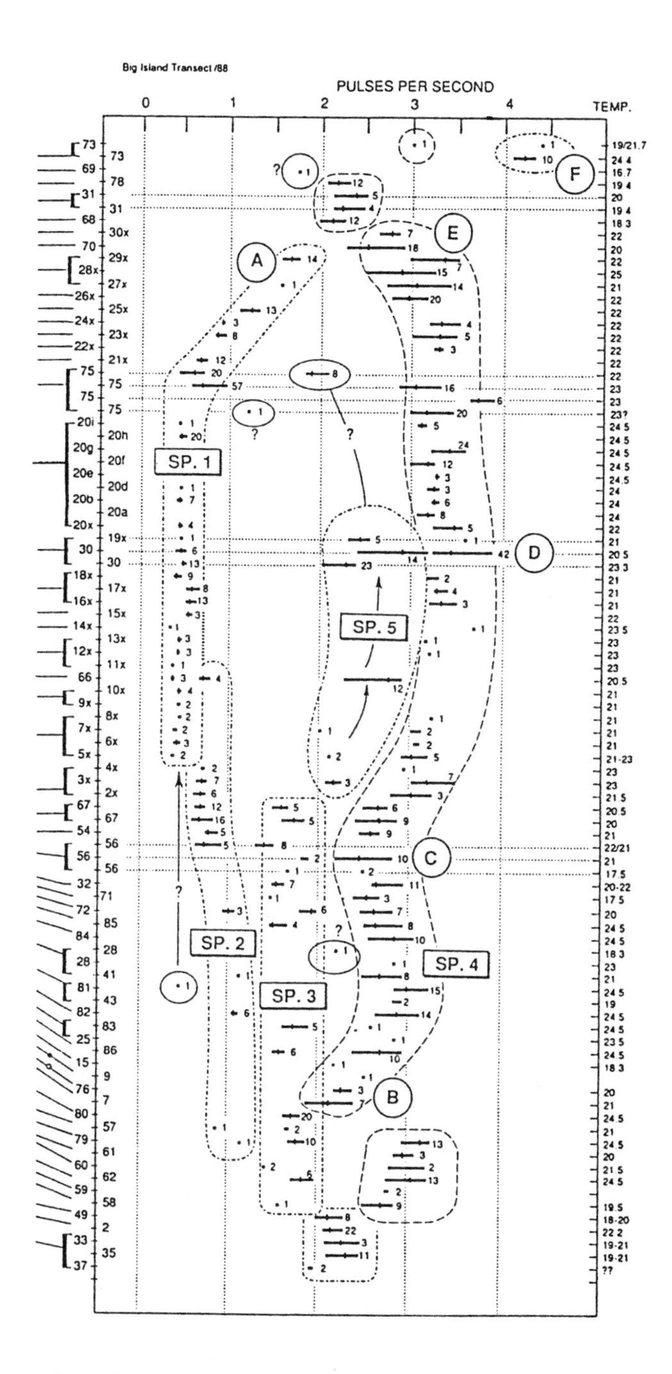

where they are syntopic. The song of sp. 4 is faster where it is sympatric with sp. 5 than at either side where sp. 5 does not occur. Some hybridization between spp. 4 and 5 appears to be occurring at D (see Figure 8).

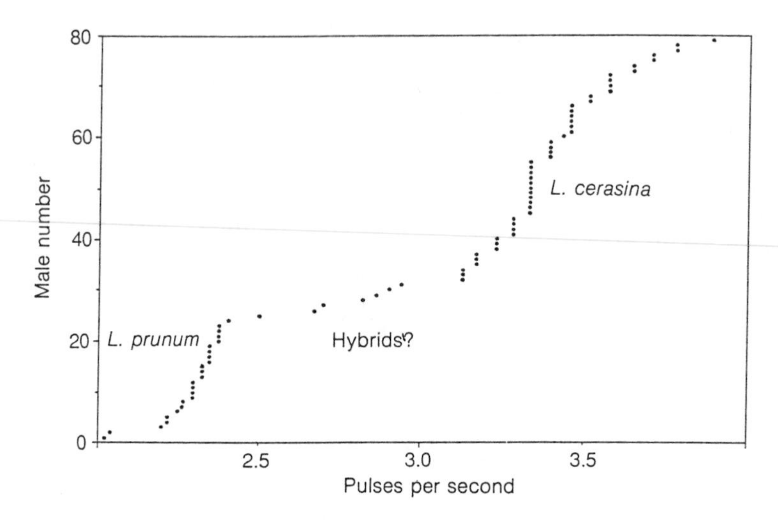

FIGURE 8. Plot of slowest to fastest males at Akaka Falls (D of Figure 7) of sp. 5 (*prunum*) and sp. 4 (*cerasina*). Intermediate songs may be of hybrids.

6. Paterson's (1978, 1982a, 1985) arguments that signal differences are due to pleiotropic effects are weakened by the fact that species that are highly divergent in songs cannot be distinguished on any other grounds—morphological, ecological, or genetic. Also, his theory does not explain why sympatric species, including siblings, are, as a rule, unambiguously distinct, whereas allopatric, geographically separate populations are not. The possibility that noncognates play an important role in molding the signals of a species is not discussed.
7. Carson's (1986) assertion that hybrid disturbances "do not serve as selective forces that elicit and strengthen premating incompatibilities, as is often assumed" probably derives from his theory that speciation is initiated and completed by founders in Hawaiian *Drosophila* (discussed under the section "Breeding Structure"). Earlier Carson and Kaneshiro (1976) had noted the importance of elaborate species-specific courtship behavior as barriers to interbreeding (see "Social Selection and Sexual Selection"). To Dobzhansky (1972b) the correlation in Hawaiian *Drosophila* between diversity in courtship and mating pattern and species diversity meant that the former was not an accidental byproduct of genetic divergence. Evidence for character displacement (and reinforcement) in this genus comes from the field (Kaneshiro 1976: "Pronounced sexual isolation is observed between the sympatric species, *heteroneura* and *sylvestris*, and further documents the concept that premating isolating mechanisms evolve as *ad hoc* products of selec-

tion between sympatric species") and has been demonstrated in laboratory studies (Knight et al. 1956; Kessler 1966).

ECOLOGICAL ISOLATION

Coexisting species usually differ in their ecological requirements if they are very closely related. This separation, though not usually believed to lead to the formation of new species, is often thought to be essential if species are to coexist syntopically. Lack (1971) in particular stressed the importance of acquiring ecological differences as a contrivance to permit coexistence: "Two species of animals can coexist in the same area only if they differ in ecology. Such ecological isolation, brought about through competitive exclusion, is of basic importance in the origin of new species, adaptive radiation, species diversity, and the composition of faunas." The notion that ecological differentiation must take place mostly during allopatry is also noted by Littlejohn (1988), who supposes that "adequate ecological differentiation in cognate populations during a disjunct allopatric phase is an essential preadaptation for coexistence" and Dobzhansky (1972b) who noted that "ecological isolation, genetically conditioned preference for different habitats, is in a sense intermediate between geographic and reproductive isolation."

Our data on Hawaiian crickets are not yet exact enough to make firm statements concerning the point at which ecological divergence takes place, relative to song divergence, but they are suggestive (Otte and Rice 1989). For the most part, the species believed to be particularly closely related often have what appear to us to be virtually identical ecologies. Carson and Stalker (1969), Heed (1971), and Carson and Kaneshiro (1976) note similar trends in *Drosophila* where subspeciation does not appear to be based primarily on ecological differences of neighboring regions, and most related species have closely similar ecologies.

In *Laupala* comingling species often show no seasonal, daily, microhabitat, or foraging differences, although slight morphological differences are often evident. Consequently we are inclined to the view that ecological differences often arise after species have come into contact. The songs and life cycles of the three *Laupala* species on Kauai (Figures 4 and 5) are noteworthy, however, in showing pronounced seasonal differences. Diurnal differences between species having nearly identical songs are seen in pairs of *Anaxipha*. Along Wahiawa Stream on Kauai two species have nearly identical songs: one species sings mainly in the daytime from under the bark of *Metrosideros* trees; the other species sings at dusk and at night in underbrush and ferns beneath the same trees. At Kaulalewelewe (West Maui) one species sings in the ferns and leaf litter close to the ground in the daytime, and is replaced by another species that sings from ferns and low tree foliage at night. The members of each pair are not cognates of each other. In both instances we cannot know without further geographic and ecological evidence whether the ecological differences coevolved in sympatry or whether preexisting differences allowed the similar song types to coexist.

How might ecological separation between cognates be achieved? We suggest the following sequence: Nearly identical species with confusingly similar signals are prone to interact and to suffer from communication interference; those insufficiently differentiated cannot coexist and displace one another to form parapatric distributions (sometimes in a mosaic pattern corresponding to their initial numbers or to differential competitiveness in a mosaic habitat). From this parapatric situation they could begin to adjust to one another ecologically, behaviorally, or both, at the zone of contact and slowly penetrate one another's ranges. Species with preexisting differences (visible to selection) can coexist at the start, but may continue to diverge if differences facilitate coexistence.

SOCIAL SELECTION AND SEXUAL SELECTION

The role of sexual selection in Hawaiian speciation has been discussed in several recent papers (Ringo 1977; Kaneshiro 1983; Carson 1986). Perhaps it was the amazing array of morphological and behavioral traits in Hawaiian insects that led *Drosophila* biologists to suspect that sexual selection has played an important role in speciation. West-Eberhard (1983) argues that social behavior may have played an important role in speciation in many groups: "Many species-specific signals heretofore attributed to selection for species recognition . . . are probably instead products of social selection."

Ringo (1977) speculated that directional intrasexual selection in Hawaiian *Drosophila* during isolation causes the sender–receiver system to evolve away from the ancestral condition through a Fisherian runaway process. Postmating barriers then evolve as pleiotropic effects of this evolution. The problem of how the runaway process is initiated, particularly given the usual tendency of signaling systems to be highly uniform within populations, is not discussed. Also, the obvious association of the great divergence of displays in Hawaiian *Drosophila* with high diversity of species (more than 60 in the Waikemoi area of Maui) suggests an interspecific cause and effect relationship. Carson and Kaneshiro (1976) noted that *Drosophila* "exhibit elaborate, species-specific courtship behavior which evolved as a major isolating mechanism among related species." Later, however, Carson (1986) suggests that the difference arose mainly during geographical isolation of the cognates.

I will attempt to evaluate the role of intraspecific relations in molding cricket acoustical signals. Because it is not always clear to me what form of sexual selection is referred to, or when sexual selection should be invoked in accounting for evolution of cricket song, I shall begin with a clarification of what I believe is meant by this term.

The sexual selection concept has been the source of much confusion (O'Donald 1980) perhaps because "the evolution of differential adaptation of the sexes is an exceedingly complex subject" whose analysis "demands integration of information and theory from several biological disciplines" (Selander 1972). Confusion existed in Darwin's own mind (although O'Donald claims otherwise). His contemporary, Wallace (1889:296) thought it was "really a form of natural selection" whose "results are as clearly deductible as those of any of the other modes in which selection acts." Huxley (1938) remarked: "None of Darwin's theories has been so heavily attacked as that of sexual selection" and he suggested that the concept be replaced by two terms: "epigamic selection" in which one sex exerts an influence on the other, and "intrasexual selection," which involves competition between members of the same sex. Mayr (1972) also reviewed this topic and concluded that many traits whose origin was ascribed to sexual selection were better explained through natural selection, whereas Selander (1972) maintained that "Strictly speaking all characters directly affecting reproduction should be attributed to sexual selection, including those that facilitate copulation and fertilization and those involved in parental care." Ehrman (1972) defined sexual selection as "all mechanisms which cause deviations from panmixia." O'Donald (1980) may not have clarified the matter by insisting that "sexual selection must be distinguished from assortative mating" and speaking of a separate "theory of assortative mating." Furthermore, in his discussion of the evolution of mating preferences he comes close to maintaining that natural and sexual selection are different phenomena: "Natural selection must now be incorporated in the model of the evolution of mating preferences in order to analyze the sequence of events in Fisher's general theory of sexual selection."

Sexual selection in its pure form was defined by Ghiselin (1974) as selection in which "the females receive no benefit from mating with one male rather than another." This definition derives from Fisher's (1930) notion that there may be situations in which it is advantageous to display in a certain way only because the opposite sex prefers individuals with such displays—even though there is no measurable benefit to preferring one trait over its alternative. This is a special mode of selection that comes about after a population is brought into a particular circumstance through ordinary natural selection. It was this mode of selection that caused Fisher to suggest his now famous (but unproven) runaway selection process.

The impression is given, therefore, that there are two kinds of selection in nature: natural and sexual. I believe this is a false dichotomy, and that the use of the sexual selection label may actually obscure the precise nature of selection pressures in any given situation. Natural selection should, of course, be distinguished from artificial selection and cultural selection (learned behavior that could affect heritable traits). Subsumed under natural selection are as many components as there are selective agents or as the imagination will allow. Temperature, light, predators, parasites, food, related species, and conspecifics are all components of selection. Each component is further divisible; conspecifics include offspring, members of the same sex, and members of the opposite sex, and each of these is further divisible by age category, by relationship, by mode of interaction, etc. Furthermore, the various components interact with one another in multifarious ways. They can act, simultaneously or sequentially, to produce or maintain adaptiveness. Since different forces tend to push populations toward different optima they often have opposing effects. The result is that all organisms have compromised phenotypes. "Runaway" sexual selection is not unique in this—predatory pressures also force organisms into adaptations that reduce their adaptedness with respect to other traits (e.g., feeding, mating, camouflage).

The term "social selection" (West-Eberhard 1983) was coined to collect the various intraspecific interactive effects under one umbrella, including those components usually referred to as sexual selection, and including culturally transmitted traits. The concept is very useful as long as its noncultural parts are not thought of as anything more than components of natural selection. A difficulty with the concept of social selection as well as with sexual selection, as given by West-Eberhard, is its emphasis on competition: "*sexual selection* refers to the subset of social competition in which the resource at stake is mates. And *social selection* is differential reproductive success due to differential success in social competition, whatever the resource at stake." All symbiotic relations are a potential source of selection.

Searcy and Andersson (1986) have correctly emphasized that selection pressures from within a species (or deme) and those exerted by other species on a signal system are not mutually exclusive. It had been proposed, for example, that speciation may initiate the evolution of sexually selected traits (Trivers 1972; Emlen 1974) in this way: When two populations that have recently speciated begin to interact, selection is expected to favor females that can distinguish between the two kinds of males. According to one such scheme, females are least likely to make mistakes *by preferring the appropriate extreme* of an available sample. The process is initiated by ordinary natural selection, with the ultimate selection pressure for change coming from the presence of another species; since female preference leads to increased female fitness, there is no need to invoke sexual selection. However, the stage is now set for the operation of sexual selection in its pure form, for a point is reached at which females no longer make mistakes, but because the mode of choosing males (preferring the most extreme males) persists, male signals continue to evolve along the same track. In the absence of opposing selection pressures, such an "overshoot" could cause a much greater divergence between species than what is necessary for females to discriminate.

The role of social selection (noncultural variety) in signal divergence in crickets, as I see it, is depicted in Figure 9. Prior to a change in the species mix, the signaling system is at equilibrium with only small mutual adjustments occurring between the sexes. The influence of one sex on the other is mostly of a stabilizing kind, with major changes in one sex being inhibited when there is a failure in communication. (Rapid changes in the species mix such that a communicative system does not have enough time to adjust to a given mix would have the same effect.) A relatively permanent change in the species mix, however, can suddenly change the signaling environment. Species with closely similar signals must suddenly cope

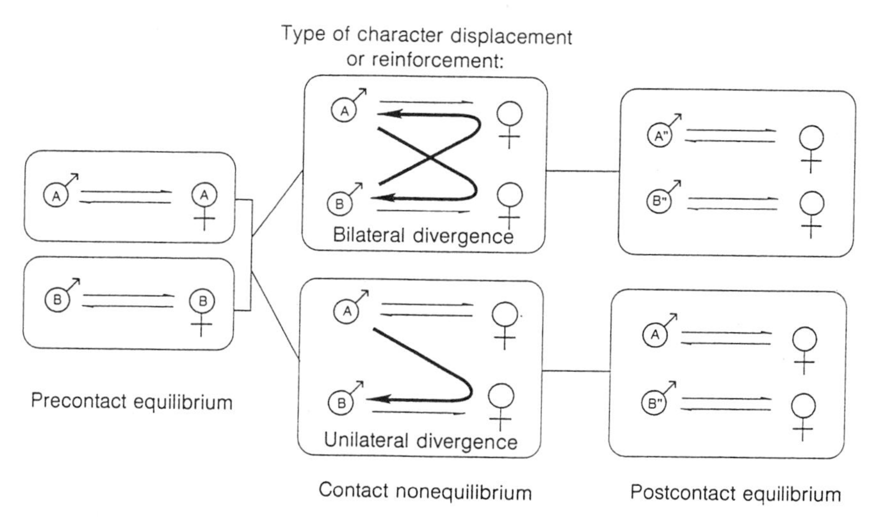

FIGURE 9. *Precontact equilibrium*: Under stable conditions male-female signal systems are likely to be at equilibrium and generally stabilizing. *Contact nonequilibrium*: As two previously separated species begin to mix (interact) the initial influence is the presence of males of another species; females, in adjusting to the new acoustic environment, then bring about a change in the signal code of their own species. Both species may adjust to one another (top center) or, if species A is common and species B rare, B may evolve much more than A. *Postcontact equilibrium*: In the upper case A has evolved into A′, B has evolved into B′. In the lower case A remained unchanged while B has evolved into B′.

with signal noise, signal interference, and the risk of fruitless reproductive interactions. As species begin to mix ecologically, the initial influence is simply the presence of males of another species. Now, if the females adjust to the new acoustic environment by changing the rules of mate selection, they will also bring about a change in the signal code of their own species. Thus, we see that social selection can be seen to be a consequence of the disturbed equilibrium brought on by the introduction of extraneous males.

Whether one or both species change (bilateral divergence versus unilateral divergence) depends on the relative numbers of the two species, with females of the minority species being under greater pressure to change than females of the majority. Sawyer and Hartl's (1981) model predicts that natural selection for mating aversion is more likely to evolve first in the population that is rarer in the sympatric zone.

Our data on crickets do not indicate that sexual selection plays an important part in signal *divergence* between species. This mode of selection is not expected to have the same outcome as does signal coevolution. For example, it predicts that diverging taxa (populations, demes, or neighborhoods) will evolve largely independently of other taxa; it does not predict hyperdispersion of signals in the signal space; nor does it predict that coexisting populations of two species will differ more than allopatric populations or that clines will be parallel or divergent, or, in the absence of other species, that the signaling will become more variable.

Finally, the social/sexual selection theory of divergence cannot readily be distinguished from adaptive peak shifts predicted by Sewell Wright's (1932) shifting balance process. Under that process, small, semiisolated populations can diverge from one another and settle on different adaptive peaks, neither one necessarily superior to the other. Intraspecific sexual selection predicts the same outcome from a different mechanism.

The feasibility of a runaway process in crickets (in accounting for large differences between males) depends on how females perceive and respond to males. There are several alternatives: First, a female could choose males on the basis of an absolute standard (neural template of Hoy et al. 1977) regardless of its distribution in the population (*absolute preference model* of Lande 1982). Under this model we would expect female preferences to evolve to a point at which females can no longer perceive those males whose signals overlap with another species. Concomitantly, male signals will shift in accordance with the new female preferences, and cause sending–receiving systems to evolve nonoverlapping distributions, but then stop.

Second, a female could choose a male after comparing him to others or in relation to the population mean (*relative preference model*). Under this model the reception of signals is less constrained, and, provided it is consistent in direction with respect to the mean, could cause signals to diverge well beyond the point of overlap. Whether the choice is absolute or relative depends in part on what kind of signal is transmitted. Signal intensity would be particularly subject to change through a relative preference and evolve continuously in one direction. But female crickets choose males principally on the basis of pulse repetition rate. The mechanism is probably one in which the females perceive (or are wired to respond to) only a narrow range of pulse rates. The alternative to having a narrow perception/response window is having a wide window within which the female must "decide" among the items she perceives. The latter capability would probably require a more complicated neural apparatus that is able to compute a variety of rates and then make a decision among them. Is it possible for female crickets to pay attention to more than one song at a time or be able to remember one song as she listens to another? Such feats would seem to be required under the relative preference model.

If females do not choose by comparison, there would seem to be less opportunity for Fisherian runaway selection and songs will not diverge much beyond what is necessary for recognition. However, I noted previously the occurrence of species whose signals have diverged *well beyond* the point of overlap (these can also be seen in Figures 6 and 7). The possible explanations for such divergence are (1) the signals diverged by the runaway process after an initial episode of ordinary selection; (2) signaling and receiving could be structurally and physiologically coupled, allowing large changes in males to be tracked by females; large changes would be favored relative to small changes if they made a greater improvement in communication; (3) if signals always interfere with one another in some degree, no matter how different they are, then selection for divergence may continue indefinitely toward some asymptote; (4) the differences were acquired in a selective regime that no longer exists.

OTHER FACTORS INFLUENCING SPECIATION

Distance from source areas

The most conspicuous feature of Hawaii's biota is not its overall diversity, which is low by tropical forest standards, but the fact that some of its genera are enormously rich in species. This genus-poor and species-rich balance can be traced to one overriding factor—its great distance from, and tenuous connection with, its source areas. The distance causes few colonists to contribute to the eventual makeup of a saturated community of species. In islands closer to continents a larger number of colonists contribute to the final biota and result in the lower ratio. The high ratio in Hawaii is the single most important factor that contributes to the impression of a high speciation rate. Consequently it may be premature to conclude that speciation in the Hawaiian biota is faster than tropical forests elsewhere.

Island barriers

The usual kinds of impediments to gene flow have probably operated to fragment crickets. But since most speciation seems to occur within islands, we are naturally interested in just what kinds of within-island barriers are involved. Typically, Hawaiian islands pass through a number of developmental stages that affect the kind, number, and duration of geographic and ecological barriers and therefore must affect speciation in an important way. But the effect of topography on differentiating Hawaiian biotas has never been explored in detail. Here I will mention several stages in this evolution, and suggest how spatial heterogeneity changes and how habitat configuration might affect the opportunities for differentiation.

Lava fields, kipukas, and succession. In the early island-building stages relatively gentle eruptions produce extensive lava beds underlaid by lava tubes (tube-like caves varying in diameter from centimeters to tens of meters). Even before ecological succession produces forest on the surface, these lava fields can be colonized by *Caconemobius*. Within decades of an eruption, forest begins to cover the lava on windward slopes, thereby altering *Caconemobius* habitat and becoming habitable to a variety of forest-inhabiting species (such as *Laupala* and *Anaxipha*). Meanwhile, the lava tubes also become optimal for *Caconemobius* habitation as roots from the overlying forest penetrate into them to provide nourishment (Howarth 1973; Gurney and Rentz 1978). These lava caves descend the slopes in parallel and are largely isolated from one another. Cavernicolous populations of *Caconemobius* may be established mainly through founder events, and, being highly linear, may be particularly susceptible to isolation by distance.

During this phase forests may be continuous over large sections and be intersected here and there by a set of parallel lava beds originating as separate streams descending from an erupting rift zone. These flows incinerate the forest and leave behind isolated forest patches of various sizes and shapes. These are, therefore, classic vicariance events that may (1) divide a large forest patch into two, (2) divide a section of forest into a series of narrow, parallel forest strips, and (3) flow around higher spots, leaving behind various habitat islands (kipukas) that can vary in area from many hectares to a few square meters. Although the forest is maximally subdivided during this stage, subdivision of the biota lasts only for as long as it takes succession to restore the forest (decades on windward slopes, tens of decades or more on drier leeward slopes) and there may not be sufficient time for speciation. Phenotypically, crickets isolated in kipukas cannot be seen to be differentiating. Indeed, an important role for kipukas in Hawaiian speciation has not yet been demonstrated.

Forest and gulch. Following cessation of the eruptive phases, volcanoes pass through a long period (tens to hundreds of thousands of years) during which their lower slopes are covered by a continuous cloak of rain forest. Impediments to lateral movement (along the slopes) would seem to be minimal, but increase gradually as numerous parallel gulches develop. Our data show that although crickets readily cross gulches in the early stages, speciation is already much in evidence during this phase (see Character Displacement and Reinforcement); this must mean either that there are imperceptible extrinsic barriers, or that speciation takes place in their absence.

Plateau and valley. As gulches deepen and become united through erosion, the land comes to consist of plateaus and deep V-shaped valleys. Habitats then consist of forest and bog on the plateaus, valley bottom forest, and whatever vegetation can grow on the steep valley walls. The transitions between habitats constitute major ecological barriers. The habitat in the valley floor is very narrow, causing cricket populations to have a linear configuration.

Ridge and valley. Erosion eventually removes many of the plateaus and produces a set of narrow, parallel ridges. Upland forest and valley bottom forest now both force crickets to assume a strongly linear configuration. Later, as the ridges are worn down, forest trees return to the slopes and the linearity of distributions is diminished.

Ridge and alluvial plain. In the final stages as ridges erode away, alluvium collects on the valley floor and produces flat plains along the rivers. Thus, ridge top species continue to be linearly distributed whereas valley bottom species assume a less linear configuration again.

These various geological and phytogeographic features may act in combination, particularly with climatic stability, to facilitate speciation. The role of forests in restricting movement of insects that have become flightless is probably important.

Climatic stability

The stability and persistence of Hawaiian habitats permit small populations to survive for long periods; small effective population sizes, in combination with the linear distributions produced by the ridge and valley system in turn could influence the fine-tuning of isolated and semiisolated populations to local conditions. Extreme humidity and temperature sensitivity in crickets (in comparison to continental crickets) (Otte and Rice 1989) are probably a result of an unchanging environment that no longer selects for animals that can cope with climatic fluctuations, but instead selects for animals that fit into very particular ecological situations. Migration in such species must be selected against because the animals are either unlikely to survive conditions during or after migration, or to be competitive in new species mixes. The higher tolerance of continental crickets can also be seen in the much greater ecological amplitude of the introduced crickets compared to the Hawaiian endemic crickets. Because of the evolution of low mobility, the Hawaiian habitats become coarse-grained, with successive generations all inhabiting the parental patch type. Such constancy facilitates the development of precise relationships among syntopic species, on a scale not possible in areas in which migration is constantly mixing the species. An even spacing of songs is most likely to be seen in such a situation.

Have the continental crickets introduced to Hawaii begun to respond to the stable climate? In the case of *Teleogryllus oceanicus* we suspect so. Although long-winged individuals are produced, no migration by night-flying individuals (of the sort that is common in Australia) has ever been detected by entomologists who pay attention to such things. Whether the most recent immigrant, *Gryllus bimaculatus* (discussed in Background), now loses its tremendous migratory tendencies should be watched carefully over a period of decades. We suspect that countless individuals must already have been lost at sea as this species made its way down the archipelago. Similar net losses every generation are bound to select against migration and constitute the first stage in selecting for a more speciation-prone organism.

Reduced tendency to migrate may already have evolved in the grasshopper *Schistocerca nitens*, a highly migratory species in the continental United States and Central America. This species, introduced within the past 50 years, is found in large numbers at south point on Hawaii Island where it inhabits grassland and shrubs immediately adjacent to sea cliffs. At this locality a constant wind blows over the land and out to sea. Unlike the normal condition in *Schistocerca*, flushed grasshoppers here fly for short distances and only into the wind. I observed two individuals near the sea cliff that were being blown out over the sea; both attempted to return to land for at least 10 minutes before they were blown away to sea.

Breeding structure

Evolution within lineages and splitting of lineages are probably strongly influenced by the breeding structure of a species (Wright 1940, 1978; Barton and Charlesworth 1984). The breeding structure is defined by several population parameters that can influence the rate at which evolution occurs and the probability that a species will become fractionated. It is influenced by effective population size, proximity of neighboring demes, migration rates among demes, and the configuration of groups of demes. The breeding structure in Hawaiian crickets must be profoundly influenced by various geological events that affect habitat shape (see Island Barriers), as well as the abundance and strength of extrinsic barriers. It also is influenced by the mobility of the animals, which is probably a result of habitat stability (see Climatic Stability).

Founder effects

Remote isolates are given great evolutionary importance by proponents of speciation through founder effects (Mayr 1963; Carson 1971; Templeton 1981) but far less importance by proponents of the shifting balance process (Wright 1940, 1978; Barton and Charlesworth 1984; Barton, this volume) or by supporters of vicariance speciation.

The mode of divergence leading to speciation in Hawaiian *Drosophila* has been the subject of much thought and modeling. What proportion of speciation events is initiated by founder events as compared to vicariance events? From the fact that cognate species of *Drosophila* often live on different islands, Carson proposed a mode of speciation in which single foundresses play the most important role in speciation in this group. But vicariance events would seem to be a viable alternative in some Hawaiian groups, such as *Achatinella* (snails), *Cyrtandra* (plants), and crickets. Vicariance events in *Drosophila* cannot be ruled out; phylogenetic data based on chromosome banding patterns show that cognates or near-cognates frequently live on one island (Carson et al. 1970; Carson and Kaneshiro 1976).

The existence of sharp clines (differentiation with distance) in crickets raises the possibility that disruptive selection is an additional mechanism leading to speciation. The operation of such selection in the Hawaiian fauna remains unexplored, for it requires extremely detailed examination of geographic variation in phenotypic and genotypic traits.

There can be little doubt that founders have been important in speciation, but controversy surrounds the question: How important are the properties of the founders themselves in causing speciation? Carson (1971, 1973, 1982) proposed that a new species is founded by a single gravid female of an aberrant genotype that manages to reach another island. Such an unusual female is the product of a population flush arising from particularly favorable conditions that allow all sorts of unusual genotypes to survive and propagate, which, under normal conditions, would not survive. The theory assumes that the important genotypic change has already occurred in the founder on the island of origin. Templeton (1980b) supports Carson's theory, stating that the founder effect may be important in Hawaiian *Drosophila*, but does not believe that the process is important in *Drosophila* generally.

Wright (1978) criticized Carson's "flush-founder" theory on the grounds that it requires a double miracle: first, the essential change leading to speciation is an extremely improbable event, and second, this recombinant must then be isolated on a new island by another event of extremely low probability. As an alternative, Wright (1978) proposed the following:

> Under a more typical operation of the shifting balance process, the founder may be a fairly ordinary representative of its species that gives rise to a new species on reaching another island by the accumulation of new interaction systems under the exceptionally favorable conditions for the process there, especially while its local populations are sparse. Under this view, speciation by isolation in different valleys on the same island is not ruled out . . . The trial and error process among numerous small local populations by which control continually shifts from old selective peaks in sets of gene frequencies to new ones, each clinched by local mass selection and spread through the species by interdeme selection, does not require the double miracle of origin of a specific difference by recombination, and isolation of the same individual by being carried to another island. This is not to deny that if the founder is a wide deviant it may give a good start toward speciation in producing a population that comes at once under control of a new selective peak.

Time effects

Older areas are more likely to have more species than younger ones, other things being equal. But in Hawaii other things are not equal: in crickets and other groups, diversity seems to depend greatly on the order of colonization (on time since colonization). The effect of time is confounded by the fact that habitat heterogeneity also changes as islands age. Also, some components of heterogeneity (i.e., subdivision of forest by lava) do not decrease as islands age. *Caconemobius* and *Laupala* diversity is inversely related to island age and geological heterogeneity (being highest in the youngest islands); in the former the greater diversity is probably the result of the disappearance of the lava habitats that the members of this genus inhabit; in the latter case the diversity is probably related to where the genus originated in the islands.

SUMMARY

1. Hawaii's enormous species-to-genus ratio is a consequence of its great distance from its source areas.
2. The endemic cricket fauna appears to be derived from four original colonizing species, belonging to three subfamilies.
3. Cladistic and distribution patterns suggest that most speciation takes place within islands.
4. Hyperdispersion of songs among syntopic *Laupala* species indicate competitive interactions among species having similar songs.
5. *Laupala* systematics indicates that speciation can occur in continuous forest, before the development of major barriers (ridges and valleys).
6. Patterns of geographic variation in song among sets of species (greater divergence in signals between coexisting populations than between allopatric populations) suggest the existence of signal coevolution among syntopic species.
7. Hyperdispersion, character displacement, and increased variability of songs in isolated species are not predicted by the social or sexual selection models that emphasize within-species selective effects.
8. The high variability seen in *Prognathogryllus* species isolated from all congeners (along with the absence of variable songs in species that live in multispecies communities) supports the notion that the presence of other species causes songs of a species to fall within fairly narrow limits.
9. The influence of one species on another probably extends well beyond the point at which they no longer interbreed; a selective influence of one species on another may cease only when the songs no longer interfere with one another; therefore theories of reproductive character displacement should take into account a somewhat broader network of interactive effects and not just relations between cognates.
10. In the signal coevolution model the influence of one sex on the other is probably mostly of a stabilizing kind, with major changes in one sex being inhibited when there is a reduction in communication. But a change in the species mix introduces a fundamental new element in the signaling environment: the presence of males of another species. Changes in female responses to the new environment may cause reverberations through an entire intraspecific signal system.

11. Character displacement is probably extremely common in Hawaii and elsewhere, but would be difficult to detect if its principal clue—character displacement in a zone of overlap—is short lived. If character displacement is common, then reinforcement must also be common, for there must be many instances of interaction between taxa that have not completed speciation.
12. The possibility that signal differences in Hawaiian crickets are due to pleiotropic effects is weakened somewhat by the fact that species that are highly divergent in songs are barely distinguishable or not distinguishable on other grounds—morphological, ecological, or chromosomal.
13. Closely related Hawaiian species tend to have nearly identical ecologies. Ecological differentiation among syntopic species may often come long after behavioral reproductive barriers have been perfected.
14. The stability and persistence of Hawaiian habitats probably permit local populations to survive for long periods and adapt to local conditions, and selects against emigration. Forests may have a large effect in restricting the movement of flightless crickets, breaking a species up into many local demes and thus affecting the population breeding structure.
15. Eruptive and erosive forces in Hawaii promote the development of linear habitats (lava islands, lava tube, ridges, valleys) that could increase the opportunities for isolation by distance.
16. Because of low mobility, habitats tend to be course grained, such that successive generations all inhabit the parental patch type. Such constancy may facilitate development of precise relationships among syntopic species, on a scale not possible in areas in which migration is constantly mixing species.
17. The existence of sharp clines (differentiation with distance) in *Laupala* raises the possibility of disruptive selection as an additional speciation mechanism.

ACKNOWLEDGMENTS

This research was initiated with grants from the American Philosophical Society and the National Geographic Society and was supported by a National Science Foundation grant (NSF 022264). Without the help of Robin Rice, my colleague and co-author of *The Hawaiian Crickets*, this research would not have been possible. I am also grateful for the assistance of the Bishop Museum, Honolulu, and for the help given to me by Gordon Nishida, Frank Howarth, John Strazanac and David Funk. The many helpful suggestions made by Richard Alexander, Thomas Walker, William

Cade, Nick Barton, Godfrey Hewitt and Dan Howard have improved the manuscript considerably.

LITERATURE CITED

Alexander, R. D. 1957. The taxonomy of the field crickets of the eastern United States (Orthoptera:Gryllidae:Acheta). Ann. Entomol. Soc. Am. 50:584–602.

Alexander, R. D. 1960. Sound communication in Orthoptera and Cicadidae. Pp. 38–92 in: W. Lanyon and R. Tavolga (eds.), *Animals Sounds and Communication*. AIBS Publication 7.

Alexander, R. D. 1962. Evolutionary change in cricket acoustical communication. Evolution 16:443–467.

Alexander, R. D. 1968. Life cycle origins, speciation, and related phenomena in crickets. Quart. Rev. Biol. 43(1):1–41.

Barton, N. H., and B. Charlesworth. 1984. Genetic revolutions, founder effects, and speciation. Annu. Rev. Ecol. System. 15:133–164.

Bentley, D. R., and R. R. Hoy. 1972. Genetic control of the neuronal network generating cricket (*Teleogryllus, Gryllus*) song patterns. Anim. Behav. 20:478–492.

Bigelow, R. S. 1960. Interspecific hybrids and speciation in the genus *Acheta* (Orthoptera, Gryllidae). Can. J. Zool. 38:509–524.

Bigelow, R. S. 1964. Song differences in closely related cricket species and their significance. Aust. J. Sci. 27(4):99–102.

Brown, W. L., and E. O. Wilson. 1956. Character displacement. System. Zool. 5:49–64.

Butlin, R. K. 1985. Speciation by reinforcement. Pp. 84–113 in: J. Gosalvez, C. Lopez-Fernandez, and C. Garcia de la Vega (eds.), *Orthoptera 1*. Fundacion Ramon Areces, Madrid.

Carlquist, S. 1980. *Hawaii: A Natural History*. Pacific Tropical Botanical Garden, SB Printers, Inc., Honolulu.

Carson, H. L. 1971. Speciation and the founder principle. Stadler Symp. 3:51–70.

Carson, H. L. 1973. Reorganization of the gene pool during speciation. Pp. 274–80 in: N. E. Morton (ed.), *Genetic Structure of Populations*. University of Hawaii Press.

Carson, H. L. 1982. Speciation as a major reorganization of polygenic balances. Pp. 411–433 in: C. Barigozzi (ed.), *Mechanisms of Speciation*. Liss, New York.

Carson, H. L. 1986. Sexual selection and speciation. Evol. Process Theory:391–409.

Carson, H. L., D. E. Hardy, H. T. Spieth, and W. S. Stone. 1970. In: M. K. Hecht and W. C. Steere (eds.), *Essays in Evolution and Genetics*. Appleton-Century-Crofts, New York.

Carson, H. L., and K. Y. Kaneshiro. 1976. *Drosophila* of Hawaii: Systematics and ecological genetics. Annu. Rev. Ecol. System. 7:311–346.

Carson, H. L., and H. D. Stalker. 1968. Polytene chromosomes relationships in Hawaiian species of *Drosophila*. Pp. 335–380 in: M. R. Wheeler (ed.), *Studies in Genetics: IV*. Publication 6918, University of Texas, Austin.

Carson, H. L., and H. D. Stalker. 1969. Polytene chromosomes relationships in Hawaiian species of *Drosophila*. Pp. 87–94 in: M. R. Wheeler (ed.), *Studies in Genetics: IV*. Publication 6918, University of Texas, Austin.

Dobzhansky, Th. 1937. *Genetics and the Origin of Species*. Columbia University Press, New York.

Dobzhansky, Th. 1970. *The Genetics of the Evolutionary Process*. Columbia University Press, New York.

Dobzhansky, Th. 1972a. Genetics and the races of man. Pp. 59–86 in: B. Campbell (ed.), *Sexual Selection and the Descent of Man, 1871–1971*. Aldine, Chicago.

Dobzhansky, Th. 1972b. Species of *Drosophila*. Science 177:664–669.

Doherty, J. A., and R. Hoy. 1985. Auditory behavior in crickets: Some views of genetic coupling, song recognition and predator detection. Quart. Rev. Biol. 60:457–472.

Ehrman, L. 1972. Genetics and sexual selection. Pp. 104–135 in: B. Campbell (ed.), *Sexual Selection and the Descent of Man, 1871–1971*. Aldine, Chicago.

Elsner, N. 1974. Neuroethology of sound production in gomophocerine grasshoppers. I. song patterns and stridulatory movements. J. Comp. Physiol. 88:67–102.
Elsner, N., and A. V. Popov. 1978. Neuroethology of acoustic communication. Adv. Insect Physiol. 13:229–275.
Emlen, J. M. 1973. *Ecology: An Evolutionary Approach*. Addison-Wesley, Reading, MA.
Fisher, R. A. 1930. *The Genetical Theory of Natural Selection*. Clarendon, Oxford.
Fulton, B. B. 1937. Experimental crossing of subspecies of *Nemobius* (Orthoptera:Gryllidae). Ann. Entom. Soc. Am. 30:201–207.
Fulton, B. B. 1933. Inheritance of song in hybrids of two subspecies of *Nemobius fasciatus* (Orthoptera). Ann. Entom. Soc. Am. 26:368–376.
Fulton, B. B. 1952. Speciation in the field cricket. Evolution 6:283–295.
Ghiselin, M. T. 1974. *The Economy of Nature and Selection in Relation to Sex*. University of California Press, Berkeley.
Grant, P. 1965. Plumage and the evolution of birds on islands. System. Zool. 14:47–52.
Grant, P. 1966. The coexistence of two wren species of the genus *Thryothorax*. Wilson Bulletin 78:266–278.
Grant, V. 1966. The selective origin of incompatibility barriers in the plant genus *Gilia*. Am. Nat. 100:99–118.
Grant, V. 1971. *Plant Speciation*. Columbia University Press, New York.
Gwynn, D. T., and G. K. Morris. 1986. Heterospecific recognition and behavioral isolation in acoustic Orthoptera (Insecta). Evolutionary Theory 8:33–38.
Helversen, D. von and O. von Helversen. 1975a. Verhaltensgenetische Untersuchungen am akustichen Kommunikationssystem der Feldheuschrecken (Orthoptera, Acrididae). I. Der gesang von Arbastarden zwischen *Chorthippus biguttulus* und *Ch. mollis*. J. Comp. Physiol. 104:273–299.
Helversen, D. von and O. von Helversen. 1975b. Verhaltensgenetische Untersuchungen am akustichen Kommunikationssystem der Feldheuschrecken (Orthoptera, Acrididae). II. Das Lautschema von Arbastarden zwischen *Chorthippus biguttulus* und *Ch. mollis*. J. Comp. Physiol. 104:301–323.
Hill, K. G., J. J. Loftus-Hills, and D. F. Gartside. 1972. Pre-mating isolation between the Australian field crickets *Teleogryllus commodus* and *T. oceanicus* (Orthoptera:Gryllidae). Aust. J. Zool. 20:153–163.
Hörmann-Heck, S. 1957. Untersuchungen über den Erbgang einiger Verhaltensweisen bei Grillenbarstarden (*Gryllus campestris* L. × *Gryllus bimaculatus* De Geer). Z. Tierpsychol. 14:137–183.
Hoy, R. R. 1974. Genetic control of acoustic behavior in crickets. Am. Zool. 14:1067–1080.
Hoy, R. R. 1978. Acoustic communication in crickets: A model system for the study of feature detection. Fed. Proc. 37:2316–2323.
Hoy, R. R., J. Hahn, and R. C. Paul. 1977. Hybrid cricket auditory behavior: Evidence for genetic coupling in animal communication. Science 195:82–84.
Hoy, R. R., and R. C. Paul. 1973. Genetic control of song specificity in crickets. Science 180:82–83.
Howarth, F. G. 1973. The cavernicolous fauna of Hawaiian lava tubes, I. Introduction. Pacific Insects 15:139–151.
Huber, F. 1962. The central nervous control of sound production in crickets and some speculations on its evolution. Evolution 16:468.
Hutchinson. 1958. Concluding remarks. Cold Spring Harbor Symp. Quant. Biol. 22:415–427.
Huxley, J. 1938. Darwin's theory of sexual selection and the data subsumed by it, in the light of recent research. Am. Natur. 72:416–433.
Kaneshiro, K. Y. 1976. Ethological isolation and phylogeny in the plantibia subgroup of Hawaiian *Drosophila*. Evolution 34:437–444.
Kaneshiro, K. Y. 1983. Sexual selection and direction of evolution in the biosystematics of Hawaiian Drosophilidae. Annu. Rev. Entomol. 28:161–78.
Kessler, S. 1966. Selection for and against ethological isolation between *Drosophila pseudoobscura* and *Drosophila persimilis*. Evolution 20:634–645.

Knight, G., A. Robertson, and C. Waddingon. 1956. Selection for sexual isolation within a species. Evolution 10:14–22.
Lack, D. 1947. *Darwin's Finches*. Harper, New York.
Lack, D. 1971. *Ecological Isolation in Birds*. Harvard University Press, Cambridge, MA.
Lande, R. 1980. Genetic variation and phenotypic evolution during allopatric speciation. Am. Natur. 116:463–79.
Lande, R. 1982. Rapid origin of sexual isolation and character divergence in a cline. Evolution 36:213–23.
Littlejohn, M. J. 1988. The retrograde evolution of homogamic acoustic signaling systems in hybrid zones. Pp. 613–635 in: B. Fritzsch (ed.), *The Evolution of the Amphibian Auditory System*. John Wiley, New York.
Loftus-Hills, J. J., M. J. Littlejohn, and K. G. Hill, 1971. Auditory sensitivity of the crickets *Teleogryllus commodus* and *T. oceanicus*. Nature New Biol. 233:184–185.
Mayr, E. 1963. *Animal Species and Evolution*. Harvard University Press, Cambridge, MA.
Mayr, E. 1972. Sexual selection and natural selection. Pp. 87–104 in: B. Campbell (ed.), *Sexual Selection and the Descent of Man, 1871–1971*. Aldine, Chicago.
Muller, H. J. 1942. Pp. 185–268 in: J. S. Huxley (ed.), *The New Systematics*. Syst. Assoc., London.
Nei, M. 1976. Models of speciation and genetic distance. Pp. 723– 65 in: S. Karlin and E. Nevo (eds.), *Population Genetics and Ecology*. Academic Press, New York.
Nei, M., T. Maruyama, and C. I. Wu. 1983. Models of evolution of reproductive isolation. Genetics 103:557–79.
O'Donald, P. 1980. *Genetic Models of Sexual Selection*. Cambridge University Press, Cambridge.
Otte, D. 1979a. Chorusing in Syrbula. Cooperation, interference competition or concealment? Entomol. News 90:159–165.
Otte, D. 1979b. Historical development of sexual selection theory. Pp. 1–18 in: M. S. Blum and N. A. Blum (eds.), *Sexual Selection and Reproductive competition*. Academic Press, New York.
Otte, D., and R. D. Alexander. 1983. The Australian crickets (Orthoptera:Gryllidae). Acad. Nat. Sci. Philadelphia Monogr. 22:1–477.
Otte, D. and R. C. A. Rice. 1989. *The Hawaiian Crickets: Systematics, Biogeography and Speciation*. Special Publications, Academy of Natural Sciences of Philadelphia, in press.
Paterson, H. E. H. 1978. More evidence against speciation by reinforcement. South African J. Sci. 74:369–371.
Paterson, H. E. H. 1982a. Perspective on speciation by reinforcement. South African J. Sci. 78:53–57.
Paterson, H. E. H. 1982b. Darwin and the origin of species. South African J. Sci. 78: 272–275.
Paterson, H. E. H. 1985. The recognition concept of species. Pp. 21–29 in: E. S. Vrba (ed.), *Species and Speciation*. Transvaal Museum Monograph No. 4, Transvaal Museum, Pretoria.
Regen, J. 1913. Über die Anlockung des Weibchens von *Gryllus campestris* L. durch telephonisch übertragene Stridulationslaute des Mänchens. Pflügers Arch. Ges. Physiol. 155:193–200.
Ringo, J. M. 1977. Why 300 species of Hawaiian *Drosophila*? The sexual selection hypothesis. Evolution 31:694–96.
Rivas, L. R. 1964. A reinterpretation of the concepts "sympatric" and "allopatric" with a proposal of the additional terms "syntopic" and "allotopic." System. Zool. 13(1):42–43.
Ryan, M. J., and E. A. Brenowitz. 1985. The role of body size, phylogeny, and ambient noise in the evolution of bird song. Am. Natur. 126:87–100.
Sawyer, S., and D. Hartl. 1981. On the evolution of behavioral reproductive isolation: The Wallace effect.Theoret. Pop. Biol. 19:261–273.
Searcy, W. A., and M. Andersson. 1986. Sexual selection and the evolution of song. Annu. Rev. Ecol. System. 17:507–533.
Selander, R. K. 1972. Sexual selection and dimorphism in birds. Pp. 180–230 in: B. Campbell (ed.), *Sexual Selection and the Descent of Man, 1871–1971*. Aldine, Chicago.

Templeton, A. R. 1980a. Modes of speciation and inferences based on genetic distances. Evolution 37:317–319.
Templeton, A. R. 1980b. The theory of speciation via the founder principle. Genetics 94:1011–1038.
Templeton, A. R. 1981. Mechanisms of speciation—a population genetic approach. Annu. Rev. Ecol. System. 12:23–48.
Trivers, R. 1972. Parental investment and sexual selection. Pp. 136–179 in: B. Campbell (ed.), *Sexual Selection and the Descent of Man, 1871–1971*. Aldine, Chicago.
Wallace, A. R. 1889. *Darwinism*. Macmillan, London.
Wallace, B. 1968. *Topics in Population Genetics*. W. W. Norton, New York.
Walker, T. J. 1957. Specificity in the response of female tree crickets (Orthoptera, Gryllidae, Oecanthinae) to calling songs of the males. Ann. Entomol. soc. Am. 50:626–636.
Walker, T. J. 1962. Factors responsible for intraspecific variation in the calling songs of crickets. Evolution 16:407–428.
Walker, T. J. 1974. Character displacement and acoustic insects. Am. Zool. 14:1137–1150.
West-Eberhard, M. J. 1983. Sexual selection, social competition, and speciation. Quart. Rev. Biol. 58:155–183.
Wilson, E. O. 1971. *The Insect Societies*. Belknap Press, Harvard University Press, Cambridge, MA.
Wilson, E. O., and W. H. Bossert. 1963. Chemical communication among animals. Recent Prog. Hormone Res. 19:673–716.
Wright, S. 1932. The roles of mutation, inbreeding, crossbreeding and selection in evolution. Proc. 6th Int. Congr. Genetics 1:356–366.
Wright, S. 1940. Breeding structure of populations in relation to speciation. Am. Natur. 74:232–248.
Wright, S. 1978. *Evolution and the Genetics of Populations*. University of Chicago Press, Chicago.

CHAPTER 21

THE GAUGE OF SPECIATION: On the Frequencies of Modes of Speciation

John D. Lynch

INTRODUCTION

Although no one has proposed explicitly the frequencies of different types of speciation in any organism group, many biologists would agree to the consensus view provided by Bush (1975), to the effect that the most common mode of speciation is the peripheral isolates model (also called founder, model 1B, model II allopatric, and peripatric), whereas a less common but nevertheless significant mode is the vicariant model (also called dichopatric, large subdivisions, model 1A, and model I allopatric) and the least common is sympatric speciation. Bush suggested that sympatric speciation might be restricted to phytophagous and zoophagous parasites and parasitoids, particularly insects, and, by implication of evoking numbers of insect species, might be relatively common (Bush 1975:352). Bush's paper preceded the current view of parapatric speciation (Endler 1977; Barton and Hewitt 1985). Parapatric speciation is not discussed here for reasons given below.

Most biologists also would concur with E. Mayr in arguing that sympatric speciation may be so rare as to not occur, at least among sexually reproducing vertebrates. Most students of speciation have directed their attention to the problems of genetics under each proposed model as an indirect effort to assess biological reality or have evaluated models by studying putative cases and attempting to corroborate ecological predictions.

In the recent literature of biology, there are several models of speciation (see Endler 1977; White 1978; Wiley 1981), but those models are of rather different applicabilities.

Paterson (1981) made a distinction between models of speciation with his characterization of them as either Class I or Class II models where Class I models were those having some empirical support whereas Class II models lack "reliable, critical, observational data . . . which compel acceptance of the view that [such] speciation . . . has actually occurred" (Paterson 1981:113). Paterson's position essentially is that only the peripheral isolates model is a Class I model (Macnamara and Paterson 1984:315–316).

Although nearly all observations under any particular model can be explained, we need to keep in mind what are the observations at our disposal. Distributions of organisms are objective observational data (sizes of distribution areas, distances separating them, amounts of geographic overlap between areas) as are systematic data [distributions of diagnostic features (apomorphies) of populations, species, and supraspecific clades] whereas speciation models and hypotheses of relationships are inferences based on various bodies of evidence.

Distributions span a continuum from sympatric to parapatric to allopatric (dichopatric) and there is a tendency to associate each distributional mode with a speciation mode bearing the same adjective. My failure to discuss explicitly parapatric speciation here derives from ambiguous claims as to the expected patterns of relationships for a group of organisms undergoing parapatric speciation. Wiley (1981:55) argues that its phylogenetic pattern is indistinguishable from that of what I here refer to as vicariant speciation (Wiley's allopatric model I). Cracraft (1982:415) argues that the phylogenetic pattern for parapatric speciation is identical to that for what I here refer to as peripheral isolates speciation. As long as contradictory positions are derivable from the model, there is no merit in distinguishing it for my purposes. Parapatric speciation has much support (Endler 1977; Barton and Hewitt 1985) and my neglect of it is a consequence of finding Cracraft's (1982:415) arguments persuasive.

This exploration grew out of a dissatisfaction with apparent positions adopted by major spokespersons in discussions about speciation. People appear to be arguing the merits of positions from perspectives that require that we ignore a large subset of the relevant data. Few observers find the footnote acknowledgment that dispersal happens satisfactory in some of the more aggressively argued papers of vicariance biogeographers. By the same token, the division of speciation models into Class I and Class II models by Paterson (1981, 1982) seems benign but carries an equal aggressiveness borne of a conviction that he knows what is correct and what is not. Paterson's (1981:113) view is explicitly and exclusively promoting peripheral isolates speciation. He assigns models "to Class I if the following question elicits an unconditionally positive answer: Do reliable, critical, observational data exist which compel acceptance of the view that speciation in accordance with this mode has actually occurred?"

Many of the questions posed in ecology, evolutionary biology, and systematics depend directly or indirectly on our assumptions about models of speciation. Some questions are rational only under certain models. Yet, aside from arguments from authority or from genetics, little light has been brought to bear on how frequent such models might have been (or are) in generating the few to several millions of species that populate this planet. Although more than three modes of speciation are posited (Bush 1975; Endler 1977; White 1978), I will focus on only three, inasmuch as those three have occupied more attention than all of the others and because the remaining variants may not be distinguishable, except in theory, from these three (Mayr 1982:1121; Wiley 1981:51–56).

Much of the early debate is summarized in the various works of Mayr (1942, 1954, 1963, 1982) whose major effect was to promote the idea that allopatric speciation accounts for all (or nearly all) speciation events. It must be emphasized that the bulk of Mayr's argument applies equally to peripheral isolates and vicariant modes of allopatric speciation; only in recent years has his writing shifted to a selective defense of the peripheral isolates model. Nonetheless, there has been a persistent argument that sympatric speciation is not only possible (in an unspecified way), but is possibly common (deriving in part from observations of numerous sympatric species difficult to distinguish from one another). The two allopatric modes discussed here may be simply extremes of a continuum rather than discrete processes (Mayr 1982; Wiley and Mayden 1985), but the distinction is important as it focuses on modes of evolution. The recent growth of vicariance biogeography (Nelson and Platnick 1981) has promoted a different model of allopatric speciation (vicariant) than that currently favored by Mayr and most other biologists (peripheral isolates). Implicitly (Wiley 1981), or explicitly (Eldredge and Cracraft 1980:307; Paterson 1981, 1985; Masters et al. 1984), the evolutionary model termed punctuated equilibrium is bound to the peripheral isolates model of Mayr (1954). Mayr's speciation model was never seen as contrary to gradualism, but the vicariance model may have served as the gradualist's model (at least heuristically—see most textbook graphic representations of speciation). Although it is not my intent to force any one-for-one correspondences between speciation models and modes of evolution, the two are almost certainly intertwined.

This chapter explores a method of identifying putative cases of vicariant, peripheral isolates, and sympatric speciations using data for vertebrate animals. The necessary information is (1) reasonably accurate knowledge of the distribution areas of the taxa under study and (2) the identification of the sister elements separated by speciation. The most rational sources of such information are corroborated hypotheses of relationships (cladograms). Most available statements of relationships are phenetic or syncretic and as such are unusable because the dendrograms, even if explicitly stated, are unstable because of the influence of autapomorphies (Farris 1977, 1978, 1979a, 1979b).

MATERIALS AND METHODS

The organisms used consist of three genera of cyprinodontiform fishes (*Fundulus*, *Heterandria*, and *Xiphophorus*), three genera of frogs (*Ceratophrys*, *Eleutherodactylus*, and *Rana*), and one genus of passeriform bird (*Poephila*).

Distribution areas were computed by using published maps (or preparing maps) and using a planimeter to calculate areas as well as areas of overlap. These maps were also used to compute geographic separations between sister taxa (unless overlapping). Distribution areas are expressed as square kilometers; distances separating sister elements are expressed in kilometers.

Sister species are either species sharing a terminal bifurcation or are equal to sister groups (one of which may be a single species or both may be sets of genealogically related species). All sister elements are strictly monophyletic, sensu Wiley (1981).

Speciation level was calculated by working down a cladogram to discover the number of necessary levels through which one must pass to reach the common ancestor of a pair of sympatric species (Figure 1). This yields only *minimum* level because the method is not without risks—chief among them being the underestimate of level because "ghost species" are not recorded (see Simberloff et al. 1981:54–55). This caveat is particularly germane when large geographic separations occur between sister elements (Wiley and Mayden 1985).

TERMS AND HYPOTHEICAL EXPECTATIONS

Sympatry is taken to be the result of one of two processes—sympatric speciation or dispersal. The degree of sympatry between daughter species in the data sets discussed below varies from 0 to 100%. Most (71.2%) daughter species pairs (sister species or elements) are allo/parapatrically distributed (a value in agreement with the data on intuitively selected geminate pairs reported by Anderson and Evensen 1978:424) and another sizable group (13.6%) exhibits only peripheral or trivial sympatry (along the edges of distributional areas, involving less than 10% of the distributional area of the less widely distributed daughter species). The cases

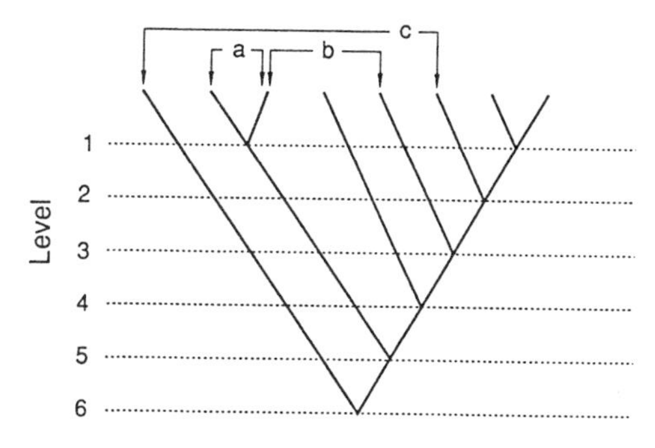

FIGURE 1. Speciation level is calculated by seeking the most recent common ancestor. Thus, for example a, the level is scored as level 1, whereas for a case of sympatry involving the two taxa in case b, the level is 5, and for case c, it is 6. When there are no geographic gaps, there is no need to conjure up "ghost species." This allows only relative "ages" that may not be comparable to the relative ages arrived at from another cladogram unless there is repeated pattern of endemism (as reported by Cracraft 1982) to date a particular speciation event (the separation of *leucotis* and *personata* is simultaneous with that separating two other subclades of *Poephila*).

of trivial sympatry include some that are the results of mapping compromises (i.e., ranges actually interdigitate) although others (more carefully studied) appear genuine. The remaining cases involve substantial sympatries (20–100% geographic overlap).

Vicariance biogeographers are often viewed with disdain because they argue against assumptions of dispersal. They see dispersal as a possible conclusion but as an unnecessary and undesirable assumption (Nelson and Platnick 1981). However, they also view the biota as experiencing only allo/parapatric speciation (as a consequence of biota fragmentation). If it is assumed that speciation is always by allopatric processes and that distributions are informative of history (Rosen 1978:161), then two conclusions follow (these are sometimes stated as assumptions): (1) sympatry is the product of postspeciational dispersal, and (2) sympatric speciation is not possible (tautological).

In the course of discussing ecological sympatric speciation among Hawaiian *Drosophila*, Wiley (1981) suggested that we could use distributional data in conjunction with phylogenetic hypotheses to falsify proposed claims of sympatric speciation. Wiley (1981:57) did not have a proposed claim, rather he had an implicit claim concerning two species of *Drosophila*. Richardson's (1974) suggestion that sympatric speciation was involved was tempered by his admission that there might be other species more closely related to one of the siblings than the other. Wiley's major argument was that when species C is considered to have arisen from species B by postulated sympatric speciation, then *if and only if a particular distribution and phylogenetic relationship are obtained, could sympatric speciation be accepted as a viable alternative* (Figure 2a). Wiley argued that if the phylogenetic relationships were otherwise (that is, species B had as a nearest relative A rather than C), then the sympatric speciation option was falsified (Figure 2b).

Wiley's (1981) argument is a rational one, but we must be careful how much generality we apply to it. If we assume that we have a complete data set (three species, with A and B being nearest relatives), and if we treat the distributional data as informative, then the most parsimonious summary is to argue that the speciation event separating A and B is an allopatric (or parapatric) one but that the event separating C and [A + B] was sympatric (Figure 2c). We must accept this conclusion because the distribution of one sister element (C) is fully sympatric with the other (A + B). However, in every case we can achieve an unparsimonious explanation by advocating allo/parapatric speciation followed by such dispersal as necessary to result in the level of sympatry observed. On the other hand, if our intent is to test models of speciation with distributional data, then we must use the available distributional data rather than evocations of dispersal.

Among the necessary components of any theory of sympatric speciation is the requirement that the daughter species be sympatric (Mayr 1963:449). As explained in the preceding paragraph, species C is entirely

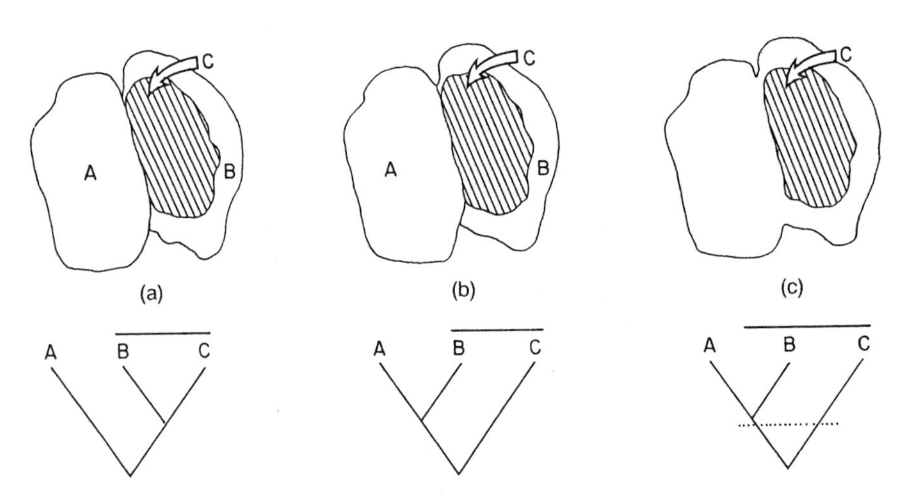

FIGURE 2. Distribution and cladogram where sympatric speciation is (a) accepted (C from B) and (b) falsified (C from B). Aside from the special case (i.e., C is derived from B), each provides evidence for sympatric speciation. (c) Distribution of species C and common ancestor AB at level c in part b.

sympatric with its sister species in each case (species B in Figure 2a and species AB in Figures 2b and c). In Figure 2a, part of the distributional area of species B is parapatric to species C (in the diagrammed case, about one-half). However, if our attention (Figure 2b) is focused on species AB rather than species C, then we might conclude that it is significant that about 75% of the distribution area of AB is parapatric to the distribution area of C. Our concern must be with the (sister) species having the smaller distributional area because it is not relevant to the question that one daughter species occurs where the other does not, only that one occurs *only* in the presence of the other. If we have a pair of daughter species (A and B or AB and C, Figure 2) and if the distribution area of one entirely encloses the other, then our first-order explanation is sympatric speciation (unless we are prepared to argue that distributional data are insufficient or are unnecessary).

A necessary condition of first-order explanations for the other modes of speciation is that the daughter species are distributed allopatrically or parapatrically. The two allopatric models considered here differ in the necessity of dispersal and in explanation of distributional disjunctions (Wiley and Mayden 1985). Vicariant speciation requires that the smaller daughter species have a "relatively large population" (Bush 1975:341) where population size is defined in terms of peripheral isolates or in terms of being polydemic. Peripheral isolates speciation requires that the smaller daughter species have a population composed of a small number of individuals (as few as a single pregnant female), but generally a single deme (Mayr 1982). Inasmuch as few data exist relative to the effective population sizes for species of organisms (and none so far as I know for the sample cases discussed below), I selected an arbitrary size (defined in terms of the area of the larger daughter species) of 5%. If the distributional area of the smaller daughter species is no more than 5% the size of the larger daughter species, then I treated it as a potential peripheral isolates case. If this value is too high, then I would overestimate the frequency of peripheral isolates speciation [now estimated as at least "most common" (Bush 1975) to considerably greater than that]. Mayr (1982:1122) noted that his theory (peripheral isolates speciation) is based on "empirical fact that when in a superspecies or species group there is a *highly divergent population or taxon*, it is *invariably* found in a peripherally isolated location" (emphases added). This certainly would be a useful parameter if it were invariate. However, I know of one case (Lynch 1982, *Ceratophrys cornuta*) in which it simply is not true. Furthermore, the pattern (if it exists) may be an artifact in that biologists are more likely to name the highly differentiated peripheral isolate than a lesser differentiated one. I do not disagree that Mayr can identify highly divergent taxa, only that peripheral (and divergent) may be in the eyes of the beholder (or be a consequence of taxonomic bias). The illustrated cases in Hapgood's (1984) semipopular article belie the objectivity of both divergence and peripherality.

Distinctions between assumptions of allopatric models

The distinctions between the two allopatric modes are best seen in recording what assumptions must be evoked to explain specific cases as vicariant or peripheral isolates speciation. Consider three cases (Figure 3). In case 1, we have three species having distribution areas approximately equal in size and having no discontinuities between them. In case 2, one species (D) is geographically disjunct from the other two (the intervening area is enclosed by dashed lines). In case 3, two species have large distribution areas but the third (J) is a small area peripheral to one of the larger species.

Under vicariant speciation, no additional explanations are required for data sets 1 or 3. For data set 2, we must account for the discontinuity by positing either extinction(s) of species or extirpation(s) of populations in the area identified as "X." *Assumptions* summary: case 1 (none), case 2 (extinctions), case 3 (none).

Under peripheral isolates speciation, we must posit dispersal(s) for all cases because, under that model, one species of a pair is the descendant of the other. Also, for all cases except 3 (species J), we must account for the enlarged distributions of the daughter species originating as peripherally isolated elements. *Assumptions* summary: case 1 (dispersals, range expansions), case 2 (dispersals, range expansions), case 3 (dispersals, one range expansion). Taking each assumption as equally unlikely, vicariant specia-

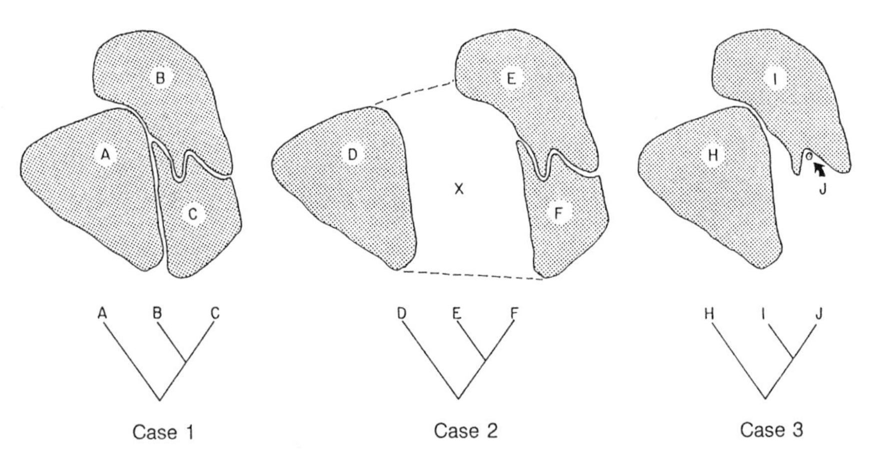

FIGURE 3. Hypothetical distributions and cladograms. Under case 1, no additional assumptions are required. Under case 2, it is necessary to account for the absence of the clade in the area denoted "X" and assume dispersals or extinctions, and possibly "ghost species" as well (Simberloff et al. 1981). Under case 3, no additional assumptions are required.

tion is preferred over peripheral isolates because the vicariant model is more parsimonious (Table 1).

Eldredge and Cracaft (1980) and Wiley (1981) have argued that the isolates in peripheral isolates speciation, if surviving, ought to be highly autapomorphous. Mayr (1982:1122) accepted this view only in part. He argued that "in the vast majority of them [founder populations] only minor genetic reorganizations occur." This view is somewhat at variance with the remark quoted earlier. The relative frequencies of autapomorphies of sister species are data rarely collected and remain expected results (although one contrary data set is available, Lynch 1982). No one has explained why, if surviving, such isolates should expand their distribution areas at disproportionately higher rates than does the larger daughter species of such a pair; such a necessity seems required if peripheral isolates speciation accounts for even the majority of cases of speciation. As pointed out by Anderson and Evensen (1978:426), the relative sizes of distribution areas of geminate (daughter) species are not distinct from that expected under the null hypothesis (but also see Diamond and Gilpin 1982, who disagree about the propriety of such null models). Although the concern of Anderson and Evensen was with the argument that new species must originate in very small populations (advanced by Eldredge and Gould 1972), their analysis applies equally well to a test of the efficacy of assumed dispersals. The variation in sizes of distribution areas for the *Rana* data (~200 km^2 to nearly 6.5 million km^2) suggests that Anderson and Evensen (1978:426) are correct in viewing the sizes of ranges of new species as random ("unless more and better data can be found that do not fit the null hypothesis").

TABLE 1. Summary of assumptions for speciation modes.

Vicariant[a]: ancestral distribution cosmopolitan; distribution areas for daughter species do not overlap, are juxtaposed (no extinctions required), and are not appreciably different in size. Distributional gaps require assumed extinctions. Sympatry requires assumed dispersals
Peripheral isolates[a]: ancestral distribution not cosmopolitan; distribution areas for daughter species do not overlap, are juxtaposed, and differ markedly in size [the "parental" species (from which daughters are derived) has a large distribution area]. Distributional gaps are expected as results of dispersals. Sympatry requires assumed dispersals
Sympatric: ancestral distribution not cosmopolitan; distribution areas for daughter species overlap; no expectations about sizes of distributional areas. No expectations about distributional gaps or dispersal

[a]Parapatric speciation may be indistinguishable from vicariant speciation (Wiley 1981:55) or may appear as a mixture of vicariant and peripheral isolates speciation (see Cracraft 1982:415).

REAL DATA SETS

Against this background, I examined seven data sets. Three sets are for freshwater fishes (*Fundulus*, *Heterandria*, and *Xiphophorus*), three are for frogs (*Ceratophrys*, *Eleutherodactylus*, and *Rana*), and one is for birds (*Poephila*). The *Ceratophrys* data set (Lynch 1982) is for lowland South America; that for *Poephila* (Cracraft 1982, 1983) is for northern Australia; those for *Eleutherodactylus* (Miyamoto 1983; Savage 1975), *Heterandria* (Rosen 1979), and *Xiphophorus* (Rosen 1960, 1978, 1979) are for Mexico and Central America; that for *Fundulus* (Wiley 1977; Lee et al. 1980) is for the eastern United States; and that for *Rana* (Hillis et al. 1983; Pace 1974) covers most of the United States and Mexico.

My principal data set is that reported by Hillis et al. (1983). Those authors provided a cladogram for 23 species of the leopard frog species complex. The cladogram (Figure 4) is fully resolved and 22 speciation events are required to generate the 23 species. The species having the smallest distribution area is *Rana dunni* (~200 km^2) and that having the largest is *R. pipiens* (~6,443,000 km^2). Hillis et al. recognized four species

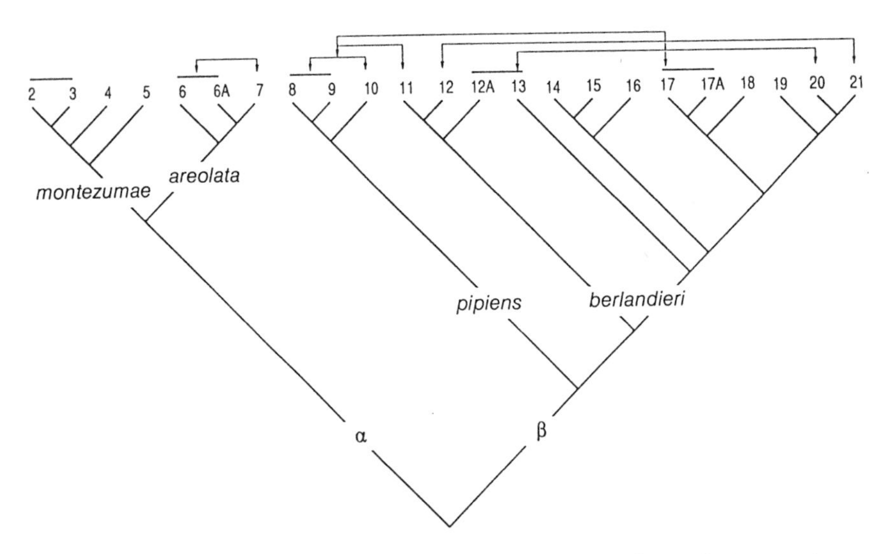

FIGURE 4. Cladogram of leopard frogs (after Hillis et al 1983). Sympatric occurrences are indicated by horizontal lines above adjacent pairs (e.g., 2 and 3) or by using horizontal lines bearing arrows pointing to sympatric species (e.g., 12 and 21). A bracket embraces allopatric species whose distributions are each overlapped (e.g., 6 and 6A by 7). The major U.S. species are *Rana areolata* (6), *R. berlandieri* (17), *R. blairi* (9), *R. capito* (6A), *R. palustris* (7), *R. pipiens* (8), and *R. spenocephala* (10).

TABLE 2. Leopard frog data set.[a]

Sister elements	Distribution areas	Separation	Overlap (%)	Speciation level	Mode[b]
(2)(3)	130,337:44,944	C[c]	60.0	1	S
(2–3)(4)	148,315:200	10			II
(2–4)(5)	148,515:238,202	15			I
(6)(6A)	330,337:539,326	30			I
(6–6A)(7)	869,663:2,492,135	C	43.7	2	?
(8)(9)	6,442,697:961,798	C	6.5	1	I
(10)(8–9)	1,525,843:7,341,574	C	6.7	2	I
(11)(12)	81,331:184,843	50			I
(12A)(11–12)	22,181:266,174	325			I
(14)(15)	5,545:33,272	10			I
(16)(14–15)	5,545:38,817	10			I
(14–16)(17–21)	44,362:870,609	50			I
(17A)(17)	79,482:576,710	C	4.6	1	I
(18)(17–17A)	40,665:652,495	170			I
(21)(20)	29,575:107,209	60			I
(19)(20–21)	40,665:136,784	175			I
(19–21)(17–18)	177,449:693,160	70			I
(13)(14–21)	22,181:914,971	C	4.5	5	(I)
(2–5)(6–7)	386,517:2,982,023	700			I
(11–21)(8–10)	1,117,298:8,765,170	C	7.6	7	I
(2–7)(8–21)	3,368,540:9,797,440	C	80.0	8	?
(11–12A)(13–21)	288,355:936,152	C	37.2	6	?

[a]For distribution maps see Hillis et al. (1983) and Pace (1974).
[b]I, vicariant speciation; II, peripheral isolates speciation; S, sympatric speciation; ?, indeterminate case (see text); (I), cases that would be called peripheral isolates by relative sizes alone, but by virtue of the actual size of the smaller, are best called vicariant cases.
[c]C, overlapping distributions.

groups (*areolata*, *berlandieri*, *montezumae*, and *pipiens*). The first (most basal) bifurcation in the cladogram separates the *alpha* and *beta* divisions of the complex. The two divisions are largely sympatric (~80% overlap). In contrast, the two species groups of the *alpha* division (*areolata* and *montezumae*) are fully allopatric and those of the *beta* division are essentially allopatric (~8% of the area of the *berlandieri* group is overlapped by species of the *pipiens* group). Within species groups, there is little geographic overlap (Table 2). In 13 cases (a case involves looking for overlap between sister elements), there is no overlap. In five cases, the overlaps are trivial (4–8%), involving only distributional range peripheries. In three cases, distributional overlaps are substantial (37–60%; species 2 and 3, 6–6A and 7, and 11–12A and 13–21).

In the leopard frog data, there is one possible case (out of 22 total, 4.6%) for sympatric speciation (species 2 and 3, *Rana megapoda* and *R. montezumae*). These species are the product of a terminal bifurcation and are substantially sympatric (60%). The three other cases of substantial geographic overlap between sister elements in the leopard frog data set are for more remote speciation events (presumably events older than that separating *R. megapoda* and *R. montezumae*). If we ignore dispersal as an obfuscating component, these four cases (out of 22) provide an upper limit to the frequency of sympatric speciation (<18.2%). If we concern ourselves only with cases within species groups, sympatric speciation is a possibility in 3 cases out of 19 (15.8%). All of these values (4.6–18.2%) are considerably in excess of the frequency that might be predicted from the literature. However, we can also build a contrary argument by emphasizing that in no case is the sympatry complete. Dispersal is an alternative explanation (and one which, intuitively, seems rather likely for older speciation events).

The first-order distinction between vicariant and peripheral isolates allopatric speciation is the relative sizes of the distributional areas (both "large" in vicariant, one "very small" in peripheral isolates). Using 5% as the limit for "very small," only two of the sister element pairs qualify as peripheral isolates [*Rana dunni*, whose sister element is *R. megapoda* + *R. montezumae* (species 4, 2 + 3), and species 13, whose sister element is the set of species numbered 14–21]. The distribution area for *R. dunni* is approximately 200 km^2 (about 0.1% the size of that of the sister element). If we also consider those cases in which "trivial" sympatry occurs, the second case surfaces (species 13/species 14–21; the distribution area for species 13 is 2.6% the size of that for the sister element). However, percentages alone provide little correspondence to effective population sizes. The first case (*R. dunni*) is for a species with a small distributional area (Zweifel 1957; Duellman 1961) whereas the second is "small" only in comparison with the sister element (the distribution area for species 13 is approximately 22,000 km^2). The remaining 16 cases [strict allopatry plus essentially allopatric (some overlap along edges of distribution areas)] involve large geographic areas (and presumably large population sizes) and larger percentages (5.1–62.3%) and qualify as vicariant speciation cases.

Using the leopard frog data set, the frequencies for the three speciation modes are as follows: peripheral isolates = 4.9–9.2%, sympatric = 4.6–18.2%, and vicariant = 72.7%. These results are contrary to conventional wisdom and to discover whether they might be merely idiographic, I examined other data sets. The other data sets (Figures 5 and 6) vary from 1 of 5 species (*Fundulus*) to 1 of 16 (*Xiphophorus*). Each is fully resolved with the exception of a single trichotomy in *Xiphophorus*. Forty-one pairs of sister elements are available in these data sets. Thirty-two are strictly allopatric, four involve "trivial" sympatry, and five involve substantial sympatry (approximately the expected distribution if the *Rana* data set is taken as a model).

The smallest data set is that for the topminnows (*Fundulus notti* group). Only vicariant speciation is called for to explain all the cases of speciation. The sizes of the distributional areas of sister species vary from 48.7 to 97.8% (Table 3). The frog genus *Ceratophrys* is only slightly larger but is more complicated. Comparing relative sizes of distributional areas, three cases appear to be peripheral isolates (*C. joazeirensis*, sister species of *C. aurita*; *C. stolzmanni*, the sister to *C. calcarata* + *C. cornuta*; and *C. calcarata*, sister species to *C. cornuta*). In the first case, the known distribution area of the peripheral isolate is small—a single locality (Mercadal 1987), whereas in the second and third cases, the distributional areas are large [*C. stolzmanni* (~35,000 km^2) is small only in comparison to the sister element (~5.2 million km^2), and that for *C. calcarata* (~200,000 km^2) is small only compared to that of *C. cornuta* (~5 million km^2)]. The three

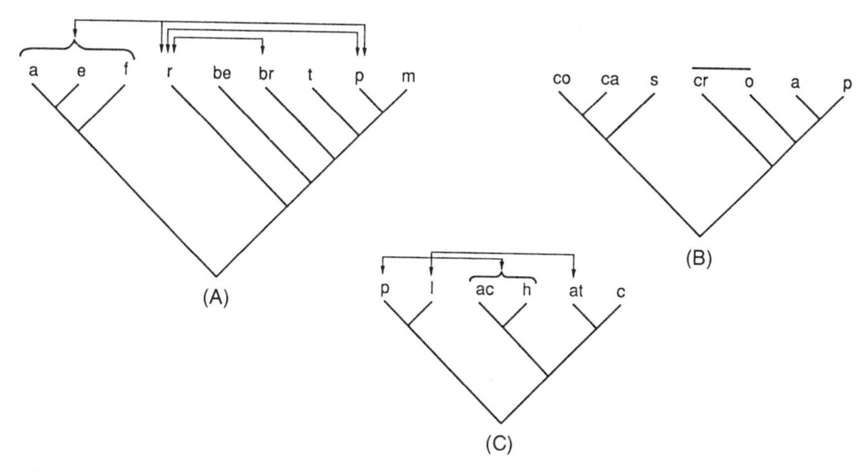

FIGURE 5. Cladograms of frog and bird groups discussed in text. (A) *Eleutherodactylus*, after Miyamoto (1983); *fleischmanni* subgroup (f, e, a), *merendonensis* (m), *punctariolus* (p), and *rugulosus* (r) mentioned in text. (B) *Ceratophrys*, after Lynch (1982); *aurita* (a), *calcarata* (ca), *cornuta* (co), *joazeirensis* (p), and *stolzmanni* (s) mentioned in text. (C) *Poephila*, after Cracraft (1982); *acuticauda (ac)*, *atropygialis* (at), *cincta* (c), *hecki* (h), *leucotis* (l), and *personata* (p). Sympatry is indicated as in Figure 4 (see legend).

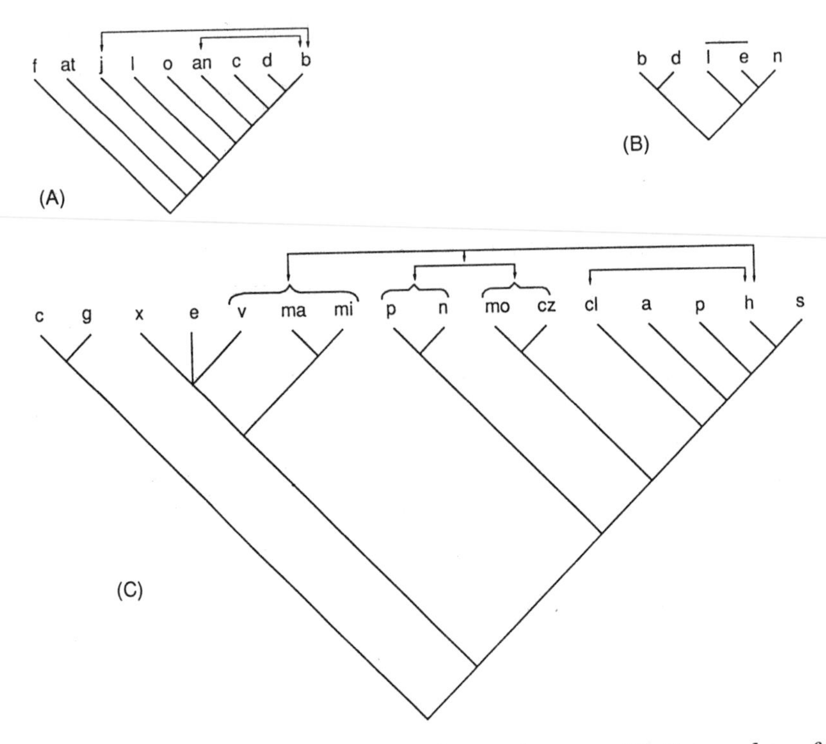

FIGURE 6. Cladograms of fish groups discussed in text. (A) *Heterandria*, after Rosen (1979); *attenuata* (at), *cataractae* (c), *dirempta* (d), *litoperas* (l), and *obliqua* (o) mentioned in text. (B) *Fundulus*, after Wiley (1977). (C) *Xiphophorus*, after Rosen (1978, 1979); *clemenciae* (cl), *cortezi* (cz), *couchianus* (c), *evelynae* (e), *gordoni* (g), *helleri* (h), *maculatus* (ma), *milleri* (mi), *montezumae* (mo), *nigrensis* (n), *pygmaeus* (p), *signum* (s), and *variegatus* (v) mentioned in text. Sympatry is indicated as in Figure 4.

remaining speciation events are apparently vicariant cases (distribution areas vary from 28.5 to 84.7% of the size of that of the sister element).

The *Eleutherodactylus* data set (*rugulosus* group sensu Miyamoto 1983) includes one peripheral isolates case (*E. merendonensis*), one case of sympatric (*fleischmanni* subgroup is entirely sympatric with its sister element), and five cases of vicariant speciation (distributional areas are 23.8–76.9% of the size of the sister element distributional area). One additional case (*E. rugulosus* and its sister element) is indeterminate—the sympatry level is 20.5%. The single peripheral isolates case in this set is anomalous in that the sister elements are separated by a distance of 760 km (Table 3). Furthermore, the sympatric case may not be convincing in that the sister elements are separated by a basal dichotomy in the cladogram (Figure 4).

TABLE 3. Data sets for *Eleutherodactylus*, *Ceratophrys*, and *Fundulus*.[a]

Sister elements	Distribution areas	Separation	Overlap (%)	Speciation level	Mode
		Eleutherodactylus			
(p)(m)	10,558:50	760			II
(pm)(t)	10,608:2,514	5			I
(pmt)(*br*)	13,072:8,547	185			I
(be)(pmt*br*	47,637:21,619	400			I
(r)(pmt*br*be)	192,936:69,256	C	20.5	5	?
(e)(a)	1,634:1,257	1			I
(f)(ea)	4,022:2,891	1			I
(pmt*br*ber)(fea)	247,989:6,913	C	100.0	6	S
		Ceratophrys			
(a)(P)	298,063:50	450			II
(o)(Pa)	351,714:298,113	540			I
(*cr*)(Pao)	909,091:649,777	C	8.0	3	I
(ca)(*co*)	202,682:5,044,709	230			(I)
(ca*co*)(s)	5,247,391:35,768	100			(I)
(cacos)(Pao*cr*)	5,283,159:1,506,707	640			I
		Fundulus			
(n)(e)	36,137:66,351	5			I
(ne)(l)	102,488:210,308	C	11.0	2	I
(d)(b)	188,981:193,128	10–20			I
(nel)(db)	301,540:382,109	1–5			I

[a]For maps see Savage (1975) for *Eleutherodactylus*, Lynch (1982) for *Ceratophrys*, and Lee et al. (1980) for *Fundulus*.

The *Heterandria* data set (Table 4) includes two cases of trivial sympatry (Miller 1974; Rosen 1979:332). Three cases of peripheral isolates speciation seem obvious (*H. attenuata*, *H. cataractae*, and *H. dirempta*). In two other cases (*H. litoperas* and *H. obliqua*), the small percentages reflect the size of the larger sister element. The other three cases are vicariant speciation cases (distribution areas are 19.2–82.7% of the sizes of the sister element distribution areas).

The *Xiphophorus* data set (Table 4) is of particular interest in that it was

TABLE 4. Data sets for *Heterandria* and *Xiphophorus*.[a]

Sister elements	Distribution areas	Separation	Overlap (%)	Speciation level	Mode
		Heterandria			
(b)(d)	102,941:251	1			II
(c)(bd)	103,192:20	40			II
(*an*)(bdc)	53,167:103,212	C	2.4	3	I
(o)(bdc*an*)	3,268:155,122	1			(I)
(l)(bdc*ano*)	1,885:158,390	1			(I)
(j)(bdc*anol*)	30,669:160,275	C	11.1	6	I
(at)(bdc*an*olj)	20:187,550	1			II
(f)(bdc*an*olj*at*)	226,896:187,570	1400			I
		Xiphophorus			
(s)(h)	20:66,993	5			II
(sh)(P)	13,325:67,013	30			I
(a)(shP)	5,405:80,336	2			I
(*cl*)(ashP)	20:85,741	C	100.0	4	S
(cz)(*mo*)	2,137:4,650	5–10			I
(cz*mo*)(ashP*cl*)	6,787:85,761	200			I
(p)(n)	126:251	45			I
(np)(cz*mo*ashP*cl*)	377:92,548	C	100.0	6	S
(mi)(*ma*)	20:50,402	5			II
(e)(v)	20:22,750				
(e)(x)	20:5,028	Unresolved trichotomy			II
(v)(x)	5,028:22,750				
(mi*ma*)(evx)	27,798:50,422	95			I
(mi*ma*evx) (cz*mo*ashP*cl*)	78,220:92,548	C	62.8	7	?
(c)(g)	251:251	230			I
(cg)(others)	502:121,623	145			(I)

[a]For maps see Rosen (1979) for *Heterandria* and Rosen (1978, 1979) for *Xiphophorus*.

used in conjunction with the *Heterandria* data set to illustrate vicariance biogeography (Rosen 1978). Using the criteria given, four cases of peripheral isolates speciation are evident (*X. signum*, *X. milleri*, *X. evelynae*, and the *couchianus–gordoni* pair from the remainder of the genus). In the case of the last example, each member of the pair is widely separated from the other *Xiphophorus* and from each other. In the other three cases, the peripheral isolate is distributed just peripheral to its sister element. Two cases of apparent sympatric speciation are evident in the data set. *X. clemenciae* is fully sympatric (Rosen 1960:169) with *X. helleri* (a member of the sister element for *X. clemenciae*). *X. nigrensis* and *X. pygamaeus* are completely sympatric with *X. cortezi* and *X. montezumae* (members of the sister element for *nigrensis–pygmaeus*). Extensive overlap also occurs between the *variegatus–maculatus* clade (mmevx) and the *helleri* clade (cmcashPMH). The overlap is 62.8% of the distributional area of the *variegatus* clade and may reflect sympatric speciation as well—however, the dichotomy is deep within the cladogram. Seven other speciation events within the genus are vicariant cases (distributional areas are 6.7–100.0% of the sizes of the sister element distributional areas).

The *Poephila* data are for a set of six species of finches found in northern Australia (Cracraft 1982, 1983). These data are particularly interesting because birds are usually thought to exhibit dispersal capacities dramatically different from fish or frogs. Four species (*P. acuticauda*, *P. atropygialis*, *P. cincta*, and *P. hecki*) are fully allopatric—these three speciation events are vicariant speciation cases, as is that between *P. leucotis* and *P. personata*. Cracraft (1982, 1983) provided maps and came to identical conclusions. However, the *leucotis–personata* clade is fully sympatric (Figure 5) with its sister element and a conclusion of sympatric speciation is most parsimonious. No case of peripheral isolates speciation can be made from these distributional and phylogenetic data (as noted earlier by Cracraft 1982:421).

Summary of the data sets

Although there is some variation among the data sets concerning the frequencies of different modes of speciation (Table 5), the similarities outweigh the differences. In spite of having apparently different vagilities, fishes and frogs and birds show the same pattern, i.e., vicariant speciation accounts for 71% of the speciation cases (range 54–100%), peripheral isolates speciation accounts for 15% (range 0–38%), and sympatric speciation accounts of 6% (range 0–12%). Those cases judged *indeterminate* (moderate levels of sympatry) above account for 7.5% (range 0–14%). The preponderance of the vicariant model of speciation is at variance with the popular notion but is in agreement with the findings of other workers (Cracraft 1982: Wiley and Mayden 1985).

The estimate of 15% for cases of peripheral isolates speciation is probably too high, because in vicariance biogeography theory, peripheral isolates (and dispersals) are random phenomena—yet in the cases for *Heterandria* and *Xiphophorus*, those cases identified as peripheral isolates are in general congruent between genera (Rosen 1978). This finding sug-

TABLE 5. Frequencies of modes of speciation in different groups of vertebrates.

Group	*N*	Vicariant	Peripheral isolates	Sympatric	Indeterminate
Rana	22	16 (+1)[a]	1	1	3[b]
Ceratophrys	6	3 (+2)[a]	1	0	0
Eleutherodactylus	8	5	1	1	1[b]
Fundulus	4	4	0	0	0
Heterandria	8	3 (+2)[a]	3	0	0
Xiphophorus	13	7	3	1	1[b]
Poephila	5	4	0	1	0

[a]The actual sizes of the distribution areas argue against assigning the cases in parentheses to peripheral isolates, even though the percentages would compel such assignment.
[b]These are cases of either sympatric speciation (because significant geographic overlap occurs) or allopatric speciation followed by extensive dispersal. In most cases, these cases are for dichotomies deep within the cladograms (putatively old events).

gests that those speciation cases are due to common causes, i.e., represent cases of microvicariance. In other cases (e.g., 12A/11–12, *Rana*; *merendonensis/punctariolus*, *Eleutherodactylus*; *calcarata*, *cornuta*, *stolzmanni*, *Ceratophyrs*; *couchianus–gordoni*/other *Xiphophorus*), the distances between sister elements suggest that the distributional areas reflect range contractions rather than dispersals.

DISCUSSION AND CONCLUSIONS

These data call for the discussion of two points: (1) testing of the "assumption" of nondispersal, and (2) determining why peripheral isolates cases are so rare.

Assumption of nondispersal

It can be argued that dispersal is the assumption and I have chosen not to make such an assumption. However, such an argument avoids the issue and concedes Rosen's (1978) assumption that sympatry is evidence of postspeciational dispersal. A discussion of dispersal is warranted inasmuch as dispersal and models of speciation are closely interwoven. For example, Brown and Gibson (1983:529) predict that "recently derived sister species would have ranges that overlapped less than those of species that are products of a more ancient splitting of phyletic lineages." Endler (1984:249) noted that they could not find data relevant to testing their prediction, but

Anderson and Evensen (1978) provided such data (albeit using intuitive phylogenetic hypotheses). Testing the assumption of nondispersal focuses directly on this point because those data sought by Brown and Gibson emerge from the analyses previously presented.

In describing the data sets, I not infrequently noted that some cases of substantial sympatries between daughter species were for species whose divisions were deep within the cladogram. How can we distinguish cases of postspeciation dispersal from cases of sympatric speciation? The theoretical expectation (Figure 7) that the level of sympatry is a function of the time since speciation follows if we accept two assumptions: (1) all speciation is allopatric (or parapatric), and (2) sympatry is evidence of dispersal. These are essentially the assumptions made by Anderson and Evensen (1978), Rosen (1978), and Wiley (1981). Of the 66 pairs of daughter species in the seven data sets, 19 pairs show some level of sympatry, but there is no correlation between an approximation of time since divergence and geographic overlap (Figure 8). These data are also are variance with the data presented by Anderson and Evensen (1978:424–25); however, that disagreement probably derives from differences in precision of identifications of daughter species (here equal sister elements in many cases, whereas Anderson and Evensen's arrangements equated phenetically similar with sister species).

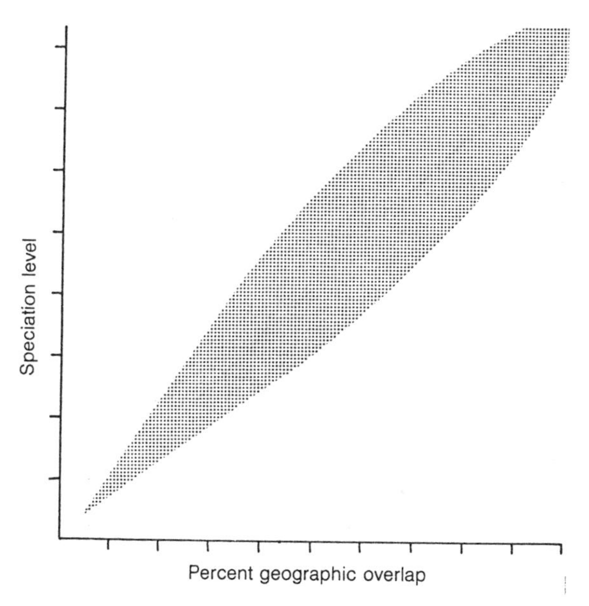

FIGURE 7. Theoretical expectation when vicariant speciation is assumed and when sympatry is assumed to result from postspeciational dispersal.

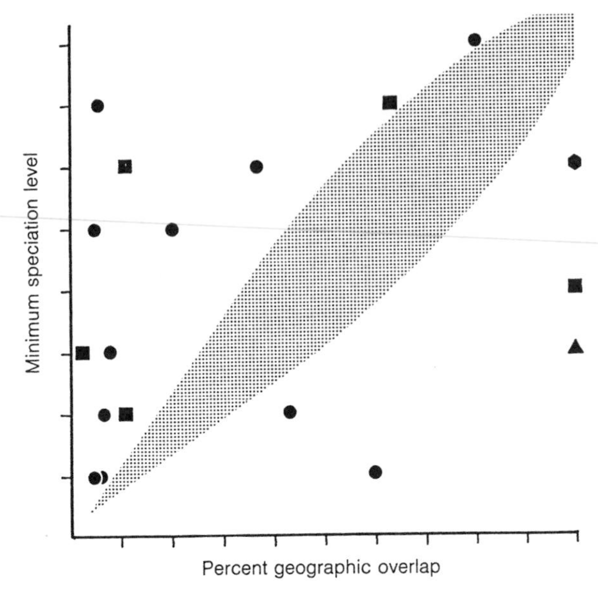

FIGURE 8. Observed results in 19 pairs of sympatric sister pairs (theoretical expectation stippled). (●) Frogs, (■) fishes, (▲) birds, (⬢) a point for one frog case and one fish case. R = +0.33.

In view of the disparity between expected and observed results, we must reject one or both of the assumptions under which the expectations were derived. Nearly 50% of all cases of sympatric occurrences involve what I earlier termed "trivial" sympatry. In these cases, the vast majority of all distributional areas for both daughter species is allopatric and the argument of postspeciational dispersal seems obvious. In these pairs as well as in the cases of the 47 pairs of allopatric daughters, there is little need to evoke dispersal. These 56 pairs include terminal pairs (the strict case for daughter species) as well as pairs whose dichotomies lie deep within the cladogram. The remaining 10 cases are widely scattered (Figure 9)—in fact, the scatter looks suspiciously like overdispersion. If so, then the values at the top of the figure may simply reflect the long-term effects of dispersal, whereas the values at the bottom of the figure may reflect sympatric speciation. However, if only 3–5 cases out of 66 possibilities reflect appreciable dispersal, then significant dispersal should not be envisioned as an important hypothesis.

The lack of fit between the expected (Figure 7) and the observed (Figure 8) falsifies the hypotheses and one or more of its assumptions. These data as well as those of Anderson and Evensen (1978) show vicariant speciation sufficient to explain fully more than 70% of the cases of

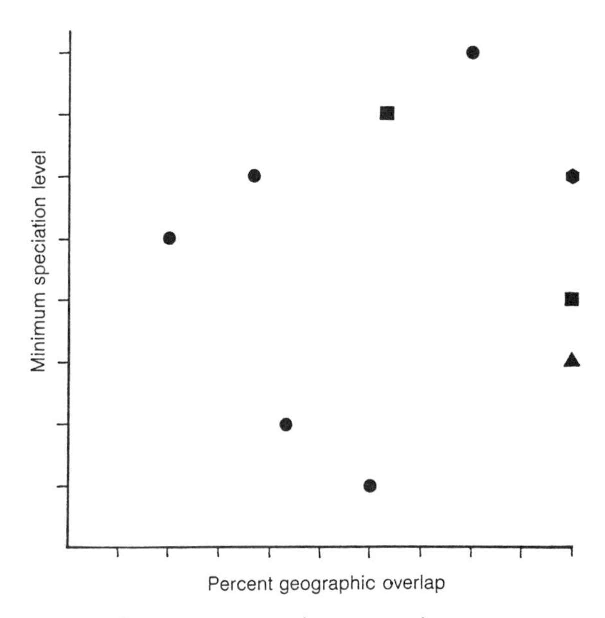

FIGURE 9. Nontrivial sympatry cases ($R = +0.12$) in Figure 8.

speciation in vertebrates [Cracraft (1982) also makes this point and points out that the massive data set of Haffer (1974) is an agreement with this conclusion]. Therefore, if we reject assumption 1 (all speciation is allopatric), *substantial sympatry is no longer necessarily due to dispersal* (that is, nontrivial sympatry *could be* due to sympatric speciation) and we may simultaneously reject assumption 2 (sympatry is evidence of post-speciational dispersal) as unnecessary.

Although sufficient to explain any observation, dispersal appears unnecessary. If that is so, why has it been so widely used and defended by biologists? Dispersal is cited frequently as a logically necessary consequence of the postulated habitat alterations of the Pleistocene (both in northern areas occupied by glaciers as well as in tropical regions in which forested areas waxed and waned). For only a limited subset of the biota is dispersal a logical necessity (i.e., those areas entirely submerged or covered by glaciers), but fossil evidence of considerable range expansion is available. At least part of the answer, or an answer, is available from historical records—the spread of introduced plant and animal species or those native species able to take advantage of the alterations of the environment associated with "normal" human activities. Biologists have, in effect, argued that because some species can, then all species do, and compounded the error of failing to discriminate between sufficiency and necessity of explanations.

Rates of dispersal (and establishment) are calculated readily for several species initially viewed as beneficial (or neutral) and later viewed as pests. For example, the European starling (Table 6) spread at an annual rate of 13 km over the first 25 years of residence in North America but at a rate of 60 km/year between 1935 and 1940 (Wing 1943). By 1873, the English sparrow had spread at a rate of 43 km/year and by 1883 it had spread at a rate of 72 km/year (Wing 1943); its rate of spreading by the time it reached far into the continent (Johnston and Selander 1965) averaged 80 km/year. The rate of spread for the European hare in Australia is comparable (Table 6), but its rate possibly was augmented by secondary introductions (Rolls 1969:20–39). The slower muskrat (Table 6) may have been aided by secondary introductions as well. Native organisms (Table 7) show slightly slower, but nevertheless appreciable, rates of spread.

For none of these selected examples would ranges have appeared constant. Against a background of easily detectable rates, we can examine the distributions of most native species and, with a few exceptions, discover

TABLE 6. Rates of dispersal by successfully introduced organisms.

Taxon	Place	Time	Distance dispersed (km)	Dispersal rate (km/year)
Passer domesticus (sparrow)[a]	North America	1852–1933	3500	43
		1852–1914	3650	59
		1852–1900	3850	80
Passer domesticus (sparrow)[b]	North America	1852–1873	900	43
		1852–1878	1470	57
		1852–1883	2220	72
Sturnus vulgaris (starling)[b]	North America	1890–1915	320	13
		1890–1925	900	26
		1890–1935	1900	42
		1890–1940	3000	60
Oryctolagus cuniculus (rabbit)[c]	Australia	1859–1877	245	13
		1877–1886	755	75
Ondatra zibethicus (muskrat)[d]	Europe	1907–1917	50–100	5–10
		1907–1927	140–150	7–8
		1907–1937	150–340	5–12
Gambusia affinis (mosquitofish)	Nebraska	1975–1987[e]	41–114	7–10
		1975–1987[f]	261	20

[a]Johnston and Selander (1965).
[b]Wing (1943).
[c]Rolls (1969:20–39).
[d]Udvardy (1969:137).
[e]Upstream dispersal.
[f]Downstream dispersal.

TABLE 7. Rates of dispersal by indigenous species.

Taxon	Place	Time	Distance dispersed (km)	Dispersal rate (km/year)
Eupithecia sinuosaria (butterfly)[a]	Europe	1890–1960	1100–1400	16–20
	Finland	1890–1895	125	25
	Finland	1895–1900	50–240	30–48
Dendrocopos syriacus (woodpecker)[b]	Hungary	1937–1960	250	12
Dasypus novemcinctus (armadillo)[c]	United States	1854–1880	240	9
		1880–1905	39	1.6
		1905–1914	330	33
		1914–1953	360	9

[a]Udvardy (1969:340).
[b]Udvardy (1969:337).
[c]Udvardy (1969:29).

that ranges are essentially constant (or are declining) over periods of observation approaching a century. Hence, we err if we generalize from the *successful* introductions (or rare mobile indigenous species) to most autochthonous faunal elements. The earlier conclusion, that dispersal is an unnecessary assumption, remain unchallenged by examining cases of known (as contrasted with logically "necessary") dispersal. Nevertheless, given the rates of dispersal observed in introduced organisms, distributional equilibria would have been achieved long ago given what we know of the timing of the last glacial retreat (12,000 years BP).

What are cases of peripheral isolates speciation so rare? Three possible answers are:

1. such speciation does not happen.
2. nature abhors a vacuum and immediately (following such speciation) ranges enlarge by dispersal, and
3. following peripheral isolates speciation, the new isolate goes extinct.

Answer (3) requires no discussion because such speciation would not be detected or contribute to pattern formation. Answer (2) was dealt with, in part, in the discussion about dispersal (see previous section). It is also counterindicated by the observation that small distribution areas appear in some taxa whose origins are deep within cladograms (Rosen 1978, 1979) and whose ages must be considerable. Answer (1) is more surprising in light of the extensive literature that developed to support the view of peripheral isolates speciation. If the method used here (see introduction) has merit, a set of maps can be scanned for a presumably monophyletic group (even in the absence of an explicit species-level phylogenetic hypothesis) to see if *any* species has a very small distribution area; if not, then there is no necessity to evoke peripheral isolates speciation [It can even be argued that the answer to Paterson's (1981) question would be "no," relegating this model to his Class II.] And, if one or more species appears to be highly divergent from the remaining, several explanations compete to explain why it is so divergent. Cracraft (1982, see also 1986:989–90) used this approach to scan Australian birds for cases of peripheral isolates speciation and found the absence of such cases remarkable. Last, if putative cases of peripheral isolates speciation are repeated in different monophyletic groups (e.g., *H. dirempta* and *X. signum*; see Rosen 1978, 1979; Wiley 1980), then a microvicariance (vicariant) explanation is more parsimonious than evoking two (or more) simultaneous but independent events (i.e., dispersals into "marginal" habitats).

For *H. dirempta* and *X. signum*, peripheral isolates that appear to have originated by way of (micro)vicariance, there is one further germane point. The habitat is not obviously "marginal" relative to that of the sister species (Rosen 1979:272–273, 330–332). The habitat of these peripheral isolates is karst formation and thus "different" from that of the sister species (but not qualitatively "marginal").

The peripheral isolates model employs, perhaps only heuristically, the notion that the habitat at the edge of the distribution area is qualitatively marginal. Such a notion seemingly explains the rapid genetic divergence sometimes postulated for such isolates. It remains to be seen if peripheral isolates exist, and, if so, whether such isolates exhibit enhanced genetic divergence. Such questions could be profitably explored by examining some or all of the nine cases initially termed peripheral isolates herein (Tables 2–5).

If there are different genetic bases of allopatric (one or both models) and sympatric speciation (beyond polyploidy), then sampling the putative cases here identified might contribute to the resolution of what have been intractable problems to date (notions of cohesion, genetic revolutions, and decoupling). Alternatively, we may discover that there are no such differences and that species are merely "victims of history" (Wiley and Brooks 1982). No resolution is available as to how sympatric speciation is possible (aside from polyploidy, which also occurs in allopatric speciation); that certainly appears to be a class II (sensu Paterson 1981) issue.

SUMMARY

In order to assess the relative frequencies of several modes of speciation, it is proposed that distributional data be used simultaneously with explicit hypotheses of relationships. Such a protocol allows the identification of probable cases of dispersal as well as other phenomena not now considered likely (e.g., sympatric speciation). Vertebrate (fish, frogs, and birds) speciation is primarily due to vicariant allopatric speciation but cases of peripheral isolates allopatric (peripatric) and sympatric speciation appear to be identified. By the method employed here for 66 cases of speciation, the vicariant model explains 71%, the peripheral isolates model initially explains 15%, and sympatric speciation explains 6%. Dispersal, long used in biogeographic and evolutionary arguments, appears to be an unnecessary assumption in speciation studies. This conclusion, when coupled with the more parsimonious argument that apparent cases of peripheral isolates speciation are cases of microvicariance, leads to the conclusion that peripheral isolates speciation may be even rarer than sympatric speciation among veterbrates.

ACKNOWLEDGMENTS

This chapter has benefitted from discussions with students since its conception in 1983. Valuable criticisms were received from the Ecology Lunch seminar, the Evolutionary Biology graduate seminar, and the 1983-84 zoogeography class at the University of Nebraska as well as from students in my biogeography course (1985) at the Universidad Nacional de Colombia (Bogotá, Colombia). Especially useful insights were provided by Tony Joern, Michal Kaspari, Cliff Lemen, and David Woodman in the course of their useful criticisms. John Endler, Richard Mayden, the late Donn Rosen, Hobart Smith, and Edward Wiley provided criticisms of earlier drafts of the earlier manuscript. I thank these colleagues for the criticisms but claim any remaining errors as my own.

LITERATURE CITED

Anderson, S., and M. K. Evensen. 1978. Randomness in allopatric speciation. System. Zool. 27:421–430.

Barton, N.H., and G. M. Hewitt. 1985. Analysis of hybrid zones. Annu. Rev. Ecol. System. 16:113–148.

Brown, J. H., and A. C. Gibson. 1983. *Biogeography*. C. V. Mosby Co., St. Louis.

Bush, G. L. 1975. Modes of animal speciation. Annu. Rev. Ecol. System. 6:339-364.

Cracraft, J. 1982. Geographic differentiation, cladistics, and vicariance biogeography: Reconstructing the tempo and mode of evolution. Am. Zool. 22:411–424.

Cracraft, J. 1983. Cladistic analysis and vicariance biogeography. Am. Sci. 71:273–291.

Cracraft, J. 1986. Origin and evolution of continental biotas: Speciation and historical congruence within the Australian avifauna. Evolution 40:977–996.

Diamond, J. M. and M. E. Gilpin. 1982. Examination of the "null" model of Connor and Simberloff for species co-occurrences on islands. Oecologia 52:64–74.

Duellman, W. E. 1961. The amphibians and reptiles of Michoacan, Mexico. Univ. Kansas Pub., Mus. Nat. Hist. 15:1–148.

Eldredge, N., and J. Cracraft. 1980. *Phylogenetic patterns and the Evolutionary Process: Method and Theory in Comparative Biology*. Columbia University Press, New York.

Eldredge, N. and S. J. Gould. 1972. Punctuated equilibria: An alternative to phyletic gradualism. Pp. 82–115 in: T. J. M. Schopf (ed.), *Models in Paleobiology*. Freeman, Cooper, San Francisco.
Endler, J.. A. 1977. Geographic variation, speciation, and clines. Monogr. Pop. Biol. (10).
Endler, J. A. 1984. [Review of] *Biogeography* by J. H. Brown and A. C. Gibson. System. Zool. 33:249–250.
Farris, J. S. 1977. On the phenetic approach to vertebrate classification. Pp. 823–850 in M. K. Hecht, D. C. Goody, and B. M. Hecht, (eds.), *Major Patterns in Vertebrate Evolution*. Plenum Press, New York.
Farris, J. S. 1978. The 11th annual numerical taxonomy conference and part of the 10th. System. Zool. 27:229–238.
Farris, J. A. 1979a. On the naturalness of phylogenetic classification. System. Zool. 28: 200–214.
Farris, J. A. 1979b. The information content of the phylogenetic system. System. Zool. 28:483–519.
Haffer, J. 1974. Avian speciation in tropical South America with a systematic survey of the toucans (Ramphastidae) and jacamars (Galbulidae). Pub. Nuttall Ornithol. Club (14):1–390.
Hapgood, F. 1984. The importance of being Ernest. Science 84 5:40–46.
Hillis, D. M., J. S. Frost, and D. A. Wright. 1983. Phylogeny and biogeography of the *Rana pipiens* complex: A biochemical evaluation. System. Zool. 32:132–143.
Johnston, R. F., and R. K. Selander. 1965. House sparrows: Rapid evolution of races in North America. Science 144:548–550.
Lee, D. S., C. R. Gilbert, C. H. Hocutt, R. E. Jenkins, D. E. McAllister, and J. R. Stauffer, Jr. 1980. *Atlas of North American Freshwater Fishes*. North Carolina State Museum of Natural History.
Lynch, J. D. 1982. Relationships of the frogs of the genus *Ceratophrys* (Leptodactylidae) and their bearing on hypotheses of Pleistocene forest refugia in South America and punctuated equilibria. System. Zool. 31:166–179.
Macnamara, M. and H. E. H. Paterson. 1984. The recognition concept of species. South African J. Sci. 80:312–318.
Masters, J., N. Caithness, and R. Rayner. 1984. Steve Gould, stasis and SMRS. South African J. Sci. 80:496–497.
Mayr, E. 1942. *Systematics and the Origin of Species from the Viewpoint of a Zoologist*. Columbia University Press, New York.
Mayr, E. 1954. Change of genetic environment and evolution. Pp. 157–180 in J. Huxley, A. C. Hardy, and E. B. Ford (eds.), *Evolution as a Process*. Allen & Unwin, London.
Mayr, E. 1963. *Animal Species and Evolution*. Belknap Press of Harvard University Press, Cambridge, MA.
Mayr, E. 1982. Speciation and macroevolution. Evolution 35:1119–1132.
Mercadal, I. T. 1987. *Ceratophrys joazeirensis* sp. n. (Ceratophryidae, Anura) del noreste de Brasil. Amphibia-Reptilia 7:313–334.
Miller, R. R. 1974. Mexican species of the genus *Heterandria*, subgenus *Pseudoxiphophorus* (Pisces: Poeciliidae). Transact. the San Diego Soc. Nat. His. 17:235–250.
Miyamoto, M. M. 1983. Frogs of the *Eleutherodactylus rugulosus* group: A cladistic study of allozyme, morphological, and karyological data. System. Zool. 32:109–124.
Nelson, G. and N. Platnick. 1981. *Systematics and Biogeography: Cladistics and Vicariance*. Columbia University Press, New York.
Pace, A. E. 1974. Systematics and biological studies of the leopard frogs (*Rana pipiens* complex) of the United States. Misc. Publ., Mus. Zool., Univ. Michigan (148):1–140.
Paterson, H. E. H. 1981. The continuing search for the unknown and unknowable: A critique of contemporary ideas on speciation. South African J. Sci. 77:113–119.
Paterson, H. E. H. 1982. Perspective on speciation by reinforcement. South African J. Sci. 78:53–57.
Paterson, H. E. H. 1985. The recognition concept of species. Pp. 21–29 in: E. S. Vrba (ed.), *Species and Speciation*. Transvaal Museum Monograph (4).

Richardson, R. H. 1974. Effects of dispersal, habitat selection, and competition on a speciation pattern of *Drosophila* endemic to Hawaii. Pp. 140–164 in M. J. D. White (ed.), *Genetic Mechanisms of Speciation in Insects*. Australia and New Zealand Book Co., Sidney.
Rolls, E. C. 1969. *They All Ran Wild: The Story of Pests on the Land in Australia*. Angus & Robertson, Sidney.
Rosen, D. E. 1960. Middle-American poeciliid fishes of the genus *Xiphophorus*. Bull., Florida State Mus., Biol. Sci. 5:57–242.
Rosen, D. E. 1978. Vicariant patterns and historical explantion in biogeography. System. Zool. 27:159–188.
Rosen, D. E. 1979. Fishes from the uplands and intermontane basins of Guatemala: Revisionary studies and comparative biogeography. Bull. Am. Mus. Nat. His. 162: 267–376.
Savage, J. M. 1975. Systematics and distribution of the Mexican and Central American stream frogs related to *Eleutherodactylus rugulosus*. Copeia 1975:254–306.
Simberloff, D., K. L. Heck, E. D. McCoy, and E. F. Connor. 1981. There have been no statistical tests of cladistic biogeographical hypothesis, Pp. 40–63 in G. Nelson and D. E. Rosen (eds.), *Vicariance Biogeography: A Critique*. Columbia University Press, New York.
Udvardy, M. D. F. 1969. *Dynamic Zoogeography with Special Reference to Land Animals*. Van Nostrand Reinhold, New York.
Wiley, E. O. 1977. The phylogeny and systematics of the *Fundulus notti* species group (Teleostei: Cyprinodontidae). Occas. Papers. Mus. Nat. Hist. Univ. Kansas (66):1–31.
Wiley, E. O. 1980. Phylogenetic systematics and vicariance biogeography. System. Bot. 5:194–220.
Wiley, E. O. 1981. *Phylogenetics: The Theory and Practice of Phylogenetic Systematics*. John Wiley, New York.
Wiley, E. O., and D. R. Brooks. 1982. Victims of history—a nonequilibrium approach to evolution. System. Zool. 31:1–24.
Wiley, E. O., and R. L. Mayden. 1985. Species and speciation in phylogenetic systematics, with examples from the North American fish fauna. Ann. Missouri Bot. Garden 72: 596–635.
Wing. L. 1943. Spread of the starling and English sparrow. Auk 60:74–87.
Zweifel, R. G. 1957. A new frog of the genus *Rana* from Michoacan, Mexico. Copeia 1957:78–83.

PART SIX

EVOLUTIONARY CONSEQUENCES OF SPECIATION

CHAPTER 22

MACROEVOLUTIONARY CONSEQUENCES OF SPECIATION: Inferences from Phytophagous Insects

Douglas J. Futuyma

Punctuated equilibrium is both a description of pattern in the fossil record and a hypothesis of evolutionary process. The pattern claimed by Eldredge and Gould (1972) is one of abrupt change in morphological characters that both before and after the "punctuation" change little over long periods ("stasis"). Although such a pattern could represent changes in evolutionary rate in a single, nondividing lineage, the process Eldredge and Gould hypothesized is that rapid changes occur in localized populations, in association with the acquisition of reproductive isolation (i.e., with the origin of a new biological species). This hypothesis was, explicitly, an application to the fossil record of Mayr's (1954) theory of peripatric speciation. Indeed, Mayr himself (1954) had described the implications of his theory for interpretation of discontinuities in the fossil record. If it were true that lineages do not evolve except in association with speciation events, long-term trends would very likely be the consequence of the differential proliferation or extinction of species with different character states, a process that was termed species selection (Stanley 1975, 1979, but see Vrba and Eldredge 1984).

Surely the most controversial aspect of punctuated equilibrium has been the proposition, again foreshadowed by Mayr (e.g., Mayr 1963:542, 615), that speciation (in the sense of bifurcation of a lineage) is a necessary condition for morphological evolution. In proposing this, Eldredge and Gould (1972) invoked Mayr's (1963) argument that coadaptation and other sources of genetic homeostasis (Lerner 1954) tend to prevent substantial evolutionary change except when genetic drift in a small population initiates a "genetic revolution" leading to both morphological change and the acquisition of reproductive isolation. In developing the genetic basis of his argument, Mayr (1963) relied heavily on the work of Dobzhansky and his students; Dobzhansky, in turn, was strongly influenced by Wright (Provine 1986), and found in Wright's work a theoretical justification for his emphasis on coadaptation. Punctuated equilibrium appears, then, to be an extreme but natural descendant of one school of thought that from the very beginning of the evolutionary synthesis has emphasized gene interaction, homeostasis, and rapid shifts between stable states: the same interactive view of nature that underlies Wright's (1931, 1982) shifting balance theory.

But throughout and since the evolutionary synthesis, a competing point of view, for which Fisher was perhaps the most prominent spokesman (Provine 1986), has emphasized the relatively unconstrained behavior of single genes and additively inherited polygenic traits, and consequently the ready response of populations to mass selection. Wherever reality may lie on the spectrum between Fisher's and Wright's views, considerable evidence—the abundance of additive genetic variation, the ready response to artificial selection, the ubiquity of geographic variation, the strength of selection relative to gene flow—argues against the supposition that speciation is a necessary condition for morphological change (Lande 1980; Templeton 1980; Turner 1981; Charlesworth et al. 1982; Barton and Charlesworth 1984). From a genetic point of view, the hypothesis that Eldredge and Gould (1972) proposed to explain the patterns claimed for the fossil record appears untenable. The evidence from population biology does not, of course, deny the possibility of a historical *pattern* of stasis and punctuation.

As a nonpaleontologist, I will not venture to judge whether or not the fossil record offers a pattern of stasis and punctuation, on which there exists some argument (e.g., Gingerich 1976; Hoffman 1982; Levinton 1988). In this chapter, rather, I will attempt to integrate several themes, in elaborating a hypothesis (Futuyma 1987) based on genetic assumptions that should be acceptable even to a Fisherian. In particular, I assume that selectable genetic variation with a local population is generally sufficient to enable selected characters to near their equilibria very rapidly, from a paleontological perspective. I suggest that although the mechanism of punctuated equilibrium may be untenable as originally formulated, speciation may in some instances nevertheless account for some punctuated patterns, but for different reasons than those postulated by Eldredge and Gould (1972). Further, speciation may in some instances facilitate anagenesis, that is, a change of many standard deviations from an ancestral character state. Mayr (1963:621) may well have been right in claiming that "speciation, the production of new gene complexes capable of ecological shifts, is the method by which evolution advances. Without speciation there would be . . . very little evolutionary progress."

Before proceeding further, an important semantic problem must be considered. Eldredge and Gould's (1972) argument that morphological change is associated with speciation has been criticized as tautological (e.g., Levinton and Simon 1980), in that paleontologists usually designate distinguishable chronologically sequential forms ("chronospecies") as different species. Advocates of punctuated equilibrium, however, have recognized that chronospecies do not provide evidence for the hypothesis that a derived morphology characterizes one of two "daughter species" of a bifurcating lineage; the necessary evidence for their hypothesis, they say, would be temporal and spatial overlap of ancestral and derived forms (Gould 1982; Eldredge and Gould 1988). In this chapter, I address not the evidence for or testability of punctuated equilibrium as a hypothesized process, but the possibility that speciation could generate morphological punctuations in lineages that, were they not to speciate, would appear relatively static. Throughout, I use the biological species concept (Mayr 1963), and use "speciation" to mean the evolution of a genetically based barrier to gene exchange. The argument that follows exemplifies the greater utility of the biological species concept than of competing species concepts in discussing macroevolution in sexually reproducing organisms.

POLYMORPHISM, RECOMBINATION, AND ADAPTATION

My argument begins with the observation that in animals, it is rare to find discrete genetically different phenotypes (i.e., bimodal or multimodal trait distributions) associated with different resources or microhabitats within a single panmictic population (Futuyma and Peterson 1985; Futuyma and Moreno 1988). That the rarity of such a distribution is attributable to recombination is indicated by the comparatively high frequency of sympatric genotypes associated with different resources, and differing in one or more traits, in parthenogenetically reproducing forms, e.g., of fish (Vrijenhoek 1984) and insects (Mitter et al. 1979; Futuyma et al. 1984; Futuyma and Peterson 1985). The variation in many, if not most, traits is polygenic; because of recombination, disruptive selection on a polygenic trait in a sexual population will seldom maintain a bimodal distribution (Felsenstein 1979). Moreover, heterogeneous selection on a single locus with discrete phenotypes will maintain polymorphism only under rather stringent conditions (Maynard Smith and Hoekstra 1980). These conditions are considerably broader if the genotypes choose habitats in which they are most fit; but if habitat selection is governed by loci different from the locus affecting the morphological character, the likelihood of polymorphism decreases as the recombination fraction increases (García-Dorado 1985). Presumably the likelihood of polymorphism decreases further if habitat selection itself is controlled by several loci, as appears to be the case in some insects (Jaenike 1986, 1988; van der Pers 1978; van Drongelen and van Loon 1980; Futuyma 1983a). At least in phytophagous insects, host preference and performance on different hosts appear to be largely under independent genetic control (Wiklund 1975; Futuyma and Peterson 1985; Via 1986; Thompson 1988), although in some vertebrates and bumblebees it appears likely that resource preference may be correlated with morphology within populations by virtue of learning (e.g., Grant 1986; Morse 1978; Smith 1987).

The conclusion of this passage is that except for nongenetic developmental polymorphisms (West-Eberhard 1986), discretely different phenotypes adapted to different resources or microhabitats will rarely coexist within populations, because of recombination. For single-locus differences (Hedrick 1986), and even more forcefully for polygenic traits, we can conclude that discrete phenotypes generally persist only if there is some interruption to random mating among genotypes favored by different environments—such as that provided by geographic segregation. In particular, an association between alleles governing resource selection and other resource-related traits will exist primarily among geographically segregated populations faced with different relative abundances of resources, rather than within populations.

THE CONSISTENCY OF SELECTION

My second theme draws on Lewontin's (1983) point that the organism is not merely the object, but the subject, of selection. *By its behavior—its genetically determined use of habitats and resources—an animal largely defines the selective pressures to which it is subject.* (In plants, evolved responses to cues for seed germination may play this role to some extent.) It is widely believed that the evolution of habitat or resource preference sets the stage for the later evolution of other adaptations to the habitat or resource. For example, this appears likely in phytophagous insects, in which closely related populations or species with different host associations often display less pronounced morphological or physiological adaptations to their hosts than do more distantly related species (Futuyma 1983b; Futuyma and Peterson 1985).

Adaptive features span a spectrum from narrowly context-dependent "special" adaptations to "general" adaptations that confer an advantage over a broad range of habitats, resources, or interspecific interactions (Wright 1941; Brown 1959). I propose as a hypothesis (to be explored further below) that genetic changes conferring general adaptation are a small minority of those that are brought about by selection. For example, in a widespread species that occupies quite different habitats in different parts of its range, most advantageous mutations that approach fixation will do so in populations that inhabit one or another of the habitats; only a minority of mutations will sweep through and be fixed throughout the species. Thus, highly general adaptations, such as the plumose hairs of bees that facilitate collection of pollen from a broad range of flowers, are most evident as synapomorphies of higher taxa; it is rare to find a species that differs from its congeners by possession of an autapomorphic general adaptation. Many, if not most, of the adaptive differences among closely related species are, instead, special adaptations to differences in the ecological niches that the organisms' behavior defines. For example, except for sexually selected traits, most of the morphological diversity among hummingbirds (Colwell 1988) can be related to their resources (presumably adopted in some instances under pressure of interspecific competition). The selection on many of the characters in which related species have diverged by natural selection has been imposed, directly or indirectly, by interactions with other species; these interactions are determined very largely by a species' habitat and resource utilization, hence by the behavior that establishes its niche. The choice of host plant by a phytophagous insect, for instance, affects the course of selection not only on digestive physiology, phenology, and morphological adaptation to the plant, but on responses to the particular predators and parasites associated with one or another plant (Jeffries and Lawton 1984). The extent to which a special adaptation evolves in response to an interacting species depends on the consistency and exclusivity of the interaction (Wheelwright and Orians 1982; Thompson 1982; Schemske 1983; Futuyma and Slatkin 1983; Howe 1984; Futuyma 1986).

EPHEMERAL ADAPTATION AND PUNCTUATED EQUILIBRIUM

The constellation of species with which a broadly distributed species interacts varies geographically and affects geographic variation in characters. In all environments, populations fluctuate and move about from one locality to another. Local populations of phytophagous insects last no longer than those of their hosts. Among both plants and animals, associations of species are so variable in time and space that the concept of "community" is hard to apply (Järvinen et al. 1986; Underwood 1986) and is rejected by some ecologists. Thus, on a local scale, the pattern of spatial variation under selection imposed by interacting species must be quite ephemeral, because (1) the spatial pattern of selection fluctuates; (2) local populations become extinct, and some of their sites are recolonized; and (3) changes in the spatial location of populations and outbreaks foster gene flow among sites, partly as a consequence of the extinction/colonization process (Slatkin 1977). Parker (1985) has found very local, possibly coevolved, differences in the responses of a legume and its fungal parasite to each other, but we should be surprised if the same geographic pattern were evident 50 years from now. The incipient divergence among the populations will have been erased by fluctuating selection and gene flow, and new geographic patterns will have arisen.

On a broader geographical scale, geographic variation lasts longer, but is nonetheless also ephemeral. As climates change, the location of a geographic variant's typical habitat and resources shifts, and with it the distribution of the populations associated with that resource. The evidence from Pleistocene distributions illustrates this clearly, and shows, moreover, that species do not move about as coherent associations, but largely independently (Davis 1976; Cushing 1965). The members of associations recognized by phytosociologists (e.g., spruce and fir) have had rather independent geographic histories. Coope (1979) describes rapid changes in the Quaternary beetle fauna of Britain, which at some times included species limited today to southern Europe or tropical Asia or Africa, and at other times included species variously distributed today in Tibet, eastern Siberia, and other parts of arctic Russia. All Quaternary beetles are indistinguishable, even in fine morphological detail, from living species.

Drawing these several themes together into the thesis of this chapter, we may recognize two effects of such shifts in geographic distributions. Consider a local regional variant of a species, possessing adaptations to a locally abundant resource or microhabitat, or to resource-associated parasites or competitors. Elsewhere, in a more widespread habitat type, the species has different characters. If a climatic change induces a change in the distribution of interacting species, some of the differences in selective regime that the variants have heretofore experienced will be lessened or cease to exist. The more similar the habitats and resources with which the isolates are associated, the more pronounced the convergence of their character optima, because the exchange of predators, parasites, etc. between two similar microhabitats occupied by the diverent populations will be greater than between two dissimilar microhabitats. Reversals of selection stemming from changes in "community" composition are thus a possible source of the apparent stasis that, carefully examined, may really be fluctuation about a long-term mean (Charlesworth 1984; Bell et al. 1985).

A second effect of climate-induced shifts in the distribution of a species' habitats or resources (e.g., in the distribution of a phytophagous insect's hosts) is that spatially segregated resources, and with them their associated populations, may be brought into sympatry. If the populations have not yet evolved reproductive isolation, recombination reduces the linkage disequilibrium both among loci governing each of the morphological or physiological traits conferring adaptation to one or the other resource, and between loci governing these traits and those affecting resource or habitat selection. Unless there is tight linkage, the physiological or morphological adaptations to the different resources become dissociated from the habitat selection that determines where they will be exercised. That is, the factors that militate against the maintenance of multiple-niche polymorphism for groups of loci again come into play when geographic isolation between divergent gene pools breaks down. Thus, for example, the divergent adaptations of geographically remote populations of Colorado potato beetle (*Leptinotarsa decemlineata*) to different species of *Solanum* (Hsiao 1978), or of the tiger swallowtail butterfly (*Papilio glaucus*) to *Populus* versus *Liriodendron* (Scriber 1983) would not persist if their hosts became broadly sympatric. As a consequence of interbreeding, the distinctive features of the more localized variant, or the one associated with the rare resource, will be lost, swamped by gene flow from the larger gene pool.

If, in contrast, geographically isolated resource- or habitat-associated populations have acquired reproductive isolation, their divergent characters can persist even as the species move about in space in concert with their resources; even if their resources become sympatric or mosaically distributed, the populations can become broadly sympatric and retain their identity. Moreover, by retaining their association with different resources or microhabitats (niches), they remain subject to many of the divergent selection pressures they experienced while allopatric. The consequence for paleontology and macroevolution is that the characters are more likely to persist long enough to be revealed in the fossil record—indeed, long enough to persist through the subsequent evolution and diversification of the clades to which the species may give rise. But if the geographic distribution of resources and habitats shifts on a time scale shorter than the time scale of speciation, divergent characters acquired in allopatry are lost to recombination, and may have so ephemeral a life as not to be recorded in the fossil record, or to be recorded as brief excursions from a long-term unchanging mean. Thus, speciation may facilitate anagenetic change, not by revolutionizing a refractory gene pool or by otherwise enabling responses to selection, but by conferring enough permanence on evolutionary changes to enable their recovery in the fossil record and by rescuing them from evolutionary oblivion.

The discussion so far has emphasized the important role of habitat selection in maintaining a population's association with a resource, and so maintaining the consistency of selection on phenotypic characters as long as they remain in linkage disequilibrium with alleles governing habitat selection. Habitat selection is not essential to the argument, however. A localized ecotypic variant of a plant, for example, remains distinct only insofar as the strength of selection exceeds that of gene flow. If its microhabitat shifts geographically in the course of time, the ecotype might persist if the new sites are colonized by seed from the old, but this process would almost certainly be diluted by gene exchange with intervening populations in different habitats. It is more likely that the populations of the ecotype that existed before the habitat shift would lose their distinctive character, and that populations with similar phenotypic properties would arise anew in the new sites, molded by selection from the local gene pool. That this is likely is indicated by the different genetic bases for similar phenotypic characters in different populations of the "same" ecotype (reviewed by Cohan 1984). Were genetic variation insufficient for the repeated origin of the ectotype where the appropriate habitat comes into existence, its appearance on the evolutionary stage would be brief, whereas a reproductively isolated, habitat-specialized species could persist as long as it were capable of dispersing from one ephemeral site to another.

ANAGENETIC TRENDS

Let us consider a character that, in part of a clade, has diverged greatly from the ancestral condition, and ask by what history it has arrived at its state. I will again focus on adaptations to resources, and will shift illustrative scenarios from phytophagous insects to feeding structures in birds, which are more easily visualized and better understood. A particularly dramatic case is the South American hummingbird *Ensifera ensifera* (Figure 1), which feeds primarily in flowers with very long, tubular corollas, and has a

FIGURE 1. Left, a "typical" hummingbird (species unidentified); right, *Ensifera ensifera*, the sword-billed hummingbird. (Photographs courtesy of VIREO/C. H. Greenewalt.)

beak both absolutely and relatively longer than that of other hummingbirds. Three historical scenarios for the evolution of such a trait are illustrated in Figure 2, in which "fitness" refers to the fitness accrued by an individual of phenotpye z_j from a unit of resource i.

1. In the most traditional scenario (Figure 2A), a single anagenetically evolving population tracks a gradual change in the optimum (i.e., a shift in the fitness function from 1 toward k). For *Ensifera*, this could result from a gradual change in the species composition of the plant community, a coevolution of beak length and corolla length in a plant species with which the bird is intimately associated, or pressure from interspecific competition toward increasingly extreme resources (flower types). In any case, anagenesis does not require speciation.
2. An ecologically generalized species feeds on both an abundant resource (1) and a very different resource (2), such as a long-corolla plant, to which it is not well adapted. Since the range of morphological variation in beak length overlaps the fitness function associated with resource 2, evolution toward a new optimum z_2 could occur if resource 1 becomes rare or extinct (Milligan 1986). If this occurs only in certain local populations of the species, then by the preceding argument the evolutionary excursion will be only temporary if (more abundant) ancestral and (less abundant) derived populations later become sympatric and interbreed. Thus, speciation can facilitate anagenesis in this scenario (Figure 2B).
3. The population is near the optimum (z_1) for resource 1, and its variation in beak length does not overlap the distant fitness function associated

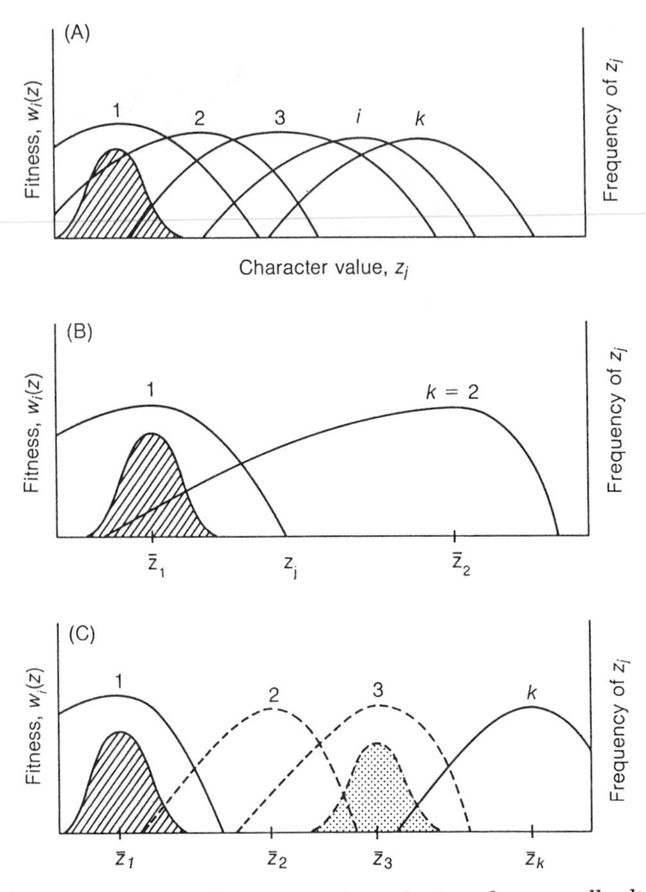

FIGURE 2. Three scenarios for anagenetic evolution of a normally distributed trait z from mean $\bar{z}_1$ to $\bar{z}_k$. In (A), the fitness function shifts gradually from $i = 1$ to k as the environment slowly changes. In (B), the character evolves in response to an increase in the abundance of environment 2, with associated fitness function w_2. In (C), the fitness function w_k does not overlap the frequency distribution of z, and evolution from $\bar{z}_1$ to $\bar{z}_k$ requires transitional adaptation to intermediate environments.

with resource k; the resource is entirely unusable. If resource 1 becomes rare or extinct so that only resource k is available, there can be no directional selection for a longer beak, and the population's fate is probably extinction. Attainment of beak length z_k from z_1 is possible only by stepwise adaptation to a succession of intermediate resources. But by the gene flow argument that I previously presented, each shift in resource (1 to 2, 2 to 3, 3 to . . . k) brings about a change in $\bar{z}$ that is ephemeral unless it is "protected" by speciation from the reversal consequent on interbreeding with populations that have retained the more ancestral association and character state. In this scenario, then, the anagenetic trend requires a succession of speciation events associated with adaptation to ever more extreme resources; speciation events prevent the "slippage" consequent on interbreeding, and act, so to speak, like pitons on an adaptive slope (Figure 2C).

Moreover, because the rate of recombination limits the variance of a polygenic character, the variance within a population will generally be less than that among species or among spatially segregated populations. Therefore the likelihood that an extreme phenotype will evolve in a clade is greater, the greater the number and phenotypic diversity of species within the clade (cf. Mayr 1963; Arnold and Fristrup 1982). Spatially segregated populations that have not acquired reproductive isolation may diverge to any degree under selection, but by the argument of this chapter, the anagenetic difference achieved by an isolate is only as long lived as its isolation from more abundant populations.

PATTERNS IN PHYTOPHAGOUS INSECTS

To date, unfortunately, I have not thought of definitive tests of the argument previously developed. In general, its predictions respecting evolutionary *patterns* are similar to those of punctuated equilibrium, although based on different mechanisms. The prediction that morphological divergence may be greater in clades in which speciation occurs frequently than those in which it occurs less often is similar to the prediction from punctuated equilibrium that Douglas and Avise (1982) explored (and, in this instance, rejected). By my argument, stasis and punctuation might more frequently be observed in "specially" adaptive than "generally" adaptive characters, if such can be discriminated in the fossil record. "Specially" adaptive characters may include, in particular, those that are geographically variable in widespread extant species. Substantial anagenetic change may well be more likely in regions with long periods of environmental stability than in less constant environments. Coope (1979) made the same suggestion, based on the extraordinary morphological stability of beetles throughout the Quaternary in the North Temperate zone, in which the movement of species "must have continually broken down the geographical barriers that separated populations, permitting genetic mixing and keeping the gene pools well stirred."

In the absence of definitive tests, we may ask if any evidence is at least consistent with the hypothesis. The hypothesis proposes that speciation "captures" and provides permanence to adaptations to particular resources. In phytophagous insects, for example, a "parent" species might be able to use several host plants. Specialization for one of these may evolve in a local population, but be ephemeral unless the population achieves reproductive isolation. If this commonly occurred, we might expect specialized species that have polyphagous relatives to feed on a subset of the polyphagous relatives' hosts. It is not necessary, a priori, that this prediction hold. In some cases the food plants of specialists are not included in the diet of related generalists; for example, in *Incisalia* (Lepidoptera: Lycaenidae), *I. henrici* feeds on both Ericaceae and Rosaceae, but although two other species feed on Ericaceae, one (*I. irus*) is apparently restricted to legumes and two others to conifers (*I. lanoraieensis* on *Picea*, and *I. niphon* on *Pinus* and perhaps *Juniperus*) (Klots 1951). This example illustrates, moreover, that closely related insects need not have similar host affiliations. For example, sibling species of *Calligrapha* (Coleoptera: Chrysomelidae) are variously specialized on plants on the Cornaceae, Salicaceae, Tiliaceae, Betulaceae, Malvaceae, and Rosaceae (Brown 1945). Some of the cerambycid genera included in the following analysis similarly include species associated with plants that are quite unrelated and have no evident commonality in their secondary chemistry; the hosts of *Phymatodes*, for example, include plants in the Pinaceae, Fagaceae, Rosaceae, and Salicaceae.

The predicted pattern appears to hold, however, in the Cerambycidae (long-horned beetles, with wood- or stem-boring larvae) of America north of Mexico (Linsley 1962a, 1962b, 1963, 1964; Linsley and Chemsak 1976, 1984). The data, presented in Figure 3, are incomplete because I lacked access to the treatments of parts of the Lepturinae and Lamiinae; the volumes I examined provide larval host records for at least some species in 177 genera (81 of them monotypic within the region covered). I designated as specialized a species for which all host records are from confamilial plants, and as generalized a species recorded from more than one plant family. My criterion was the number of families used, rather than lower taxa of plants, because few host records are given at the level of plant species and because the relatively few specialists recorded from two or more confamilial genera often occur on the same groups of genera (e.g., *Salix* and *Populus*, and *Picea* and *Abies*), suggesting that these plants are seldom "viewed" by cerambycids as different.

Figure 3 plots, for those genera that include at least one generalized and at least one specialized species, the frequency distribution, over genera, of the related ratios *A*/*B* and *C*/*D*, where *A* is the number of specialists recorded from at least one of the plant genera included in the diet of congeneric generalists, *B* is the number of specialized species in the genus, *C* is the number of plant genera used by specialists that are included

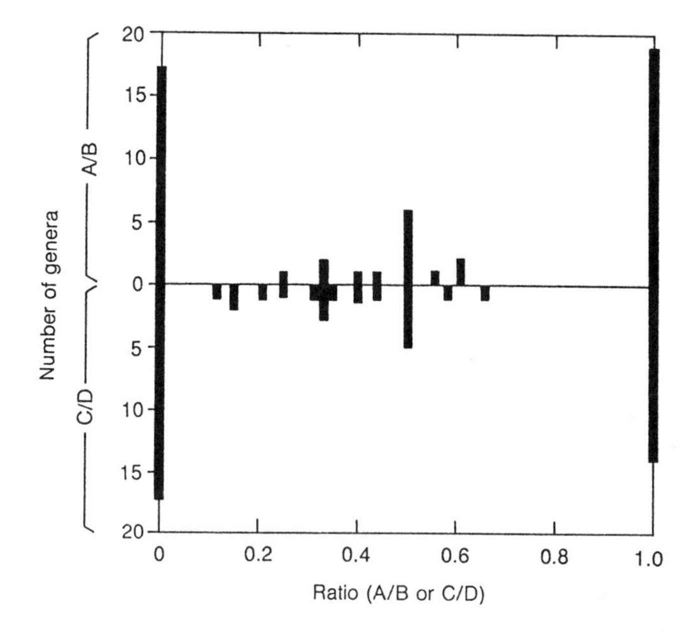

FIGURE 3. The number of North American cerambycid genera, containing both host plant specialists and generalists, with ratios *A/B* (above the horizontal axis) and *C/D* (below). The ratios are explained in the text. *A/B* measures, in effect, the proportion of specialized species that feed on the hosts of congeneric generalists. *C/D* measures the proportion of hosts used by the specialists in a genus that are also used by generalists.

in the host list of at least one congeneric generalist, and *D* is the number of plant genera recorded for the aggregate diets of the specialists in the genus. The values 0 and 1 predominate because of small numbers of species in many of the genera. Taken at face value, the plot suggests that in about half the genera, the diets of specialists are subsets of those of congeneric generalists.

I compared the observed number of specialized species recorded from the hosts of congeneric generalists with the number expected if specialists were allocated at random among all the plant genera (x_T) from which the cerambycids in these genera have been recorded (both generalists and specialists). The expected proportion of specialists feeding on congeneric generalists' hosts was calculated as

$$p_e = \frac{1}{n_s} \sum_i \frac{n_i x_i}{x_T}$$

where n_s is the number of specialized species (summed over genera), n_i is the number of specialists in cerambycid genus i, x_i is the number of plant genera from which generalists in genus i have been recorded, and x_T is the total number of plant g^nera as previously described. For the full data set, $x_T = 116$, $n_s = 137$, and $p_e = 0.094$; the difference between the observed (56) and expected (12.9) number of specialists on congeneric generalists' hosts is highly significant ($\chi^2 = 167.3$, df = 1). Because p_e is highly sensitive to x_T, I reanalyzed the data by deleting plant genera (and associated insects) from which only one cerambycid species has been recorded, on the supposition that these plants have been little surveyed for cerambycids. This reduces x_T to 74 and n_s to 130; $p_e = 0.132$, and the difference between observed (56) and expected (17.2) numbers of specialists on congeneric generalists' hosts is still highly significant ($\chi^2 = 47.324$, df = 1). When the analysis was redone using a species' association with one versus more than one plant genus as the criterion of specialization versus generalization, the difference between the observed (66) and expected (12.5) number of specialists feeding on hosts of polyphagous congeners was again highly significant.

These data are consistent with the pattern expected if restriction and (presumably) adaptation to one of a parent generalist's host were frequently accompanied and stabilized by reproductive isolation. I offer this observation only for speculation; the data base is surely weak for at least three reasons. First, host records are very incomplete, and many of the apparent specialists may have broader host ranges than the records indicate. Second, many of the genera are probably paraphyletic; if a phylogeny were available, inclusion of many other, often small, genera into larger genera might alter the pattern. Third, there is no phylogenetic information on the direction of evolution between generalized and specialized feeding habits.

Some very preliminary information (Futuyma, in press) from my current work is compatible with the hypothesis that divergence may be facilitated by intervening speciation events. *Ophraella* (Coleoptera: Chrysomelidae, Galerucinae) is a clearly monophyletic North American genus of leaf-eating beetles. Each of most of the 13 currently recognized species (LeSage 1986) feeds only on certain species in a single genus of Asteraceae (Compositae); in two cases a species feeds on two closely related host genera (*O. communa* on *Ambrosia* and *Iva*; *O. pilosa* on *Solidago* and *Aster*). For a chrysomelid to maintain a population on a plant, at least five criteria must be met: the plant must elicit feeding in both larvae and adults, it must, if eaten, support larval growth and adult survival, and it must elicit oviposition. *Ophraella communa* and *O. notulata* appear to be close relatives, and feed on related plants, but when reared on *O. notulata's* host, *O. communa* feeds readily but shows low larval growth and survival. Moreover, few females will oviposit on the plant. Thus adoption of even a closely related host seems to entail change in several physiological traits.

I have performed a number of feeding trials of adult *Ophraella* on the hosts of their congeners, as the availability of beetles and plants has permitted. A positive feeding response by at least some individuals is probably the minimal requirement for founding a colony on a new host; the likelihood of oviposition and larval growth is even lower than that of adult feeding. Tests on nonhosts consisted of confining beetles individually with foliage for at least 24 hours, and in some cases up to 72 hours, and scoring the leaf area consumed; typically, a beetle will attack its own host immediately if starved for 24 hours. Figure 4 schematically illustrates the results to date; the plants are divided into tribes and the beetles into the four "species groups" recognized by LeSage. I am presently engaged in a phylogenetic analysis that will alter this scheme; preliminary data suggest that *O. conferta* and *O. cribrata* are more closely related to each other, and *O. americana*/*O. pilosa* less closely related to *O. cribrata*, than in LeSage's arrangement. Anything more than a minimal (trace) score indicates that at least some individuals consumed the test plant. Granted that the matrix is incompletely filled, some indication of pattern nevertheless is apparent. For the most part, if beetles responded positively to nonhosts at all, they responded to plants in the same tribe as their natural host. In most of the instances in which a species responded to a plant in a different tribe, the plant is the host of a member of the same species group of *Ophraella* (cf. responses of *O. artica* and of *O. communa* and *O. ?californiana*[1] to *Ambrosia* and *Artemisia*).

Considering only adult feeding, a positive response to a congener's host implies a potential (yet to be tested) for genetic variation in the threshold of acceptance, hence the potential for selection to alter the response. It would take large samples to show that there is no variation in response to a distant relative's host, but in some instances little variation is evident. For example, none of 323 *O. communa*, which is associated with *Ambrosia*, ate *Solidago altissima* after 48 hours; there may well be little scope for selection for a shift between *Ambrosia* and *Solidago altissima*. Evolution of the capacity to use these very different hosts, then, may have required transition through other hosts. That all species of *Ophraella* are highly host specific in nature suggests that these transitions have occurred via a history of speciation events.

POPULATIONS AND SPECIES

Modern evolutionary theory has focused on the population as the locus of evolutionary change, to the extent that some authors have cited the pre-

[1]The *Artemisia*-associated population from Texas, here referred to as *?californiana*, will be described as a new species in a forthcoming publication (Futuyma, in preparation).

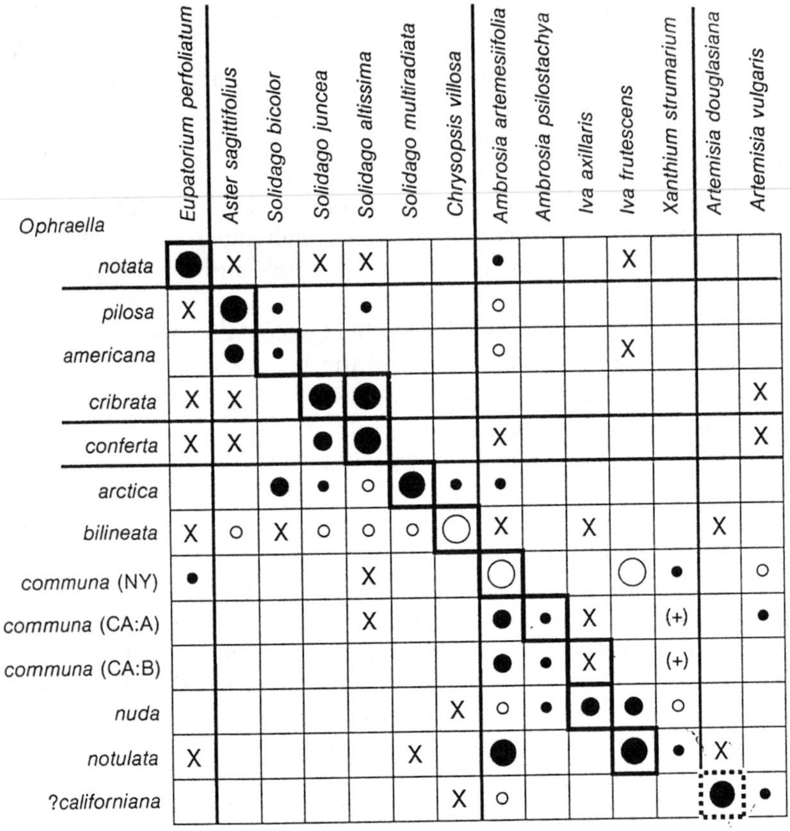

Ophraella	*Eupatorium perfoliatum*	*Aster sagittifolius*	*Solidago bicolor*	*Solidago juncea*	*Solidago altissima*	*Solidago multiradiata*	*Chrysopsis villosa*	*Ambrosia artemesiifolia*	*Ambrosia psilostachya*	*Iva axillaris*	*Iva frutescens*	*Xanthium strumarium*	*Artemisia douglasiana*	*Artemisia vulgaris*
notata	⬤	X		X	X			•			X			
pilosa	X	⬤	•		•			○						
americana		●	•					○			X			
cribrata	X	X		⬤	⬤									X
conferta	X	X		●	⬤			X						X
arctica			●	•	○	⬤	•	•						
bilineata	X	○	X	○	○	○	◯	X		X			X	
communa (NY)	•				X			◯			◯	•		○
communa (CA:A)					X			●	•	X		(+)		•
communa (CA:B)								●	•	X		(+)		
nuda							X	○	•	●	●	○		
notulata	X					X		⬤			⬤	•	X	
?californiana							X	○					⬤	•

FIGURE 4. A highly schematic figure of the feeding responses of adult beetles in the genus *Ophraella* (rows) to host plants (columns) used by members of the genus. Heavy vertical lines separate tribes of Asteraceae; heavy horizontal lines divide *Ophraella* into the species groups recognized by LeSage (1986). Each symbol represents mean consumption over 24 hours, the test in most cases extending 2–3 days. X represents no feeding, a small open circle trace feeding, and increasingly large filled circles slight, moderate, and heavy feeding. Large open circles represent heavy feeding in other, not readily comparable, experiments. (+) represents a nonquantified observation. The heavily outlined cells are naturally occurring associations; the broken border for ?*californiana*/*Artemisia douglasiana* represents a Texas population collected on *Artemisia* cf. *Carruthii* and resembling in its host association *O. californiana*, which has been recorded on *A. douglasiana* in California. Two Californian populations of *O. communa* (A and B) are included. *Xanthium*, a relative of *Iva*, has been recorded only as an incidental host of *O. communa* in California; *Artemisia vulgaris*, a European species that is abundant in New York, is not a natural host of any species. The data on which the diagram is based will appear in Futuyma (in press).

eminence of populations over species in evolutionary theory as an argument against the utility of the biological species concept (Sokal and Crovello 1970). It is of course true that gene frequency change within populations in the sine qua non of evolution. Moreover, populations commonly are so responsive to selection that one may well argue that there exist few constraints on their evolution other than ecological forces of stabilizing selection.

To focus on the population as the exclusive locus of evolution, however, is in a way to take as parochial and reductionist a view as to suppose that a single-locus model is an adequate representation of the evolutionary process. As single loci are embedded in genomes, and contribute to characters that function in the larger context of the organism, so a population is embedded in a network of populations—a species— and its fate is not independent of theirs. This realization is central to Wright's shifting balance theory (Wright 1931, 1982), a theory of the long-term evolution of a species (and which, incidentally, suggests few if any definitive tests). In Wright's conception, a low level of gene flow enhances local genetic variability, small population size in concert with selection permits occasional shifts to higher adaptive peaks, and the greater productivity of these populations spreads the favorable gene combinations throughout the species: "the competition between subgroups will bring about rapid changes in the gene frequencies of the species as a whole" (Wright 1931:151). Thus "an evolutionary advance through the field of possible combinations of the genes . . . should be relatively rapid and practically unlimited," and "the direction of evolution of the species as a whole will be closely responsive to the prevailing conditions, orthogenetic as long as these are constant, but changing with sufficiently long continued environmental change" (Wright 1931:151, 152).

In Wright's theory, speciation plays no special role in macroevolution, but neither is the local population the sole locus of evolution; its fate affects and is affected by that of other populations, and the evolving unit is ultimately the species. The impact of gene flow among populations is to enhance adaptation. The theory describes, however, only character modifications that are advantageous in all populations of the species, in whatever environments they occupy (i.e., general adaptations).

It is extremely difficult to say how commonly improvements of the kind described by Wright occur. If we may assume that adaptive features held in common by numerous members of a clade in spite of their very different ecological circumstances exemplify Wright's general improvements (a debatable assumption), then the relative rapidity with which they become stabilized after their origin suggests that Wright's "practically unlimited," "orthogenetic" advances in generally adaptive characters account for a small fraction of the evolutionary events in a lineage. For example, acquisition of the distinctive tarsal structure that is an adaptation for running in perissodactyls occurred in less than five million years, a speed of transition indicating "a considerably higher rate of evolution in late Paleocene proto-perissodactyls than occurred during most of the subsequent 55 million years of perissodactyl evolution" (Radinsky 1966), during which a considerable adaptive radiation occurred.

If, then, most adaptive evolutionary events within populations involve adaptations to the population's peculiar ecological circumstances (of which there is no dearth of examples), the imprint of these events on evolution in the long term may be lessened by gene exchange with other, larger populations. In the perspective of geological time, the particular environment of a population may persist, but its location is impermanent. Spatial movement of a population with that of its environment brings about gene exchange with other populations. As parthenogenesis within a population can rescue an advantageous gene combination from destruction by recombination, so reproductive isolation of a population from other populations can preserve its distinctive adaptive character indefinitely through evolutionary time. In a macroevolutionary perspective, the biological species may play an irreducibly important role.

SUMMARY

In many animals, genetically determined habitat or resource selection maintains consistency of selection pressures and, consequently, stability in many phenotypic characters. Spatially segregated populations can diverge to any degree both in habitat or resource selection and in morphological or physiological adaptations. These differences rapidly break down under recombination and fluctuating selection if populations associated with different habitats or resources become sympatric and interbreed, as is likely when the geographic distribution of resources is altered. Such alteration is inevitable in the fullness of time. Acquisition of reproductive isolation enables retention of niche-associated differences in habitat selection and other phenotypic characters, even as a population moves about geographically in concert with the movement of its resources. Thus, reproductive isolation provides evolutionary changes with a degree of permanence that may cause a pattern of punctuation and stasis in the fossil record. A succession of such speciation events may under some circumstances be necessary for substantial anagenetic change. Patterns of host specificity in some phytophagous insects are consistent with (but by no means compel) the hypothesis that speciation facilitates association with and adaptation to special resources.

ACKNOWLEDGMENTS

I am grateful to H. R. Ackacaya, M. Burgman, and L. R. Ginzburg for the formulation used to analyze cerambycid host associations, to J. Coyne and C. Mitter for insightful comments on the manuscript, and to the reviewers and to members of the Department of Ecology and Evolution at Stony Brook, whose comments improved the manuscript's clarity. The research on which this chapter draws has been supported by the National Science Foundation (BSR8516316). This is Contribution No. 677 in Ecology and Evolution from the State University of New York at Stony Brook.

LITERATURE CITED

Arnold, A. J., and K. Fristrup. 1982. The theory of evolution by natural selection: A hierarchical expansion. Paleobiology 8:113–129.

Barton, N. H., and B. Charlesworth. 1984. Genetic revolutions, founder effects, and speciation. Annu. Rev. Ecol. System. 15:133–164.

Bell, M. A., J. V. Baumgartner, and E. C. Olson. 1985. Patterns of temporal change in single morphological characters of a Miocene stickleback fish. Paleobiology 11: 258–271.

Brown, W. J. 1945. Food-plants and distribution of the species of *Calligrapha* in Canada, with descriptions of new species (Coleoptera, Chrysomelidae). Can. Entomol. 77: 117–133.

Brown, W. L., Jr. 1959. General adaptation and evolution. System. Zool. 5:157–168.

Charlesworth, B. 1984. Some quantitative methods for studying evolutionary patterns in single characters. Paleobiology 10:308–318.

Charlesworth, B., R. Lande, and M. Slatkin. 1982. A neo-Darwinian commentary on macroevolution. Evolution 36:474–498.

Cohan, F. M. 1984. Can uniform selection retard random genetic divergence between isolated conspecific populations? Evolution 38:495–504.

Colwell, R. K. In press. A morphometric study of adaptive radiation in hummingbirds. Fieldiana.

Coope, G. R. 1979. Late Cenozoic fossil Coleoptera: Evolution, biogeography, and ecology. Annu. Rev. Ecol. System. 10;249–267.

Cushing, E. J. 1965. Problems in the Quaternary phytogeography of the Great Lakes region. Pp. 403–416 in: H. L. Wright, Jr. and D. G. Frey (eds.), *The Quaternary of the United States*. Princeton University Press, Princeton, NJ.

Davis, M. A. 1976. Pleistocene biogeography of temperate deciduous forests. Geosci. Man 13:13–26.

Douglas, M. E., and J. C. Avise. 1982. Speciation rates and morphological divergence in fishes: Tests of gradual versus rectangular modes of evolutionary change. Evolution 36:224–232.

Eldredge, N., and S. J. Gould. 1972. Punctuated equilbria: An alternative to phyletic gradualism. Pp. 82–115 in: T. J. M. Schopf (ed.), *Models in Paleobiology*. Freeman, Cooper, San Francisco.

Eldredge, N., and S. J. Gould. 1988. Punctuated equilibrium prevails. Nature (London) 332:211–212.

Felsenstein, J. 1979. Excursions along the interface between disruptive and stabilizing selection. Genetics 93:773–795.

Futuyma, D.J. 1983a. Selective factors in the evolution of host choice by insects. Pp. 227–244 in: S. Ahmad (ed.). *Herbivorous Insects: Host-Seeking Behavior and Mechanisms*. Academic Press, New York.

Futuyma, D. J. 1983b. Evolutionary interactions among herbivorous insects and plants. Pp. 207–231 in: D. J. Futuyma and M. Slatkin (eds.), *Coevolution*. Sinauer Associates, Sunderland, MA.

Futuyma, D. J. 1986. Evolution and coevolution in communities. Pp. 369–381 in: D. M. Raup and D. Jablonski (eds.), *Patterns and Processes in the History of Life*. Dahlem Konferenzen, Springer-Verlag, Berlin.

Futuyma, D. J. 1987. On the role of species in anagenesis. Am. Natur. 130:465–473.

Futuyma, D. J. In press. The evolution of host specificity in herbivorous insects: Genetic, ecological, and phylogenetic aspects. *In* P. W. Price, W. Benson, T. Lewinsohn and G. W. Fernandez (eds.), *Herbivory: Tropical and Temperate Perspectives*. Wiley, New York.

Futuyma, D. J., R. P. Cort, and I. van Noordwijk. 1984. Adaptation to host plants in the fall cankerworm (*Alsophila pometaria*) and its bearing on the evolution of host affiliation in phytophagous insects. Am. Natur. 123:287–296.

Futuyma, D. J., and S. C. Peterson. 1985. Genetic variation in the use of resources by insects. Annu. Rev. Entomol. 30:217–238.

Futuyma, D. J., and G. Moreno. 1988. The evolution of ecological specialization. Annu. Rev. Ecol. System. 19:207–233.

Futuyma, D. J., and M. Slatkin. 1983. Introduction. Pp. 1–13 in: D. J. Futuyma and M. Slatkin (eds.). *Coevolution*. Sinauer Associates, Sunderland, MA.

García-Dorado, A. 1986. The effect of niche preference on polymorphism protection in a heterogeneous environment. Evolution 40:936–945.

Gingerich, P. D. 1976. Paleontology and phylogeny: Patterns of evolution at the species level in early Tertiary mammals. Am. J. Sci. 276:1–28.

Gould, S. J. 1982. The meaning of punctuated equilibrium and its role in validating a hierarchical approach to macroevolution. Pp. 83–104 in: R.Milkman (ed.), *Perspectives on Evolution*. Sinauer Associates, Sunderland, MA.

Gould, S. J., and N. Eldredge. 1977. Punctuated eqilibria: The tempo and mode of evolution reconsidered. Paleobiology 3:115–151.

Grant, P. R. 1985. *Ecology and Evolution of Darwin's Finches*. Princeton University Press, Princeton, NJ.

Hedrick, P. W. 1986. Genetic polymorphism in heterogeneous environments: A decade later. Annu. Rev. Ecol. System. 17:535–566.

Hoffman, A. 1982. Punctuated versus gradual mode of evolution, a reconsideration. Evol. Biol. 15:411–436.

Howe, H. F. 1984. Constraints on the evolution of mutualisms. Am. Natur. 123:764–777.

Hsiao, T. H. 1978. Host plant adaptation among geographic populations of the Colorado potato beetle. Entomol. Exp. Appl. 24:237–247.

Jaenike, J. 1986. Genetic complexity of host-selection behavior in *Drosophila*. Proc. Natl. Acad. Sci. U. S. A. 83:2148–2151.

Jaenike, J. 1988. Genetics of oviposition-site preference in *Drosophila tripunctata*. Heredity 59:363–369.

Järvinen, O., and 10 others. 1986. The neontologico-paleontological interface of community evolution: How do the pieces in the kaleidoscopic biosphere move? Pp. 331–350 in: D. M. Raup and D. Jablonski (eds.). *Patterns and Processes in the History of Life*. Dahlem Konferenzen, Springer-Verlag, Berlin.

Jeffries, M. J., and J. H. Lawton. 1984. Enemy free space and the structure of ecological communities. Biol. J. Linnaean Soc. 23:269–286.

Klots, A. B. 1951. *A Field Guild to the Butterflies*. Houghton Mifflin, Boston.

Lande, R. 1980. Microevolution in relation to macroevolution. Paleobiology 6:402–416.

Lerner, I. M. 1954. *Genetic Homeostasis*. Oliver & Boyd, Edinburgh.

LeSage, L. 1986. A taxonomic monograph of the Nearctic galerucine genus *Orphraella* cox (Coleoptera:Chrysomelidae). Mem. Entomol. Soc. Can. No. 133.

Levinton, J. 1988. *Genetics, Paleontology, and Macroevolution*. Cambridge University Press, Cambridge and New York.

Levinton, J. S., and C. M. Simon. 1980. A critique of the punctuated equilibria model and implications for the detection of speciation in the fossil record. System. Zool. 29:130–142.

Lewontin, R. C. 1983. The organism as the subject and object of evolution. Scientia (Milan) 118:65–82.

Linsley, E. G. 1962a. *The Cerambycidae of North America. Part II. The Taxonomy and Classification of the Parandrinae, Prioninae, Spondylinae, and Aseminae*. University of California Publications in Entomology 19. University of California Press, Berkeley.

Linsley, E. G. 1962b. *The Cerambycidae of North America. Part III. Taxonomy and Classification of the Subfamily Cerambycinae, Tribes Opsimini through Megaderini*. University of California Publications in Entomology 20, University of California Press, Berkeley.

Linsley, E. G. 1963. *The Cerambycidae of North America. Part IV. Taxonomy and Classification of the Subfamily Cerambycinae, Tribes Elaphidionini through Rhinotragini*. University of California Publications in Entomology 21, University of California Press, Berkeley.

Linsley, E. G. 1964. *The Cerambycidae of North America. Part V. Taxonomy and Classification of the Subfamily Cerambycinae, Tribes Callichromini through Ancylocerini*. University of California Publications in Entomology 22, University of California Press, Berkeley.

Linsley, E. G., and J. A. Chemsak. 1976. *The Cerambycidae of North America. Part VI, No. 2. Taxonomy and Classification of the Subfamily Lepturinae*. University of California Publications in Entomology 80, University of California Press, Berkeley.

Linsley, E. G., and J. A. Chemsak. 1984. *The Cerambycidae of North America. Part VII, No. 1. Taxonomy and Classification of the Subfamily Lamiinae, Tribes Parmenini through Acanthoderini*. University of California Publications in Entomology 102, University of California Press, Berkeley.

Maynard Smith, J., and R. Hoekstra. 1980. Polymorphism in a varied environment: How robust are the models? Genet. Res. 35:45–57.

Mayr, E. 1954. Change of genetic environment and evolution. Pp. 157–180 in: J. Huxley, A. C. Hardy, and E. B. Ford (eds.), *Evolution as a Process*. Allen & Unwin, London.

Mayr, E. 1963. *Animal Species and Evolution*. Harvard University Press, Cambridge, MA.

Milligan, B. G. 1985. Punctuated equilibrium induced by ecological change. Am. Natur. 127:522–532.

Mitter, C., D. J. Futuyma, J. C. Schneider, and J. D. Hare. 1979. Genetic variation and host plant relations in a parthenogenetic moth. Evolution 33:777–790.

Morse, D. H. 1978. Size-related foraging differences of bumblebee workers. Ecol. Entomol. 3:189–192.

Parker, M. A. 1985. Local population differentiation for compatibility in an annual legume and its host-specific pathogen. Evolution 35:713–723.

Provine, W. B. 1986. *Sewall Wright and Evolutionary Biology*. University of Chicago Press, Chicago.

Radinsky, L. B. 1966. The adaptive radiation of the phenacodontid condylarths and the origin of the Perisssodactyla. Evolution 20:408–417.

Schemske, D. W. 1983. Limits to specialization and coevolution in plant-animal mutualisms. Pp. 67–111 in: M. H. Nitecki (ed.), *Coevolution*. University of Chicago Press, Chicago.

Scriber, J. M. 1983. The evolution of feeding specialization, physiological efficiency and host races in selected Papilionidae and Saturniidae. Pp. 373–412 in: R. F. Denno and M. S. McClure (eds.), *Variable Plants and Herbivores in Natural and Managed Systems*. Academic Press, New York.

Slatkin, M. 1977. Gene flow and genetic drift in a species subject to frequent local extinctions. Theoret. Pop. Biol. 12:253–262.

Smith, T. B. 1987. Bill size polymorphism and intraspecific niche utilization in the African finch *Pyrenestes*. Nature (London) 325:717–719.

Sokal, R. R., and T. J. Crovello. 1970. The biological species concept: A critical evaluation. Am. Natur. 104:127–154.

Stanley, S. M. 1975. A theory of evolution above the species level. Proc. Natl. Acad. Sci. U. S. A. 72:646–650.

Stanley, S. M. 1979. *Macroevolution: Pattern and Process*. Freeman, San Francisco.

Templeton, A. R. 1980. Macroevolution. Evolution 34:1224–1227.

Thompson, J. N. 1982. *Interaction and Coevolution*. John Wiley, New York.
Thompson, J. N. 1988. Evolutionary ecology of the relationship between oviposition preference and performance of offspring in phytophagous insects. Entomol. Exp. Appl. 47:3–14.
Turner, J. R. G. 1981. Adaptation and evolution in *Heliconius*: A defense of neoDarwinism. Annu. Rev. Ecol. System. 12:99–121.
Underwood, A. J. 1986. What is a community? Pp. 351–367 in: D. M. Raup and D. Jablonski (eds.), *Patterns and Processes in the History of Life*. Dahlem Konferenzen, Springer-Verlag, Berlin.
van der Pers, J. C. N. 1978. Responses from olfactory receptors in females of three species of small ermine moths (Lepidoptera: Yponomeutidae) to plant odours. Entomol. Exp. Appl. 28:199–203.
van Drongelen, W., and J. A. van Loon. 1980. Inheritance of gustatory sensitivity in F_1 progeny of crosses between *Yponomeuta cagnagellus* and *Y. malinellus* (Lepidoptera). Entomol. Exp. Appl. 28:199–203.
Via, S. 1986. Genetic covariance between oviposition preference and larval performance in an insect herbivore. Evolution 40:778–785.
Vrba, E. S., and N. Eldredge. 1984. Individuals, hierarchies, and processes: Towards a more complete evolutionary theory. Paleobiology 10:146–171.
Vrijenhoek, R. C. 1984. Ecological differentiation among clones: The frozen niche variation model. Pp. 217–231 in: K. Wohrmann and V. Loeschke (eds.), *Population Biology and Evolution*. Springer-Verlag, Berlin.
West-Eberhard, M. J. 1986. Alternative adaptations, speciation, and phylogeny (a review). Proc. Natl. Acad. Sci. U. S. A. 83:1388–1392.
Wheelwright, N. T., and G. H. Orians. 1982. Seed dispersal by animals: Contrasts with pollen dispersal, problems of terminology, and constraints on coevolution. Am. Natur. 119:402–413.
Wiklund, C. 1975. The evolutionary relationship between adult oviposition preferences and larval host plant range in *Papilio machaon* L. Oecologia 18:185–197.
Wright, S. 1931. Evolution in Mendelian populations. Genetics 16:97–159.
Wright, S. 1941. The "age and area" concept extended. (Review of *The course of evolution by differentiation or divergent mutation rather than by selection*, by J. C. Willis). Ecology 22:345–347.
Wright. S. 1982. Character change, speciation, and the higher taxa. Evolution 36:427–443.

CHAPTER 23

THE RELATIONSHIP BETWEEN SPECIATION AND MORPHOLOGICAL EVOLUTION

Allan Larson

INTRODUCTION

The study of speciation frequently conflates two conceptually distinct evolutionary phenomena: evolutionary change in organismal morphology and development versus evolutionary change in the reproductive properties of populations. Although speciation is formally conceived as an evolutionary splitting of lineages featuring intrinsic genetic and reproductive closure of populations (Mayr 1982), many studies of "speciation" have been largely studies of morphological evolution. Two recent developments have stimulated more precise investigation of speciation and morphological change as conceptually distinct historical phenomena. One of these is the theory of punctuated equilibrium, which postulates that speciation and morphological change are congruent in geological time (Eldredge and Gould 1972; Gould and Eldredge 1977). By formally postulating this association, the theory requires that speciation and morphological evolution be recognizable empirically as distinct historical events whose coincidence in time can be tested. The recent, widespread use of biochemical polymorphisms to study geographic variation constitutes the second development that has promoted a more precise analysis of the congruence between speciation and morphological evolution. Biochemical polymorphisms provide the first widely applicable means of evaluating the genetic structure and continuity of populations in a manner independent of morphological variation.

Punctuated equilibrium is a theory about the tempo and mode of evolution; the tempo of morphological change through time is postulated to be discontinuous or "punctuated" rather than gradual, and the mode of change is a speciational rather than phyletic ("anagenetic") one (Gould 1982). The different combinations of the alternative postulated tempos and modes of evolution generate four hypothetical evolutionary "processes," all of which have been invoked to explain empirical evolutionary observations (Table 1). The punctuated tempo combined with the speciational mode of change generates the process of punctuated equilibrium. This process has been invoked in a number of paleontological studies; two good examples are the evolution of Cenozoic molluscs from Turkana Basin (Williamson 1981) and the evolution of alcelaphine antelopes (Vrba 1980). The gradualist tempo combined with the speciational mode generates the process called "cladistic gradualism" by Greenwood (1984), who invoked it to explain the evolution of the haplochromine cichlid fishes of Lake Victoria. The gradualist tempo combined with the phyletic mode of change constitutes "phyletic gradualism," the opposite of punctuated equilibrium and the viewpoint against which the theory of punctuated equilibrium was formed; this process is invoked, for example, by Bock (1979) in his evolutionary study of the drepanidid honeycreepers. The punctuated tempo combined with the phyletic mode generates the process called "punctuated phyletic evolution, " which is presented in detail and invoked by Turner (1986) to explain the evolution of *Heliconius* butterflies. Identification of the evolutionary processes deployed in particular cases may be controversial, as illustrated by the fact that Sheldon (1987) and Eldredge and Gould (1988) invoke opposite processes (phyletic gradualism and punctuated equilibrium, respectively) to explain the morphological evolutionary patterns of mid-Ordovician trilobites. In recent theoretical and empirical evolutionary studies, the process of punctuated phyletic evolution appears to be the most frequently invoked of these alternatives although, as I will discuss, there are major differences in the way that this process is perceived to occur.

Paleontological investigation of these alternative evolutionary processes has been criticized primarily for the lack of independent criteria for identifying speciation and morphological change (Levinton and Simon 1980; Wake et al. 1983; Turner 1986). Turner (1986) notes that if the genetic splitting of a lineage produces morphologically uniform, sibling species, and one lineage subsequently incorporates morphological changes rapidly in the phyletic mode, the paleontological outcome would be indistinguishable from the case in which morphological change is incorporated in a branching event with one branch retaining the ancestral morphology. The issue of sibling speciation thus becomes an important one for the investigation of evolutionary process. If sibling speciation can

TABLE 1. Four postulated evolutionary processes differing by tempo and/or mode of change.

	Tempo	
Mode	**Gradual**	**Punctuated**
Phyletic evolution	Phyletic gradualism Bock (1979): drepanidid honeycreepers	Punctuated phyletic evolution Turner (1986): *Heliconius* butterflies
Speciation	Cladistic gradualism Greenwood (1984): cichlid fishes	Punctuated equilibrium Vrba (1980): alcelaphine antelopes Williamson (1981): molluscs

be demonstrated to be common in a particular group, the observation of punctuated morphological change through time and an accompanying persistence of the ancestral morphology does not necessarily demonstrate punctuated equilibrium, because punctuated phyletic evolution is equally consistent with this observation (Turner 1986). Turner (1986) notes that a surprising amount of the past can be inferred from neontological data (facts collected from extant species) by using cladistic techniques to reconstruct phylogeny; hypotheses of the historical relationship between speciation and morphological evolution can be tested from the neontological perspective using a combination of morphological and molecular data.

NEONTOLOGICAL INVESTIGATIONS OF EVOLUTIONARY PROCESS

If morphological homogeneity or "stasis" through time represents the continuity of species rather than undetected sibling speciation, the morphological stasis observed among extant populations on a geographical scale should represent reproductive continuity of populations rather than undetected sibling species. Comparative molecular and morphological studies of geographic variation among populations can be used to test the postulated relationship between speciation and morphological change.

Speciation (in bisexually reproducing organisms) is an historical event that imposes intrinsic reproductive closure on an evolving population.

Through this event, the population becomes genetically independent of all others with which it had formerly been compatible, and an evolutionary branching of the lineage results. Species produced in this manner may be viewed as containment devices that prevent the flow of genetic information across their boundaries. Because natural populations are of finite size and new genetic variants are produced regularly by mutation, a continuous process of fixation of new molecular variants will occur (Kimura 1983; Nei 1987). Analysis of geographic variation in molecular features can be used to identify the boundaries of species; these boundaries are demonstrated by the maintenance of fixed genetic differences at the molecular level where reproductively incompatible populations geographically overlap (Larson 1984). If a pair of evolutionarily divergent populations retains reproductive compatibility, at least a small amount of genetic mixing should be observed in regions of geographic contact. This approach permits evaluation of the species status of paired populations regardless of their morphological similarities or differences. Because molecular systems demonstrating a variety of characteristic evolutionary rates are now available for study, paired populations representing a large range of divergence times can be analyzed in this manner.

This approach is illustrated using the lungless salamanders of the family Plethodontidae, which include many instances of morphological homogeneity or stasis among populations in geographic dimension (Wake et al. 1983). Many genera of this family have been studied extensively at both the molecular and morphological levels. Abundant variation interpretable as genetically segregating polymorphism has been detected among plethodontid populations at the protein level using electrophoretic techniques (Larson 1984). Geographic surveys of presumably conspecific populations reveal a recurring pattern of strong fragmentation; multiple fixed differences among such populations are the rule rather than the exception (Larson 1984; Larson et al. 1984b). The multiple fixed differences that separate populations are analyzed on a fine geographic scale to test for reproductive closure of populations that make geographic contact. The molecular information is used to estimate the relative recency of common ancestry of the populations studied, and populations are then analyzed in the pairs that constitute the most recent evolutionary branching events identified.

I previously presented an analysis of 22 independent comparisons of populations conducted in this manner (Larson 1984). These comparisons identified a minimum of 15 speciational events, only three of which could have been accompanied by substantial morphological change. For the other cases, morphological differentiation was not identified even on the basis of living specimens, and any fossil material from the paired lineages almost certainly would be attributed to a single species. These results indicate that the formation of sibling species is a regular feature of plethodontid evolution, and increased resolution of geographic molecular variation can be expected to increase rather than to decrease the estimated frequency of sibling speciation. Following my study, Sessions and Kezer (1987) showed on the basis of chromosomal variation that two species of *Aneides* that I had analyzed are each divisible into at least two reproductively distinct subsets. The *Aneides* lineage was one of only two in my original study for which the most recent events of branching speciation appeared to have featured notable morphological divergence. Extensive molecular analysis has shown also that *Plethodon glutinosus*, treated as a single species in my earlier review, comprises 16 groups of populations that appear to have achieved the species level of divergence (Highton 1984; Highton et al. 1989); three of these are confirmed as reproductively closed units by the maintenance of genetic distinctness in sympatry. These results strengthen my original conclusion that recent events of branching speciation in the plethodontids appear to feature minimal divergence at the morphological level.

Plethodontid sibling species are listed in Table 2. This list includes pairs of morphologically similar or identical populations that have been demonstrated to be reproductively closed units by comparative cytological and molecular studies of sympatric populations. The criteria of morphological similarity follow the precedent of Wake et al. (1983) and emphasize osteological structure. Some pairs of species have not been distinguished consistently on the basis of any phenotypic criteria; others display differences in pigmentation or minor proportional differences that can be used to sort them, but are not separated by osteological novelties (for a detailed review of these comparisons, see Larson 1984). In a few cases, character displacement is observed in areas of sympatry (Highton 1972, 1985). In the *Plethodon cinereus* group (#10–14), a single geographically widespread species makes contact throughout its range with numerous, closely related endemic species. It is likely that widespread geographic fragmentation of a once continuous species has produced numerous geographically isolated populations, some of which have evolved reproductive closure and subsequently expanded their ranges to become sympatric with neighboring populations.

It is likely that continued investigation of these groups on a finer scale will reveal additional cryptic species. Protein variation demonstrates high levels of genetic fragmentation among conspecific populations of most plethodontid species; fixed differences at the protein level are frequent and fixation indices show that on average more than half of the total genetic variance of the species is partitioned among populations (Larson 1984; Larson et al. 1984b). The magnitude of differentiation among populations at the protein level indicates that genetic isolation among the pop-

TABLE 2. Pairs of morphologically similar plethodontid species whose genetic distinctness has been demonstrated by comparative biochemical or cytological studies.

Species pair	Number of fixed differences[a]	Genetic distance[b]	Source[c]
1. *Aneides aeneus* sp. 1 and 2	N/A	N/A	14
2. *Aneides ferreus* sp. 1 and 2	N/A	N/A	7, 14
3. *Batrachoseps attenuatus* and *B. pacificus*	13	1.53	6
4. *Batrachoseps nigriventris* and *B. pacificus*	13	0.99	6
5. *Desmognathus imitator* and *D. ochrophaeus*	8	0.59	5
6. *Plethodon aureolus* and *P. glutinosus*	2	0.29	12, 15
7. *Plethodon aureolus* and *P. teyahalee*	6	0.42	12, 15
8. *Plethodon glutinosus* and *P. teyahalee*	6	0.36	12, 15
9. *Plethodon glutinosus* and *P. kentucki*	1	0.45	11, 15
10. *Plethodon cinereus* and *P. shenandoah*	7	0.26	1, 9
11. *Plethodon cinereus* and *P. hubrichti*	8	0.32	1, 9
12. *Plethodon cinereus* and *P. hoffmani*	7	0.36	1, 9
13. *Plethodon cinereus* and *P. nettingi*	11	0.47	1, 9
14. *Plethodon cinereus* and *P. richmondi*	9	0.46	1, 9
15. *Plethodon dorsalis* and *P. websteri*	14	1.70	4, 8, 13
16. *Plethodon dunni* and *P. vehiculum*	16	1.20	3
17. *Plethodon jordani* and *P. yonahlossee*	8	0.32–0.39	9, 11
18. *Pseudoeurycea smithi* and *P. unguidentis*	10	0.94	2
19. *Thorius dubitus* and *T. troglodytes*	3	0.41	10
20. *Thorius dubitus* and *T.* sp. A	5	0.56	10
21. *Thorius troglodytes* and *T.* sp. A	7	0.91	10

[a]Number of proteins for which mutually exclusive allelic composition is observed. (Numbers in italics include cases in which the predominant allele of one species appears as a rare variant in the other.)
[b]Genetic distances calculated according to Nei (1972, 1978).
[c]References are (1) Highton (1972), (2) Lynch et al. (1977), (3) Feder et al. (1978), (4) Larson and Highton (1978), (5) Tilley et al. (1978), (6) Yanev (1978), (7) Beatty (1979), (8) Highton (1979), (9) Highton and Larson (1979), (10) Hanken (1983a), (11) Highton and MacGregor (1983), (12) Highton (1984), (13) Highton (1985), (14) Sessions and Kezer (1987), (15) Highton et al. (1989).

ulations is ancient, regardless of whether the genetic barrier represents biological speciation or strictly geographic isolation. Temporal calibrations of protein evolution against paleontological standards indicate that the ages of these branching events are multiple millions of years (Maxson and Maxson 1979; Larson 1984). This observation indicates strong morphological evolutionary stasis (Wake et al. 1983).

Speciation in this group appears to proceed characteristically with little or no accompanying morphological change. Although this does not constitute a demonstration that morphological change proceeds by a phyletic rather than speciational mode when it does occur, it shows that morphological evolutionary stasis transcends the species level and therefore cannot be explained adequately as a species-level phenomenon. It confirms the importance of Turner's observation that the persistence of an ancestral morphology after the sudden appearance of a novel one does not necessarily imply that morphological change was accomplished in a speciational mode (Turner 1986); a macroevolutionary pattern of this kind would be expected whenever one of the sibling species subsequently incorporates morphological change in the phyletic mode long after the phylogenetic branching event. Another very important observation relevant to the issue of macroevolutionary pattern is the presence of skeletal polymorphisms within many extant populations of plethodontids. These polymorphisms involve variation in the number and shape of skull and limb bones and cartilages, with variant forms being present usually in low but occasionally in substantial frequencies. The alternative states of these polymorphisms are separated by qualitative differences that resemble very closely important skeletal differences that distinguish higher taxa in plethodontids (Alberch 1983; Hanken 1983b; Hanken and Dinsmore 1986; Wake and Larson 1987). The status of these polymorphic populations relative to long-term branching events cannot be determined; however, the presence of these polymorphisms indicates that morphological transitions can be accomplished within the phyletic mode. Considering the high frequency of sibling speciation, the most reasonable interpretation of both observations is that branching speciation often occurs without substantial morphological change, and that novel morphologies are fixed subsequently at low frequency within the phyletic mode. This describes a process of punctuated phyletic evolution.

Speciation and morphological change are viewed as distinct and decoupled phenomena in the plethodontids, with speciation occurring at a higher rate than the origins of morphological novelties. Although it departs from the original statement of punctuated equilibrium, this evolutionary pattern nonetheless confirms two important features of the theory of punctuated equilibrium: the importance of a punctuated tempo of evolution featuring long periods of morphological evolutionary stasis, and the importance of hierarchical structure in evolutionary processes. The finding of numerous, reproductively closed species within the boundaries of what was formerly considered to be a genetically continuous unit increases the relative macroevolutionary importance of interspecific processes of sorting (Vrba and Gould 1986). This observation is consistent also with the suggestion that the production of species may be a largely nonadaptive event. If a novel morphological feature is fixed within a species of limited geographic scope, its chance of long-term survival may be small unless it confers a greater resistance to extinction or enhances speciation; this is an example of the effect hypothesis of macroevolution (Vrba 1980).

HIERARCHY AND THE DESCRIPTION OF EVOLUTIONARY PROCESSES

Few issues in evolutionary biology elicit more controversy than the suggestion that evolutionary processes are hierarchical in their structure (Grene 1987). A common pattern is for an empirical observation to be interpreted as demonstrating hierarchical evolutionary forces, and for subsequent criticism to claim either that reductionistic explanations (with an emphasis on genic-level properties) are adequate to explain the phenomenon, or that it is too artificial to be representative of "real" evolutionary processes. A consequence is that two separate theoretical frameworks guide the study of evolution, and the conflict between these frameworks is evident at many levels of evolutionary investigation. In each case, a major controversy centers on whether a given level of biological organization contains emergent properties that exert control on the direction or timing of evolution.

The response to the theory of punctuated equilibrium has not been an exception to this pattern. To analyze this response, it is necessary to consider four levels at which the conflict between hierarchical and nonhierarchical perspectives is evident. Perhaps the most obvious of these is the controversy concerning the reality and importance of historical processes that transcend the species level (Stanley 1979; Gould 1985); the major issue here is whether species contain irreducible properties that can serve as the basis for a higher order process of sorting (Vrba and Gould 1986). A second major controversy addresses the issue of whether the structuring of populations in space or time can induce adaptive evolution that would not occur in a large, continuous population (Barton and Charlesworth 1984; Carson and Templeton 1984); this issue is closely associated with a third controversy regarding the genetic architectures of speciation and morphological evolution (Templeton 1982; Gottlieb 1984, 1985; Gottlieb and Ford 1987; Coyne and Lande 1985; Goodnight 1987, 1988). A genetic architecture featuring a major locus combined with epistatic modifiers of the penetrance and expressivity of its pleiotropic effects can be constructed most effectively in the context of a hierarchically subdivided population structure (Wright 1982). This hierarchical genetic architecture also has the potential for inducing rapid phenotypic transitions and for having its additive variance increased by a founder event, properties that are lacking in nonhierarchical genetic architectures (Carson and Templeton 1984). The importance of this genetic architecture relative to nonhierarchical ones in which all loci have strictly additive effects on the phenotype, remains controversial. The question of whether emergent, epigenetic properties of organismal development impose an additional level of causality in evolutionary pattern and direction constitutes a fourth major controversy that separates hierarchical and nonhierarchical viewpoints (Alberch 1980; Goodwin 1984; Wake and Larson 1987).

Although punctuated phyletic evolution appears to be the prevailing view of the tempo and mode of evolution, the issue of hierarchy raised by punctuated equilibrium remains unresolved; it is simply deflected to levels of explanation other than the emergent properties of species. Phyletic explanations of punctuated change generally invoke "shifting balance," "stabilizing selection," and "adaptive zones" in the explanation of this pattern of change (Charlesworth et al. 1982; Kirkpatrick 1982; Petry 1982; Wright 1982; Lande 1985, 1986). Careful inspection of these terms reveals, however, that each one is used to describe both a hierarchical and a nonhierarchical process of evolution. Resolution of the issue of hierarchy requires discrimination of the alternative hypothetical evolutionary processes encompassed by each term.

Shifting balance

The shifting balance process is invoked frequently as a model that accounts for a punctuated tempo of evolution within a phyletic rather than speciational mode. This process describes the evolution of adaptive gene complexes in large populations that are subdivided into many small, semiisolated demes (Wright 1982). The issue of hierarchy enters the discussion of this process in three contexts: the structure of populations, the structure of genetic systems, and the occurrence of selective processes at both the organismal and populational levels. Theoretical work on shifting balance has generated two somewhat disparate descriptions of the process that differ in their use of hierarchy.

Gould (1982) described Wright's version of shifting balance (Wright 1982) as a friendly challenge to punctuated equilibrium because it confirms the importance of both a punctuated evolutionary tempo and hierarchical processes. Evolution is driven by the internal dynamics of a biological system that is hierarchically organized both in genetic architecture and population structure, in the absence of environmental change.

Polygenic inheritance, pleiotropy, and epistasis characterize the relationship between genotype and phenotype. Nonadditive genic interactions generate alternative, potentially adaptive genetic combinations that are separated by relatively maladaptive ones; this is represented graphically as an "adaptive landscape" containing adaptive "peaks" and "valleys" corresponding to favorable and unfavorable genetic compositions of the population, respectively. The action of mass selection in a large population will maintain a particular, adaptive polygenic complex, but will inhibit the attainment of alternative, possibly superior ones if this requires moving the population through an adaptive valley. The action of random genetic drift in a small population may cause the population to move into an adaptive valley, but by doing so may permit a selective transition to a novel adaptive peak. A hierarchical population structure featuring a large population subdivided into many relatively small demes that exchange migrants will promote the evolution of novel adaptive gene complexes; populations with this hierarchical structure combine the properties of large and small populations with respect to the action of selection and drift. The combination of genes having a major effect on the phenotype with other genes that modify the penetrance and expressivity of their pleiotropic effects is promoted by this process (Wright 1982). Adaptive gene complexes arise initially in a particular deme, and are then spread throughout the entire population by interdemic selection resulting from the differential fertility of the demes.

An alternative formulation of shifting balance that has been used to challenge punctuated equilibrium (Lande 1980, 1985, 1986) lacks many of the hierarchical elements of the process previously described. In this alternative version, additive polygenic inheritance is emphasized almost exclusively, and the hierarchical adaptive gene complexes in which a gene of major effect is combined with many polygenic modifiers are not invoked (Barton and Charlesworth 1984; Lande 1980). Genetic changes occurring in small, completely isolated populations are viewed as having greater importance than those occurring in large subdivided populations (Lande 1980, 1985); variation is maintained by recurring mutation, rather than by migration from neighboring populations. A process of interdemic selection may still operate, but it involves differential extinction and colonization among completely isolated populations rather than differential exchange of migrants (Lande 1980). These two versions of shifting balance are quite divergent, sharing little more than the adaptive landscape and the joint action of selection and drift. To a large degree, the conflict between these two versions preserves the general conflict between hierarchical and nonhierarchical perspectives that Gould (1982) considered to be an essential point of the punctuated equilibrium controversy.

Stabilizing selection

One of the most commonly invoked explanations of morphological evolutionary stasis is stabilizing selection (Charlesworth et al. 1982; Williamson 1987), which attributes morphological evolutionary stasis to the selective elimination of any individuals that deviate from a standard phenotype. The theory of stabilizing selection places a fairly minimal restriction on the exact mechanisms that underlie phenotypic stability. Although no version of stabilizing selection postulates that phenotypic stability is achieved by a strict constancy of genotype through time, the genetics of phenotypic stability are highly controversial. Schmalhausen's statement of stabilizing selection invokes the generation of novel genotypes that improve phenotypic stability and produce an autonomous and self-regulatory organismal development (Schmalhausen 1949). This version of stabilizing selection is explicitly hierarchical, and represents an intermediate position relative to two major explanations of stasis that have developed subsequently. One of these is a highly reductionistic version of stabilizing selection, which attributes phenotypic stability to additive polygenic systems whose variation is constantly depleted by selection against phenodeviant individuals and then restored by recurring mutation (Lande 1980); the other attributes stability largely to an intrinsic resilience of epigenetic systems (Rachootin and Thompson 1981; Wake et al. 1983). The former perspective carries the requirement that the perceived environment of the organism impose a constant optimum phenotype throughout the evolutionary period during which stasis persists (Lande 1986), and it has been criticized largely on the perceived weakness of this position (Wake et al. 1983). According to the epigenetic perspective, phenotype stability reflects at least partly the limitations on diversity imposed by the structure of morphogenetic systems (Oster and Alberch 1982); this proposal has been criticized on the basis of artificial selection experiments which demonstrate that populations may contain heritable variation that permits departure from the standard phenotype (Charlesworth et al 1982; Williamson 1987).

Williamson (1987) proposed a mechanism of stabilizing selection that incorporates both constraint and genetic variation. His argument is based on the observation that departures from a standard phenotype achieved by artificial selection are rarely stable; novel phenotypes tend to be accompanied by high levels of developmental instability, preventing their persistence when in selective competition with individuals demonstrating the standard developmental pattern. Following Carson's notion that selective elimination of phenodeviant individuals through intraspecific competition will be greatest when population density is high, Williamson (1987) proposes that morphological evolutionary stasis characterizes periods of high population density, and that transitions to novel forms occur when selection is relaxed, such as during a population flush (Carson and Templeton 1984). This explanation is consistent with Schmalhausen's notion of genotypic stabilization of development and with the notion of adaptive gene complexes. This proposal also explains how a particular stabilized phenotype can persist through the action of "internal" forces as the environment varies, in spite of the presence of heritable variation that would permit its potential disruption by selection. Because it invokes hierarchical processes at the levels of the genotype and population structure and also potentially at the organismal level, it is bound to be controversial (Barton and Charlesworth 1984).

The issue of stabilizing selection may be viewed, therefore, as preserving rather than resolving the conflict between hierarchical and nonhierarchical perspectives. Williamson's formulation of stabilizing selection shares many features with the theory of punctuated equilibrium, with perhaps a somewhat relaxed requirement for speciation and morphological change to be coupled (Williamson 1987).

Hybridization phenomena may prove to be important in understanding the kinds of genetic systems that underlie the stabilization of alternative phenotypes. Destabilization of normal development and of resistance to disease is observed repeatedly in hybrid populations (Sage et al 1986; Schwab 1987), indicating that genetic mechanisms of phenotypic stabilization are disrupted. Incongruence between patterns of introgression observed at the molecular and morphological levels can be used to identify phenotypic features whose expression is potentially subject to genetic stabilization and its disruption. At least one clear example of this phenomenon is observed in plethodontids. The effects of introgression on body proportions and coloration in hybrids of *Bolitoglossa franklini* and *B. lincolni* are asymmetrical; introgression has a pervasive effect on the morphology of *B. lincolni*, but *B. franklini* can accept a fair amount of introgression without apparent morphological change (Wake et al. 1980; Wake and Lynch 1988). The resilience of the phenotype of *B. franklini* suggests that it has been subjected to genetic stabilization that resists even a fairly substantial alteration of genetic background. The structure of gene complexes that can confer phenotypic stability has been reviewed by Carson and Templeton (1984).

Adaptive zones

The concept of adaptive zones is frequently invoked along with shifting balance and stabilizing selection in discussions of the tempo and mode of evolution. Usage of this term is quite variable, particular regarding the hierarchical level to which it applies. Simpson (1953) used the term adaptive zone to refer to an evolutionary group possessing a characteristic pattern of organismal/environmental interaction. The evolutionary origin of an adaptive zone features the derivation of a novel biological characteristic that redefines the nature of organismal/environmental interaction, and the subsequent evolutionary diversification that occurs in the context of novel selective forces. The novel feature is called a "key innovation" (Miller 1949) and can be viewed as imposing simultaneously a constraint and an opportunity for diversity on the adaptive evolution of a lineage (Wake and Larson 1987). Used in this manner, the adaptive zone may include a large number of species and different morphologies; it is equivalent to an adaptive landscape with many alternative adaptive peaks, each representing a different solution to a given set of selective forces. Following the usage of Simpson (1953), a change of adaptive zone is a stimulus for morphological and ecological diversification but does not restrict such change to a gradual versus punctuated tempo or to a phyletic versus speciational mode. A more restrictive usage is one in which adaptive zone refers specifically to the occupation of a particular peak on an adaptive landscape (Petry 1982). This latter usage is apparently the one intended in many criticisms of punctuated equilibrium.

In his defense of a phyletic mode of punctuated evolution, Petry (1982) views the key innovation as "a genetically controlled trait that determines the selective surface but does not necessarily contribute to the observed phenotype." If a novel feature of this kind is fixed in a population, the accompanying alteration of the selective surface may induce selectively driven change within the phyletic mode, and the change may proceed rapidly enough to appear to be instantaneous on a geological time scale. The morphological differentiation stimulated by a change of adaptive zone is not expected to occur all at one time or in a single lineage, however. The model of diversification encompassed by the traditional usage of adaptive zone is one that transcends both the species level and the general issue of punctuated equilibrium.

In an analysis of morphological evolution in the plethodontid genus *Aneides*, I invoked this model of change in a transspecific context (Larson et al. 1981). Two morphological novelties shared by all members of this genus (modifications of the jaw and mesopodial elements) were interpreted to be "key innovations" that served to stimulate subsequent modifications occurring differentially among the extant lineages of *Aneides*. I disagree with Petry's conclusion that the order of mutational events in one character and the selective events elsewhere cannot be ascertained (Petry 1982). In the analysis of *Aneides*, the morphological evolutionary events were ordered using a cladistic analysis in conjunction with molecular analyses of phylogenetic relationships. The features hypothesized to be "key innovations" can be inferred from this analysis to have evolved prior to the other morphological novelties observed in *Aneides*, consistent with the hypothesized changes of adaptive landscape. Comparisons of the distribution of morphological and molecular changes in the phylogeny of *Aneides* indicate that major morphological change is highly episodic relative to molecular evolution (Larson et al. 1981; Larson 1984).

The concept of adaptive zone therefore also leaves open the question of the role of hierarchical processes in setting the tempo of evolution. In the more traditional usage, this concept transcends the issue of punctuated equilibrium and identifies perhaps an even higher level of causality in evolutionary change (for example, two lineages demonstrating punctuated equilibrium may nonetheless differ greatly in the duration of periods of stasis); in the more restricted usage, it refers to occupation of peaks on an adaptive landscape and therefore encompasses the same hierarchical conflicts that characterize shifting balance and stabilizing selection.

HISTORICAL ANALYSIS OF THE ROLES OF HIERARCHICAL EVOLUTIONARY PROCESSES IN SPECIATION AND MORPHOLOGICAL EVOLUTION

The conflicting evolutionary processes represented by the alternative statements of shifting balance, stabilizing selection, and adaptive zone make potentially testable predictions regarding evolutionary tempo and mode. Differences among groups in the expected historical relationship between speciation and morphological evolution provide one possible test.

As previously discussed, the expected role of population structure in speciation and morphological evolution differs between the hierarchical and nonhierarchical formulations of these evolutionary mechanisms. The testing of these hypotheses on a macroevolutionary scale may utilize the observation that different evolutionary groups may possess characteristic population structures that transcend the species level and impose particular patterns of organization on genetic variation. This is evident, for example, among major vertebrate groups. Protein variation reveals that conspecific populations of birds demonstrate less genetic differentiation on average than do conspecific populations of amphibians or mammals (Barrowclough 1983); rates of macroevolutionary change also differ among these groups (Larson et al. 1984a). Characteristic differences in macroevolutionary pattern are observed also among marine molluscs that demonstrate two alternative modes of larval ecology that impose different genetic structures on populations (Jablonski 1986). Rates of speciation and extinction are higher in molluscs with nonplanktrotrophic larvae, in which the genetic structure of populations is highly subdivided, than in planktrotrophic molluscs whose populations tend to be more homogeneous genetically; the planktrotrophic and nonplanktrotrophic groups differ also in the mode of speciation most frequently deployed (Jablonski 1986).

Population genetic structures that are conducive to disruption of coadapted gene complexes may induce morphological change and speciation simultaneously (Templeton 1986), increasing the expected coincidence of these events in geological time. Lineages that are prone to experiencing founder-flush events contain the highest potential for repeated disruption of coadapted gene complexes (Carson and Templeton 1984). Lineages whose population structure is not conducive to disruption of stabilized genetic systems may be expected alternatively to maintain morphological stability; speciation may occur gradually by selective alteration of the mate recognition system without altering the genetic stabilization of phenotype in such lineages (Paterson 1985). Available data suggest that morphological novelties may arise more frequently in lineages that are subject to repeated founder events than in sister lineages that do not regularly experience such events (Jablonski et al. 1983; Templeton 1986). The historical association between speciation and morphological change in a particular lineage is predicted to be contingent upon the population structures characteristic of the lineage if hierarchical structure is important to evolutionary processes; the nonhierarchical perspective predicts alternatively that the historical association between speciation and morphological evolution will not depend on the population genetic structures characteristic of a lineage.

To address this issue analysis of population genetic structure must be conducted on a very fine scale. The extreme genetic fragmentation observed among plethodontid populations on a large geographical scale does not indicate that they are subjected to either the spatial or temporal structuring that is expected to induce disruption of stabilized genetic systems. Ecological and behavioral data suggest that local plethodontid populations often are dense, and that lack of long-distance dispersal produces the geographic fragmentation observed among them; the long-term histories of these populations appear to feature gradual expansion of dense populations into geographically contiguous and ecologically favorable regions, with geographical fragmentation being imposed subsequently by climatic changes (Highton and Webster 1976; Larson 1984). Under this population structure, genetically stabilized morphologies would not regularly be subjected to disruption, although speciation could be achieved gradually in geographically isolated populations by selective alteration of the mate recognition system (Paterson 1985).

To examine the detailed genetic structure of these and other populations, highly variable genetic systems that can resolve relationships among individual organisms must be used; molecular genetic systems with high resolution have been identified in other vertebrates (Wong et al.

1986), and this methodology hopefully will become generally applicable for evolutionary studies in many groups of organisms. This analysis will be most effective where characteristic population genetic structures differ between pairs of sister lineages for which the coincidence of speciation and morphological evolution can be evaluated. The macroevolutionary significance of hierarchical evolutionary processes can be investigated also by examining the historical association between population genetic structure and rates of speciation and morphological evolution in sister groups. The use of molecular phylogenies to evaluate evolutionary rates (Wilson et al. 1987) makes this approach feasible even in groups that lack a good fossil history. The relationship between speciation and morphological evolution and its dependence upon particular genetic architectures and population structures is potentially subject also to experimental investigation in *Drosophila* and other genera that offer good population genetic study systems (Powell 1978; Carson and Templeton 1984; Bryant et al. 1986a, 1986b).

The issue of hierarchical structure in the historical processes underlying speciation and morphological evolution is viewed by Gould (1982) as being more fundamental than the strict historical coincidence of these events. The historical coincidence between speciation and morphological evolution studied in conjunction with the evolution of population structure and genetic architecture should clarify the role of hierarchical evolutionary processes.

SUMMARY

Evolutionary studies often fail to make adequate distinction between morphological evolution and speciation, the latter defined as the evolutionary splitting of lineages featuring intrinsic reproductive and genetic closure of populations. By formally postulating an historical association between speciation and morphological evolution, the theory of punctuated equilibrium requires that these phenomena be recognizable as distinct evolutionary events whose historical coincidence can be evaluated empirically. Comparative morphological and molecular studies of the plethodontid salamanders indicate that speciation and morphological evolution are historically decoupled in this group, and that speciation proceeds at a higher rate than does the origin of morphological novelties. Morphological evolutionary stasis is prevalent and requires for its explanation a mechanism that transcends speciation. Phyletic explanations of punctuated morphological evolution commonly invoke the processes of shifting balance and stabilizing selection, although each of these comprises alternative hierarchical and nonhierarchical formulations that preserve much of the controversy surrounding the challenge of punctuated equilibrium. Comparative studies of the historical relationship between speciation and morphological evolution in different groups may clarify the role of hierarchical processes in evolution.

ACKNOWLEDGMENTS

I thank Daniel Dykhuizen, John Endler, David Wake, and anonymous reviewers for their reviews of the manuscript. Support was provided by NSF Grant BSR-8708393.

LITERATURE CITED

Alberch, P. 1980. Ontogenesis and morphological diversification. Am. Zool. 20:653–667.

Alberch, P. 1983. Morphological variation in the neotropical salamander genus, *Bolitoglossa*. Evolution 37:906–919.

Barrowclough, G. F. 1983. Biochemical studies of microevolutionary processes. Pp. 223–270 in: A. H. Brush and G. A. Clark, Jr. (eds.), *Perspectives in Ornithology*. Cambridge University Press, Cambridge.

Barton, N. H., and B. Charlesworth. 1984. Genetic revolutions, founder events and speciation. Annu. Rev. Ecol. System. 15:133–164.

Beatty, J. J. 1979. Morphological variation in the clouded salamander, *Aneides ferreus* (Cope) (Amphibia: Caudata: Plethodontidae). Ph.D. dissertation. Oregon State University, Corvallis.

Bock, W. J. 1979. The synthetic explanation of macroevolutionary change—A reductionist approach. Bull. Carnegie Mus. Nat. Hist. 13:20–69.

Bryant, E. H., S. A. McCommas, and L. M. Combs. 1986a. The effect of an experimental bottleneck upon quantitative genetic variation in the housefly. Genetics 114: 1191–1211.

Bryant, E. H., L. M. Combs, and S. A. McCommas. 1986b. Morphometric differentiation among experimental lines of the housefly in relation to a bottleneck. Genetics 114: 1213–1223.

Carson, H. L., and A. R. Templeton. 1984. Genetic revolutions in relation to speciation phenomena: The founding of new populations. Annu. Rev. Ecol. System. 15:97–131.

Charlesworth, B., R. Lande, and M. Slatkin. 1982. A neo-Darwinian commentary on macroevolution. Evolution 36:474–498.

Coyne, J. A., and R. Lande. 1985. The genetic basis of species differences in plants. Am. Natur. 126:141–145.

Eldredge, N., and S. J. Gould. 1972. Punctuated equilibria: An alternative to phyletic gradualism. Pp. 82–115 in: T. J. M. Schopf (ed.), *Models in Paleobiology*. Freeman, Cooper, San Francisco.

Eldredge, N., and S. J. Gould. 1988. Punctuated equilibrium prevails. Nature (London) 332:211–212.

Feder, J. H., G. Z. Wurst, and D. B. Wake. 1978. Genetic variation in western salamanders of the genus *Plethodon*, and the status of *Plethodon gordoni*. Herpetologica 34:64–69.

Goodnight, C. J. 1987. On the effect of founder events on epistatic genetic variance. Evolution 41:80–91.

Goodnight, C. J. 1988. Epistasis and the effect of founder events on the additive genetic variance. Evolution 42:441–454.

Goodwin, B. C. 1984. Changing from an evolutionary to a generative paradigm in biology. Pp. 99–120 in: J. W. Pollard (ed.), *Evolutionary Theory: Paths into the Future*. John Wiley, New York.

Gottlieb, L. D. 1984. Genetics and morphological evolution in plants. Am. Natur. 123: 681–709.

Gottlieb, L. D. 1985. Reply to Coyne and Lande. Am. Natur. 126:146–150.

Gottlieb, L. D., and V. S. Ford. 1987. Genetic and developmental studies of the absence of ray florets in *Layia discoidea*. Pp. 1–17 in: H. Thomas and D. Grierson (eds.), *Developmental Mutants in Higher Plants*. Cambridge Unversity Press, Cambridge.

Gould, S. J. 1982. The meaning of punctuated equilibrium and its role in validating a hierarchical approach to macroevolution. Pp. 83–104 in: R. Milkman (ed.), *Perspectives on Evolution*. Sinauer Associates, Sunderland, MA.

Gould, S. J. 1985. The paradox of the first tier: An agenda for paleobiology. Paleobiology 11:2–12.

Gould, S. J., and N. Eldredge. 1977. Punctuated equilibria: The tempo and mode of evolution reconsidered. Paleobiology 3:115–151.

Greenwood, P. H. 1984. African cichlids and evolutionary theories. Pp. 141–154 in: A. A. Echelle and I. Kornfield (eds.), *Evolution of Fish Species Flocks*. University of Maine Press, Orono.

Grene, M. 1987. Hierarchies in biology. Am. Sci. 75:504–510.

Hanken, J. 1983a. Genetic variation in a dwarfed lineage, the Mexican salamander genus *Thorius* (Amphibia: Plethodontidae): Taxonomic, ecologic and evolutionary implications. Copeia 1983:1051–1073.

Hanken, J. 1983b. High incidence of limb skeletal variants in a peripheral population of the red-backed salamander, *Plethodon cinereus* (Amphibia: Plethodontidae), from Nova Scotia. Can. J. Zool. 61:1925–1931.

Hanken, J., and C. E. Dinsmore. 1986. Geographic variation in the limb skeleton of the red-backed salamander, *Plethodon cinereus*. J. Herpetol. 20:97–101.

Highton, R. 1972. Distributional interactions among eastern North American salamanders of the genus *Plethodon*. Virginia Polytech. Inst. Res. Div. Monogr. 4:139–188.

Highton, R. 1979. A new cryptic species of salamander of the genus *Plethodon* from the southeastern United States (Amphibia: Plethodontidae). Brimleyana 1:31–36.

Highton, R. 1984. A new species of woodland salamander of the *Plethodon glutinosus* group from the southern Appalachian Mountains. Brimleyana 9:1–20.

Highton, R. 1985. The width of the contact zone between *Plethodon dorsalis* and *P. websteri* in Jefferson County, Alabama. J. Herpetol. 19:544–546.

Highton, R., and A. Larson. 1979. The genetic relationships of the salamanders of the genus *Plethodon*. System. Zool. 28:579–599.

Highton, R., and J. R. MacGregor. 1983. *Plethodon kentucki* Mittleman: A valid species of Cumberland Plateau woodland salamander. Herpetologica 39:189–200.

Highton, R., G. C. Maha, and L. R. Maxson. 1989. Biochemical evolution in the slimy salamanders of the *Plethodon glutinosus* complex in the eastern United States. Univ. Illinois Biol. Monogr., in press.

Highton, R., and T. P. Webster. 1976. Geographic protein variation and divergence in populations of the salamander *Plethodon cinereus*. Evolution 30:33–45.

Jablonski, D. 1986. Larval ecology and macroevolution in marine invertebrates. Bull. Marine Sci. 39:565–587.

Jablonski, D., J. J. Sepkoski, Jr., D. J. Bottjer, and P. M. Sheehan. 1983. Onshore-offshore patterns in the evolution of phanerozoic shelf communities. Science 222:1123–1125.

Kimura, M. 1983. *The Neutral Theory of Molecular Evolution*. Cambridge University Press, Cambridge.

Kirkpatrick, M. 1982. Quantum evolution and punctuated equilibria in continuous genetic characters. Am. Natur. 119:833–848.

Lande, R. 1980. Genetic variation and phenotypic evolution during allopatric speciation. Am. Natur. 116:463–479.

Lande, R. 1985. Expected time for random genetic drift of a population between stable phenotypic states. Proc. Natl. Acad. Sci. U.S.A. 82:7641–7645.

Lande, R. 1986. The dynamics of peak shifts and the pattern of morphological evolution. Paleobiology 12:343–354.

Larson, A. 1984. Neontological inferences of evolutionary pattern and process in the salamander family Plethodontidae. Evol. Biol. 17:119–217.

Larson, A., and R. Highton. 1978. Geographic protein variation and divergence in the salamanders of the *Plethodon welleri* group (Amphibia, Plethodontidae). System. Zool. 27:431-448.
Larson, A., D. B. Wake, L. R. Maxson, and R. Highton. 1981. A molecular phylogenetic perspective on the origins of morphological novelties in the salamanders of the tribe Plethodontini (Amphibia, Plethodontidae). Evolution 35:405-422.
Larson, A., E. M. Prager, and A. C. Wilson. 1984a. Chromosomal evolution, speciation and morphological change in vertebrates: The role of social behaviour. Chromosom. Today 8:215-228.
Larson, A., D. B. Wake, and K. P. Yanev. 1984b. Measuring gene flow among populations having high levels of genetic fragmentation. Genetics 106:293-308.
Levinton, J. S., and C. M. Simon. 1980. A critique of the punctuated equilibria model and implications for the detection of speciation in the fossil record. System. Zool. 29:130-142.
Lynch, J. F., S. Y. Yang, and T. J. Papenfuss. 1977. Studies of neotropical salamanders of the genus *Pseudoeurycea*, I. Systematic status of *Pseudoeurycea unguidentis*. Herpetologica 33:46-52.
Maxson, L. R., and R. D. Maxson. 1979. Comparative albumin and biochemical evolution in plethodontid salamanders. Evolution 33:1057-1062.
Mayr, E. 1982. *The Growth of Biological Thought*. Belknap Press of Harvard University Press, Cambridge.
Miller, A. H. 1949. Some ecologic and morphologic considerations in the evolution of higher taxonomic categories. Pp. 84-88 in: E. Mayr and E. Schuz (eds.), *Ornithologie als Biologische Wissenschaft*. Carl Winter/Universitatsverlag, Heidelberg.
Nei, M. 1972. Genetic distance between populations. Am. Natur. 106:283-292.
Nei, M. 1978. Estimation of average heterozygosity and genetic distance from a small number of individuals. Genetics 89:583-590.
Nei, M. 1987. *Molecular Evolutionary Genetics*. Columbia University Press, New York.
Oster, G., and P. Alberch. 1982. Evolution and bifurcation of developmental programs. Evolution 36:444-459.
Paterson, H. E. H. 1985. The recognition concept of species. Pp. 21-29 in: E. S. Vrba (ed.), *Species and Speciation. Transval Museum Monograph No. 4*. Transval Museum, Pretoria.
Petry, D. 1982. The pattern of phyletic speciation. Paleobiology 8:56-66.
Powell, J. R. 1978. The founder-flush speciation theory: An experimental approach. Evolution 32:465-474.
Rachootin, S. P., and K. S. Thomson. 1981. Epigenetics, paleontology, and evolution. Evol. Today 2:181-193.
Sage, R. D., D. Heyneman, K. -C. Lim, and A. C. Wilson. 1986. Wormy mice in a hybrid zone. Nature (London) 342:60-63.
Schmalhausen, I. I. 1949. *Factors of Evolution*. Blakiston, Philadelphia.
Schwab, M. 1987. Oncogenes and tumor suppressor genes in *Xiphophorus*. Trends Genet. 3:38-42.
Sessions, S. K., and J. Kezer. 1987. Cytogenetic evolution in the plethodontid salamander genus *Aneides*. Chromosoma 95:17-30.
Sheldon, P. R. 1987. Parallel gradualistic evolution of Ordovician trilobites. Nature (London) 330:561-563.
Simpson, G. G. 1953. *The Major Features of Evolution*. Columbia University Press, New York.
Stanley, S. M. 1979. *Macroevolution*. Freeman, San Francisco.
Templeton, A. R. 1982. Genetic architectures of speciation. Pp. 105-121 in: C. Barigozzi (ed.), *Mechanisms of Speciation*. Liss, New York.
Templeton, A. R. 1986. The relation between speciation mechanisms and macroevolutionary patterns. Pp. 497-512 in: S. Karlin and E. Nevo (eds.), *Evolutionary Processes and Theory*. Academic Press, Orlando.

Tilley, S. G., R. B. Merritt, B. Wu, and R. Highton. 1978. Genetic differentiation in salamanders of the *Desmognathus ochrophaeus* complex (Plethodontidae). Evolution 32:93–115.

Turner, J. R. G. 1986. The genetics of adaptive radiation: A neo-Darwinian theory of punctuated equilibrium. Pp. 183–207 in: D. M. Raup and D. Jablonski (eds.), *Patterns and Processes in the History of Life.* Springer-Verlag, Heidelberg.

Vrba, E. S. 1980. Evolution, species and fossils: How does life evolve? South African J. Sci. 76:61–84.

Vrba, E. S., and S. J. Gould. 1986. The hierarchical expansion of sorting and selection: Sorting and selection cannot be equated. Paleobiology 12:217–228.

Wake, D. B., and A. Larson. 1987. Multidimensional analysis of an evolving lineage. Science 238:42–48.

Wake, D. B., and J. F. Lynch. 1988. The taxonomic status of *Bolitoglossa resplendens* (Amphibia: Caudata). Herpetologica 44:105–108.

Wake, D. B., G. Roth, and M. H. Wake. 1983. On the problem of stasis in organismal evolution. J. Theoret. Biol. 101:211–224.

Wake, D. B., S. Y. Yang, and T. J. Papenfuss. 1980. Natural hybridization and its evolutionary implications in Guatemalan plethodontid salamanders of the genus *Bolitoglossa* Herpetologica 36:335–345.

Williamson, P. G. 1981. Paleontological documentation of speciation in Cenozoic molluscs from Turkana Basin. Nature (London) 293: 437–443.

Williamson, P. G. 1987. Selection or constraint: A proposal on the mechanism for stasis. Pp. 129–142 in: K. S. W. Campbell and M. F. Day (eds.), *Rates of Evolution.* Allen & Unwin, London.

Wilson, A. C., H. Ochman, and E. M. Prager. 1987. Molecular time scale for evolution. Trends Genet. 3:241–247.

Wong, Z., V. Wilson, A. J. Jeffries, and S. L. Thein. 1986. Cloning a selected fragment from a human DNA "fingerprint": Isolation of an extremely polymorphic minisatellite. Nucleic Acids Res. 14:4605–4616.

Wright, S. 1982. Character change, speciation, and the higher taxa. Evolution 36:227–443.

Yanev, K. P. 1978. Evolutionary studies of the plethodontid salamander genus *Batrachoseps.* Ph.D. dissertation. University of California, Berkeley.

CHAPTER 24

SPECIATION AND DIVERSITY: The Integration of Local and Regional Processes

Robert E. Ricklefs

INTRODUCTION

Most of the chapters in this volume have addressed the mechanisms of speciation. My purpose, however, is to examine the consequences of variation in rates of species formation and other long-term processes for patterns of biological diversity on local and regional scales. Local diversity, by which I mean the number of species within a small area of homogenous habitat, reflects the interaction of local processes, which tend to reduce diversity, and regional processes, which tend to enhance diversity. Local processes include various physical disturbances and catastrophes that reduce the number of individuals in populations, stochastic variation, whose force increases as population size decreases, and such interactions between species as competition and predation, which may lead to exclusion of one or more species (MacArthur 1972). These local processes, which tend to reduce diversity, are balanced by long-term, regional processes that either facilitate the coexistence of species or bring new species upon the local scene. Movement of individuals between habitats and through the geographic range of the population lessens the probability of extinction or exclusion. As a consequence of such movements individuals may recolonize an area of local extinction or augment a population in a habitat unproductive for the species (the "mass effect" of Schmida and Wilson 1985). The evolution of ecological diversification and mutualistic interactions also may facilitate local coexistence. Both these processes may pave the way for species to enter a new locality and thereby increase local diversity. In general, however, species are added both to regions and to local communities by the processes of speciation, which may involve vicariant isolation of entire biotas, and interregional biotal dispersal [i.e., involving a large portion of the flora and fauna; what has been called, for example, faunal interchange (Webb 1976)]. Only these mechanisms can increase diversity within a region and, of course, speciation alone increases the total number of species on earth.

In this chapter, I argue that speciation and biotal dispersal strongly influence both regional and local diversity. Although most biologists accept historical and geographic components to regional diversity (see Darlington 1957; Udvardy 1969), many believe that local diversity is regulated solely by local processes whose influence is so powerful as to override that of regional processes (see reviews by Pianka 1966; MacArthur 1969). I shall demonstrate the falseness of this notion, and argue that ecologists must, therefore, place studies of diversity in a larger context. Comparisons of diversity among ecologically similar, but geographically removed localities clearly demonstrate the impact of regional processes on local as well as regional diversity (for a review, see Orians and Paine 1983). Contrary to common sense, these studies imply that rapid, local processes cannot fully compensate the effects of slow, regional processes. I suggest that this paradox, contrasting the apparent asymmetry between rates of regional and local processes with the relative symmetry of their effects, can be resolved in two ways. First, many population traits exhibit long-term persistence, generally attributable to evolved properties. These persistent qualities control, to a large extent, the outcome of local interactions between species and they change only over evolutionary time scales. Second, population processes, including habitat specialization and within-habitat specialization, tend to bring competitive and other interactions into balance and greatly extend the time scales over which they act. Both factors prolong the exclusion of species to evolutionary rather than ecological time scales, and, thus, to scales that approach those of speciation and biotal dispersal.

WHY MUST WE INVOKE REGIONAL PROCESSES?

During the past 25 years, ecologists have directed most of their efforts concerning variation in local species diversity toward understanding the influence of local factors. This perspective grew out of the tradition of population biology, whose origins date to the 1920s. Prior to that time, most explanations of variation in diversity over the surface of the earth relied on regional processes of long duration. Alfred Russel Wallace (1878) believed that tropical environments were older than more polar environments; the processes of proliferation had therefore resulted in the accumulation of species over longer periods in the tropics than at higher latitudes.

The "time" hypothesis reached its acme in the ideas of the botanist J. C. Willis. In his book *Age and Area*, published in 1922, Willis argued that history and geography were the primary influences on the accumulation of species within local floras. Willis observed that localized species were often rare within their limited geographic ranges. He supposed that these were new species, formed by "macromutations." Accordingly, because speciation proceeded at random with respect to the environment, new species were poorly adapted, hence rare and locally distributed. As these new populations evolved, they became better adapted to their environments and their populations began to increase in density and geographic range.

According to Willis's logic, species accumulate within a region in direct proportion to age and area (hence in proportion to the opportunity for chance "mutation" to produce new species). In hindsight, some of Willis's ideas verge on silliness, but he did provide a logically consistent mechanism for the generation of diversity in the spirit of the genetics of his day.

The historic and geographic perspectives that inspired Willis largely faded from ecology and evolutionary biology during the 1920s, when both population biology and population genetics became firmly established as powerful and successful paradigms. Elsewhere (Ricklefs 1987) I have traced what Kingsland (1985) referred to as the "eclipse of history," from the development of theoretical models of population interactions by Lotka (1925) and Volterra (1926), through experimental confirmation of the theory by Gause (1934) and others, to the extrapolation of population biology from theory and the laboratory microcosm to natural systems by Lack (1944, 1947) and Hutchinson (1957, 1959). The Lotka–Volterra view of the world was largely deterministic and homogeneous. Laboratory experiments created similarly homogeneous microcosms. By extrapolation, coexistence of species within a community depended on local processes of competition and predator–prey relations. Theory suggested and experiment confirmed that coexistence becomes more difficult as ecological similarity between competitors increases. This soon gave rise to a notion that species could coexist only if they differed by more than some limiting level of similarity (MacArthur and Levins 1967; May 1975). Although this level varied according to local conditions, such as environmental variability and predation pressure, it nonetheless set an upper limit to the density of species packing in ecological niche space, hence an upper limit to the number of species that can coexist locally.

Most ecologists have accepted the influence of historical factors on diversity of large regions (MacArthur 1969, 1972). Discrepancies between regional diversity (gamma diversity, in the sense of Whittaker 1972), determined by regional and historical factors, and local (alpha) diversity, determined by local factors, could be accommodated by replacement of species between habitats (Table 1). Where regional diversity greatly exceeds local diversity, species occupy relatively few habitats; thus, species composition turns over rapidly between habitats (beta diversity) (e.g., Cody 1975).

The population biological perspective took the phenomenon of local diversity from systematists and biogeographers and placed it in the hands of ecologists, who rushed to explore this problem during the 1960s and 1970s. Their primary objective was to test the hypothesis that local interactions, principally competition, modified by predation, determined local diversity in concert with local conditions of the environment. In doing so, ecologists concentrated on demonstrating that species compete for resources within local communities and that their interaction manifests itself evolutionarily in the even dispersion of species within morphological space. The first is necessary to the local-interaction hypothesis but not sufficient to validate it. The second was pursued to demonstrate competition within communities (Brown and Wilson 1956; Hutchinson 1959; Grant 1975; Schluter et al. 1985), but is not otherwise relevant to the local-interaction hypothesis.

Field experiments have demonstrated that interspecific competition is

TABLE 1. Comparison of two ecologically similar regions with similar levels of local (alpha) diversity, but different levels of beta diversity and regional (gamma) diversity.

Habitat A	Habitat B	Habitat C	Habitat D
	Region I (7 species): low beta diversity		
Species a	a	a	b
b	b	b	d
	c	d	e
		e	f
			g
	Region II (11 species): high beta diversity		
Species h	i	k	m
i	j	l	o
	k	m	p
		n	q
			r

pervasive, but by no means universal, in natural communities (Schoener 1983). Claims that competition results in regularity in the distribution of species within ecological or morphological space have been vigorously debated recently (see chapters in Strong et al. 1984; Diamond and Case 1986). But the outcome of the debate has little bearing on the local-interaction hypothesis (Walter et al. 1984; Maurer 1985): competition could regulate local diversity without producing evolutionary manifestations on a local scale. Conversely, demonstrating competition is not equivalent to demonstrating a role in regulating number of species.

More direct tests of the local-interaction hypothesis involve the principles of convergence and saturation. If local interactions determined diversity, communities developed under similar local conditions would have similar numbers of species, regardless of the taxonomic origin of the biota or the total diversity of the region (Recher 1969; MacArthur 1972; Cody 1975). Corollary to this expectation, local diversity should be independent of regional diversity when the latter exceeds the level set by limiting similarity in the local community (Terborgh and Faaborg 1980).

Schluter (1986) has pointed out that, as a process, convergence can be identified only by the tendency of local assemblages of species in similar habitats to resemble each other more closely than do the regional source pools of species from which they are derived. Hence the process of convergence does not imply similarity. But, for practical reasons, ecologists usually analyze similarity rather than convergence with reference to regional sources. Similarity, or lack thereof, nevertheless pertains to the local-interaction hypothesis: even if convergence can be demonstrated, lack of similarity indicates that local assemblages have not achieved local equilibria and continue to bear the imprint of regional/historical processes (tempered by whatever convergence has occurred).

Few ecologists have tested the notion of convergence/similarity, partly owing to the logistical difficulties of collecting comparable data in many historically independent regions of the world. Exceptions include Pianka (1973,1975; desert lizards), Blair et al. (1976; desert vertebrates), Mares (1985; desert mammals), Cody (1975, 1983; Mediterranean-climate birds), Recher (1969; temperate-zone birds), Karr (1976; tropical forest birds), Pearson (1977; tropical dry-forest birds), Lawton (1984; herbivores of bracken), Schmida (1981; Mediterranean-climate vegetation), and Whitmore (1984; rain forest trees). Orians and Paine (1983) summarized several of these studies and could find little evidence for convergence on regional scales (which would be expected on the basis of differing histories and geographic positions of the major geographic regions). But they also found little evidence for convergence (similarity) on local scales.

For many systems and groups of organisms, similarity may be rejected outright. The fivefold, or greater, difference in diversity of mangrove species of plants between Malaysia and the Caribbean (Richards 1952), similarly large differences in the diversity of corals and reef inhabitants between the same regions (e.g., Ehrlich 1975), and smaller but persistent difference on both regional and local scales in floral diversity of Mediterranean-climate habitats (Schmida 1981) are oft-cited examples.

The relatively low regional diversity of tropical rain forests in Africa compared to Amazonia and, particularly, Malesia (generally, the Malay Peninsula and islands between it and Australia, including New Guinea and the Phillipines; Whitmore 1984) is well known. Both Richards (1952) and Whitmore (1984) attribute these differences in diversity to regional/historical factors. Richards (1952:231) commented that "whatever the true explanation [for the poverty of the African rain forest], it can hardly be due to any ecological factor operating at the present day." Whitmore (1984:12) states that "Contrary to widespread belief it can now be seen that the extreme floristic richness of the Malay archipelago is not due to the flowering plants having evolved in the region somewhere between Assam and Fiji. It arises in part because there are two floristic elements of different origin [Asian and Australian] now considerably intermingled." Whitmore demonstrates quite convincingly that differences in regional species richness of trees are translated into differences in local species richness, as well. Hubbell and Foster (1987:226), referring to species richness in a forest plot on Barro Colorado Island, Panama, state the case explicitly: "We suspect that the maintenance of high tree species diversity has more to do with regional tree species richness and availability of potential immigrants, which in turn are dictated by the interaction of climate, the historical dynamic biogeography of particular tree taxa, and speciation processes on a regional and subcontinental scale."

Even mobile organisms such as birds show the imprint of history on the diversity of local assemblages. Karr (1976) remarked on the relative poverty of forest bird communities in West Africa compared to Panama. Based on comparisons of bird communities in closely matched habitats in South America and Malaysia, Pearson (1977) stated, "One major generalization, as earlier suggested by Haffer (1969), Karr (1976), and others, involves the influence of historical factors on these communities. From theorized geological history alone one can predict the rank order of the number of bird species on the six plots."

Finally, the regional process of colonization has been accorded a prominent role in the equilibrium theory of island biogeography (MacArthur and Wilson 1967) as a determinant of diversity on islands. Cox and Ricklefs (1977) and others have demonstrated a correlation between local (within-habitat) and regional (island-wide) diversity on islands (Ricklefs 1987). Schluter's (1986) analysis protocol partitions diversity into habitat (local) and regional components by two-way analysis of variance. Although his

TABLE 2. Analysis of variance in number of species recorded in small areas of homogeneous habitat with respect to locality within the Caribbean basin and habitat type.[a]

Source	df	SS	MS	F	p
Locality	4	2335	584	13.7	<0.0001
Habitat	8	2417	302	7.1	<0.0001
Error	32	1365	43		
Total	44	6117			

[a]Data from Ricklefs and Cox (1978).

comparison of finch communities on five continents revealed no significant regional component, similar analysis of passerine birds in nine matched habitats in five areas around the Caribbean basin demonstrates a strong regional component to diversity (Table 2).

A second test of the local interaction hypothesis involves the identification of local saturation of species richness by the leveling off of local diversity as regional diversity rises. This test was first proposed by Terborgh and Faaborg (1980), who observed that local communities of landbirds on islands in the West Indies reached diversity plateaus of between 10 and 30 species, depending on habitat, when the total species pool on the island reached about 40 species. Using a different data set on passerine birds from the West Indies and two Caribbean mainland localities, Cox and Ricklefs (1977) found that the local diversity continued to increase even as the regional pool of species approached 100 species (Ricklefs 1987). Cornell (1985) also failed to find saturation in the cynipine gall wasps on oaks (*Quercus*) in California. Unfortunately, few other data sets permit testing of the saturation hypothesis.

Although convergence and saturation cannot be rejected as important patterns, striking counterexamples strengthen the complementary viewpoint that history influences the composition of present-day, local communities.

THE TIME SCALES OF LOCAL AND REGIONAL PROCESSES

If comparative community ecology strongly implies that events and circumstances reach down through time, then this implication creates a major paradox. That is, how can processes that work on vastly different scales of time and space interact to determine patterns of local diversity? The study of hierarchical systems tells us that processes must be expressed on similar scales if they are to interact strongly (Allen and Starr 1982; O'Neill et al. 1986). Conventional wisdom places demographic processes, including species interactions, on scales of tens to hundreds of generations. In contrast, regional events such as speciation and biotal dispersal occur with frequencies of one in thousands to millions of generations.

Table 3 contains some wild guesses and estimates for several categories of ecological, biogeographic, and evolutionary processes: (1) individual responses to environmental change, (2) demographic responses, including movements between habitats and interactions between populations, (3) evolutionary change within a population, and (4) speciation, transspecific evolution, and biotal dispersal. Individual responses, and birth and death processes, clearly have a time course of a single generation or less. Local movements between habitats may occur within a generation or may reflect density-dependent population responses to longer-term variation in habitat quality. Estimates of the time scales of density-dependent regula-

TABLE 3. Estimates of characteristic time periods of processes that influence local diversity.

	Generations								
	1	10	100	10^3	10^4	10^5	10^6	10^7	10^8
Phenotypic response	—								
Birth and death	—								
Local movement	—								
Density dependence	—	—							
Competitive exclusion		—	—						
Dispersion			—	—					
Quantitative selection		—	—						
Allele substitution			—	—					
Genetic drift, time					—	—	—	—	
Genetic drift, rate of fixation							—	—	
Reproductive isolation						—	—		
Speciation							—	—	
Taxonomic evolution								—	—
Trans-specific evolution								—	—
Biotal dispersal									—

tion and population interactions are based on theoretical considerations and the results of laboratory experiments (Miller 1967). Gene flow across the range of a population may require longer periods (Endler 1977), even though individuals may be capable of moving long distances in short periods.

Response to artificial selection upon quantitative traits may run its course within ten or so generations (Falconer 1981). Allele substitution takes longer depending on the initial and final frequencies considered and the strength of selection (Crow and Kimura 1970; Hartl 1980). The time required for the fixation of a mutant by drift is on the order of four times the effective population size, and the rate of fixation of mutations in a population is on the order of the mutation rate (Futuyma 1986). Characteristic durations of regional processes and intervals between regional events probably are on the order of 10^5 generations and longer.

I arrived at figures for speciation events by guessing at the ages of several adaptive radiations. For example, the 13 species of Darwin's finches (Geospizinae) on the Galapagos Islands could have been produced by 3.7 serial bifurcations (speciation events) during a history of perhaps three million years, or a rate of 1.23×10^{-6}. Forty species of Hawaiian honeycreepers might have been similarly produced over 10 million years at a rate of 0.53×10^{-6}, 600 species of *Eucalyptus* over 40,000,000 years at a rate of 0.23×10^{-6}, and 170 species of *Haplochromis* (Cichlidae) in 0.75×10^6 years at a rate of 9.9×10^{-6} years. Even though some mechanisms may produce new species in one generation, such events may occur very infrequently. During the Pleistocene epoch major relocation of species and intermixing of faunas and floras occurred over periods of 10^5 to 10^6 years; most of the major events of biotal dispersal, involving separation and rejoining of continents and seas, have occurred with much lower frequency (Valentine 1973; Hallam 1973; Eldredge and Cracraft 1980; Stehli and Webb 1985).

No reasonable revision of these estimates will bring local demographic processes and regional processes of species production into the same ballpark.

THE PERSISTENCE OF INDIVIDUAL SPECIES, AND COMMUNITY TRAITS

The idea of historical factors reaching down through time presupposes that events and circumstances create traits that persist far beyond the duration of their creative forces. Such persistence is the basis of systematics, which classifies species into a hierarchy of groupings based on traits inherited from more and more distant ancestors. The persistence of populations, species, and even community characters has, however, only recently begun to receive attention (Brooks 1985). I provide several examples to demonstrate the general principle and illustrate its potential dimensions.

Jablonski (1987) has suggested that the geographic ranges of species may persist over millions of years. He observed that closely rated taxa of late Cretaceous bivalves and gastropods tended to occupy ranges of similar area. He further suggested that temporal duration in the fossil record is positively associated with geographic range. Combining these observations, Jablonski claimed to have demonstrated heritability for species-level traits that affected probability of extinction—the prerequisites for species level selection (Stanley 1975). It could be argued that ecological relationships determine geographic range (Brown 1984) and that similar species retain similar ecologies. If this were true, persistence would apply to individual-level traits rather than species-level traits. Species duration itself might be ecologically constrained by factors related to those responsible for geographic range. But regardless of whether the traits involved reside at the species or individual level, Jablonski's data indicate long-term persistence of traits involved in ecological interactions. In contrast, many contemporary genera contain species with vastly different geographic ranges (e.g., Rapoport 1982; Bock and Ricklefs 1983), suggesting that ecological interactions are perhaps governed more by locally adaptive traits.

Roger Latham and I (unpublished) have examined the ranges of contemporary species belonging to genera of plants endemic to both eastern Asia and eastern North America, with objectives similar to Jablonski's. These genera are remnants of a widespread, temperate, "arctotertiary" flora that at one time stretched in a continuous band across northern Europe, Asia, and North America (Fernald 1929; Berry 1937; Graham 1972). As polar climates cooled, elements of this flora retreated to more southerly latitudes and became fragmented and relictual. Species endemic to mesophytic forests of eastern North America and eastern Asia represent a small fraction of the total extant flora, which has been discussed in detail by Li (1952).

If the ecological relations of species were evolutionarily conservative, a correlation might be observed between the ranges of representatives of various genera in each hemisphere. Indeed, for species of herbaceous perennials the ranges of congenerics in the two hemispheres were strongly correlated for geographic area ($r = 0.58, p = 0.008$; Figure 1); for woody species, the correlations were not significant. Among all species, the latitude of the center of the range was strongly correlated between Asia and North America [$F(1,54) = 13.0, p = 0.0007, r^2 = 0.19$].

The 58 genera listed by Li contain more species in Asia (total of 366 species) than in North America (132 species), paralleling the general pattern in mesophytic forest (Latham and Ricklefs, unpublished). Further-

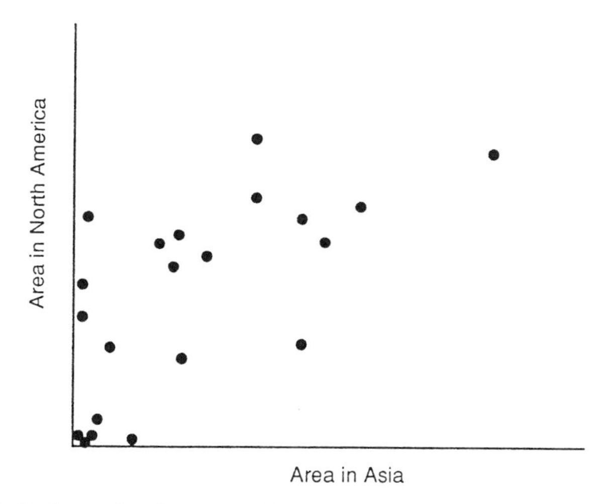

FIGURE 1. Relationship between the geographic area occupied by a genus of herbaceous perennial plants in eastern North America and the area occupied by the same genus in eastern Asia. The relationship is highly significant $F(1{,}20) = 8.8$, $p = 0.008$, $r^2 = 0.316$.

more, the number of species in a genus does not appear to covary between the two regions (Figure 2). Thus, although certain aspects of ecological relationships appear to persist, being constrained at least at the generic level, species production, species duration, or both appear to be more labile; diversification and the maintenance of species richness may be related to characteristics of the region.

If ecological relationships determined the areal extent of species ranges, the two ought to be correlated. Ecological factors associated with local rarity have received attention recently from ecologists (e.g., Karr 1977; Hubbell and Foster 1987; Rabinowitz et al. 1987). Furthermore, local abundances of species have been found to bear a direct relationship to distribution and geographic range on both local and regional scales of resolution (Hanski 1982; Bock and Ricklefs 1983; Brown 1984). However, patterns revealed by such studies demonstrate that local and regional population characteristics are linked, but do not permit us to estimate their persistence.

The studies of Ricklefs and Cox (1972, 1978) and Cox and Ricklefs (1977) on passerine birds of the West Indies are unique in demonstrating that local ecological characteristics, expressed as habitat distribution and local abundance of populations, bear a detectable relationship to regional and historic characteristics of the species. We assigned each species to one of four categories according to its geographic range and degree of tax-

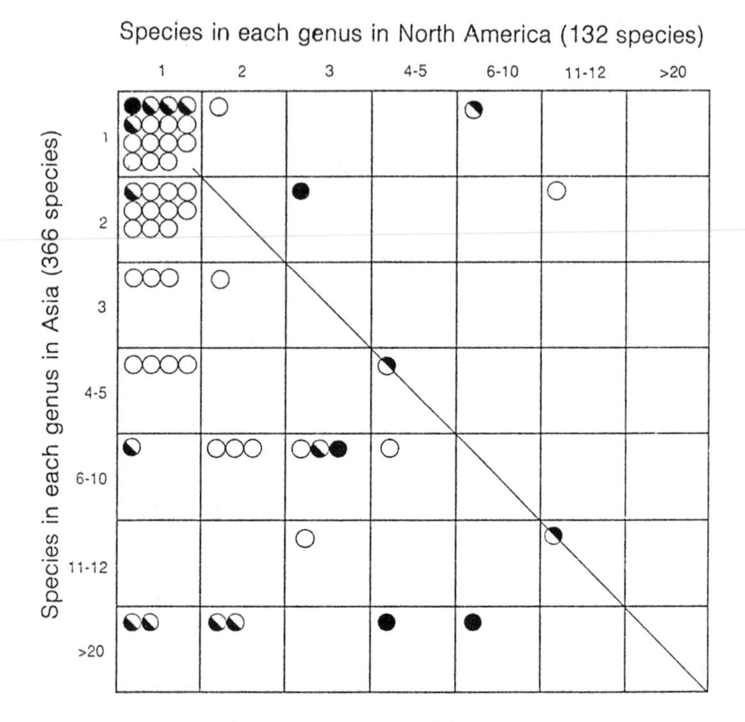

FIGURE 2. Relationship between number of species of each genus of plant in eastern Asia and eastern North America, demonstrating a general lack of correlation. Circles are filled according to representation of genus in tropical latitudes in each region (Asia: lower left; North America: upper right).

onomic differentiation at the subspecies level within the West Indies (Table 4). We proposed that stages I through IV constitute an historic sequence of expansion and contraction, similar to models proposed by Darlington (1943,1957), Dillon (1966), Erwin (1981), and Rapoport (1982), and called the "taxon cycle" by Wilson (1961). Regardless of the historic validity of the taxon cycle (Pregill and Olson 1981), these categories derive solely from regional patterns that reflect dispersal through the island chain, degree of evolutionary divergence between populations, and extinction of populations from islands within the West Indies, hence from regional/historic factors (Ricklefs and Cox 1972). For each of three islands (Jamaica, St. Lucia, and St. Kitts), we tabulated the abundances of each species in nine matched habitats. From these counts, we obtained, for each species present on the island, measures of habitat diversity (number of habitats occupied: 0–9) and average local abundance (scaled from 0 to 10). The three islands differ in area; they also differ in

TABLE 4. Characteristics of distribution of species in the stages of the taxon cycle.[a]

Stage of cycle	Distribution among islands	Differentiation between island populations
I	Expanding or widespread	Island populations similar to each other
II	Widespread over many neighboring islands	Widespread differentiation of populations on different islands
III	Range fragmented due to extinction	Widespread differentiation
IV	Endemic to one island	—

[a]From Ricklefs and Cox (1972).

number of species, both in total and in each taxon cycle category (Table 5).

Ecologists generally believe that among ecologically comparable areas, the habitat ranges and local densities of individual species vary in inverse relation to the number of species. Thus, the total density of individuals of all species is conserved. This principle, called density compensation, has been widely reported (e.g., Crowell 1962; MacArthur et al. 1972; Cody 1983); it is cited as evidence of local interaction between species.

The West Indian data allow us to partition the relative influences of

TABLE 5. Number of species of passerine bird on each of three islands in the West Indies according to stage of the taxon cycle.[a]

		Jamaica	St. Lucia	St. Kitts
Area (square miles)		4450	233	65
Elevation (feet)		7402	3117	4314
Stage	I	5	8	6
	II	10	7	6
	III	8	9	2
	IV	12	2	0
Total		35	26	14

[a]From Ricklefs and Cox (1978).

local interactions and regional/historic processes on the local ecological distributions and abundances of birds. The results of a two-way analysis of variance, in which the main effects are island (local interactions) and distribution/differentiation category (reflecting regional and historic processes), are presented in Table 6. Although the ANOVA explains only 15 and 23% of the total variance, taxon cycle stage is a significant effect for habitat breadth and local population density, whereas island influences neither. I interpret these data to indicate the primacy of regional/historic factors over local interactions in determining local ecological amplitude.

Ricklefs and Cox (1972) proposed a mechanism, involving a notion similar to Brown's (1959) "general adaptation," to account for the taxon cycle. Briefly, we postulated that species expanding their ranges within the West Indies have somehow escaped limitations on their populations, possibly through the evolution of disease resistance, the acquisition of a well-adapted general phenotype (whatever that means), or a pervasive and favorable change in the environment. New colonists to islands initially occupy broad arrays of habitats at high density. Subsequently, while they adapt to local conditions, the local biota also adapts to them. We suggested that because each newcomer is a major selective force in the environment of local species, the evolutionary responses of the latter together reduce the general adaptiveness of the newcomer, thereby making it vulnerable to competitive exclusion from succeeding invaders. Regardless of the mech-

TABLE 6. Analysis of variance in population characteristics with respect to island (local conditions) and stage of taxon cycle (regional and historical attributes of the species).[a]

	Habitat breadth (1)	Local density (2)	Overall abundance (1) × (2)
Island (2 df)			
Type III SS (%)	3.8	0.0	3.2
F	1.8	0.0	1.5
p	0.17	0.99	0.23
Stage (3 df)			
Type III SS (%)	14.4	22.5	16.5
F	4.5	7.0	5.3
p	0.006	0.0004	0.0026

[a]Seventy-five species of passerine birds were distributed among islands as Jamaica 35, St. Lucia 26, and St. Kitts 14, and among taxon cycle stages as I 19, II 23, III 19, IV 14 (data from Ricklefs and Cox 1978).

anism, however, species in categories III and IV undoubtedly represent evolutionarily older invasions of islands or reexpansions from within the islands than do species in categories I and II, which represent either recent expansions or species with high levels of gene flow between islands. Compared to type I and II species, type III and IV species have lower vagility and reduced ecological amplitude, and they have suffered greater historic extinction resulting from anthropogenic changes in their environments (Ricklefs and Cox 1972). In this example, change in local ecology from one category to the next represents not the persistence of traits but the predictable change in traits over time. Thus, species characteristics nonetheless reveal the imprint of history.

THE HISTORICAL DIMENSION OF DIVERSITY

Regional diversity reflects the accumulation and disappearance of species over long periods. Although the historic record of diversity within a region cannot be reconstructed, the contribution to total diversity of each level of the taxonomic hierarchy can be used as an index to the history of diversification. In particular, in comparisons between regions, it is possible to determine whether differences in diversity have arisen within the recent or more distant evolutionary past.

A prevalent explanation for the high diversity of plants and other groups in Amazonia rests on the idea that repeated fragmentation of forest habitat during the Pleistocene caused a flourish of species production (Prance 1982; Haffer 1969, 1974). If this were true, differences in number of species between Amazonia and less diverse areas should appear at the level of species within genera rather than at higher taxonomic levels.

To test this notion it is necessary to accept that taxonomic designations are homogeneous between regions; the results are nonetheless revealing (Table 7). The roughly twofold difference in number of species of forest trees between a small area of Costa Rica and the forested area of eastern North America arises at the family level and, particularly at the level of genera within families. Genera in Central America are, by comparison, rather depauperate in species. The passerine birds of Amazonian Peru and the United States reveal a similar pattern. Samples of similar numbers of trees in forests in Central America and Malesia provide a closer comparison; these indicate similar richness at the generic level, but much greater richness in Malesia at the level of species within genera. Comparisons of birds in forest habitats in Amazonia and West Africa also indicate relatively greater generic richness in the New World tropics and relatively greater species richness within genera in the Old World. None of the data presented here supports the hypothesis that forest fragmentation within Amazonia during the Pleistocene has resulted in an unusual pro-

TABLE 7. Taxonomic levels of diversity among forest trees and passerine birds in tropical America compared to other regions.

	Families	Genera	Species	G/F	S/G
FOREST TREES					
Rain forests of Golfo Dulce, Costa Rica (Allen 1956)	64	260	417	4.1	1.6
Deciduous trees of eastern North America (Preston 1961)	33	59	205	1.8	3.5
Malesia (Whitmore 1984)					
Malaya (1.6 ha, 791 trees)		113	210		1.9
Sabah (1.9 ha, 1215 trees)		99	198		2.0
Central America (Thorington et al. 1982, Barro Colorado Island, 5 ha, 856 trees)		80	112		1.4
PASSERINE BIRDS					
Tambopata Reserved Zone, Peru (Pearson 1980)	18	172	279	9.6	1.6
All of North America (various field guides)	24	114	293	4.8	2.6
Moist forest, Panama (Karr 1976)	15	75	86	5.0	1.2
Moist forest, Liberia (Karr 1976)	14	38	57	2.7	1.5

liferation of species compared to other regions of the world. Of course, the data may also indicate differences in taxonomic practices between New World and Old World specialists. But, in either case, the patterns merit attention.

If we accept data on taxonomic levels of diversity at face value, and if each taxonomic level represents a similar passage of time and evolutionary diversification in different regions, it would appear that the high diversity of Amazonia has persisted for long periods; contemporary high generic and familial diversities reflect processes well in the evolutionary past. With respect to the relatively high diversity of species within genera in Malesia, recent geographic conditions or perhaps characteristics of the species themselves may have promoted speciation. Other explanations could account for the data, as well. If generic characters of the trees of Malesia were more persistent than those of other areas, divergences within genera could be as old as contemporary generic level differences elsewhere. Finally, exclusion of some genera by a small number of recent, speciose competitively dominant genera may have obliterated patterns of diversity established earlier.

Taxonomic data themselves are open to divergent interpretations. But molecular techniques can now assign approximate ages to divergence points between species, opening the way to interpreting diversification with respect to the history of climate and geography.

RESOLVING THE TIME SCALE PARADOX

Biological systems retain the effects of large-scale processes. Yet those processes presumably are weak compared to local interactions, which operate on time scales three to five orders of magnitude shorter. This discrepancy between historical/regional and contemporary/local processes recalls the contrast between group and individual selection: the potential strength of individual selection is proportional to the rate of turnover of individuals within a population, whereas that of group selection is proportional to the lower rate of extinction of subpopulations. Partly for this reason, few evolutionary biologists believe in the efficacy of group selection when counteracting individual selection (Wiens 1966). The latter would always predominate, it seemed, because of its speed. Realistic models of group selection depend on small group size and correspondingly rapid turnover of groups (Wade 1978; Wilson 1983).

The influence of regional processes on diversity is measured in rates of species production, dispersal, and extinction, all of which require evolutionary time. The outcomes of local processes such as species interactions are thought of as occurring within a few tens of generations—that is, on ecological scales of time (Udvardy 1981). Theoretical and experimental studies appear to bear this out. How, then, may we reconcile the relative sluggishness of regional/historical processes and their clear imprint on local diversity?

Allow me a few statements of faith. In part, the paradox stems from a mistaken perception of the confinement and speed of local processes. Theoretical systems are simple and homogenous, or nearly so. Laboratory microcosms also simplify nature to the barest necessities. In theoretical systems, the outcomes of species interactions depend on the strengths and asymmetries of interaction coefficients, higher values generally leading more rapidly to exclusion (May 1975). Experimental systems are deliberately concocted to emphasize these same qualities to obtain results within short periods (e.g., Gause 1934).

In natural systems, which comprise much more complexity, both locally and regionally, species interactions are ameliorated by density-dependent responses, which involve specialization with respect to habitat and to niche space within habitat. To some degree, such responses may require evolutionary change within the population, but most populations exhibit phenotypic plasticity with respect to within- and between-habitat specialization, so that responses may occur within one to a few generations. In this way, strong interactions and asymmetries are reduced to levels that permit coexistence.

Furthermore, incompatible use of ecological niches probably arises infrequently in natural situations compared to the laboratory. Laboratory studies throw together species, often not naturally coexisting, into contrived, unnatural situations, which may emphasize competitive or predator–prey interactions. Taxa produced by speciation are ecological copies of a successful mode of life. In the case of vicariant speciation, community patterns are maintained more or less intact. Mechanisms for reducing interactive effects of other populations have evolved over long periods in ancestral lines. Speciation is therefore unlikely to produce strong interactions, except perhaps with parent or sister taxa following secondary sympatry. In these cases, small differences in habitat or niche within habitat would seem sufficient to permit coexistence. I suggest, therefore, that the accumulation of diversity by speciation does not generate the strong interactions that lead to exclusion over short periods, as observed in the laboratory. Theoretical and laboratory systems are qualitatively inadequate models of natural communities.

Biotal dispersal, as well as the introduction of alien species by man, may have consequences different than speciation within regions. The outcome will depend on the particular niche requirements of the mixing faunas and floras. If they are similar, then species may interact strongly, and increased rates of local exclusion and extinction will follow mixing. If mixing biotas are organized differently ecologically, they may be mutually compatible and diversity may remain elevated for considerable periods, as Whitmore (1984) has suggested for the rain forests of Malesia.

CONCLUSIONS

Variation in both regional and local diversities in ecologically comparable areas having long-independent evolutionary histories indicate that regional/historic processes, principally speciation and biotal dispersal, are important determinants of species diversity. This conclusion is by no means novel—with respect to regional diversity, it has been widely held among systematists and biogeographers for more than a century—but it does contrast strongly with the prevalent attitude among ecologists that local diversity is regulated by local ecological interactions, particularly competition and predation, within the context of the local physical environment.

I believe that we stand at the threshold of a major integration of ecological and evolutionary perspectives that will ultimately bridge the gap in temporal and spatial scales that separates these disciplines. For the ecologist, this synthesis will require abandoning parochial limits on the scales of relevant processes (5 years and a few hectares or less being typical), broadening ecological hypotheses concerning community diversity to include movements between habitats, dispersal between regions, and speciation, and incorporating data of biogeography, systematics, and paleontology into analyses of ecological patterns. At the same time, evolutionists should appreciate the relevance of large-scale processes to ecological problems, and emphasize the ecological contexts of evolutionary change and speciation to the extent that they now do their genetic bases.

At this point, I can visualize several fruitful avenues of investigation. First, to establish the validity of the regional/historic point of view, comparative studies of local and regional diversity should be undertaken through a systematic sampling scheme in physically matched environments in selected geographic areas. It will be important to survey groups with different dispersal capabilities and different relative ages. Such studies could assess the relative contributions of historic processes to contemporary patterns of diversity. Variation in these patterns among groups of organisms may also indicate the roles of particular processes or events.

Second, comparisons of individual and population traits of sister taxa will allow the degree of persistence or evolutionary flexibility to be assessed. Determining whether persistence represents evolutionary or ecological conservativeness would require experimental approaches, but persistence itself should provide clues to the ecological context of evolution and, conversely, the evolutionary basis of community assemblage. Conservatism of ecological relationships would suggest that populations can select environments so as to minimize detrimental (that is, selective or extinction-causing) interactions, thereby extending ecological relationships to the scale of evolutionary and regional/historic processes.

Third, ecologists should use systematic data to assess the buildup of diversity at different taxonomic levels. Molecular techniques may yield a less ambiguous assessment of the time dimension of local and regional assemblages of communities, perhaps tying certain components of diversity to specific geological events. Required are analytical methods of partitioning diversity into different taxonomic levels and distinguishing among biotal dispersal, species proliferation, and extinction as causes of variation in diversity. Recent advances in cladistics and vicariance biogeography would appear to be good starting places.

Fourth, ecologists should explore the geographic and habitat extent of populations and investigate the ecological basis of species turnover between communities. For the ecologist, the link between regional and local diversity is essentially important, as it ties together the local/contemporary and regional/historic perspectives. Such studies would also elucidate the relative roles of habitat selection versus evolutionary response to environment in determining the geographic (and perhaps temporal) extent of the population.

Each of these suggestions advocates large-scale comparative approaches at a time when ecology is becoming more reductionist and experimental. The validity and general acceptance of exploring such approaches will depend entirely on their success, which can be judged only in terms of the expectations one holds. No foreseeable analysis is likely to produce an historic account of community development. However, the simple analyses presented in this chapter demonstrate the inadequacy of models of species diversity based entirely on local processes of short period, and the reasonableness of the point of view that ecological interactions may be played out on evolutionary time scales. I believe that, in time, related approaches will produce a synthetic concept of ecology and evolution in a greatly elaborated context of habitat and geography.

SUMMARY

This chapter explores the influence of regional and historic factors, including rate of speciation, on patterns of biological diversity on both local and regional scales. Comparative studies reveal large discrepancies in the diversity of some groups between ecologically similar areas on different continents having long-distinct evolutionary histories. These discrepancies strongly imply that local communities are not regulated solely by local interactions among species, but that historic differences in species production, migration, and extinction leave their imprint on local species diversity.

Analysis of population traits of sister taxa demonstrates that characteristics of individuals influencing ecological relationships, perhaps including interspecific interactions, have persistences that can be measured on evolutionary time scales. Geographic range, habitat preferences, and local abundance appear to be conserved in some groups over periods that include the time required for speciation and for the extinction of populations on large islands.

Comparisons of diversity at different taxonomic levels suggest that the establishment of global patterns of variation in diversity may have been influenced by processes and events occurring in the distant past.

I suggest that local/contemporary processes do not completely dominate regional/historic processes in determining local diversity because habitat selection in a complex environment tends to balance and equalize interactions among species, thereby prolonging the outcome of competitive exclusion to evolutionary time scales.

I believe that we are on the threshold of a major synthesis of ecology and evolution in which ecological investigations are expanded by incorporating systematic, biogeographic, and paleontological data and insights, and evolution, placed in an expanded ecological context, addresses the development of biological communities.

LITERATURE CITED

Allen, P. H. 1956. *The Rain Forests of Golfo Dulce.* University of Florida Press, Gainesville.

Allen, T. F. H., and T. B. Starr. 1982. *Hierarchy. Persepctives for Ecological Complexity.* University of Chicago Press, Chicago.

Berry, E. W. 1937. Tertiary floras of eastern North America. Bot. Rev. 3:31–46.

Blair, W. F., A. C. Hulse, and M. A. Mares. 1976. Origins and affinities of vertebrates of the North American Sonoran Desert and the Monte Desert of Northwestern Argentina. J. Biogeog. 3:1–18.

Bock, C. E., and R. E. Ricklefs. 1983. Range size and local abundance of some North American songbirds: A positive correlation. Am. Natur. 122:295–299.

Brooks, D. R. 1985. Historical ecology: A new approach to studying the evolution of ecological associations. Ann. Missouri Bot. Garden 72:660–680.

Brown, J. H. 1984. On the relationship between abundance and distribution of species. Am. Natur. 124:255–279.

Brown, W. L., Jr. 1959. General adaptation and evolution. System. Zool. 7:157–168.

Brown, W. L., Jr., and E. O. Wilson. 1956. Character displacement. System. Zool. 5:49–64.

Cody, M. L. 1975. Towards a theory of continental species diversities: Bird distributions over Mediterranean habitat gradients. Pp. 214–257 in: M. L. Cody and J. M. Diamond (eds.), *Ecology and Evolution of Communities.* Harvard University Press, Cambridge, MA.

Cody, M. L. 1983. Bird diversity and density in South African forests. Oecologia 59:201–215.

Cornell, H. V. 1985. Species assemblages of cynipid gall wasps are not saturated. Am. Natur. 126:565–569.

Cox, G. W., and R. E. Ricklefs. 1977. Species diversity, ecological release, and community structuring in Caribbean land bird faunas. Oikos 29:60–66.

Crow, J. F., and M. Kimura. 1970. *An Introduction to Population Genetics Theory.* Harper & Row, New York.

Crowell, K. L. 1962. Reduced interspecific competition among the birds of Bermuda. Ecology 43:75–88.

Darlington, P. J., Jr. 1943. Carabidae of mountains and islands: Data on the evolution of isolated faunas, and on atrophy of wings. Ecol. Monogr. 13:37–61.

Darlington, P. J., Jr. 1957. *Zoogeography: The Geographical Distribution of Animals.* John Wiley, New York.

Diamond, J. M., and T. J. Case (eds.). 1986. *Community Ecology.* Harper & Row, New York.

Dillon, S. L. 1966. The life cycle of the species: An extension of current concepts. System. Zool. 15:112–126.

Endler, J. A. 1977. *Geographic Variation, Speciation, and Clines.* Princeton University Press, Pinceton, NJ.

Ehrlich, P. R. 1975. The population biology of coral reef fishes. Annu. Rev. Ecol. System. 6:211–247.

Eldredge, N., and J. Cracraft. 1980. *Phylogenetic Patterns and the Evolutionary Process.* Columbia University Press, New York.

Erwin, T. L. 1981. Taxon pulses, vicariance, and dispersal: An evolutionary synthesis illustrated by carabid beetles. Pp. 159–196 in: G. Nelson and D. E. Rosen (eds.), *Vicariance Biogeography. A Critique.* Columbia University Press, New York.

Falconer, D. S. 1981. *Introduction to Quantitative Genetics,* 2nd ed. Ronald Press, New York.

Fernald, M. L. 1929. Some relationships of the floras of the Northern Hemisphere. Proc. Int. Cong. Plant Sci. Ithaca 2:1487–1507.

Futuyma, D. J. 1986. *Evolutionary Biology*, 2nd ed. Sinauer Associates, Sunderland, MA.

Gause, G. J. 1934. *The Struggle for Existence*. Williams & Wilkins, Baltimore.

Graham, A. (ed.). 1972. *Floristics and Paleoflorisitics of Asia and Eastern Northern America.* Elsevier, Amsterdam.

Grant, P. R. 1975. The classical case of character displacement. Pp. 237–337 in: Th. Dobzhansky, M. K. Hecht, and W. C. Steere (eds.), *Evolutionary Biology*, Vol. 8. Plenum, New York.

Haffer, J. 1969. Speciation in Amazonian forest birds. Science 165:131–137.

Haffer, J. 1974. *Avian Speciation in Tropical South America*. Publications of the Nuttall Ornithological Club, Number 14.

Hallam, A. (ed.). 1973. *Atlas of Palaeobiogeography*. Elsevier, London.

Hanski, I. 1982. Dynamics of regional distribution: The core and satellite hypothesis. Oikos 38:210–221.

Hartl, D. 1980. *Principles of Population Genetics*. Sinauer Associates, Sunderland, MA.

Hubbell, S. P., and R. B. Foster. 1987. Commonness and rarity in a neotropical forest: Implications for tropical tree conservation. Pp. 205–231 in: M. E. Soulé (ed.), *Conservation Biology*. Sinauer Associates, Sunderland, MA.

Hutchinson, G. E. 1957. Concluding remarks. Cold Spring Harbor Symp. Quant. Biol. 22:415–427.

Hutchinson, G. E. 1959. Homage to Santa Rosalia, or why are there so many kinds of animals? Am. Natur. 93:145–159.

Jablonski, D. 1987. Heritability at the species level: Analysis of geographic ranges of Cretaceous mollusks. Science 238:360–363.

Karr, J. R. 1976. Within-and between-habitat avian diversity in African and Neotropical lowland habitats. Ecol. Monogr. 46:457–481.

Karr, J. R. 1977. Ecological correlates of rarity in a tropical forest bird community. Auk 94:240–247.

Kingsland. S. E. 1985. *Modeling Nature*. University of Chicago Press, Chicago.

Lack, D. 1944. Ecological aspects of species formation in passerine birds. Ibis 86: 260–286.

Lack, D. 1947. *Darwin's Finches*. Cambridge University Press, Cambridge.

Lawton, J. H. 1984. Non-competitive populations, nonconvergent communitites, and vacant niches: The herbivores of bracken. Pp. 67–95 in: D. R. Strong, Jr., D. Simberloff, L. G. Abele, and A. B. Thistle (eds.), *Ecological Communities. Conceptual Issues and the Evidence*. Princeton University Press, Princeton, NJ.

Li, H. L. 1952. Floristic relationships between eastern Asia and eastern North America. Transact. Am. Philos. Soc., New Ser. 42:371–429.

Lotka, A. J. 1925. *Elements of Physical Biology*. Williams & Wilkins, Baltimore.

MacArthur, R. H. 1969. Patterns of communities in the tropics. Biol. J. Linnean Soc. 1:19–30.

MacArthur, R. H. 1972. *Geographical Ecology*. Harper & Row, New York.

MacArthur, R. H., J. M. Diamond, and J. R. Karr. 1972. Density compensation in island faunas. Ecology 53:330–342.

MacArthur, R. H., and R. Levins. 1967. The limiting similarity, convergence, and divergence of coexisting species. Am. Natur. 101:377–385.

MacArthur, R. H., and E. O. Wilson. 1967. *The Theory of Island Biogeography*. Princeton University Press, Princeton, NJ.

Mares, M. A. 1985. Mammal faunas of xeric habitats and the Great American Interchange. Pp. 489–520 in: F. G. Stehli and S. D. Webb (eds.), *The Great American Interchange*. Plenum, New York.

Maurer, B. A. 1985. On the ecological and evolutionary roles of interspecific competition. Oikos 45:300–302.

May, R. M. 1975. *Stability and Complexity in Model Ecosystems*, 2nd. ed. Princeton University Press, Princeton, NJ.

Miller, R. S. 1967. Pattern and process in competition. Adv. Ecol. Res. 4:1–74.

O'Neill, R. V., D. L. De Angelis, J. B. Waide, and T. F. H. Allen. 1986. *A Hierarchical Concept of Ecosystems*. Princeton, NJ.
Orians, G. H., and R. T. Paine. 1983. Convergent evolution at the community level. Pp. 431–458 in: D. J. Futuyma and M. Slatkin (eds.), *Coevolution*. Sinauer Associates, Sunderland, MA.
Pearson, D. L. 1977. A pantropical comparison of bird community structure on six lowland forest sites. Condor 79:232–244.
Pearson, D. L. (ed.). 1980. Preliminary Floral and Faunal Survey, Tambopata Reserved Zone, Madre de Dios, Peru, 1979. Mimeo, 35 pp.
Pianka, E. R. 1966. Latitudinal gradients in species diversity: A review of concepts. Am. Natur. 100:33–46.
Pianka, E. R. 1973. The structure of lizard communities. Annu. Rev. Ecol. System. 4:53–74.
Pianka, E. R. 1975. Niche relations of desert lizards. Pp. 292–314 in: M. L. Cody and J. M. Diamond (eds.), *Ecology and Evolution of Communities*. Harvard University Press, Cambridge, MA.
Prance, G. T. (ed.). 1982. *The Biological Model of Diversification in the Tropics*. Columbia University Press, New York.
Pregill, G. K., and S. L. Olson. 1981. Zoogeography of West Indian vertebrates in relation to Pleistocene climatic cycles. Annu. Rev. Ecol. System. 12:75–98.
Preston, R. J., Jr. 1961. *North American Trees*, The M.I.T. Press, Cambridge, MA.
Rabinowitz, D., S. Cairns, and T. Dillon. 1987. Seven forms of rarity and their frequency in the flora of the British Isles. Pp. 182–204 in: M. E. Soulé (ed.), *Conservation Biology*. Sinauer Associates, Sunderland, MA.
Rapoport, E. H. 1982. *Areography. Geographical Strategies of Species*. Pergamon Press, Oxford.
Recher, H. 1969. Bird species diversity and habitat diversity in Australia and North America. Am. Natur. 103:75–80.
Richards, P. W. 1952. *The Tropical Rainforest*. Cambridge University Press, London.
Ricklefs, R. E. 1987. Community diversity: Relative roles of local and regional processes. Science 235:167–171.
Ricklefs, R. E., and G. W. Cox. 1972. Taxon cycles in the West Indian avifauna. Am. Natur. 106:195–219.
Ricklefs, R. E., and G. W. Cox. 1978. Stage of taxon cycle, habitat distribution, and population density in the avifauna of the West Indies. Am. Natur. 112:875–895.
Schluter, D. 1986. Tests for similarity and convergence of finch communities. Ecology 67:1073–1085.
Schluter, D., T. D. Price, and P. R. Grant. 1985. Ecological character displacement in Darwin's finches. Science 227:1056–1059.
Schoener, T. W. 1983. Field experiments on interspecific competition. Am. Natur. 122:240–285.
Shmida, H. 1981. Mediterranean vegetation in California and Israel: similarities and differences. Israel J. Bot. 30:105–123.
Schmida, A., and M. V. Wilson. 1985. Biological determinants of species diversity. J. Biogeog. 12:1–20.
Stanley, S. M. 1975. A theory of evolution above the species level. Proc. Natl. Acad. Sci. 72:646–650.
Stehli, F. G., and S. D. Webb (eds.). 1985. *The Great American Biotic Interchange*. Plenum, New York.
Strong, D. L., Jr., D. Simberloff, L. G. Abele, and A. Thistle (eds.) 1984. *Ecological Communities. Conceptual Issues and Evidence*. Princeton University Press, Princeton, NJ.
Terborgh, J. W., and J. Faaborg. 1980. Saturation of bird communities in the West Indies. Am. Natur. 116:178–195.
Udvardy, M. D. F. 1969. *Dynamic Zoogeography with Special Reference to Land Animals*. Van Nostrand Reinhold, New York.

Udvardy, M. D. F. 1981. The riddle of dispersal: Dispersal theories and how they affect vicariance biogeography. Pp. 6–38 in: G. Nelson and D. E. Rosen (eds.), *Vicariance Biogeography. A Critique*. Columbia University Press, New York.

Valentine, J. W. 1973. *Evolutionary Paleoecology of the Marine Biosphere*. Prentice-Hall, Englewood Cliffs, NJ.

Volterra, V. 1926. Variations and fluctuations of the numbers of individuals in animal species living together. Translation reprinted in: R. N. Chapman (ed.), *Animal Ecology*. McGraw-Hill, New York.

Wade, M. J. 1978. A critical review of the models of group selection. Quart. Rev. Biol. 53:101–114.

Wallace, A. R. 1878. *Tropical Nature and Other Essays*. Macmillan, New York and London.

Walter, G. H., P. E. Hulley, and A. J. F. K. Craig. 1984. Speciation, adaptation and interspecific competition. Oikos 43:246–248.

Webb, S. D. 1976. Mammalian faunal dynamics of the great American interchange. Paleobiology 2:220–234.

Whitmore, T. C. 1984. *Tropical Rain Forests of the Far East*. Clarendon Press, Oxford.

Whittaker, R. H. 1967. Gradient analysis of vegetation. Biol. Rev. 42:207–264.

Whittaker, R. H. 1972. Evolution and measurement of species diversity. Taxon 21: 213–251.

Wiens, J. A. 1966. On group selection and Wynne-Edwards' hypothesis. Am. Sci. 54: 273–287.

Willis, J. C. 1922. *Age and Area. A Study of Geographical Distribution and Origin of Species*. Cambridge University Press, Cambridge.

Wilson, D. S. 1983. The group selection controversy: History and current status. Annu. Rev. Ecol. System. 14:159–187.

Wilson, E. O. 1961. Nature of the taxon cycle in the Melanesian ant fauna. Am. Natur. 95:169–193.

PART SEVEN

CONCLUSION

CHAPTER 25

CONCEPTUAL AND OTHER PROBLEMS IN SPECIATION

John A. Endler

INTRODUCTION

The chapters in this volume bring out and clarify a number of conceptual issues as well as other issues about speciation that were not initially obvious. Among the most important are just what do we mean by species and what are we trying to explain? I shall attempt to focus on these and related problems arising from the preceding chapters. This is made more difficult, and more interesting, by the great diversity of viewpoints.

WHAT IS A SPECIES?

Species are "tools that are fashioned for characterizing organic diversity" (Levin 1979). Just as there are a variety of chisels made for different purposes, different species concepts are best for different purposes; and just as it is inadvisable to use a carving chisel to cut a mortise, problems arise when one species concept is used when it is inappropriate. Confusion and controversy have often resulted because different people working with different groups of organisms mean different things by "species." There are at least four major differences in aims of species concepts: (1) taxonomic versus evolutionary, (2) theoretical versus operational, (3) contemporaneous versus clade, and (4) reproductive versus cohesive. These contrasts are somewhat arbitrary, grade into each other, and give only a flavor of the various meanings of the word "species" in the literature and in this book.

Taxonomic and evolutionary species

Species may be defined so that they may be identified and classified into convenient groups, and so that there is a standard nomenclature that all types of biologists may use. This was the original use of "species," and its most extreme development is found in the phenetic approach to taxonomy (Sneath and Sokal 1973). This is also the aim of "transformed" or "pattern" cladists, who, in addition, wish to order species into hierarchies of shared characters with no assumptions about the processes that generated them (Nelson, this volume; Ridley 1986). Thus, the taxonomic species is purely empirical.

The term "evolutionary species" has several related meanings (Mayr 1963; Wiley 1981; Templeton, this volume; Futuyma, this volume; and see below), but all emphasize the species as an evolutionary unit. In other words, an evolutionary species evolves (roughly) as a unit, and evolves independently of other such units. Thus, evolutionary species concepts make assumptions about the processes that result in the pattern that we observe, and are a mixture of empirical and theoretical concepts. As a result of the assumption of evolutionary independence, the evolutionary species concepts are most frequently used by taxonomists interested in phylogeny reconstruction, as well as by population and evolutionary biologists. There are several kinds of evolutionary species concepts, and these will be discussed briefly later in this chapter. In summary, the major difference between taxonomic and evolutionary species concepts is whether or not assumptions are made about processes that give rise to patterns of variation among taxa.

Theoretical and operational species

Practicing taxonomists and other biologists need to be able to reliably identify species; the species must be diagnosable (Cracraft, this volume). These operational species definitions serve a practical purpose that does not necessarily correspond to theoretical species concepts, and theoretical species concepts are often impractical. For example, although the biological species concept is popular among zoologists, its reproductive isolation criterion is not practical; for most species it is not possible to do breeding tests or observe natural breeding. As a result, the majority of classifications use operational definitions based on unique and on hierarchies of shared morphological traits. Birds are an exception, and it is probably no coincidence that ornithologists invented and are strong proponents of the biological species concept. Although it is often the case that the theoretical and evolutionary species correspond, and the taxonomic and operational definitions correspond, this is by no means always true. For example, phylogenetic systematists (cladists) have a good theoretical notion of species that they try to make operational (Wiley 1981; Cracraft, this volume).

We hope that the operational definitions are good approximations to the theoretical definitions. If this were not the case, our interpretations of evolution based on incorrectly identified patterns of variation could be misleading or completely wrong. This is more than an academic problem; in the conservation of endangered species in zoos and botanical gardens, it is essential not to try to interbreed individuals from different but unidentified (sibling) species, which can result in sterility and other problems, further reducing their effective population size, increasing the rate of genetic drift and of inbreeding depression (Ryder et al., this volume).

Contemporaneous and clade species

There are two major kinds of evolutionary species concepts, and these roughly correspond to the fields of evolutionary biology and phylogenetic systematics, with paleontology falling somewhere in between. I am not sure that these have been named before, so for brevity I have called them the *contemporaneous* and *clade species concepts.* Evolutionary biologists (as well as population biologists and ecologists) consider species as defined with respect to groups of contemporaneous organisms, either currently living, or living together in the past. This "snapshot" approach to species is implicit in the biological species concept with its emphasis on reproductive isolation, although we can make assumptions about potential interbreeding with immediate ancestors and descendants (Mayr 1963). Relative to evolutionary biologists, phylogenetic systematists are more concerned with patterns of ancestral-descendant relationships among characters and among species, and in the branching process of evolution. Cladists use several different concepts. Cracraft's (this volume, pp. 34–35) is "an irreducible (basal) cluster of organisms, diagnosably distinct from other such clusters, and within which there is a parental pattern of ancestry and descent." Putting this in a slightly different way, a clade species is a basal, differentiated, monophyletic evolutionary taxon as well as the smallest evolutionary unit (see Cracraft, this volume and Nelson, this volume for discussions). This is similar to Ghiselin's (1974, 1981) suggestion that species are logical individuals with origins (speciation), continued existence as recognizable units, and ends (extinction). Cracraft (this volume) discusses the variable interpretations and some of the different kinds of problems that arise when the contemporaneous and clade species concepts are applied independently to the same organisms. In summary, the contemporary species concept takes an evolutionarily instantaneous view of species, whereas the clade concept treats a species as a group connected by common descent as well as by common biology.

There is another important difference between the contemporaneous and clade species concepts, and that is in their application. Most evolutionary biologists are more interested in anagenesis (morphological or genetic change within species, and perhaps similar directional change among species within clades), whereas most phylogenetic systematists are more interested in cladogenesis (patterns of clade branching or speciation). It is clear that the contemporaneous definition is more useful for and indeed emphasizes anagenesis while the clade definition is more useful for and emphasizes speciation. Thus, it is no surprise that the two concepts are differentially popular with (though not exclusively used by) evolutionary biologists and cladists. This has given rise to rather different viewpoints about evolution and speciation along with heated controversies, as hinted in Cracraft's and Nelson's chapters in this volume. This has also affected the matching operational definitions. If the primary concern is speciation, the relationships among species, and phylogenetic reconstruction, then the operational definition must stress the presence of shared derived traits or character states (synapomorphies, Wiley 1981; Ridley 1986; Cracraft, this volume; Nelson, this volume), whereas if the primary concern is with what happens within species, then criteria that ensure evolutionary independence are more important, as in the biological species concept (Mayr 1963).

Reproductive and cohesive species

Templeton's chapter in this volume presents an admirable discussion of four kinds of evolutionary species concepts: evolutionary, isolation, recognition, and cohesion. His evolutionary species concept is simply a population (or group of populations) that shares a common evolutionary fate through time. It thus recognizes only the pattern of evolution and makes no assumptions about evolutionary processes or causes (see also Wiley 1981). This species concept can be made operational, and is most similar to concepts used by many paleontologists and some systematists. The other three concepts make assumptions about the processes that give rise to evolutionarily independent entities.

The *biological species concept* (Mayr 1963; Futuyma 1987, this volume) defines species in terms of their being reproductively isolated from other such entities, by premating, postmating, or both kinds of mechanisms. Such reproductive isolation yields evolutionary independence from the other isolates. This has also been renamed the *isolation species concept* by Paterson (1985; Paterson and Macnamara 1984) because there are other kinds of biological species concepts (see below). Paterson (1985; Paterson and Macnamara 1984) notes that this concept is relational; "isolation species" must be defined or recognized relative to other such species. This is fine for a simple evolutionary species concept, but it usually leads to impractical operational definitions. In addition, as Paterson (1985; Paterson and Macnamara 1984) and Templeton (this volume) have pointed out, reproductive isolation and speciation can be evolutionarily independent. On the one hand, speciation can occur in the absence of the evolution of reproductive isolation if the new species evolved in allopatry and remain so. On the other hand, there are also many organisms that clearly exist as evolutionarily independent entities as well as being very distinct biologically and ecologically, yet they have little or no reproductive isolation (Templeton, this volume; this is particularly common in plants: Grant 1957, 1981; Levin 1979; Jonsell 1984). Therefore the major problem with the isolation species concept is that the process (reproductive isolation) is neither necessary nor sufficient for speciation or the maintenance of evolutionarily independent entities, although it can maintain evolutionarily independent entities if there is secondary contact (Endler 1977).

As a result of these and other problems with the biological (isolation) species concept, the *recognition species concept* was derived independently by Ghiselin (1974) and Paterson (1985), and aptly named by Paterson. *Recognition species* are (1) "the most extensive units in the natural economy such that reproductive competition occurs among their parts" (Ghiselin 1974), or (2) "the most inclusive population of individual, biparental organisms which are a common fertilization system" (Paterson 1985). Templeton (this volume) points out that, like isolation mechanisms, fertilization systems may evolve independently of species, so we must include the entire reproductive process and life cycle in the definition; this is essentially Ghiselin's (1974) point. The recognition species concept emphasizes internal properties rather than properties relating one species to another; recognition facilitates reproduction, and isolation can be irrelevant to speciation. Another advantage of the recognition concept is that it easily accommodates sexual selection as a method promoting divergence of recognition systems between speciating groups (Lande 1981), and the notion that species are fields for gene recombination (Templeton, this volume). The fact that members of the same species recognize each other for reproduction yields the pattern of evolutionary independence. The two concepts emphasize different aspects of the same phenomenon for sympatric or parapatric species, but the recognition concept is more biologically meaningful for allopatric species and populations.

The isolation and recognition species concepts both place an emphasis on reproduction, and will be collectively called the *reproductive species concept*. But, as already mentioned, there are many organisms that clearly exist as evolutionarily, biologically, and ecologically independent entities, yet they share a common mating system with other such units (Grant 1957, 1981; Levin 1979; Jonsell 1984; Templeton, this volume). Botanists were conscious enough of this problem 30 years ago to coin the term *syngameon*

(Grant 1957; Beaudry 1960) for groups of naturally hybridizing but distinct species. At the same time, zoologists used the terms *rassenkreis* and *semi-species* for similar groups of taxa (Rensch 1959; Mayr 1963), and the term *species complex* is still used for these groups. How similar must a mating (or fertilization) system be (or how much reproductive isolation must there be) for a taxon to be recognized as a species? As botanists and protozoologists have long known, there is an even more serious problem with reproductive species concepts for asexually reproducing species (Levin 1979; Jonsell 1984; Templeton, this volume)—it is unreasonable to regard every clone as a separate species, yet they are sexually isolated and have separate fertilization systems. Of course, on a larger scale they share the same reproductive system, and are in a sense in reproductive competition with each other (Ghiselin 1974), but these are matters of degree. These are more than operational species problems because they get at the very basic question of the *causes* of speciation. There seem to be real biological and evolutionarily independent entities present in nature (as in Templeton, this volume), or, as Ghiselin (1981) suggests, species are logical individuals with beginnings and ends, yet reproduction may have little to do with their boundaries. What causes them and what maintains them if sexual reproduction is neither necessary nor sufficient?

For these reasons botanists (such as Levin 1979; Jonsell 1984) as well as others (Templeton, this volume) have argued that we should discard the reproductive species concept and concentrate on what factors cause groups of organisms to be similar to one another, and on the processes that maintain this similarity of morphology, biology, ecology, behavior, and genetics. Templeton (this volume) aptly calls this the *cohesion species concept*, and there are similarities to what Wiley (1981) calls the evolutionary species concept. There have been several proposed mechanisms promoting the cohesion of species. One of the oldest is that gene flow holds species together because it homogenizes allele frequencies and coadapted gene complexes (for example, Mayr 1963). This mechanism has come under increased attack because of the widespread occurrence of isolation by distance, and because gene flow (no matter what its scale) will homogenize a widespread species gene pool only if the genes are selectively neutral (reviewed by Endler 1977; Jonsell 1984). Another idea is that stabilizing selection and common ecological factors keep species homogeneous, and the degree of homogeneity is related to the degree of spatial homogeneity in environmental conditions (Raven 1976; Van Valen 1976; Endler 1977; Jonsell 1984). Stability and changes in time can also give rise to stasis and punctuated equilibrium patterns (Van Valen 1975; Jonsell 1984), and this connects this cohesive (or ecological) species concept with the phylogenetic concept. Still another cohesive concept is one found in paleontology (Eldredge and Cracraft 1980), phylogenetic systematics (Eldredge and Cracraft 1980; Wiley 1981), quantitative genetics (Cheverud 1984; Cheverud et al. 1983, 1985; Atchley 1984; Schluter 1984), and in evolutionary genetics (Templeton, this volume). Species are cohesive units because they share the same developmental genetic systems. A cohesive species has genetic and phenotypic cohesion (Templeton, this volume), minimal geographic and temporal variation in its phenotypic and genetic variance-covariance matrices, or the same set of developmental and phylogenetic constraints. These are shared because of ancestral-descendant relationships (Templeton, this volume; Cracraft, this volume; Eldredge and Cracraft 1980). By Templeton's definition, a cohesion species is "the most inclusive population of individuals having the potential for phenotypic cohesion through intrinsic cohesion mechanisms" (Templeton, p. 12). We could expand this to include both parts of the quantitative genetic evolutionary mechanisms: a cohesion species is the most inclusive group of organisms whose range of phenotypic variation is limited by genetically and environmentally based cohesion mechanisms. Both versions of the definition allow us to include ancestral-descendant continuity (phylogenetic species concept) because common developmental genetic systems are unlikely to have arisen by convergence (e.g., Kauffman 1985), and for the same reason allows us to include asexually reproducing species.

But the cohesive species concept is poor when we try to make it operational. How similar is similar? This can degenerate into the phenetic species concept. Even if we aim to use cladistic methods, we still have to recognize species before we work out their evolutionary relationships. Worse, if the concern is with ancestral-descendant relationships in an asexual species complex, where do we draw the line? And how do we distinguish such species from higher taxa (Templeton, this volume; Nelson, this volume)? And how would we work out the genetic variance- covariance matrix of a fossil assemblage? (It would be possible to work out the phenotypic variance-covariance matrix of fossil assemblages and investigate how they change through time, but this would potentially confuse genetic and environmental factors that cause anagenesis and cladogenesis.) It could be argued that once we obtain the complete DNA sequences for taxa in a lineage we could compare them and thus know what happened, but this can also degenerate into the phenetic species concept applied to DNA sequences. Unfortunately, the same problem applies to the recognition concept: how similar must recognition systems be to define two groups as being members of the same species? This is a practical problem to potential mates as well, which is precisely the point of the reproductive species concept.

In summary, the reproductive species concept emphasizes isolation between entities (isolation or biological concept) or recognition within entities (recognition concept) during reproduction, whereas the cohesion concept emphasizes genetic and phenotypic cohesion within an entity, whether caused by genetics or environment or both. They differ in the assumed causes of origin and maintenance of the species produced: reproductive specificity or common developmental genetics. Both concepts are reasonable theoretically, but both have problems when translated to operational definitions.

It is clear that species concepts vary radically depending on their purpose, be it theoretical or operational, taxonomic or evolutionary, contemporaneous or clade, reproductive or cohesive. It is unproductive, and often positively misleading, to apply one species concept to all species or to answer all questions. Their use and utility depend on what is to be investigated and explained.

WHAT ARE WE TRYING TO EXPLAIN?

A perusal of the chapters in this volume, as well as the literature on speciation in general, suggests that we are not all trying to explain the same thing when we discuss speciation. The differences follow from our explicit or implicit use of a particular species definition. For example, Mayr (1963:426) defines speciation as "the splitting of an originally uniform species into several daughter species," whereas Templeton (this volume, p. 24) defines it as "the process by which new genetic systems of cohesion mechanisms evolve within a population." The use of the isolation species concept (Mayr) leads to an interest in cladogenesis, whereas the use of the cohesion concept (Templeton) leads to either an interest in anagenesis and morphological (and genetic) divergence, or an interest in cladogenesis, or both.

There are at least three different evolutionary processes that we are trying to explain when we discuss speciation: (1) the evolution of reproductive isolation (or recognition) and the resulting cladogenesis, (2) the evolution of biological diversity—morphology as well as genetics, development, physiology, and behavior, and (3) the evolution of DNA sequence diversity.

Evolution of reproductive isolation and cladogenesis

Given the reproductive or clade species concept, there are two central questions of pattern and process: (1) pattern: what is the phylogenetic history, or more specifically the pattern of species multiplication (cladogenesis)? and (2) process: what causes the evolution of reproductive systems that ensure the integrity of the evolutionary unit and its isolation from other such species? The first question is thoroughly discussed by Eldredge and Cracraft (1980) and Wiley (1981), and is briefly considered in this volume by Cracraft, Nelson, and Lynch.

The evolution of reproductive isolation is a contentious issue because we know very little about it. It is usually divided into two components, premating and postmating isolation, depending on what stage the isolation (or recognition) takes place (Mayr 1963; Templeton, this volume). Both pre- and postmating isolation appear to increase with time in sister species of *Drosophila* (Coyne and Orr 1989).

There seems to be general agreement that postmating isolation can evolve as a by-product of genetic divergence, through selection or genetic drift (or both), aided by allopatry or isolation by distance. It is often sex linked (Coyne and Orr, this volume) and may even be involved in the evolution of sex chromosomes. Various aspects of divergence associated with founder events (Barton) or hybrid zones (Harrison and Rand; Hewitt; Patton and Smith) are discussed in detail in this volume. Particularly novel is Hewitt's discussion of how hybrid zones break up otherwise continuously distributed species. Hybrid zones are clines with some degree of pre- and postmating isolation at least in the center of the zone, and can develop either from primary or secondary contact. Barton and others (reviewed in Barton and Hewitt 1985) have previously shown how the presence of hybrid zones can restrict gene flow across the zone. Hewitt (this volume) suggests that if the hybrid zone is spatially extensive enough to enclose or subdivide a species range as well as restricting gene flow, then it results in more isolation between the hybridizing population groups than we would suspect merely by mapping them. This will accelerate the divergence between the groups, yielding progressively more postmating isolation, which itself makes the hybrid zone more of a gene flow barrier. Hybrid zones can occur anywhere and subdivide otherwise continuous populations into two or more diverging sections that may diverge in this way to species status (Hewitt, this volume). This is a new and interesting form of parapatric speciation that has not heretofore been discussed.

The evolution of premating isolation is more problematic. There are at least six processes that can favor the evolution of premating isolation: (1) the evolution of premating isolation as a by-product of genetic divergence among populations (random or selected), (2) microhabitat or habitat choice and genetic predisposition, (3) reinforcement of premating isolation by postmating isolation, (4) reproductive character displacement, (5) sexual selection, and (6) natural selection due to sensory biases caused by the environment. Very little is known about any of these processes.

All other things being equal, if populations diverge at random, or diverge as a result of natural selection from slightly (or strongly) different environments among the isolates, then premating isolation could evolve as a simple by-product of this divergence. The divergence could be either directly in whatever genes affect mating behavior, or could arise through pleiotropy of other diverging traits. This is a similar argument to one of those for the evolution of postmating isolation; as for postmating isolation, it does not necessarily actively promote isolation, but is permissive of it.

Microhabitat or habitat choice may provide isolation if the mate recognition system depends on location-dependent cues, and breeding sites are chosen on the basis of cues based on similarity to the mate's birthplace. This may be true for many insects, and can potentially lead to sympatric speciation (Bush 1975; Diehl and Bush, this volume; Wilson, this volume). The genetic system, combined with frequency- and density-dependent selection, may directly result in isolation (Wilson, this volume). Either process may lead to sympatric, parapatric, or allopatric speciation, depending upon the biology of the species concerned. Wilson (this volume) gives an example of microhabitat selection and divergence in a sunfish, although we do not know whether the divergence took place in the lake, or was an effect of two different human introductions.

Reinforcement is hypothesized to result from individual selection for positive assortative mating (like × like) within species or hybridizing taxa, as a result of reduced fitness in the offspring of unlike matings (postmating isolation). Unfortunately, this has been shown in only a few laboratory experiments; evidence for it in the field is weak or ambiguous (Butlin, this volume), and no laboratory experiments have gone to 100% premating isolation. Perhaps the strongest evidence against it was given by Coyne and Orr (1989). In presently allopatric species both pre- and postmating isolation appear to evolve at roughly the same rate (as estimated by allozymic genetic distances), whereas in species that are at least partially sympatric with each other premating isolation appears to evolve faster than postmating isolation (Coyne and Orr 1989). This is the opposite of that expected from the hypothesis that postmating isolation favors the evolution of premating isolation. These conclusions depend upon the validity of the "molecular clock" calibrated by allozymes. However, the pattern is suggestive even if the "clock" were highly variable. If postmating isolation antedates and favors premating isolation, then we would not expect pre- and postmating isolation to be equally common in allopatric species and premating isolation to be more common in sympatric or nearly sympatric species.

Unlike reinforcement, reproductive character displacement occurs between good reproductive ("biological") species (Butlin, this volume), and results in an exaggeration of differences in premating traits. There are three possible causes, all with the same effects: (1) Individual selection against gamete wastage. This is the same argument used for reinforcement, and there is more laboratory support, but still no selection to 100% isolation. (2) Divergence before contact coupled with poor sampling, which gives a false appearance of divergence in sympatry but not in allopatry

(Grant 1975). (3) Individual selection to prevent confusion with sounds from the sister species (Littlejohn 1960), to prevent active jamming of signals from conspecifics, or to maximize the signal-to-background noise ratio during intersexual and intrasexual communication (Brush and Narins 1989). Selection to prevent confusion is also a mechanism to improve efficiency of the recognition system.

Sexual selection is capable of generating or amplifying premating isolation. If there is heritable variation for female choice criteria and heritable variation for male traits chosen by females, then sexual selection will occur. Under certain conditions the system will "run away," leading to rapid evolution of both the male trait and the female choice criteria (Lande 1981). It is quite possible for allopatric populations to diverge and even run away in different directions, leading to divergent reproduction recognition systems, hence the formation of new species (Lande 1981). Indeed, this can happen even if the two populations are isolated by distance and there is initially only clinal variation for the male trait (Lande 1982). Since runaway sexual selection can proceed very rapidly, and postmating isolation can evolve as a by-product of genetic divergence between populations, premating isolation may evolve more rapidly than postmating isolation. Thus, the observations of Coyne and Orr (1989) may be explained by sexual selection. However, this does not explain why the frequency or rates are apparently equal when the populations are allopatric; perhaps that is merely a function of random divergence of both suites of traits.

A possible example of the early stages of divergence is found in guppies (*Poecilia reticulata*): Breden and Stoner (1987) found that females from a high predation locality prefer male models with duller color patterns, whereas females from a low predation locality in the same stream prefer brighter color patterns. Two potentially confounding problems here are (1) females from all stream localities prefer males from their own populations, and (2) guppies from high predation localities school more tightly than those from low predation and there is a strongly female-biased sex ratio, so females from high predation localities will be attracted to duller males that appear more like the most common females. Hopefully these confounding factors and others (Endler 1988; Houde 1988b) will be eliminated in future experiments. A second study by Houde (1988a) found that females from some populations strongly favor males with more orange and females from other populations have only a weak preference for orange. These two guppy studies suggest the possibility that populations isolated by distance may be in the early stages of running away in different directions, as in Lande's (1982) model. This example also gives some idea of the difficulty of working with sexual selection problems, and indeed any problems dealing with sexual reproduction.

Because the female choice criteria evolves through covariance with the male trait (Lande 1981), the male trait may evolve and diverge faster than the female choice criteria, leading to the kinds of mating asymmetries found between hypothesized ancestral and descendant species such as in the Hawaiian *Drosophila* (Kaneshiro 1989): females of the new species still recognize males of the old species, but not vice versa. If secondary contact occurs between the ancestor and the recently diverged descendant species, then such asymmetries may actually result in the species' fusing. Fusion might result because the females of the new species may actually prefer males of the old species to their own males (Ryan and Wagner 1987). Sexual selection does not guarantee divergence, isolation, and speciation, even if it proceeds in different directions in different places.

Sexual selection can proceed at any time until zygote formation, so elaboration and differentiation of male traits such as clasping and intromittent organs may be involved, even after mating per se, and many species are isolated (or recognized!) only be differences in genitalia (Eberhard 1985). In summary, sexual selection is capable of rapidly generating premating isolation if it proceeds, or runs away, in different directions in different places.

Another process that can result in premating isolation is natural selection due to sensory biases caused by the environment. During courtship males send a signal (visual, auditory, olfactory, taste, electrical, pressure, etc.) to the female, and the female receives signals from one or more males, possibly males from congeneric species, and noise from the environment. In addition, the medium through which the signal is sent (air, water, soil, plant surfaces, silk, etc.) will attenuate the signal in various ways (examples for vision in Endler 1986a). In different environments the background noise will be different, and for many sensory modes, the attenuation properties will also be different in different places. Such differences may also change seasonally. For this reason there will be sensory biases that will vary geographically, making different kinds of male signals most efficient in different places. This will bias the direction of evolution of efficiency in sexual signals even in the absence of sexual selection, and cause divergence among signals in different places; for brevity I will call this "sensory drive." Given sufficient environmental differences or time, sensory drive could lead to premating isolation between populations. If sexual selection operates as well as sensory drive, and conditions are suitable for the runaway process, then sensory drive may set the direction of the runaway process. If so, and there is sufficient spatial variation in signal transmission properties, then different populations may run away in different directions set by sensory drive, leading to isolation (or specific recognition). In addition, sexual selection may directly affect the female sensory system rather than just female behavior. Bimodal and unimodal variation in photoreceptor spectral sensitivity is known in guppies (Archer et al. 1987), ground squirrels (Jacobs et al. 1985), several monkeys (Jacobs 1986; Jacobs and Neitz 1987a, 1987b; Jacobs et al. 1987) and humans (Neitz and Jacobs 1986). As a result of this variation, different females will perceive the same male differently, and the "brighter" of one male to one female may be the "duller" of the two to another female. This variation is known to be heritable in some primates (Jacobs and Neitz 1987; Neitz and Jacobs 1986; Jacobs et al. 1987). Given genetic polymorphism for cone pigments (or other sensory mechanisms), there will be heritable differences in female preferences, even if the females are "hard wired" to prefer the "brighter" male, and the female receptors will evolve under sexual selection. Combined with sensory drive, and conditions for runaway sexual selection, strong isolation or very specific mate recognition could evolve quite rapidly.

For whatever reason premating isolation (or specific recognition) evolves, considerations of signal transmission and reception will not only bias the direction of evolution (sensory drive), but may also constrain it. Ryan (1986) showed that the structure of the inner ear in anurans (frogs and toads) affects which sounds can be heard most effectively. Anurans have one or two inner ear structures, the amphibian papilla (AP) and the basilar papilla (BP), and the AP may be greatly reduced in some lineages. These two structures have different ranges of detectable sound frequencies or "windows," so when the AP is reduced or absent, one sound window is absent, restricting the potential sound variation usable in mate recognition in evolutionary time as long as the AP does not change. Indeed, lineages with the AP reduced or absent appear to have evolved more slowly and have more restricted call types than do those with a complex AP. Fossil and cladistic evidence suggests that the rate of speciation and call complexity evolved in parallel with the complexity of the amphibian papilla (Ryan 1986). The cone pigment variation in ground squirrels is only about 4 nm (Jacobs et al. 1985), suggesting that they too may have limits to how rapidly sexual selection for male traits may run. In summary, any consideration of premating isolation must consider the evolution of the signal transmission and signal reception systems.

The evolution of premating isolation is complex, and very little is known about it. There are six main questions particularly worthy of further study:

1. What causes the origin of reproductive isolation or divergence in recognition mechanisms? Which is more important, recognition of mates within species or isolation from other species?
2. What causes the maintenance of recognition systems? The causes of maintenance may not be the same as the causes of origins.
3. What factors cause geographic variation in recognition systems and which are the most important?

4. What is the relative importance of the five possible causes of premating isolation, and what conditions favor each cause?
5. If Coyne and Orr's (1989) observation is correct and general, why do premating systems evolve faster than postmating systems, and why particularly in sympatric species and not in allopatric species?
6. How do these factors affect the rates and modes of speciation? (See also Templeton 1981, this volume; Lynch, this volume.)

The evolution of biological diversity

Many of us went into biology (particularly evolution and systematics) because we were fascinated by biological diversity. For many of us the explanation of this diversity in morphology, genetics, development, physiology, and behavior remains *the* basic question. For those of us who do systematics of organisms that are impractical to breed, or do ecology or functional morphology in the broad sense, the study of sexual recognition or isolation seems rather academic, except insofar as it tells us when we have sibling species. Our interest is perhaps weakened as we discover that there is very frequently a lack of correspondence between morphology and sexual isolation (Larson, this volume; Patton and Smith, this volume), as has been known for a long time in perennial plants (see previous section; Raven 1976; Levin 1979; Jonsell 1984; Barrett, this volume). Is reproductive isolation (or recognition) a "red herring?"

There are at least three arguments against isolation being a "red herring":

1. Where examined, most species are indeed isolated from their sister species, and, even in plants, mating system shifts are frequently associated with speciation (Barrett, this volume).
2. Groups of populations that are isolated from each other can diverge more rapidly than those with gene flow (see section on reproductive species; Mayr 1963; Lynch, this volume; Futuyma, this volume). This is the classic argument for the importance of evolutionary independence.
3. In geological time, local selection may change in direction, and adjacent areas may have selection working in opposite directions, so local adaptations are ephemeral. As a result, the long-term average selection over the entire species range may be weak or absent, and little anagenesis can occur. On the other hand, if isolation restricts gene flow to smaller areas, they are relatively less likely to experience fluctuation of selection in different directions, and are therefore more likely to diverge (Futuyma 1987, this volume). These are all arguments for the reproductive species concept.

On the other hand there are strong arguments suggesting that isolation is a "red herring," at least in some organisms. The counterargument for (1) is the apparent irrelevance of isolation in many groups of taxa (see the cohesion species section): there are too many sibling species as well as too many extremely distinct taxa with no or little reproductive isolation. Mating system shifts are often associated with speciation, but speciation is not always associated with mating system changes (Barrett, this volume).

The counterargument for the gene flow argument (2) is that it ignores the power of isolation by distance. It is true that totally isolated populations will diverge faster than those isolated by distance, but this generality is only qualitatively true if the divergence is due to genetic drift. If there are selective differences between populations isolated by distance, strong differentiation can occur very rapidly indeed (Endler 1977; Lande 1982; Jonsell 1984; Barton and Hewitt 1985; Hewitt, this volume). Reproductive isolation matters only if the geographic scale of gene flow is large. It is the selective differences driving the divergence; spatially restricted or totally absent gene flow is *permissive* only of anagenesis and divergence. There is an additional counterargument for (2). Hybridization is widespread in plants and animals, and even a low level of hybridization is sufficient to increase the genetic variation of the introgressing species (a fine example is found in Grant 1986). Increased genetic variation means more rapid evolution under natural selection or genetic drift, so isolation may actually be a hindrance to anagenesis and to divergence caused by geographically varying selection. Since too much gene flow will also be a hindrance, there is probably some intermediate rate of introgression that results in maximum divergence, but this has not been investigated.

The counterargument for (3) is that local adaptation will be ephemeral only if the *sign* of the selection differential or coefficient changes direction frequently and if the gene flow scale is large enough for the spatial and temporal average of selection to have a magnitude close to zero. It is unlikely to be close to zero over long periods of time, even if it fluctuates a lot. In fact, the assumption of roughly constant sign of natural selection is a fundamental assumption of the Pleistocene forest refuge hypothesis used to explain the extraordinarily high species diversity in tropical forests (Prance 1982); in spite of major climatic changes, conditions were constant enough in the refugia to support forest animals and plants and allow them to diverge between forest refugia. Once again, total isolation is not necessary; the conditions within the proposed refugia areas may have been constant even if the refugia were not totally isolated (Endler in Prance 1982).

But no matter how weakly coupled speciation and isolation appear to be in some groups (Larson, this volume; Patton and Smith, this volume), it is clear that gene flow can have an effect on speciation and extinction rates (Lynch, this volume). For many marine benthic invertebrates, particularly gastropods, species with planktotrophic larval stages, which can move long distances, are less differentiated, have a lower speciation rate, and last longer in the fossil record than do species with nonplanktotrophic larvae, which can move only short distances (Jackson 1974; Shuto 1974; Hanson 1978; Kauffman 1978; Jablonski and Lutz 1983; Jablonski 1987). Short-distance larvae yield small-scale gene flow distances, making both isolation by distance and physical isolation more likely than it would be for species with long-distance larvae. As a result, species with nonplanktotrophic larvae are more likely to differentiate in response to local ecological conditions than species with planktotrophic larvae, yielding more anagenesis and cladogenesis. But they are also more likely to go extinct because their smaller geographic ranges are more likely to be wholly within an area whose climate and ecology have become inhospitable. This results in higher turnover of species with small-scale gene flow, further encouraging cladogenesis, and anagenesis when extirpated areas are recolonized. This is essentially an expansion of Futuyma's (1987, this volume) points, adding isolation by distance and reducing the importance of complete isolation. It is clear that more work needs to be done on the interplay between geographic scale of gene flow and the scales of geographic and temporal variation in ecological factors that lead to natural selection (as in Lynch, this volume). In the meantime the decision as to whether the origin of reproductive isolation/recognition or morphological divergence is more important or more interesting to study is largely a matter of taste.

If we decide that morphological divergence is more important or interesting than reproductive and evolutionary independence, then there are a different set of questions to be addressed. The major questions are (1) what are the mechanisms and causes of the *origins* of new variants? and (2) what are the mechanisms and causes of the *replacement* of old by new variants? This distinction is important, but has been relatively neglected by evolutionists in the past 20 years (Endler 1986b, 1986c; Endler and McLellan 1988).

New variants may enter a population or species by gene flow, mutation, recombination, hybridization, and hybrid dysgenesis. Since variation is a fundamental requirement for evolution, the rate of appearance of new variants can set limits to both the rate and direction of evolution, although it does not have to do so (Endler 1986b; Endler and McLellan 1988). There are two approaches to the study of origins: (1) detailed studies of molecular genetics and development and (2) comparative speciology and phylogeny reconstruction. The former elucidates the physical mechanisms promoting and determining new forms and the latter traces the patterns of appearance of the new forms during speciation. Unfortunately these have been the almost exclusive provinces of allopatric fields until very recently—molecular biology and development as distinct from systematics. Quantitative genetics has fallen somewhere in between in its attempts to quantify genetic structure (embodied in the variance–covariance matrix) in the absence of detailed knowledge of molecular biology of morphogenesis. From these fields we have the impression that mutations may not be completely random and there may be constraints or biases in the directions and rates of evolution caused by the developmental genetic system (Endler 1986c; Endler and McLellan 1988). We know very little about these phenomena, but the data are now being actively collected by molecular evolutionists and systematists alike. Eventually we should know something about the effect of the developmental genetic system on the direction and rate of evolution, as well as the evolution of the developmental genetic system itself, since it is unlikely to have remained constant since the origin of eukaryotes.

Once a new variant appears it must then spread and replace the old variant, and this can occur by genetic drift, natural selection, or genetic biasing mechanisms such as meiotic drive. These are rather different mechanisms from those involving origination, and have also had a largely separate history in allopatric fields (population genetics and ecology). Both the mechanisms and resulting patterns are largely irrelevant to systematics and phylogeny reconstruction, except insofar as they (for example, selection) can cause convergence and homeoplasy. Fortunately the multiple gaps between population genetics, molecular genetics, developmental genetics, and systematics are now being perceived and are beginning to be closed.

One gap that is not being filled as rapidly as the others is that between ecology and population genetics. Most population geneticists and many other evolutionary biologists treat organisms as machines that act independently of the environment. This is largely a historical artifact of the background of the early leaders of the field: Mayr worked on museum specimens, Dobzhansky worked on flies with unknowable ecology and largely with laboratory populations, Simpson worked on fossil mammals, Wright worked on domestic animals, and Haldane was perhaps empirically most interested in humans—only Fisher made excursions into more ecological problems, but he was most interested in theory. So in spite of a hundred years of study we know almost nothing about *why* natural selection occurs or what conditions favor the process, although there is no shortage of "informed speculation" (Endler 1986b, 1986c). This is rather like trying to do chemistry with no knowledge of the solutes in which the reactions of interest are taking place! However, this is changing with the encouragement of new ecological genetic methods (reviewed in Endler 1986b). Excellent approaches to incorporating ecology into speciation are found in Grant (1986), Grant and Grant (this volume), and Roughgarden and Pacala (this volume).

There are some major unanswered questions regarding the replacement of variants:

1. What is the relationship between variation in form and variation in function? How does this change with microenvironment and habitat within and between species?

2. What is the relationship between variation in function and variation in fitness? How does this change with microenvironment and habitat within and between species?
3. What are the ecological consequences of variation and how does this feed back on the evolution of variation?
4. When and how should phenotypic plasticity evolve, when and how should developmental colonization evolve, and when should polymorphisms for genetically "fixed" functions evolve?
5. What is the effect of various time scales of fluctuation in the environment relative to generation length and relative to the time scale's ecological phenomena?
6. Is morphology (in the broad sense) a spatial average of the effects of natural selection in conditions over many adjacent genetic neighborhoods or is it more affected by temporal variation?
7. What causes replacement of variants and what causes "stasis" or equilibrium of form and function?
8. How much "stasis" is due to ecological equilibrium and how much is due to genetic homeostasis and other internal constraints?
9. Why do some clades speciate frequently whereas others do not, and what makes a species more prone to speciation?
10. What makes populations or species more or less prone to extirpation and extinction?
11. Does divergence (anagenesis and cladogenesis) occur more frequently as the result of expansion into new niches after climatic change, or as the result of expansion into niches or adaptive zones vacated by continual local extinction of other species?

There is clearly much to be learned, and many of these questions also involve the question of origins of variation.

In summary, in studying the biological diversity of speciation rather than the evolutionary independence of species, there are an entirely different set of questions to be asked, and these involve the integration of all possible fields of biology. The study of the causes of evolutionary independence is easier and more circumscribed, but it may not tell us much about speciation since evolutionary independence does not guarantee evolutionary divergence.

The evolution of DNA sequence diversity

A third phenomenon we are trying to explain in speciation studies is the evolution of DNA sequence diversity. This is the most basic level of explanation because it addresses the information used to generate genetic systems and organisms. Unfortunately, at present we know virtually nothing about the relationship between the DNA sequence of an organism and its morphology, form, and function. Until the functional significance of sequence differences is known, DNA sequence information tells us little about speciation. As molecular biology progresses beyond the natural history stage it will help us make the connection and probably will offer profound insights into the causes of divergence and speciation. About all that we can say now is that genome evolution can affect the rate of formation of evolutionarily independent entities if it affects the physical arrangement of the genome and its regulation in such a way that genetic compatibility is affected (Krieber and Rose 1986; see also papers in Dover and Flavell 1982 and in Karlin and Nevo 1986). So genome evolution may directly affect isolation or recognition systems. Presumably it will affect the rate and direction of the origin of new variants too, but we know very little about this process (reviewed in Endler and McLellan 1988). A major problem is the widespread occurrence in various eukaryotic genomes of large amounts of DNA with unknown function even at the molecular level. It is questionable whether this material is relevant to the organism or whether it is merely parasitic or "selfish" DNA. This makes it difficult to go from sequence to function, and largely restricts the use of sequence data to reconstructing phylogenies (for which it is invaluable).

OTHER PHENOMENA THAT MAY BE ASSOCIATED WITH SPECIATION

There are various phenomena that ought to be integrated more with speciation studies. As already mentioned in the last section, we need to know why cladogenesis may be more important in some lineages, whereas anagenesis may be more important in others (apart from taxonomic artifacts). What are the fundamental causes of both, and what long-term patterns of environmental conditions favor one over the other? Could we address any of these questions by long-term artificial selection experiments in which we specifically try to alter recognition/isolation systems or specifically try to affect anagenesis? Most previous artificial selection experiments are anagenesis oriented, and have not tried to change traits thought to be important at the specific or even generic levels. Some artificial selection experiments specifically attempted to change the degree of assortative mating or other measures of recognition/isolation (reviewed in Templeton 1981), but very few of them have investigated causes or looked for morphological correlates. It would be very interesting to artificially select in two directions for a trait used in mate recognition to try to produce isolation as a by-product. It would also be very interesting to artificially select for traits thought to be important at the generic level.

What is the relationship between cladogenesis at the DNA, molecular, biochemical, trait, and species levels of organization? Do the reconstructed phylogenies at each level necessarily correspond? Should they? (See Figure 1.) This is an important practical as well as theoretical question because if the phylogenies do not correspond, it is not possible to infer species phylogenies from (say) molecular phylogenies, and vice versa. Avise (Avise et al. 1983, 1987; Neigel and Avise 1986), Brother (1985), and Hillis (1987) have presented some excellent discussions of this problem.

There are also a number of curious phenomena that may bear directly on speciation processes. Mention has already been made of the possible effects of the gene flow distance on rates of speciation and extinction. Another interesting pattern is the observation that perennial or woody plants seem to differentiate as rapidly and extensively as animals with respect to morphology and ecology, yet do not seem to form what zoologists would like to call good biological species. On the other hand, annual plants do seem to behave like animals and form "standard" biological species (Barrett, this volume; Gill, this volume). Why should this be true? An answer might tell us something about the mechanisms of anagenesis and cladogenesis, and which is favored under which conditions.

Breeding system evolution may also have profound implications for speciation, particularly cladogenesis. Barrett (this volume) provides some fascinating insights on the effects of changes in breeding system on the evolution of floral morphology, rates of inbreeding, and its feedback to evolution of breeding systems in higher plants. If there is no loss of fitness on inbreeding, then floral morphology can change rapidly, and if outcrossing starts again, then speciation could go quite rapidly. If inbreeding depression is common, then both cladogenesis and anagenesis may be slower. Barrett makes the following empirical observations: (1) shifts in mating systems are often associated with speciation, (2) mating systems are often changed through polyploidy or through major mutations affecting floral structure, and (3) the importance of sympatric speciation and small population size is manifest in plants, even though contentious in animals. The interplay between the evolution of breeding systems and morphological evolution needs more investigation in animals too, and is part of the general question of the interplay between anagenesis and cladogenesis.

CONCLUSION

Species concepts vary radically depending on their purpose. It is unproductive, and often positively misleading, to apply one species concept to all species or to answer all questions. To some (Cracraft 1987) this suggests that there cannot be a general theory of evolution that uses species as its currency. This may indeed be correct when considering anagenesis, and it may be correct for cladogenesis if, for example, perennial plants have dif-

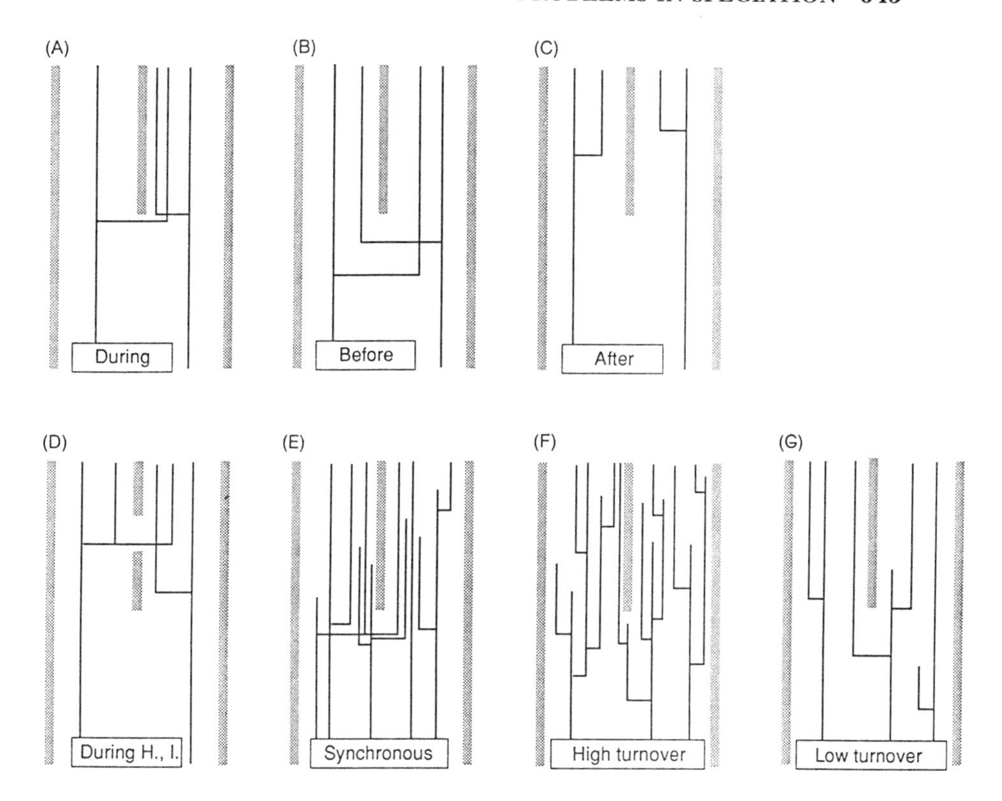

FIGURE 1. The lack of necessary correspondence between the origination of evolutionary independence (formation of isolation or recognition species) and divergence of the genome. The heavy vertical bars indicate the limits to interbreeding (time going from bottom to top) as the single reproductive species splits into two species at the bottom of the middle bar (cladogenesis). The fine lines indicate the ancestral–descendant relationships of old and new mutants within the species. Cladogenesis of the species can correspond in time to cladogenesis of the genome (A), but different parts of the genome may also exhibit cladogenesis before (B) or after (C) the species splits, and, in fact, it can even happen as a result of hybridization (D). Cladogenesis of the genome may be synchronous (E) or asynchronous (F, G) with cladogenesis of the species. The rate of turnover of the genome and its variants will affect the rate of divergence of species, and if asynchronous with species cladogenesis it may be difficult to deduce what happened, particularly if there is low turnover in the genome (G). Because most studies of cladogenesis are based on parts of the genome that have little or nothing to do with reproduction, there is a tendency to believe that (E) is most common, but we actually have no idea which pattern is prevalent, and the effects of hybridization (D) are virtually never considered. For further discussion see Avise et al. (1983, 1987), Neigel and Avise (1986), Brothers (1985), and Hillis (1987).

ferent mechanisms and patterns of speciation (and hybridization) than annuals and most animals. What one finds interesting and important in speciation depends on which species concept is being used implicitly or explicitly and whether one is interested in the phenomena of anagenesis (differentiation within a lineage) or in cladogenesis (splitting of lineages). If one concentrates on the biological species concept (isolation or recognition), then mechanisms associated with reproduction and evolutionary independence are most important, and cladogenesis is the most interesting topic. If one concentrates on the cohesion species concept (common developmental genetic systems, physiology, and ecology), then mechanisms associated with form and function are most important and anagenesis is the most interesting topic, although cladogenesis may also be addressable. Even among those people studying anagenesis, there is a difference in emphasis depending on whether the origins of new variants or the processes that cause new variants to replace old ones are being studied (Endler 1986b). These approaches have been characteristic of separate fields for quite some time, much to the impoverishment and fragmentation of speciation theory and evolutionary biology in general. It is hoped that this volume has shown the need to bring the diversity of fields and aspects of biology together again.

LITERATURE CITED

Archer, S. N., J. A. Endler, J. N. Lythgoe, and J. C. Partridge. 1987. Visual pigment polymorphism in the guppy, *Poecilia reticulata*. Vision Res. 27:1243–1252.

Atchley, W. R. 1984. Ontogeny, timing of development, and genetic variance–covariance structure. Am. Natur. 123:519–540.

Avise, J. C., J. F. Shapira, S. W. Daniel, C. F. Aquadro, and R. A. Lansman. 1983. Mitochondrial DNA differentiation during the speciation process in *Peromyscus*. Mol. Biol. Evol. 1:38–56.

Avise, J. C., J. Arnold, R. M. Ball, E. Bermingham, T. J. Lamb, J. E. Neigel, C. A. Reeb, and N. C. Saunders. 1987. Intraspecific phylogeography: The mitochondrial bridge between population genetics and systematics. Annu. Rev. Ecol. System. 18:489–522.

Barton, N. H., and B. Charlesworth. 1984. Genetic revolutions, founder effects, and speciation. Annu. Rev. Ecol. System. 15:133–164.

Barton, N. H., and G. M. Hewitt. 1985. Analysis of hybrid zones. Annu. Rev. Ecol. System. 16:113–148.

Beaudry, J. R. 1960. The species concept: Its evolution and present status. Rev. Can. Biol. 19:219–240.

Breden, F., and G. Stoner. 1987. Male predation risk determines female preference in the Trinidad guppy. Nature (London) 329:831–833.

Brothers, D. J. 1985. Species concepts, speciation, and higher taxa. Pp. 35–42 in: E. Vrba (ed.), *Species and Speciation*. Transvaal Museum Monograph 4. Pretoria, South Africa.

Brush, J. S., and P. M. Narins. 1989. Chorus dynamics of a neotropical amphibian assemblage: Comparison of computer simulation and natural behavior. Anim. Behav. 37:33–44.

Bush, G. L. 1975. Modes of animal speciation. Annu. Rev. Ecol. System. 6:339–364.

Cheverud, J. M. 1984. Quantitative genetics and developmental constraints on evolution by selection. J. Theoret. Biol. 110:155–171.

Cheverud, J. M., J. J. Rutledge, and W. R. Atchley. 1983. Quantitative genetics of development: Genetic correlations among age-specific trait values and the evolution of ontogeny. Evolution 317:896–905.

Cheverud, J. M., M. M. Dow, and W. Leutenegger. 1985. The quantitative assessment of phylogenetic constraints in comparative analyses: Sexual dimorphism in body weight among primates. Evolution 39:1335–1351.

Coyne, J. A., and H. A. Orr. 1989. Patterns of speciation in *Drosophila*. Evolution 43:362–381.

Cracraft, J. 1987. Species concepts and the ontology of evolution. Biol. Philos. 2:63–80.

Dover, G. A., and R. B. Flavell. 1982. *Genome Evolution*. Academic Press, New York.

Eberhard, W. G. 1985. *Sexual Selection and Animal Genitalia*. Harvard University Press, Cambridge, MA.

Eldredge, N., and J. Cracraft. 1980. *Phylogenetic Patterns and the Evolutionary Process*. Columbia University Press, New York.

Endler, J. A. 1977. *Geographic Variation, Speciation, and Clines*. Princeton University Press, Princeton, NJ.

Endler, J. A. 1986a. Defense against predation. Pp. 109–134 in: M. E. Feder and G. V. Lauder (eds.), *Predator-Prey Relationships, Perspectives and Approaches from the Study of Lower Vertebrates*. University of Chicago Press, Chicago.

Endler, J. A. 1986b. *Natural Selection in the Wild*. Princeton University Press, Princeton, NJ.

Endler, J. A. 1986c. The newer synthesis? Some conceptual problems in evolutionary biology. Oxford Rev. Evol. Biol. 3:224–243.

Endler, J. A. 1988 Sexual selection and predation risk in guppies. Nature (London) 332: 593–594.

Endler, J. A., and T. McLellan. 1988. The processes of evolution: Toward a newer synthesis. Annu. Rev. Ecol. Sytem. 19:395–421.

Futuyma, D. 1987. On the role of species in anagenesis. Am. Natur. 130:465–473.

Ghislein, M. T. 1974. A radical solution to the species problem. System. Zool. 23:536–544.

Ghiselin, M. T. 1981. Categories, life, and thinking. Behav. Brain Sci. 4:269–313.

Grant, P. R. 1975. The classical case of character displacement. Evol. Biol. 8:237–337.

Grant, P. R. 1986. *Ecology and Evolution of Darwin's Finches*. Princeton University Press, Princeton, NJ.

Grant, V. 1957. The plant species in theory and practice. In: E. Mayr (ed.), *The Species Problem*. A.A.A.S., Washington, D.C.

Grant, V. 1981. *Plant Speciation*, 2nd ed. Columbia University Press, New York.

Hansen, T. A. 1978. Larval dispersal and species longevity in lower tertiary gastropods. Science 199:885–887.

Hillis, D. 1987. Molecular versus morphological approaches to systematics. Annu. Rev. Ecol. System. 18:23–42.

Houde, A. E. 1988a. Genetic difference in female choice between two guppy populations. Anim. Behav. 36:510–516.

Houde, A. E. 1988b. Sexual selection in guppies called into question. Nature (London) 333:711.

Jablonski, D. 1987. Heritability at the species level: Analysis of geographic ranges of Cretaceous mollusks. Science 238:360–363.

Jablonski, D., and R. A. Lutz. 1983. Larval ecology of marine benthic invertebrates: Paleobiological implications. Biol. Rev., Cambridge Philos. Soc. 58:21–89.

Jackson, J. B. C. 1974. Biogeographic consequences of eurytopy and stenotopy among marine bivalves and their evolutionary significance. Am. Natur. 108:541–560.

Jacobs, G. H. 1986. Color vision variations in non-human primates. Trends Neuro. Sci. 9:320–323.

Jacobs, G. H., and J. Neitz. 1987a. Polymorphism in the middle wavelength cone in two species of South American monkey: *Cebus apella* and *Callicebus moloch*. Vision Res. 27:1263–1268.

Jacobs, G. H., and J. Neitz. 1987b. Inheritance of color vision in a New World monkey (*Saimiri sciureus*). Proc. Natl. Acad. Sci. U.S.A. 84:2545–2549.

Jacobs, G. H., J. Neitz, and M. Crognale. 1985. Spectral sensitivity of ground squirrel cones measured with ERG flicker photometry. J. Comp. Physiol. A 156:503–509.

Jacobs, G. H., J. Neitz, and M. Crognale. 1987. Color vision polymorphism and its photopigment basis in a callitrichid monkey (*Saguinus fuscicollis*). Vision Res. 27:2089–2100.

Jonsell, B. 1984. The biological species concept reexamined. Pp. 159–169 in: W. F. Grant (ed.), *Plant Biosystematics*. Academic Press, New York.

Kauffman, E. G. 1978. Evolutionary rates and patterns among cretaceous Bivalvia. Phil. Trans. R. Soc. London Ser. B 284:277–304.

Kauffman, S. A. 1985. Self-organization, selective adaptation, and its limits. Pp. 169–207 in: D. J. Depew and B. H. Weber (eds.), *Evolution at a Crossroads: The New Biology and the New Philosophy of Science*. M.I.T. Press, Cambridge, MA.

Kaneshiro, K. Y. 1989. The dynamics of sexual selection and founder effects in species formation. In: L. V. Giddings, K. Y. Kaneshiro, and W. W. Anderson (eds.), *Genetics, Speciation, and the Founder Principle*. Oxford University Press, New York and London.

Karlin, S., and E. Nevo. 1986. *Evolutionary Processes and Theory*. Academic Press, New York.

Krieber, M., and M. Rose. 1986. Molecular aspects of the species barrier. Annu. Rev. Ecol. System. 17:456–485.

Lande, R. 1981. Models of speciation by sexual selection on polygenic traits. Proc. Natl. Acad. Sci. U.S.A. 78:3721–3725.

Lande, R. 1982. Rapid origin of sexual isolation and character divergence in a cline. Evolution 36:213–223.

Levin, D. A. 1979. The nature of plant species. Science 204:381–384.

Littlejohn, M. J. 1960. Call discrimination and potential reproductive isolation in *Pseudacris triseriata* females from Oklahoma. Copeia 1960:370–371.

Mayr, E. 1963. *Animal Species and Evolution*. Belknap Press, Cambridge, MA.

Neigel, J. E., and J. C. Avise. 1986. Phylogenetic relationships of mitochondrial DNA under various demographic models of speciation. Pp. 515–534 in: S. Karlin and E. Nevo (eds.), *Evolutionary Processes and Theory*. Academic Press, New York.

Neitz, J., and G. H. Jacobs. 1986. Polymorphism of the long-wavelength cone in normal human color vision. Nature (London) 323:623–625.

Paterson, H. E. H. 1985. The recognition concept of species. Pp. 21–29 in: E. Vrba (ed.), *Species and Speciation*. Transvaal Museum Monograph 4, Pretoria, South Africa.

Paterson, H. E. H., and M. Macnamara. 1984. The recognition concept of species. South African J. Sci. 80:312–318.

Prance, G. (ed.) 1982. *Biological Diversification in the Tropics*. Columbia University Press, New York.

Raven, P. H. 1976. Systematics and plant population biology. System. Bot. 1:284–316.

Rensch, B. 1959. *Evolution above the Species Level*. Methuen & Co., London.

Ridley, M. 1986. *Evolution and Classification, the Reformation of Cladism*. Longman, London.

Ryan, M. J. 1986. Neuroanatomy influences speciation rates among anurans. Proc. Natl. Acad. Sci. U.S.A. 83:1379–1382.

Ryan, M. J., and W. E. Wagner, J. 1987. Asymmetries in mating preferences between species: Female swordtails prefer heterospecific males. Science 236:595–597.

Schluter, D. 1984. Morphological and phylogenetic relations among the Darwin's finches. Evolution 38:921–930.

Shuto, T. 1974. Larval ecology of prosobranch gastropods and its bearing on biogeography and paleontology. Lethaia 7:239–256.

Sneath, P. H. A.,and R. R. Sokal. 1973. *Numerical Taxonomy, the Principles and Practice of Numerical Classification*. W.H. Freeman, San Francisco.

Templeton, A. 1981. Mechanisms of speciation—a population genetic approach. Annu. Rev. Ecol. System. 12:23–48.

Van Valen, L. 1976. Ecological species, multispecies and oaks. Taxon 25:233–239.

Wiley, E. O. 1981. *Phylogenetics, the Theory and Practice of Phylogenetic Systematics*. Wiley-Interscience, New York.

Author Index

Numbers in *italic* indicate the bibliography page(s) on which the citation can be found.

Subject Index